SOC 6

INTRODUCTION TO SOCIOLOGY

NIJOLE V. BENOKRAITIS
University of Baltimore

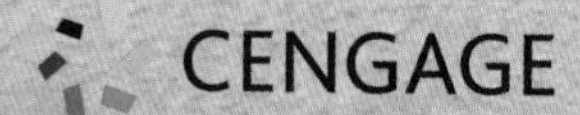

Australia • Brazil • Canada • Mexico • Singapore • United Kingdom • United States

***SOC* 6**
Nijole V. Benokraitis

Senior Vice President, Higher Ed Product, Content, and Market Development: Erin Joyner

Product Manager: Elizabeth Beiting-Lipps

Content/Media Developer: Sarah Keeling

Product Assistant: Allison Balchunas

Marketing Manager: Christopher Walz

Marketing Coordinator: Quynton Johnson

Sr. Content Project Manager: Tim Bailey

Sr. Art Director: Bethany Bourgeois

Text Designer: Lou Ann Thesing

Cover Designer: Lisa Kuhn, Curio Press, LLC / Chris Miller, Cmiller Design

Cover Image: 7 pips/Shutterstock.com

Intellectual Property Analyst: Alexandra Ricciardi

Intellectual Property Project Manager: Nick Barrows

Production Service: MPS Limited

Library of Congress Control Number: 2017957884

Student Edition ISBN: 978-1-337-40521-8
Student Edition with MindTap ISBN: 978-1-337-40516-4

Cengage
200 Pier 4 Boulevard
Boston, MA 02210
USA

ACKNOWLEDGEMENTS

We thank the following faculty members for their valuable feedback in revising this edition:

Megan Allen, *Blue Ridge Community and Technical College*
David Briscoe, *University of Arkansas at Little Rock*
Ashley Chambers, *Blue Ridge Community and Technical College*
Rose De Luca, *Emmanuel College*
Sarah Deward, *Eastern Michigan University*
Steven Fulks, *Barton College*
Jennifer Kunz, *West Texas A&M University*
Diane Levy, *University of North Carolina at Wilmington*
Kathleen Lowney, *Valdosta State University*
Timothy McLean, *Herkimer College*
Amanda Miller, *University of Indianapolis*
Ken Muir, *Appalachian State University*
Jessica Oladapo, *Rock Valley College*
Jodie Simon, *Wichita State University*
Viviene Wood, *University of Western Georgia*
Rochelle Zaranek, *Macomb Community College*

Printed in the United States of America
Print Number: 07 Print Year: 2021

BENOKRAITIS

SOC 6

BRIEF CONTENTS

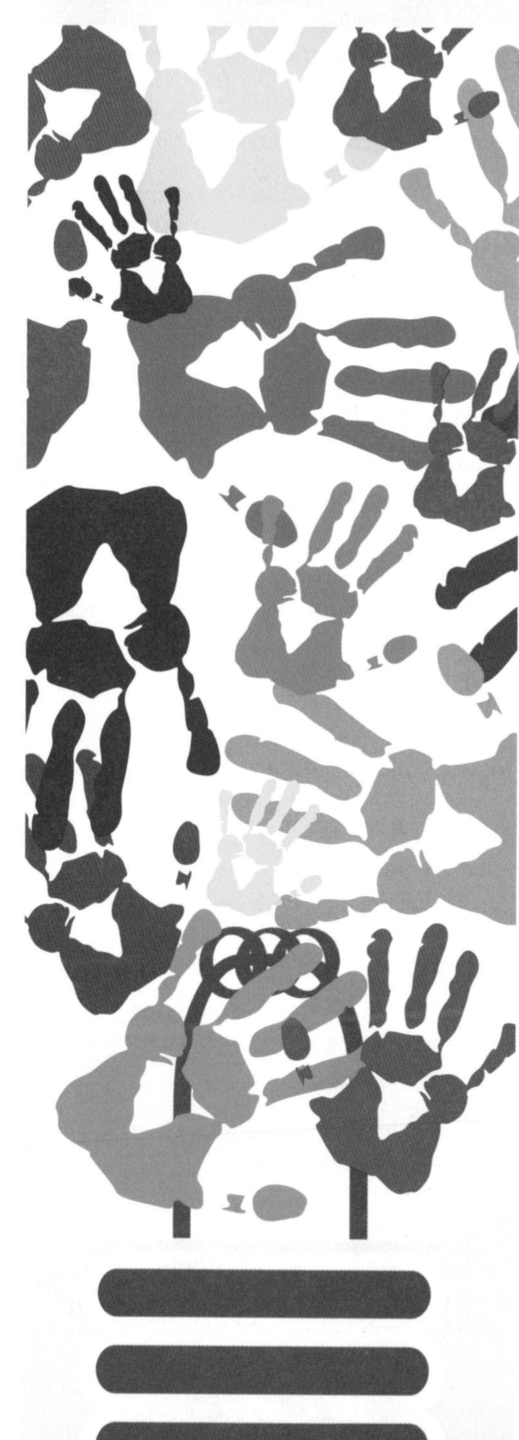

7 pips/Shutterstock.com

CONTENTS

Michele Burgess/Alamy Stock Photo; Jeremy Woodhouse/AGE Fotostock; BSIP/Newscom; vgajic/E+/Getty Images; Jim West/Alamy Stock Photo; John Lund/Getty Images

Chip Somodevilla/Getty Images News/Getty Images

PG Arphexad/Alamy Stock Photo

Henryk T. Kaiser/Getty Images

Andrew Lichtenstein/Corbis News/Getty Images

Masterfile

14 Health and Medicine 282

Lisa S./Shutterstock.com

US Army Photo/Alamy Stock Photo

15 Population, Urbanization, and the Environment 302

16 Social Change: Collective Behavior, Social Movements, and Technology 326

1 Thinking Like a Sociologist

LEARNING OBJECTIVES

After studying this chapter, you will be able to...

1-1 Explain what sociology is and how it differs from other social sciences and common sense.

1-2 Explain how and why a sociological imagination helps us understand society.

1-3 Identify and illustrate why it's worthwhile to study sociology.

1-4 Describe and explain the origins of sociology, why sociology developed, and its most influential early theorists.

1-5 Compare, illustrate, and evaluate the four contemporary sociological perspectives.

After finishing this chapter go to **PAGE 19** for **STUDY TOOLS**

Texting and emailing are associated with the highest risk of car crashes, and headset cell phones aren't much safer than handheld cell phones. Almost 90 percent of U.S. drivers say that distracted driving is a serious safety threat. However, 70 percent talk on a cell phone, text, surf the Internet, and even video chat while driving. In 2014 alone, such distractions contributed to 18 percent of all crashes that resulted in death or severe injury (AAA Foundation for Traffic Safety, 2015; AT&T Newsroom, 2015; National Center for Statistics and Analysis, 2016).

WHAT DO YOU THINK?

Sociology is basically common sense.

1	2	3	4	5	6	7
strongly agree				strongly disagree		

Why is there such a disconnection between many Americans' attitudes and behavior? This chapter examines these and other questions. Let's begin by considering what sociology is (and isn't) and how a "sociological imagination" can give us more control over our lives. We'll then look at how sociologists grapple with complex theoretical issues in explaining social life. Before reading further, take the True or False?

1-1 WHAT IS SOCIOLOGY?

Stated simply, **sociology** is the scientific study of human behavior in society. Sociologists study behavior patterns that occur between individuals, among small groups (e.g., families), large organizations (e.g., Apple), and entire societies (e.g., the United States). But, you might protest, "I'm unique."

1-1a Are You Unique?

Yes and no. Each of us is unique in the sense that you and I are like no one else on earth. Even identical twins, who have the same physical characteristics and genetic matter, often differ in personality and interests. One of my colleagues likes to tell the story about his 3-year-old twin girls who received the same doll. One twin chattered that the doll's name was Lori, that she loved Lori, and would take good care of her. The second twin muttered, "Her name is Stupid," and flung the doll into a corner.

Despite some individual differences, identical twins, you, and I are like other people in many ways. Around the world, we experience grief when a loved one dies, participate in rituals that celebrate marriage or the birth of a child, and want to have healthy and happy lives. Some actions, such as terrorist attacks, are unpredictable. For the most part, however, people conform to expected and acceptable behavior. From the time that we get up until we go to bed, we follow a variety of rules and customs about what we eat, how we drive, how

True *or* False?

EVERYBODY KNOWS THAT...

1. The death penalty reduces crime.
2. Women's earnings are now similar to men's, especially in high-income occupations.
3. People age 65 and older make up the largest group of those who are poor.
4. There are more married than unmarried U.S. adults.
5. Divorce rates are higher today than in the past.
6. Latinos are the fastest-growing racial-ethic group in the United States.
7. The best way to get an accurate measure of public opinion is to poll as many people as possible.
8. Illegal drugs are the biggest health hazard.

The answers are at the end of 1-1.

sociology the scientific study of human behavior in society.

Marriage without Love? No Way!

When I ask my students, "Would you marry someone you're not in love with?" most laugh, raise an eyebrow, or stare at me in disbelief. "Of course not!" they exclaim. In fact, the "open" courtship and dating systems common in Western nations, including the United States, are foreign to much of the world. In many African, Asian, Mediterranean, and Middle Eastern countries, marriages are arranged: They forge bonds between families rather than individuals, and preserve family continuity along religious and socioeconomic lines. Love isn't a prerequisite for marriage in societies that value kin groups rather than individual choices (see Chapters 9 and 12).

we act in different social situations, and how we dress for work, classes, and leisure activities.

So what? you might shrug. Isn't it "obvious" that we dress differently for classes than for job interviews? Isn't all of this just plain old common sense?

1-1b Isn't Sociology Just Common Sense?

No. Sociology goes well beyond conventional wisdom, what we call common sense, in several ways:

- **Common sense is subjective.** If a woman crashes into my car, I might conclude, according to conventional wisdom, statements that we've heard over the years, that "women are terrible drivers." In fact, most drivers involved in crashes are men—especially teenagers and those age 70 and older (Insurance Institute for Highway Safety, 2013). Thus, *objective* data show that, overall, men are worse drivers than women.
- **Common sense ignores facts.** Because common sense is subjective, it ignores facts that challenge cherished beliefs. For example, many Americans are most concerned about street crimes, such as robbery or homicide. FBI and sociological data show, however, that we're much more likely to be assaulted or murdered by someone we know or live with (see Chapters 7 and 12).
- **Common sense varies across groups and cultures.** Many Americans believe that working harder decreases poverty. In contrast, Europeans tend to think that poverty is due to forces outside people's control (see Chapters 3 and 8). Thus, common sense notions about economic success vary considerably across countries.
- **Much of our common sense is based on myths and misconceptions.** A common myth is that living together is a good way to find out whether partners will get along after marriage. Generally, however, couples who live together before marriage have higher divorce rates than those who don't (see Chapter 12).

Sociology, in contrast to conventional wisdom, examines claims and beliefs critically, considers many points of view, and enables us to move beyond established ways of thinking. The *sociological perspective* analyzes how social context influences people's lives. The "sociological imagination" is at the center of the sociological perspective.

True *or* False?

EVERYBODY KNOWS THAT...

All of the answers are false.

1. States without the death penalty have had consistently lower homicide rates than those with death penalties (see Chapter 7).
2. Regardless of education or occupation, women's earnings are lower than men's (see Chapters 9 and 11).
3. Children ages 5 and under make up the largest group of Americans who are poor (see Chapter 8).
4. The number of unmarried U.S. adults outnumbers those who are married (see Chapter 12).
5. Divorce rates are lower today than they were between 1975 and 1990 (see Chapter 12).
6. Latinos are the largest ethnic group in the United States, but Asian Americans are the fastest growing (see Chapter 10).
7. What matters in polling is not the number of people polled, but their representativeness in the population studied (see Chapter 2).
8. In the United States and worldwide, tobacco use is the leading cause of preventable death and disability (see Chapter 14).

1-2 WHAT IS A SOCIOLOGICAL IMAGINATION?

According to sociologist C. Wright Mills (1916–1962), social factors such as religion, ethnicity, and politics affect our behavior. Mills (1959) called this ability to see the relationship between individual experiences and larger social influences the **sociological imagination**. The sociological imagination emphasizes the connection between personal troubles (biography) and structural (public and historical) issues.

Consider unemployment. If only a small group of people can't find a job, it's a *personal trouble* that may be due, in part, to an individual's low educational attainment, lack of specific skills that employers want, not searching for work, and so on. If unemployment is widespread, it's a *public issue* because economic problems are also the result of structural factors such as mass layoffs, sending jobs overseas, technological changes, and restrictive hiring policies (see Chapter 11). Thus, people may be unemployed regardless of skills, a college degree, and job searches.

A sociological imagination helps us understand how larger social forces impact our everyday lives. It identifies why our personal troubles are often due to larger public issues and policies over which we have little, if any, control. A sociological imagination relies on both micro- and macro-level approaches to understand our social world.

1-2a Microsociology: How People Affect Our Everyday Lives

Microsociology examines the patterns of individuals' social interaction in specific settings. In most of our relationships, we interact with others on a micro, or "small," level (e.g., members of a work group discussing who will perform which tasks). These everyday interactions involve what people think, say, or do on a daily basis.

1-2b Macrosociology: How Social Structure Affects Our Everyday Lives

Macrosociology focuses on large-scale patterns and processes that characterize society as a whole. Macro, or "large," approaches are especially useful in understanding some of the constraints—such as economic forces and public policies.

Microsociology and macrosociology differ conceptually, but are interrelated. Consider the reasons for divorce. On a micro level, sociologists study factors like extramarital affairs, substance abuse, arguments about money, and other everyday interactions that fuel marital tension and unhappiness, leading to divorce. On a macro level, sociologists look at how the economy, laws, cultural values, and technology affect divorce rates (see Chapter 12). Examining micro, macro, and micro–macro forces is one of the reasons why sociology is a powerful tool in understanding (and changing) our behavior and society at large (Ritzer, 1992).

Jessica Miglio/Netflix/Everett Collection

Netflix's *Orange Is the New Black* revolves around a White, upper-middle-class woman who's in prison for drug smuggling. For sociologists, the series illustrates the connection between micro-level individual behavior (the inmates' and prison guards' experiences) and macro-level factors (social class, family structures, racial discrimination, corruption, and prison overcrowding).

1-3 WHY STUDY SOCIOLOGY?

Sociology offers explanations that can greatly improve the quality of your everyday life. These explanations can influence choices that range from your personal decisions to expanding your career opportunities.

1-3a Making Informed Decisions

Sociology can help us make more informed decisions. We often hear that grief counseling is essential after the death

sociological imagination seeing the relationship between individual experiences and larger social influences.

microsociology examines the patterns of individuals' social interaction in specific settings.

macrosociology examines the large-scale patterns and processes that characterize society as a whole.

of a loved one. In fact, 4 in 10 Americans are better off without it. Grief is normal, and most people work through their losses on their own, whereas counseling sometimes prolongs depression and anxiety (Stroebe et al., 2000).

1-3b Understanding Diversity

The racial and ethnic composition of the United States is changing. By 2025, only 58 percent of the U.S. population is projected to be White, down from 76 percent in 1990 and 86 percent in 1950 (Passel et al., 2011; U.S. Census Bureau, 2012). As you'll see in later chapters, this racial/ethnic shift has already affected interpersonal relationships as well as education, politics, religion, and other spheres of social life.

Recognizing and understanding diversity is one of sociology's central themes. Our gender, social class, marital status, ethnicity, sexual orientation, and age—among other factors—shape our beliefs, behavior, and experiences. If, for example, you're a White middle-class male who attends a private college, your experiences are very different from those of a female Vietnamese immigrant who is struggling to pay expenses at a community college.

Increasingly, nations around the world are intertwined through political and economic ties. What happens in other societies often has a direct or indirect impact on contemporary U.S. life. Decisions in oil-producing countries, for example, affect gas prices, spur the development of hybrid cars that are less dependent on oil, and stimulate research on alternative sources of energy.

Suzanne Tucker/Shutterstock.com

1-3c Shaping Social and Public Policies and Practices

Sociology is valuable in applied, clinical, and policy settings because many jobs require understanding society and research to create social change. According to a director of a research institute, sociology increased her professional contributions: "I can look at problems of concern to the National Institutes of Health and say 'here's a different way to solve this problem'" (Nyseth et al., 2011: 48).

1-3d Thinking Critically

We develop a sociological imagination not only when we understand and can apply the concepts, but also when we can think, speak, and write critically. Much of our thinking and decision making is often impulsive and emotional. In contrast, critical thinking involves knowledge and problem solving (Paul and Elder, 2007).

Critical sociological thinking goes even further because we begin to understand how our individual lives, choices, and troubles are shaped by race, gender, social class, and social institutions like the economy, politics, and education (Eckstein et al., 1995; Grauerholz and Bouma-Holtrop, 2003). *Table 1.1* summarizes some of the basic elements of critical sociological thinking.

Snapshots

"I love our lunches out here, but I always get the feeling that we're being watched."

Jason Love/Cartoonstock.com

Table 1.1 What Is Critical Sociological Thinking?

Critical sociological thinking requires a combination of skills. Some of the basic elements include the ability to:
• rely on reason rather than emotion
• ask questions, avoid snap judgments, and examine popular and unpopular beliefs
• recognize one's own and others' assumptions, prejudices, and points of view
• remain open to alternative explanations and theories
• examine competing evidence (see Chapter 2)
• understand how public policies affect private troubles

Some well-known people who were sociology majors: Rev. Martin Luther King, Ronald Reagan, Michelle Obama, Robin Williams, and Joe Theismann. Library of Congress Prints and Photographs Division[Leffler, Warren K/ LC-DIG-ds-00836]; U.S. National Archives and Records Administration (NARA); Official White House Photo; Everett Collection/Shutterstock.com; Nate Fine/Getty Images Sport/Getty Images; iStock.com/belterz

1-3e Expanding Your Career Opportunities

A degree in sociology is a springboard for many jobs and professions. A national survey of under-graduate sociology majors found that 44 percent were in administrative or management positions, 22 percent were employed in social service and counseling, 18 percent were in sales and marketing, and 12 percent were teachers (Senter et al., 2014; see also Senter and Spalter-Roth, 2016).

What specific skills do sociology majors learn that are useful in their jobs? Some of the most important are being better able to work with people (71 percent), to organize information (69 percent), to write reports that nonsociologists understand (61 percent), and to interpret research findings (56 percent) (Van Vooren and Spalter-Roth, 2010).

In other cases, students major in sociology because it provides a broad liberal arts foundation for professions such as law, education, and social work. The Medical College Admission Test (MCAT) now includes material from sociology because "Being a good physician is about more than scientific knowledge. It's about understanding people—how they think, interact, and make decisions" (Olsen, 2016: 72).

Even if you don't major in sociology, developing your sociological imagination can enrich your job skills. Sociology courses help you learn to think abstractly and critically, formulate problems, ask incisive questions, search for data in the most reliable and up-to-date sources, organize material, and improve your oral presentations (ASA Research Department, 2013; Spalter-Roth et al., 2013).

1-4 SOME ORIGINS OF SOCIOLOGICAL THEORY

During college, most of my classmates and I postponed taking theory courses (regardless of our major) as long as possible. "This stuff is boring, boring, boring," we'd grump, "and has nothing to do with the real world." In fact, theorizing is part of our everyday lives. Every time you try to explain why your family and friends behave as they do, for example, you're theorizing.

As people struggle to understand human behavior, they develop theories. A **theory** is a set of statements that explains why a phenomenon occurs. Theories produce knowledge, guide our research, help us analyze our findings, and, ideally, offer solutions for social problems.

Sociologist James White (2005: 170–171) describes theories as "tools" that don't profess to know "the truth" but "may need replacing" over time as our understanding of society becomes more sophisticated. In effect, theories evolve over time because of cultural and technological changes. You'll see shortly, for example, that sociological theories changed considerably after the rise of feminist scholarship during the late 1960s.

Sociological theories didn't emerge overnight. Nineteenth-century thinkers grappled with some of the same questions that sociologists try to answer today: Why do people behave as they do? What holds society together? What pulls it apart? Of the many early sociological theorists, some of the most influential were Auguste Comte, Harriet Martineau, Émile Durkheim, Karl Marx, Max Weber, Jane Addams, and W. E. B. Du Bois.

theory a set of statements that explains why a phenomenon occurs.

1-4a Auguste Comte

Creatas Images/Jupiter Images

Auguste Comte (1798–1857) coined the term *sociology* and is often described as the "father of sociology." Comte maintained that the study of society must be **empirical**. That is, information should be based on observations, experiments, or other data collection rather than on ideology, religion, intuition, or conventional wisdom.

He saw sociology as the scientific study of two aspects of society: social statics and social dynamics. *Social statics* investigates how principles of social order explain a particular society, as well as the interconnections between institutions. *Social dynamics* explores how individuals and societies change over time. Comte's emphasis on social order and change within and across societies is still useful today because many sociologists examine the relationships between education and politics (social statics), as well as how such interconnections change over time (social dynamics).

Hulton Archive/Getty Images

The Father of Sociology—Auguste Comte

1-4b Harriet Martineau

Harriet Martineau (1802–1876), an English author, published several dozen books on a wide range of topics in social science, politics, literature, and history. Her translation and condensation of Auguste Comte's difficult material for popular consumption was largely responsible for the dissemination of Comte's work. "We might say, then, that sociology had parents of both sexes" (Adams and Sydie, 2001: 32). She emphasized the importance of systematic data collection through observation and interviews, and an objective analysis of data to explain events and behavior. She also published the first sociology research methods textbook.

Martineau, a feminist and strong opponent of slavery, denounced many aspects of capitalism as alienating and degrading, and criticized dangerous workplaces that often led to injury and death. Martineau promoted improving women's positions in the workforce through education, nondiscriminatory employment, and training programs. She advocated women's admission into medical schools and emphasized issues such as infant care, the rights of the aged, suicide prevention, and other social problems (Hoecker-Drysdale, 1992).

After a long tour of the United States, Martineau described American women as being socialized to be subservient and dependent rather than equal marriage partners. She also criticized American and European religious institutions for expecting women to be pious and passive rather than educating them in philosophy and politics. Most scholars, including sociologists, ridiculed and dismissed such ideas as too radical.

1-4c Émile Durkheim

Émile Durkheim (1858–1917), a French sociologist and writer, agreed with Comte that societies are characterized by unity and cohesion because their members are bound together by common interests and attitudes. Whereas Comte acknowledged the importance of using scientific methods to study society, Durkheim actually did so by poring over official statistics to test a theory about suicide (Adams and Sydie, 2001).

SOCIAL FACTS

To be scientific, Durkheim maintained, sociology must study **social facts**—aspects of social life, external to the individual, that can be measured. Sociologists can determine *material facts* by examining demographic characteristics such as age, place of residence, and population size. They can gauge *nonmaterial facts*, like communication processes, by observing everyday behavior and how people relate to each other (see Chapters 3 to 6). For contemporary sociologists, social facts

Spencer Arnold/Hulton Archive/Getty Images

Harriet Martineau

empirical information that is based on observations, experiments, or other data collection rather than on ideology, religion, intuition, or conventional wisdom.

social facts aspects of social life, external to the individual, that can be measured.

also include collecting and analyzing data on *social currents* such as collective behavior and social movements (see Chapter 16).

DIVISION OF LABOR

One of Durkheim's central questions was how people can be autonomous and individualistic while being integrated in society. **Social solidarity**, or social cohesiveness and harmony, according to Durkheim, is maintained by a **division of labor**—an interdependence of different tasks and occupations, characteristic of industrialized societies, that produces social unity and facilitates change.

As the division of labor becomes more specialized, people become increasingly dependent on others for specific goods and services. Today, for example, many couples who marry often contract "experts" (e.g., photographers, florists, deejays, caterers, bartenders, travel agents, and even "wedding planners").

Figure 1.1 U.S. Suicide Rates, by Sex and Age

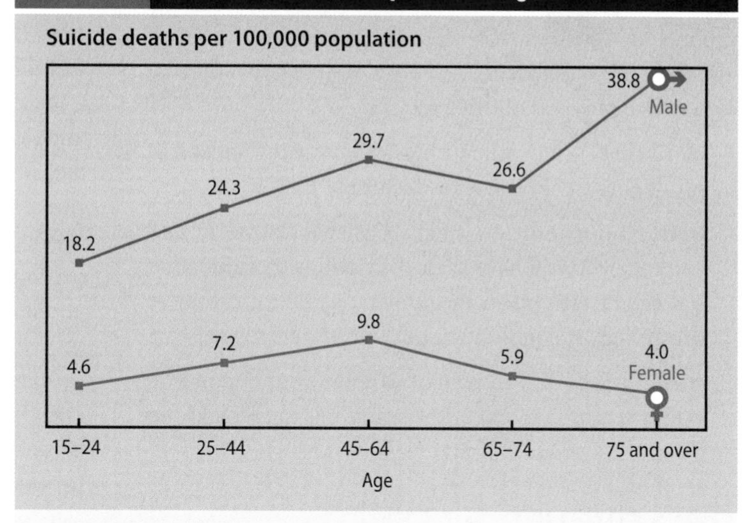

Source: Based on Curtin et al., 2016, Figures 2 and 3.

SOCIAL INTEGRATION

Durkheim was one of the first sociologists to test a theory using data. In his classic study, *Suicide*, Durkheim (1897/1951) relied on extensive data collection to test his theory that suicide is associated with social integration. He concluded that people who experience meaningful social relationships in families, social groups, and communities are less likely to commit suicide than those who feel alone, helpless, or hopeless. Thus, many seemingly isolated individual acts, including suicide, are often the result of structural arrangements, such as weak social ties.

We typically hear about high teenage suicide rates, but they're much higher at later ages. As in Durkheim's day, men have higher suicide rates than women across all age groups (*Figure 1.1*), and White males age 85 and older are the most likely to take their own lives (National Center for Health Statistics, 2015).

Pictorial Press Ltd/Alamy Stock Photo

Émile Durkheim

Durkheim's connection of social integration to the suicide rate is still relevant today. The high suicide rates of older White men are due to a complex interplay of depression, substance abuse, access to guns, hopelessness because of terminal illnesses, and not being "connected" to family, friends, community groups, and support systems that women tend to develop throughout their lives (American Association of Suicidology, 2009; see also Chapters 9 and 12).

1-4d Karl Marx

Karl Marx (1818–1883), a German social philosopher, is often described as the most influential social scientist who ever lived. Marx, like Comte and Durkheim, tried to explain the societal changes that were taking place during the Industrial Revolution.

The Industrial Revolution began in England around 1780 and spread throughout Western Europe and the United States during the nineteenth century. A number of technological inventions—like the spinning wheel, the steam engine, and large weaving looms—enabled the development of large-scale manufacturing and mining industries over a relatively short period. The extensive mechanization shifted agricultural and home-based work to factories in cities. As masses of people migrated from small farms to factories to find jobs, urbanization and capitalism grew rapidly.

CAPITALISM

Unlike his predecessors and contemporaries, Marx (1867/1967, 1964) maintained that economic issues produce divisiveness rather than social solidarity. For Marx, the most important

social solidarity social cohesiveness and harmony.

division of labor an interdependence of different tasks and occupations, characteristic of industrialized societies, that produces social unity and facilitates change.

social change was the development of **capitalism**, an economic system in which the ownership of the means of production—such as land, factories, large sums of money, and machines—is private. As a result, Marx saw industrial society as composed of three social classes:

- **capitalists**—the ruling elite who own the means of producing wealth (such as factories)
- **petit bourgeoisie**—small business owners and workers who still have their own means of production but might end up in the proletariat because they're driven out by competition or their businesses fail
- **proletariat**—the masses of workers who depend on wages to survive, have few resources, and make up the working class

CLASS CONFLICT

Marx believed that society is divided into the haves (capitalists) and the have-nots (proletariat). For Marx, capitalism is a class system in which conflict between the classes is common and society is anything but cohesive. Instead, class antagonisms revolve around struggles between the capitalists, who increase their profits by exploiting workers, and workers, who resist but give in because they depend on capitalists for jobs.

Marx argued that there's a close relationship between inequality, social conflict, and social class. History, he maintained, is a series of class struggles between capitalists and workers. As wealth becomes more concentrated in the hands of a few capitalists, Marx predicted, the ranks of an increasingly dissatisfied proletariat would swell, leading to bloody revolution and eventually a classless society. The Occupy Wall Street movement showed that thousands of Americans are very unhappy about the growing inequality between the haves and the have-nots, but there hasn't been a "bloody revolution" in the United States, unlike some countries in the Middle East.

ALIENATION

In industrial capitalist systems, Marx (1844/1964) contended, **alienation**—feeling separated from one's group or society—is common across all social classes. Workers feel alienated because they don't own or control either the means of production or the product. Because meaningful labor is what makes us human, Marx maintained, our workplace has alienated us "from the essence of our humanness." Instead of collaboration, a capitalistic society encourages competition, backstabbing, and "looking out for number one."

According to Marx, capitalists are also alienated. They regard goods and services as important simply because they're sources of profit. Capitalists don't care who buys or sells their products, how the workers feel about the products they make, or whether buyers value the products. The major focus, for capitalists, is on increasing profits as much as possible rather than feeling "connected" to the products or services they sell. Every year, for example, companies must recall cars, pharmaceutical items, toys, and food that cause injuries, illness, or death.

Comstock/Stockbyte/Getty Images

Roger Viollet Collection/Getty Images

Karl Marx

Daryl Lang/Shutterstock.com

Occupy Wall Street (OWS) was a protest movement against corporate greed, corruption, and influence on government. It began in mid-September, 2011, in New York City's Wall Street financial district. The OWS slogan, "We are the 99%," referred to U.S. income and wealth inequality between the wealthiest 1 percent and the rest of the population. OWS received global attention and spawned similar movements worldwide but was short-lived (see Chapter 16).

capitalism an economic system based on the private ownership of property and the means of production.

alienation feeling separated from one's group or society.

1-4e Max Weber

Max Weber (pronounced VAY-ber; 1864–1920) was a German sociologist, economist, legal scholar, historian, and politician. Unlike Marx's emphasis on economics as a major factor in explaining society, Weber focused on social organization, a subjective understanding of behavior, and a value-free sociology.

SOCIAL ORGANIZATION

For Weber, economic factors were important, but ideas, religious values, ideologies, and charismatic leaders were just as crucial in shaping and changing societies. He maintained that a complete understanding of society requires an analysis of the social organization and interrelationships among economic, political, and cultural institutions. In his *Protestant Ethic and the Spirit of Capitalism*, for example, Weber (1920/1958) argued that the self-denial fostered by Calvinism supported the rise of capitalism and shaped many of our current values about working hard (see Chapters 3 and 6).

SUBJECTIVE UNDERSTANDING

Max Weber

Weber posited that an understanding of society requires a "subjective" understanding of behavior. Such understanding, or *verstehen* (pronounced fer-SHTAY-en), involves knowing how people perceive the world in which they live. Weber described two types of *verstehen*. In *direct observational understanding*, the social scientist observes a person's facial expressions, gestures, and listens to his/her words. In *explanatory understanding*, the social scientist tries to grasp the intention and context of the behavior.

Is It Possible to Be a Value-Free Sociologist?

Max Weber was concerned about popular professors who took political positions that pleased many of their students. He felt that these professors were behaving improperly because science, including sociology, must be "value free." Faculty must set their personal values aside to make a contribution to society. According to a sociologist who agrees with Weber, sociology's weakness is its tendency toward moralism and ideology:

> *Many people become sociologists out of an impulse to reform society, fight injustice, and help people. Those sentiments are noble, but unless they are tempered by skepticism, discipline, and scientific detachment, they can be destructive. Especially when you are morally outraged and burning with a desire for action, you need to be cautious (Massey, 2007: B12).*

Some argue that being value free is a myth because it's impossible for a scholar's attitudes and opinions to be totally divorced from her or his scholarship (Gouldner, 1962). Many sociologists, after all, do research on topics that they consider significant and about which they have strong views.

Others maintain that one's values should be passionately partisan, should frame research issues, and should improve society (Feagin, 2001). Sociologists shouldn't apologize for being subjective in their teaching and research. By staying silent, social scientists "cede the conversation to those with the loudest voices or deepest pockets . . . people with megaphones who spread sensational misinformation" that deprives the public of the best available data (Wang, 2015: A48).

Can sociologists *really* be value free—especially when they have strong feelings about many societal issues? Should they be?

CAN SOCIOLOGISTS BE VALUE FREE—ESPECIALLY WHEN THEY HAVE STRONG FEELINGS ABOUT MANY SOCIETAL ISSUES? *SHOULD THEY BE?*

If a person bursts into tears (direct observational understanding), the observer knows what the person may be feeling (anger, sorrow, and so on). An explanatory understanding goes a step further by spelling out the reason for the behavior (rejection by a loved one, frustration if you lose your smartphone, humiliation if a boss yells at you in public).

VALUE-FREE SOCIOLOGY

One of Weber's most lasting and controversial views was the notion that sociologists must be as objective, or "value free," as possible in analyzing society. A researcher who is **value free** is one who separates her or his personal values, opinions, ideology, and beliefs from scientific research.

During Weber's time, the government and other organizations demanded that university faculty teach the "right" ideas. Weber encouraged everyone to be involved as citizens, but he maintained that educators and scholars should be as dispassionate as possible politically and ideologically. The task of the teacher, Weber argued, was to provide students with knowledge and scientific experience, not to "imprint" the teacher's personal political views and value judgments (Gerth and Mills, 1946). "Is It Possible to Be a Value-Free Sociologist?" examines this issue further.

Wallace Kirkland//Time Life Pictures/Getty Images

Jane Addams with a child at Hull House

1-4f Jane Addams

Jane Addams (1860–1935) was a social worker who cofounded Hull House, one of the first settlement houses in Chicago that served the neighborhood poor. An active reformer throughout her life, Jane Addams was a leader in the women's suffrage movement and, in 1931, was the first American woman to be awarded the Nobel Peace Prize for her advocacy of negotiating, rather than waging war, to settle disputes.

Sociologist Mary Jo Deegan (1986) describes Jane Addams as "the greatest woman sociologist of her day." However, she was ignored by her colleagues at the University of Chicago (the first sociology department established in the United States in 1892) because discrimination against women sociologists was "rampant" (p. 8).

value free separating one's personal values, opinions, ideology, and beliefs from scientific research.

Despite such discrimination, Addams published articles in many popular and scholarly journals, as well as many books on the everyday life of urban neighborhoods, especially the effects of social disorganization and immigration. Much of her work contributed to symbolic interaction, an emerging school of thought that you'll read about shortly. One of Addams' greatest intellectual legacies was her emphasis on applying knowledge to everyday problems. Her pioneering work in criminology included ecological maps of Chicago that were later credited to men (Moyer, 2003).

1-4g W. E. B. Du Bois

W. E. B. Du Bois (pronounced Do-BOICE; 1868–1963) was a prominent Black sociologist, writer, editor, social reformer, and orator. The author of almost two dozen books on Africans and Black Americans, Du Bois spent most of his life responding to the critics and detractors of Black life. He was the first Black individual to receive a Ph.D. from Harvard University, but once remarked, "I was in Harvard but not of it."

Du Bois helped found the National Association for the Advancement of Colored People (NAACP) and edited its journal, *Crisis.* The problem of the twentieth century, he wrote, is the problem of the color line. Du Bois believed that the race problem was one of ignorance, and advocated a "cure" for prejudice and discrimination. Such cures included promoting Black political power and civil rights and providing Black individuals with a higher education rather than funneling them into technical schools.

These and other writings were unpopular at a time when Booker T. Washington, a well-known Black educator, encouraged Black people to be patient instead of demanding equal rights. As a result, Du Bois was dismissed as a radical by his contemporaries but was rediscovered by a new generation of Black scholars during the 1970s and 1980s. Among his many contributions, Du Bois examined the oppressive effects of race and social class, advocated women's rights, and played a key role in reshaping Black–White relations in America (Du Bois, 1986; Lewis, 1993).

All of these and other early thinkers agreed that people are transformed by each other's actions, social patterns, and historical changes. They and other scholars shaped contemporary sociological theories.

1-5 CONTEMPORARY SOCIOLOGICAL THEORIES

How one defines "contemporary sociological theory" is somewhat arbitrary. The mid-twentieth century is a good starting point because "the late 1950s and 1960s have, in historical hindsight, been regarded as significant years of momentous changes in the social and cultural life of most Western societies" (Adams and Sydie, 2001: 479). Some of the sociological perspectives had earlier origins, but all matured during this period.

Sociologists typically use more than one theory to explain behavior. The theories view our social world somewhat differently, but all of them analyze why society is organized the way it is and why we behave as we do. Four of the most influential theoretical perspectives are functionalism, conflict theory, feminist theories, and symbolic interaction.

W. E. B. Du Bois

Hulton Archive/Hulton Archive/Getty Images

1-5a Functionalism

Functionalism (also known as *structural functionalism*) maintains that society is a complex system of interdependent parts that work together to ensure a society's survival. Much of contemporary functionalism grew out of the work of Auguste Comte and Émile Durkheim, both of whom believed that human behavior is a result of social structures that promote order and integration in society.

One of their contemporaries, English philosopher Herbert Spencer (1820–1903), used an organic analogy to explain the evolution of societies. To survive, Spencer (1862/1901) wrote, our vital organs—like the heart, lungs, kidneys, liver, and so on—must function together. Similarly, the parts of a society, like the parts of a body, work together to maintain the whole structure.

SOCIETY IS A SOCIAL SYSTEM

Prominent American sociologists, especially Talcott Parsons (1902–1979) and Robert K. Merton (1910–2003), developed the earlier ideas of structure and function. For these and other functionalists, a society is a system that is composed of major institutions such as government, religion, the economy, education, medicine, and the family.

Each institution or other social group has *structures*, or organized units, that are connected to each other and within which behavior occurs. Education structures like colleges, for instance, aren't only organized internally in terms of who does what and when, but depend on other structures like government (to provide funding), business (to produce textbooks and construct buildings), and medicine (to ensure that students, staff, and faculty are healthy).

FUNCTIONS AND DYSFUNCTIONS

Each structure fulfills certain *functions*, or purposes and activities, to meet different needs that contribute to a society's stability and survival (Merton, 1938). The purpose of education, for instance, is to transmit knowledge to the young, to teach them to be good citizens, and to prepare them for jobs (see Chapter 13).

Dysfunctions are social patterns that have a negative impact on a group or society. When one part of society isn't working, it affects other parts, generating conflict, divisiveness, and social problems. Consider religion. In the United States, the Catholic Church's stance on issues such as not ordaining women to be priests and denouncing abortion and non-heterosexuality has produced a rift between those who embrace or question papal edicts. In other countries, religious intolerance has led to wars and terrorism (see Chapter 13).

MANIFEST AND LATENT FUNCTIONS

There are two kinds of functions. **Manifest functions** are intended and recognized; they're present and clearly evident. **Latent functions** are unintended and unrecognized; they're present but not immediately obvious.

Consider the marriage ceremony. Its primary manifest function is to publicize the formation of a new family unit and to legitimize sexual intercourse and childbirth (even though both might occur outside of

functionalism (*structural functionalism*) maintains that society is a complex system of interdependent parts that work together to ensure a society's survival.

dysfunctions social patterns that have a negative impact on a group or society.

manifest functions purposes and activities that are intended and recognized; they're present and clearly evident.

latent functions purposes and activities that are unintended and unrecognized; they're present but not immediately obvious.

Sociology and Other Social Sciences: *What's the Difference?*

Mitchell Funk/Photographer's Choice/ Getty Images

How would different social scientists study the same phenomenon, such as homelessness? Criminologists might examine whether crime rates are higher among homeless people than in the general population. Economists might measure the financial impact of programs for the homeless. Political scientists might study whether and how government officials respond to homelessness. Psychologists might be interested in how homelessness affects individuals' emotional and mental health. Social workers are most likely to try to provide needed services (e.g., food, shelter, medical care, and jobs). Sociologists have been most interested in examining homelessness across gender, age, and social class, and explaining how this social problem devastates families and communities.

According to sociologist Herbert Gans (2005), sociologists "study everything." There are currently 43 different subfields in sociology, and the number continues to increase, because sociologists' interests range across many areas.

marriage). The latent functions of a marriage ceremony include communicating a "hands-off" message to suitors, providing the new couple with household goods and products through bridal showers and wedding gifts, and redefining family boundaries to include in-laws or stepfamily members.

CRITICAL EVALUATION

You'll see in later chapters that functionalism is useful in seeing the "big picture" of interrelated structures and functions. Its influence waned during the 1960s and 1970s, however, because functionalism was so focused on order and stability that it often ignored social change. For example, functionalism couldn't explain the many rapid changes sparked by the civil rights, women's, and gay movements.

A second and related criticism is that functionalism often glosses over the widespread inequality that a handful of powerful people create and maintain. Conflict theorists, especially, have pointed out that what's functional for some privileged groups is dysfunctional for many others.

conflict theory examines how and why groups disagree, struggle over power, and compete for scarce resources.

1-5b Conflict Theory

In contrast to functionalism—which emphasizes order, stability, cohesion, and consensus—**conflict theory** examines how and why groups disagree, struggle over power, and compete for scarce resources (like property, wealth, and prestige). Conflict theorists see disagreement and the resulting changes in society as natural, inevitable, and even desirable.

SOURCES OF CONFLICT

The conflict perspective has a long history. As you saw earlier, Karl Marx predicted that conflict would result from widespread economic inequality, and W. E. B. Du Bois criticized U.S. society for its ongoing and divisive racial discrimination. Since the 1960s, many sociologists—especially feminist and minority scholars—have emphasized that the key sources of economic inequity in any society include race, ethnicity, gender, age, and sexual orientation.

Conflict theorists agree with functionalists that many societal arrangements are functional. But, conflict theorists ask, who benefits? And who loses? When corporations merge, workers in lower-end jobs are often laid off while the salaries and benefits of corporate executives soar and the value of stocks (usually held by higher

SOCIOLOGISTS TYPICALLY USE MORE THAN ONE THEORY TO EXPLAIN BEHAVIOR AND WHY SOCIETY IS ORGANIZED THE WAY IT IS.

skynesher/E+/Getty Images

Among other manifest functions, schools transmit knowledge and prepare children for adult economic roles. Among their latent functions, schools provide matchmaking opportunities. What are some other examples of education's manifest and latent functions?

social classes) rise. Thus, mergers might be functional for people at the upper end of the socioeconomic ladder, but dysfunctional for those on the lower rungs.

SOCIAL INEQUALITY

Unlike functionalists, conflict theorists see society not as cooperative and harmonious, but as a system of widespread inequality. For conflict theorists, there's a continuous tension between the haves and the have-nots, most of whom are children, women, minorities, people with low incomes, and the poor.

Many conflict theorists focus on how those in power—typically wealthy White Anglo-Saxon Protestant males (WASPs)—dominate political and economic decision making in U.S. society. This group controls a variety of institutions—like education, criminal justice, and the media—and passes laws that benefit primarily people like themselves (see Chapters 8 and 11).

CRITICAL EVALUATION

Conflict theory explains how societies create and cope with disagreements. However, some have criticized conflict theorists for overemphasizing competition and coercion at the expense of order and stability. Inequality exists and struggles over scarce resources occur, critics agree, but conflict theorists often ignore cooperation and harmony. Voters, for example, can boot dominant White males out of office and replace them with women and minority group members. Critics also point out that the have-nots can increase their power through negotiation, bargaining, lawsuits, and strikes.

"Sometimes the best man for the job isn't."

Author Unknown

1-5c Feminist Theories

You'll recall that influential male theorists generally overlooked or marginalized early female sociologists' contributions. Until the feminist activism of the 1960s and 1970s, men—who dominated universities and scholarship—were largely "blind to the importance of gender" (Kramer and Beutel, 2015: 17).

Feminist scholars agree with contemporary conflict theorists that much of society is characterized by tension and struggle, but **feminist theories** go a step further by focusing on women's social, economic, and political inequality. The theories maintain that women often suffer injustice primarily because of their gender, rather than personal inadequacies like low educational levels or not caring about success. Feminist scholars assert that people should be treated fairly and equally regardless not only of their sex but also of other characteristics such as their race, ethnicity, national origin, age, religion, class, sexual orientation, or disability. They emphasize that women should be freed from traditionally oppressive expectations, constraints, roles, and behavior (see Reger, 2012).

FOCUSING ON GENDER

Feminist scholars have documented women's historical exclusion from most sociological analyses (see, for example, Smith, 1987, and Adams and Sydie, 2001). Before the 1960s women's movement in the United States, very few sociologists published anything about gender roles, women's sexuality, fathers, or intimate partner violence. According to sociologist Myra Ferree (2005: B10), during the 1970s, "the Harvard social-science library could fit all its books on gender inequalities onto a single half-shelf." Because of feminist scholars, many researchers—both women and men—now routinely include gender as an important research variable on both micro and macro levels.

Globally, except for some predominantly Muslim countries, solid majorities of both women and men support gender equality and agree that women should be able to work outside the home. When jobs are scarce, however, many women and men believe

feminist theories examine women's social, economic, and political inequality.

that men should be given preferential treatment ("Gender Equality...," 2010). Thus, even equal rights proponents place a higher priority on men's economic rights.

LISTENING TO MANY VOICES

Feminist scholars contend that gender inequality is central to *all* behavior, ranging from everyday interactions to political and economic institutions, but feminist theories encompass many perspectives. For example, *liberal feminism* endorses social and legal reform to create equal opportunities for women. *Radical feminism* sees male dominance in social institutions (e.g., as the economy and politics) as the major cause of women's inequality. *Global feminism* focuses on how the intersection of gender with race, social class, and colonization has exploited women in the developing world (see Lengermann and Niebrugge-Brantley, 1992). Most of us are feminists because we endorse equal opportunities for women and men in the economy, politics, education, and other institutions.

> "I myself have never been able to find out precisely what feminism is; I only know that people call me a feminist whenever I express sentiments that *differentiate me from a doormat.*"
>
> Rebecca West, British journalist

CRITICAL EVALUATION

Feminist scholars have challenged employment discrimination, particularly practices that routinely exclude women who aren't part of the "old boy network" (Wenneras and Wold, 1997). One criticism, however, is that many feminists are part of an "old girl network" that hasn't always welcomed different points of view from Black, Asian American, Native American, Muslim, Latina, lesbian, working-class, and disabled women (Lynn and Todoroff, 1995; Jackson, 1998; Sánchez, 2013).

A second criticism is that feminist perspectives often overlook gender, social class, and generational gaps. Shortly before the 2016 presidential election, 69 percent of women voters said that Donald Trump, the Republican nominee, didn't respect women. However, 42 percent of those women voted for him anyway (Hartig et al., 2016). At least 90 percent of people vote for their party's candidate, but Trump appealed to many White women, particularly those without a college degree and those living in rural areas (*Table 1.2*).

symbolic interaction theory (*interactionism*) examines people's everyday behavior through the communication of knowledge, ideas, beliefs, and attitudes.

There are many reasons for a presidential candidate's victory, but some observers have attributed Clinton's defeat to many voters'—particularly working-class White women's—frustrations about diminished possibilities for their husbands and sons to provide for their families, fears about downward mobility and poverty, concerns about a growing number of immigrants, and a scarcity of jobs in small towns and rural areas (Featherstone, 2016; Morin, 2016; Roberts and Ely, 2016). In contrast, Trump's slogan to "Make America Great Again" resonated with millions of voters, especially those without college degrees, who feel economically disenfranchised.

During the 2016 presidential race, a large majority of millennials (people born after 1980) supported Bernie Sanders—a 73-year-old senator from Vermont—over Clinton or Trump. Sanders' platform called for the most progressive and drastic changes to the U.S. political and economic structures (e.g., free tuition, changes in energy policies, and greater equality of wealth). Some analysts believe that millennial enthusiasm for Sanders is an example of a "feminist generation gap" that has increased because younger and older feminists have different values, convictions, and goals (Norman, 2016; Rosen, 2016).

Some critics, including feminists, also question whether feminist scholars have lost their bearings by concentrating on personal issues like greater sexual freedom rather than broader social issues, particularly wage inequality (Chesler, 2006; Shteir, 2013; Rosen, 2016).

1-5d Symbolic Interaction

Symbolic interaction theory (sometimes called *interactionism*) is a micro-level perspective that examines people's everyday behavior through the communication

Table 1.2 How Women Voted in the 2016 Presidential Election, by Selected Characteristics			
	PERCENTAGE WHO VOTED FOR...		
	HILLARY CLINTON (DEMOCRAT)	DONALD TRUMP (REPUBLICAN)	OTHER/NO ANSWER
White women	42	53	5
Black women	94	4	2
Latinas	68	26	6
College-educated White women	51	45	4
White women without a college degree	34	62	4
Rural White women	34	62	4

Sources: Based on Huang et al., 2016; Levinson, 2016; Malone, 2016; Mohdin, 2016; Morin, 2016.

of knowledge, ideas, beliefs, and attitudes. Whereas functionalists, conflict theorists, and some feminist theories emphasize structures and large (macro) systems, symbolic interactionists focus on *process* and keep the *person* at the center of their analysis.

There have been many influential symbolic interactionists, whom we'll cover in later chapters. In brief, George Herbert Mead's (1863–1931) assertion that the human mind and self arise in the process of social communication became the foundation of the symbolic interaction schools of thought in sociology and social psychology. Herbert Blumer (1900–1987) coined the term *symbolic interactionism* in 1937, developed Mead's ideas, and proposed that people interpret or "define" each other's actions, especially through symbols, instead of merely reacting to them.

Erving Goffman (1922–1982) enriched these earlier theories by examining human interaction in everyday situations ranging from jobs to funerals. Among his other contributions, Goffman used "dramaturgical analysis" to compare everyday social interaction to a theatrical presentation (see Chapter 5).

CONSTRUCTING MEANING

Our actions are based on **social interaction** in the sense that people take each other into account in their own behavior. Thus, we act differently in different social settings and continuously adjust our behavior, including our body language, as we interact (Goffman, 1959; Blumer, 1969). A woman's interactions with her husband differ from those with her children. And she will interact still differently when she is teaching, talking to a colleague in the hall, or addressing an audience at a professional conference.

For symbolic interactionists, society is *socially constructed* through human interpretation (O'Brien and Kollock, 2001). That is, meanings aren't inherent but are created and modified through interaction with others. For example, a daughter who has batting practice with her dad will probably interpret her father's behavior as loving and involved. In contrast, she'll see batting practice with her baseball coach as less personal and more goal-oriented. In this sense, our interpretations of even the same behavior, such as batting practice, vary across situations and depend on the people with whom we interact.

SYMBOLS AND SHARED MEANINGS

Symbolic interaction looks at subjective, interpersonal meanings and how we interact with and influence each other by communicating through *symbols*—words, gestures, or pictures that stand for something and that can have different meanings for different individuals.

After the 9/11 terrorist attacks, many Americans displayed the flag on buildings, bridges, homes, and cars to show their solidarity and pride in the United States. In contrast, some groups in the Middle East burned the U.S. flag to show their contempt for U.S. culture and policies. Thus, symbols are powerful forms of communication that show how people feel and interpret a situation.

To interact effectively, our symbols must have *shared meanings*, or agreed-on definitions. One

social interaction a process in which people take each other into account in their own behavior.

of the most important of these shared meanings is the *definition of the situation*, or the way we perceive reality and react to it. Relationships often end, for example, because people view emotional closeness differently ("We broke up because my partner wanted more sex. I wanted more communication."). We typically learn our definitions of the situation through interaction with *significant others*—especially parents, friends, relatives, and teachers—who play an important role in our socialization (as you'll see in Chapters 4 and 5).

CRITICAL EVALUATION

Unlike other theorists, symbolic interactionists show how people play an active role in shaping their lives on a micro level. One of the most common criticisms is that symbolic interaction overlooks the widespread impact of macro-level factors (e.g., economic forces, social movements, and public policies) on our everyday behavior and relationships. During economic downturns, for example, unemployment and ensuing financial problems create considerable interpersonal conflict among couples and families (see Chapters 11 and 12). Symbolic interaction rarely considers such macro-level changes in explaining everyday behavior.

A related criticism is that interactionists sometimes have an optimistic and unrealistic view of people's everyday choices. Most of us enjoy little flexibility in our daily lives because deeply embedded social arrangements and practices benefit those in power. For instance, people are usually powerless when corporations transfer jobs overseas or cut the pension funds of retired employees.

Some also believe that interaction theory is flawed because it ignores the irrational and unconscious aspects of human behavior (LaRossa and Reitzes, 1993). People don't always consider the meaning of their actions or behave as reflectively as interactionists assume. Instead, we often act impulsively or say hurtful things without weighing the consequences of our actions or words.

1-5e Other Theoretical Approaches

Table 1.3 summarizes the major sociological perspectives that you've just read about. However, new theoretical perspectives arise because society is always changing. For example, *postmodern theory* analyzes contemporary societies that are characterized by postindustrialization, consumerism, and global communications.

Sociology, like other social sciences, has subfields. The subfields—such as socialization, deviance, and social stratification—offer specific theories that reinforce and illustrate functionalist, conflict, feminist, and interactionist approaches. No single theory explains social life completely. Each theory, however, provides different insights that guide sociological research, the topic of Chapter 2.

Lane Oatey/Blue Jean Images/Getty Images

José Nicolas/Sygma/Getty Images

For many people, a diamond, especially in an engagement ring, signifies love and commitment. For others, diamonds represent Western exploitation of poor people in Africa who are paid next to nothing for their backbreaking labor in mining these stones.

Table 1.3 Leading Contemporary Perspectives in Sociology

THEORETICAL PERSPECTIVE	FUNCTIONALISM	CONFLICT	FEMINIST	SYMBOLIC INTERACTION
Level of Analysis	**Macro**	**Macro**	**Macro and Micro**	**Micro**
Key Points	• Society is composed of interrelated, mutually dependent parts. • Structures and functions maintain a society's or group's stability, cohesion, and continuity. • Dysfunctional activities that threaten a society's or group's survival are controlled or eliminated.	• Life is a continuous struggle between the haves and the have-nots. • People compete for limited resources that are controlled by a small number of powerful groups. • Society is based on inequality in terms of ethnicity, race, social class, and gender.	• Women experience widespread inequality in society because, as a group, they have little power. • Gender, ethnicity, race, age, sexual orientation, and social class—rather than a person's intelligence and ability—explain many of our social interactions and lack of access to resources. • Social change is possible only if we change our institutional structures and our day-to-day interactions.	• People act on the basis of the meaning they attribute to others. Meaning grows out of the social interaction that we have with others. • People continuously reinterpret and reevaluate their knowledge and information in their everyday encounters.
Key Questions	• What holds society together? How does it work? • What is the structure of society? • What functions does society perform? • How do structures and functions contribute to social stability?	• How are resources distributed in a society? • Who benefits when resources are limited? Who loses? • How do those in power protect their privileges? • When does conflict lead to social change?	• Do men and women experience social situations in the same way? • How does our everyday behavior reflect our gender, social class, age, race, ethnicity, sexual orientation, and other factors? • How do macro structures (such as the economy and the political system) shape our opportunities? • How can we change current structures through social activism?	• How does social interaction influence our behavior? • How do social interactions change across situations and between people? • Why does our behavior change because of our beliefs, attitudes, values, and roles? • How is "right" and "wrong" behavior defined, interpreted, reinforced, or discouraged?
Example	• A college education increases one's job opportunities and income.	• Most low-income families can't afford to pay for a college education.	• Gender affects decisions about a major and which college to attend.	• College students succeed or fail based on their degree of academic engagement.

STUDY TOOLS 1

READY TO STUDY? IN THE BOOK, YOU CAN:

- ☐ Check your understanding of what you've read with the Test Your Learning Questions provided on the Chapter Review Card at the back of the book.
- ☐ Tear out the Chapter Review Card for a handy summary of the chapter and key terms.

ONLINE AT CENGAGEBRAIN.COM WITHIN MINDTAP YOU CAN:

- ☐ Explore: Develop your sociological imagination by considering the experiences of others. Make critical decisions and evaluate the data that shape this social experience.
- ☐ Analyze: Critically examine your basic assumptions and compare your views on social phenomena to those of your classmates and other MindTap users. Assess your ability to draw connections between social data and theoretical concepts.
- ☐ Create: Produce a video demonstrating connections between your own life and larger sociological concepts.
- ☐ Collaborate: Join your classmates to create a capstone project.

2 Examining Our Social World

Franck Boston/Shutterstock.com

LEARNING OBJECTIVES

After studying this chapter, you will be able to...

2-1 Compare knowledge based on tradition, authority, and research.

2-2 Explain why sociological research is important in our everyday lives.

2-3 Describe the scientific method.

2-4 Describe the basic steps of the research process.

2-5 Compare and illustrate the five most common sociological data collection methods, including their strengths and limitations.

2-6 Explain why ethics are important in scientific research.

After finishing this chapter go to **PAGE 37** for **STUDY TOOLS**

Spring break is all about beer fests, wet T-shirt contests, frolicking on the beach, and hooking up, right? Maybe not. A national survey found that 70 percent of college students stay home with their parents, and 84 percent of those who throng to vacation spots report consuming alcohol in moderation (The Nielsen Company, 2008). If you suspect that these numbers are too high or too low and wonder how the survey was done, you're thinking like a researcher, the focus of this chapter.

WHAT DO **YOU** THINK?

People can find data to support any opinion they have.

1	2	3	4	5	6	7
strongly agree				strongly disagree		

2-1 HOW DO WE KNOW WHAT WE KNOW?

Much of our knowledge is based on *tradition*, a handing down of statements, beliefs, and customs from generation to generation ("The groom's parents should pay for the wedding rehearsal dinner"). Another common source of knowledge is *authority*, a socially accepted source of information that includes "experts," parents, government officials, police, judges, and religious leaders ("My mom says that . . ." or "According to the American Heart Association . . .").

Knowledge based on tradition and authority simplifies our lives because it provides us with basic rules about socially and legally acceptable behavior. The information can be misleading or wrong, however. Suppose a 2-year-old throws a temper tantrum at a family barbecue. One adult comments, "What that kid needs is a smack on the behind." Someone else immediately disagrees: "All kids go through this stage. Just ignore it."

Who's right? To answer this and other questions, sociologists rely on **research methods**, organized and systematic procedures to gain knowledge about a particular topic. Much research shows, for example, that neither ignoring a problem nor inflicting physical punishment (like spanking) stops a toddler's bad behavior. Instead, most young children's misbehavior can be curbed by having simple rules, being consistent in disciplining misbehavior, praising good behavior, and setting a good example (see Benokraitis, 2015).

2-2 WHY IS SOCIOLOGICAL RESEARCH IMPORTANT IN OUR EVERYDAY LIVES?

In contrast to knowledge based on tradition and authority, sociological research is important in our everyday lives for several reasons:

1. **It counteracts misinformation.** Blatant dishonesty and misinformation spread rapidly through digital communication. In 2015, violent crimes were 77 percent below their 1993 level, but 70 percent of Americans believe that the rate has increased (Gramlich, 2016; Truman and Morgan, 2016). Such unfounded fears—fueled by mass shootings, the media's focus on crime, powerful lobby groups such as the National Rifle Association, and Donald Trump's false claim that "inner city crime is reaching record levels"—are partly responsible for the growth of gun ownership in the past 20 years (Cohn et al., 2013; see also Chapter 7).
2. **It exposes myths.** According to many newspapers and television shows, suicide rates are highest during the Christmas holidays. In fact, suicide rates are lowest in December and highest in the spring and fall (but the reasons for these peaks are unclear). Another myth is that more women are victims of domestic violence on Super Bowl Sunday than on any other day of the year, presumably because men become intoxicated and abusive. In fact, intimate partner violence rates are high, and consistent, throughout the year (Romer, 2011; "Super Bull Sunday," 2015).
3. **It helps explain *why* people behave as they do.** A recent study predicted that older drivers, particularly those age 70 and older, would be more likely than younger drivers to have fatal crashes,

research methods organized and systematic procedures to gain knowledge about a particular topic.

"Facebook Causes 20 Percent of Today's Divorces." What?!

The founder and self-described leader of the United Kingdom's online divorce site (Divorce-Online) sent out a press release titled "Facebook Is Bad for Your Marriage—Research Finds," and claimed that Facebook causes 20 percent of today's divorces. News media around the world ran stories about this press release with headlines such as "Facebook to Blame for Divorce Boom." You'll see in Chapter 12 that there's no divorce boom, so where did the 20 percent number come from?

In 2009, the managing director of Divorce-Online scanned its online divorce petition database for use of the word *Facebook*, and found 989 instances in about 5,000 petitions. Divorce-Online never said that the petitions were only those filed by members of the American Academy of Matrimonial Lawyers, who comprise a very small percentage of all divorce attorneys. Two years later, many Internet sites and blogs were still spreading the fiction that "Facebook Causes Divorce" (see Bialik, 2011). In reality, as you'll see in Chapter 12, there are a number of interrelated macro- and micro-level reasons for divorce; there's no single "cause," much less Facebook.

but found the opposite (Braitman et al., 2011). The researchers couldn't explain why their prediction turned out to be false. Sociologists posit that older Americans are less likely to have fatal crashes because many avoid driving at night or during bad weather, and they're much less likely than younger drivers to use cell phones or text while driving (Halsey, 2010).

4. **It affects social policies.** According to the captain of a large North Carolina Police Department, "Research dictates everything that officers do, whether we realize it or not," in reducing crime. Examples include tracking criminal activity in high-risk locations, implementing research-based policies in training patrol officers and detectives, and managing limited resources more efficiently (Nolette, 2015).
5. **It sharpens critical thinking skills.** Many Americans, particularly women, rely on talk shows for information on a number of topics. Oprah Winfrey has featured and applauded guests who maintained, among other things, that children contract autism from the measles, mumps, and rubella (MMR) vaccinations they receive as babies; that fortune cards can help people diagnose their illnesses; and that people can wish away cancer (Kosova and Wingert, 2009; Clemmons et al., 2015; see also "Clueless," 2014, for other recent examples of "celebrity bogus science").

RTimages/Shutterstock.com

All of these claims are false, and as you'll see in Chapter 14, endanger our health.

A fact-checking website found that, during the 2016 presidential campaign, only 25 percent of Hillary Clinton's and 4 percent of Donald Trump's statements were true (PolitiFact, 2016). *Fake news*—misinformation that deliberately misleads people for financial, political, or other gain—has been around for a long time. About 84 percent of Americans are confident that they can identify false news (Barthel et al., 2016).

Many of us, however, are susceptible to *confirmation bias*, a tendency to embrace and recall information that confirms our beliefs and ignores or downplays contrary evidence. The scientific method, which requires critical thinking skills that you read about in Chapter 1, strengthens our ability to separate fact from fiction, but do people always believe scientific findings?

THE SCIENTIFIC METHOD

Sociologists rely on the **scientific method**, a body of objective and systematic techniques used to investigate phenomena, acquire knowledge, and test hypotheses and theories. The techniques include careful data collection, exact measurement, accurate recording and analysis of the findings, thoughtful interpretation of results, and, when appropriate, generalization of the findings to a larger

scientific method a body of objective and systematic techniques used to investige phenomena, acquire knowledge, and test hypotheses and theories.

group. Before collecting any data, however, social scientists must grapple with a number of research-related issues. Let's begin with concepts, variables, and hypotheses.

2-3a Concepts, Variables, and Hypotheses

A basic element of the scientific method is a **concept**—an abstract idea, mental image, or general notion that represents some aspect of the world. Some examples of concepts are "blood pressure," "religion," and "marriage."

Because concepts are abstract, scientists use variables to measure (*operationalize*) concepts. A **variable** is a characteristic that can change in value or magnitude under different conditions. Variables can be attitudes, behaviors, or traits (e.g., ethnicity, age, and social class).

An **independent variable** is a characteristic that has an effect on the **dependent variable**, the outcome. A **control variable** is a characteristic that is constant and unchanged during the research process.

Scientists can simply ask a research question ("Why are people poor?"), but they usually begin with a **hypothesis**—a statement of the expected relationship between two or more variables—such as "Unemployment increases poverty." In this example, "unemployment" is the independent variable and "poverty" is the dependent variable.

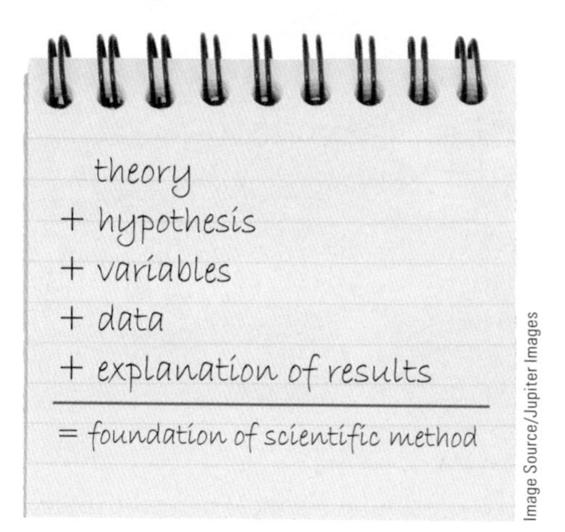

Image Source/Jupiter Images

Figure 2.1 Deductive and Inductive Approaches

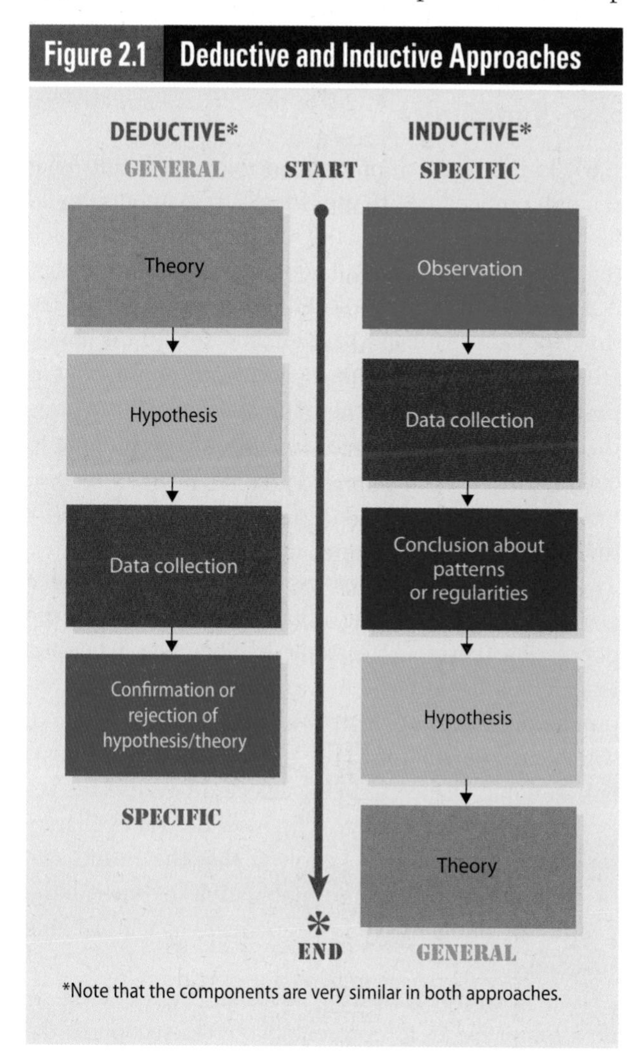

*Note that the components are very similar in both approaches.

Researchers might also use control variables, like education, to explain the relationship between unemployment and poverty. For example, people with at least a college degree generally have lower poverty rates than those with lower educational levels because the former are less likely to experience long periods of unemployment.

2-3b Deductive and Inductive Reasoning

Deduction and induction are two different but equally valuable approaches in examining the relationship between variables. Generally, **deductive reasoning** begins with a theory, prediction, or general principle that is then tested through data collection. An alternative mode of inquiry, **inductive reasoning**, begins with specific observations, followed by data collection, a conclusion about patterns or regularities, and the formulation of hypotheses that can lead to theory construction (*Figure 2.1*).

concept an abstract idea, mental image, or general notion that represents some aspect of the world.

variable a characteristic that can change in value or magnitude under different conditions.

independent variable a characteristic that has an effect on the dependent variable.

dependent variable the outcome that may be affected by the independent variable.

control variable a characteristic that is constant and unchanged during the research process.

hypothesis a statement of the expected relationship between two or more variables.

deductive reasoning begins with a theory, prediction, or general principle that is then tested through data collection.

inductive reasoning begins with a specific observation, followed by data collection, a conclusion about patterns or regularities, and the formulation of hypotheses that can lead to theory construction.

Monica Schipper/WireImage/Getty Images

Are *The Voice* voters an example of a probability or nonprobability sample of the show's fans?

Taking a deductive approach, you might decide to test a theory of academic success using the following hypothesis: "Students who study in groups perform better on exams than those who study alone." You would collect the data, ultimately confirming or rejecting your hypothesis (or theory).

Alternatively, you might notice that your classmates who participate in study groups seem to get higher grades on exams than those who study alone. Using an inductive approach, you would collect the data systematically and formulate hypotheses (or suggest a theory) that could then be tested deductively. Most social science research involves both inductive and deductive reasoning.

reliability the consistency with which the same measure produces similar results time after time.

validity the degree to which a measure is accurate and really measures what it claims to measure.

population any well-defined group of people (or things) that researchers want to know something about.

sample a group of people (or things) that's representative of the population researchers wish to study.

probability sample each person (or thing) has an equal chance of being selected because the selection is random.

nonprobability sample there is little or no attempt to get a representative cross section of a population.

2-3c Reliability and Validity

Sociologists are always concerned about reliability and validity. **Reliability** is the *consistency* with which the same measure produces similar results time after time. If, for example, you ask "How old are you?" on two subsequent days and a respondent gives two different answers, such as 25 and 30, there's either something wrong with the question or the respondent is lying. Respondents might lie, but scientists must make sure that their measures are as reliable as possible.

Validity is the degree to which a measure is accurate and *really* measures what it claims to measure. Consider student course evaluations. The measures of a "good" professor often include items like whether she or he was "interesting," "fair," or "knowledgeable about the course content."

Because we don't know what students mean by "interesting" and "fair," and students don't know how "knowledgeable" an instructor is, how accurate are such measures in differentiating between "good" and "bad" professors? A study at two large public universities found that a third of the students admitted being dishonest in end-of-semester course evaluations. Some fibbed to make their instructors look good, but most lied to "punish" professors they didn't like, especially when they received lower grades than they thought they deserved (Clayson and Haley, 2011). Such research findings raise questions about the accuracy and usefulness of student course evaluations in measuring an instructor's actual performance.

2-3d Sampling

Early in the research process, sociologists decide what sampling procedures to use. Ideally, researchers would like to study all the units of the population in which they're interested—say, all adolescents who use drugs. A **population** is any well-defined group of people (or things) that researchers want to know something about. Obtaining information about and from populations is problematic, however. The population may be so large that it would be too expensive and time consuming to conduct the research. In other cases—such as all adolescents who use drugs—it's impossible even to identify the population.

As a result, researchers typically select a **sample**, a group of people (or things) that's representative of the population they wish to study. In obtaining a sample, researchers decide whether to use probability or nonprobability sampling. A **probability sample** is one in which each person (or thing, such as an email address) has an equal chance of being selected because the selection process is *random*. The most desirable characteristic of a probability sample is that the results can be generalized to the larger population because all the people (or things) have had an equal chance of being selected.

In a **nonprobability sample**, there's little or no attempt to get a representative cross section of the population. Instead, researchers use sampling criteria such as convenience or the availability of respondents or information. Nonprobability samples are especially useful when sociologists are exploring a new topic or want to get people's insights on a particular topic before launching a larger study (Babbie, 2013).

Television news programs, newsmagazines, and entertainment shows often provide a toll-free number, a texting number, or a website and encourage viewers to vote on an issue (such as whether marijuana should be legal in all states). How representative are these voters of the general population? And how many enthusiasts skew the results by voting more than once?

But, you might think, if as many as 100,000 people respond, doesn't such a large number indicate what most people think? No. Because the respondents are self-selected and don't comprise a random sample, they're not representative of a population.

A recent study that looked at data from almost 209,000 people found that drinking one to five cups of coffee a day was associated with lower mortality and lower risks of death from heart and neurological diseases. Although the sample size was huge, it was composed primarily of White medical and health professionals. Thus, the results aren't generalizable to other populations (Ding et al., 2015).

2-3e The Time Dimension

Researchers compare variables in two ways: cross-sectional studies and longitudinal studies. The data can be *longitudinal* (collected at two or more points in time from the same or different samples of respondents) or *cross-sectional* (collected at one point in time). *Figure 2.2* shows a change over time in Americans' attitudes toward same-sex marriage; this is an example of a longitudinal study. If the researchers had collected data at only one point in time (2017, 2000, or 1996), this would have been a cross-sectional study. Cross-sectional studies provide valuable information, but longitudinal studies are especially useful in examining trends in behavior or attitudes; a researcher can compare similar populations across different years or follow a particular group of people over time.

2-3f Qualitative and Quantitative Approaches

In **qualitative research**, sociologists examine and interpret nonnumerical material. In a study of grandfathers who were raising their grandchildren, for example, the researcher tape-recorded in-depth interviews and then analyzed the responses to

Figure 2.2 Acceptance of Same-Sex Marriage Has Increased

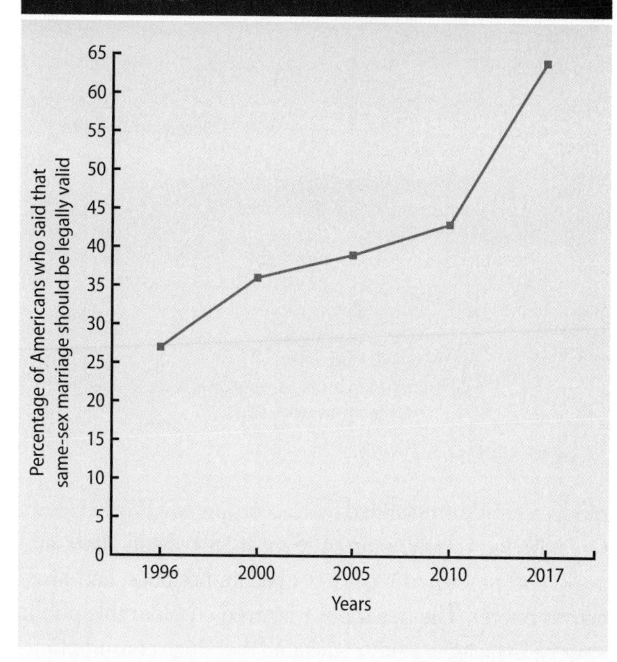

Source: Based on McCarthy, 2017.

questions about financial issues and daily parenting tasks (Bullock, 2005).

In **quantitative research**, sociologists focus on a numerical analysis of people's responses or specific characteristics, studying a wide range of attitudes, behaviors, and traits (e.g., homeowners versus renters). In one national probability study, for example, the researchers surveyed almost 7,000 respondents to understand the influence of grandparents who live with their children and grandchildren (Dunifon and Kowaleski-Jones, 2007).

Which approach should a researcher use? It depends on her or his purpose. Consider college attrition. Quantitative data provide information on characteristics such as national college graduation rates. Qualitative data, in contrast, yield in-depth descriptions of why some college students drop out whereas others graduate. In many studies, sociologists use both approaches.

2-3g Correlation Is Not Causation

Ideally, researchers would like to determine **causation**, a relationship in which one variable is the direct consequence of another. The

qualitative research examines and interprets nonnumerical material.

quantitative research focuses on a numerical analysis of people's responses or specific characteristics.

causation a relationship in which one variable is the direct consequence of another.

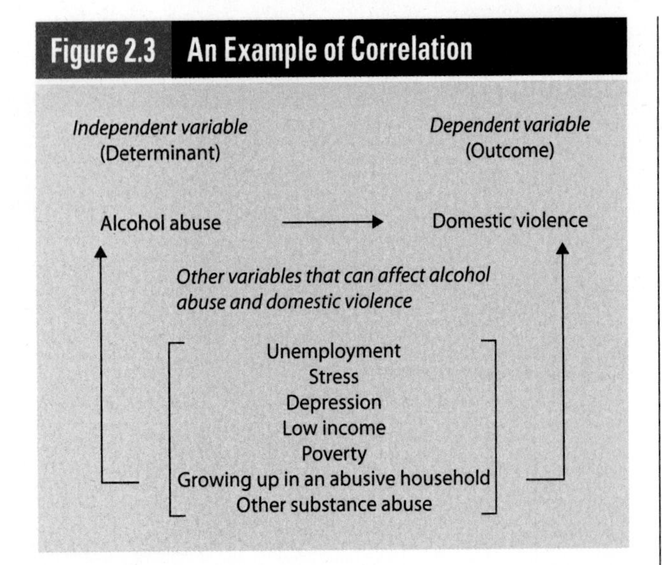

media recently proclaimed that, according to a World Health Organization report, eating red meat (e.g., beef, lamb, and pork) and processed meat (e.g., bacon, hot dogs, and ham) causes cancer. The headlines triggered considerable public anxiety. In fact, the study concluded that "high consumption" of red and processed meat "may be associated with a small risk" of several types of cancer (Bouvard et al., 2015).

Because it's difficult to determine causation, sociologists and other scientists calculate a **correlation**, the relationship between two or more variables. For example, much research shows that there's an association between alcohol abuse and domestic violence, and the more frequent the alcohol abuse, the greater the likelihood of domestic violence (see Chapter 12).

Alcohol abuse and domestic violence often occur together, but this doesn't mean that one *causes* the other. Domestic violence also occurs when people don't drink, and not all people who abuse alcohol become violent or aggressive. Instead, there may be other factors that affect alcohol abuse, domestic violence, or both variables (*Figure 2.3*). Sociologists rarely use the term *cause* because they can't *prove* that there's a cause-and-effect relationship. Instead, a researcher might conclude that alcohol abuse is "associated (or correlated) with," "contributes to," or "increases the likelihood of" rather than "causes" domestic violence.

2-4 BASIC STEPS IN THE RESEARCH PROCESS

Hypotheses construction, deductive and inductive reasoning, establishing reliability and validity, and sampling are some of the preliminary and often most challenging steps in the research process. *Figure 2.4* outlines the scientific method, using a deductive approach that begins with an idea and ends with writing up (and sometimes publishing) the results. Later in this chapter, we'll examine some studies that use an inductive approach.

correlation the relationship between two or more variables.

1. **Choose a topic to study.** The topic can be general or very specific. Some sociologists begin with a new question or idea; others extend or refine previous research findings. A topic can generate new information, replicate a previous study, or propose an intervention (e.g., a new substance abuse program).
2. **Summarize the related research.** In what is often called a *literature review*, a sociologist summarizes the pertinent research, shows how her or his topic

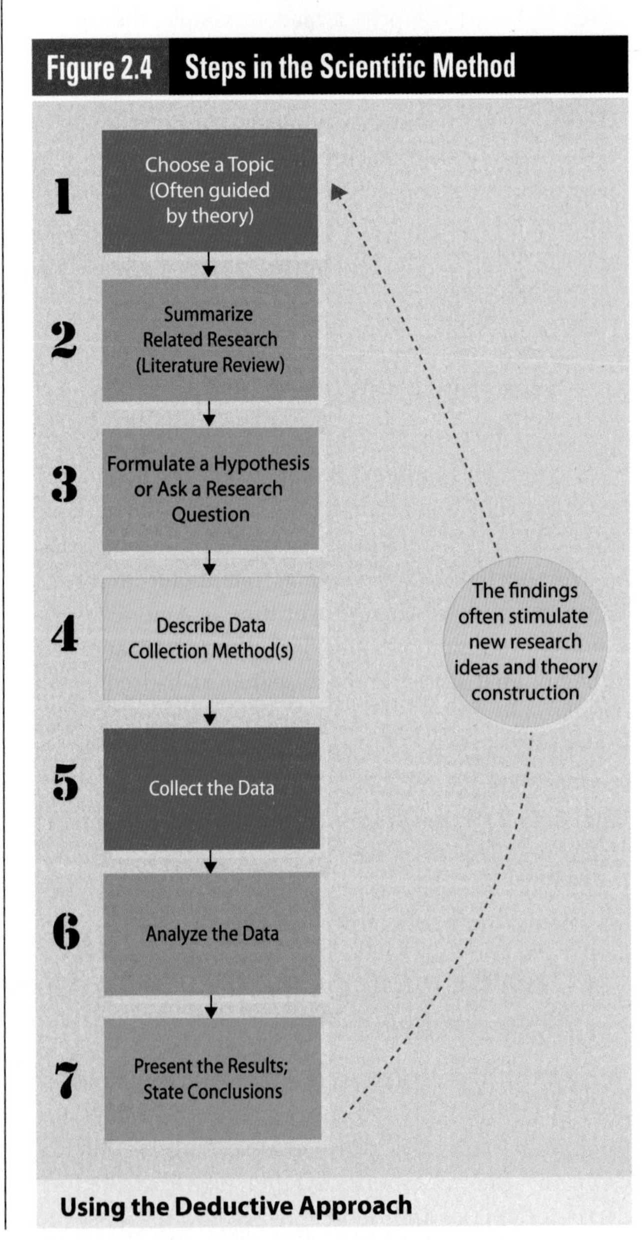

is related to previous and current research, and indicates how the study will extend the body of knowledge. If the research is applied, a sociologist also explains how the proposed service or program will improve people's lives.

3. **Formulate a hypothesis or ask a research question.** A sociologist next states a hypothesis or asks a research question. In either case, she or he has to be sure that the measures of the variables are as reliable and valid as possible.
4. **Describe the data collection method(s).** A sociologist describes which method or combination of methods (sometimes called *methodology, procedure*, or *research design*) is best for testing a hypothesis or answering a research question. This step also describes sampling, the sample size, and the respondents' characteristics.
5. **Collect the data.** The actual data collection might rely on fieldwork, surveys, experiments, or existing sources of information like Census Bureau statistics.
6. **Analyze the data.** After coding (tabulating the results) and running statistical tests, a sociologist presents the findings as clearly as possible.
7. **Present and explain the results.** After analyzing the data, a sociologist explains why the findings are important. This can be done in many ways. She or he might show how the results provide new information, enrich our understanding of behavior or attitudes that researchers have examined previously, or refine existing theories or research approaches.

In drawing conclusions about the study, sociologists typically discuss its implications. For instance, does a study of juvenile arrests suggest that new policies should be implemented, that existing ones should be changed, or that current police practices may be affecting the arrest rates? That is, the researcher answers the question "So what?" by showing the importance and usefulness of the study.

2-5 SOME MAJOR DATA COLLECTION METHODS

Sociologists typically use one or more of the following major data collection methods: surveys, field research, content analysis, experiments, and secondary analysis of existing data. Because each method has strengths and weaknesses, researchers must decide which will provide the most accurate information, given time and budget constraints.

"How long have you been dead? Do you have any complaints about your treatment here? Do you have any suggestions?"

Farris, Joseph/Cartoonstock.com

2-5a Surveys

Many sociologists use **surveys** that include questionnaires, face-to-face or telephone interviews, or a combination of these techniques. Two important elements in survey research are sampling and constructing a series of questions for *respondents*, the people who answer the questions.

SELECTING A SAMPLE

Random sample surveys are preferred because the results can be generalized to a larger population. Researchers can obtain representative samples through *random digit dialing*, which involves selecting area codes and exchanges (the next three numbers) followed by four random digits. In the procedure called *computer-assisted telephone interviewing* (CATI), the interviewer uses a computer to select random telephone numbers, reads the questions to the respondent from a computer screen, and then enters the answers in precoded spaces, saving time and expense by not having to reenter the data after the interview.

QUESTIONNAIRES AND INTERVIEWS

A survey must have a specific plan for asking questions and recording answers. The most common way to do this is to give respondents a **questionnaire**, a series of written questions that ask for

survey a data collection method that includes questionnaires, face-to-face or telephone interviews, or a combination.

questionnaire a series of written questions that ask for information.

information. The questions can be *closed-ended* (the researcher provides a list of answers that a respondent chooses), *open-ended* (the researcher asks respondents to answer questions in their own words) (*Table 2.1*), or a combination. Questionnaires can be mailed, used during an interview, or self-administered (e.g., student course evaluations, Web surveys).

The **interview**, in which a researcher directly asks respondents a series of questions, is another way to collect survey data. Interviews can be conducted face-to-face or by telephone. *Structured interviews* use closed-ended questions, whereas *unstructured interviews* use open-ended questions that allow respondents to answer as they wish.

Table 2.1 One Question–Two Formats

Open-Ended Version		
How would you describe your current financial situation?		
Closed-Ended Version		
How would you describe your current financial situation?		
[] Excellent	[] Fair	[] Terrible
[] Very good	[] Poor	[] Not sure
[] Good		

Suppose that you're the researcher on an income study. Do you think that you'd use the open- or closed-ended version? Why?

Now, suppose that you're the respondent. Do you think that you'd be more likely to answer the open- or closed-ended version? Why?

ONLINE SURVEYS

Two types of online surveys are becoming increasingly popular. One is sent via email, either as text in the body of the message or as an attachment. The second type is the more familiar survey that's posted on a website. Like the more traditional surveys that use self-administered questionnaires, telephones, and face-to-face interviews, online surveys have strengths and weaknesses.

STRENGTHS

Surveys are usually inexpensive, simple to administer, and have a fast turnaround. Self-administered questionnaires can be mailed to thousands of respondents. Having a large number of cases is important in describing, analyzing, and explaining many variables simultaneously.

Telephone interviews are popular because they're a relatively cost-effective way to collect data. Face-to-face interviews have high response rates (often up to 85 percent) because they involve personal contact. People are more likely to discuss sensitive issues (e.g., drug use, sexual behavior) in an interview than via a mailed questionnaire, a phone survey, or an electronic survey. If respondents don't understand a question, are reluctant to answer, or give incomplete answers, interviewers can clarify, keep respondents from digressing, or *probe* by asking respondents to elaborate on an answer (Babbie, 2013).

interview a researcher directly asks respondents a series of questions.

With the innovation of "robo-polls," the entire interview is conducted by a programmed recording that interprets the respondent's spoken answers, records them, and determines how to continue the interview. This method cuts out the cost of hiring people, but it's unclear whether respondents are more reluctant or more willing to express socially unacceptable views to a machine than to a person (Barrow, 2015).

Online surveys, like mail surveys, are relatively inexpensive to administer, and the responses can be tabulated quickly. Online surveys are particularly appropriate for certain targeted groups (e.g., people who access health sites), and can increase the pool of people who are increasingly difficult to reach by phone. Online surveys can also provide respondents with visual material, including videos, to look at and respond to (Keeter, 2010; DeSilver, 2014).

LIMITATIONS

Unlike questionnaires and telephone surveys, face-to-face interviews can be very expensive. A major limitation of surveys that use mailed questionnaires is a low response rate, often only about 10 percent (Higgins, 2015). If the questions are unclear, complicated, or seen as offensive, respondents may simply throw the questionnaire away or give answers on subjects they know nothing about (Babbie, 2013).

A survey's wording can also substantially affect people's answers. According to a recent Gallup poll, for instance, 51 percent of Americans agreed that doctors should be allowed to "assist the patient to commit suicide," but a whopping 70 percent said that doctors should be allowed to "end the patient's life by some painless means." Thus, not using the word *suicide* was much more acceptable even though the patient's outcome was the same (Saad, 2013).

Another concern is *social desirability bias*—the tendency of respondents to give the answer that they think they "should" give or that will cast them in a favorable light. For example, the proportion of Americans who say they voted in a given election is always much higher than the actual number of votes cast (Taylor and Lopez, 2013).

People also underreport (or lie about) socially undesirable behaviors and traits (e.g., using illicit drugs or mental health problems), but exaggerate attendance at religious services, donating to charity, and washing their hands after using a restroom. In a recent survey, 28 percent of the respondents admitted that they had lied to their doctors about smoking, drinking alcohol, using illicit drugs, not exercising, having unsafe sex with multiple partners, and other risky behavior (Reddy, 2013; Bharadwaj et al., 2015; Keeter et al., 2015).

Because most online surveys are based on nonprobability samples, they underrepresent groups such as Black individuals, Latinos, and older, poorer, and less-educated people who don't use the Internet. During the 2016 election, polls underestimated Donald Trump's level of support because, among other reasons, less-educated voters—who comprised a strong pro-Trump segment of the population—refused to respond to online surveys (Kennedy et al., 2016; Mercer et al., 2016).

Because the survey is the research approach you'll encounter most often in sociology and many other disciplines, it's important to be an informed consumer. You can't simply assume that a survey, including a public opinion poll, is accurate or representative of a larger population. Asking a few basic questions about a survey, such as those in *Table 2.2*, will help you assess its credibility.

Table 2.2 Can I Trust These Numbers?

To determine a survey's credibility, ask:
• Who sponsored the survey? A government agency, a political organization, a business, or a group that's lobbying for change?
• What's the purpose of the survey? To provide objective information, to promote an idea or a political candidate, or to get attention through sensationalism?
• How was the sample drawn? Randomly? Or was it a "self-selected opinion poll" (a SLOP, according to Tanur, 1994)?
• How were the questions worded? Were they clear and objective, or biased? If the survey questions aren't provided, why not?
• How did the researchers report their findings? Were they objective, or did they make value judgments?

2-5b Field Research

In **field research**, sociologists collect data by systematically observing people in their natural surroundings. In *participant observation*, researchers interact with the people they're studying; they may or may not reveal their identities as researchers. If you recorded interaction patterns between students and professors during your classes, you would be engaging in participant observation. In *nonparticipant observation*, researchers study phenomena without being part of the situation (observing young children in classrooms through one-way mirrors). Researchers sometimes combine participant and nonparticipant observation.

Some field research studies are short term (e.g., observing, over a few weeks or months, whether and how parents discipline their unruly children in grocery stores). Others, called *ethnographies*, require a considerable amount of time in the field. Sudhir Venkatesh (2008), while a graduate student at the University of Chicago, spent more than 6 years studying the culture and members of the Black Kings, a crack-selling gang in Chicago's inner city.

Observational studies are usually highly structured and carefully designed projects in which data are recorded and then converted to quantitative summaries. These studies may examine complex communication patterns, measure the frequency of acts (e.g., the number of head nods or angry statements), or note the duration of a particular behavior (e.g., the length of eye contact) (Stillars, 1991). Thus, observational studies are much more complex and sophisticated than they appear to the general public.

STRENGTHS

Unlike other data collection methods, field research provides rich detail in describing and understanding attitudes and behavior in the "real world." The researcher can explore new topics from the respondents' viewpoint using observation or in-depth interviews. Because observation usually doesn't disrupt the natural surroundings, the researcher doesn't directly influence the subjects. Field research is also more flexible than some other methods. The researcher can

field research data collected by observing people in their natural surroundings.

Jon Feingersh/AGE Fotostock

Field researchers study a variety of topics, including how customers, cocktail waitresses, bouncers, and paid consultants help people hook up at a nightclub (see Grazian, 2008, for a description of this study).

modify the research design, for example, by deciding to interview (rather than just observe) key people after the research has started.

Validity is usually a strength of field research because researchers can determine how participants define and interpret concepts such as "success," "inequality," or "marital infidelity." Compared with other data collection methods, field research is also more flexible if a situation changes. Studying crime in poor neighborhoods, for example, may require shifting the focus of the analysis if the key participants are killed or imprisoned (see Goffman, 2014). Finally, field research is a powerful data collection method because it describes a phenomenon using the subjects' own words.

LIMITATIONS

Observation can be expensive if a researcher needs elaborate recording equipment, must travel far or often, or has to live in a different society or community for an extended period. Researchers who study other cultures must often learn a new language, a time-consuming task.

A second limitation is the problem of *reactivity*: If people know they're being observed, they might behave differently. A study of 392 restaurants from five different chains (e.g., the Olive Garden) found that employee theft decreased by 22 percent after management installed surveillance cameras (Pierce et al., 2013). In the case of participant observation, anything that the researcher does or doesn't do can affect what's being observed because the researcher's presence changes a group's dynamics, behavior, and outcomes (see Chapter 6 on the Hawthorne effect).

A field researcher may encounter other data collection challenges. Homeless and battered women's shelters, for example, are usually—and understandably—wary of researchers intruding on their residents' privacy. Even if the researcher has access to such groups, it's often difficult to be objective while collecting and interpreting the data because poverty and domestic violence can evoke strong emotional reactions (e.g., anxiety, anger against perpetrators, sympathy for victims). Finally, the findings can't be generalized because the data come from nonprobability samples.

AP Images/Mel Evans

Counting the homeless is an ongoing research problem. Communities must provide accurate numbers to qualify for federal funds, but census takers have difficulty getting such counts because safety rules prohibit them from entering private property (such as warehouses) or dark alleys. In one approach, the YWCA in Trenton, New Jersey, gave free haircuts at a fair for the homeless to attract the city's homeless people to a place where they could be counted (Jonsson, 2007). Is such field research unethical because it violates people's privacy? Or innovative in helping a community get more federal funding for the homeless?

2-5c Content Analysis

Content analysis systematically examines some form of communication. Researchers can apply this unobtrusive data collection method to almost any form of written or oral communication: speeches, TV programs, online blogs, advertisements, office emails, songs, tweets, advice columns, poems, or Facebook chatter, to name just a few.

The researcher develops categories for coding the material, sorts and analyzes the data in terms of frequency, intensity, or other characteristics, and draws conclusions about the results. Look at the two baby cards in *Figure 2.5*. What messages do they send about gender roles? These are only two cards, but we could do a content analysis of all birth congratulations at a particular store or from a number of stores or online sites to determine whether such cards reinforce stereotypical gender role expectations.

Sociologists have used content analysis to examine a number of topics. A few examples are images of women and men in video games and music videos, changes in child-rearing advice in popular parenting magazines, gender and ethnic differences in yearbook photographs, and gender biases in job advertisements (Downs and Smith, 2010; Zhang et al., 2010; Wallis, 2011; Lindner, 2012; Wondergem and Friedlmeier, 2012).

Figure 2.5 How Do Views of Female and Male Babies Differ?

As these cards illustrate, girls and boys are viewed differently from the time they're born. In these and other baby cards, girls, but not boys, are typically described as "sweet" or "precious." Also, the images usually show boys as active but girls as passive.

STRENGTHS

Content analysis is usually inexpensive and often less time consuming than other data collection methods, particularly field research. If, for example, you wanted to examine the content of television commercials during football games, you wouldn't need fancy equipment, a travel budget, or a research staff.

A second advantage is that researchers can recode errors fairly easily. This isn't the case with surveys. If you mail a questionnaire with poorly constructed items, it's too late to change anything.

Third, content analysis is unobtrusive. Because researchers aren't interacting with human subjects, they don't need permission to do the research, and don't have to worry about influencing the respondents' attitudes or behavior.

Finally, researchers can use content analysis in both cross-sectional and longitudinal studies. In one study, the researchers analyzed gender portrayals in the 101 top-grossing G-rated films in the United States and Canada between 1990 and 2005 (Smith et al., 2010; see also Joshi et al., 2011). It would be very difficult, using most other data collection methods, to analyze such material over a 15-year period.

LIMITATIONS

Content analysis can be very labor intensive, especially if a project is ambitious. In one study, the researchers examined the amount and intensity of violence in children's animated movies that were released between 1938 and 1999 (Yokota and Thompson, 2000). It took several years to code one or more of the major characters' words, expressions, and actions.

A second disadvantage is that the coding may be subjective. Having several researchers on a project can increase coding objectivity, but only one researcher often codes the content.

A third limitation is that content analysis often reflects social class bias. Because most books, articles, speeches, films, and so forth are produced by people in upper socioeconomic levels, content analysis rarely captures the behavior or attitudes of working-class people and the poor. Even when documents created by lower-class individuals or groups are available, it's difficult to determine whether the coding reflects a researcher's social class prejudices.

Finally, content analyses can't always tell us *why* people behave as they do. We'd have to turn to studies that use different data

content analysis a data collection method that systematically examines some form of communication.

An Experiment That Went Awry

In a well-known 1971 experiment, psychologist Philip Zimbardo created a mock prison, using 24 undergraduate volunteers as prisoners and guards. He stopped the experiment after 6 days because some of the participants experienced intense negative reactions. For more information, see Stanford Prison Experiment, prisonexp.org. We'll examine this study, including the ethical issues it raised, in more detail in Chapter 6.

Gary Friedman/Los Angeles Times/Getty Images

collection methods—like surveys and field research—to understand why, for example, people buy stereotypically feminine or masculine baby cards or produce video games that portray men, but not women, as competitive and aggressive (see Lindner, 2012).

2-5d Experiments

An **experiment** is a controlled artificial situation that allows researchers to manipulate variables and measure the effects. The classic experimental design includes two equal-sized groups that are similar on characteristics such as gender age, ethnicity, race, and education.

Suppose our hypothesis is "Watching a film on race discrimination reduces prejudice." In the **experimental group**, the participants are exposed to the independent variable (they watch the film on race discrimination). In the **control group**, participants don't watch the film because they're not exposed to the independent variable. Before the experiment, we'll measure the dependent variable (prejudice) in both groups using a *pretest* (a prejudice scale). After the experimental group watches the film, we'll measure both groups again using a *posttest* (the same prejudice scale). If we find a difference in the scores of the dependent variable (prejudice), we might conclude that the independent variable affects the dependent variable (e.g., watching the film on race discrimination reduced or increased prejudice). *Figure 2.6* illustrates this basic experimental design.

experiment a controlled artificial situation that allows researchers to manipulate variables and measure the effects.

experimental group the participants who are exposed to the independent variable.

control group the participants who aren't exposed to the independent variable.

STRENGTHS

Experiments come closer than other data collection methods in suggesting a possible cause-and-effect relationship. They're also usually less expensive and time consuming than other data collection techniques (especially large surveys and multiyear field research), and there's often no need to purchase special equipment. A second and related strength is that because researchers recruit students or other volunteers (as in medical studies), participants are usually readily available and don't expect much, if any, monetary compensation.

A third advantage is that experiments can be *replicated* (repeated) many times with different participants. Replication strengthens the researchers' confidence in the validity and reliability of the measures and the study's results. For example, doctors stopped prescribing hormone pills for menopausal women because better-designed experiments that replicated earlier studies found that the pills increased the risk for breast cancer and strokes (Ioannidis, 2005).

LIMITATIONS

One disadvantage of laboratory experiments is their reliance on student volunteers or paid respondents. Students often feel obligated to participate as part of their grade, or they may fear antagonizing an instructor who's conducting a study. Participants might also give the answers that they think the researcher expects (the social desirability bias described earlier). In the case of paid subjects, those who are the busiest, don't need the cash, move, or get sick may not participate fully or drop out of the study.

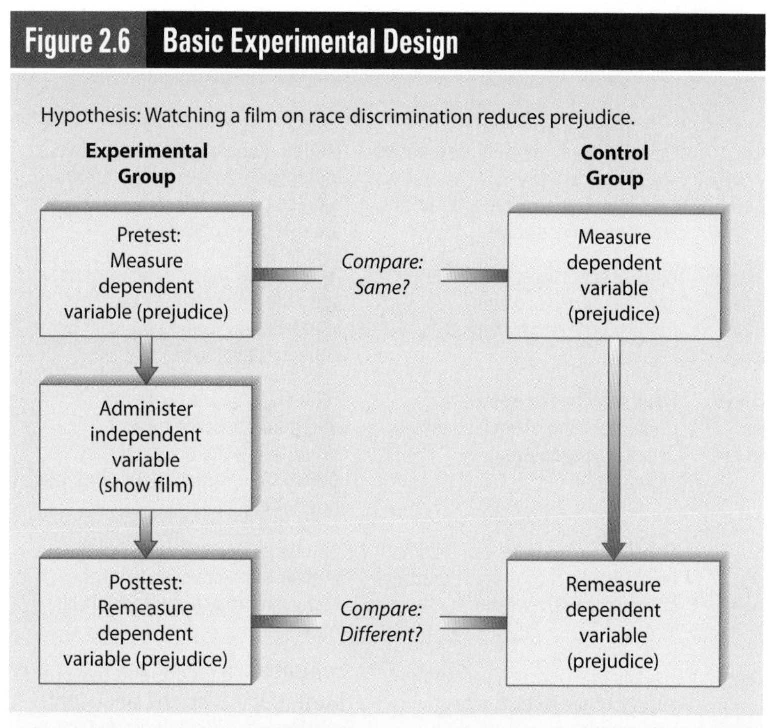

Figure 2.6 Basic Experimental Design

Source: Based on Babbie, 2013, Figure 5.1.

Another limitation is that experiments, especially those conducted in laboratories, are artificial. People *know* that they're being observed and may behave very differently than they would in a natural setting (Gray et al., 2007; see also Chapter 6 on the Hawthorne effect).

Third, the results of experimental studies can't be generalized to a larger population because they come from small, nonrepresentative, or self-selected samples. For example, college students who participate in experiments aren't necessarily representative of other college students, much less of people who aren't in college.

Fourth, experiments aren't suitable for studying large groups of people, a major focus in sociology. As you saw in Chapter 1, many sociologists examine macro-level issues (like employment trends) that can't be observed in a laboratory or other experimental setting.

Finally, even if a researcher finds an association between variables (e.g., watching a film reduces prejudice), it doesn't mean that the former "causes" the latter. The control group's responses in the posttest may have been influenced by other factors, including communicating with the members of the experimental group (Cook and Campbell, 1979) or living in a racially mixed neighborhood, which tends to diminish racial and ethnic prejudice (see Chapter 10). That is, experiments have few control variables that might be explaining the relationship between the independent and dependent variables.

Because of such limitations, consumers are often confused and frustrated by conflicting advice. Recently, different experiments have concluded that caffeine can relieve *or* lead to chronic headaches, inhibit *or* improve memory, increase *or* reduce depression, and that some high-fat foods are healthier than low-fat foods (Bartlett, 2011; Mozzafarian and Ludwig, 2015).

2-5e Secondary Analysis

Sociologists rely heavily on **secondary analysis**, an unobtrusive data collection method that examines information collected by someone else. The data may be historical materials (e.g., court proceedings), personal documents (e.g., letters and diaries), public records (e.g., state archives on births, marriages, and deaths), or official statistics (e.g., U.S. Bureau of Labor reports). Researchers with slightly different interests could analyze or reanalyze the data or examine only specific variables. For instance, the federal government compiles information on a large number of variables—including population, age, and education—that social scientists, marketing analysts, health practitioners, and other researchers analyze for their own purposes.

STRENGTHS

Secondary analysis is usually accessible, convenient, and inexpensive. A staggering amount of data on topics like income and employment—for the United States and worldwide—are readily available at libraries, colleges, and online.

donatas1205/Shutterstock.com

Second, there's a huge savings in time. Because many of the data sets are stored electronically, researchers can spend most of their time analyzing, rather than collecting, the data.

A third advantage of secondary analysis is the data's high quality. The federal government and nationally known survey organizations have budgets and well-trained staff equipped to address any data collection problems. Because the samples are representative of national populations, researchers can be more confident about generalizing the findings.

Fourth, data sets often contain hundreds of variables. Thus, researchers can access national, international, and historical

secondary analysis examination of data that have been collected by someone else.

Table 2.3 Some Data Collection Methods in Sociological Research

Method	Example	Advantages	Disadvantages
Surveys	Sending questionnaires and/or interviewing students on why they succeeded in college or dropped out	Questionnaires are fairly inexpensive and simple to administer; interviews have high response rates; findings are often generalizable	Mailed questionnaires may have low response rates; respondents tend to be self-selected; interviews are usually expensive
Secondary analysis	Using data from the National Center for Education Statistics (or similar organizations) to examine why students drop out of college	Usually accessible, convenient, and inexpensive; often longitudinal and historical	Information may be incomplete; some documents may be inaccessible; some data can't be collected over time
Field research	Observing classroom participation and other activities of first-year college students with high and low grade-point averages (GPAs)	Flexible; offers deeper understanding of social behavior; usually inexpensive	Difficult to quantify and to maintain observer/subject boundaries; the observer may be biased or judgmental; findings are not generalizable
Content analysis	Comparing the transcripts of college graduates and dropouts on variables such as gender, race/ethnicity, and social class	Usually inexpensive; can recode errors easily; unobtrusive; permits comparisons over time	Can be labor intensive; coding is often subjective (and may be distorted); may reflect social class bias
Experiments	Providing tutors to some students with low GPAs to find out if such resources increase college graduation rates	Usually inexpensive; plentiful supply of subjects; can be replicated	Subjects aren't representative of a larger population; the laboratory setting is artificial; findings can't be generalized

information that they themselves can't collect (Boslough, 2007). Finally, because many of the data sets are longitudinal, researchers can look at trends and changes over time.

LIMITATIONS

Secondary analysis has several drawbacks. First, it may be challenging to access historical materials because the documents are deteriorating, housed in only a few libraries in the country, or part of private collections. It may also be difficult to determine the accuracy and authenticity of historical data.

Second, existing sources may not have the exact data a researcher needs. In 2000, the Census Bureau changed how it counted race and ethnicity to include the growing numbers of recent immigrants and mixed-race Americans. Doing so increased the accuracy of gauging racial and ethnic diversity, but created 63 possible racial-ethnic combinations. As a result, today's numbers are different from those before 2000, making comparisons on race and ethnicity across time problematic (Saulny, 2011; see also Pratt et al., 2015).

A third and related limitation is that even the same source may modify its measures of variables because of legal and social changes. Regarding marital and family status, for example, the Census Bureau added "separated" only in 1950, "unmarried partner" in 1990, and started categorizing same-sex married couples as families only in 2013 (Bates and DeMaio, 2013; Morello, 2014). Such changes are necessary. They can be frustrating, however, if, for instance, you wanted to compare the number of separated Americans between 1940 and 2015 or to determine the number of same-sex married couples since 2004, when Massachusetts became the first state to legalize same-sex marriage.

evaluation research examines whether a social intervention has produced the intended result.

Finally, because the researcher didn't collect the data, she or he has no control over the specific variables used or how they were measured, and may not know how seriously problems such as low response rates, validity, and reliability affect the data. Most federal and other established data sets provide such information, but researchers can't simply assume that all data sets are problem-free.

Table 2.3 summarizes the data collection methods described in this chapter. Often, researchers use a combination of methods because "the social world is a multifaceted and multi-layered reality that reveals itself only in part with any single method" (Jacobs, 2005: 4). You should also be familiar with evaluation research, an increasingly important research tool.

2-5f Evaluation Research: Pulling It All Together

Evaluation research examines whether a social intervention has produced the intended result. *Social* *interventions* are typically programs and strategies that seek to prevent or change negative outcomes (e.g., teenage pregnancy, substance abuse).

Evaluation research "refers to a research purpose rather than a specific method" (Babbie, 2013: 192). Thus, sociologists use a variety of data collection methods, such as secondary analysis and content analysis to examine an organization's records, surveys to gauge employee and client satisfaction, and interviews with the staff and program recipients for in-depth information on an organization's processes.

Evaluation research is *applied* because it's intended to have some real-world effect (Weiss, 1998). Examples include *needs assessment studies* (e.g., to improve or initiate a new service like an after-school program), *cost-benefit studies* (e.g., to determine whether funding teen abstinence programs reduces pregnancies), and *monitoring studies* (e.g., to examine the effect of new policing policies on crime rates). The findings are generally used to improve the efficiency and effectiveness of a particular policy or group.

STRENGTHS

Evaluation research is important for several reasons. First, because local and state governments have been cutting their budgets, social service agencies often rely on evaluation research to streamline their programs, and to achieve the best results at the lowest possible cost (Kettner et al., 1999).

Second, evaluation research is versatile because it includes qualitative and quantitative approaches. It can address almost any topic such as the effectiveness of driver education programs, foreign aid policies, job training programs, and premarital counseling, to name just a few. The research costs can also be low if, for example, researchers can use secondary analysis rather than collect new data.

Third, evaluation research addresses real-life problems that confront many families and communities. Because state and local resources are shrinking and there are hundreds of intervention strategies, evaluation research can be invaluable to program directors or agency heads in deciding which programs to keep, improve, or cut (Peterson et al., 1994).

LIMITATIONS

Evaluations can usually address only one or a few of the many factors that affect behavior. For instance, adolescent substance abuse is not only due to individual high-risk decisions, but also to multiple interrelated variables that include peers, school policies, and cultural values (Bandy, 2012).

Another limitation is that the social context affects evaluation research because of politics, vested interests, and conflicts of interest. Groups that fund the evaluation may pressure researchers to present only positive findings. Also, the results may not be well received if they contradict deeply held beliefs, challenge politicians' pet projects, or conflict with official points of view (Olson, 2010; Babbie, 2013).

Kristopher Skinner/ZUMA Press/Newscom

DARE programs have been popular with schools, police departments, parents, and politicians across the country even though evaluation research has found that the curriculum doesn't reduce students' drug use.

Consider the popular DARE (Drug Abuse Resistance Education) program, introduced in 1983, to teach elementary school children about the dangers of drug use. Social scientists evaluated DARE in 1994, 1998, and 1999, and found no significant difference in the incidence of drug use among students who had and hadn't completed the DARE curriculum. When DARE funding was threatened, its promoters dismissed the research as "voodoo science" (Miller, 2001).

2-6 ETHICS AND SOCIAL RESEARCH

Researchers today operate under much stricter guidelines than they did in the past. In conducting their research, what ethical and political dilemmas do sociologists (and other social scientists) encounter?

2-6a What Is Ethical Research?

Any agency (such as a university or hospital) wishing to receive federal research support must establish an institutional review board (IRB)—a panel of faculty (and sometimes others) who review all proposals involving human subjects. The chief responsibility of an IRB is to ensure that the subjects' rights will be protected (Office for Human Research Protections, 2016).

Because so much research relies on human participants, the federal government, IRBs, and many professional organizations have formulated codes of ethics to protect participants. Regardless of the discipline or the research methods used, all ethical standards have at least three golden rules:

- First, *do no harm* by causing participants physical, psychological, or emotional pain.
- Second, the researcher must get the participant's *informed consent* to be in a study. This includes the participant's knowing what the study is about and how the results will be used.
- Third, researchers must always protect a participant's *confidentiality* even if the participant has broken a law that she or he discloses to the researcher. *Table 2.4* lists some of the basic ethical principles in sociological research.

2-6b Scientific Misconduct

Ethics violations are most common when the research is funded by for-profit companies and organizations. Medical researchers, particularly, have been accused of considerable scientific dishonesty. Some of the alleged violations have included the following: changing research results to please the corporation (usually tobacco or pharmaceutical companies) that sponsored the research; being paid by companies to deliver speeches to health practitioners that endorse specific drugs, even if the medications don't reduce health problems; allowing drug manufacturers to ghostwrite articles that are published in prestigious medical journals (and even draft textbooks); and falsifying data. A study of more than 2,000 articles that were published in some of the most prestigious medical journals found that almost 68 percent were withdrawn because of misconduct and fraud rather than errors (Project on Government Oversight, 2010; Fang et al., 2012; McNeill, 2012).

In the social sciences, some data collection methods are more prone to ethical breaches than others. Surveys, secondary analysis, and content analysis are less vulnerable than field research and experiments because the researchers typically don't interact directly with or affect participants. In contrast, experiments and field research can raise ethical questions because of deception or influencing the participants' feelings, attitudes, or behavior.

Table 2.4 Some Basic Principles of Ethical Sociological Research

In its most recent update, the American Sociological Association (1999) has reinforced its previous ethical codes and guidelines for researchers. Researchers must:
• Obtain all subjects' consent to participate in the research and their permission to quote from their responses and comments.
• Not exploit subjects or research assistants involved in the research for personal gain.
• Never harm, humiliate, abuse, or coerce the participants in their studies, either physically or psychologically.
• Honor all guarantees to participants of privacy, anonymity, and confidentiality.
• Use the highest methodological standards and be as accurate as possible.
• Describe the limitations and shortcomings of the research in their published reports.
• Identify the sponsors who funded the research.
• Acknowledge the contributions of research assistants (usually underpaid and overworked graduate students) who participate in the research project.

2-6c External Pressures on Researchers

Social science research is likely to be challenged when it focuses on sensitive or controversial issues. Federal agencies have avoided funding nearly all research on gun violence partly in response to powerful lobbyists, like the National Rifle Association, that pressure Congress to prevent any studies that deal with firearms (Kelderman, 2016).

Some religious groups and school administrators have successfully derailed research on adolescent sexual behavior because it undermines "traditional family values" and might make a school district look bad (e.g., if a study reports a high incidence of drug use or sexual activity) (Kempner et al., 2005).

In the first days of the Trump administration in 2017, the White House "stifled" federal scientists who were working on projects (e.g., pollution, food safety, climate) that some of the most powerful corporations oppose (Jackson, 2017; Mann, 2017). The CEO of a powerful energy company wanted the University of Oklahoma to dismiss scientists who found a possible link between fracking for oil and gas and earthquakes, and offered to sit on a search committee to fill the vacancies. The CEO is like many other politicians and corporate leaders "who embrace science when it's convenient but work to undermine, discredit, and inhibit science when it conflicts with their narrow self-interests" (Ganem, 2016: 25).

2-6d Do People Believe Scientific Findings?

Not always. For example, 98 percent of scientists, compared with 33 percent of U.S. adults, say that humans

Rzucamokiem/Shutterstock.com

Wong Sze Yuen/Shutterstock.com

Fisher-Price and other toy companies claim that, based on their research, their iPhone and iPad apps can teach infants as young as 9 months old to read and count (Singer, 2013). In fact, studies have found that such products do more harm than good because infants who watch the apps learn fewer words than those whose parents talk or read to them (see Chapter 12).

and other living things have evolved over time. And although there has been no scientific evidence of paranormal events, 60 percent of Americans believe that people have seen or been in the presence of ghosts (Speigel, 2013; Masci, 2017).

Why do many people reject scientific findings? One reason may be that when social media publicizes unscientific polls and falsehoods that people already believe, they're unlikely to change their minds about an issue, even after being presented with scientific data (Taub and Nyhan, 2017).

Second, research findings often challenge personal attitudes and beliefs that people cherish. For example, a number of conservative Christians believe that global warming is a myth because only God, not humans, can affect the climate (Sheppard, 2011; Smith and Leiserowitz, 2013).

Third, an explosion of information can create confusion, especially when online searches oversimplify complex phenomena or present misleading or unreliable content. Average users "don't know how to work through all the information [and misinformation] that's thrown at them on a daily basis" (Dreid, 2016: A24).

Finally, many scientists, including social scientists, spend little or no time engaging in policy debates or explaining their research to the public and journalists (Basken, 2016). As a result, "lobbyists and activists can promote their ideological agendas . . . over those of good science and public policy" (Lynas, 2015).

STUDY TOOLS 2

READY TO STUDY? IN THE BOOK, YOU CAN:

- ☐ Check your understanding of what you've read with the Test Your Learning Questions provided on the Chapter Review Card at the back of the book.
- ☐ Tear out the Chapter Review Card for a handy summary of the chapter and key terms.

ONLINE AT CENGAGEBRAIN.COM WITHIN MINDTAP YOU CAN:

- ☐ Explore: Develop your sociological imagination by considering the experiences of others. Make critical decisions and evaluate the data that shape this social experience.
- ☐ Analyze: Critically examine your basic assumptions and compare your views on social phenomena to those of your classmates and other MindTap users. Assess your ability to draw connections between social data and theoretical concepts.
- ☐ Create: Produce a video demonstrating connections between your own life and larger sociological concepts.
- ☐ Collaborate: Join your classmates to create a capstone project.

3 Culture

Bruce Yuanyue Bi/Lonely Planet Images/Getty Images

LEARNING OBJECTIVES

After studying this chapter, you will be able to...

- 3-1 Describe and illustrate a culture's characteristics.
- 3-2 Explain the significance of symbols, language, values, and norms.
- 3-3 Discuss and illustrate cultural similarities.
- 3-4 Discuss and illustrate cultural variations.
- 3-5 Differentiate between high culture and popular culture.
- 3-6 Explain how and why technology affects cultural change.
- 3-7 Compare and evaluate the theoretical explanations of culture.

After finishing this chapter go to **PAGE 59** for **STUDY TOOLS**

Once when I returned a set of exams, a student who was unhappy with his grade blurted out an obscenity. A voice from the back of the classroom snapped, "You ain't got no culture, man!" Is it true that people who use vulgar language "ain't got no culture"?

WHAT DO **YOU** THINK?

I shape my culture; my culture doesn't shape me.

1	2	3	4	5	6	7
strongly agree						strongly disagree

3-1 CULTURE AND SOCIETY

As popularly used, *culture* often means an appreciation of the finer things in life, such as Shakespeare's sonnets, the opera, and using civil language. In contrast, sociologists use the term in a much broader sense: **Culture** refers to the learned and shared behaviors, beliefs, attitudes, values, and material objects that characterize a particular group or society. Thus, culture shapes a people's total way of life.

Most human behavior isn't random or haphazard. Among other things, culture influences what you eat; how you were raised and will raise your own children; if, when, and whom you will marry; how you make and spend money; and what you read. Even people who pride themselves on their individualism conform to most cultural rules. The next time you're in class, for example, count how many students are *not* wearing jeans, T-shirts, sweatshirts, or sneakers—clothes that are the prevalent uniform of adolescents, college students, and many adults.

A **society** is a group of people who share a culture and a defined territory. Society and culture go hand in hand; neither can exist without the other. Because of this interdependence, social scientists sometimes use the terms *culture* and *society* interchangeably.

3-1a Some Characteristics of Culture

All societies, despite their diversity, share some cultural characteristics and functions (Murdock, 1940; see also Aguilar and Stokes, 1995). We don't see culture directly, but it shapes our attitudes and behaviors.

1. **Culture is learned.** Culture isn't innate but learned, and affects how we think, feel, and behave. If a child is born in one region of the world but raised in another, she or he will learn the customs, attitudes, and beliefs of the host culture.
2. **Culture is transmitted from one generation to the next.** We learn many customs, habits, and attitudes informally through parents, relatives, friends, and the media. We also learn culture formally in such settings as schools, workplaces, and community organizations. Whether our learning is formal or informal, culture is cumulative because each generation transmits cultural information to the next one.
3. **Culture is shared.** Culture brings members of a society together. We have a sense of belonging because we share similar beliefs, values, and attitudes about what's right and wrong. Imagine the chaos if we did what we wanted (as physically assaulting an annoying neighbor) or if we couldn't make numerous daily assumptions about other people's behavior (as stopping at red traffic lights).
4. **Culture is adaptive and always changing.** Culture changes over time. New generations discard technological aspects of culture that are no longer practical, such as replacing typewriters with personal computers, and landlines with cell phones. Attitudes can also change over time. Compared with several generations ago, for example, many Americans now say that nonmarital sex is acceptable (*Figure 3.1*).

Culture shapes who we are, but remember that it's people who create culture. As a result, culture changes as people adapt to their surroundings. Since the attacks on September 11, 2001, most people worldwide, including Americans, don't complain at airports when we have to pass through a metal detector, when our baggage is X-rayed or searched, when items are confiscated, and when

culture the learned and shared behaviors, beliefs, attitudes, values, and material objects that characterize a particular group or society.

society a group of people who share a culture and defined territory.

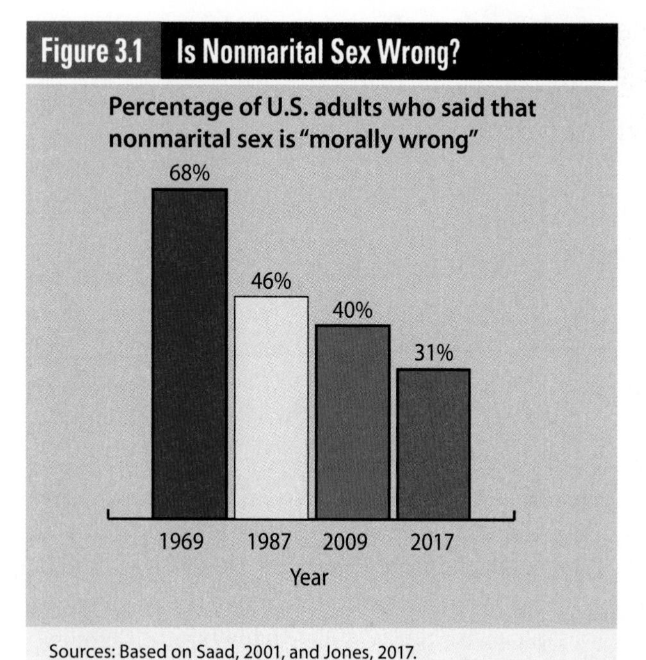

Sources: Based on Saad, 2001, and Jones, 2017.

airports use full-body scanners. Thus, as the dangers in society have grown, people have passed laws or implemented rules that have increased their security but decreased their privacy.

3-1b Material and Nonmaterial Culture

Cultures that people construct are both material and nonmaterial (Ogburn, 1922). **Material culture** consists of the physical objects that people make, use, and share. These objects include buildings, furniture, music, weapons, jewelry, hairstyles, and the Internet. **Nonmaterial culture** consists of the ideas that people create to interpret and understand the world. Beliefs about the supernatural, customs, and rules of behavior are examples of nonmaterial culture.

Because material and nonmaterial culture are created by people, physical objects and ideas can vary considerably across societies (e.g., the farming tools people create, how they form governments, or raise their children). Frequently, as you'll see shortly, there's a "cultural lag" because material culture (invention of the automobile and the cell phone) changes faster than nonmaterial culture (laws about pollution and texting while driving).

material culture the physical objects that people make, use, and share.

nonmaterial culture the ideas that people create to interpret and understand the world.

symbol anything that stands for something else and has a particular meaning for people who share a culture.

World History Archive/Alamy Stock Photo

iStock.com/Duckycards

Communication symbols vary across societies and change over time. The earliest known map of the world (700–500 B.C.E.) used cuneiform script; modern tablets are loaded with icons that provide a variety of information, including maps.

3-2 THE BUILDING BLOCKS OF CULTURE

Recently, 74-year-old James Davis granted his dying wife's wish to be buried in their front yard. The Alabama Supreme Court agreed with a lower court judge that family members can't bury people anywhere they like. Davis refused to budge because the town doesn't have an ordinance governing burials, and his family members and neighbors supported his decision. After four years of court hearings and appeals, a city crew exhumed the body (Reeves, 2013).

Who's right—Davis, his family and neighbors, or the judges? To answer such questions, we must understand the building blocks of culture, particularly symbols, language, values, norms, and rituals.

3-2a Symbols

A **symbol** is anything that stands for something else and has a particular meaning for people who share a culture. In most societies, for example, a handshake communicates friendship or courtesy, a wedding ring signals that a person isn't a potential dating partner, and a siren denotes an emergency. People influence each other through the use of symbols. A smile and a frown communicate different information and elicit different responses. Through symbols, we engage in *symbolic interaction* (see Chapter 1).

SYMBOLS TAKE MANY FORMS

Written words are the most common symbols, but we also communicate by tattooing our bodies or not, celebrating or disregarding wedding anniversaries, and purchasing or avoiding goods and services that indicate our social status. Gestures (e.g., raised fists, hugs, and stares) convey important messages about people's feelings and attitudes.

The U.S. Flag Code describes the flag as a "living thing" and forbids its display on clothing, bedding, drapery, or in advertising (Luckey, 2008). Is this man disrespecting a cherished symbol? Or showing his patriotism?

Tim Macpherson/Stone/Getty Images

SYMBOLS DISTINGUISH ONE CULTURE FROM ANOTHER

Every nation's flag, including its colors and emblems, is an important symbol of national identity. A third of the world's flags have religious symbols (Theodorou, 2014). Countries usually create or change their flags after becoming independent to create and reinforce national consciousness and pride.

In Islamic societies, many women wear a head scarf (*hijab* or *hejab*) that covers their hair and neck, a veil that hides part of the face from just below the eyes (*nijab*), or a garment that covers the entire body (*chador* or *burqa*). Whatever form it takes, the veil is just a piece of cloth, but it means different things to people across and within cultures.

SYMBOLS CAN UNIFY OR DIVIDE A SOCIETY

Symbols unify people. About 75 percent of Americans say that they display the nation's flag in their home, workplace, car, or clothing ("75%—A Nation of Flag Wavers," 2011). Every Fourth of July, many Americans celebrate independence from Britain with parades, firecrackers, barbecues, and speeches by local and national politicians. Because such behavior signifies freedom and democracy, some immigrants purposely choose July 4th to be naturalized.

Naturalization ceremonies themselves teem with symbolic objects and behavior—flags, national anthems, oaths and pledges of allegiance, uniformed military members, and photographs of past immigrants. Such images and rituals "work together to elicit an emotional sense of belonging, pride, and patriotism" that reinforce a sense of community (Aptekar, 2016: 49).

Symbols can also be divisive. Recently, South Carolina's governor, with the legislature's approval, decided to stop flying the Confederate flag on statehouse grounds because it represents slavery and white domination of African Americans. Protesters denounced the flag's removal because they see it as a proud emblem of their Southern heritage.

SYMBOLS CAN CHANGE OVER TIME

Societies create new symbols all the time. We now have recycling symbols on many everyday items, restroom symbols that include transgender people, red ribbons that communicate support for people with HIV and AIDS, ever-changing emoticons that express a person's feelings or mood, and hundreds of text and online chat abbreviations (e.g., afaik, "as far as I know") that vary across groups depending on the user's age, occupation, and other characteristics.

Symbols communicate information that varies across societies and may change over time. In 1986, the International Red Cross changed its name to the International Red Cross and the Red Crescent Movement to encompass a number of Arab branches, and adopted a crescent emblem in addition to the well-known cross. Israel wanted to use a red Star of David, rejecting the cross as a Christian symbol and the crescent as Islamic.

Left: The two emblems of the International Movement of the Red Cross and the Red Crescent. Right: The proposed red diamond emblem.

AP Images/Keystone/Salvatore Di Nolfi; Fabrice Coffrini/AFP/Getty Images

Red Cross officials offered a red diamond as the new shape, but some countries rejected the diamond because it represents bloody conflicts in many African countries that mine diamonds (Whitelaw, 2000). The red cross and crescent continue to be the organization's emblems because the issue hasn't been resolved.

The meaning of even simple symbols varies across groups and can change over time. The # is commonly called the pound and number sign (#2 pencil). For musicians, # denotes a note that's sharp. For proofreaders, # means "insert a space." We're used to seeing # on telephone pads, but since the ascent of social media, # is a hashtag used to mark keywords or topics in a tweet (#culture).

Source: Hallmark Cards, Inc.

Every year, Hallmark produces "Keepsake" Christmas ornaments and apparel, many of which are collectibles. In 2013, one of the new pieces paid tribute to "Deck the Halls," a traditional Christmas carol. Hallmark changed the lyrics from "Don we now our gay apparel" to "Don we now our fun apparel" because the meaning of gay has changed since the song's publication in the 1800s (Wong, 2013). Because of some negative reactions, Hallmark apologized for the change (Murphy, 2013). Why do you think one word generated so much controversy?

3-2b Language

Perhaps the most powerful of all human symbols is **language**, a system of shared symbols that enables people to communicate with one another. Language is a human invention that communities of people have endowed with meaning. In every society, children begin to grasp the essential structure of their language at a very early age and without any instruction. Babbling rapidly leads to uttering words and combinations of words that represent an idea, feeling, or physical object.

Anthropologists Edward Sapir (1929) and Benjamin Whorf (1956) theorized that thoughts and behavior are determined by language. Their theory, known as the *Sapir-Whorf hypothesis,* posited that language provides people with a framework for interpreting social reality and the world around them. Thus, as languages vary, so do interpretations of social reality.

Subsequent research found little support for the Sapir-Whorf hypothesis. Some critics maintained that language affects but does not "determine" culture. Others argued that language influences some thinking processes, but not all (e.g., Berlin and Kay, 1969; Heider and Olivier, 1972; Schlesinger, 1991). Still, the Sapir-Whorf hypothesis has generated considerable research across several disciplines, including sociology, in exploring the relationship between language, thought, and culture.

language a system of shared symbols that enables people to communicate with one another.

WHY LANGUAGE IS IMPORTANT

Language makes us human: It conveys our ideas, transmits information, and influences people's attitudes and behavior. Language also directs our thinking, controls our actions, and shapes our expression of emotions. The meaning of words change over time, but some changes are more controversial than others. Both birds and humans now tweet. Spam is a canned meat product, but also an unsolicited electronic message. Hallmark generated considerable controversy, however, when it changed "gay" to "fun" on a line of Christmas decorations and accessories, including a sweater (see photo).

In a recent survey of 14 developed countries, including the United States, large majorities of people said that speaking the country's language is very important in being "one of us." Language was far more important in being "part of the nation" than birthplace, religious affiliation, education, clothing preferences, and celebrating the host country's holidays (Stokes, 2017). For many nations, then, language is the cornerstone of acceptance into a group.

LANGUAGE AND GENDER

Many professors routinely use phrases such as "Okay, guys, in class today..." and no one objects. One of my colleagues illustrates the linkage between language and gender, and how it affects our thinking, by saying "Okay, gals, in class today...." "I always get a reaction of gaping mouths, laughs, and bewildered looks," he says. The students react differently to "guys" and "gals" because we've internalized male terms (e.g., *guys*, *policeman*, and *maintenance man*) as normal and acceptable.

Language has a profound influence on how we think about and act toward people. Those who adhere to traditional language contend that nouns such as *fireman*, *chairman*, *mailman*, and *mankind* and pronouns like *he* refer to both women and men, and that women who object to such

Wanted: Someone Who Can Translate Federal Gibberish into Plain English

It's almost impossible to find a U.S. government document that people can understand. In 2010, President Obama signed the Plain Writing Act, which requires federal documents to be rewritten in plain English. Here's a "before" and "after" example from the Department of Health and Human Services Handbook that assures health care for low-income HIV/AIDS patients:

BEFORE: Title I of the CARE Act creates a program of formula and supplemental competitive grants to help metropolitan areas with 2,000 or more reported AIDS cases meet emergency care needs of low-income HIV patients. Title II of the Ryan White Act provides formula grants to States and territories for operation of HIV service consortia in the localities most affected by the epidemic, provision of home and community-based care, continuation of insurance coverage for persons with HIV infection, and treatments that prolong life and prevent serious deterioration of health. Up to 10 percent of the funds for this program can be used to support Special Projects of National Significance.

AFTER: Low income people living with HIV/AIDS gain, literally, years, through the advanced drug treatments and ongoing care supported by HRSA's Ryan White Comprehensive AIDS Resources Emergency (CARE) Act.

Language is supposed to help people communicate. However, the Plain Writing Act isn't enforced, and many federal agencies, including the Internal Revenue Service, are exempted. Does a government have more control if people don't understand its rules and regulations?

Sources: Based on Health Resources and Services Administration (2010), and "Government Resolves...," (2011).

usage are too sensitive. Suppose, however, that all of your professors used only *she, her*, and *women* when referring to all people. Would the men in the class feel excluded? One solution is to use sentences with plural, gender-neutral pronouns (e.g., *they, them*) and gender-neutral nouns (e.g., *firefighter, police officer*, and *mail carrier*).

LANGUAGE, RACE, AND ETHNICITY

Words—written and spoken—create and reinforce both positive and negative images about race and ethnicity. Someone might receive a *black mark*, and a *white lie* isn't "really" a lie. We *blackball* someone, *blacken* someone's reputation, denounce *blackguards* (villains), and prosecute people in the *black market.* In contrast, a *white knight* rescues people in distress, the *white hope* brings glory to a group, and the good guys wear *white hats.*

Racist or ethnic slurs, labels, and stereotypes demean and stigmatize people. Derogatory ethnic words abound: *honky, hebe, kike, spic, chink, jap, polack, wetback,* and many others. Self-ascribed racial epithets are as harmful as those imposed by outsiders. When Italians refer to themselves as *dagos* or African Americans call each other *nigger*, they tacitly accept stereotypes about themselves and legitimize the general usage of such derogatory labels (Attinasi, 1994).

Through language, children learn about their cultural heritage and develop a sense of group identity. Spanish is the fastest-growing and most spoken non-English language in the United States today. Some demographers predict, however, that about 34 percent of Latinos will speak only English at home by 2020, up from 25 percent in 2010 (Lopez and Gonzalez-Barrera, 2013). In many Latino and other immigrant families, the native language begins to disappear after just one generation. Assimilation has positive effects, but can also lead to communication problems between Americanized children and their non-English-speaking neighbors and kin in the homeland (Tobar, 2009; see also Chapter 10).

LANGUAGE AND SOCIAL CHANGE

Language is dynamic and changes over time. The most recent edition of the New American Bible, an annual best seller, has changed the word "booty," which now has a sexual connotation, to "spoils of war." "Cereal," which many think of as breakfast food, is now "grain" to reference loads of wheat (*Huffington Post*, 2011).

Jiri Hera/Shutterstock.com; Ispace/Shutterstock.com

U.S. English is composed of hundreds of

Table 3.1 U.S. English Is a Mixed Salad

Here are a few examples of English words borrowed from other languages, showing our mixture of heritages. What words would you add to the list?

☐ Africa: apartheid, Kwanzaa, safari	☐ Spain: anchovy, bizarre
☐ Alaska and Siberia: husky, igloo, kayak	☐ Thailand: Siamese
☐ Bangladesh: bungalow, dinghy	☐ Turkey: baklava, caviar, kebob
☐ Hungary: coach, goulash, paprika	☐ France: bacon, police, ballet
☐ India: bandanna, cheetah, shampoo	☐ Japan: geisha, judo, sushi
☐ Iran and Afghanistan: bazaar, caravan, tiger	☐ Norway: iceberg, rig, walrus
☐ Israel: kosher, rabbi, Sabbath	☐ Mexico: avocado, chocolate, coyote
☐ Italy: fresco, spaghetti, piano	☐ Germany: strudel, vitamin, sauerkraut, kindergarten

thousands of words borrowed from many countries and groups that were in the Americas before the colonists arrived (Carney, 1997). *Table 3.1* provides a few examples of English words borrowed from other languages.

In response to cultural and technological changes, our vocabulary now includes *sexting*, *ebook*, *tweet*, *staycation*, *facetime*, *unfriend*, and *selfie*, among other new words. And to some people's dismay and others' delight, some writers are now substituting *partner* for the traditional *spouse*, *wife*, or *husband*.

3-2c Values

Values are the standards by which people define what is good or bad, moral or immoral, proper or improper, desirable or undesirable, beautiful or ugly. These widely shared standards provide *general guidelines* for everyday behavior rather than concrete rules for specific situations. Honesty, for example, is an important but abstract U.S. value. Thus, faculty typically define in course syllabi what they mean by cheating, such as using material from the Internet or other sources without proper acknowledgment, signing an attendance sheet for another student, or getting help on an assignment.

MAJOR U.S. VALUES

Sociologist Robin Williams (1970: 452–500) has identified a number of core U.S. values. All are central to the American way of life: They're widespread, have endured over time, and indicate what we believe is important in life.

values the standards by which people define what is good or bad, moral or immoral, proper or improper, desirable or undesirable, beautiful or ugly.

1. **Achievement and success.** U.S. culture stresses personal achievement, especially occupational success. Many Americans believe that if they work hard, apply themselves, and save their money, they'll be successful in the future.
2. **Activity and Work.** Americans want to "make things happen." They respect people who are focused and disciplined, and assume that hard work will be rewarded. Journalists and others often praise those who work past their retirement age.
3. **Humanitarianism.** U.S. society emphasizes concern for others, helpfulness, kindness, and support. During natural disasters—like earthquakes, floods, fires, and tornadoes—at home and abroad, many Americans are enormously generous (see Chapter 5).
4. **Efficiency and practicality.** Americans emphasize technological innovation, up-to-dateness, practicality, and getting things done. Many American colleges now tout their programs or courses as being practical instead of emphasizing knowledge and intellectual growth.
5. **Progress.** Americans focus on the future rather than the present or the past. The next time you walk down the aisle of a grocery store, note how many products are "new," "improved," and "better than ever."
6. **Material possessions.** Americans "have more stuff now than any society in history" (even if they can't afford the products). When the stuff accumulates, we buy stuff to stuff it into. Consequently, an entire industry, self-storage, has emerged to house our extra belongings (Sanburn, 2015: 46).
7. **Freedom and equality.** Countless U.S. documents affirm freedom of speech, freedom of the press, and freedom of worship. U.S. laws also tell Americans that they're "created equal" and have the same legal rights regardless of race, ethnicity, gender, religion, disability, age, or social class.
8. **Conformity.** Most people don't want to be labeled as "strange," "peculiar," or "different." They conform because they want to be accepted, liked, hired, and promoted. As a result, we "toe the line," "go along to get along," and try to "fit in."
9. **Democracy.** Democracy promises the average citizen equal political rights. It emphasizes equality, freedom,

and faith in the people rather than giving power to a monarch, dictator, or emperor.

10. **Individualism.** U.S. culture values each person's development. Thus, we often encourage children to be independent, creative, and self-motivated.

VALUES SOMETIMES CONFLICT

Williams acknowledged that these core American values are sometimes contradictory and even hypocritical. For example,

- We believe in religious freedom, but 82 percent of Americans say that it's important that Christians be able to worship freely, compared with 61 percent who say the same for Muslims (Zoll and Swanson, 2015).
- We cherish freedom of speech, but protesters physically attacked political candidates' supporters during the 2016 presidential campaign. At many colleges, campus speakers have been shouted down or disinvited because a small number of vocal students "demand the right to be free from speech that they find offensive or upsetting" (Stone, 2016: B10; see also Kueppers, 2016).
- We prize individualism, but reward people who conform to a group's expectations (see Chapter 6).
- We encourage responsibility, but often blame popular culture rather than parenting for children's bad behavior (see Chapter 12).
- We rail against waste, rising food prices, and the growing number of landfills, but throw away up to 40 percent of food every year (Gunders, 2012).

VALUES VARY ACROSS CULTURES AND CHANGE OVER TIME

Cultural values can change because of technological advances, immigration, and contact with outsiders. For example, the Japanese parliament passed a law making love of country a compulsory part of school curricula. The lawmakers and their numerous supporters hope that teaching patriotism will counteract the American-style emphasis on individualism and self-expression that they believe has undermined Japanese values of cooperation, self-discipline, responsibility, and respect for others (Wallace, 2006).

In the United States, surveys of first-year college students show a shift in values. Between 1968 and 2013, for example, "developing a meaningful philosophy of life" plummeted in importance whereas "being very well off financially" surged (Eagan et al., 2014). Consistent with our humanitarian values, however, a sizeable majority (63 percent) of the students said that "helping others who are in difficulty" is very important, but women were more likely to feel this way (*Figure 3.2*).

iStock.com/David Henderson

Every year, Americans throw out more food than plastic, paper, metal, and glass. The French government fines large supermarkets up to $82,000 for throwing away food that's considered edible (Ferdman, 2015).

The share of U.S. adults ages 25 and older who have never been married is at an historic high—20 percent in 2016 compared with only 9 percent in 1960. You'll see in Chapter 12 that marriage rates vary by social class, among other factors, but values also affect marital status. For example, 50 percent of Americans, another historic high, say society is just as well off if people have priorities other than marriage and children (Wang and Parker, 2014; U.S. Census Bureau News, 2016).

3-2d Norms

Values are general standards, whereas **norms** are a society's specific rules of

norms specific rules of right and wrong behavior.

Figure 3.2 What Really Matters in Life?

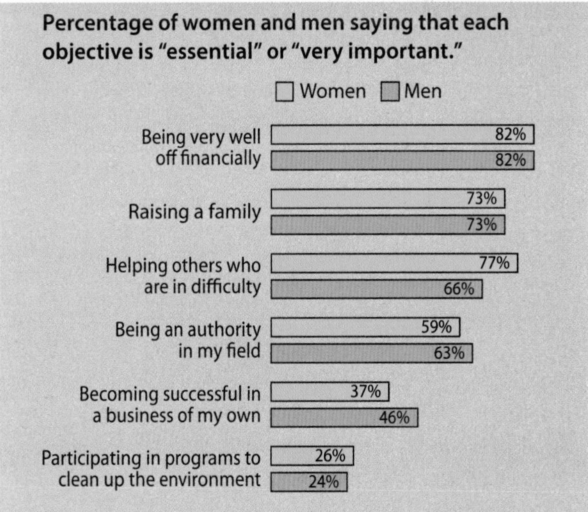

These objectives of first-year college students reflect U.S. values. How are women and men similar? How do they differ? How might you explain the differences?

Source: Based on Eagan et al., 2014: 66, 92.

right and wrong behavior. Norms tell us what we *should*, *ought*, and *must* do, as well as what we *should not, ought not*, and *must not* do: don't talk in church, stand in line, and so on. We may not like many of the rules, but they make our everyday lives more orderly and predictable. Here are some general characteristics of norms:

- Most are *unwritten,* passed down orally from generation to generation (use the good dishes and tablecloth on special occasions).
- They're *instrumental* because they serve a specific purpose (get rid of garbage because it attracts roaches and rats).
- Some are *explicit* (save your money "for a rainy day"), whereas others are *implicit* (be respectful at a wake or funeral).
- They *change* over time (it's now more acceptable to have a child out of wedlock but much less acceptable to smoke or drink alcohol while pregnant).
- Most are *conditional* because they apply in specific situations (slipping out of your smelly shoes may be fine at home, but not on an airplane).
- Because they're situational, norms can be *rigid* ("You *must* turn in your term paper at the beginning of class on May 5") or *flexible* ("Turn in your term papers by the end of the semester").

Norms regulate our behavior, but can differ because of characteristics such as age, social class, race, and gender. For example, a female professor's students sometimes comment on her clothes in their course evaluations and want her to "dress like a professor." In contrast, her husband, also a professor, "has yet to hear a single student comment about his wardrobe." So she changes her businesslike outfits every day while her husband usually wears khaki pants (or jeans) and a polo shirt day in and day out (Johnston, 2005). Thus, "her" and "his" norms differ.

folkways norms that involve everyday customs, practices, and interaction.

Sociologists differentiate three types of norms—folkways, mores, and laws—that vary because some rules are more important than others. As a result, a society punishes some wrongdoers more severely than others.

FOLKWAYS

Folkways are norms that involve everyday customs, practices, and interaction. Etiquette rules are good examples of folkways: Cover your mouth when you cough or sneeze, say "please" and "thank you," knock before entering someone's office, and don't pick your ear wax or toenails in public.

We automatically practice numerous folkways because we've internalized them since birth. We often don't realize that we conform to folkways until someone violates one, such as picking one's nose or popping pimples in public, or talking loudly on a cell phone.

Folkways vary from one country to another. Punctuality is more important in Western countries than in parts of Latin America and the Mediterranean. Tipping is practically nonexistent in Australia, China, and Japan, but expected in the United States. Austrians, Germans, and the Swiss consider chewing gum in public vulgar. There are also many differences in table manners: Europeans keep their hands above the table at all times, not in their laps; "cleaning your plate" may be insulting in the Philippines, North Africa, and some regions of China because an empty plate suggests that the host didn't serve enough; and in some countries, like India, it's customary to eat with one's hands instead of using knives and forks (Becker, 2015; Louie, 2015; eDiplomat, 2016).

Yonhap/AFP/Getty Images/Newscom

South Korea's national media criticized Microsoft founder Bill Gates for his "rude" handshake when meeting President Park Geun-Hye in 2013. As one person tweeted, "How can he put his hand in his pocket when meeting a leader of a state?!" A one-hand shake is disrespectful in South Korea and other parts of Asia, and is usually reserved for someone younger or a good friend (Agence France-Presse, 2013; Evans, 2013). Which types of norms did Gates violate—folkways, mores, or laws?

Folkways can change in response to macro-level changes. The growth of technology has altered many campus folkways. In the past, a syllabus was typically a few pages long, outlining basic course requirements and deadlines. Now, many syllabi—some as long as 10 pages—look more like legal documents. They describe, often in great detail, a variety of rules ranging from attendance to laptop and cell phone use during class, even forbidding videos of professors that may appear on YouTube (Wasley, 2008).

MORES

Mores (pronounced "MORE-ayz") are norms that people consider very important because they maintain moral and ethical behavior. According to U.S. cultural mores, people must be sexually faithful to their spouse or partner, loyal to their country, and must not kill another person (except during war or in self-defense).

Folkways emphasize *ought to* behavior whereas mores define *must* behavior. The Ten Commandments are a good example of both civil and religious mores. Other mores include ethical guidelines (don't cheat, don't lie), expectations about behavior (don't use illicit drugs), and rules about sexual partners (don't have sexual intercourse with children or family members).

Mores, like folkways, can change. Until about the late 1960s, 80 percent of U.S. nonmarital babies were given up for adoption. This rate has dropped to about 1 percent because most unwed mothers, who are no longer stigmatized for having nonmarital babies, keep their infants (Benokraitis, 2015).

The strictest mores are *taboos*, which forbid acts that violate social customs, religious or moral beliefs, or laws. Cannibalism, incest, and infanticide (killing infants) are examples of tabooed behavior.

Taboos define and reinforce mores, but there are many violations. For instance, 92 percent of Americans say that having an extramarital affair is morally wrong. Over a lifetime, however, 21 percent of husbands and 15 percent of wives admit to having had extramarital sex ("Adultery in New England . . .," 2014; Dugan, 2015).

LAWS

Laws are formally defined norms about what is legal or illegal. Unlike folkways and most mores, laws are deliberate, formal, "precisely specified in written texts," and "enforced by a specialized bureaucracy," usually police and courts, which have the power to punish violators (Hechter and Opp, 2001: xi).

Laws change over time. Every year, states pass thousands of bills, ranging from drones to school sports. So far, 29 states and the District of Columbia have legalized medical marijuana, and 8 states and the District of Columbia have legalized marijuana for recreational use. Every state now permits gambling (e.g., lotteries, casinos, racetrack betting). Some people believe that legalizing marijuana and gambling violates mores, but taxing both of these activities has greatly augmented state coffers.

Laws also vary across societies. You'll see shortly and throughout this textbook that countries' laws differ considerably regarding, for example, alcohol, adultery, elections, health care, religious worship, freedom of speech, and adoption.

SANCTIONS

Most people conform to norms because of **sanctions**, rewards for good or appropriate behavior and/or penalties for bad or inappropriate behavior. Children learn norms through both *positive* sanctions (praise, hugs, smiles, new toys) and *negative*

mores norms that people consider very important because they maintain moral and ethical behavior.

laws formally defined norms about what is legal or illegal.

sanctions rewards for good or appropriate behavior and/or penalties for bad or inappropriate behavior.

China's new "Elderly Rights Law" allows aging parents to sue children who don't visit them "often." The judge ordered the daughter of this 77-year-old woman to visit her mother every two months and on at least two of the country's major holidays (Makinen, 2013).

AFP/Getty Images

sanctions (frowning, scolding, spanking, withdrawing love) (see Chapter 4).

Negative sanctions vary in the degree of punishment. When we violate folkways, the sanctions are relatively mild (e.g., gossip, raised eyebrows, ridicule) because folkways aren't critical for a society's survival. If you don't bathe or brush your teeth, you may not be invited to parties because others will see you as crude, but not sinful or evil.

The sanctions for violating mores and some laws can be severe: loss of employment, expulsion from college, whipping, torture, banishment, imprisonment, and even execution. The punishment is usually harsh because the behavior threatens a society's moral foundations. For Hindus, who make up more than 80 percent of India's 1.2 billion population, the cow is a revered religious symbol. Much of India bans its slaughter, sale, and consumption. Hindu mobs have beaten to death Muslims that they suspected of killing, stealing, or smuggling cows. In one case, and as the family had maintained, a forensic test found that the murdered victim had eaten goat meat (Raj, 2015).

Sanctions aren't always consistent, despite universally held norms. Someone who can hire a good attorney often receives no penalty or a light one for a serious crime. In some cultures, young girls and women may be lashed for engaging in premarital sex, but men aren't punished at all for the same behavior, even though it violates Islamic law (Peter, 2012). Also, laws are sometimes enforced selectively or not at all. In the United States and elsewhere, only a fraction of offenders who violate rape, sexual harassment, and domestic violence laws are prosecuted (Jewkes et al., 2013; see also Chapters 7 and 12).

3-2e Rituals

Rituals are formal and repeated behaviors that unite people. They include giving gifts during Christmas, oaths of allegiance, wedding and funeral ceremonies, worship rites, wearing Halloween costumes, graduations, veterans' parades, and even shaking hands. What all rituals have in common is the transmission and reinforcement of norms that unite us and strengthen relationships (Rossano, 2012).

Rituals are also outward symbols of values. During the playing of our national anthem, for example, everyone stands and faces the flag. Many people place their hands over their hearts, and men remove their caps and hats. Such ritualistic behavior demonstrates respect and reinforces shared values about patriotism and democracy.

rituals formal and repeated behaviors that unite people.

cultural universals customs and practices that are common to all societies.

Noah Seelam/Getty Images

In some Indian states, violators who possess or sell beef can be punished by up to five years in prison (Bagri and Najar, 2015).

3-3 SOME CULTURAL SIMILARITIES

You've seen that there's considerable diversity across societies in symbols, language, values, and norms. There are also some striking similarities because of cultural universals, ideal versus real culture, ethnocentrism, and cultural relativism.

3-3a Cultural Universals

Cultural universals are customs and practices that are common to all societies. Anthropologist George Murdock and his associates studied hundreds of societies and compiled a list of 88 activities that they found among all cultures (*Table 3.2* provides some examples).

There are many cultural universals, but specific behaviors vary across cultures, from one group to another in the same society, and over time. For example, all societies have food taboos, but they differ across societies. For about 75 percent of the world's people, locusts, crickets, silkworms, grasshoppers, ants, and other insects are a big part of the diet (Huis et al., 2013). Vietnamese restaurants offer cat on the menu, and many poor people in rural China eat dog meat because they can't afford poultry, pork, and beef. Increasingly, however, urban, middle-class Chinese households "have learned to love dogs at the end of a leash rather than on a skewer" ("Dog-lovers . . . ," 2015: 46; see also Wan, 2011).

Table 3.2 Some Cultural Universals

Athletics	Food taboos	Housing	Medicine
Cooking	Funeral rites	Inheritance rules	Music
Courtship	Games	Kin terminology	Property rights
Dancing	Gift giving	Language	Religious rituals
Division of labor	Greetings	Magic	Sexual restrictions
Etiquette	Hairstyles	Marriage	Status differentiation

Source: Based on Murdock, 1940.

3-3b Ideal versus Real Culture

The **ideal culture** comprises the beliefs, values, and norms that people say they hold or follow. In every society, however, ideal culture differs from **real culture**, or people's actual everyday behavior.

Consider parent-child relationships. Americans say that they love and cherish their children, but every year hundreds of thousands of children experience abuse and neglect on a daily basis. Indeed, 92 percent of the offenders are parents (U.S. Department of Health and Human Services, 2017; see also Chapter 12).

In China, the 2,500-year-old Confucian ideal of *filial piety* (respecting and taking care of one's parents) still runs deep. However, nearly half of the country's older people now live apart from their children, most of whom live and work in urban areas—"a phenomenon unheard of a generation ago"—leaving elders to fend for themselves (Mencher, 2013). Thus, ideal culture and actual behavior are often inconsistent.

About 75 percent of the world's people consume insects, which are high in protein, vitamins, and fiber, and usually low in fat. More than two out of every three American adults are overweight or obese, the highest percentage on the planet (see Chapter 14). Would we be healthier if we ate insects instead of gulping down hamburgers and french fries?

Kevin Foy/Alamy Stock Photo

3-3c Ethnocentrism and Cultural Relativism

Ethnocentrism is the belief that one's culture, society, or group is inherently superior to others. Thus, "*our* customs, *our* laws, *our* food, *our* traditions, *our* music, *our* religion, *our* beliefs and values, and so forth, are somehow better than those of other societies" (Smedley, 2007: 32). Because people internalize their culture and tend to see their way of life as the best and the most natural, they often disparage those with differing attitudes and behavior as inferior, wrong, or backward. How many of you, for example, are repelled at the thought of eating insects, cats, or dogs?

Some of my Black students argue that it's impossible for African Americans to be ethnocentric because they experience much prejudice and discrimination. *Any* group can be ethnocentric, however (Rose, 1997). An immigrant from Nigeria who believes that native-born African Americans are lazy and criminal is just as ethnocentric as a native-born African American who describes Nigerians as arrogant and "uppity."

Ethnocentrism can be functional. Appreciating one's own country and heritage promotes loyalty and cultural unity, reinforces conformity, and maintains stability. Members of a society become committed to their particular values and customs, and transmit them to the next generation. As a result, life is (generally) orderly and predictable.

Ethnocentrism is also dysfunctional because viewing others as inferior generates hatred, discrimination, and conflict. Many current wars and battles—as those in some African nations and between Israelis and Palestinians—are due to religious, ethnic, or political intolerance (see Chapter 13). Thus, ethnocentrism thwarts intergroup understanding and cooperation.

ideal culture the beliefs, values, and norms that people say they hold or follow.

real culture people's actual everyday behavior.

ethnocentrism the belief that one's own culture, society, or group is inherently superior to others.

The opposite of ethnocentrism is **cultural relativism**, the belief that no culture is better than another and should be judged by its own standards. Most Japanese mothers stay home with their children, whereas many American mothers work outside the home. Is one practice better than another? No. Because Japanese fathers are expected to be the breadwinners, it's common for many Japanese women to be homemakers. In the United States, some mothers choose to stay home, but economic recessions and the loss of many high-paying jobs have catapulted many married women into the labor force to help support their families (see Chapters 10, 11, and 12). Thus, Japanese and American parenting may differ, but one culture isn't better or worse than the other.

Vatican Pool-Corbis/Corbis News/Getty Images

Dozens of rules and customs govern what an American first lady should and shouldn't wear while visiting other countries. In keeping with Vatican protocol, Melania Trump wore long sleeves, formal black clothing, and a black veil in a meeting with Pope Francis.

3-4 SOME CULTURAL VARIATIONS

There's considerable cultural variation not only *across* societies but also *within* the same society. *Subcultures* and *countercultures* account for some of the complexity within a society.

3-4a Subcultures

A **subculture** is a group within society that has distinctive norms, values, beliefs, lifestyle, or language. A subculture is part of the larger, dominant culture but has its own particular ways of thinking, feeling, and behaving. U.S. society contains thousands of subcultures based on a variety of characteristics, interests, or activities:

- **Ethnicity** (Irish, Mexican, Vietnamese)
- **Religion** (evangelical Christians, atheists, Mormons)
- **Politics** (Maine Republicans, Southern Democrats, independents)
- **Sex and gender** (heterosexual, lesbian, transgender)
- **Age** (older widows, middle schoolers, college students)
- **Occupation** (surgeons, prostitutes, truck drivers)
- **Music and art** (jazz aficionados, country music buffs, art lovers)
- **Social class** (billionaires, working poor, middle class)
- **Recreation** (mountain bikers, poker players, dancers)

As you read this list, you'll realize that most of us are members of numerous subcultures. We usually participate in subcultures without having much commitment to any of them (e.g., basketball fans, guitar players, collectors of ceramic frogs). On the other hand, members of racial or ethnic subcultures may intentionally live in the same neighborhoods, associate with each other, have close personal relationships, and marry others who are similar to themselves.

To fit in, members of most subcultures adapt to the larger society, but they may also maintain some of their traditional customs. Among Ghanaian immigrants, for instance, funerals are lavish celebrations rather than sad gatherings. The funerals are "all-night affairs with open bars and window-rattling music." As in Ghana, the guests need not know the deceased or even the family. Many people attend simply to meet other Ghanaians and to "cut loose on the dance floor" (Dolnick, 2011: A1; see also Bax, 2013).

In many instances, subcultures arise because of technological or other societal changes. After the Internet emerged, subcultures appeared that identified themselves as hackers, techies, and computer geeks.

3-4b Countercultures

A **counterculture** is a group within society that openly opposes and/or rejects some of the dominant culture's norms, values, or laws. Countercultures usually emerge when people believe they can't achieve their goals within the existing society. As a result, the groups develop values and practices that run counter to those of the

cultural relativism the belief that no culture is better than another and should be judged by its own standards.

subculture a group within society that has distinctive norms, values, beliefs, lifestyle, or language.

counterculture a group within society that openly opposes and/or rejects some of the dominant culture's norms, values, or laws.

Figure 3.3 Active U.S. Hate Groups, 2016

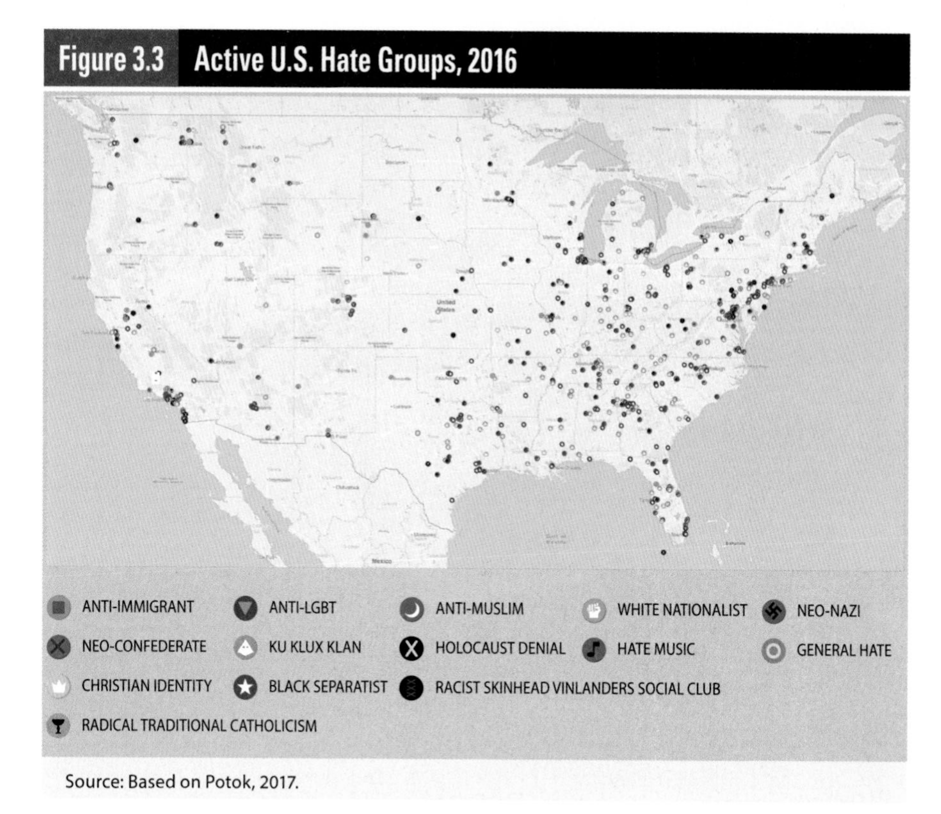

Source: Based on Potok, 2017.

dominant society. Some countercultures are small and informal, but others have millions of members and are highly organized.

Most countercultures are law-abiding. Some, however, are violent and extremist. There are active hate groups in every state (*Figure 3.3*), but their number has increased from 784 in 2014 to 917 in 2016. Hate groups include antigovernment militia, white supremacists, abortion foes, domestic Islamist radicals, neo-Nazis, and a small but growing number of Black separatists. The groups target people because of their race, ethnicity, religion, or sexual orientation (Potok, 2017).

The most dramatic change has been the growth of anti-Muslim organizations—from 34 in 2015 to 101 in 2016. Some analysts have attributed the near-tripling of these groups to international terrorism, a growing fear of domestic attacks, and President Trump's campaign rhetoric that characterized refugees and immigrants from predominantly Muslim nations as terrorists (Johnson, 2016; Eversley, 2017; Potok, 2017). The FBI is particularly concerned about violence carried out by individuals who aren't affiliated with any countercultural movement and are, therefore, hard to detect (Kurzman and Schanzer, 2015).

3-4c Multiculturalism

Multiculturalism (sometimes called *cultural pluralism*) refers to the coexistence of several cultures in the same geographic area, without one culture dominating another.

Many applaud multiculturalism because it encourages intercultural dialogue (e.g., schools offering programs in African American, Latino, Arabic, and Asian studies). Supporters hope that emphasizing multiculturalism—especially in schools and the workplace—will decrease ethnocentrism, racism, sexism, and other forms of discrimination.

Despite its benefits, not everyone is enthusiastic about multiculturalism. Not learning the language or following the customs of the country where one lives and works can spur tension and conflict. European political leaders in Germany, Great Britain, and France have described multiculturalism, particularly integrating Muslims, as a "failed policy" and a "nightmare" that's had "disastrous results." Instead of adopting the dominant culture's values and beliefs, according to critics, some ethnic and religious subcultures "build tiny nations" within the host nation, thereby jeopardizing the host country's national identity (Marquand, 2011; Kymlicka, 2012; Tharoor, 2015; see also Chapter 13).

3-4d Culture Shock

People who travel to other countries often experience **culture shock**—confusion, disorientation, or anxiety that accompanies exposure to an unfamiliar way of life. Conventional cues about how to behave are missing or have a different meaning. Culture shock affects some people more than others, but the most stressful changes involve differences in food, clothes, etiquette, values, language, the general pace of life, and a lack of privacy (Spradley and Phillips, 1972; Pedersen, 1995).

To some degree, everyone is *culture bound* because we've internalized cultural norms and values. During a recent tour to North Korea, an American college student stole a political poster as a souvenir. North Korea's Supreme Court sentenced him to

multiculturalism (sometimes called *cultural pluralism*) the coexistence of several cultures in the same geographic area, without one culture dominating another.

culture shock confusion, disorientation, or anxiety that accompanies exposure to an unfamiliar way of life.

Does Pop Culture Make Kids Fat?

The prevalence of obesity among American children ages 2 to 19 years has more than tripled, increasing from 5 percent in the 1960s to 17 percent in 2014 (Fryar et al., 2016). There are many reasons for the increase, but physicians and researchers lay much of the blame on popular culture, especially the advertising industry. The advertising, particularly on television, uses cartoon characters like SpongeBob SquarePants and Scooby-Doo to sell sugary cereals, cookies, candy, and other high-calorie snacks. In contrast, many European countries forbid advertising on children's television programs. After considerable pressure from the federal government, the nation's largest food companies agreed to curb advertising, but children are watching more fast-food ads on TV than ever before. Only six companies are responsible for more than 70 percent of all TV fast-food ads viewed by children age 18 and younger ("Food Companies Propose...," 2011; Harris et al., 2013).

iStock.com/Juanmonino

15 years of prison and hard labor. Many Americans were outraged: They dismissed the theft as a small and harmless college-style prank, and attributed the harsh punishment to the escalating tensions with the United States over nuclear weapons. But in North Korea, where the leader is treated as a deity, "political slogans are sacrosanct," harming slogans bearing leaders' names is "one of the most serious crimes" and is considered an "act of hostility against the state" (Sang-Hun, 2016: A4; Dawson, 2017).

3-5 HIGH CULTURE AND POPULAR CULTURE

High culture refers to the cultural expression of a society's highest social classes. Examples include opera, ballet, theatre, and classical music. In contrast, **popular culture** refers to the beliefs, practices, activities, and products that are widespread among a population. Examples include television, music, social media, advertising, sports, hobbies, fads, fashions, movies, the food we eat, the gossip we share, and the jokes we tell (Levine, 1998; Gans, 1999).

high culture the cultural expression of a society's highest social classes.

popular culture the beliefs, practices, activities, and products that are widespread among a population.

cultural capital resources and assets that give a group advantages.

Social class affects our participation in high and popular culture. Members of the upper class can pursue the fine arts and similar activities because they have **cultural capital**—resources such as knowledge, verbal and social skills, education, and other assets that give a group advantages (Bourdieu, 1986). Resources alone, however, don't determine our cultural tastes and interests. For example, in Baltimore (as in many other cities), tickets for the symphony, opera, or ballet are usually much less expensive than those for a Ravens football game or rock concert.

Cultural capital sets up boundaries between social classes, but our cultural expressions and products often contain elements of both popular and high culture. For instance, Black jazz players have inspired European classical music composers and themselves perform classical music (Salamone, 2005; Lawn, 2013).

The Queen of England has conferred knighthood—one of the highest honors given to an individual in the United Kingdom—on pop singers like Elton John and Paul McCartney. Since 1998, the John F. Kennedy Center for the Performing Arts in Washington, D.C., has presented an annual Mark Twain Prize for American Humor to popular culture comedians like Richard Pryor, Tina Fey, and Bill Murray.

American fast-food restaurants are popular in many Asian countries. Because of greater competition from local fast-food chains, U.S. companies now offer a large number of items geared specifically to Asian tastes. At one of China's Pizza Huts, the menu includes shrimp pizza, fried squid, and green tea ice cream. In Latin America, McDonald's sells "McMollettes," English muffins filled with refried beans, cheese, and salsa. In India, McDonald's offers an entirely meat-free menu.

iStock.com/Holger Mette

Millions of Americans who think of themselves as "cultured" watch "trashy TV" like *Hoarders*, *The Bachelor/The Bachelorette*, *The Real Wives of . . .*, and *Keeping Up with the Kardashians*. Doing so allows them to feel superior to the shows' conventional "lowbrow" viewers ("Thank God I'm not like that!") (McCoy and Scarborough, 2015). Thus, the divide between popular and high culture isn't as rigid as we might think.

3-5a The Influence of Mass Media

People don't always believe everything they read or see on television or online. Nonetheless, many are highly influenced by a popular culture that's largely controlled and manipulated by newspapers and magazines, television, movies, music, and ads (see Chapter 4). These **mass media**, or forms of communication designed to reach large numbers of people, have enormous power in shaping public attitudes and behavior.

Many of my students who claim that they "don't pay any attention to ads" come to class wearing branded apparel: Budweiser caps, Old Navy t-shirts, and Nike footwear. Advertising affects our self-image and self-esteem. As you'll see in later chapters, many people spend considerable time and money to meet ideal standards of femininity and masculinity—generated by ads—that are unrealistic and impossible to attain. (We'll examine fashion and fads, two important components of popular culture, in Chapter 16.)

Because companies are driven by the pursuit of profit, considerable mass media content is basically marketing. Many television shows and MTV have become "a kind of sophisticated infomercial" (O'Donnell, 2007: 30). A third of the content on morning shows is essentially selling something (a book, music, a movie, or another television program) that the corporation owns. The *Dr. Phil Show* is crowded with infomercials that promote the host's numerous products, including his diet and relationship books, his wife's skin care line, and his and his son's TV show, publishing company, and "Doctor on Demand" app. U.S. mass media also exert a powerful influence abroad.

3-5b The Globalization of Popular Culture

Because the United States is the biggest producer of popular culture, many countries have become inundated with America's movies, music, television shows, satellite broadcasts, fast food, clothing, and other consumer goods. Overseas corporations contribute to the spread of U.S. popular culture because they profit by distributing the products of popular musicians, authors, and filmmakers (The Levin Institute, 2013).

Some observers have described the spread of U.S. popular culture as **cultural imperialism**, a process by which the cultural values and products

mass media forms of communication designed to reach large numbers of people.

cultural imperialism the cultural values and products of one society influence or dominate those of another.

of one society influence or dominate those of another. A great deal of cultural imperialism is voluntary. For example, much of England's population has embraced American pop music, films, television programs, architecture, and is now experiencing an "obesity epidemic" because of the "bread-potatoes-and-lard-based fast food on which America nourishes itself" (Heffer, 2010).

Other countries complain that U.S. cultural imperialism displaces authentic local culture and results in cultural loss. Iran's government has denounced Batman, Spider-Man, and Harry Potter toys as a "cultural invasion" that challenges the country's conservative and religious values. The curvaceous and often scantily clad Barbie dolls with peroxide-blond hair have been especially singled out as "destructive culturally and a social danger" (Peterson, 2008: 4).

Only 42 percent of Americans, compared with 79 percent in 2002, believe that the world at large sees the United States favorably. This belief may be a major reason for many Americans' wariness of U.S. global involvement. A large majority (70 percent) say that the president should focus on domestic rather than foreign policy, 57 percent believe that other countries should deal with their own problems, and 43 percent think that the United States should "mind its own business" internationally ("Public Uncertain...," 2016; Swift, 2017).

3-6 CULTURAL CHANGE AND TECHNOLOGY

Why do cultures persist? How and why do they change? And what happens when technology changes faster than cultural norms, values, and laws?

3-6a Cultural Persistence: Why Cultures Are Stable

In many ways, culture is a conservative force. As you saw earlier, values, norms, and language are transmitted from generation to generation. Such **cultural integration**, or the consistency of various aspects of society, promotes order and stability. Even when new behaviors and beliefs emerge, they commonly adapt to existing ones. Recent immigrants may speak their native language at home and celebrate their own holy days, but they're expected to gradually absorb the host country's values, to obey its civil and criminal laws, and to adopt its national language. Life would be chaotic and unpredictable without such cultural integration.

cultural integration the consistency of various aspects of society that promotes order and stability.

3-6b Cultural Dynamics: Why Cultures Change

Cultural stability is important, but all societies change over time. Some of the major reasons for cultural change include diffusion, invention and innovation, discovery, and external pressures.

DIFFUSION

A society may change because of *diffusion*, the spread of cultural beliefs and activities from one group to another. The influences may have occurred so long ago that the members of a society consider their culture to be entirely their own creation. However, anthropologist Ralph Linton (1964) has estimated that 90 percent of the elements of any culture are a result of diffusion (*Figure 3.4*).

Diffusion can be direct and interpersonal, occurring through trade, tourism, immigration, intermarriage, or the invasion of one country by another. Diffusion can also be indirect and largely impersonal, as in the Internet transmissions that zip around the world.

INVENTION AND INNOVATION

Cultures change because people are continually finding new ways of doing things. *Invention*, the process of creating new things, brought about products such as toothpaste (invented in 3000 B.C.E.), eyeglasses (262 C.E.), flushable toilets (the sixteenth century), fax machines (1843—that's right, invented in 1843!), credit cards (1920s), computer mouses (1964), Post-it™ notes (1980), and DVDs (1995).

Innovation—turning inventions into mass-market products—also sparks cultural changes. An innovator is someone who markets an invention, even if it's someone else's idea. For example, Henry Ford invented nothing new but "assembled into a car the discoveries of other men behind whom were centuries of work," an innovation that changed people's lives (Evans et al., 2006: 465).

Inventors and innovators are rarely isolated geniuses who work alone. Instead, they depend on collaborative relationships and social networks for ideas and help. Thomas Edison's laboratory, which had a staff of 35 inventors, produced the light bulb. The Human Genome Project involved 110 scientists around the world from 20 universities, research centers, institutes, and hospitals. Even Michelangelo used 12 assistants to paint the renowned Sistine Chapel in Rome (Dahlin, 2011).

DISCOVERY

Like invention, *discovery* involves exploration and investigation, and results in new products, insights, ideas,

Figure 3.4 The 100 Percent American

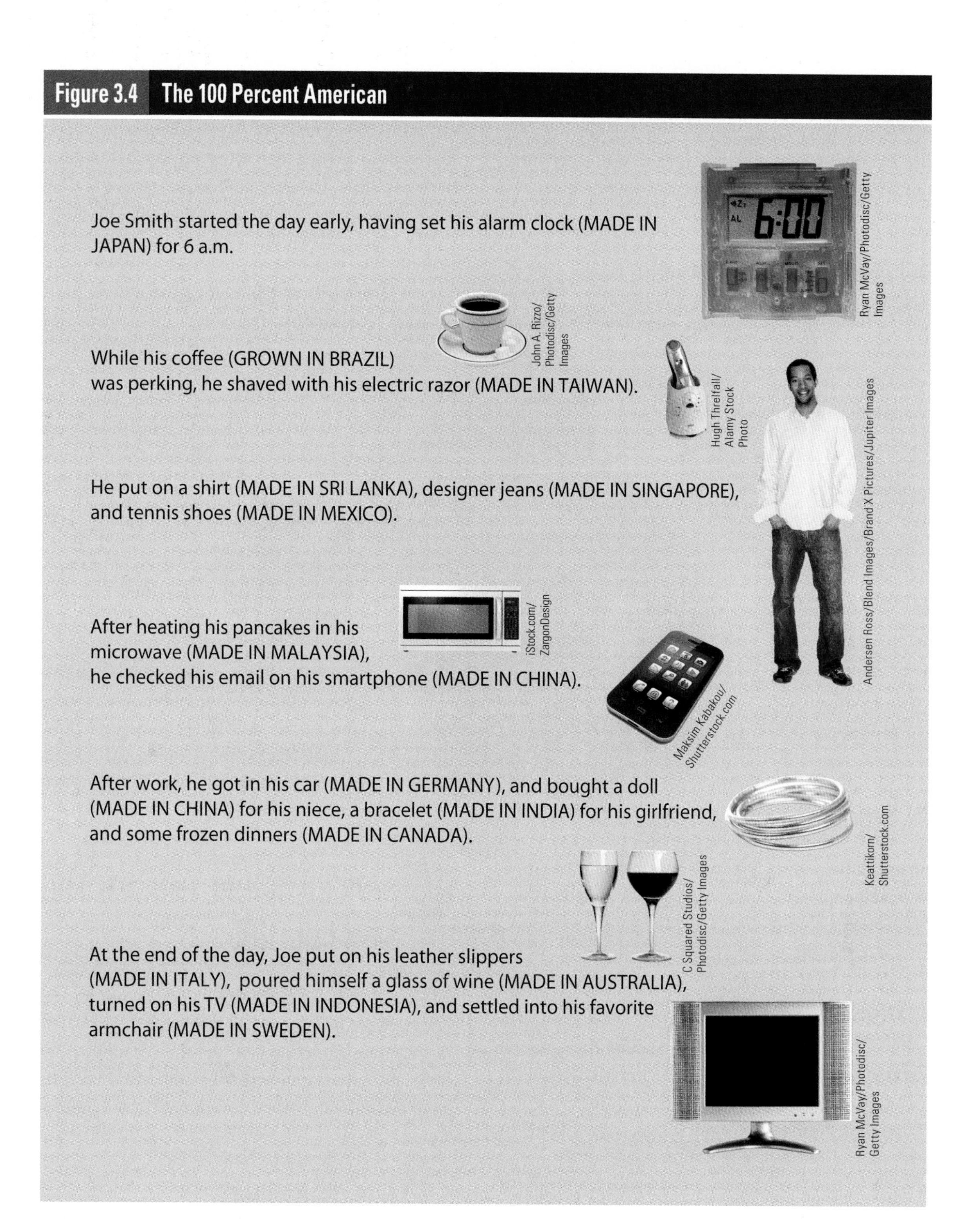

or behavior. The discovery of penicillin prolonged lives, which, in turn, means that more grandparents (as well as great-grandparents) and grandchildren get to know each other. However, longer life spans also mean that many adult children and grandchildren now care for older family members over many years (see Chapter 12).

Discovery usually requires dedicated work and years of commitment, but some discoveries occur by chance, called the *serendipity effect.* George de Mestral, a Swiss electrical engineer, was hiking through the woods and was annoyed by burrs that clung to his clothing. Why were they so difficult to remove? A closer examination showed that the burrs had hooklike arms that locked into the open weave of his clothes. The discovery led de Mestral to invent a hook-and-loop fastener. His invention, Velcro—derived from the French words *velour* (velvet) and *crochet* (hooks)—can now be found on everything from wallets to spacecraft.

EXTERNAL PRESSURES

External pressure for cultural change can take various forms. In its most direct form—war, conquest, or colonization—the dominant group uses force or the threat of force to change an existing culture. When the Soviet Union invaded and occupied many small countries (e.g., Lithuania, Latvia, Estonia, Ukraine, and Armenia) after World War II, it forbade citizens to speak their native languages, banned traditions and customs, and turned churches into warehouses.

Pressures for change can also be indirect. Some countries (e.g., Thailand, Vietnam, China, and Russia) have reduced their prostitution and international sex trafficking because of widespread criticism by the United Nations and some European countries. The United Nations has no power to intervene in a country's internal affairs but can embarrass nations by publicizing human rights violations (Farley, 2001).

Henrik Lehnerer/Shutterstock.com

3-6c Technology and Cultural Lag

Some parts of culture change more rapidly than others. **Cultural lag** refers to the gap that occurs when material culture changes faster than nonmaterial culture.

There are numerous examples of cultural lag in modern society because our nonmaterial culture (e.g., norms, values, and laws) hasn't kept up with material culture (e.g., technological advances). Texting while driving has resulted in thousands of fatal car crashes (see Chapters 1 and 2), but many states have neither banned this practice nor enforced the laws they've passed.

Because technology is moving faster than laws to regulate and ensure Internet-based communication privacy, there are many breaches. The National Security Agency (NSA), one of the largest U.S. intelligence organizations, has secretly collected phone records of millions of Americans, and spied on leading U.S. technology companies (e.g., Facebook, Google, Apple, Yahoo) as well as European political heads of state who are allies (Timberg and Nakashima, 2013).

cultural lag the gap that occurs when material culture changes faster than nonmaterial culture.

Cultural lags can create confusion, ambiguity about what's right and wrong, conflict, and a feeling of helplessness. They also expose contradictory values and behavior. U.S. technology executives have railed against the NSA's mass compilation of data, but the companies themselves collect information on users that they sell to advertisers (Miller, 2013; Sengupta, 2013). Some consumers complain that retailers track their behavior, but have no problem with cookies, profiles, apps, and other online tools that let e-commerce sites know who they are, how they shop, and what they purchase (Clifford and Hardy, 2013). And as you'll see in several later chapters, even when mores, values, and laws catch up with technological innovations and inventions, they can create unexpected problems such as pollution, drug shortages, and high medical costs.

3-7 SOCIOLOGICAL PERSPECTIVES ON CULTURE

What is the role of culture in modern society? And how does culture help us understand ourselves and the world around us? Functionalist, conflict, feminist, and symbolic interaction scholars offer important insights in answering these and other questions. *Table 3.3* summarizes the perspectives.

3-7a Functionalism

Functionalists focus on society as a system of interrelated parts (see Chapter 1). Similarly, in their analysis of culture, functionalists emphasize the social bonds that attach people to society.

KEY POINTS

For functionalists, culture is the cement that binds people together. As you saw earlier, norms and values shape our lives, provide guideposts for our everyday behavior,

Table 3.3 Sociological Perspectives of Culture

THEORETICAL PERSPECTIVE	FUNCTIONALIST	CONFLICT	FEMINIST	SYMBOLIC INTERACTIONIST
Level of Analysis	Macro	Macro	Macro and Micro	Micro
Key Points	• Similar beliefs bind people together and create stability. • Sharing core values unifies a society and promotes cultural solidarity.	• Culture benefits some groups at the expense of others. • As powerful economic monopolies increase worldwide, the rich get richer and the rest of us get poorer.	• Women and men often experience culture differently. • Cultural values and norms can increase inequality because of sex, race/ethnicity, and social class.	• Cultural symbols forge identities (that change over time). • Culture (like norms and values) helps people merge into a society despite their differences.
Examples	• Speaking the same language (English in the United States) binds people together because they can communicate with one another, express their feelings, and influence one another's attitudes and behaviors.	• Much of the English language reinforces negative images about gender ("slut"), race ("honky"), ethnicity ("jap"), and age ("old geezer") that create inequality and foster ethnocentrism.	• Using male language (e.g., "congressman," "fireman," and "chairman") conveys the idea that men are superior to and dominant over women, even when women have the same jobs.	• People can change the language they create as they interact with others. Many Americans now use "police officer" instead of "policeman," and "single person" instead of "bachelor" or "old maid."

and promote cultural integration and societal stability. Especially in developed countries like the United States that have high immigration rates, cultural norms and values help newcomers adjust to the host society.

All societies have similar strategies for meeting human needs. Cultural universals, like religious rituals, may play out differently in different countries, but all known societies have religious rituals. Critics often blame popular culture for societal problems, including violence and crime, but popular culture also unites people with similar interests and produces new technologies (e.g., LinkedIn, YouTube, Instagram).

Functionalists also note that culture can be dysfunctional. Some countercultures, particularly right-wing extremist groups, can create chaos by bombing federal buildings and killing or injuring hundreds of people.

CRITICAL EVALUATION

In emphasizing culture's role in meeting people's daily needs, functionalism often overlooks diversity and social change. A number of influential functionalists have proposed, for instance, that immigration should be restricted because it dilutes shared U.S. values. Such proposals overlook the many contributions that newcomers make to society (see Chapter 10).

Another issue is how much culture really binds people together. Only 52 percent of adults say that they're "extremely proud" to be Americans, down from 70 percent in 2003. Some of the reasons for the declining patriotism include frustration with political leaders, a belief that moral values have deteriorated, and pessimism about the economy (McCarthy, 2015; Jones, 2016).

3-7b Conflict Theory

Unlike functionalists, conflict theorists argue that culture can generate considerable inequality instead of unifying society. Because the rich and powerful determine economic, political, educational, and legal policies for their own benefit and control the mass media, the average American has little power to change the culture—whether it's low wages, unpopular wars, or corporate corruption (see Chapters 8 and 11).

KEY POINTS

Conflict theorists maintain that many cultural values and norms benefit some members of society more than others. We're taught to work hard, for example, but who profits from the average worker's efforts? U.S. taxpayers paid billions of dollars to bail out financial industries that collapsed because of greedy top executives. The executives then used the money to give themselves and their staff higher salaries and bonuses than ever before. Very rarely, moreover, are wealthy people incarcerated for crime (see Chapters 7 and 8). Thus, those at the top of the socioeconomic ladder can violate laws and mores, including honesty, that other Americans are expected to observe.

FotoFlirt/Alamy Stock Photo

Anne-Marie Palmer/Alamy Stock Photo

India's high-income families usually have lavish weddings that follow centuries-long traditions. For functionalists, the customs and rituals reinforce cultural identity and bind people. For conflict theorists, India's inequality also creates a large population of poor people who are excluded from the traditions and ceremonies that reinforce social bonds.

Conflict theorists also point out that technology serves primarily the rich. Desperate low-income women use online sites to sell their hair, breast milk, and ovarian eggs. Selling kidneys is illegal in the United States, but one kidney sells for more than $100,000 on the Internet black market (Weller, 2015).

CRITICAL EVALUATION

According to some critics, conflict theorists place too much emphasis on societal discord and downplay a culture's benefits. For instance, cultural values integrate members of a society and decrease divisiveness.

Conflict theorists maintain that mass media conglomerates control popular culture and promote goods and services that most people don't need and/or can't afford (e.g., big screen TVs, the most recent smartphones, and designer clothes). Critics counter that people aren't mindless puppets but can make choices, including not living on credit. Thus, according to some conflict theory critics, conglomerates shouldn't be blamed for people's irrational behavior.

3-7c Feminist Theories

Feminist scholars agree with conflict theorists that culture creates considerable inequality, but they focus on gender differences. Feminists are also more likely than other theorists to examine multicultural variations across groups.

KEY POINTS

Culture affects gender roles. When media portrayals of women are absent or stereotypical, we get a distorted view of reality (Neuhaus, 2010). Feminist scholars also emphasize that subcultures (e.g., female students or single mothers) may experience culture differently than their male counterparts do because women typically have fewer resources, particularly income and power (see Chapters 9 and 11).

Unequal access to resources often results in women having fewer choices than men, and living under laws and customs that subordinate women. In the United States, a nonmarital birth typically impoverishes women but not men. Only 10 states or jurisdictions have specific laws that prevent or punish forced marriages of girls under age 16 (Reiss, 2015). In most Islamic societies, men dictate women's attire and behavior (Heath, 2008; Zahedi, 2008). In many Latin American countries, violence against women is not only "firmly entrenched" but has recently spiked (Llana and Brodzinsky, 2012).

In India, gang rapes are a "routine and a largely invisible crime." Young women who move to a city are policed by a network of males. If the women violate village moral codes, including owning a cell phone, they disappear or are quickly married to village men of their parents' choosing (Barry, 2013; Barry and Choksi, 2013; Srivastava and Mehrotra, 2013).

ANIMATED HEALTHCARE LTD/Science Photo Library/Getty Images

CRITICAL EVALUATION

Feminist analyses expand our understanding of cultural components that other theoretical perspectives ignore or gloss over. Like conflict theorists, however, feminist theorists often emphasize discord rather than how culture integrates women and men into society. Another weakness is that

Hurst Photo/Shutterstock.com

Which sociological perspectives help explain how children's birthday parties reproduce culture?

feminist scholars tend to focus on the cultural experiences of low-income and minority women but not of their male counterparts.

3-7d Symbolic Interaction

Interactionists examine culture through micro lenses. They're most interested in understanding how individuals—in contrast to groups, organizations, or societies—create, maintain, and modify culture.

KEY POINTS

Interactionists study how culture influences our everyday lives. Language, you'll recall, shapes our views and behavior. People within and across societies create a variety of symbols that may change over time. Technology has also changed many people's communication patterns. Some complain that texting dominates our lives. A much viewed YouTube clip shows the bride pulling out her cell phone from under her veil and texting with one hand as she walks down the aisle with her father (Pflaumer, 2011). On the other hand, students become more engaged when they're required to use technology, like tweeting weekly observations that illustrate their reading assignments (Lang, 2013).

For interactionists, many cultural components are socially constructed. Consider food consumption. Americans may think it's odd to eat insects, but cultural factors affect what we define as good to eat and how food is prepared. Thus, some visitors consider many U.S. foods strange or "nasty" (e.g., Cheez Whiz, Velveeta, corn dogs, supermarket white bread and rolls, super-sweet iced tea, grits). According to some international travelers, many of our sugary breakfast cereals, particularly those with different colors and marshmallows, are "disgusting" (Martin, 2013; Govender, 2014).

Symbolic interactionists also note that our values and norms, like other components of culture, aren't superimposed by some unknown external force. Instead, as people construct their perception of reality, they create, change, and reinterpret values and norms through interaction with others. Peer pressure, for example, can either encourage or discourage inappropriate language or behavior.

CRITICAL EVALUATION

Micro approaches help us understand what culture means to people and how these meanings differ across societies. However, symbolic interactionists don't offer a systematic framework that explains *how* people create and shape culture or develop shared meanings of reality. Why, for instance, are some of us less polite than others even though we share the same cultural values and norms? Another weakness is that interactionists don't address the linkages between culture and subcultures. Language bonds people, but interactionists say little about how organized groups (like the English-only movement) try to maintain control over language to promote their own beliefs.

STUDY TOOLS 3

READY TO STUDY? IN THE BOOK, YOU CAN:

- ☐ Check your understanding of what you've read with the Test Your Learning Questions provided on the Chapter Review Card at the back of the book.
- ☐ Tear out the Chapter Review Card for a handy summary of the chapter and key terms.

ONLINE AT CENGAGEBRAIN.COM WITHIN MINDTAP YOU CAN:

- ☐ Explore: Develop your sociological imagination by considering the experiences of others. Make critical decisions and evaluate the data that shape this social experience.
- ☐ Analyze: Critically examine your basic assumptions and compare your views on social phenomena to those of your classmates and other MindTap users. Assess your ability to draw connections between social data and theoretical concepts.
- ☐ Create: Produce a video demonstrating connections between your own life and larger sociological concepts.
- ☐ Collaborate: Join your classmates to create a capstone project.

4 Socialization

Comstock Images/Getty Images

LEARNING OBJECTIVES

After studying this chapter, you will be able to...

4-1 Define and illustrate socialization and explain its importance.

4-2 Describe the nature versus nurture debate.

4-3 Compare social learning and symbolic interaction theories of socialization.

4-4 Describe and illustrate five socialization agents.

4-5 Explain how socialization changes throughout life.

4-6 Explain when and how resocialization occurs.

After finishing this chapter go to **PAGE 81** for **STUDY TOOLS**

16 and Pregnant, a popular reality show that aired between 2009 and 2014, spawned three equally popular *Teen Mom* series. The producers say that the goal of the programs is to decrease teen pregnancy by showing the harmful effects of not using protection or birth control, and the resulting struggles in raising a baby. Critics argue that the shows glamorize teen pregnancy, make the teen moms instant celebrities, and ignore the long-term negative impact on the mothers' and children's healthy social development (Thompson, 2010). Both sides are talking about **socialization**, the lifelong process through which people learn culture and become functioning members of society.

WHAT DO YOU THINK?

I'm always the same person, no matter where I am or who's around.

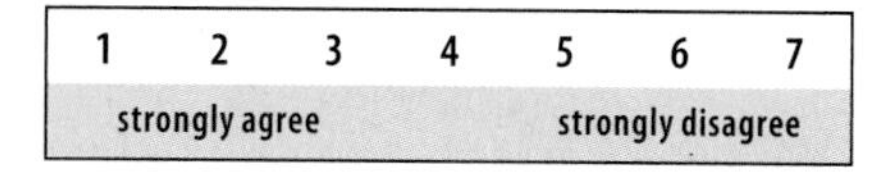

SOCIALIZATION: ITS PURPOSE AND IMPORTANCE

Socialization is critical in all societies. To understand why, let's begin by looking at the purpose of socialization.

4-1a What Is the Purpose of Socialization?

Socialization—from childhood to old age—can be relatively smooth or bumpy, depending on factors such as age, gender, race/ethnicity, and social class. Generally, however, socialization has four key functions that range from providing us with a social identity to transmitting culture to the next generation.

SOCIALIZATION ESTABLISHES OUR SOCIAL IDENTITY

Have you ever thought about how you became the person you are today? Sociology professors sometimes ask their students to give 20 answers to the question "Who am I?" (Kuhn and McPartland, 1954). How would you respond? You would probably include a variety of descriptions like college student, single or married, female or male, and son or daughter. Your answers would show a sense of your *self* (a concept we'll examine shortly). You are who you are largely because of socialization.

SOCIALIZATION TEACHES US ROLE TAKING

Why do you act differently in class than with your friends? Because we play different roles in different settings. A *role* is the behavior expected of a person in a particular social position (see Chapter 5). The way we interact with a parent is typically very different from the way we talk to an employer, a child, or a professor. We all learn appropriate roles through socialization.

SOCIALIZATION CONTROLS OUR BEHAVIOR

In learning appropriate roles, we absorb values and a variety of rules about how we should (and shouldn't) interact in everyday situations. If we follow the rules, we're usually rewarded. If we break the rules, we may be punished. Socialization maintains social order and controls our behavior by teaching us to conform to social expectations.

We conform because we've internalized societal norms and values. **Internalization** is the process of learning cultural behaviors and expectations so deeply that we accept them without question. Obeying laws, paying bills on time, and respecting teachers are examples of internalized behaviors.

SOCIALIZATION TRANSMITS CULTURE TO THE NEXT GENERATION

Each generation passes its culture on to the next generation. The culture that is transmitted, as you saw in Chapter 3, includes language, beliefs, values, norms, and symbols.

socialization the lifelong process through which people learn culture and become functioning members of society.

internalization the process of learning cultural behaviors and expectations so deeply that we accept them without question.

The Harlow Studies and Emotional Attachment

In the early 1960s, psychologists Margaret and Harry Harlow (1962) conducted several studies on infant monkeys. In one group, a "mother" made of terry cloth provided no food, while a "mother" made of wire did so through an attached baby bottle containing milk. In another group, the cloth mother provided food but the wire mother didn't. Regardless of which mother provided milk, when both groups of monkeys were frightened, they clung to the cloth mother. The Harlows concluded that physical contact and comfort were more important to the infant monkeys than nourishment.

Since then, some sociologists have cited the Harlow studies to argue that emotional attachment may be even more critical than food for human infants. Do you see any problems with sociologists' generalizing the results of animal studies to humans?

Nina Leen/Time Life Pictures/Getty Images

4-1b Why Is Socialization Important?

Social isolation can be devastating. An example of its negative effects is Genie. When Genie was 20 months old, her father decided she was "retarded," put her in a wire mesh cage, and locked her away in a back room with the curtains drawn and the door shut. Genie's father frequently beat her with a wooden stick and no one in the house was allowed to speak to her. When she was discovered in 1970 at age 13, a psychiatrist described Genie as "unsocialized, primitive, and hardly human." Except for high-pitched whimpers, she never spoke. Genie had little bowel control, experienced rages, and tried to hurt herself. After living in a rehabilitation ward and a foster home, she learned to eat normally, was toilet-trained, and gradually developed a vocabulary, but her language use never progressed beyond that of a 3- or 4-year-old (Curtiss, 1977).

The research on children who are isolated (Genie) or institutionalized (orphans) demonstrates that socialization is critical to our development. Behaviors such as talking, eating with utensils, and controlling our bowel movements don't come naturally. Instead, we learn to do all of these things beginning in infancy. When children are deprived of social interaction, they don't develop the characteristics that most of us see as normal and human.

4-2 NATURE AND NURTURE

Biologists focus on the role of heredity (or genetics) in human development. In contrast, most social scientists, including sociologists, underscore the importance of learning, socialization, and culture. This difference of opinion is often called the *nature–nurture debate* (*Table 4.1*).

4-2a How Important Is Nature?

Those who argue that nature (biology) shapes behavior point to two kinds of evidence—developmental and health differences between males and females, and unsuccessful sex reassignment cases.

DEVELOPMENTAL AND HEALTH DIFFERENCES

Boys mature more slowly than girls, get sick more often, and are less likely to have mastered the self-control and fine-motor skills necessary for a successful start in school. Boys are also at greater risk than girls for most of the major learning and developmental disorders—as much as four times more likely to suffer from autism, attention deficit disorder, and dyslexia. Girls, however, are at least twice as likely as boys to suffer from depression, anxiety, and eating disorders (Eliot, 2012).

Table 4.1 The Nature–Nurture Debate

Nature	Nurture
Human development is . . .	**Human development is . . .**
Innate	Learned
Biological, physiological	Psychological, social, cultural
Due largely to heredity	Due largely to environment
Fairly fixed	Fairly changeable

Among adults, the senses of smell and taste are more acute in women than in men, and hearing is better and lasts longer in women than in men. Women, however, have a higher risk of developing diabetes. Some conditions (migraine headaches and breast cancer) are more common in women, whereas others (hemophilia and skin cancer) are more common in men (McDonald, 1999; Kreeger, 2002).

UNSUCCESSFUL SEX REASSIGNMENT

Scientists who believe that nature (not nurture) molds behavior also point to unsuccessful sex reassignment attempts. In the 1960s, John Money, a highly respected psychologist, published numerous articles and books in which he maintained that gender identity is determined as much by culture and nurture as by hormones (see, for example, Money and Ehrhardt, 1972). His views were based on the case of David Reimer, a child who underwent sex reassignment after his penis was mutilated in a botched circumcision. His parents, upon Money's recommendation, raised David as a girl (Brenda). Money maintained that Brenda was happy and healthy. A biologist and a psychiatrist who followed up on David/Brenda's case in the 1990s, however, concluded that the sex reassignment had failed (Diamond and Sigmundson, 1997).

Lichtmeister/Shutterstock.com
graphixmania/Shutterstock.com
Denys Kurbatov/Shutterstock.com

Nature or nurture?

4-2b How Important Is Nurture?

Most sociologists maintain that nurture is more significant than nature because socialization and culture shape human behavior. They point to successful sex reassignment cases (see Vitello, 2006) and two types of data to support their argument: cross-cultural variations in male violence, and the environment's effect on biology.

CROSS-CULTURAL VARIATIONS IN MALE VIOLENCE

We often hear that males are "naturally" more aggressive than females because their glands produce more testosterone, the dominant male hormone. If men were innately aggressive due to biology, they would be equally violent across all societies. This isn't the case. The proportion of women who have ever suffered physical violence by a male partner varies considerably: 16 percent in East Asia, 21 percent in North America, 41 percent in western Latin America, and 66 percent in central sub-Saharan Africa (World Health Organization, 2013). All mass murderers (those who have killed a large number of people during one incident) have been men, but most such murders have occurred in the United States (Christakis, 2012; Kluger, 2014). Such variations reflect cultural laws and practices and other environmental factors (nurture) rather than biology or genetics (nature) (Chesney-Lind and Pasko, 2004).

HOW ENVIRONMENT AFFECTS BIOLOGY

Much research shows that environment (nurture) influences children's genetic makeup (nature). For example, birth defects associated with prenatal alcohol exposure can occur in the first 3 to 8 weeks of pregnancy, before a woman even knows she's pregnant. A woman's single drinking "binge"—lasting four hours or more—can permanently damage an unborn child's brain. A man who drinks heavily may have genetically damaged sperm that also leads to birth defects. Thus, alcohol abuse (an environmental factor) by either a woman or a man can have a devastating, irreversible, and lifelong negative biological impact on a child (Denny et al., 2009; Warren, 2012).

In a growing field known as "fetal origins," scientists are finding that the nine months of pregnancy constitute the most consequential period of our lives, permanently influencing the wiring of the brain, the functioning of organs like the heart and liver, and behavior. That is, the kind and quantity of nutrition you received in the womb; the pollutants, drugs, and infections you were exposed to during gestation; your parents' health and eating habits; and your mother's stress level during pregnancy—all these factors have lasting effects in infancy, childhood, and adulthood (Begley, 2010; Paul, 2010).

Adult behavior can influence a child's biological makeup in other ways. Physical, psychological, or sexual abuse can blunt a child's biological development and lead to behavioral and emotional problems throughout life (see Chapter 12). Siblings share much genetic makeup, but the quality of the schools they attend and the teachers they have affect their interests, social skills, academic outcomes, and earnings in adulthood (Whitehurst, 2016).

Is children's obesity due to nature, nurture, or both?

Fotoluminate/Shutterstock.com

4-2c What Can We Conclude About the Nature–Nurture Debate?

Humans have hundreds of traits—like height, metabolism, aggression, leadership traits, and cognitive ability—that are partly inherited (Freese, 2008). A team of researchers analyzed 50 years of twin studies and more than 300 characteristics. They found that, on average, 49 percent of the individual differences were genetic and 51 percent were environmental, but some traits (e.g., a cleft lip) were more likely to be inherited than others (e.g., antisocial behavior) (Polderman et al., 2015).

Our social environment can enhance or dampen biological characteristics. Children who are genetically predisposed to obesity don't always become overweight if parents discourage overeating and encourage physical recreational activities (Martin, 2008).

The Case of Brenda/David

In 1963, twin boys were being circumcised when the penis of one of the infants, David Reimer, was accidentally burned off. Encouraged by John Money, a highly respected psychologist, the parents agreed to raise David as "Brenda." The child's testicles were removed, and surgery to construct a vagina was planned. Money reported that the twins were growing into happy, well-adjusted children, setting a precedent for sex reassignment as the standard treatment for 15,000 newborns with similarly injured genitals (Colapinto, 1997, 2001).

In the mid-1990s, a biologist and a psychiatrist followed up on Brenda's progress and concluded that the sex reassignment had failed. Almost from the beginning, Brenda refused to be treated like a girl. When her mother dressed her in frilly clothes as a toddler, Brenda tried to rip them off. She preferred to play with boys and stereotypical boys' toys such as machine guns (Diamond and Sigmundson, 1997).

At age 14, Brenda rebelled and stopped living as a girl: She refused to wear dresses, urinated standing up, refused vaginal surgery, and decided she would either commit suicide or live as a male. When his father finally told David the true story of his birth and sex reassignment surgery, David recalls that "all of a sudden everything clicked. For the first time things made sense and I understood who and what I was" (Diamond and Sigmundson, 1997: 300).

David had a mastectomy (breast removal surgery) at age 14 and underwent several operations to reconstruct a penis. At age 25, he married an older woman and adopted her three children. At age 38, he committed suicide. Most suicides have multiple reasons, but some speculated that David committed suicide because of the "physical and mental torments he suffered in childhood that haunted him the rest of his life" (Colapinto, 2004: 96). David's experience suggests to some scientists that nature outweighs nurture in shaping a person's gender identity.

Str Old/REUTERS

David Reimer was raised as a girl, Brenda, until he was 14.

Genetic intelligence thrives in an enriching environment but is stifled in poor and disadvantaged conditions. Children of average intelligence who are adopted by rich parents tend to grow up to be rich adults; genetically talented children raised by biological parents who are poor are much less likely to enjoy economic success as adults (Bryant, 2014; Black et al., 2015; Tucker-Drob and Bates, 2016; see also Chapters 8 and 13). Biological factors play an important role in human behavior, but sociology's larger concern is how socialization affects people's development, even overriding some genetic predispositions and influences.

4-3 SOCIOLOGICAL EXPLANATIONS OF SOCIALIZATION

Functionalism provides a foundation for understanding the purposes of socialization described at the beginning of this chapter, but doesn't tell us *how* socialization works. There are well-known psychological and psychosocial theories of human development (e.g., Sigmund Freud, Erik Erikson, Jean Piaget, and Lawrence Kohlberg). In sociology, two influential micro approaches that explain socialization are *social learning theories* and *symbolic interaction theories* (*Table 4.2* summarizes these perspectives).

4-3a Social Learning Theories

The central notion of **social learning theories** is that people learn new attitudes, beliefs, and behaviors through social interaction, especially during childhood. The learning is direct and indirect, and a result of observation, reinforcement, and imitation (Bandura and Walters, 1963; Mischel, 1966; Lynn, 1969).

DIRECT AND INDIRECT LEARNING

Reinforcement refers to direct or indirect rewards or punishments for particular behaviors. Consider gender roles. A little girl who puts on her mother's makeup may be told that she's cute, but her brother will be scolded ("boys don't wear makeup"). Children also learn through indirect reinforcement. If a little boy's male friends are punished for crying, he'll learn that "boys don't cry."

Children also learn through *observation* and *imitation*. Even when children aren't directly rewarded or punished for "behaving like a boy" or "behaving like a girl," they learn about gender by watching who does what in their families. A father who is rarely at home because he's always working sends the message that men are supposed to earn money. A mother who always complains about being overweight or old sends the message that women are supposed to be thin and young.

SelectStock/Vetta/Getty Images

Because parents are emotionally important to their children, they're typically a child's most powerful **role models**, people we admire and whose behavior we imitate. Other role models, as you'll see shortly, include siblings, grandparents, teachers, friends, and even celebrities.

Parents and other role models reinforce particular behaviors, but much of our learning is informal and occurs in a variety of social contexts. A recent study analyzed nearly 6,000 children's picture books published from 1900 to 2000. Females were central characters in no more than 33 percent of the books published in any given year (McCabe et al., 2011). Such underrepresentation teaches children, though informally, that males occupy more important roles in society than do women.

LEARNING AND PERFORMING

Social learning theorists also distinguish between *learning* and *performing* behavior. Children and adults learn norms and roles through observation, but they don't always imitate the behavior. For example, children may see their friends cheat in school but don't do so themselves. Adults, similarly, may see their coworkers steal office supplies but buy their own. We behave as we do, then, because our society teaches us what's appropriate and inappropriate.

CRITICAL EVALUATION

Social learning theories help us understand why we behave as we do, but much of the emphasis is on early socialization rather than what occurs throughout life. Social learning theories don't explain why reinforcement and modeling affect some children more than others, especially those in the same family. There may be personality

social learning theories people learn new attitudes, beliefs, and behaviors through social interaction.

role models people we admire and whose behavior we imitate.

Table 4.2 Key Elements of Socialization Theories

Social Learning Theories	Symbolic Interaction Theories
• Social interaction is important in learning appropriate and inappropriate behavior.	• The self emerges through social interaction with significant others.
• Socialization relies on direct and indirect reinforcement.	• Socialization includes role taking and controlling the impression we give to others.
Example: Children learn how to behave when they are scolded or praised for specific behaviors.	*Example:* Children who are praised are more likely to develop a strong self-image than those who are always criticized.

differences, but if learning is as effective as social learning theorists maintain, siblings' attitudes and behavior should be more similar than different. This is often not the case, even with identical twins.

Social learning theories often ignore factors such as birth order, which brings different advantages and disadvantages. Except for affluent families, larger families have fewer resources for the second, third, and later children. The first child may be disciplined and monitored more, but also has higher grades than her or his younger siblings, and is more likely to attend college. In contrast, later-born children often have fewer rules, less parental supervision, lower grades, and are less likely to receive financial support to attend college (Conley, 2004; Hotz and Pantano, 2013).

Another limitation is that social learning theories disregard the interconnections between social structures and socialization. As family forms become more complex, children's socialization becomes more intricate. In the last 50 years or so, children have experienced greater family instability due to a biological parent's divorce, remarriage, cohabitations, and having children with more than one partner. One outcome is that by age 9, children who have undergone such disruptions are more likely to demonstrate behavioral problems that include delinquency (e.g., stealing, vandalizing property) and classroom disruptions (e.g., fighting, bullying, having temper tantrums) (Fomby and Osborne, 2017).

self an awareness of one's social identity.

looking-glass self a self-image based on how we think others see us.

4-3b Symbolic Interaction Theories

Symbolic interaction theories have had a major impact in explaining social development. Sociologists Charles Horton Cooley (1864–1929), George Herbert Mead (1863–1931), and Erving Goffman (1922–1982) were especially influential in showing how social interaction shapes socialization.

CHARLES HORTON COOLEY: EMERGENCE OF THE SELF AND THE LOOKING-GLASS SELF

Newborn infants lack a sense of **self**, an awareness of their social identity. Gradually, they begin to differentiate themselves from their environment and to develop a sense of self.

After carefully observing the development of his young daughter, Charles Horton Cooley (1909/1983) concluded that children acquire a sense of who they are through their interactions with others, especially by imagining how others view them. The sense of self, then, isn't innate but emerges out of social relationships. Cooley called this social self the *reflected self*, or the **looking-glass self**, a self-image based on how we think others see us. He proposed that the looking-glass self develops in an ongoing process of three phases:

Pat Byrnes The New Yorker Collection/The Cartoon Bank

- **Phase 1: Perception.** We imagine how we appear to other people and how they *perceive* us ("She thinks I'm attractive" or "I bet he thinks I'm fat").
- **Phase 2: Interpretation of the perception.** We imagine how others *judge* us ("She's impressed with me" or "He's disgusted with the way I look").
- **Phase 3: Response.** We experience *self-feelings* based on how we think other people judge us. If we think they see us in a favorable light, we may feel proud, happy, or self-confident ("I'm terrific"). If we think others see us in a negative light, we may feel angry, embarrassed, or insecure ("I'm pathetic").

Our interpretation of others' perceptions (phase 2) may be totally wrong. The looking-glass self, remember, refers to how we *think* others see us rather than how they *actually* see us. Our perceptions of other people's views—whether we're right or wrong—affect our self-image, which then affects our behavior.

Cooley focused on how children acquire a sense of who they are through their interactions with others, but he noted that the process of forming a looking-glass self doesn't end in childhood. Instead, our self-concept may change over time because we reimagine ourselves as we think others see us—attractive or ugly, interesting or boring, intelligent or stupid, graceful or awkward, and so on.

GEORGE HERBERT MEAD: DEVELOPMENT OF THE SELF AND ROLE TAKING

Cooley described how an individual's sense of self emerges, but not how it develops. George Herbert Mead, one of Cooley's colleagues, took up this task. For Mead (1934), the most critical social interaction occurs in the family, the foundation of socialization.

Our self develops, according to Mead, when we learn to differentiate the *me* from the *I*—two parts of the self. The *I* is creative, imaginative, impulsive, spontaneous, self-centered, and sometimes unpredictable. The *me* that has been successfully socialized is aware of other people's attitudes, has self-control, and has internalized social roles. Instead of impulsively and selfishly grabbing another child's toys, for example, as the *I* would do, the *me* asks for permission to use someone's toys and shares them with others.

Photodisc/Getty Images

Parents who encourage their children to express themselves help them develop a positive self-image.

For Mead, the *me* forms as children engage in **role taking**, learning to take the perspective of others. Children gradually acquire this ability early in the socialization process through three sequential stages:

1. **Preparatory stage (roughly birth to 2 years).** An infant doesn't distinguish between the self and others. The *I* is dominant, while the *me* is forming in the background. In this stage, children learn through imitation. They may mimic daddy's shaving or mommy's angry tone of voice without really understanding the parent's behavior. In this exploratory stage, children engage in behavior that they rarely associate with words or symbols, but they begin to understand cause and effect (e.g., crying leads to being picked up). Gradually, as the child begins to recognize others' reactions and forms a looking-glass self, he or she develops a self.

2. **Play stage (roughly 2 to 6 years).** The child begins to use language and to understand that words (like *dog* and *cat*) have a shared cultural meaning. Through play, children begin to learn role taking in two ways. First, they emulate significant others. A **significant other** is someone whose opinions we value and who influences our thinking, especially about ourselves. A significant other can be anyone: usually a parent, grandparent, or sibling in early childhood; a teacher, girlfriend/boyfriend, spouse, best friend, or employer in later childhood and adulthood. The child learns that he or she has a self that's distinct from others, that others behave in many different ways, and that others expect her or him to behave in specific ways. In other words, the child learns social norms (see Chapter 3).

   In the play stage, the child moves beyond imitation and acts out imagined roles ("I'll be the mommy and you be the daddy"). The play stage involves relatively simple role taking because the child plays one role at a time but doesn't yet understand the relationships between roles. This stage is crucial, according to Mead, because the child *is learning to take the role of the other*. For the first time, the child tries to imagine how others behave or feel. The *me* grows stronger because the child is concerned about the judgments of significant others.

role taking learning to take the perspective of others.

significant other someone whose opinions we value and who influences our thinking, especially about ourselves.

Also in the play stage, children experience **anticipatory socialization**, learning how to perform a role that they'll occupy in the future. By playing "mommy" or "daddy," children prepare themselves for eventually becoming parents. Anticipatory socialization continues in later years when, for instance, expectant parents attend childbirth classes, job seekers practice their skills in mock interviews, and high school students visit campuses and attend orientation meetings to prepare for college life.

3. **Game stage (roughly 6 years and older).** This stage involves acquiring the ability to understand connections between roles. The child must "not only take the role of the other . . . but must assume the various roles of all participants in the game, and govern his action accordingly" (Mead, 1964: 285).

Figure 4.1 Mead's Three Stages in Developing a Sense of Self

STAGE 1: **Preparatory Stage** (under age 2)
No distinction between self and others; the child is self-centered and self-absorbed
Learns through observation

STAGE 2: **Play Stage** (aged 2 to about 6)
Distinguishes between self and others
Imitates **significant others** (usually parents)
Learns role taking, assuming one role at a time in "let's pretend" and other play that teaches **anticipatory socialization**

STAGE 3: **Game Stage** (aged 6 and older)
Understands and anticipates multiple roles
Connects to societal roles through the **generalized other**

Mead used baseball to illustrate this stage. In baseball (or other organized games and activities), the child plays one role at a time (batter) but understands and anticipates the actions of other players (pitcher, shortstop, runners on bases) on both teams. To successfully participate in a game of baseball, the individual must be able to anticipate the roles of others. The same is true of participating in society. Thus, as children grow older and interact with a wider range of people, they learn to anticipate, respond to, and fulfill a variety of social roles.

At the game stage, children connect to societal roles through the **generalized other**, the norms, values, and expectations of society that affect a person's behavior.

This concept suggests that when an individual acts, she or he takes into account people in general. For example, a child is honest not merely to please a particular parent, but because parents, teachers, and others endorse honesty as an important social value. For sociologists, the development of the generalized other is a central feature of socialization because the *me* becomes an integral part of the self (*Figure 4.1*).

anticipatory socialization learning how to perform a role that a person will occupy in the future.

generalized other the norms, values, and expectations of society that affect a person's behavior.

impression management providing information and cues to others to present oneself in a favorable light while downplaying or concealing one's less appealing characteristics.

After the generalized other has developed, the *me* never fully controls the *I*, even in adulthood. We sometimes break rules or act impulsively, even though we know better. When we're frustrated, sick, tired, or angry, we may lash out against other people even though they haven't done anything wrong.

ERVING GOFFMAN: STAGING THE SELF IN EVERYDAY LIFE

Cooley and Mead described how the self and role taking emerge and develop during early socialization. Erving Goffman (1959, 1969) extended these analyses by showing that we interact differently in different settings throughout adulthood. Goffman proposed that social life mirrors the theater because we're like actors: We engage in "role performances," want to influence an "audience," and can have considerable control over the image that we project while we're "on stage." (We'll explore these concepts in greater detail in Chapter 5.)

In a process that Goffman called **impression management**, we provide information and cues to others to present ourselves in a favorable light while downplaying or concealing our less appealing characteristics. Being successful in this presentation of the self requires managing three types of expressive resources. First, we try to control the *setting*—the physical space, or "scene," where the interaction takes place. In the classroom, "a professor may use items such as chalk, lecture notes, computers, videos, and desks to facilitate a class and show that he or she is an excellent teacher" (Sandstrom et al., 2006: 105).

Who's engaging in impression management? How? Why?

A second expressive resource is controlling *appearance,* such as clothing and titles that convey information about our social status. When physicians or professors use the title "Doctor," for example, they're telling the "audience" that they expect respect.

The third expressive resource is *manner*—the mood or style of behavior we display that sends important messages to the audience. Faculty members regularly manage their manner when interacting with students. When you email or text a professor a question about your course paper, the response is usually encouraging ("Have you also thought about...") even though the professor may actually be annoyed ("You haven't thought about this very much; stop wasting my time").

As we move from one situation to another, according to Goffman, we maintain self-control, for example, by avoiding emotional outbursts and altering our facial expressions and verbal tones. Thus, all of us engage in impression management practically every day.

CRITICAL EVALUATION

Symbolic interactionists have provided major insights about socialization, particularly of young children, and have shown that fitting into our social world is a complex process that continues through adulthood. Like other theories, however, symbolic interaction has its limitations.

First, it's not clear why some children have a positive looking-glass self and are successful later in life even when the cues are consistently negative. Over the years, some of my best students have included those who grew up in abusive households or attended low-quality schools.

Second, essential concepts such as *self, me,* and *I* are vague. Because the concepts are imprecise, it's difficult, if not impossible, to measure them.

Third, stage theories, like Mead's, maintain that children automatically pass through play and game stages as they get older. Some critics argue that the stages aren't as rigid as symbolic interactionists claim. The stages may overlap or vary, for example, depending on whether parents and other primary caregivers are actively involved in the child's upbringing and provide enriching interaction. Even then, some children never develop the role of the generalized other (see Ritzer, 1992).

Fourth, some have questioned the value of the concept of the generalized other in understanding early socialization processes. Because children interact in many social contexts (home, preschool, play groups), they don't simply assume the role of one generalized other, but many. That is, as children get older, they may have several **reference groups**—people who shape an individual's self-image, behavior, values, and attitudes in different contexts (Merton and Rossi, 1950; Shibutani, 1986).

Fifth, interactionists credit people with more free will than they have. Individuals don't always have the power or ability to affect others' reactions or their own lives. For example, impression management is less common among lower than higher socioeconomic groups, whose members have internalized such skills to be successful in jobs. Also, how people react to us may be due to characteristics (e.g., gender, age, race/ethnicity) that we can't change (Powers, 2004).

Finally, a major criticism of Cooley, Mead, Goffman, and other interactionists is that they tend to downplay or ignore macro-level factors that affect people's development. As the next two sections show, institutions such as the family, education, economy, and popular culture have a major impact on socialization. Let's begin by looking at socialization agents who play important roles in our lives.

4-4 PRIMARY SOCIALIZATION AGENTS

By fretting, crying, and whining, infants teach adults when to feed, change, and pick them up. Babies are actively engaged in the socialization process, but family, friends, peer groups, teachers, and the media are some of the primary **socialization agents**—the individuals, groups, or institutions that teach us how to participate effectively in society.

reference groups people who shape an individual's self-image, behavior, values, and attitudes in different contexts.

socialization agents the individuals, groups, or institutions that teach us how to participate effectively in society.

4-4a Family

Parents, siblings, grandparents, and other family members play a critical role in our socialization. Parents, however, are the first and most influential socialization agents.

HOW PARENTS SOCIALIZE CHILDREN

The purpose of socialization is to enable children to regulate their behavior and to make responsible decisions. In teaching their children social rules and roles, parents rely on several learning techniques, such as reinforcement, that you've just read about. Parents also manage many aspects of the environment that influence a child's social development: They choose the neighborhood (which then determines what school a child attends), decorate the child's room in a masculine or feminine style, provide her or him with particular toys and books, and arrange enriching social activities (e.g., sports, art, music).

Gelpi JM/Shutterstock.com

PARENTING STYLES

Four common parenting styles affect socialization (*Table 4.3*). In *authoritarian parenting*, parents tend to be harsh, unresponsive, and rigid, using their power to control a child's behavior. *Authoritative parenting* is warm, responsive, and involved, but parents set limits and expect appropriately mature behavior from their children. *Permissive parenting* is lax: Parents set few rules but are usually warm and responsive. *Uninvolved parenting* is indifferent and neglectful: Parents focus on their own needs, spend little time interacting with the children, and know little about their interests or whereabouts (Baumrind, 1968, 1989; Maccoby and Martin, 1983; Aunola and Nurmi, 2005).

Table 4.3 Parenting Styles

Parenting Style	Characteristics	Example
Authoritarian	Very demanding, controlling, punitive	"You can't borrow the car because I said so."
Authoritative	Demanding, controlling, warm, supportive	"You can borrow the car, but be home by curfew."
Permissive	Not demanding, warm, indulgent, set few rules	"Borrow the car whenever you want."
Uninvolved	Neither supportive nor controlling	"I don't care what you do; I'm busy."

Healthy child development is most likely in authoritative homes because parents are consistent in combining warmth, monitoring, and discipline. Authoritative parenting tends to produce children who are self-reliant, achievement-oriented, and successful in school. Adolescents in households with authoritarian, permissive, or uninvolved parents tend to have poorer psychosocial development, lower school grades, lower self-reliance, higher levels of delinquent behavior, and are more likely to be swayed by harmful peer pressure (e.g., to use drugs and alcohol) (Hillaker et al., 2008; Meteyer and Perry-Jenkins, 2009).

Such findings vary by social class and race/ethnicity, however. For many Latino and Asian immigrants, authoritarian parenting produces positive outcomes, like respectful behavior and academic success. This parenting style is also effective in safeguarding children who are growing up in low-income neighborhoods with high crime rates and drug peddling (Pong et al., 2005).

Across all social classes, Black parents—particularly mothers—engage in *racial socialization*, a process that teaches children to be proud of their race and how to recognize and diminish the negative effects of discrimination. For example, many middle- and upper-middle-class Black parents spend considerable time teaching their sons how to navigate the "thug" image. Some of the strategies include suppressing their anger, frustration, or excitement when interacting with authority figures, like teachers and police officers, and monitoring how they dress (Dow, 2016).

SIBLINGS

Siblings, like parents, are important socialization agents. Older siblings may transmit beliefs that having sex at an early age, smoking, drinking, and marijuana use "aren't a big deal" (Altonji et al., 2010). On the other hand, older siblings are positive role models when they encourage their younger brothers and sisters to do well in school, to stay away from friends who get in trouble, and to get along with people. Siblings can also have a positive impact by helping their younger brothers and sisters with homework and protecting them from neighborhood bullies, as well as reducing

AP Images/Yonhap/Shin Young-kuyn

During the last decade, numerous summer camps have sprung up in South Korea where children ages 7 to 19 live in military-style barracks, undergo drills and exercise from dawn to dusk, eat very simple meals, and have no access to computers, television, or cell phones. Most are sent by parents "who realize that their pampered offspring need more discipline to become better students and grow into conscientious adults" (Glionna, 2009: A1). Some of the kids' offenses have included getting low grades, accidentally breaking a window, and, most commonly, talking back to their parents.

the chances of risky sexual behavior and teen pregnancy. During adulthood, siblings often act as confidantes and offer financial and emotional assistance during stressful times (McHale et al., 2012; Killoren et al., 2014; Killoren and Roach, 2014).

GRANDPARENTS

In many ways, grandparents are the glue that keeps a family close. Some grandparents and grandchildren live far apart and rarely see each other, but the majority see their grandchildren often, frequently do things with them, and offer them emotional and instrumental support (e.g., listening and giving them money). Many recent Asian immigrants live in extended families and are more likely to do so than any other group, including Latinos. In such co-residence, grandparents are often "historians" who transmit values and cultural traditions to their grandchildren even if there are language barriers (Benokraitis, 2015; see also Chapter 12).

Grandparents are sometimes *surrogates* because they provide regular care or raise the grandchildren. Whether custody is legal or informal, about 10 percent of all U.S. children live with a grandparent, up from 5 percent in 1990. About 2.6 million grandparents are raising their grandchildren; 33 percent of these households have no parent present (Ellis and Simmons, 2014; "National Grandparents Day…," 2016).

The most common pattern is *living-with* grandparents who have the grandchild in their own home or, less often, live in the home of a grandchild's parents. Living-with grandparents take on child-rearing responsibilities either because their children haven't moved out of the house or because teenage or adult parents can't afford to live on their own. These living arrangements have increased the number of **multigenerational households**, homes in which three or more generations live together. In effect, then, grandparents are important socialization agents.

4-4b Play, Peer Groups, and Friends

As you saw earlier, Mead believed that play and games are important in children's development. They help us become more skilled in using language and symbols, learn role playing, and internalize roles that we don't necessarily enact (the generalized other). A **peer group** consists of people who are similar in age, social status, and interests. All of us are members of peer groups, but such groups are especially influential until about our mid-20s. After that, coworkers, spouses, children, and close friends are usually more important than peer groups in our everyday lives.

PLAY

Play serves several important functions. First, it promotes cognitive development. Whether it's doing simple five-piece puzzles or tackling complex video games like *SimCity* and *RollerCoaster Tycoon*, play encourages children to think, formulate strategies, and budget and manage resources. From an early age, however, play is generally gender typed. Girls' sections of online and toy stores are swamped with cosmetics, dolls and accessories, arts and crafts kits, and housekeeping toys. Boys' sections feature sports supplies, building sets, workbenches, construction equipment, and toy guns. Such gender stereotyping sends girls the message that their most important goals should be looking pretty, nurturing others, and developing cooking and housework skills, often to the exclusion of exploration and invention. Boys are getting the message that they should be adventurous,

multigenerational households homes in which three or more generations live together.

peer group people who are similar in age, social status, and interests.

Pressmaster/Shutterstock.com

Grandparents often provide stability in family relationships and a continuity of family rituals and values. When a teacher asked her class of 8-year-olds what they liked best about a grandparent, the essays included comments like "They don't say 'Hurry up'" and "When they take us for a walk, they slow down for things like pretty leaves and caterpillars."

aggressive, competitive, and athletic, and that "caring for a baby or the home is girly and ... undesirable" (Filipovic, 2016: 51).

About 92 percent of American girls ages 3 to 12 own at least one Barbie doll. After Barbie sales plummeted 20 percent because of greater competition by other toy manufacturers, in 2016 Mattel introduced three new Barbie body types—tall, short, and curvy—in a greater number of skin tones, eye colors, and hairstyles (Abrams, 2016; Dockterman, 2016). A year later, popular doll company American Girl released Logan Everett, its first-ever boy doll.

Some toymakers have responded to parents' demands for more gender-neutral toys. For example, GoldieBlox and Roominate, created by female engineers, offer construction toys aimed at girls that encourage interest in math, science, and engineering, and Amazon has stopped using gender-based categories for children's toys (Li, 2015; Tabuchi, 2015).

The second function of play—especially when it's structured—is to enhance children's social development. Children who spend more of their free time in structured and supervised activities—like hobbies and sports—rather than just hanging out with their friends, perform better academically, are emotionally better adjusted, and have fewer problems at school and at home. In effect, sports and hobbies provide children with constructive ways to channel their energy and intelligence (McHale, 2001).

Third, play can strengthen peer relationships. Beginning in elementary school, few things are more important to most children than being accepted by their peers. Even if children aren't popular, belonging to a friendship group enhances their psychological well-being and ability to cope with stress (Rubin et al., 1998; Scarlett et al., 2005).

PEERS AND FRIENDS

Peer influence usually increases as young children get older. Especially during the early teen years, friends often reinforce desirable behavior or skills that enhance a child's self-image ("Wow, you're really good in math!"). Thus, to apply Cooley's concept, peers can help each other develop a positive looking-glass self.

Peers also serve as good role models. Children acquire a wide array of information and knowledge

AP Images/Nick Ut

Godong/BSIP SA/Alamy Stock Photo

When Target announced that it would stop labeling girls' and boys' toys in their stores, hundreds of parents flocked to its website to complain because "Boys will always be boys and girls will always be girls!" (Steinmetz, 2015: 25). Others argue that toys are more gender typed than in the past (see Tabuchi, 2015, and Chapter 9). What do *you* think?

by observing their peers. Even during the first days of school, children learn to imitate their peers at standing in line, raising their hands in class, and being quiet while the teacher is speaking.

Among teens and young adults who are lesbian or gay, heterosexual friends can be especially supportive in accepting one's gay or lesbian sexual orientation and disclosing it to family. Peers can also have a significant impact in teen dating violence by encouraging their friends to leave a relationship, to report the abuse, and to seek protection orders (Shilo and Savaya, 2011; Oudekerk et al., 2014). In college, close friends often offer academic support and motivation to succeed: They study together, quiz each other before exams, and proofread each other's papers. Close college friends can also provide social support, a sense of belonging, and relationships that last beyond college (McCabe, 2016).

Not all peer or friend influence is positive, however. Among seventh- to twelfth-graders, having a best friend who engages in sexual intercourse, is truant, joins a gang, and uses tobacco and marijuana increases the probability of imitating such behavior. If close college friends are low achievers and unmotivated, they can interfere in studying, discourage class attendance, and offer little, if any, emotional or academic support (Haas and Schaefer, 2014; McCabe, 2016).

4-4c Teachers and Schools

Oprah Winfrey often praised her fourth-grade teacher, who recognized her abilities, encouraged her to read, and inspired her to excel academically. Like family and peer groups, teachers and schools are important socialization agents.

THE SCHOOL'S ROLE IN SOCIALIZATION

By the time children are 4 or 5, school fills an increasingly large portion of their lives. The school's primary purpose is to instruct children and enhance their cognitive development. Schools don't simply transmit knowledge; they also teach children to think about the world in different ways. Because of the emphasis on multiculturalism, for example, children often learn about other societies and customs. Even outside of classes, schools affect children's daily activities through homework assignments and participation in clubs and other extracurricular activities.

Because many parents are employed, schools have had to devote more time and resources to topics—such as sex education, abusive adolescent relationships, and drug abuse prevention—that were once the sole responsibility of families (Miller et al., 2015). In many ways, then, schools play an increasingly important role in socialization.

Siede Preis/Getty Images

TEACHERS' IMPACT ON CHILDREN'S DEVELOPMENT

Teachers are among the most important socialization agents. From kindergarten through high school, teachers play numerous roles in the classroom—instructor, role model, evaluator, moral guide, and disciplinarian, to name just a few. Well-trained teachers can motivate students and increase their academic achievement regardless of their innate ability, talents, behaviors, or home circumstances (Gershenson, 2015; see also Chapter 13).

Many educators have criticized students' increasing disrespect toward adults and each other, bad manners, and bullying. Instead of complaining, some Maryland elementary school teachers and guidance counselors have started after-school clubs called "Guys with Ties, Girls with Pearls." Once a week, the students dress up and learn etiquette (e.g., being polite, not talking with a mouth full), how to write thank-you notes, and what not to say on Facebook, Twitter, and Snapchat (Bowie, 2016). According to some critics, teachers are wasting valuable time and resources on behavior that should be taught at home, but such programs are growing in popularity across the country.

4-4d Popular Culture and the Media

Because of tablets, smartphones, YouTube, and social networking sites, young people are constantly connected to electronic media (see Chapter 5). How does such technology affect socialization?

ELECTRONIC MEDIA

The American Academy of Pediatrics (AAP) advises parents to avoid use of screen media, other than video-chatting, for children younger than 18 months. It also recommends limiting screen use for children ages 2 to 5 to 1 hour per day, and only of high-quality programs. The controls would encourage more interactive activities such as talking, playing, singing, and reading together "that will promote proper brain development" (American Academy of Pediatrics, 2016). Despite the AAP's advice, 97 percent of children

Moodboard-Mike Watson Images/Getty Images

Class field trips enhance children's socialization. Among other benefits, the field trips expand learning beyond the classroom, enrich students' understanding of subject matter, and encourage them to think about occupational choices and careers that they haven't considered.

under age 4—regardless of family income—use mobile devices like smartphones and tablets. Parents are especially likely to let their youngsters use mobile devices when the adults are doing house chores (70 percent), keeping a child calm in public places (65 percent), running errands (58 percent), and putting children to sleep (28 percent) (Kabali et al., 2015).

The average young American now spends more time with electronic media than in school. The typical 8- to 10-year-old spends nearly 8 hours a day using some type of electronic media (which is more time than many adults spend in a full-time job), and 13- to 18-year-olds spend almost 9 hours a day (Rideout, 2015). Too much time on the Internet has been linked with violence, cyberbullying, trouble in school, obesity, lack of sleep, and lower grades (Strasburger and Hogan, 2013). Screen media usage itself isn't necessarily harmful, but it typically involves "passive consumption," such as watching and listening to content produced by someone else, rather than writing and creating one's own art and music or interacting face-to-face with family and friends (Rideout, 2015).

There's also a growing concern that violent video games make violence seem normal. Playing violent video games (e.g., *Grand Theft Auto, Call of Duty,* and *Halo*) can increase a person's aggressive thoughts, feelings, and behavior both in laboratory settings and in real life. Violent video games also encourage male-to-female aggression because much of the violence is directed at women (Carnagey and Anderson, 2005; Brody, 2015).

Still, it's not clear why violent video games affect people differently. Many young males who enjoy playing such games aren't any more aggressive, vicious, or destructive than those who aren't violent video game enthusiasts (Ferguson and Olson, 2013). Instead, factors such as angry and aggressive personalities, indifference to other people's feelings, and the quantity of violent gaming increase the likelihood of violent behavior (see Markey et al., 2014; Przybylski and Mishkin, 2015).

What? Adolescents Can Buy Ultraviolent Video Games, but Not a Magazine with an Image of a Nude Woman?!

In 2011, the Supreme Court ruled (7–2 in *Brown v. Entertainment Merchants Association*) that video games, even ultraviolent ones, that are sold to minors are protected by the First Amendment's guarantee of free speech. The majority said that none of the scientific studies *prove* that violent games *cause* minors to act aggressively, and that parents, not the government, should decide what is appropriate for their children (Schiesel, 2011; Walls, 2011).

Hot Property/Shutterstock.com

Not surprisingly, the video game industry was jubilant over the decision. Many parents and lawmakers, on the other hand, agreed with one of the dissenting justices that it makes no sense to forbid selling a 13-year-old boy a magazine with an image of a nude woman, but not an interactive video game in which the same boy "actively . . . binds and gags the woman, then tortures and kills her" (Barnes, 2011: A1).

Gillian Charters-Barnes/Shutterstock.com

Many men's health and fitness magazines routinely feature models who have undergone several months of extreme regimens, including starvation and dehydration, to tighten their skin and make their muscles "pop." The magazines also use lighting tricks and Photoshop to project an idealized image of hypermasculinity that, in reality, is impossible to attain (Christina, 2011; see also Ricciardelli et al., 2010).

ADVERTISING

In 2006, dozens of companies promised to limit child-directed TV ads that used cartoons and other characters to promote unhealthy food. Some companies have run shorter commercials, but the content hasn't changed very much (Kunkel et al., 2015; see also Chapter 3).

Advertisers are targeting children at increasingly younger ages. A new form of advertising called *advergaming* combines free online games with advertising. Advergaming is growing rapidly. Sites like Wonka.com, operated by Nestlé, and Barbie.com, operated by Mattel, attract millions of young children and provide marketers with an inexpensive way to "draw attention to their brand in a playful way, and for an extended period of time" (Moore, 2006: 5).

Young people see 45 percent more beer ads and 27 percent more ads for hard liquor in teen magazines than adults do in their magazines (Jernigan, 2010). Girls ages 11 to 14 are subjected to about 500 advertisements a day on the Internet, billboards, and magazines in which the majority of models are "nipped, tucked, and airbrushed to perfection" (Bennett, 2009: 43).

What effect do such ads have on girls' and women's self-image? About 43 percent of 6- to 9-year-old girls use lipstick or lip gloss, and 38 percent use hairstyling products. In addition, 8- to 12-year-old girls spend more than $40 million a month on beauty products, 80 percent of 10-year-old girls have been on a diet, and 80 percent of girls ages 13 to 18 list shopping as their favorite activity (Hanes, 2011; Seltzer, 2012). Many girls and young women believe that they have to be gorgeous, thin, and almost perfect to be loved by their parents and boyfriends (Schwyzer, 2011).

Many women, especially White women, who are unhappy with their bodies, turn to cosmetic surgery. In 2016 women had 92 percent of all cosmetic procedures; the number (almost 14.2 million) has increased 118 percent since 2000. Men had 1.3 million procedures, a 27 percent increase since 2000. Seventy percent of all patients were White. The most common cosmetic surgery was breast enlargement or breast lift for women and nose reshaping or eyelid surgery for men (American Society of Plastic Surgeons, 2017).

4-5 SOCIALIZATION THROUGHOUT LIFE

Biological aging comes naturally, but social aging is a different matter. As we progress through the life course—from infancy to death—we are expected to learn culturally approved norms, values, and roles.

4-5a Infancy and Toddlerhood

Infancy (between birth and 12 months) and toddlerhood (ages 1 to 3) encompass only a small fraction of the average person's life span, but are periods of both extreme helplessness and considerable physical and cognitive growth. Some scientists describe healthy infants' brains as "small computers" because of their enormous capacity for learning (Gopnik et al., 2001: 142).

The quality of relationships with adults and other caregivers has a profound effect on infants' and toddlers' development. Engaging infants in talk increases their vocabulary. By age 2, toddlers who know more words have better language skills that, in turn, help them control their behavior at age 3 and later (Vallotton and Ayoub, 2011).

That's Not All ...

There are many socialization agents other than family, play and peer groups, teachers and schools, and the media. If you think about the last few years, which groups or organizations have influenced who you are? What about the military, a company you've worked for, college and athletic clubs, religious organizations, support groups, athletic memberships, community groups, or others? All affect our socialization throughout life.

Lewis Hine/Archive Farms/Getty Images

Childhood—viewed as a distinct stage of development—is a fairly recent phenomenon (Ariès, 1962). Until 1938, children made up a large segment of the U.S. labor force, especially in factories. They worked 6 days a week, 12 or more hours a day, received very low wages, and were often injured because of dangerous equipment.

On average, family members spend about 30 minutes a day reading to children under age 2 (Rideout, 2014). The biggest change is children's media use. In 2014, 99 percent of all children under age 2 were using mobile devices (e.g., smartphone, tablet), up from only 10 percent in 2011. They watched TV almost an hour a day, used apps up to 20 minutes a day, and played video games another 13 minutes a day. About 20 percent of 1-year-olds have their own tablets, and 28 percent of 2-year-olds can navigate a mobile device without help (Kabali et al., 2015).

You saw earlier that the AAP has advised parents to limit their infants' and toddlers' screen time. Only 28 percent of parents say that media decreases face-to-face family time, but they often underestimate the amount of time that children under age 2 spend with electronic devices (Rideout, 2013). Electronic media usage increases digital literacy. According to a pediatrics professor, however, young children "need laps more than apps" (quoted in Louis, 2015: A16).

4-5b **Childhood Through Adolescence**

Physical, emotional, and other development continues through childhood and adolescence (roughly ages 4 to 12). Children acquire cognitive skills that include processing thoughts, speaking, reasoning, memorizing and recalling information, analyzing problems, and evaluating options. The quantity and quality of time that parents and teachers invest in developing cognitive development affect children's long-term outcomes.

Social class enhances or limits cognitive growth. Middle-class parents, regardless of race or ethnicity, engage in "concerted cultivation" child-rearing processes that deliberately try to improve their children's cognitive and social skills. In contrast, working-class and poor parents—because of a lack of free time or resources—facilitate "natural growth" by granting their children more autonomy in everyday life (Lareau, 2011). Especially since the early 1990s, the gap between high- and low-educated parents' investment in children's development has increased (Altintas, 2016).

Besides reading to their children, helping them with homework, and attending their children's school-based events, high-income parents are much more likely than lower-income parents to enroll children in extracurricular activities (*Figure 4.2*). Extracurricular activities hone cognitive skills, decrease the likelihood of antisocial behavior, boost self-esteem, teach time management and collaborating to achieve a common goal, and uncover leadership potential.

Self-help books and self-proclaimed child experts routinely instruct parents to praise their kids to bolster their self-esteem. Is such advice misguided? A recent study of parents with children ranging from 7 to 12 years old found that constantly praising kids as "special" or "exceptional" and applauding every little achievement results in narcissism. The children often demand special treatment, can't tolerate setbacks, are more violent

Figure 4.2 Higher Income Children Are More Likely to Be in Extracurricular Activities

	$75,000+	$30,000–$74,999	<$30,000
Participated in sports or athletic activities	84%	69%	59%
Done volunteer work	64%	49%	37%
Taken lessons in music, dance, or art	62%	53%	41%

Note: The data are from a nationally representative sample of parents who have at least one child ages 6 to 17.

Source: Based on "Parenting in America…," 2015, p. 11.

than other people, and are at high risk for drug addiction and depression (Brummelman et al., 2015). A recent review of 41 studies concluded that, in the long run, consistent and well-deserved (but not constant) praise strengthens the parent–child relationship, teaches children to overcome setbacks, and improves their problem-solving skills (Owen et al., 2012; Petersen, 2012).

Ryan McVay/Getty Images

Most U.S. children enjoy happy and healthy lives, but many don't. More than 5 million children younger than 18 (7 percent of all children in this age group) have an incarcerated parent. Another 416,000 are in the foster care system. Almost 13 percent of all children live in households where a parent or other adult uses, manufactures, or distributes illicit drugs. And, every year, at least 684,000 children experience neglect or physical and sexual abuse (U.S. Department of Justice, 2011; Murphey and Cooper, 2015; Child Information Gateway, 2016; U.S. Department of Health and Human Services, 2017). Thus, for millions of U.S. children, the socialization process is shaky at best.

4-5c The Teenage Years

Ages 13 to 19 is another period of tremendous change. Teenagers establish their own identity and become more independent as they mature and break away from parental supervision, a healthy development process. The most difficult part of raising adolescents, according to some parents, is dealing with their changing moods and behavior.

Tim Pannell/Corbis

Nationally, 57 percent of 6- to 17-year-olds participate in at least one after-school extracurricular activity. The benefits include learning time management, exploring diverse interests, and building social skills, but some family practitioners believe that today's children are too overbooked and overscheduled (Laughlin, 2014; Lawrence, 2015). Do you agree?

For many years, people attributed such dramatic changes to "raging hormones." Scientists now believe that there's a link between a teen's baffling behavior and the fact that his or her brain may be changing far more than was thought previously. During adolescence, the brain matures at different rates. Areas involved in basic functions such as processing information and controlling physical movement mature first. The parts of the brain responsible for controlling impulses, avoiding risky behavior, and planning ahead—the hallmarks of adult behavior—are among the last to mature (National Institute of Mental Health, 2011).

Parental involvement is usually beneficial in a child's development, but "helicopter parents," who hover over their kids, are hyperinvolved, intrusive, and overcontrolling. Anecdotal examples include parents verbally attacking teachers about their children's low grades, demanding that their child be moved to another class before the school year has even begun, completing their homework assignments, showing up in the guidance counselor's office with college applications that they have filled out for their children, and haggling with college professors over the student's grade on an exam or paper (Krache, 2008; Weintraub, 2010; Rochman, 2013).

Helicopter parenting diminishes teens' and young adults' ability to develop decision-making and problem-solving skills. However well-intentioned, helicopter parents increase their own stress and decrease their children's feelings of autonomy and competence. In turn, not believing in one's ability to achieve goals on one's own leads to teenagers' and young adults' feeling anxious, depressed, and more dissatisfied with life (Shellenbarger, 2013; Schiffrin et al., 2014).

4-5d Adulthood

Socialization continues throughout adulthood. People adopt a series of new roles that may include work, marriage, parenthood, divorce, remarriage, buying a house, and experiencing the death of a child, parent, or grandparent. We'll examine these transitions in later chapters. Two of the most important roles in adulthood are work and parenthood, but let's begin with young adults who are reluctant to leave their parents' nest.

THE CROWDED EMPTY NEST

During the 1960s and 1970s, sociologists almost always included the "empty nest" in describing the family life course. This is the stage in which parents, typically in their

Figure 4.3 Who's Not Leaving the Nest?

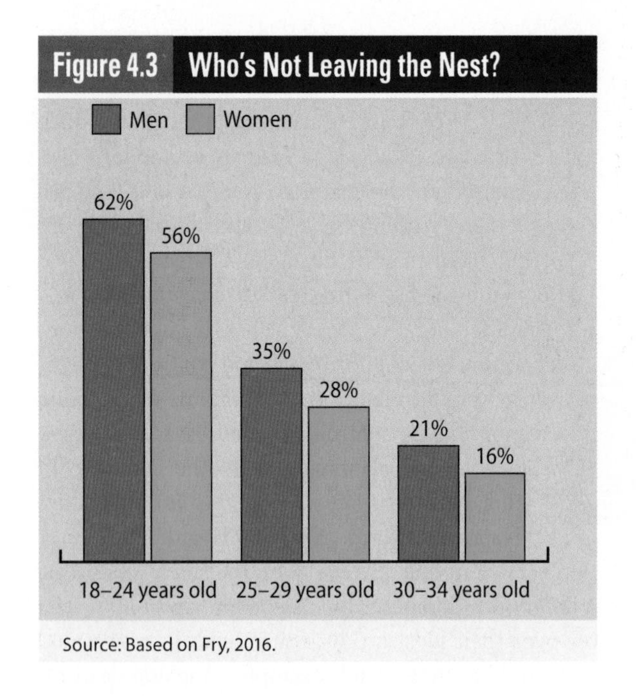

Source: Based on Fry, 2016.

50s, find themselves alone after their children have married, gone to college, or found jobs and moved out.

Today, young adults are living at home longer than they did in the past. The terms *boomerang children* and *boomerang generation* refer to young adults who move back in with their parents, often with a spouse, girlfriend or boyfriend, and children in tow.

Most young adults leave the parental nest by age 23, but the proportion ages 18 to 34 who are living with parents increased from 20 percent in 1960 to 32 percent in 2014 (Fry, 2016). Among young adults ages 18 to 34, men are more likely than women to live with their parents (*Figure 4.3*).

Macro-level factors have contributed to many Americans' delayed transition to adulthood, especially living independently. Financial insecurity, student loan debt, low wages, divorce, credit card debt, and going from job to job until they find work they enjoy have made it harder for young middle-class adults to maintain the lifestyles that their parents created. And the transition to independence gets tougher the lower a person's occupation and education. In 2014, for instance, only 19 percent of 18- to 34-year-olds who were college graduates were living with their parents, compared with 36 percent who had not completed a college degree (Fry, 2016).

Howard McWilliam

On a micro level, almost 33 percent of today's parents, compared with 19 percent in 1993, say that children shouldn't be expected to be on their own financially until age 25 or later (Taylor et al., 2012). Among those ages 18 to 24, 87 percent live at home because parents "make it easy for me to stay" (Payne and Copp, 2013). They enjoy the comforts of the pampering parental nest. According to one of my male students in his late 20s, for example, "My mom enjoys cooking, cleaning my room, and just having me around. I don't pitch in for any of the expenses, but we get along great because she doesn't hassle me about my comings and goings" (Benokraitis, 2015: 354).

Many young adults, as you'll see in Chapter 12, are postponing marriage and don't feel a need to establish their own homes. Perhaps most important, the stigma traditionally linked to young adults' living at home has faded. Among those ages 25 to 34 who live at home, 61 percent say they have friends or family members who have moved back in with their parents (Parker, 2012). The implication (and perhaps justification) of moving back home or not leaving in the first place is that "a lot of people are doing it."

WORK ROLES

The average American who was born between 1957 and 1964 has held 11 jobs from ages 18 to 46 alone; 26 percent have held 15 jobs or more ("Number of Jobs Held . . . ," 2012). Such job changing means that our occupational socialization is an ongoing process.

Learning work roles can be difficult because each job has different demands and expectations. Even when there's formal job training, we must also learn the subtle rules that are implicit in many job settings. A supervisor who says, proudly and loudly, "I have an open door policy. Come in and chat whenever you want," may *really* mean "I have an open door policy if you're not going to complain about something." Even if we remain in the same job for many years, we must often acquire new skills, especially technological ones, or risk being laid off.

Being laid off can be stressful at any age. In the case of midlife men (generally defined as those ages 45 to 60), such changes are especially traumatic: They must learn new job-hunting skills and often adjust to as much as a 70 percent decrease in their earnings (see Chapter 11). Instead of enjoying the security of a job in their midlife years, these men must undergo occupational socialization that assaults their self-image as good employees and family providers.

PARENTING ROLES

Like workplace roles, parenting does *not* come naturally. Most first-time parents muddle through by trial and error. Family sociologists often point out that we get more training for driving a car or a fishing license than for parenting.

Most couples don't realize that children are expensive. Middle-income couples, with an average income of $81,700 a year, spend about 16 percent of their earnings on a child during the first 2 years (*Figure 4.4*). These couples will spend almost $234,000 for each child from birth through high school. Child-rearing costs are much higher for single parents and especially for low-income families if a child is disabled, chronically ill, or needs specialized care that welfare benefits don't cover. Expenses are also much higher for parents whose children attend private schools (Lino et al., 2017).

Figure 4.4 How Much Does It Cost to Raise a Child?

In 2015, middle-income married couples—those with before-tax income between $59,200 and $107,400—spent $12,680 per year for a child under age 2. This amount doesn't include the costs of prenatal care or delivery.

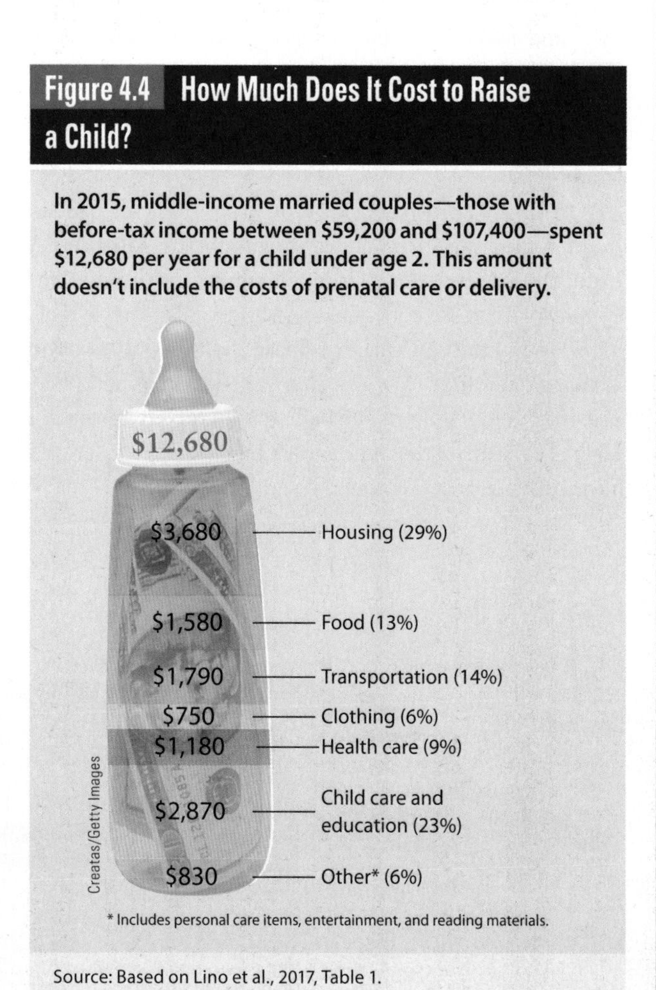

* Includes personal care items, entertainment, and reading materials.

Source: Based on Lino et al., 2017, Table 1.

About 23 percent of Americans are "sandwiched" between their children and their parents. That is, they're caring for at least one parent age 65 or older, raising a child under age 18, and/or providing economic assistance to a grown child. Pulled in many directions, these midlife adults must learn to deal with the stress of juggling a full-time job and caregiving responsibilities (Parker and Horowitz, 2015; see also Chapter 12).

4-5e Later Life

Socialization continues in later life. People age 65 and older often remarry and must learn to play new roles such as stepparent or step-grandparent. Many others undergo unwanted rites of passage such as widowhood, divorce or re-divorce, becoming a custodial parent, and dealing with the death of an adult child or grandchild.

Because we live longer, we may spend 20 percent of our adult life in retirement. When retired people are unhappy, it's usually because of health or income problems rather than loss of the worker role (see Chapter 14). If retirement benefits don't keep up with inflation or if a retiree isn't covered by a pension plan, people can plunge into poverty soon after retirement. Poverty is devastating at any age, but sliding down the economic ladder during one's 60s and 70s is especially traumatic because there's little chance of recouping one's income (see Chapters 12 and 14).

Older people who are in poor health and experience continuous pain sometimes welcome death. Among those who are healthy, some reenter the labor force or volunteer in various organizations, many are involved with their grandchildren, and others forge new relationships after widowhood (see Chapter 12). Thus, even in the later years, many people continue to learn new roles.

4-6 RESOCIALIZATION AND TOTAL INSTITUTIONS

Throughout life, many people undergo **resocialization**, the process of unlearning old ways of doing things and adopting new attitudes, values, norms, and behavior. Resocialization ranges from mild to intense, and can be voluntary or involuntary.

resocialization unlearning old ways of doing things and adopting new attitudes, values, norms, and behavior.

4-6a Resocialization

Much resocialization is voluntary, as when an American wife moves to the Middle East with her Iranian husband and follows local customs about veiling, women's submissive roles, and staying out of public life. Other examples include entering a religious order, seeking treatment in a drug rehabilitation facility, or serving in the military.

Voluntary resocialization is usually mild. First-year college students, new employees, and members of support groups, for instance, must learn new rules and live up to an individual's, a group's, or an organization's expectations. Such changes, however, are usually short term and may even be enjoyable.

On the other hand, some voluntary resocialization can be long, difficult, and intense. New parents' lives change dramatically as they learn parenting roles. In divorce, people may experience a sense of relief and liberation, but must also cope with emotional distress, financial changes, and child custody disputes. People who experience a permanent physical disability must cope with physical limitations, psychological issues, and stigma. Soldiers who return from wars, even if they haven't suffered physical injuries, describe coming home and living a normal life as daunting. They must reconnect with their children, adjust to spouses who have become more independent, and deal with posttraumatic stress disorders (PTSD) that include depression and nightmares (Jones, 2013; van Agtmael, 2013).

Resocialization can also be involuntary when it occurs against a person's will. Examples include children sent to a foster home or older people persuaded to move to an assisted-living facility. Whether voluntary or not, total institutions are the most intense forms of resocialization.

total institutions isolated and enclosed social systems that control most aspects of the participants' lives.

4-6b Total Institutions

A **total institution** is an isolated and enclosed social system that controls most aspects of its participants' lives. Prisons, mental hospitals, nursing homes, juvenile detention camps, convents and monasteries, boarding schools, and military training centers are all total institutions to varying degrees. Because a total institution's goal is resocialization, people are isolated from the rest of society, stripped of their former identities, and required to conform to new rules and behavior (Goffman, 1961).

Resocialization in total institutions is an intense two-step process. First, the staff uses *degradation ceremonies*, humiliating rituals that publicly stigmatize people and try to erode their identities and independence (Garfinkel, 1956). Inmates are strip-searched, deloused, and fingerprinted; their heads are shaved; they're issued a uniform and are given a serial number; they have no privacy, limited access to family and friends, and little communication with the outside world; and they're told what to do and when. In the military, soldiers get short haircuts, wear matching uniforms, follow rigid schedules, and make few personal decisions. The purpose of these practices and restrictions is to destroy any sense of individualism, autonomy, and past identity, and to produce a more compliant person (Goffman, 1961; see also Chapter 7).

In the second resocialization step, the staff tries to build a new identity that conforms to the institution's expectations. Inmates who conform are rewarded with simple privileges (e.g., making phone calls, more visits with family) and an earlier release. Soldiers often bond with one another, become more self-disciplined, may use their new skills in civilian jobs, and are proud of having served in the military ("Once a Marine, always a Marine").

iStock.com/Davincidig; iStock.com/Bridget McGill

Most resocialization occurs in institutions—ranging from monastic orders to the military.

Total institutions don't always succeed in resocializing people. Inmates, for instance, find ways to make money, to barter, to smuggle in banned items (including drugs and smartphones), and to make and enforce their own rules through a system of rewards and punishments (Kilgore, 2015; Biunno, 2016). Netflix's *Orange Is the New Black* series dramatizes female inmates' making and breaking formal and informal rules to cope with imprisonment.

As this chapter shows, socialization is a powerful force in shaping who we are. It doesn't produce robots, however, because people are creative, adapt to new environments, and change as they interact with others.

STUDY TOOLS 4

READY TO STUDY? IN THE BOOK, YOU CAN:

- ☐ Check your understanding of what you've read with the Test Your Learning Questions provided on the Chapter Review Card at the back of the book.
- ☐ Tear out the Chapter Review Card for a handy summary of the chapter and key terms.

ONLINE AT CENGAGEBRAIN.COM WITHIN MINDTAP YOU CAN:

- ☐ Explore: Develop your sociological imagination by considering the experiences of others. Make critical decisions and evaluate the data that shape this social experience.
- ☐ Analyze: Critically examine your basic assumptions and compare your views on social phenomena to those of your classmates and other MindTap users. Assess your ability to draw connections between social data and theoretical concepts.
- ☐ Create: Produce a video demonstrating connections between your own life and larger sociological concepts.
- ☐ Collaborate: Join your classmates to create a capstone project.

5 Social Interaction in Everyday Life

Chip Somodevilla/Getty Images News/Getty Images

LEARNING OBJECTIVES

After studying this chapter, you will be able to...

5-1 Explain the importance of social interaction and its relationship to social structure.

5-2 Define and illustrate status set, ascribed and achieved statuses, master status, and status inconsistency.

5-3 Explain how and why social roles differ, and how people cope with role conflict and role strain.

5-4 Compare and illustrate symbolic interaction, social exchange, and feminist explanations of social interaction.

5-5 Describe and illustrate nonverbal communication, its importance, and cross-cultural variations.

5-6 Summarize the benefits and costs of online interaction.

After finishing this chapter go to **PAGE 99** for **STUDY TOOLS**

What, if anything, annoys you about airplane passengers? Among U.S. adults, the top offenders are parents who don't control their whining or misbehaving children, people behind them who kick the seat, and travelers who talk loudly or listen to loud music, usurp several overhead bins, become drunk or disruptive, and never stop talking (Expedia, 2017).

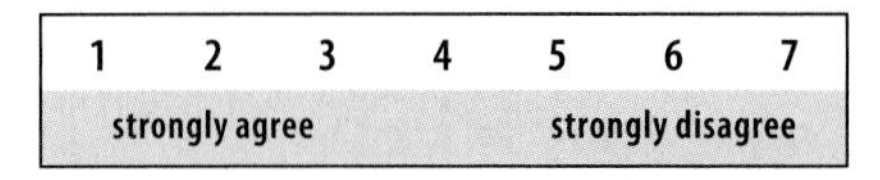

I spend more time on social media than I do talking face-to-face with my family and friends.

1	2	3	4	5	6	7
strongly agree						strongly disagree

Such complaints illustrate **social interaction**, the process of acting toward and reacting to people around us. Social interaction—verbal, nonverbal, face-to-face, or on social media—is central to all human social activity, can be cooperative or competitive, and can be interpreted in various ways across cultures and groups over time. As people interact, they create social structure.

5-1 SOCIAL STRUCTURE

Social structure is an organized pattern of behavior that governs people's relationships (Smelser, 1988). Because social structure shapes our actions, life is typically orderly and predictable rather than chaotic or random. We're often not aware of the impact of social structure until we violate cultural rules, formal or informal, that dictate our daily behavior—whether in class, at home, at work, or in an airplane.

Every society has a social structure that encompasses statuses, roles, groups, organizations, and institutions (Smelser, 1988). We'll examine groups, organizations, and institutions in later chapters. Let's take a closer look here at statuses and roles, two building blocks of social structure.

5-2 STATUS

For most people, the word *status* signifies prestige. An executive, for example, has more status than a receptionist, and a physician has a higher status than a nurse. For sociologists, **status** is a social position that a person occupies in a society (Linton, 1936). Thus, executive, secretary, physician, and nurse are all social statuses. Other examples of statuses are musician, voter, sister, parent, police officer, and friend.

Sociologists don't assume that one position is more important than another. A mother, for example, isn't more important than a father, and an adult isn't more important than a child. Instead, all statuses are significant because they determine social identity, or who we are.

5-2a Status Set

Every person has many statuses (*Figure 5.1*). Together, they form her or his **status set**, a collection of social statuses that a person occupies at a given time (Merton, 1968). Dionne, one of my students, is female, African American, 42 years old, divorced, mother of two, daughter, aunt, Baptist, voter, bank supervisor, volunteer at a soup kitchen, and country music fan. All of these socially defined positions (and others) make up Dionne's status set.

Status sets change throughout the life course. Because she'll graduate next year and is considering remarrying and starting an after-school program, Dionne will add at least three more statuses to her status set and will also lose the statuses of divorced and college student. As Dionne ages, she'll continue to gain new statuses (grandmother, retiree) and lose others (supervisor at a bank, or wife, if she is widowed).

Statuses are *relational*, or complementary, because they're connected to other statuses: A *husband* has a *wife*, a *real-estate agent* has *customers*, and a *teacher* has *students*. No matter how many

social interaction acting toward and reacting to people around us.

social structure an organized pattern of behavior that governs people's relationships.

status a social position that a person occupies in a society.

status set a collection of social statuses that a person occupies at a given time.

Figure 5.1 Is This Your Status Set?

Being a college student is only one of your current statuses. What other statuses comprise your status set?

statuses you occupy, every status is linked to those of other people. These connections between statuses influence our behavior and relationships.

5-2b Ascribed and Achieved Status

Status sets include both ascribed and achieved statuses. An **ascribed status** is a social position that a person is born into. We can't control, change, or choose our ascribed statuses, which include sex (male or female), age, race, ethnicity, and family relationships. Your ascribed statuses, for example, might include *male*, *Latino*, and *brother*.

An **achieved status**, in contrast, is a social position that a person attains through personal effort or assumes voluntarily. Your achieved statuses might include college graduate, mother, and employee. Unlike our ascribed statuses, our achieved statuses can be controlled and changed. We have no choice about being a son or daughter (an ascribed status), but we have an option to become a parent (an achieved status).

Students sometimes think that religion and social class are ascribed rather than achieved statuses. It's true that someone may be born into a family that practices a certain religion or one that is poor, middle class, or wealthy. Because we can change these statuses through our own actions, however, neither religion nor social class is an ascribed status. A Catholic might convert to Judaism (or vice versa), and thousands of Americans born into poor or working-class families have become millionaires (see Chapter 8).

ascribed status a social position that a person is born into.

achieved status a social position that a person attains through personal effort or assumes voluntarily.

master status overrides other statuses and forms an important part of a person's social identity.

status inconsistency the conflict that arises from occupying social positions that are ranked differently.

5-2c Master Status

A **master status** overrides other statuses and forms an important part of a person's social identity (Hughes, 1945; Becker, 1963). A master status is usually immediately apparent, makes the biggest impression, affects others' perceptions, and, consequently, often shapes a person's entire life.

Some master statuses—like sex, age, and race—are ascribed. Whatever you do, for example, people will see you as male or female and react accordingly. Other master statuses are achieved. For many people, their occupation is their master status: Occupation tends to be the first thing we ask about when we meet someone; conveys much information about income, education, skills, and achievements; and lingers after we've left our jobs (e.g., someone is introduced as a former CEO or ex-military officer). Other achieved master statuses can be based on religion, community leadership, family wealth, or deviant behavior.

5-2d Status Inconsistency

Because we hold many statuses, some clash. **Status inconsistency** refers to the conflict that arises from

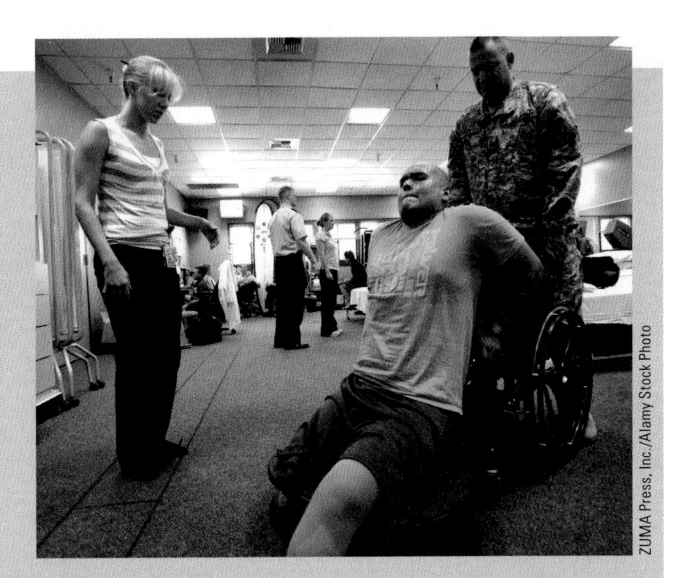

A physical disability can become a master status. One of the military's major tasks is helping severely injured soldiers become as independent as possible.

occupying social positions that are ranked differently. Examples include a teacher who works as a bartender or a skilled welder who stocks shelves at a grocery store because neither can find a better job in a stagnant economy. We'll cover status inconsistency in more detail in Chapter 7. For now, you should be aware that you occupy many statuses, and some of them may clash now and in the future.

5-3 ROLE

Each status is associated with one or more roles. A **role** is the behavior expected of a person who has a particular status. We *occupy* a status but *play* a role. In this sense, a role is the dynamic aspect of a status (Linton, 1936).

College student is a status, but the role of a college student requires many *formal* behaviors such as going to class, reading, thinking, completing weekly assignments, writing papers, and taking exams. *Informal* behaviors may include joining a student club, befriending classmates, attending football games, and even abusing alcohol on weekends.

Like statuses, roles are relational. Playing the role of professor requires teaching, advising students, answering email and text messages, and grading exams and assignments. Most professors are also expected to serve on committees, do research, publish articles and/or books, and perform services like giving talks to community groups. Thus, the status of college student or professor involves numerous role requirements that govern who does what, where, when, and how.

Roles can be rigid or flexible. A secretary typically plays a role that's defined by rules about when to come to work, how to answer the phone, when to submit the necessary work, and in what format. A boss, on the other hand, usually enjoys considerable flexibility: She or he has more freedom to come in late or leave early, to determine which projects should be completed first, and to decide when to hire or fire employees.

Because roles are based on mutual obligations, they ensure that social relations are fairly orderly. We know what we're supposed to do and what others expect of us. If professors fail to meet their role obligations by missing classes or coming to classes unprepared, students may respond by studying very little, cutting classes, and submitting negative course evaluations. If students miss classes, don't turn in the required work, or cheat, professors can fail them or assign low grades.

5-3a Role Performance

Roles define how we're *expected* to behave in a particular status, but people vary considerably in fulfilling the responsibilities associated with their roles. Many college students succeed, whereas others fail; some professors inspire their students, whereas others put them to sleep. These differences reflect **role performance**, the *actual* behavior of a person who occupies a status. For example, a professor may demand more of graduate students than of undergraduates. An instructor may also interact differently with a 19-year-old student than a 40-year-old student who's anxious about returning to college.

5-3b Role Set

We occupy many statuses and play many roles associated with each status. A **role set** is the array of roles attached to a particular status. The status of a college student, for example, may include at least six different roles (*Figure 5.2*). Each of these

role the behavior expected of a person who has a particular status.

role performance the actual behavior of a person who occupies a status.

role set array of roles attached to a particular status.

Figure 5.2 Role Set of a Typical College Student

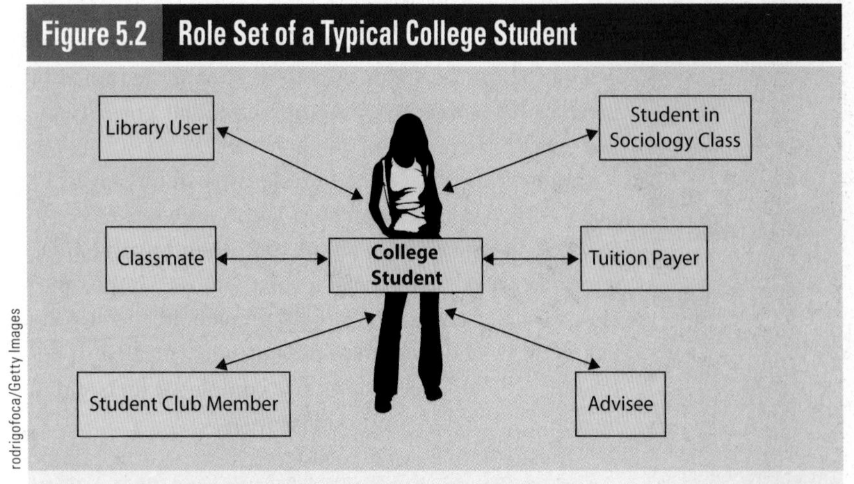

What other roles comprise your college student role set?

roles is linked to people with different statuses (e.g., professor, registrar, librarian). Because a different set of norms governs each of these relationships, the student interacts differently with a classmate than with a reference librarian or a professor.

Figure 5.2 shows only one status—that of college student. If you think about a college student's other statuses (employee, son or daughter, parent, girlfriend or boyfriend, husband or wife), you can see that meeting the expectations of numerous role sets can create considerable role conflict and role strain.

5-3c Role Conflict and Role Strain

Playing many roles often leads to **role conflict**, difficulties in playing two or more contradictory roles. College students who have a job, especially if it's full-time, often experience role conflict because both employers and professors expect them to excel, but it's very difficult to meet these expectations. The role conflict increases if the student has young children or cares for an aging parent or grandparent.

Whereas role conflict arises because of the expectations of contradictory roles, **role strain** is due to the conflicting demands within the same role. Students experience role strain when several exams are scheduled on the same day and they're also involved in time-consuming extracurricular activities. Faculty members experience role strain when the requirements of being a professor—teaching, research, and community service—sap their energy and time. Military chaplains report role strain in preaching about peace while blessing those about to go to war.

role conflict difficulties in playing two or more contradictory roles.

role strain difficulties due to conflicting demands within the same role.

Almost all of us experience role strain because many inconsistencies are built into our roles. *Table 5.1* gives examples of some of the reasons for role conflict and role strain.

5-3d Coping with Role Conflict and Role Strain

Role conflict and role strain can produce tension, hostility, aggression, and stress-related physical problems that include insomnia, headaches, ulcers, eating disorders, anxiety attacks, chronic fatigue, nausea, weight loss or gain, and drug and alcohol abuse (Weber et al., 1997). Some role conflict and role strain may last only a few weeks, but others are long-lived (e.g., working in a stressful or low-paying job).

To deal with role conflict and role strain, some people *deny that there's a problem.* Employed mothers,

Table 5.1 Why Do We Experience Role Conflict and Role Strain?

Reason	Example
Because many people are overextended, some roles are bound to conflict with others.	Students may study less than they want because employers demand that they work overtime or on weekends.
People have little or no training for many roles.	Parents are expected to turn out "perfect" kids even though they receive training for driving a car but none for parenting.
Some role expectations are unclear or contradictory.	Some employers pride themselves on having family-friendly policies but expect employees to work 12 hours a day, travel on weekends, and use vacation days to care for a sick child.
Highly demanding jobs often create difficulties at home.	Some jobs (being in the military, policing, and firefighting) require people to be away from their families for extended periods of time or during crises.

especially those who are divorced, often become supermoms. They provide home-cooked meals, attend their children's sports activities, and volunteer for a school's fund-raising campaign even though they're exhausted, must do laundry at midnight, and neglect their own interests and well-being. Supermoms may succeed over a number of years in managing role conflict and role strain. Eventually, however, they may become angry and resentful, or experience health or emotional problems (Douglas and Michaels, 2004).

There are five more effective ways to minimize role conflict and role strain. First, we can reduce role conflict through *compromise* or *negotiation.* To decrease the conflict between work and family roles, many couples draw up schedules that require fathers to do more of the housework and child rearing (see Chapter 12).

Glenda/Shutterstock.com

Second, we can *set priorities.* If extracurricular activities interfere with studying, which is more important? Succeeding in college always requires making sacrifices, such as attending fewer parties or not seeing friends as often as we'd like.

Third, we can *compartmentalize* our roles. Many college students take courses in the morning, work part-time during the afternoon or evening, and devote part of the weekend to leisure activities. It's not always easy but usually possible to separate our various roles.

Fourth, we can decide *not to take on more roles.* One of the most effective ways to avoid role conflict is to just say "no" to requests to do volunteer work, pressure from family or friends to take on unwanted tasks (e.g., babysitting), or pleas to become involved in college or community activities.

Finally, we can *exit* a role or status. Withdrawing from community activities and club offices, for example, can decrease role conflict and role strain considerably. There's always pressure to remain in a role, but none of us is indispensable, and there are many people who are eager to replace us. Although necessary, some role exits are painful and long-lived. Divorce, for instance, isn't a quick event but a process that goes on for many years (sometimes decades), during which two people (and their children) must redefine their expectations and adjust to a different household structure (see Chapter 12).

Geo Martinez/Shutterstock.com

Like other people, college students can reduce their role conflict and role strain by setting priorities and compartmentalizing their roles.

5-4 EXPLAINING SOCIAL INTERACTION

Statuses and roles shape our everyday relationships, but social interaction provides the basis of these relationships. At the macro level, functionalists and conflict theorists study interaction among and between social structures, institutions, and societies (see, especially, Chapters 1 and 6). However, three micro perspectives—symbolic interaction, exchange theory, and some feminist theories—are especially useful in understanding why we interact as we do, and why people sometimes interpret the same words differently. The explanations offer different insights, but each shows how people interact in their daily lives (*Table 5.2* summarizes these approaches).

5-4a Symbolic Interaction

For interactionists, the most significant characteristic of all human communication is that people take each other and the context into account (Blumer, 1969). When your professors ask, "How are you?" they expect a "Fine, thanks" and probably barely look at you. A doctor, on the other hand, usually looks you in the eye when asking "How are you?" and takes notes as soon as you start to reply. Thus, "How are you?" has different meanings in different social contexts and elicits different responses ("I'm doing okay" vs. "I've been having a lot of headaches lately"). In this sense, people construct reality.

Table 5.2 Sociological Explanations of Social Interaction

Perspective	Key Points
Symbolic Interactionist	• People create and define their reality through social interaction. • Our definitions of reality, which vary according to context, can lead to self-fulfilling prophecies.
Social Exchange	• Social interaction is based on a balancing of benefits and costs. • Relationships involve trading a variety of resources, such as money, youth, and good looks.
Feminist	• Females and males act similarly in many interactions but may differ in communication styles and speech patterns. • Men are more likely to use speech that's assertive (to achieve dominance and goals), whereas women are more likely to use language that connects with others.

SOCIAL CONSTRUCTION OF REALITY

"Human reality is socially constructed reality" (Berger and Luckmann, 1966: 172). That is, we produce, interpret, and share the reality of everyday life with others. This social construction of reality typically evolves through direct, face-to-face interaction, but the interaction can also be indirect, as in watching television or participating in social media.

Businesspeople, advertisers, politicians, educators, and even social scientists use words intentionally to shape or change our perceptions of reality. For example, doublespeak is "language that pretends to communicate but really doesn't. [It] makes the bad seem good, the negative appear positive, the unpleasant appear attractive or at least tolerable" (Lutz, 1989: 1).

There are several kinds of doublespeak: *Euphemisms* are words or phrases that avoid a harsh, unpleasant, or distasteful reality; *gobbledygook* (or *bureaucratese*) overwhelms the listener with big words and long sentences; and *inflated language* makes simple everyday things seem complex (Lutz, 1989; see also Beard and Cerf, 2015). Take the Doublespeak quiz to see how much doublespeak you understand or use.

Cory Thoman/Shutterstock.com

self-fulfilling prophecy if we define something as real and act on it, it can, in fact, become real.

ethnomethodology the study of how people construct and learn to share definitions of reality that make everyday interactions possible.

SOCIAL INTERACTION AND SELF-FULFILLING PROPHECIES

Our perceptions of reality shape our behavior. In an oft-cited statement, known as the *Thomas Theorem,* sociologists W. I. Thomas and Dorothy Thomas (1928: 572) observed, "If men define situations as real, they are real in their consequences." That is, our behavior is a result of how we interpret a situation ("My mother-in-law hates me, so I go shopping when she drops in").

Carrying this idea further, sociologist Robert Merton (1948/1966) proposed that our definitions of reality can result in a **self-fulfilling prophecy**: If we define something as real and act on it, it can, in fact, become real. For example, physical education teachers who publicly humiliate students may turn them off physical fitness for good. One man said, "To this day I feel totally inadequate in team-related activities and have a natural reflex to AVOID THEM AT ALL COSTS" (Strean, 2009: 217; capitalization in original). Thus, gym teachers' negative comments make students feel inadequate, regardless of their ability, and change their behavior during adulthood.

Our perceptions of reality shape our behavior, but how do people define that reality? Interactionists use two research tools—*ethnomethodology* and *dramaturgical analysis*—to help answer this question.

ETHNOMETHODOLOGY

A term coined by sociologist Harold Garfinkel (1967), **ethnomethodology** is the study of how people construct and learn to share definitions of reality that make everyday interactions possible. That is, we base our interactions on common assumptions about what makes sense in specific situations (Schutz, 1967; Hilbert, 1992).

People make sense of their everyday lives in two ways. First, by observing conversations, people discover the general rules that we all use to interact. Second, people can understand interaction rules by breaking them. Over a number of years, Garfinkel instructed his students to purposely violate everyday interaction rules and then to analyze the results.

In these exercises, some of his students went to a grocery store and insisted on paying more than was asked for a product. Others were instructed, in the course of an ordinary conversation "to bring their faces up to the subject's until their noses were almost touching" (Garfinkel, 1967: 72).

When students violated interaction norms, they were sanctioned. Grocery clerks became hostile when the students insisted on paying more than the marked price for a product, and people backed off when their noses were almost touching: "Reports were filled with accounts of astonishment, bewilderment, shock, anxiety, embarrassment, and anger" (Garfinkel, 1967: 47).

Violating interaction rules, even unspoken ones, can trigger anger, hostility, and frustration. College students become upset if professors are sarcastic or disrespectful, are unprepared or consistently late, constantly read from the book, or discourage students' comments and questions (Berkos et al., 2001).

Doublespeak

1. automotive internist	A. a lie
2. reducing costs	B. bombing
3. alternative fact	C. undertaker
4. auto dismantler and recycler	D. cutting salaries and/or firing people
5. air support	E. fake
6. previously owned	F. car mechanic
7. genuine imitation	G. junk dealer
8. bereavement care expert	H. stock market crash
9. equity retreat	I. used
10. revenue enhancement	J. tax increase

Answers: 1-F; 2-D; 3-A; 4-G; 5-B; 6-I; 7-E; 8-C; 9-H; 10-J

DRAMATURGICAL ANALYSIS

Dramaturgical analysis is a research approach that examines social interaction as if occurring on a stage where people play different roles and act out scenes for the "audiences" with whom they interact. According to sociologist Erving Goffman (1959, 1967), life is similar to a play in which each of us is an actor, and our social interaction is much like theatre because we're always on stage and always performing. In our everyday "performances," we present different versions of ourselves to people in different settings (audiences).

Because most of us try to present a positive image of ourselves, much social interaction involves *impression management*, a process of suppressing unfavorable traits and stressing favorable ones (see Chapter 4). To control information about ourselves, we often rely on props to convey or reinforce a particular image. For example, physicians, lawyers, and college professors may line their office walls with framed diplomas, medical certificates, or community awards to give the impression that they're competent, respected, and successful.

For Goffman, a performance involves front- and back-stage behaviors. The *front stage* is an area where an actual performance takes place. In front-stage areas, such as living rooms or restaurants, the setting is clean and the servers or hosts are typically polite and deferential to guests. The *back stage,* an area concealed from the audience, is where people can relax. Bedrooms and restaurant kitchens are examples of back stages. After guests have left, the host and hostess may kick off their shoes and gossip about their company. In restaurants, cooks and servers may criticize the guests, use vulgar language, and yell at each other. Thus, the civility and decorum of the front stage may change to rudeness in the back stage.

Another example of front- and back-stage behavior involves faculty and students. Professors want to create the impression that they're well prepared, knowledgeable, and hardworking. In back-stage areas like their offices, however, faculty may complain to their colleagues that they're tired of teaching a particular course or grading terrible exams and papers, or they may confess that they don't always prepare for classes as well as they should. Students often plead with instructors for higher grades: "But I studied very hard" or "I'm an A student in my other courses." In back-stage areas like dormitories, student lounges, or libraries, however, students may admit to their friends that they barely studied or that they have a low grade-point average.

dramaturgical analysis examines social interaction as if occurring on a stage where people play different roles and act out scenes for the audiences with whom they interact.

iStock.com
Thinkstock Images/Stockbyte/Getty Images
Luca Campedri/Nonstock/Jupiter Images

The chaotic and crowded back-stage conditions in many restaurant kitchens differ markedly from the efficient and relaxed front-stage behavior of its servers. Can you think of a situation where your front- and back-stage behaviors differ dramatically?

the domestic violence research shows that women stay in abusive relationships because their self-esteem has eroded after years of criticism and ridicule from both their parents and their spouses or partners ("You'll be lucky if anyone marries you," "You're dumb," "You're ugly," and so on). In effect, the abused women (and sometimes men) believe that they have nothing to offer in a relationship, that they don't have the right to expect benefits (especially when the abuser blames the victim for provoking the anger), or that enduring an abusive relationship is better than being alone. An estimated 33 percent of unhappily married couples stay together because one or both partners believe that they have too much to lose if they divorce (e.g., not seeing children, financial insecurity) (Wong, 2014; see also Chapter 12).

5-4b Social Exchange Theory

Social exchange theory proposes that individuals seek through their interactions to maximize their rewards and minimize their costs. An interaction that elicits a reward, such as approval or a smile, is more likely to be repeated than an interaction that evokes disapproval or criticism (Thibaut and Kelley, 1959; Homans, 1974; Blau, 1986). Because interactions are most satisfying when there's a balance between giving and taking, individuals must decide whether and which resources to exchange.

People bring various tangible and intangible resources to a relationship (money, status, intelligence, good looks, youth, power, affection). Any of a person's resources can be traded for more, better, or different resources. Marriages between older men and young women often reflect an exchange of the man's power, money, and/or fame for the woman's youth, physical attractiveness, and ability to bear children.

Many of our cost–reward decisions are conscious and deliberate, but others are passive or based on long-term negative interactions. Much of

social exchange theory proposes that individuals seek through their interactions to maximize their rewards and minimize their costs.

5-4c Feminist Theories

Cultural norms and gender role expectations shape our interaction patterns. Generally, women are socialized to be more comfortable talking about their feelings,

Tyler Olson/Shutterstock.com

In predominantly White organizations, many Black professionals believe that they're held to different interaction standards than their White coworkers. Consequently, they manage their emotions to avoid being labeled as "too sensitive," "angry," "irritating," or "unpleasant" (Wingfield, 2010).

whereas men are socialized to be dominant and take charge, especially in the workplace. Women often ask questions that probe about feelings ("Were you glad it happened?"). Women are also more likely than men to do conversational "maintenance work" that encourages conversation ("What happened at the meeting?") (Lakoff, 1990; Farley et al., 2010).

Women's speech tends to be supportive ("You must've felt terrible") and tentative ("I kind of feel that . . . "). "Uptalk" is a style of speech in which every sentence rises in pitch, as if the speaker is always asking a question ("My name is Jennifer?"). Women are much more likely than men to uptalk, which shows uncertainty and a lack of self confidence. And, in many social situations, women—particularly those ages 40 to 60—often apologize for the way they look (weight, hair, wrinkles). Men don't do so, "even when they look terrible" (Wolven, 2014; Wood, 2015).

Compared with female speech, men's speech often reflects *conversational dominance*, speaking more frequently, and for longer periods. Men also show dominance by interrupting or ignoring others, reinterpreting the speaker's meaning, changing the topic, or giving unsolicited and unwanted advice (Tannen, 1990; Toth, 2011; Gurkoff and Ranieri, 2012).

Such interaction differences aren't innate because of one's sex. Instead, much interaction is contextual—that is, it varies depending on the situation. For example, fathers who do much of the parenting have communication styles that are similar to those of mothers. In the workplace, women and men who occupy high-level decision-making positions have similar interaction styles with superiors and subordinates (Cameron, 2007).

You've seen, so far, that much of our interaction is verbal and face-to-face. Nonverbal communication also affects our relationships.

5-5 NONVERBAL COMMUNICATION

Nonverbal communication refers to messages sent without using words. This silent language, a "language of behavior" that conveys our real feelings, can be more potent than our words (Hall, 1959: 15). For example, sobbing sends a much stronger message than saying "I feel very sad." Some of the most common nonverbal messages are silence, visual cues, touch, and personal space.

5-5a Silence

Silence expresses a variety of emotions: Agreement, apathy, awe, confusion, disagreement, embarrassment,

"This concludes my lecture on non-verbal communication. Any comments or questions?"

Chris Wildt/CartoonStock.com

regret, respect, sadness, thoughtfulness, and fear, to name just a few. In various contexts, and at particular points in a conversation, silence means different things to different people. Sometimes silence saves us from embarrassing ourselves. Think, for example, about the times you fired off an angry email or text message, regretted doing so an hour later, and then dreaded getting a fuming response.

Many of us have experienced the pain of getting "the silent treatment" from friends, family members, coworkers, or lovers. Not talking to people who are important to us builds up anger and hostility. Initially, the "offender" may work hard to make the closemouthed person discuss a problem. Eventually, the target of the silent treatment gets fed up, gives up, or ends a relationship. In intimate relationships, silence often signals emotional detachment and the beginning of a breakup.

5-5b Visual Cues

Visual cues, another form of nonverbal communication, include gestures, facial expressions, and eye contact. Let's consider a few examples of such body language in our daily interactions.

GESTURES

Most of us think we know what certain gestures mean—folding your arms across your chest indicates a closed, defensive attitude; leaning forward often shows interest; shrugging your shoulders signals indifference; and knitted eyebrows show worry or concentration. Constantly rolling one's eyes, which conveys contempt and superiority, can be an early predictor of divorce (Weisman, 2016).

Finger-pointing usually directs attention outward, placing blame or responsibility on someone else. The common "talk to the hand" gesture sends a stronger message: "Go away!" or

nonverbal communication messages that are sent without using words.

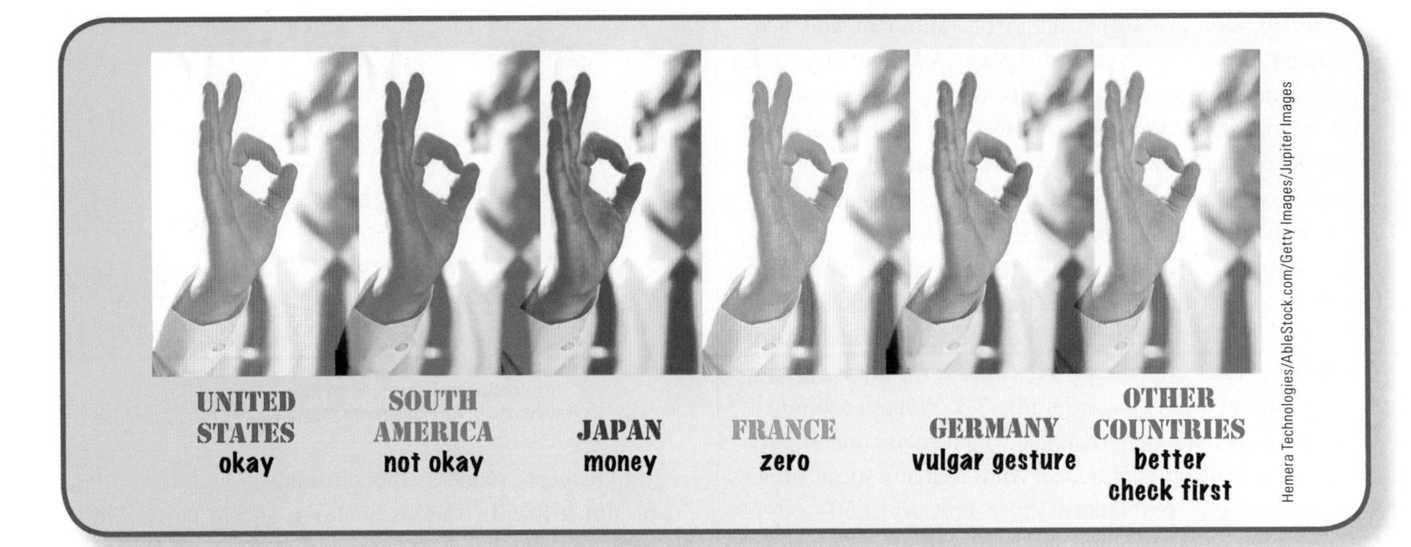

"I'm not listening to you." Gestures, however, should be interpreted in context. Because of habit or hearing problems, for instance, a coworker may always lean forward when listening, regardless of interest.

The same gesture may have different meanings in different countries:

- Tapping one's elbow several times with the palm of one's hand indicates that someone is sneaky (in Holland), stupid (in Germany and Austria), or mean or stingy (in South America).
- Twisting an index finger into one cheek means "She's beautiful" in Libya, and that something tastes good in Italy. The same gesture in southern Spain means that a man is effeminate, and "You're crazy!" in Germany.
- In many Middle Eastern countries, people view the shoes and the soles of one's feet as unclean. Thus, stretching out one's legs with the feet pointing at someone or crossing one's legs so that a sole faces another person is considered rude (Morris, 1994; Lynch and Hanson, 1999; Jandt, 2001; Donadio, 2013).

Departing jets get a wave from the ground crew in Tokyo.

In a remarkable tarmac ritual, and regardless of weather, the ground crews at Japanese airports use many gestures as a jet pushes back from the gate. First, the crew members line up, snap to attention, and then, "in perfect unison," they bow. This gesture is directed toward the passengers ("Thank you for visiting Japan"), the plane's flight crew ("We respect your expertise and dedication"), and, ultimately, as recognition of the ground crew's commitment to service ("We have fulfilled our duties to the best of our abilities"). Then they straighten up, smile broadly, and wave goodbye to show the passengers that they had been welcome to Japan (Gottlieb, 2011: 44).

FACIAL EXPRESSIONS

Facial expressions reveal emotions, but they can be deceptive. First, our facial expressions don't always show our true feelings. Parents tell their children "Don't you roll your eyes at me!" or "Look happy when Aunt Minnie hugs you." Thus, children learn that displaying their real feelings—especially when they're negative—is unacceptable.

Second, faces can lie about feelings. Parents often know when children are lying because they avoid eye contact, cry, or blink and swallow frequently. Some doctors say that they can tell if their patients are lying because of signs such as avoiding eye contact, pausing, or voice inflections, but many adults can control their facial expressions. They can deceive, successfully and over many years, because they've rehearsed the lies in their heads, are smooth talkers with a reputation for being trustworthy, or have gotten away with lying in the past.

Is she being friendly? Or flirting?

Digital Vision/Thinkstock

In nonverbal interaction, people can correctly distinguish truth from lies only about 54 percent of the time (Reddy, 2013; Shea, 2014).

Third, facial expressions can be misleading because of cultural variations. American businesspeople have grumbled that Germans are cool and aloof, whereas many German businesspeople have complained that Americans are excessively friendly and hide their true feelings with grins and smiles. The Japanese, who believe that it's rude to display negative feelings in public, smile more than Americans do to disguise embarrassment, anger, and other negative emotions (Jandt, 2001).

The only consistent research finding, across cultures, is that women smile more than men. This difference may be due to cultural norms that socialize women to hide negative feelings that might make people uncomfortable, and to show deference to men (Szarota, 2010; McDuff et al., 2017). Men are more likely than women to mistake smiles and friendliness as being flirtatious and sexually seductive (La France et al., 2009; Rutter and Schwartz, 2012).

EYE CONTACT

Eye contact serves several social purposes (Eisenberg and Smith, 1971; Ekman and Friesen, 1984; Siegman and Feldstein, 1987). First, we get much information about other people by looking at their eyes. Eyes open wide show surprise, fear, or a flicker of interest. When we're angry, we stare in an unflinching manner. When we're sad or ashamed, our eyes may be cast down.

Second, appropriate eye contact depends on the social context. Especially during job interviews, eye contact conveys attentiveness, confidence, and respect. Managers expect eye contact from subordinates, but subordinates often complain that a manager's prolonged eye contact is domineering or intimidating. Looking at a coworker when speaking shows trust and interest, but an overly intense gaze can be disturbing (Chen et al., 2013). When two people like each other, they establish eye contact more often and for longer periods than when there's tension in the relationship.

Finally, cultural norms affect eye contact. In many Asian cultures, including Japan, meeting another's eyes can be rude. Asians are more likely than Westerners to regard a person who makes frequent eye contact as angry, unapproachable, and unpleasant (Akechi et al., 2013). In effect, then, cross-cultural communication breaks down when we don't understand cultural rules about acceptable body language.

5-5c Touch

Touching, another important form of nonverbal communication, sends powerful messages about our feelings and attitudes. Parents worldwide communicate with their infants through touch—stroking, holding, patting, rubbing, and cuddling.

Touching can be positive (hugging, embracing, kissing, and holding hands) or negative (hitting, shoving, pushing, spanking). Other forms of touching are controlling.

Rod Aydelotte-Pool/Getty Images

Pictured are former President Bush and Saudi Crown Prince Abdullah at Bush's Texas ranch in 2002. Many U.S. journalists raised questions about two men holding hands. In response, Arab Americans were quick to point out that Arab society sees the outward display of affection between men as an expression of respect and trust. Thus, government officials and military officers often hold hands as they walk together or converse with one another.

One of my students left a boyfriend because "He said he loved and trusted me, but gripped my arm tightly every time I talked to another guy and pulled me away." Between intimate partners, a decline in the amount of touching may signal that feelings are cooling off.

GENDERED TOUCHING

Whether touching is viewed as positive or negative depends on the situation and one's gender. In higher education, even when the faculty member is popular, male and female students perceive touching differently. When female professors touch male students on the arm while talking to them, the students view the gesture as friendly. When a male professor touches a female student on the arm, she may get nervous because she's afraid that the touching may escalate (Lannutti et al., 2001; Fogg, 2005). Generally, women are more likely than men to initiate hugs and touches that express support, affection, and comfort. In contrast, men more often use touching to assert power or show sexual interest (Wood, 2015).

CROSS-CULTURAL VARIATIONS IN TOUCHING

As with other forms of nonverbal communication, the interpretation of touching varies from culture to culture. In some Middle Eastern countries, people don't offer anything to another with the left hand because it's used to clean oneself after using the toilet. Among many Chinese and other Asian groups, hugging, backslapping, and handshaking aren't as typical as they are in the United States because such touching is seen as too intimate (Lynch and Hanson, 1999).

According to one scholar, Americans—even strangers—seem to be hugging more. Nonetheless, she maintains, the United States is a "medium touch" culture: "More physically demonstrative than Japan, where a bow is the all-purpose hello and goodbye, but less demonstrative than Latin or Eastern European cultures, where hugs are robust and can include a kiss on both cheeks" (Drexler, 2013: C3).

5-5d Personal Space

In the example of airlines at the beginning of this chapter, you saw that the distance that people establish between themselves is an important aspect of nonverbal communication. Americans are more annoyed by a passenger who invades their seat space (47 percent) than skimpy airline leg room (39 percent), long security lines (37 percent), or weather delays (20 percent) (Aguila, 2014). Personal space plays a significant role in our everyday nonverbal interactions, reflects power and status, and varies across societies.

WHEN IS OUR PERSONAL SPACE VIOLATED?

Our space is public or private. In the public sphere, which is usually formal, we have clearly delineated spaces: "This is your locker," "That's her office," or "You're parking in my spot." We usually decorate our public spaces with businesslike artifacts such as awards or framed photos of a group's accomplishments.

After a snowstorm, people in densely populated cities who shovel parking spots "mark" their personal spaces with lawn chairs, strollers, and even a table set for two, complete with a bottle of wine. Some of the most aggressive (and locally accepted) retaliation occurs in South Boston, where residents punish violators by slashing their tires or smashing their car windows (Goodnough, 2010).

Private spaces send a different message. They convey informality and a relaxed feeling. Private spaces—homes,

dpa picture alliance/Alamy Stock Photo

Why do many men use more public space than women? Because they're larger and have longer legs? Or because men, but not women, feel "totally empowered to take up a lot public space," even if it inconveniences someone else (see Bahadur, 2013)?

apartments, and dorm rooms—are often mini-museums that reflect people's interests, hobbies, hygiene, and personalities.

People, particularly men, sometimes invade our personal space by standing too close to us in a line, leaning against us, using all of the shared armrest in tight airline seats, or plopping their feet on the chair next to us in a classroom, airport, or movie theatre. Not all space intrusions are physical encroachments. Loud cell phone conversations are a good example of an auditory intrusion into our personal space.

In intimate relationships (as between family members), our personal space is often 2 feet or less because we're at ease with close physical proximity. In contrast, in public situations (someone speaking to a large audience or faculty in large classrooms), the personal space is often 12 feet or more because the speaker, who has a higher social status than the audience, has a formal relationship with the listeners and avoids close physical contact (Hall, 1966).

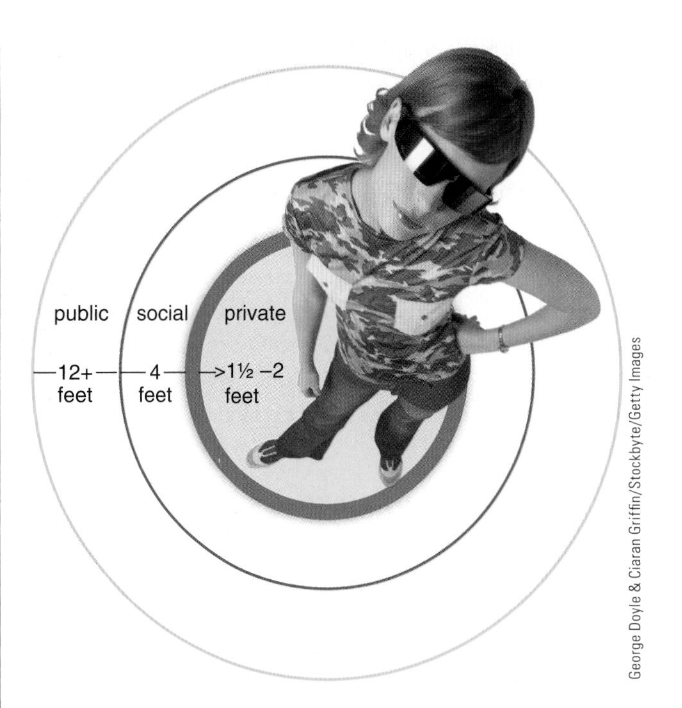

George Doyle & Ciaran Griffin/Stockbyte/Getty Images

SPACE AND POWER

Space usage signifies who has privilege, status, and power. Wealthy people can afford enormous apartments in the city or houses in the suburbs, whereas the poor are crowded into the most undesirable sections of a city or town or in trailer parks. Executives, including college presidents, usually have huge offices (even entire suites), whereas faculty members often share office space, even though many of their discussions with students are confidential. Generally, the higher the socioeconomic status, the greater the consumption of space, including large cars, reserved parking spaces, private dining areas, first-class airline seats, and luxurious skyboxes at sports stadiums.

Some wealthy men feel entitled to violate women's personal space. In a leaked recording, Donald Trump boasted about "grabbing women by the pussy." "When you're a star," he crowed, "you can do anything." Trump denied such sexual assaults, but more than a dozen women have accused him of groping them, grabbing their buttocks, or kissing them against their will. Almost a third of the men who voted for Donald Trump said that grabbing a woman by her genitalia without consent isn't a sexual assault (Nelson and Crockett, 2017; *New York Times*, 2016; PerryUndem, 2017).

CROSS-CULTURAL VARIATIONS AND SPACE

Cultural norms and values determine how we use space. Americans not only stand in line but also have strict queuing rules: "Hey, the end of the line is back there" and "That guy is trying to cut into the line." In contrast, "along with Italians and Spaniards, the French are among the least queue-conscious in Europe" (Jandt, 2001: 109). In these countries and others, people routinely push into the front of a group waiting for taxis, food, and tickets, and nobody objects.

Americans maintain personal space in an elevator by moving to the corners or to the back. In contrast, an Arab male may stand right next to another man even when no one else is in the elevator. Because most Arab men don't share American concepts of personal space in public places and in private conversations, they consider it offensive if the other man steps or leans away. In many Middle Eastern countries, however, there are strict rules about women and men separating themselves spatially during religious ceremonies and about women avoiding any physical contact with men in public places (Office of the Deputy Chief . . . , 2006; Commisceo Global, 2016; see also Chapter 3).

5-6 ONLINE INTERACTION

Many of us now interact in *cyberspace*, an online world of computer networks. In cyberspace, **social media** are websites that enable users to create, share, and/or exchange information and ideas. Social media include social networking sites (e.g., Facebook and Twitter), gaming sites (e.g., *Second Life*), video sites (e.g., YouTube), and blogs (websites

social media websites that enable users to create, share, and/or exchange information and ideas.

maintained by people who provide commentary and other material). Who's online and why? And what are the benefits and costs of online interaction?

5-6a Who's Online and Why?

The percentage of Americans age 18 and older who are connected to the Internet increased from practically zero in 1994 to 88 percent in 2016. The first iPhone was released in 2007. By 2016, 77 percent of Americans over the age of 12 owned a smartphone, giving them online access all the time ("Driven to Distraction," 2017; Smith, 2017).

DEMOGRAPHIC VARIATIONS

About 87 percent of women and men use the Internet. Only 67 percent of Americans aged 65 and older are online compared with 96 percent of those aged 30 to 49, and 99 percent of 18- to 29-year-olds (Anderson and Perrin, 2017).

Race, ethnicity, and social class also affect a person's likelihood of being online. As *Figure 5.3* shows, Asian Americans are the most connected group in the United States. Their greater "connectivity" is due primarily to high education and income levels. Many Asian American parents are professionals who can afford computers and online service, and they encourage their children to use technology for education, a major avenue of upward mobility. Asian Americans also tend to live in urban and suburban areas where high-speed Internet connections are readily available (File and Ryan, 2014; see also Chapters 10 and 13).

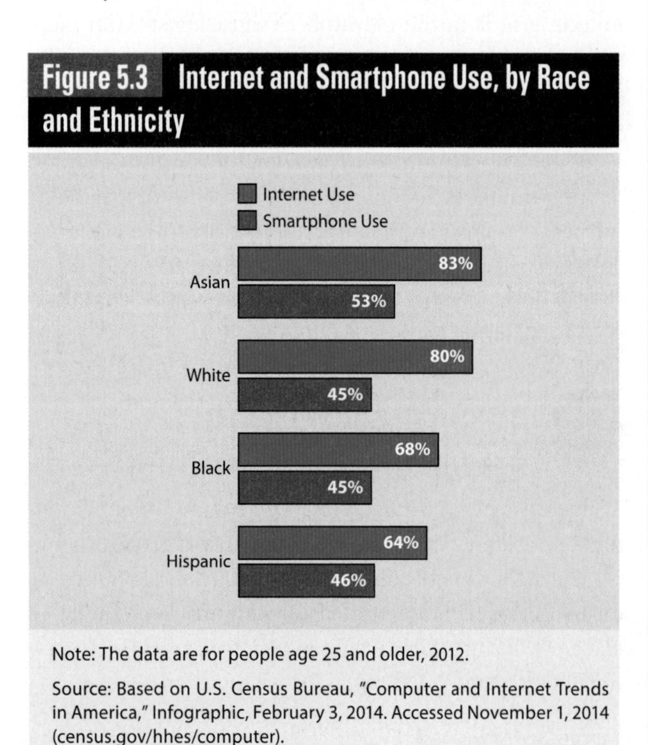

Figure 5.3 Internet and Smartphone Use, by Race and Ethnicity

Note: The data are for people age 25 and older, 2012.

Source: Based on U.S. Census Bureau, "Computer and Internet Trends in America," Infographic, February 3, 2014. Accessed November 1, 2014 (census.gov/hhes/computer).

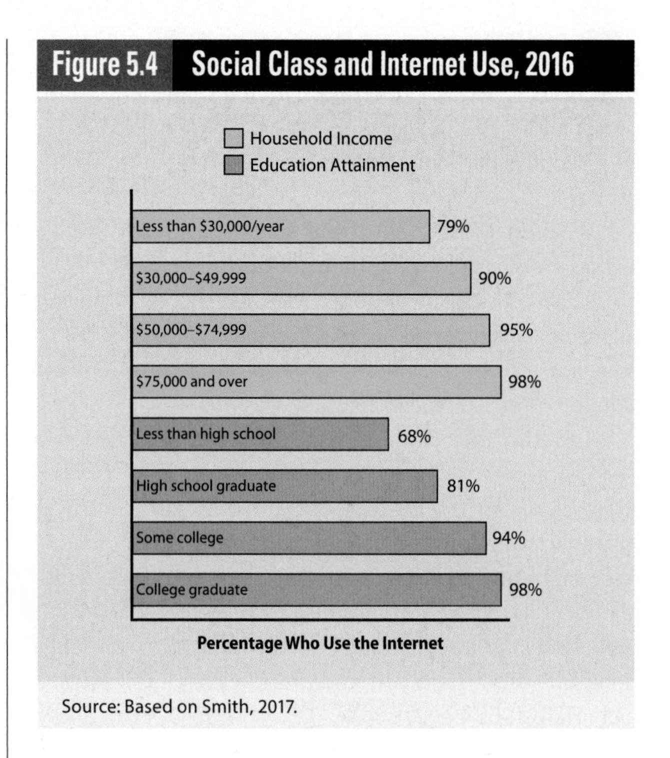

Figure 5.4 Social Class and Internet Use, 2016

Source: Based on Smith, 2017.

The U.S. offline population has declined substantially—from 48 percent in 2000 to only 13 percent in 2016 (Anderson and Perrin, 2016)—but there's still a digital divide between social classes. The higher a person's education and family income, the greater the likelihood of being online (*Figure 5.4*). Thus, children from the poorest families are the most likely to lack technological skills.

WHAT WE DO ONLINE

The average American spends almost 11 hours a day staring at a screen (Nielsen, 2016). Among our many online activities, the most popular are *social networking sites* that connect people who have similar personal or professional interests. About 70 percent of adults use social networking sites, up from only 5 percent in 2005 ("Social Media Fact Sheet," 2017). We spend more time on social media and email than on radio and news sites (see *Table 5.3*).

Among online adults ages 18 and older, 79 percent use Facebook at least once a day, more than double the share that uses Instagram (32 percent), Pinterest (31 percent), LinkedIn (29 percent), and Twitter (24 percent). Besides social networking, more than half of online users access healthcare information, seek parenting advice, do

Table 5.3 How Americans Spend Their Online Time

Average minutes per day spent on . . .	
Email	56
Social media	50
Search	26
Online radio and news	14 minutes each

Source: Based on Statista, 2016.

banking, learn about community events or activities, and use navigation devices (Greenwood et al., 2016). Thus, people go online for a variety of reasons.

5-6b What Are the Benefits and Costs of Online Interaction?

You'll see in Chapter 16 that technology brings both benefits and costs. The same is true of online interaction, ranging from family ties to privacy issues.

FAMILY TIES

Social media has had a mixed effect on families. By age 3, many children have learned technological skills (e.g., creating a password, navigating some websites, and using tablets). Video games like *Minecraft* encourage school age children to build, explore, collaborate, and improve hand-eye coordination and problem solving (Dockterman, 2013; Lewin, 2013). On the other hand, screen time erodes the quality and quantity of interaction between parents and young children: "Some parents are perpetually tuned into their own devices, responding to every ping of their cellphones and tablets" (Brody, 2015: D7).

Cell phones have increased the frequency of interaction between parents, especially if both are employed, to coordinate schedules and to chat with their children. Also, parents report spending more time with their children by playing home video games. On the other hand, some children complain that they rarely receive their parents' full attention because a parent is often immersed in email, texting, or being online even when pushing a swing, driving, or eating meals.

A team of researchers observed 55 parents and their young children at fast-food restaurants. Forty parents immediately took out mobile devices and used them throughout most of the meal (Radesky et al., 2014). By paying more attention to their electronic gadgets than their children, parents are sending their kids a powerful message: "My cell phone is more important than you are."

Among adults who are married or in committed relationships, 74 percent say that the Internet, smartphones, and social media have strengthened their communication and emotional closeness. On the other hand, 20 percent believe that the technology has had a negative effect on their relationships (e.g., a cell phone distracts a spouse or partner when the couple is together, they argue about the amount of time one of them spends online) (Lenhart and Duggan, 2014).

ONLINE RELATIONSHIPS

An increasing number of people use social media to discuss important matters, get social support, and keep up or revive dormant relationships. Some ethnic groups, especially Asian Americans, who are rarely featured on television, have become successful on YouTube. They've found "millions of eager fans" who follow their comedy sketches and discuss topics, particularly sex and race, that are taboo among older generations (Considine, 2011: ST6; see also Pipkin, 2013).

Millions of Americans turn to the Web to find romance. About 11 percent of Internet users (representing 9 percent of all American adults) have used online dating sites (e.g., Match.com, plentyoffish.com). Seventy percent of online daters believe that online dating is a good way to meet prospective marriage mates, but only 5 percent of Americans who are currently married or in a long-term relationship met their partner online. Some 54 percent of online daters, particularly women, say that people have lied about their age, height, weight, income, occupation, and marital status. And, in 2015 alone, the Federal Bureau of Investigation (2016) received nearly 13,000 complaints about romance scams from people who reported being swindled out of a total of over $204 million (Smith and Anderson, 2015; see also Benokraitis, 2015).

CONNECTIONS AND DISCONNECTIONS

Online interaction develops networking skills, helps people find jobs, and encourages political engagement. Some of the benefits of texting include faster communication and avoiding uncomfortable interactions. On the other hand, there's a "constant compulsion to connect" (Suddath, 2013: 80). For example:

- The average worker checks his or her email "an extraordinary" 74 times a day (Roberts, 2014).

- On average, 18- to 24-year-olds use their smartphones more than 80 times a day ("Driven to Distraction," 2017).
- Nearly half of adults say they couldn't live without their smartphones, 41 percent check them several times an hour, and 11 percent do so every few minutes (Anderson, 2015; Newport, 2015).

Fully 57 percent of U.S. teenagers have met a new friend online while browsing social networks or playing video games (Lenhart et al., 2015). The average Facebook-using teen has 300 "friends," 33 percent of whom she or he has never met in person. According to some critics, social networking sites, especially Facebook, are superficial, give people a false sense of connection to others, inflate egos, and diminish time to develop relationships with the few friends that we have (DiSalvo, 2010; Madden et al., 2013).

The more often young adults use Facebook, according to several studies, the unhappier they become, and regardless of gender, level of loneliness, or self-esteem. The root cause of the unhappiness is envy, even though Facebook users suspect that their "friends" Photoshop images and exaggerate their achievements, job success, vacations, and love life. Jealousy, anger, and loneliness are especially likely if Facebook users receive fewer positive comments, "likes," and general feedback than their friends do (Krasnova et al., 2013; Kross et al., 2013). Even adults admit that social media makes them feel bad about themselves when their friends describe awards, promotions, and "living a more fulfilling, fabulous life than I am" (Melton, 2013).

iStock.com/pearleye

Relying too much on technology can hurt relationships. Smartphones are displacing (or reducing) romantic interactions and creating conflict when partners don't give each other their full attention. Instead of mingling and conversing with coworkers during lunch or a break, many are texting or checking their email (Turkle, 2015; Roberts and David, 2016). The next time you go to a restaurant, watch how often—instead of chatting with each other—people check messages and social networks even before scanning the menu, and how often they check their cell phones throughout a meal.

ONLINE HARASSMENT

Nationally, 94 percent of parents say that they've talked with their children about proper online behavior, but 34 percent of 12- to 17-year-old students have experienced cyberbullying at least once during the school year. *Cyberbullying* refers to deliberately using digital media to communicate false, embarrassing, or hostile information about someone. Cyberbullying is less common than traditional bullying, but has more profound negative outcomes that include depression, academic and mental health problems, severe isolation, and, most tragically, suicide (Anderson, 2016; Cyberbullying Research Center, 2016).

Harassment—from name calling to threatening behavior—is becoming a common part of online life. Fully 72 percent of adult Internet users have seen someone harassed online, and 47 percent have experienced it personally. Half of the latter didn't know the person who had most recently attacked them. Men are more likely to experience name-calling and embarrassment, whereas women are targets of more intense and serious forms of online assaults like sexual harassment and cyberstalking (Lenhart et al., 2016).

Cyberstalking is the repeated use of electronic communications to harass, threaten, or frighten someone. Nationally, 8 percent of Internet users have been cyberstalked to the point of feeling unsafe or afraid. Young people, particularly women under 30, and LGBTQ individuals are especially likely to be targets of cyberstalking. Cyberstalking may evolve into offline stalking, including abusive or harassing phone calls, vandalism, threatening or obscene mail, trespassing, and physical assault (Lenhart et al., 2016).

Tweeters can be engaging, funny, and supportive. Many of these faceless strangers can also be racist, sexist, abusive, and hateful. Because tweets are short and easy to read, public accusations and vile, ugly things said about anyone can quickly get a huge audience's attention.

PRIVACY ISSUES

A major cost of online interaction is jeopardizing our privacy because email and text messages are neither anonymous nor confidential. Many companies monitor and preserve their employees' email messages and can

Highwaystarz-Photography/Getty Images

use them to discipline or fire people. It's also becoming increasingly common for college admissions officers and employers to search social media sites before deciding whether to accept a student or to make a job offer (Belkin and Porter, 2012; Singer, 2013).

People often text and say things in emails they'd never say in person, especially when they're flirting, angry, frustrated, or tired. Emails, text messages, Snapchat photos, and videos don't disappear after being deleted but may last indefinitely in cyberspace. Divorce lawyers have successfully retrieved deleted emails, Facebook posts, tweets, and Instagram photos to convince judges that a divorcing spouse shouldn't get financial support or custody of the children (Hillin, 2014).

Only 3 percent of American adults say they trust Facebook with their personal data, but millions willingly give out personal information on Facebook and other social media sites—photos, phone number, address, age, education and work background, political and religious views, and so on (Fleming, 2015; Beres, 2016). It takes a private investigator just a few clicks to get a composite picture of someone from public records, email messages, and social media sites.

Tech giants like Apple, Facebook, and Google know more about their users than many people realize—where they go, what they search for, what they buy, which restaurants they frequent, what they do for fun, and when they go to bed. Besides collecting data on its own users, Facebook has partnerships with several companies that collect behavioral data—from store transactions and customer email lists to divorce and Web browsing records—that it then sells to advertisers (Angwin et al., 2016). Neither state nor federal laws prohibit the collection or sharing of such data by third parties.

STUDY TOOLS 5

READY TO STUDY? IN THE BOOK, YOU CAN:

- ☐ Check your understanding of what you've read with the Test Your Learning Questions provided on the Chapter Review Card at the back of the book.
- ☐ Tear out the Chapter Review Card for a handy summary of the chapter and key terms.

ONLINE AT CENGAGEBRAIN.COM WITHIN MINDTAP YOU CAN:

- ☐ Explore: Develop your sociological imagination by considering the experiences of others. Make critical decisions and evaluate the data that shape this social experience.
- ☐ Analyze: Critically examine your basic assumptions and compare your views on social phenomena to those of your classmates and other MindTap users. Assess your ability to draw connections between social data and theoretical concepts.
- ☐ Create: Produce a video demonstrating connections between your own life and larger sociological concepts.
- ☐ Collaborate: Join your classmates to create a capstone project.

6 Groups, Organizations, and Institutions

PG Arphexad/Alamy Stock Photo

LEARNING OBJECTIVES

After studying this chapter, you will be able to...

6-1 Compare and illustrate the different types of social groups, explain why people conform to group pressure, and discuss the impact of social networks.

6-2 Describe and illustrate three types of formal organizations, summarize the strengths and shortcomings of bureaucracies, and explain how informal groups affect organizations.

6-3 Compare the theoretical explanations of groups and organizations, including their contributions and limitations.

6-4 Explain why social institutions are important and how they're interconnected.

After finishing this chapter go to **PAGE 117** for **STUDY TOOLS**

Every year, about a dozen of our neighbors, all fervent Ravens football fans, get together to buy season tickets and to plan elaborate tailgating parties. They wear Ravens caps and purple jerseys, and most of the drivers attach pennants to their car antennas. These fervent football fans are a social group.

WHAT DO YOU THINK?

It's a good idea for companies to use video surveillance, including in restrooms.

1	2	3	4	5	6	7
strongly agree						strongly disagree

6-1 SOCIAL GROUPS

A **social group** is two or more people who share some attribute and interact with one another. They have a sense of belonging or "we-ness." Friends, families, work groups, religious congregations, clubs, athletic teams, and Vietnam veterans are all examples of social groups. Each of us is a member of many social groups, but we identify more closely with some groups than others.

6-1a Types of Social Groups

Some groups are small and personal (families); others are large and impersonal (financial organizations). Some are highly organized and stable (political parties); others are fluid and temporary (high school classmates). The most basic types of social groups are primary and secondary groups.

PRIMARY AND SECONDARY GROUPS

A **primary group** is a a relatively small group of people who engage in intimate face-to-face interaction over an extended period (Cooley, 1909/1983). Primary groups, such as families and close friends, are our emotional glue. They have a powerful influence on our social identity because we interact with them on a regular basis over many years, usually throughout our lives. Because primary group members genuinely care about each other, they contribute to one another's personal development, security, and well-being. Our family and close friends, particularly, stick with us through good and bad, and we feel comfortable being ourselves with them.

Lew Robertson/Spirit/Corbis

In contrast to primary groups, a **secondary group** is a large, usually formal, impersonal, and temporary collection of people who pursue a specific goal or activity. Your sociology class is a good example of a secondary group. You might have a few friends in class, but students typically interact infrequently and impersonally. When the semester (or quarter) is over and you've accomplished your goal of passing the course, you may not see each other again (especially if you're attending a large college or university). Other examples of secondary groups include sports teams, labor unions, and a company's employees.

Unlike primary groups, secondary groups are usually highly structured: There are many rules and regulations, people know (or care) little about each other personally, relationships are formal, and members are expected to accomplish specific tasks. Whereas primary groups meet our *expressive* (emotional) needs, secondary groups fulfill *instrumental* (task-oriented) needs. Once a task or activity is completed—whether it's earning a grade, turning in a committee report, or building a bridge—secondary groups split up and become members of other secondary groups.

Table 6.1 summarizes the characteristics of primary and secondary groups. These characteristics are **ideal types**—general traits that describe a social phenomenon rather than every case. Ideal types provide composite pictures of how structures, events, and behaviors differ rather than specific descriptions of reality. Because primary and secondary groups are ideal types, their characteristics can vary. Thus, primary group members may sometimes devote themselves to meeting instrumental needs (running a family-owned business), and secondary group members (military units and athletic teams) can develop lasting ties.

social group two or more people who share some attribute and interact with one another.

primary group a small group of people who engage in intimate face-to-face interaction over an extended period.

secondary group a large, usually formal, impersonal, and temporary collection of people who pursue a specific goal or activity.

ideal types general traits that describe a social phenomenon rather than every case.

Table 6.1	Characteristics of Primary and Secondary Groups	
	Characteristics of a Primary Group	**Characteristics of a Secondary Group**
Interaction	• Face-to-face • Usually small	• Face-to-face or indirect • Usually large
Communication	• Emotional, personal, and satisfying	• Emotionally neutral and impersonal
Relationships	• Intimate, warm, and informal • Usually long-term • Valued for their own sake (expressive)	• Typically remote, cool, and formal • Usually short-term • Goal-oriented (instrumental)
Individual Conformity	• Relatively free to stray from norms and rules	• Expected to adhere to rules and regulations
Membership	• Members aren't easily replaced	• Members are easily replaced
Examples	• Family, close friends, girlfriends and boyfriends, self-help groups, street gangs	• College classes, political parties, professional associations, religious organizations

IN-GROUPS AND OUT-GROUPS

During the 2016 presidential campaign, Donald Trump accused Mexico of "sending" people who are criminals to the United States: "*They're* bringing drugs. *They're* bringing crime. *They're* rapists." Trump's comments illustrate in-groups and out-groups. Members of an **in-group** share a sense of identity and belonging that typically excludes and devalues outsiders. **Out-group** members are people who are viewed and treated negatively because they're seen as having values, beliefs, and other characteristics different from those of an in-group. For example, "we" vegetarians are healthier than "you" meat eaters, "we" computer geeks are smarter than "you" fraternity and sorority "types," and so on.

Almost everyone sees others as members of in-groups and out-groups. From ancient times to the present, people in various parts of the world have made "we" and "they" distinctions based on race, ethnicity, gender, sexual orientation, religion, age, social class, and other social and biological characteristics (Tajfel, 1982; Hinkle and Schopler, 1986). Such distinctions can promote in-group solidarity and cohesion. They can also create conflict and provoke inhumane actions, including wars, massacres of out-group members, and civil wars (see Chapter 3).

One person's in-group may be another person's out-group. The general public and criminal justice system see gangs as dangerous out-groups. Many young people who join gangs, however, experience in-group benefits. Attractions include making money, although illegally; close relationships with family members and friends who are already involved in gang life; protection against violent family or community members and rival gangs; a sense of self-worth, support, and belonging; and the status of being an "outlaw" that the entertainment industry often glamorizes as sexy and exciting (Taylor and Smith, 2013).

in-group people who share a sense of identity and belonging that typically excludes and devalues outsiders.

out-group people who are viewed and treated negatively because they're seen as having values, beliefs, and other characteristics different from those of an in-group.

reference group people who shape our behavior, values, and attitudes.

REFERENCE GROUPS

An in-group or an out-group can become a **reference group**, people who shape our behavior, values, and attitudes (Merton and Rossi, 1950). Reference groups influence who we are, what we do, and who we'd like to be

AF archive/Alamy Stock Photo

Is the cast of characters on the television series *NCIS* an example of a primary group? A secondary group? Both? Neither?

in the future. Unlike primary groups, however, reference groups rarely provide personal support or face-to-face interaction over time.

Reference groups might be people with whom we already associate (a college club or an athletic team). They can also be groups that we admire and want to be part of (teachers or doctors). Each person has many reference groups. Your sociology professor, for example, may be a member of several professional associations, a golf enthusiast, a parent, and a homeowner. Identification with each of these groups influences her or his everyday attitudes and actions.

Like in-groups and primary groups, reference groups can have a strong impact on our self-identity, self-esteem, and sense of belonging because they shape our current and future attitudes and behavior. We typically add or drop reference groups throughout the life course. If you aspire to move up the occupational ladder, your reference group may change from entry-level employees to managers, vice presidents, and CEOs.

6-1b Group Size and Structure

Two important characteristics of a social group are its size and leadership. Both affect a group's interaction and dynamics.

GROUP SIZE: DYADS, TRIADS, AND BEYOND

German sociologist Georg Simmel (1858–1918) pioneered the study of **dyads**, groups with two members, and **triads**, groups with three members. A dyad (e.g., parent–child, lovers, husband–wife, close friends) is the most cohesive of all groups because its members tend to have a personal relationship and to interact more intensely than do people in larger groups. Dyads can be unstable, however, because both persons must cooperate. If either member doesn't fulfill her or his responsibilities, the dyad collapses (Simmel, 1902; Mills, 1958).

This triad studies together for sociology exams. What might happen if one student drops out of the study group? What if three more students join the group?

Adding only one member to a dyad has important consequences because it changes a group's dynamics (behavior over time). A triad is more stable than a dyad because the group can continue if a person drops out, and a member may patch up an argument that erupts between the other two. On the other hand, a third person may intentionally create conflict to break up a relationship or to attain a dominant position. In any triad, two of the members can gang up against or leave out the third (Simmel, 1902; Mills, 1958). Triads also tend to interact less often and intensely than do dyads (e.g., think about a third roommate who moves in, another person who joins you and a coworker at lunch, or someone who tags along on a date).

As more members are added—a fourth, fifth, and so on—interaction changes even more and alliance possibilities increase. At about seven or eight members, the group may break down into dyads and triads because it's difficult to have a single conversation about the task at hand (Becker and Useem, 1942). As a result, larger groups often require some kind of leadership to keep the members focused.

GROUP LEADERSHIP

Several classic studies (Lewin et al., 1939; Bales, 1950) identified three basic types of group leaders. An **authoritarian leader** gives orders, assigns tasks, and makes all major decisions. There's a clear division between leaders and followers, and leaders focus on completing *instrumental* tasks (e.g., marketing a new smartphone). Authoritarian leadership is most effective when there's little time for group decision making and the leader is the most knowledgeable member of the group. On the other hand, authoritarian leaders tend to stifle creativity and may be viewed as controlling and bossy.

By contrast, a **democratic leader** encourages group discussion and includes everyone in the decision-making process. Democratic leaders have the final say, but tend to be concerned about meeting a group's *expressive* needs (e.g., maintaining harmony, high morale, cohesiveness). Democratic leaders engage group members and encourage their creativity, but have also been blamed for not taking charge in times of crisis, and being less productive than groups with authoritarian leaders.

A **laissez-faire** (pronounced les-ey-fair) **leader**

dyad a group with two members.

triad a group with three members.

authoritarian leader gives orders, assigns tasks, and makes all major decisions.

democratic leader encourages group discussion and includes everyone in the decision-making process.

laissez-faire leader offers little or no guidance to group members and allows them to make their own decisions.

Lisa Werner/Alamy Stock Photo

U.S. government shutdowns occur when political leaders don't compromise on legislation that funds federal agencies and operations.

offers little or no guidance to group members and allows them to make their own decisions. Laissez-faire leaders are similar to the permissive parents you met in Chapter 4: There are few rules, few demands, and people do pretty much what they want. Laissez-faire leadership is effective in situations where group members are highly qualified in an area of expertise. However, this leadership style often leads to poorly defined roles, low productivity, a lack of motivation, little cooperation between group members, and not meeting a group's goals.

It's difficult for a leader to meet both a group's instrumental and expressive needs because involving everyone in decision making may result in not accomplishing a group's objectives. Also, leadership styles are more effective in some settings than others. Imagine, for example, if your college professors were laissez-faire ("Take the exams whenever you want") or if your employers were always democratic ("Let's vote on who'll do what today").

6-1c Group Conformity

Most Americans see themselves as rugged individualists who have minds of their own (see the discussion of values in Chapter 3). A number of studies have shown, however, that many of us are profoundly influenced by group pressure. Four of the best known of these studies are by Solomon Asch, Stanley Milgram, Philip Zimbardo, and Irving Janis.

ASCH'S RESEARCH

In a now classic study of group influence, social psychologist Solomon Asch (1952) told subjects that they were taking part in an experiment on visual judgment. After seating six to eight male undergraduates around a table, Asch showed them the line drawn on card 1 and asked them to match the line to one of three lines on card 2 (*Figure 6.1*). The correct answer, clearly, is line C.

All but one of the subjects—who usually sat in the last chair—were Asch's confederates, or accomplices. In the first test, all the confederates selected the correct matching line. In the other tests, each of them, one by one, deliberately chose an incorrect line. Thus, the nonconfederates faced a situation in which seven other group members had unanimously agreed on a wrong answer. Averaged over all of the trials, 37 percent of the nonconfederate subjects ended up agreeing with the group's incorrect answers. When they were asked to judge the length of the lines alone, away from the influence of the group, they made errors only 1 percent of the time.

Asch's research demonstrated the power of groups over individuals. Even when we know that something is wrong, we may go along with the group to avoid ridicule or exclusion. Remember that these experiments were done in a laboratory and with people who didn't know each other. A group's influence on a person's attitudes and behavior can be even stronger when it's a real-life situation.

MILGRAM'S RESEARCH

In a well-known laboratory experiment on obedience, psychologist Stanley Milgram (1963, 1965) asked 40 volunteers to administer electric shocks to other study participants. In each experimental trial, one participant was a "teacher" and the other a "learner," one of Milgram's accomplices. The teachers were businessmen, professionals, and blue-collar workers.

Figure 6.1 Cards in Asch's Experiment

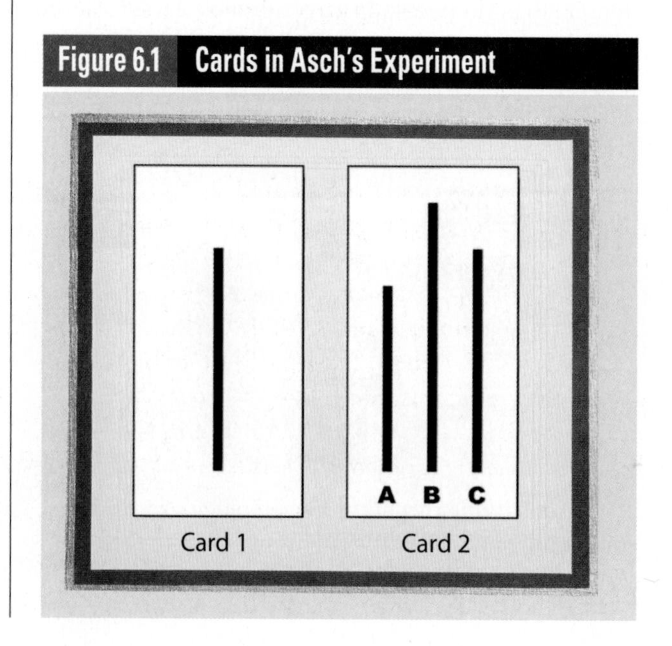

The learner was strapped into a realistic-looking chair that supposedly regulated electric currents. The teacher read aloud pairs of words that the learner had to memorize. Whenever the learner gave a wrong answer, the teacher was told to apply an electric shock from a low of 15 volts to a high of 450 volts. The learner didn't actually receive a shock, but was told to fake pain and fear. When the learners shrieked in agony, the majority of the teachers, although distressed, obeyed the study supervisor and administered the shocks when told to do so.

Alexandra Milgram

Milgram's study was controversial. Ordering electric shocks raised numerous ethical questions about the participants' suffering extreme emotional stress (see Chapter 2, and Perry, 2013, for a recent critique). However, the results showed that an astonishingly large proportion of ordinary people obeyed an authority figure's instructions to inflict pain on others. It's easy to sit back and say "I'd never do something like that," but we do. For example, many workers obey employers when told to ignore evidence that their product is unsafe (Tavris, 2013).

ZIMBARDO'S RESEARCH

The Stanford Prison Experiment conducted by social psychologist Philip Zimbardo also underscores the influence of groups on behavior (Haney et al., 1973; Zimbardo, 1975). Zimbardo recruited volunteers through a local California newspaper for an experiment on prison life. He then selected 24 young men, most of them college students.

On a Sunday morning, nine of the men were "arrested" at their homes as neighbors watched. The men were booked and transported to a mock prison that Zimbardo and his colleagues had constructed in the basement of the psychology building at Stanford University. The "prisoners" were searched, issued an identification number, and outfitted in a dresslike shirt and heavy ankle chains. Those assigned to be "guards" were given uniforms, billy clubs, whistles, and reflective sunglasses. The guards were told that their job was to maintain control of the prisoners but not to use violence.

All the young men quickly assumed the roles of either obedient and docile prisoners or autocratic and controlling guards. The guards became increasingly more cruel and demanding. The prisoners complied with dehumanizing demands (such as eating filthy sausages) to gain the guards' approval and bowed to their authority.

Zimbardo's study was scheduled to run for 2 weeks but was stopped after 6 days because the guards became increasingly aggressive. Among other things, they forced the prisoners to clean out toilet bowls with their bare hands, locked them in a closet, and made them stand at attention for hours. Instead of simply walking out or rebelling, the prisoners became withdrawn and depressed. Zimbardo ended the experiment because of the prisoners' stressed-out reactions.

The experiment raised numerous ethical questions about the harmful treatment of participants. It demonstrated, however, the powerful effect of group conformity: People exercise authority, even to the point of hurting others, or submit to authority if there's group pressure to conform (Zimbardo et al., 2000).

JANIS'S RESEARCH

Sometimes intelligent people, including those in highly responsible positions, make disastrous and irrational decisions. Why? Social psychologist Irving Janis (1972, 1982) cautioned presidents and other heads of state to

AP Images

Many social scientists have used Milgram's findings to explain the Abu Ghraib prison in Iraq, where U.S. soldiers humiliated prisoners because they were "just following orders."

> **Sometimes intelligent people, even those in highly responsible positions, make disastrous and irrational decisions.**

be wary of **groupthink**—a situation in which in-group members make faulty decisions because of group pressures, rather than critically testing, analyzing, and evaluating ideas and evidence. To demonstrate group loyalty, Janis argues, individuals don't raise controversial issues, don't question weak arguments, or probe "soft-headed thinking." As a result, influential leaders often make decisions, based on their advisors' consensus, that turn out to be political and economic disasters.

Examples of groupthink "fiascoes" that Janis studied included U.S. failure to anticipate the attack on Pearl Harbor in 1941, and the escalation of the Vietnam War during the 1960s. More recently, a U.S. Senate study (2004) concluded that U.S. leaders' decision to invade Iraq in 2003 was based on a groupthink dynamic that relied on unproven and inaccurate assumptions, inadequate or misleading sources, and a dismissal of conflicting information showing that Iraq had no weapons of mass destruction.

Janis and other researchers have focused on high-level decision making, but groupthink is common in all kinds of groups—student clubs, PTAs, search committees, juries, fraternities, religious groups, and college administrators (Hansen, 2010; Cohen and DeBenedet, 2012). People can avoid groupthink by hammering out disagreements and seeking advice from informed and objective people outside the group.

groupthink in-group members make faulty decisions because of group pressures, rather than critically testing, analyzing, and evaluating ideas and evidence.

social network a web of social ties that links individuals or groups to one another.

formal organization a complex and structured secondary group designed to achieve specific goals in an efficient manner.

6-1d Social Networks

A **social network** is a web of social ties that links individuals or groups to one another. It may involve as few as three people or millions.

Network links between people or groups can be strong or weak. Some of our social networks, like our primary and secondary groups, may be tightly knit, involve interactions on a daily basis, and have clear boundaries about who belongs and who doesn't. In other cases, our social networks connect us to large numbers of people whom we don't know personally, with whom we interact only rarely or indirectly, and the boundaries change as people come and go. Examples of distant networks include members of the American Sociological Association and LinkedIn subscribers.

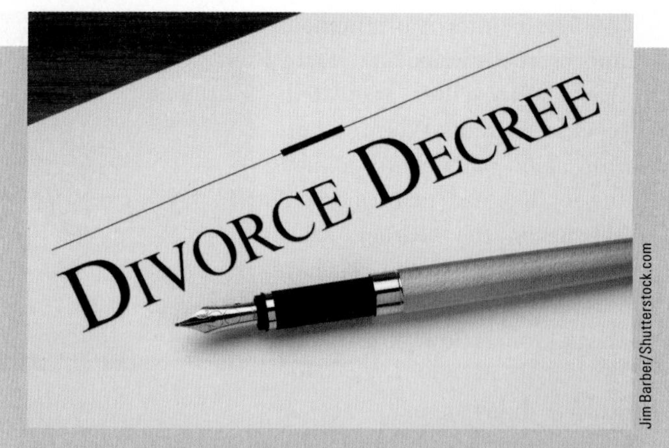

Jim Barber/Shutterstock.com

Do social networks affect divorce rates? A research team that analyzed three decades of divorce data found that people were 75 percent more likely to get divorced if a friend was divorced, and 33 percent more likely to end a marriage if a friend's friend was divorced (McDermott et al., 2013; Chapter 12 discusses some of the micro- and macro-level reasons for divorce).

Electronic communities (also called *online communities* and *virtual communities*) are social networks whose members interact via the Internet. They exchange information, give advice, and engage in public discussions. Some electronic communities—ranging from art to yoga enthusiasts—have thousands of members. Others are small, focused on a single interest, and geographically close. One of my friends used meetup.com to find a local group of equally fervent cyclists who ride on Saturday mornings.

6-2 Formal Organizations

A **formal organization** is a complex and structured secondary group designed to achieve specific goals in an efficient manner. We depend on a variety of formal organizations to provide goods and services in a stable and predictable way. Examples include water departments, food producers and grocery chains that stock our favorite items, and garment industries and retailers that manufacture and sell the clothes we wear.

6-2a Characteristics of Formal Organizations

Formal organizations share some common characteristics:

- Social statuses and roles are organized around shared expectations and goals.
- Norms governing social relationships among members specify rights, duties, and sanctions.
- A formal hierarchy includes leaders or individuals who are in charge.

Modern, complex societies rely on formal organizations, including businesses and schools, to accomplish complex tasks. Organizations differ, however, in their goals, membership, and degree of hierarchy.

6-2b Types of Formal Organizations

Sociologist Amitai Etzioni (1975) identified three types of formal organizations—utilitarian, normative, and coercive—based on their purpose and people's reasons for participating.

UTILITARIAN ORGANIZATIONS

Most people belong to a *utilitarian organization*, one that provides an income or other specific current or future material reward. Examples include government agencies, factories, corporations, and schools. Colleges are workplaces, but also offer students the opportunity to earn degrees and future income.

Stan Rohrer/Alamy Stock Photo

Is this a utilitarian, normative, or coercive organization? (If your immediate response was "coercive," try again.)

Table 6.2 Volunteering in the United States, 2015

Percentage of adults who volunteer	25
Total number of volunteers age 16 and older	63 million
Median annual hours per volunteer	52
Total dollar value of volunteer time	$184 billion

Sources: Based on BLS News Release, 2016, and Corporation for National & Community Service, 2016.

NORMATIVE ORGANIZATIONS

People join *normative organizations* (also called *voluntary organizations* or *voluntary associations*) because of shared interests and to pursue goals that they consider personally worthwhile or rewarding. Examples include political parties, civic organizations (e.g., Habitat for Humanity), religious organizations, choirs, cultural groups, occupational groups (e.g., American Accounting Association), and groups that have a recreational focus (e.g., Beer Collectors of America).

Helping others is a strong American value (see Chapter 3). Thus, every year, millions of Americans—many of whom have full-time jobs—are volunteers in a variety of local, regional, and national organizations, saving government agencies billions of dollars by providing needed services (*Table 6.2*). Charitable donations, more than $390 billion, set a record in 2016. Individual contributions made up 72 percent of all giving, compared with only 5 percent from corporations (Giving USA Foundation, 2017).

COERCIVE ORGANIZATIONS

Membership in *coercive organizations* is largely involuntary. People are pushed or forced to join these organizations because of punishment (prisons) or treatment (psychiatric hospitals, drug rehabilitation centers). Most coercive organizations are total institutions characterized by strict rules, the members' isolation, and resocialization (see Chapter 4).

A single formal organization can fall into all three categories. For example, the military is a coercive organization because of its rigid rules, resocialization, and members' isolation from the rest of the population. It's a utilitarian organization that provides millions of jobs for both soldiers and civilians. The military is also a normative organization: Some people join during wartime because of civic responsibility; many volunteer their time and resources to support soldiers and their families; and others help veterans find jobs, housing, and health care for war-related injuries.

6-2c Bureaucracies

A **bureaucracy** is a formal organization designed to accomplish goals and tasks in an efficient and rational way. Bureaucracies aren't a modern invention but existed thousands of years ago in ancient Egypt, China, and Africa. Some bureaucracies function more smoothly than others, and some formal organizations are more bureaucratic than others. Whether they're relatively small (a 100-bed hospital) or huge (the Social Security Administration), bureaucracies have some common characteristics.

IDEAL CHARACTERISTICS OF BUREAUCRACIES

Max Weber (1925/1947) identified six key characteristics of the ideal type of bureaucracy. In Weber's model, the following characteristics describe what an efficient and productive bureaucracy *should* be like:

- **High degree of division of labor and specialization.** People perform very specific tasks.
- **Hierarchy of authority.** Workers are arranged in a hierarchy in which each person is supervised by someone in a higher position. The resulting pyramids—often presented in organizational charts—show who has authority over and is responsible to whom. Thus, there's a chain of command, stretching from top to bottom, that coordinates decision making.
- **Explicit written rules and regulations.** Detailed written rules and regulations cover almost every possible kind of situation and problem that might arise, including hiring, firing, salary scales, rules for sick pay and absences, and everyday operations.
- **Impersonality.** There's no place for personal likes, dislikes, or tantrums. Workers are expected to behave professionally. An impersonal workplace in which all employees are treated equally minimizes conflict and favoritism and increases efficiency.
- **Qualifications-based employment.** People are hired based on objective criteria such as skills, education, experience, and standardized test scores. If workers perform well and have the necessary credentials and technical competence, they'll move up the career ladder.
- **Separation of work and ownership.** Neither managers nor employees own the offices they work in, the desks they sit at, the technology they use, or the products that they assemble, invent, or design.

bureaucracy a formal organization designed to accomplish goals and tasks in an efficient and rational way.

goal displacement a preoccupation with rules and regulations rather than achieving the organization's objectives.

Dave Carpenter/CartoonStock.com

A "rational matter-of-factness," Weber maintained, makes bureaucracies more productive by "eliminating from official business love, hatred, and all purely personal, irrational, and emotional elements" (Weber, 1946: 216). Weber viewed bureaucracies as superior to other forms of organization because they're more efficient and predictable. He worried, however, that bureaucracies could become "iron cages" because people become trapped in them, "their basic humanity denied."

BUREAUCRATIC DYSFUNCTIONS

Weber described the ideal characteristics of a productive and efficient bureaucracy, but what's the reality? As you read through the following list of problems, think about the ones you've experienced while working in a bureaucracy or dealing with one.

- *Weak reward systems* reduce the motivation to do a good job and are thus a major source of inefficiency and lack of innovation (Barton, 1980). Besides low wages or salaries, weak reward systems include few or no health benefits, little recognition, unsafe equipment and work environments, and few incentives to be creative.
- *Rigid rules* squelch creativity and can lead to **goal displacement**, a preoccupation with rules and regulations rather than achieving the organization's

PzAxe/Shutterstock.com

About 67 percent of U.S. employees are disengaged and emotionally detached from the job because of poor leadership, lack of upward mobility, boredom, and similar problems. Such alienation can lead to poor physical and mental health, low productivity, inferior customer service, and high absenteeism—all of which jeopardize a team's performance and the organization's financial vitality (Harter and Adkins, 2017).

objectives (Merton, 1968). A global management consultant criticizes companies for creating numerous "dumb rules" that hobble the best and most talented workers: "The more rules, the less passion, which means less motivation" (Daskal, 2016: 19).

- Rigid rules and goal displacement often lead to **alienation**, a feeling of isolation, meaninglessness, and powerlessness. Alienation—at all levels—may result in high turnover, tardiness, absenteeism, stealing, sabotage, stress, health problems, and in some cases, whistle-blowing (reporting organizational misconduct to legal authorities).
- *Communication problems* are common in bureaucracies. Because communication typically flows down rather than up the hierarchy, employees (and many managers below the highest echelons) rarely know what's going on. Supervisors and their subordinates may be reluctant to discuss problems or offer suggestions because they fear being criticized, demoted, or fired (Blau and Meyer, 1987; Patel, 2017). Communication problems also waste time and resources. An employee prepared a monthly report for nearly three years because no one ever mentioned that the company no longer needed it (Harnish, 2014).
- The **iron law of oligarchy** is the tendency of a bureaucracy to become increasingly dominated by a small group of people (Michels, 1911/1949). A handful of people can control and rule a bureaucracy because the top officials and leaders monopolize information and resources. As a result, those at the top maintain their power and privilege.
- The cumulative effect of these and other bureaucratic dysfunctions can result in a *dehumanization* that stifles organizational creativity and freedom. As a result, work becomes more automated and impersonal (see "The McDonaldization of Society").

Bureaucracies are plagued by other problems like favoritism and dishonest employee evaluations (Kilmann, 2011). They function, however, largely because of the internal development of informal groups and collaborative work teams.

Health Care on Aisle 7!

To avoid medical and health insurance bureaucracies, more people are turning to *retail clinics* (also called *convenient care clinics*). The first one opened in 2000, the number grew to 1,400 by 2012, and is expected to increase to at least 3,000 by 2018 (Bachrach and Frohlich, 2016; Statista, 2017).

The clinics—usually staffed by nurse practitioners—offer low-cost treatment for more than 25 common conditions (e.g., strep throat and ear infections), and provide health screening tests, immunizations, and physicals. The clinics are located in pharmacies, grocery stores, and "big box" stores like Target and Walmart. There's no need to make an appointment. Waiting time, if any, is minimal. Patients are in and out in about 15 minutes, and most of the clinics are open evenings and weekends ("Health Care in America," 2015).

Brand X Pictures/Getty Images

alienation a feeling of isolation, meaninglessness, and powerlessness.

iron law of oligarchy the tendency of a bureaucracy to become increasingly dominated by a small group of people.

6-2d Getting Around Bureaucratic Dysfunctions

Weber's model overlooked the many ways that workers get around rigid bureaucratic regulations and impersonality. Since around 2000, self-managing work teams have become more common. Let's begin with a brief look at some of the early studies' missteps and contributions regarding the importance of informal group networks.

HISTORIC STUDIES OF INFORMAL GROUP NETWORKS

For many years, and well into the 1930s, experts concentrated on organizational efficiency based on the principles of *scientific management* developed by Frederick Winslow Taylor, a mechanical engineer who wanted to improve industrial productivity. Taylor (1911/1967) considered workers—especially those in factories and on assembly lines—as mere adjuncts to machines, assumed that people don't like to work, maintained that employees need close supervision, and believed that management had to enforce cooperation because most workers aren't capable of handling even the simplest tasks.

Studies conducted by industrial psychologists and sociologists between 1927 and 1932 shook up Taylor's views. The research teams studied employees at the Western Electric Company's Hawthorne plant in Chicago. The *Hawthorne studies*, as they're often called, found that informal groups were critical to the organization's functioning (Roethlisberger and Dickson, 1939/1942; Mayo, 1945; Landsberger, 1958).

Informal social groups can promote an organization's goals if they collaborate, are cohesive, and motivate their members. They can also resist an organization's goals and formal rules. In one of the Hawthorne studies, Roethlisberger and Dickson (1939/1942) spent 6 months observing a group of 14 men who wired telephone switchboards in what the company called the "bank-wiring room."

The bank-wiring room work group consisted of nine wiremen, three solderers, and two inspectors. Management offered financial incentives for higher productivity, but the work group pressured people to limit output. The wiremen developed the following norms that controlled the group's behavior:

1. **You shouldn't turn out too much work.** If you do, you're a "rate-buster" and a "speed king."
2. **You shouldn't turn out too little work.** If you do, you're a "chiseler."
3. **You shouldn't tell a supervisor anything that will be detrimental to a coworker.** If you do, you're a "squealer."
4. **You shouldn't "act officious."** If you're an inspector, for example, you shouldn't act like one.

If individuals violated any of these norms, the other workers used "binging" (striking a person on the shoulder) to punish them.

Why did the men control productivity rather than take advantage of the management's promise of higher wages? They feared that if they produced at a high level, they'd be required to do so regularly. They also worried that high productivity would lead to layoffs because supervisors would conclude that fewer workers could achieve the same output. In addition, the bank-wiring room men experienced high morale and job satisfaction because they felt they had some control over their work. In contradiction to Taylor's scientific management perspective, then, the Hawthorne studies concluded that there's an important relationship between formal and informal organization.

MODERN WORK TEAMS

Since the Hawthorne studies, many companies have recognized that informal groups affect workers' productivity. As a result, numerous organizations have implemented alternative management strategies.

Today, *self-managing work teams* are the dominant model in most large organizations (Cloke and Goldsmith, 2002; Neider and Schriesheim, 2005). Contrary to Taylor's notion that workers do and managers think, self-managing work teams, sometimes referred to as *postbureaucratic organizations*, involve groups of 10 to 15 people who take on the duties of their former supervisors. They gather, interpret, and act on the information, and take collective responsibility for their actions—whether it's designing a new refrigerator or handling a university's food service. Not held back by unresponsive managers, effective self-managing groups focus on their goals and are committed to the organization's success (Colvin, 2012).

COLLABORATION

Many organizations emphasize collaboration. The benefits include a diverse pool of ideas, greater worker participation and job satisfaction, a sense of belonging, and speedier completion of tasks, because people can achieve more when they combine their skills.

On the other hand, "collaborative overload" can result in interruptions that increase the total time needed

The McDonaldization of Society

According to sociologist George Ritzer (1996: 1), the organizational principles that underlie McDonald's, the well-known fast-food chain, are beginning to dominate "more and more sectors of American society as well as of the rest of the world." McDonaldization has four components: efficiency, calculability, predictability, and control.

1. *Efficiency* means that consumers have a quick way of getting meals. In a society where people rush from one place to another and where both parents are likely to work or single parents are pressed for time, McDonald's (and similar franchises) offers an efficient way to satisfy hunger and avoid much "fuss and mess" (see also Jargon, 2013). Like their customers, McDonald's workers function efficiently: The menu is limited, the registers are automated, and employees perform their tasks rapidly and easily.
2. *Calculability* emphasizes the quantitative aspects of the products sold (portion size and cost) and the time it takes to get the products. Customers often feel that they're getting a lot of food for what appears to be a nominal sum of money and that a trip to a fast-food restaurant will take less time than eating at home. In reality, soft drinks are sold at a 600 percent markup because most of the space in the cup is taken up by ice. It costs at least twice as much and takes twice as long to get to the restaurant, stand in line, and pick up the food than to prepare a similar meal at home.
3. *Predictability* means that products and services will be the same over time and in all locales. "There's great comfort in knowing that McDonald's offers no surprises" (p. 10). The Quarter Pounder is the same regardless where customers order. Outside of the United States, "homesick American tourists in far-off countries can take comfort in the knowledge that they'll likely run into those familiar golden arches and the restaurant they have become so accustomed to" (p. 81). Workers, like customers, behave in predictable ways: "There are, for example, six steps to window service: greet the customer, take the order, assemble the order, present the order, receive payment, thank the customer and ask for repeat business" (p. 81).
4. *Control* means that technology shapes behavior. McDonald's controls customers subtly by offering limited menus and uncomfortable seats that encourage diners to eat quickly and leave. "Consumers know that they're supposed to line up, move to the counter, order their food, pay, carry the food to an available table, eat, gather up their debris, deposit it in the trash receptacle, and return to their cars. People are moved along in this system not by a conveyor belt, but by the unwritten, but universally known, norms for eating in a fast-food restaurant" (p. 105).

   McDonald's controls employees more openly and directly. Workers are trained to do a limited number of things in precisely the same way, managers and inspectors make sure that subordinates toe the line, and McDonald's has steadily replaced human beings with technologies that include soft-drink dispensers that shut off when the cup is full and a machine that rings and lifts the French fries out of the oil when they're crisp. Such technology increases the corporation's control over workers because employees don't have to use their own judgment or need many skills to prepare and serve the food.

Efficiency, calculability, predictability, and control reflect a rational system (remember Weber?) that increases a bureaucracy's efficiency. On the other hand, Ritzer contends, McDonaldization reflects the "irrationality of rationality" because the results can be harmful. The huge farms that now produce "uniform potatoes to create those predictable French fries" of the same size rely on the extensive use of chemicals that then pollute water supplies. And the enormous amount of nonbiodegradable trash that McDonald's produces wastes our money because we–and not McDonald's–pay for landfills.

McDonald's is such a powerful model that many businesses have cloned its four dimensions of efficiency, calculability, predictability, and control. Examples include "McDentists" and "McDoctors," drive-in clinics designed to deal quickly with minor dental and medical problems; "McChild" care centers like KinderCare; and "McPaper" newspapers, such as *USA Today* (Ritzer, 2008).

iStock.com/EllenMoran

Table 6.3 Sociological Perspectives on Groups and Organizations

Theoretical Perspective	Level of Analysis	Main Points	Key Questions
Functionalist	Macro	Organizations are made up of interrelated parts and rules and regulations that produce cooperation in meeting a common goal.	• Why are some organizations more effective than others? • How do dysfunctions prevent organizations from being rational and effective?
Conflict	Macro	Organizations promote inequality that benefits elites, not workers.	• Who controls an organization's resources and decision making? • How do those with power protect their interests and privileges?
Feminist	Macro and micro	Organizations tend not to recognize or reward talented women and regularly exclude them from decision-making processes.	• Why do many women hit a glass ceiling? • How do gender stereotypes affect women in groups and organizations?
Symbolic Interactionist	Micro	People aren't puppets but can affect what goes on in a group or organization.	• Why do people ignore or change an organization's rules? • How do members of social groups influence workplace behavior?

to complete a task. The most vulnerable are *knowledge workers*, people who acquire, organize, analyze, and use information (e.g., programmers, data and systems analysts, product developers). Some researchers estimate that knowledge workers "waste" 70 to 85 percent of their time attending meetings (virtual or face-to-face), answering email or phone calls, and dealing with an avalanche of requests (Drucker, 2001; "The Collaboration Curse," 2016; Cross et al., 2016).

6-3 SOCIOLOGICAL PERSPECTIVES ON SOCIAL GROUPS AND ORGANIZATIONS

How do organizations operate? And how do social groups affect massive bureaucracies? Functionalist, conflict, feminist, and interactionist perspectives offer different insights, providing a multifaceted understanding of groups and organizations (*Table 6.3* summarizes these perspectives).

6-3a Functionalism: Groups and Organizations Benefit Society

Functionalist perspectives emphasize that social groups and organizations are composed of interrelated, mutually dependent parts. As Weber pointed out, when a bureaucracy operates rationally and efficiently, workers and bosses cooperate to turn out a final product, whether it's an automobile or a can of soup.

Effective leadership is a primary factor in job satisfaction. Even in our sprawling government bureaucracies, and across all organizational levels, the happiest workers are those whose bosses reward motivation and commitment, encourage integrity, and promote the employees' professional development and creativity (Partnership for Public Service, 2011).

A recent study of a large company with technology-based service jobs (e.g., retail clerks, airline gate agents, call-center workers) found that replacing a supervisor from the bottom 10 percent of the pool with one from the top 10 percent increased output almost as much as adding a tenth person to a nine-worker team. The researchers concluded that productivity increases when the most promising workers are paired with top bosses who provide frequent feedback and teach better work methods and skills (Lazear et al., 2013). Thus, good bosses don't threaten or punish employees, but provide concrete training to get the job done.

Some companies have boosted productivity by encouraging workers' interaction. Examples include providing group rather than individual breaks, and carving out meeting spaces that accommodate three or four people instead of 12 or more (Silverman, 2013).

Groups and organizations can also be dysfunctional. Workers may be alienated because of few rewards, favoritism, and incompetent supervisors. Only about a third of U.S. workers and managers are engaged in their jobs. Among employees who feel unchallenged, bored, or dissatisfied, 75 to 80 percent spend more than 2 hours a day surfing the Net or on Twitter, Facebook, and other social media sites. Wasting time—which results in low productivity, unhappy customers, and lower profits—costs

Courtesy of King Flour Company

Since 1975, the number of employee-owned companies in the United States has grown from 1,600 to more than 11,500, and now represents about 12 percent of the private sector workforce. These firms are profitable because the workers are highly motivated: They have a stake in the business, control the decision making, contribute original ideas and cost-saving solutions, and depend on each other. Pictured here are people learning to bake bread at the King Arthur Flour Company's Baking Education Center in Norwich, Vermont, an employee-owned company that has "a ferocious focus on treating employees well" (Koba, 2013; Marikar, 2015).

organizations up to $398 billion annually (Adkins, 2015; OfficeTime, 2015).

Since 2008, business spending on employees has grown only 2 percent compared with 26 percent on equipment and software (Rampell, 2011). Many people contend that doing so increases unemployment rates. From a functionalist perspective, however, investing more in equipment than workers is a rational bureaucratic response to the rising costs of health care benefits and economic competition from other countries (see Chapter 11).

CRITICAL EVALUATION

For critics, especially conflict theorists, functionalists exaggerate cooperation and tend to gloss over dysfunctions like worker dissatisfaction and alienation. At some of the largest and most successful companies like Amazon, current and former employees have complained of 80-hour work weeks, interrupted vacations, sending negative feedback about coworkers to bosses, and little tolerance for those struggling with life-threatening illnesses or family crises (Kantor and Streitfeld, 2015). Still, many Americans' views of corporations are contradictory. About 63 percent (compared with 48 percent in 2001) are dissatisfied with the size and power of major companies, but only 20 percent want greater regulation (Riffkin, 2016).

Another issue is whether informal social networks improve worker morale and control as much as functionalists claim. After all, a tedious and monotonous job is tedious and monotonous regardless of the degree of informal coworker interaction.

6-3b Conflict Theory: Some Benefit More Than Others

Conflict theorists contend that organizations promote inequality that benefits the top of the hierarchy, not workers. In many companies, those at higher levels are more comfortable hiring and promoting people like themselves (male, White, and middle class or higher). Phrases like "fitting the mold" and "having common interests" are often code words for belonging to an in-group. In 2016, women and minorities made up only 14.4 percent of Fortune 500 corporate boards—up slightly from 12.8 percent in 2010. The negligible gains "are certainly not representative of the broad demographic changes we have seen in the United States . . ." (Deloitte and Alliance for Board Diversity, 2017: 1).

Inequality in income, status, and other rewards means that owners and managers can easily exploit workers. Those at the top dictate to those at the middle and the bottom. Because organizations serve elites, conflict theorists argue, they routinely ignore workers' needs and interests.

What about rewarding the best employees? According to a management professor, "I've examined scores of empirical studies since the early 1980s and have not found convincing evidence that performance reviews are fair, accurate or consistent across managers, or that they improve organization effectiveness." Instead, subjective evaluations "are intimidating tools that make employees too scared to speak their minds" (Culbert, 2011: A25).

There's also considerable incompetence, waste, and corruption in many government agencies and big business. For example:

- In 2014 alone, the federal government paid more than $125 billion to ineligible Medicare, Medicaid, and low-income recipients ("Government Efficiency and Effectiveness," 2015).
- Since at least 2012, the Veterans Administration has ignored the agency's competitive bidding laws, wasting at least $6 billion a year on high prices for medical goods and services (Rein and Wax-Thibodeaux, 2015).

- Between 2008 and 2012, the Agriculture Department paid $22 million to farmers who had been dead for at least two years (GAO, 2013).
- In 2016, the Pentagon buried an internal study that it had wasted $125 billion on accounting, human resources, and property management. At least 40 percent of all U.S. weapons given to the Afghan and Iraqi armies "can't be traced" (Vicens, 2015; Whitlock and Woodward, 2016).
- Since 2000, the federal government has used more than $3 billion taxpayer dollars to build or renovate 36 professional sports stadiums (e.g., for the New York Yankees and Indianapolis Colts). Dozens of studies have shown that such facilities don't increase local economic development or create jobs (Gayer et al., 2016).
- At least 110 Americans, so far, have been injured or killed by an airbag that Ford, Honda, Nissan, and Toyota automakers knew was defective. Volkswagen sold more than 600,000 diesel cars over a decade that the company equipped with illegal devices to evade U.S. emission rules (see Chapters 7 and 15).

Besides wasting billions of taxpayer dollars, according to many federal employees, government agencies hire and promote incompetent people, offer few advancement opportunities and rewards for good performance, and keep increasing the layers of unnecessary executives and managers. One employee complained, for instance, that "It takes 13 steps and five layers to get a signature from our office director" (Rein, 2011: B4; U.S. Office of Personnel Management, 2014).

Compared with government agencies, privately owned organizations have more freedom, authority, and control over activities and decision making, but at what cost? When Walmart opened its small Express stores in many rural or isolated towns, local groceries and pharmacies closed because they couldn't compete with Walmart's prices. To increase profits at its Supercenters and some Neighborhood Markets, in 2016 Walmart shut all 102 of its Express stores. Investors were pleased, but some residents had to make 50-minute roundtrip drives to the nearest grocery stores and pharmacies (Pettypiece, 2016).

Because of these and similar problems, many Americans are disillusioned with a number of institutions and organizations (*Figure 6.2*).

ZUMA Press, Inc./Alamy Stock Photo

CRITICAL EVALUATION

Conflict theory has several weaknesses. First, it assumes that greater equality leads to a more successful and productive organization. However, self-managing work teams can fail because of ineffective team leaders or management interference. Employee-owned companies have bankrupted because of inexperienced management, competition, and low worker morale (Lencioni, 2002; "When Workers Are Owners," 2015).

Second, conflict theorists rarely credit many organizations' supporting controversial social issues that promote greater equality. Regardless of motives (e.g., improving a tarnished public image, boosting profits, becoming more

Figure 6.2 Many Americans Have Little Confidence in Organizations

Percentage who said, in 2016, that they have "a great deal" or "quite a lot" of confidence in ...

Institution	Percent
The church or organized religion	41
The medical system	39
The presidency and U.S. Supreme Court	36 (each)
The public schools	30
Banks	27
The criminal justice system and organized labor	23 (each)
Television news	21
Newspapers	20
Big business	18
Congress	9

Note: A majority of Americans had high confidence only in the military (73 percent), small business (68 percent), and the police (56 percent).

Source: Based on Norman, 2016.

socially responsible), a number of the nation's most powerful CEOs and businesses—including Walmart, NASCAR, Apple, AT&T, and Southwest Airlines—have supported controversial gay marriage laws and fought discrimination against lesbian, gay, and transgender customers (Green and Higgins, 2016; see also Chapters 7, 9, and 11).

Third, do conflict theorists focus too much on organizational deficiencies? There are, after all, many organizations that are efficient, profitable, and that don't exploit their workers (Kaplan, 2010).

6-3c Feminist Theories: Men Benefit More Than Women

Feminist scholars point out that across all social classes, women (and especially minority women) consistently fare worse than men, especially in leadership roles. Women have enjoyed increased success in organizations since the 1980s, but rarely at the highest levels. For example, women hold only 21 (4.2 percent) of the CEO positions at Fortune 500 companies (Merelli, 2017), and there's a huge gender gap in S&P 500 companies between the female labor force and CEOs (*Figure 6.3*).

Figure 6.3 Women in S&P 500 Companies

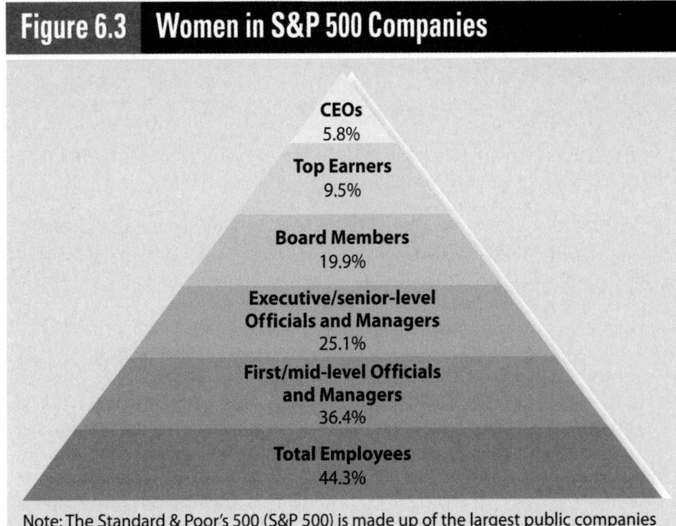

Note: The Standard & Poor's 500 (S&P 500) is made up of the largest public companies of the leading U.S. industries (as selected by economists).

Source: Catalyst, 2017.

Companies benefit from female leadership. Female CEOs are more likely than their male counterparts to mentor women workers, to match people to job skills, and to pay wages that reward productivity. On average, female managers are more engaged than male managers, more likely to encourage their subordinates' development, to recognize and praise their accomplishments, and to make them feel valued (Fitch and Agrawal, 2015; Flabbi et al., 2016).

Why, then, does the gender gap in top positions persist? Large numbers of women still hit a **glass ceiling**—workplace attitudes or organizational biases that prevent them from advancing to leadership positions. Examples of obstacles include men's negative attitudes about women in the workplace, women's placement in staff positions that aren't on the career track to the top, a lack of mentoring, biased evaluations by male supervisors, and little or no access to highly visible committees or task forces (Barreto et al., 2009; Auster and Prasad, 2016; see also Chapters 9 and 11).

In contrast, many men who enter female-dominated occupations (e.g., nursing and teaching) receive higher wages and faster promotions, a phenomenon known as a **glass escalator**. In 2016, for example, men made up only 11 percent of all full-time registered nurses. However, the average female nurse earned $59,436, about 10 percent less than the $65,572 that the average male nurse earned. Men on glass escalators also move up to supervisory positions more quickly than do females (Goudreau, 2012; U.S. Bureau of Labor Statistics, 2017).

CRITICAL EVALUATION

Feminist theories show that groups and organizations treat many talented women like outsiders. One weakness, however, is that much of the emphasis is still on White and Black women even though Latinas and Asian American women comprise a large segment of the labor force.

A second limitation is that even when feminist scholars say that both sexes suffer from organizational stereotypes, they tend to gloss over stereotypes that affect men and how female supervisors reinforce glass ceilings and escalators (see Huffman, 2013). Many women experience a double standard, but they fail for some of the same reasons as men: Female managers and executives make mistakes, may not network within and outside a company, avoid lateral moves (that can lead to a promotion), or turn down promotions and job offers that involve risk-taking, long hours, and work-related stress (Reingold, 2016).

6-3d Symbolic Interaction: People Define and Shape Their Situations

Whereas functionalist, conflict, and (some) feminist theorists examine groups and organizations on a macro level, symbolic interactionists focus on micro-level behavior. They emphasize that an individual's perception

glass ceiling workplace attitudes or organizational biases that prevent women from advancing to leadership positions.

glass escalator men who enter female-dominated occupations receive higher wages and faster promotions than women.

and definition of a situation shape group dynamics and, consequently, organizations.

Group leaders or members can create or reinforce conformity (as shown in the Asch and Milgram studies). Informal groups can also determine what goes on in an organization by refusing to obey the rules and implementing their own (as in the Hawthorne studies). Thus, according to symbolic interactionists, individuals make choices, change rules, and mold their own identities instead of being manipulated (Kivisto and Pittman, 2001).

Symbolic interactionists also note that people's outcomes depend on how coworkers and bosses interpret the *same* behavior. For example, whether women are supervisors or clerical workers, if they lose their temper, they're overwhelmingly seen as too emotional, incompetent, out of control, weak, and worth less pay. Their angry male counterparts, in contrast, are often viewed as authoritative, tough, and forceful (Brescoll and Uhlmann, 2008).

CRITICAL EVALUATION

Symbolic interaction explains how members of groups and organizations interpret the world around them and, as a result, affect what goes on. However, symbolic interaction, a micro-level theory, ignores macro-level factors that affect workers and consumers. About 51 percent of Americans seek out "all natural" products, but much of the food contains chemically processed vitamins and artificial ingredients. The average consumer has little knowledge of and control over such false advertising, a macro-level factor (Esterl, 2013).

In contrast to interactionists' claims, most people can't shape or change their situations. Formal organizations (the U.S. government and large companies) often invade the privacy of citizens, workers, or consumers by collecting information and monitoring people's behavior. Knowledge workers must share large, noisy, open office spaces and do much of their work when they get home at night. A majority (51 percent) of police officers say their job is frustrating because of factors over which they have no control, such as biased media coverage, unfair departmental disciplinary procedures, and a shortage of resources ("The Collaboration Curse," 2016; Morin et al., 2017).

We've looked at social groups and formal organizations, both of which are part of larger structures that sociologists call *social institutions.* The final section introduces you to this important concept and lays the foundation for analyses of social institutions in later chapters.

social institution an organized and established social system that meets one or more of a society's basic needs.

6-4 SOCIAL INSTITUTIONS

A **social institution** (or simply, *institution*) is an organized and established social system that meets one or more of a society's basic needs. There are five major social institutions, worldwide, that ensure a society's survival:

- The *family* replaces a society's members through procreation, socializes children, and legitimizes sexual activity between adults.
- The *economy* determines how a society produces, distributes, and consumes goods and services.
- The *government* (which includes police, military, and courts) maintains law and order, creates and enforces laws, regulates elections, and protects people's civil liberties.
- *Education* helps to socialize children, transmits values and knowledge, and teaches work-related skills.
- *Religion* offers meaning and purpose through shared beliefs, values, and practices related to the supernatural.

Besides these core institutions (which we'll examine in Chapters 7 and 11–14), people develop new institutions in response to cultural changes. In the past, for example, the family and religion addressed most health problems. Only in relatively recent times have *medicine* and *health care systems* emerged as distinct institutions that diagnose, treat, and try to prevent illness and disease. Because of technological and economic changes, the *mass media* is another relatively recent institution. It disseminates information to a large audience, creates desires for goods and services, influences our attitudes about social issues, and reinforces a society's values and beliefs.

Five major social institutions

- family
- economy
- government
- education
- religion

6-4a Why Social Institutions Are Important

People create social institutions to address basic needs or problems. Institutions have an organized purpose, weave together norms and values, and, consequently, guide behavior. They may differ across countries, but each institution carries out certain tasks in a particular society that contribute to its overall functioning and stability. No two families are exactly alike, but the family institution affirms broadly shared cultural agreements about what a family should be and do. In the same vein, no two grocery stores are exactly alike, but the economy establishes and maintains a variety of rules that make food shopping predictable.

Institutions aren't just "somewhere out there." We create, live inside, change, resist, and reproduce them, and participate simultaneously in multiple institutions. Understanding institutions, and how they're interconnected, can tell us a lot about how a society functions.

6-4b How Social Institutions Are Interconnected

Institutions are linked. Consider the connections among seven institutions that affect obesity, a serious health problem. *Medicine* tells us that obesity is a major (and preventable) reason for disease, disability, and death, and for steadily rising *health care* costs (see Chapter 14). The *mass media* publicizes such information and medical advice about prevention. It also promotes unhealthy food through persuasive marketing techniques (e.g., free gifts, celebrity endorsements, false nutrition claims) on television and other electronic devices (Jenkin et al., 2014).

In *education*, many schools now offer healthier meals, but school vending machines still sell high-fat snacks and sugary drinks. When state and local *governments* slash education budgets, schools decrease physical education and sports programs. At the federal level, the most recent food guidelines advise Americans to consume less saturated and trans fats and sugar, but don't tell people to avoid sodas, sugary drinks, and processed meat (Evich, 2016; Yeager, 2016).

Families contribute to the growing obesity problem. By the time children enter school, their parents have already shaped their eating habits and preferences (see Chapters 4 and 14). Parents and other family members can encourage physical activity, restrict certain foods, and limit exposure to mass media marketing that promotes unhealthy eating habits.

Caspar Benson/Getty Images

On average, a 12-ounce can of soda contains at least 9 teaspoons of sugar. Over the course of a year, drinking one soda a day can make you 16 pounds fatter (Loumarr, 2015). Which social institutions affect your beverage choices?

The *economy* has far-reaching effects on obesity through its linkages to other institutions. Almost all politicians rely on the food industry to help finance their elections and to hire them as lobbyists after they leave public office (see Chapter 11). Beverage giants like Coca-Cola and PepsiCo have donated millions of dollars to nearly 100 prominent health groups (e.g., National Institutes of Health, American Diabetes Association) while simultaneously spending millions to defeat public health measures aimed at curbing soda consumption. The mass media relies on food ads even though companies make unhealthy products. And, because of advertising and packaging claims, many families think that fruit and sports drinks and flavored waters—all of which have high amounts of sugar—are healthy (Munsell et al., 2015; Aaron and Siegel, 2017).

STUDY TOOLS 6

READY TO STUDY? IN THE BOOK, YOU CAN:

- ☐ Check your understanding of what you've read with the Test Your Learning Questions provided on the Chapter Review Card at the back of the book.
- ☐ Tear out the Chapter Review Card for a handy summary of the chapter and key terms.

ONLINE AT CENGAGEBRAIN.COM WITHIN MINDTAP YOU CAN:

- ☐ Explore: Develop your sociological imagination by considering the experiences of others. Make critical decisions and evaluate the data that shape this social experience.
- ☐ Analyze: Critically examine your basic assumptions and compare your views on social phenomena to those of your classmates and other MindTap users. Assess your ability to draw connections between social data and theoretical concepts.
- ☐ Create: Produce a video demonstrating connections between your own life and larger sociological concepts.
- ☐ Collaborate: Join your classmates to create a capstone project.

7 Deviance, Crime, and Social Control

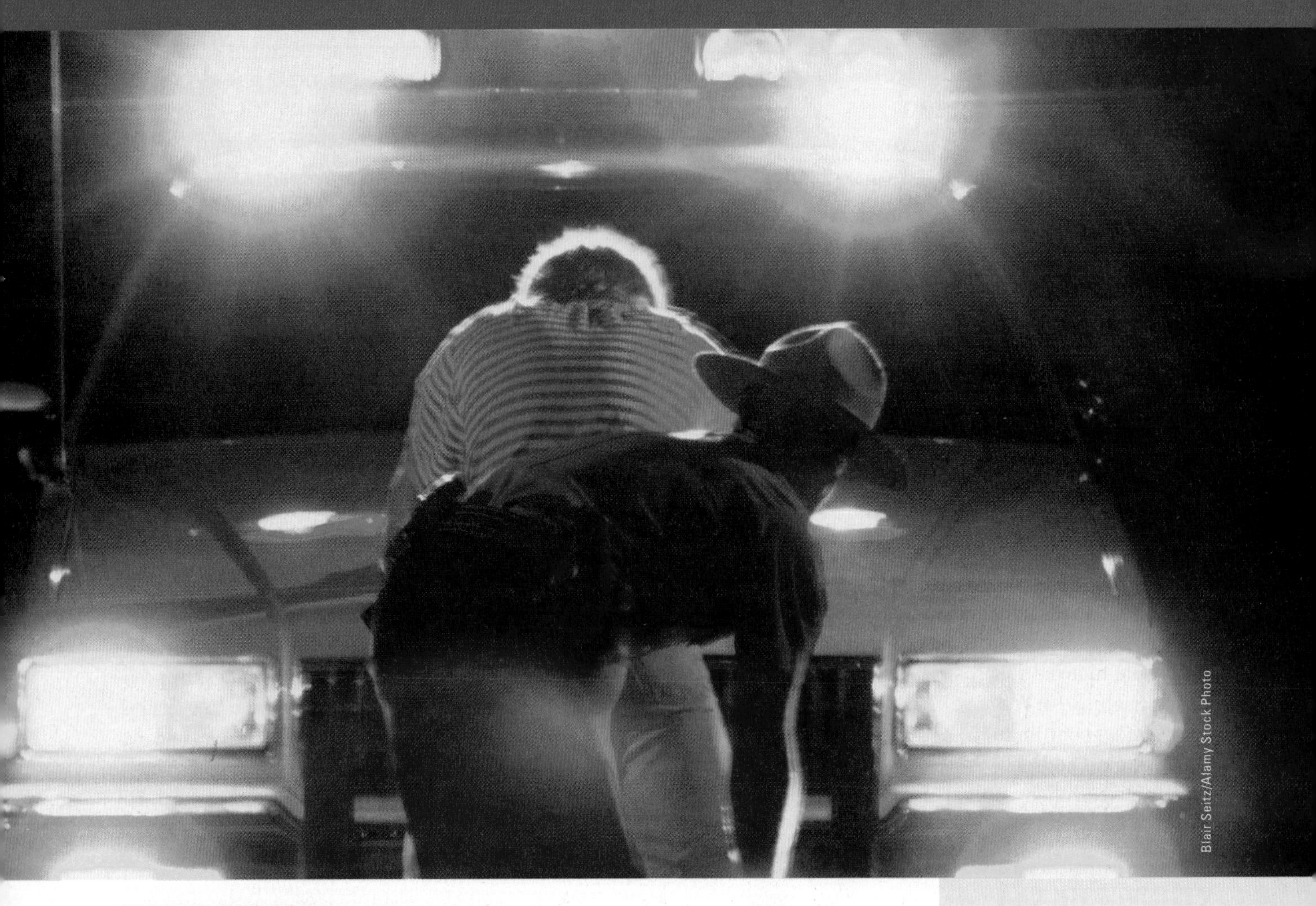

Blair Seitz/Alamy Stock Photo

LEARNING OBJECTIVES

After studying this chapter, you will be able to...

- 7-1 Differentiate between deviance and crime, and describe the key characteristics of deviance.
- 7-2 After evaluating the two major crime measures, identify and illustrate the different types of crime.
- 7-3 Describe, illustrate, and evaluate functionalist perspectives on deviance.
- 7-4 Describe, illustrate, and evaluate conflict perspectives on deviance.
- 7-5 Describe, illustrate, and evaluate feminist perspectives on deviance.
- 7-6 Describe, illustrate, and evaluate symbolic interaction perspectives on deviance.
- 7-7 Identify and evaluate the criminal justice system's social control methods.

After finishing this chapter go to **PAGE 137** for **STUDY TOOLS**

A group of hackers broke into AshleyMadison.com, a website that arranges extramarital affairs, and released more than 35 million of the subscribers' names, addresses, and personal information. The online profiles included thousands of American and European military and government personnel and high-level executives. Many people believed that the deceitful spouses got what they deserved, but hundreds of church leaders resigned, some people lost their jobs, and several subscribers committed suicide (Weise and Vanden Brook, 2015; Rosenthal, 2016).

WHAT DO YOU THINK?

There would be less crime if the punishments were more severe.

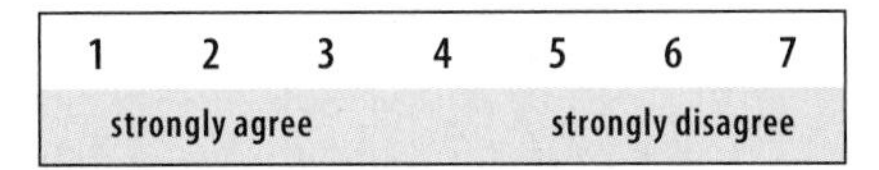

Who was deviant? The hackers? The website owners? The subscribers? And what, exactly, is deviance? Before reading further, take the True or False quiz to see how much you know about U.S. deviance and crime.

WHAT IS DEVIANCE?

Have you ever driven above the speed limit? Cheated on an exam? Engaged in underage drinking? All are examples of **deviance**—violations of social norms. Unlike the general public, sociologists use the concept *deviance* nonjudgmentally: They're interested in understanding and explaining why all of us violate norms from time to time, and why people judge some acts (hacking) more negatively than others (subscribing to an infidelity site).

Deviance becomes **crime**, a violation of society's formal laws, when it breaks rules that have been written into law and enforced by a political authority. For example, trafficking children under age 18 for prostitution, pornography, or sex is deviant. It's not a crime, however, until a country enacts laws that punish offenders.

True *or* False?

HOW MUCH DO YOU KNOW ABOUT U.S. DEVIANCE AND CRIME?

1. All deviant behavior is criminal.
2. Deviance serves a useful purpose in society.
3. Most crime victims are women.
4. Crime rates have decreased since 1990.
5. People are more likely to be arrested for a drug violation than for driving while drunk.
6. Murder and assault are more common than illegal gambling and prostitution.
7. Suburbs are safer than cities.
8. Death penalties deter crime.

The answers for #2, #4, and #5 are true; the others are false. You'll see why as you read this chapter.

7-1a Some Key Characteristics of Deviance

Deviance is universal because it exists in every society. Still, deviance can vary quite a bit over time, from situation to situation, from group to group, and from culture to culture (Sumner, 1906; Schur, 1968):

- **Deviance can be a trait, a belief, or a behavior.** People usually do something to be considered deviant. We can also be treated as outsiders simply because of our appearance, skin color, religious beliefs, or sexual orientation (Becker, 1963; see also Chapters 9, 10, and 13).
- **Deviance is accompanied by social stigmas.** A **stigma** is a negative label that devalues a person and changes her or his self-concept and social identity. Stigmatized individuals may react in many ways: They may alter their appearance (as through cosmetic surgery), associate with others like themselves who accept them (as in gangs), hide information about some aspect of their deviance (an

deviance a violation of social norms.

crime a violation of society's formal laws.

stigma a negative label that devalues a person and changes her or his self-concept and social identity.

ex-convict who doesn't reveal that status), or divert attention from a stigma by excelling in some area (as music or sports) (Goffman, 1963).

- **Deviance varies across and within societies.** What's accepted or tolerated in one society may be deviant in another. The United States comprises only 5 percent of the world's population, but accounts for 31 percent of global public *mass shootings* (incidents that result in killing four or more people outside homes and that aren't gang-related, hostage situations, or robberies). Between 1983 and 2013, the United States had 78 mass shootings, nearly twice as many as 24 other industrialized countries combined. Within the United States, 8 of the 10 deadliest mass shootings between 1982 and 2016 occurred in the South and West. Across and within countries, stricter gun control laws and lower firearm ownership rates are correlated with fewer mass shooting incidents (Florida, 2013; Lankford, 2015; Lemieux, 2015; Myers, 2016).

- **Deviance varies across situations.** What's acceptable in one context may be stigmatized in another. Among U.S. adults, between 74 and 77 percent say it's generally okay to use a cell phone on public transportation and while walking down the street or waiting in line. In contrast, only 4 to 5 percent approve of using cell phones at meetings, movie theaters, and religious services (Rainie and Zickuhr, 2015).

- **Deviance is formal or informal.** *Formal deviance* is behavior that violates laws. A leading example is crime, a topic we'll examine shortly. In contrast, *informal deviance* is behavior that disregards accepted social norms, such as picking one's nose or teeth or scratching one's private parts in public, belching loudly, and not dressing appropriately (e.g., wearing jeans to a wedding reception or job interview).

- **Perceptions of deviance can change over time.** Many behaviors that were acceptable in the past are now seen as deviant. Only during the 1980s and 1990s did U.S. laws define date rape, marital rape, stalking, and child abuse as crimes. Smoking—widely accepted in the past—has been banned in most public places, and increasing numbers of employers don't hire people who use tobacco. Texas is now one of the eight states that allows students to carry concealed guns in four-year public colleges. In 2017, Nevada became the first state to install vending machines with free needles for drug users.

On the other hand, most Americans now shrug off behaviors that were stigmatized in the past. Cohabitation, seen as sinful and immoral 40 years ago, is now widespread and considered normal. Americans' support for legalizing marijuana for recreational use increased from only 12 percent in 1969 to 60 percent in 2016 (Geiger, 2016). And a third of Americans have at least one tattoo, up from only 16 percent in 2003 (Geiger, 2016; Shannon-Missal, 2016).

7-1b Who Decides What's Deviant?

Because deviance is culturally relative and standards change over time, who decides what's right or wrong? An important group is those who have authority or power. During our early years, parents and teachers specify acceptable and unacceptable behavior (see Chapter 4). As we get older, laws also define what's deviant (e.g., being forbidden to drive until age 16 or to purchase or consume alcohol until age 21).

Many employers—especially those in businesses and federal agencies—don't hire people with visible "body art" ("Tattoos in the Workplace . . . ," 2014). Lawmakers pass numerous laws that people must obey, but often exempt themselves. Examples include having immunity from employment discrimination and sexual harassment lawsuits, and passing laws that allow guns in

More than 800 women in Ghana live in exile in witch camps. Most are poor widows or older women who are blamed for outbreaks of disease in their villages and relatives' or neighbors' illness or death (Küntzle and Blondé, 2013). Murders of women and children accused of sorcery have risen since 2009. Saudi Arabia, Nepal, India, Papua New Guinea, Tanzania, and Uganda still kill people suspected of witchcraft (Hoffman, 2014).

Markus Matzel/ullstein bild/Getty Images

Alfred Wekelo/Shutterstock.com; Jeff J Mitchell/Getty Images News/Getty Images; Dmitry Naumov/Shutterstock.com; iStock.com/belterz

schools, restaurants, public buildings, and churches, but not in state legislative chambers (Skoning, 2013; Deprez and Selway, 2014).

Public attitudes and behavior also affect definitions of deviance. Many people, including health practitioners, now use the word "overweight" instead of "fat" or "obese" to decrease the stigma or being fat. On the other hand, there has been considerable *normalization of deviance*, the gradual process through which unacceptable practices or standards become acceptable. Examples include the growing acceptance of same-sex and interracial marriage, legalized gambling, political corruption, and recreational marijuana (see Chapters 3, 10, and 11). Normalizing some types of deviance generates considerable income, particularly at the state level.

7-2 TYPES OF DEVIANCE AND CRIME

Noncriminal deviance—such as suicide, alcoholism, lying, mental illness, and adult pornography—is common. Sociologists are especially interested in *criminal deviance*, behavior that violates laws, because most of the transgressions are visible, publicized by the media, and threaten people's lives or property. Measuring criminal deviance may seem straightforward, but isn't as simple as it appears.

7-2a Measuring Crime

Two of the most important sources of crime statistics are the FBI's *Uniform Crime Report* (UCR) and the U.S. Department of Justice's *National Crime Victimization Survey* (NCVS). Each has its strengths and weaknesses.

The UCR data are crimes reported to the police and arrests made each year. The statistics are useful in examining trends over time, but the UCR doesn't include offenses such as corporate crime, kidnapping, and Internet crimes; simple assaults (those not involving weapons or serious injury); nor the 53 percent of all violent crimes (e.g., rape, robbery, and aggravated assault) and 65 percent of property crimes (e.g., burglary and motor vehicle theft) that aren't reported to the police (Gramlich, 2017). Because of such limitations, many sociologists also use victimization surveys to measure the extent of crime.

A **victimization survey** interviews people about being crime victims. Because the response rates are at least 90 percent, the NCVS offers a more accurate picture of many offenses than does the UCR. The NCVS also includes both reported and unreported crime, and isn't affected by police discretion in deciding whether to arrest an offender.

Still, many victims don't report a crime to either the police or interviewers for a variety of reasons: They fear reprisal for getting the offender in trouble, believe that the police can't or won't do anything to help, or the crime, particularly rape, is too painful to discuss. Some people don't want to admit having been victims—especially when the perpetrator is a family member or friend. Others may be too embarrassed to tell an interviewer that they were victimized while drunk or engaged in drug sales (Hagan, 2008; Truman and Langton, 2015).

No one knows the extent of U.S. crime because even the best sources (the UCR and NCVS) provide only estimates, and data for higher-income lawbreakers are more elusive than for lower-income offenders. Let's begin with the latter, often called *street crimes*, offenses committed by ordinary people, usually in public places.

7-2b Street Crimes

A majority of Americans (59 percent) say crime is an "extremely" or "very serious" problem in the United

victimization survey interviews people about being crime victims.

One of the biggest costs for retail stores is *wardrobing*, returning expensive clothes after wearing them. Such "borrowing" has also become prevalent in fine jewelry, high-priced cameras, men's clothes (like business suits for special occasions), and even household tools. Wardrobing has increased 40 percent since 2009, and cost retailers $9.1 billion in 2015 (Timberlake, 2013; Allen, 2015). Who do you think pays for wardrobing fraud?

Esline/Shutterstock.com

SCAM

Becky Stares/Shutterstock.com

States (McCarthy, 2015). They're usually referring to *violent crimes* (e.g., murder, rape, robbery, aggravated assault) that use force or the threat of force against others. *Property crimes* (e.g., burglary, motor-vehicle theft, arson) involve the destruction or theft of property, but offenders don't use force or the threat of force.

INCIDENCE OF VIOLENT AND PROPERTY CRIMES

Property crimes are more common than violent crimes, but both have been falling (*Figure 7.1*). Despite occasional crime spikes in some large cities, between 2008 and 2015 violent and property crime rates fell 26 percent and 22 percent, respectively. The average person is 433 times more likely to experience a theft than to be murdered, and the odds of being murdered or robbed are now less than half of what they were in the early 1990s (Glassner, 2010; Oppel, 2011; Truman and Morgan, 2016).

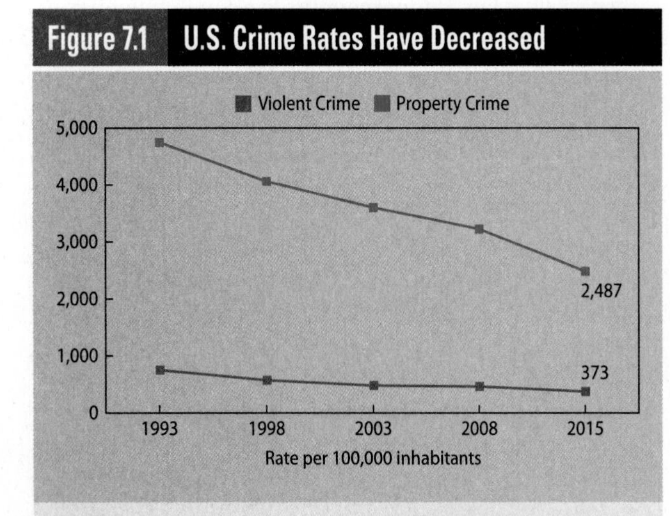

Figure 7.1 U.S. Crime Rates Have Decreased

Source: Based on Federal Bureau of Investigation, 2016, Table 1.

About 75 percent of Americans say that neither they nor anyone in their household has experienced a crime. Nonetheless, 53 percent worry "a great deal" about being a crime victim. Since 1989, up to 89 percent have said that crime had increased compared with the previous year (McCarthy, 2015; Davis, 2016).

Why is there such a disconnect between many Americans' perceptions of crime and official statistics? Among other reasons, the media inspire fear by publicizing violent crimes; television shows and movies exaggerate the prevalence of crime; and many people believe misinformation. Both during and after the 2016 presidential election, Donald Trump claimed that the U.S. murder rate "is the highest it's been in 27 years," that Philadelphia's murder rate was "just terribly increasing," and that "carnage" plagued "inner cities" (Diamond, 2017; Hee Lee, 2017; see also Chapter 2). The statements were false, but they sparked public concerns and misperceptions about crime.

OFFENDERS AND VICTIMS

Most offenders are never caught, but street crimes aren't random. Of those arrested in 2015, for example:

- 34 percent were under age 25, and 30 percent were ages 25 to 34. In contrast, only 6 percent were 55 or older, and many of the crimes involved drug abuse and drunk driving rather than violent and property crimes.
- 73 percent were males. They accounted for 80 percent of arrests for violent crime and 62 percent for property crime.
- 70 percent were White, 27 percent were Black, and 3 percent were Native American (represented as American Indian in the data), Latino, or Asian (Federal Bureau of Investigation, 2016).

Men are more likely than women to be crime victims, but the type of victimization differs. Males experience higher rates of homicide, robbery, and aggravated assault, whereas females are more likely to be victims of rape, sexual assault, and intimate partner violence.

Violent victimization rates are lower for White and Latino individuals (17 per 1,000 persons age 12 and older) than for Black, Asian, and Native American

Men are much more likely than women to be victims of violent crime.

Comstock/Stockbyte/Getty Images

individuals (25). The rates are also much higher for people under age 25 (28) than those age 65 and older (5). People in low-income households (under $10,000 a year) are almost four times more likely than high-income households ($75,000 or more a year) to experience violent victimization. Overall, the most likely victims of violent crimes are young, Black, female, poor, urban, and women who are divorced or separated (Truman and Morgan, 2016; see also Norman, 2015).

Crime rates are higher in low-income areas, but does this mean that poor people are more deviant? Because police devote more resources to poor neighborhoods, those at the lower end of the socioeconomic ladder are more likely to be caught, arrested, prosecuted, and incarcerated. *Intellectual property theft*—which includes offenses like software piracy, bootlegging musical recordings and movies, selling company trade secrets, and copyright violations—is primarily a middle-class crime, but the offenders are difficult to catch.

Figure 7.2 Reasons for Hate Crimes, 2015

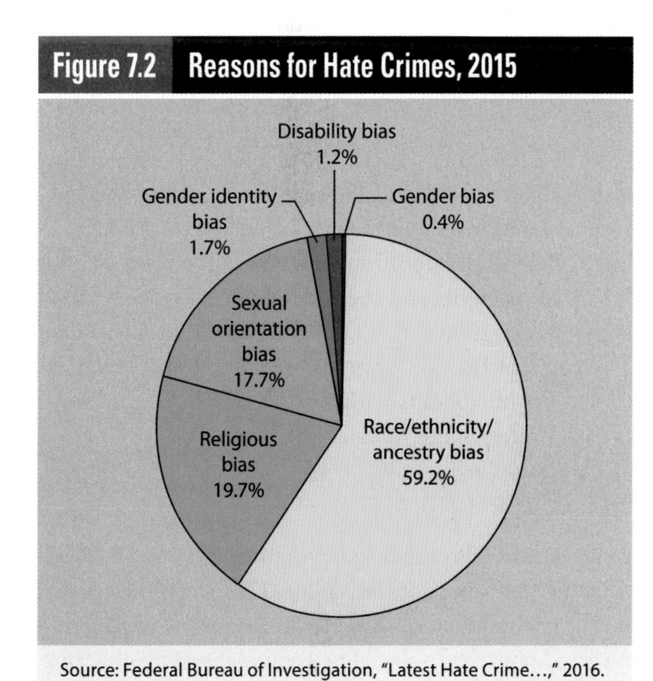

Source: Federal Bureau of Investigation, "Latest Hate Crime...," 2016.

7-2c Hate Crimes

In 2016, 21-year-old Dylann Roof killed nine African Americans in a South Carolina church. Roof said that "Blacks were taking over the world" and that he was rescuing the "White race" (Silverstein, 2015). The massacre is an example of a **hate crime** (also known as a *bias crime*), a criminal offense motivated by the fact or perception that the victims differ from the perpetrator. The attacks, often violent, are based on prejudice toward race, ethnicity, ancestry, religion, disability, sexual orientation, gender, or gender identity (Federal Bureau of Investigation, 2016; see also *Figure 3.3* and the related discussion of hate groups on page 51).

The FBI recorded 6,885 hate crimes in 2015: 83 percent were directed at individuals; the remaining 17 percent were against businesses or financial institutions, religious organizations, or the government. (The number is probably much higher because more than half of victims don't report hate crimes to police.) Almost 60 percent of the incidents were racially motivated (*Figure 7.2*). Forty-eight percent of the offenders were White, 24 percent were Black, 16 percent were of unknown races, and 9 percent were multi-racial (Federal Bureau of Investigation, 2016; Masucci and Langton, 2017).

Why do people commit hate crimes? In *thrill hate* crimes, the offenders are "just having fun" by attacking minorities, bashing gay people, or destroying property. Perpetrators of *defensive hate crimes* feel justified, even obligated, to protect their neighborhood against outsiders who are the "wrong" race, ethnicity, and so on. *Retaliatory hate crimes* are motivated by revenge for a real or imagined incident (e.g., attacking Muslims after the 2013 Boston Marathon bombing) (McDevitt et al., 2002; Levin and Nolan, 2017).

7-2d White-Collar Crime

White-collar crime refers to illegal activities committed by high-status people in the course of their occupations (Sutherland, 1949; Cressey, 1953). The offenses don't involve violence because white-collar criminals use their status and powerful positions to enrich themselves or others. Typical white-collar crimes include

hate crime (also known as *bias crime*) a criminal offense motivated by the fact or perception that the victims differ from the perpetrator.

white-collar crime illegal activities committed by high-status people in the course of their occupations.

embezzlement (stealing an employer's or client's money); identity theft; bribery; insider trading (illegal stock manipulation); and mass marketing, mortgage, and investment fraud (Federal Bureau of Investigation, 2014).

Bank tellers, who have low salaries and easy access to customer information, are particularly susceptible to white-collar crime. Between 2011 and 2015, Wells Fargo employees opened more than 2 million fake accounts to meet sales goals. The fraud resulted in $2.6 million in unnecessary fees for tens of thousands of clients. Wells Fargo fired 5,300 tellers and low-level managers. The executive who supervised the 6,000 retail branches and knew about the swindles retired with a $125 million payout (Clifford and Silver-Greenberg, 2016; Ochs, 2016).

White-collar crimes are lucrative because most offenders aren't caught. According to one estimate, there's only one FBI investigator per almost 4,000 financial services sales agents, personal financial advisers, and financial analysts. Thus, the odds of being arrested are very small (Rohrlich, 2010; see fbi.gov/wanted/wcc for annual updates of the FBI's most wanted white-collar criminals).

Even when white-collar criminals are arrested and convicted, few are incarcerated. Of the almost 3,800 suspected embezzlers, for example, only 14 percent have been arrested and sentenced; of those arrested and sentenced, only 7 percent served any prison time (Motivans, 2013). Because so few of the most affluent white-collar offenders are punished, they enjoy lavish lifestyles that include yachts, private jets, and $2 million birthday parties (Lublin, 2013; Spitznagel, 2013).

7-2e Corporate Crimes

Corporate crimes (also known as *organizational crimes*) are illegal acts committed by executives to benefit themselves and their companies. Corporate crimes include a vast array of illicit activities such as conspiracies to stifle free market competition, price-fixing, tax evasion, and false advertising. The target of the crime can be the general public, the environment, or even a company's own workers (see Chapters 11 and 15). Often, corporate offenders commit multiple crimes (fraud, insider trading, and perjury).

corporate crimes (also called *organizational crimes*) illegal acts committed by executives to benefit themselves and their companies.

cybercrime (also called *computer crime*) illegal activities that are conducted online.

Corporate crime is common. In the auto industry alone, and during just the last few years, General Motors (GM) knew about a malfunctioning ignition switch as far back as 2001, but started recalling cars only in 2014 after dozens of death and injury lawsuits. At least four automakers knew that Takata's airbags had a deadly defect linked to deaths and injuries, but installed them anyway "to save on costs." Well over a year before Volkswagen publicly admitted cheating on pollution tests, executives knew that the diesel cars were giving off more than 40 times the legal U.S. limit of nitrogen oxide, which can cause respiratory problems (Vlasic, 2015; Krisher and Biesecker, 2016; Tabuchi and Boudette, 2017: B1).

AP images/Kathy Willens

In 2009, Bernard Madoff—a highly respected stockbroker and investment advisor—pleaded guilty to defrauding thousands of investors of almost $18 billion between 1991 and 2008. Madoff was sentenced to 150 years in a federal prison. So far, it's cost almost $1 billion to recover 61 percent of the stolen money (Ross et al., 2016).

Corporate crime is common because it's profitable. In the ignition switch problems, the Department of Justice (DOJ) fined GM $900 million, which amounts to less than 1 percent of the company's annual revenue. Education Management Corporation, one of the nation's largest for-profit education companies, received $11 billion from students and the federal government by lying about its programs. The DOJ fine, again, was less than 1 percent of the corporation's revenue, but left tens of thousands of students with worthless degrees and huge debts (United States Senate, 2016).

7-2f Cybercrime

Cybercrime (also called *computer crime*) refers to illegal activities that are conducted online. These high-tech crimes include defrauding consumers with bogus financial investments, embezzling, sabotaging computer systems, hacking, and stealing confidential business and personal information.

In 2016, there were 4,149 data breaches, worldwide, that stole more than 4.2 billion records; 68 percent of the breaches were targeted at American companies, government agencies, and individuals. In the same year, U.S. internal data breaches hit a record high of 1,093, a 40 percent increase from 2015. Since 2007, the leading reasons for data breaches have been *hacking* (illegal access to a computer), *phishing* (masquerading as a legitimate person or group to obtain a login name and password), and *skimming* (stealing credit card information). Criminals can use stolen information (e.g., social security numbers, addresses, and names) to file false tax returns, order credit cards, and drain people's bank accounts (Internet Theft Resource Center, 2017; Risk Based Security, 2017).

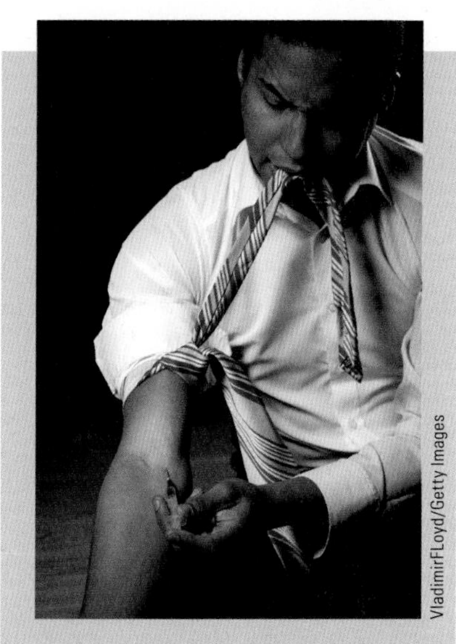
VladimirFLoyd/Getty Images

Is this a victimless crime?

In 2015, the FBI received almost 289,000 complaints that resulted in more than $1 trillion in financial losses. The most common offenses were identity theft and payments for goods and services that people never received. About 64 percent of Americans have experienced illegal credit card charges (Federal Bureau of Investigation, 2016; Olmstead and Smith, 2017). Stealing personal information has become so routine, according to one observer, "that most such breaches are quickly forgotten" and are "merely an occasional inconvenience" ("The Ashley Madison Hack," 2015: 14).

7-2g Organized Crime

Organized crime refers to activities of individuals and groups that supply illegal goods and services for profit. Organized crime includes human and drug trafficking, loan-sharking (lending money at unlawful rates), illegal gambling, pornography, theft rings, hijacking cargo, and laundering money obtained from illicit activities (e.g., prostitution) by depositing the money in legitimate enterprises (e.g., banks) (Federal Bureau of Investigation, 2017).

One of the most profitable activities, *organized retail crime*, involves stealing billions of dollars in merchandise that is then sold online, a practice known as "e-fencing." Organized crime costs the retail industry approximately $30 billion each year (National Retail Federation, 2016).

Since 2013, organized crime rings have orchestrated 80 percent of global cybercrimes, usually targeting financial institutions. In 2014, a Russian crime syndicate stole 1.2 billion U.S. usernames, passwords, and email addresses in "the largest hack in history." Criminal organizations employ highly sophisticated staff who circumvent bank security systems. Russia and China are especially likely to sponsor organized cybercrime, which makes money by stealing other countries' intellectual property and government secrets (IBM Journal Staff, 2016; Kessem, 2016; Rodgers, 2016).

7-2h Victimless Crimes

Victimless crimes (also called *public order crimes*) are illegal acts that have no direct victim. Because crimes such as illicit drug use, prostitution, illegal gambling, and pornography are voluntary, the people involved don't consider themselves victims or report the offenses. Prostitutes argue, for instance, that they're simply providing services that people want, and drug users claim that they're hurting only themselves.

Some contend that this term is misleading because victimless crimes often lead to property and violent crimes, as when addicts commit burglary, rob people at gunpoint, and engage in identity theft to get money for drugs. In a national survey, 70 percent of the respondents said that a family member's drug abuse had a negative effect on the emotional or mental health of at least one other family member, and 39 percent experienced financial problems because they went into debt to pay for an uninsured addict's treatment (Saad, 2006). In 2014, 36 percent of Americans—up from 15 percent in 1947—said that alcohol was a major cause of family problems. Drug use is also often a reason for divorce and family turmoil (White and Witkus, 2013; Jones, 2014).

The next sections examine four important sociological perspectives that help us understand why people are deviant. *Table 7.1* summarizes these theories.

organized crime activities of individuals and groups that supply illegal goods and services for profit.

victimless crimes (also called *public order crimes*) illegal acts that have no direct victim.

Table 7.1 Sociological Explanations of Deviance

Theoretical Perspective	Level of Analysis	Key Points
Functionalist	Macro	• Deviance is both functional and dysfunctional. • Anomie increases the likelihood of deviance. • People are deviant when they experience blocked opportunities to achieve the culturally approved goal of economic success.
Conflict	Macro	• There's a strong association between capitalism, social inequality, power, and deviance. • The most powerful groups define what's deviant. • Laws rarely punish the illegal activities of the powerful.
Feminist	Macro and micro	• There's a large gender gap in deviant behavior. • Women's deviance reflects their general oppression due to social, economic, and political inequality. • Many women are offenders or victims because of patriarchal beliefs and practices and living in a rape culture.
Symbolic Interactionist	Micro	• Deviance is socially constructed. • People learn deviant behavior from significant others such as parents and friends. • If people are labeled or stigmatized as deviant, they're likely to develop negative self-concepts and engage in criminal behavior.

7-3 FUNCTIONALIST PERSPECTIVES ON DEVIANCE

For functionalists, deviance is a normal part of society. They don't endorse undesirable behavior, but view deviance as both functional and dysfunctional.

7-3a Dysfunctions and Functions

Dysfunctions are the negative consequences of behavior, but what's dysfunctional for one group or individual may be functional for another. For example, gangs are functional because they provide members with a sense of belonging, identity, and protection. Gangs are also dysfunctional because they commit violent and property crimes (Egley and Howell, 2013).

Deviance is *dysfunctional* because it:

- **Creates tension and insecurity.** Any violation of norms—a babysitter who cancels at the last minute or the theft of your smartphone or tablet—makes life unpredictable and increases anxiety.
- **Erodes trust in personal and formal relationships.** Crimes like date rape and stalking make many women suspicious of men, and victims of identity theft have had problems obtaining banking services or credit cards because financial organizations don't trust them (see Harrell and Langton, 2013).
- **Decreases confidence in institutions.** In the 2008 stock market crash, taxpayers had to pay for the financial industry's corporate fraud and mismanagement. Since then, millions of people, even those who didn't lose money, worry that their retirement funds may disappear in the future (see Chapter 12).
- **Is costly.** Besides personal costs to victims (e.g., fear, emotional trauma, physical injury), deviance is expensive. All of us pay higher prices for consumer goods and services (e.g., auto and property insurance), as well as higher taxes for prosecuting criminals and for building and maintaining prisons. And as you saw earlier, white-collar, corporate, cyber, and organized crime cost the U.S. economy hundreds of billions of dollars every year.

Deviance can also be *functional* because it provides a number of societal benefits (Durkheim, 1893/1964; Erikson, 1966; Sagarin, 1975). For example, deviance

- **Affirms cultural norms and values.** Negative reactions, like expelling a college student who's caught cheating or incarcerating a robber, reinforce a society's norms and values about being honest and law-abiding.
- **Provides temporary safety valves.** Some deviance is accepted under certain conditions, enabling people to "blow off steam." Typically, a community and police will tolerate noise, underage drinking, loud music, and obnoxious behavior during college students' spring break because it's a short-lived nuisance.
- **Bolsters the economy.** Deviance can benefit a community financially. Some towns welcome new prisons because they generate jobs and stimulate the economy (Semuels, 2010). Unlike a manufacturing plant that might close down or move to another country, prisons are rarely dismantled.

- **Triggers social change.** When auto collisions increased, states started passing laws that banned distracted driving. Since 1999, drunk driving and deadly crashes among persons aged 16 to 25 have declined primarily because of drinking age laws, mass media campaigns, and sobriety checkpoints (Governors Highway Safety Association, 2014; Azofeifa et al., 2015).

7-3b Anomie and Social Strain Theory

Functionalists offer many explanations of deviance, but two of the most influential are anomie and strain theories. Both analyze why so many people engage in deviant behavior even though they share many of the same goals and values as those who conform to social norms.

DURKHEIM'S CONCEPT OF ANOMIE

Émile Durkheim (1893/1964, 1897/1951) introduced the term **anomie** to describe the condition in which people are unsure how to behave because of absent, conflicting, or confusing social norms. During periods of rapid social change, such as industrialization in Durkheim's time, societal rules may break down. As many young people moved to the city to look for jobs in the nineteenth century, norms about proper behavior that existed in the countryside crumbled. Even today, as you'll see in Chapter 15, many urban newcomers experience anomie and miss the neighborliness that was common at home.

MERTON'S CONCEPT OF SOCIAL STRAIN

Robert Merton (1938) expanded the concept of anomie to explain how social structure helps to create deviance. According to Merton, Americans are socialized to believe that anyone can realize the American dream of accumulating wealth and being successful economically. To achieve the *cultural goal* of economic success, society emphasizes legitimate and *institutionalized means* such as education, hard work, saving, starting at the bottom and working one's way up, and making sacrifices instead of seeking instant gratification.

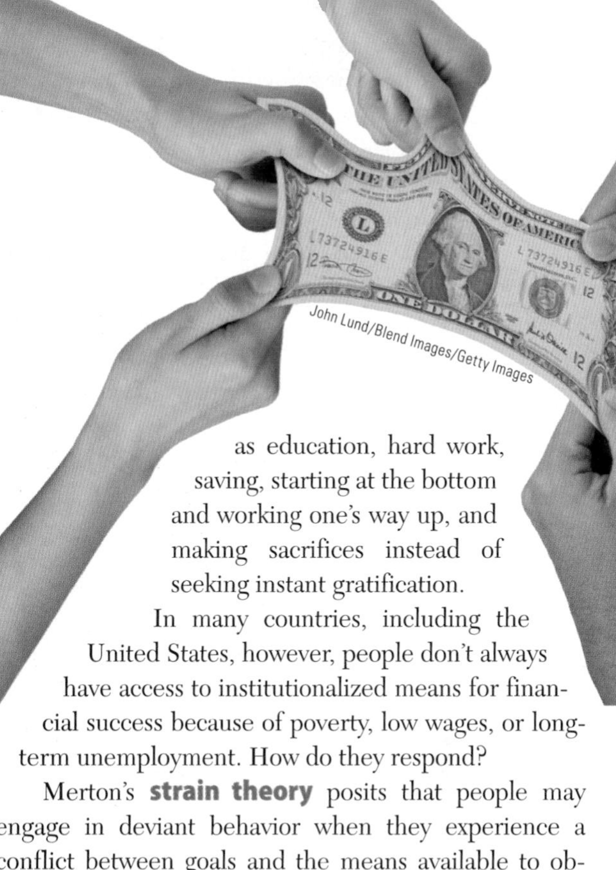

John Lund/Blend Images/Getty Images

In many countries, including the United States, however, people don't always have access to institutionalized means for financial success because of poverty, low wages, or long-term unemployment. How do they respond?

Merton's **strain theory** posits that people may engage in deviant behavior when they experience a conflict between goals and the means available to obtain the goals. Not all people turn to deviance to resolve social strain (*Table 7.2*). Most of us *conform* by working harder and longer to become successful. You're reading this textbook, to illustrate Merton's typology, because you've chosen conformity to achieve success through an institutionalized means (a college education).

Merton's other four modes of adaptation signify

anomie the condition in which people are unsure how to behave because of absent, conflicting, or confusing social norms.

strain theory people may engage in deviant behavior when they experience a conflict between goals and the means available to obtain the goals.

Table 7.2 Merton's Strain Theory of Deviance

Mode of Adaptation	Accept Cultural Goals?	Accept Institutionalized Means to Achieve Goals?	Examples
Conformity	Yes	Yes	College graduate
Innovation	Yes	No	Thief
Ritualism	No	Yes	Low-level bureaucrat
Retreatism	No	No	Alcoholic
Rebellion	No—seeks to replace goals	No—seeks to replace means for achieving goals	Antigovernment militia member

Source: Based on Merton, 1938.

deviance. *Innovation* occurs when people endorse the cultural goal of economic success but turn to illegitimate means, especially crime, to achieve the goal. Resorting to crime is functional for people who believe that they can't achieve success through hard work, education, and deferred gratification.

In *ritualism,* people have modest aspirations, but follow customs and rules because they enjoy doing so or they want to keep their jobs. Ritualists aren't criminals, but they're deviant because they've given up on becoming financially successful. Instead, they do what they're told and "go along to get along."

In *retreatism,* people reject both the goals and the means for success. They try to escape their failure by withdrawing—mentally, physically, or both—through alcohol and other drugs, becoming drifters, living in the streets, or otherwise dropping out of society.

In *rebellion*, people feel so alienated that they want to change the social structure entirely by substituting new goals and means for the current ones. Contemporary examples are terrorists and U.S. militia groups that oppose the federal government (see Chapter 3).

7-3c Critical Evaluation

A major contribution of functionalism is showing how social structure, not just individual attitudes and behavior, produces or reinforces deviance and crime. Functionalist theories also have weaknesses: (1) The concepts of *anomie* and *strain theory* are limited because they overlook the fact that not everyone in the United States embraces financial success as a major goal in life; (2) the theories don't explain why women's crime rates are much lower than men's, especially because women generally have fewer legitimate opportunities for financial success than men; (3) why crime rates have declined since 2000 despite much poverty and unemployment; and (4) why people commit some crimes (setting fires just for kicks or murdering an intimate partner) that have nothing to do with being successful (Anderson and Dyson, 2002; Williams and McShane, 2004).

The most consistent criticism is that functionalism typically focuses on lower-class deviance and crime. Conflict theorists have filled this gap by examining middle- and upper-class crime.

ORLANDO POLICE/The New York Time/Redux

In 2016, 29-year-old Omar Mateen killed 49 people and wounded 53 others at a gay nightclub in Orlando, Florida. Do you think that Merton's strain theory explains Mateen's crime?

7-4 CONFLICT PERSPECTIVES ON DEVIANCE

Functionalists ask, "Why do some people commit crimes and others don't?" Most conflict theorists ask, "Why are some acts defined as criminal while others aren't?" (Akers, 1997).

7-4a Capitalism and Social Inequality

For conflict theorists, there's a strong association between capitalism, social inequality, and deviance. People and groups with economic and political power view deviance as any behavior that threatens their own interests (Vold, 1958; Turk, 1969, 1976; Kraska, 2004). As a result, the mass media, owned by the affluent, routinely focus on "street crimes" rather than "suite crimes."

The United States, like many other countries, is a capitalist economy based on the private ownership of property (see Chapter 11). Characterized by personal gain rather than by collective well-being, capitalistic goals include continually maximizing profits, accumulating wealth, and promoting economic self-interest.

From a conflict perspective, capitalism produces crime for several interrelated reasons (Chambliss, 1969; Quinney, 1980; Snider, 1993). First, greed and self-interest perpetuate deviance. A study of 36 countries found that 42 percent of board directors and senior managers—pressured to increase growth and fearing personal pay cuts—"cooked the books" by overstating revenue, encouraging customers to buy unnecessary stock, and bribing powerful people to begin or continue their business (Ernst & Young, 2013).

Second, capitalism is a competitive system that encourages corporate crime. When only a few firms dominate an industry, they can easily eliminate competitors, conspire to fix prices, falsify financial statements, and commit fraud (see Eavis, 2015).

Third, because capitalism creates considerable social inequality, those at the top can control and exploit lower-level workers. In the financial services/banking industry, almost half

Investigation Discovery, which debuted in 2008, is one of the country's fastest growing cable television networks, airing in 157 countries and territories (Steel, 2015). The shows re-enact true crime incidents, primarily homicides, but also kidnappings, stalkings, sexual assaults, and intimate partner violence (e.g., *Scorned: Love Kills, Deadly Affairs, Fatal Vows, Wives with Knives*). Why do you think this network is so popular?

of the workers said that illegal or unethical activity was "part and parcel of succeeding in this highly competitive field," and that their company's confidentiality policies and procedures bar them from reporting wrongdoing (Tenbrunsel and Thomas, 2015).

The current U.S. tax system allows companies to avoid paying $135 billion each year in taxes. Such "tax dodging" prevents crucial investments in healthcare, infrastructure, education, family homelessness, and poverty reduction (Oxfam America, 2017).

7-4b Laws, Power, and Social Control

From a conflict perspective, criminal laws serve the interests of the capitalist ruling class. The most powerful groups in society control the law, which defines what's deviant, and who will be punished (Chambliss and Seidman,1982; Reiman and Leighton, 2010).

Why do so many high-status people commit crimes? Because they can, according to conflict theorists. First, most white-collar and corporate crimes are *not criminalized* (Turk, 1969; Lilly et al., 1995). Generally, the wealthy and powerful make and enforce laws that protect their property. Laws against higher-status criminals are relatively lenient and seldom enforced, whereas laws against lower-status offenders, particularly those who commit property crimes, are harsher and enforced more often (Thio et al., 2012).

Second, the powerful have more *opportunities for deviance*. As you saw earlier, white-collar and corporate criminals have access to resources and the necessary occupational skills to defraud customers and employees.

Third, powerful individuals and corporations are *rarely prosecuted*. Prosperous corporations pay millions, or billions, "to make cases disappear before any public hearing." Even in "flagrant corporate law breaking," and despite repeated promises by the Obama Administration and the DOJ, federal law enforcement agencies rarely prosecute corporate criminals (United States Senate, 2016).

Fourth, there are *few penalties* for high-status crimes because lawmakers and defendants share a common cultural background. Of the senior Wall Street bank executives who played leading roles in the 2008 financial crisis, only one has been convicted of fraud. Offshore banks knowingly shelter U.S. tax evaders because there are no punishments for doing so. Fraud by the rich wipes out 40 percent of the world's wealth, but no one goes to jail (Taibbi, 2014; "Financial Crime," 2015).

When U.S. prosecutors bring civil charges, the fines are typically less than half of 1 percent of annual revenue and are passed on to shareholders. Recently, top officials at Goldman Sachs agreed to pay a $550 million fine to settle accusations of securities fraud that made almost $52 billion in profits. The fine represented four days of the bank's revenues (Summers, 2010). If, using a comparable percentage, you made $100,000 a year illegally and someone paid your $1,000 fine, would there be a financial incentive for honesty?

7-4c Critical Evaluation

Conflict theories explain the linkages between capitalism, laws, power, and social class that benefit those at the top. According to some critics, however, conflict theories (1) exaggerate the importance of capitalism because other industrialized societies (e.g., Japan, Scandinavian countries), which enforce fraud laws, have much lower white-collar and corporate crime rates; (2) downplay street crimes even though victim injuries, incarceration, and crime prevention programs are costly to society; (3) overlook the fact that some affluent people, including executives, don't always get away with crimes; and (4) ignore the ways that crime is functional for society (e.g., provides jobs, affirms law-abiding cultural norms and values) (Leonardsen, 2004; Ernst & Young, 2013; "Jail Bait," 2016).

Finally, some contend, the most influential conflict theories focus almost entirely on men as both victims and offenders (Moyer, 2001; Belknap, 2007). Feminist theories have been filling this gap.

Rich Criminals, Lenient Sentences Ryan LeVin (pictured here), a super-rich heir to a family costume jewelry fortune, killed two men while hurtling down a street at more than 100 miles an hour, then fled the scene. He pleaded guilty to vehicular homicide and faced 30 years in prison. LeVin was sentenced to only 2 years of house arrest, which he served at one of his parents' luxury seaside condos. LeVin never apologized to the victims' families. His major concern was getting back his Porsche (Shepherd, 2011; Mayo, 2012).

In another recent case, Ethan Couch, 18, killed four people while driving drunk. The defense argued for leniency because Couch suffered from "affluenza," being so rich and spoiled that he was totally unaware of consequences and, thus, not responsible for his actions. The judge sentenced Couch to only 10 years of probation, which he violated 2 years later (Garza and Williams, 2016).

Sun Sentinel/Getty Images

7-5 FEMINIST PERSPECTIVES ON DEVIANCE

Until recently, most sociological research on deviance focused almost entirely on males. Nearly all sociologists were men who saw women as not worthy of much analytical attention or assumed that explanations of male behavior were equally applicable to females (Flavin, 2001; Belknap, 2007). Feminist scholars' analyses of female offenders and victims have greatly enriched our understanding of deviance.

Siede Preis/Getty Images

7-5a Women as Offenders

Crime is still generally a male activity. Men constitute about half of the U.S. population, but 73 percent of all arrests. Overall, and across 28 offenses, men commit more crimes and more serious crimes than women (Federal Bureau of Investigation, 2016). Among the top arrests, embezzlement is the only crime with no gender gap, and prostitution is the only female-dominated crime (*Table 7.3*).

Socialization explains some of the gender gaps in arrest rates. Compared with males, females have more responsibilities from an earlier age and are supervised more closely, expected to behave, and discouraged from risk-taking behavior. When women take risks, they're likely to do so to protect loved ones rather than to gain status or financial profits (England, 2005; Steffensmeier et al., 2013; see also Chapter 4).

Men are much more likely than women to commit violent crimes (*Table 7.3*), but women's arrest rates have increased for some offenses. Between 2006 and 2016, male arrests dropped for all offenses except larceny-theft; female arrests rose 26 percent for larceny-theft and 20 percent for property crimes (Federal Bureau of Investigation, 2016).

Why? More women have white-collar occupations (e.g., bank tellers, accountants,

Table 7.3 Top Six Reasons for Arrests, by Sex, 2015

Percentage of females arrested for ...		Percentage of males arrested for ...	
Prostitution	64	Sex offenses (except rape and prostitution)*	92
Embezzlement	50	Weapons	91
Larceny-theft	43	Murder or non-negligent manslaughter	89
Fraud	39	Robbery	86
Property crime	38	Burglary, arson, or drunkenness	81 (each)
Forgery or counterfeiting	35	Violent crime	80

*97 percent of rape arrestees were male.

Source: Based on Federal Bureau of Investigation, 2016, Table 42.

sales managers) that present opportunities for fraud, theft, and embezzlement. Because women, on average, are in low-income jobs, some resort to crime—especially shoplifting, petty theft, and prostitution—to survive financially or to support a family. Even at managerial levels, women are much more likely than men to commit fraud because of financial difficulties due to divorce or family problems rather than "generally unscrupulous or shrewd behavior" (Abrahams, 2015; Association of Certified Fraud Examiners, 2016).

Almost 9 percent of all individuals arrested are adolescents. Like adults, boys account for more arrests (70 percent) than girls, but girls' arrests have increased dramatically—from 17 percent in 1980 to 30 percent in 2015 (Federal Bureau of Investigation, 2016).

The higher arrest rates are due to a combination of factors: Policy changes, including more aggressive policing; parents' and school officials' greater likelihood of calling the police to deal with girls' (especially Black girls') disruptive behavior; and a substantial increase in juvenile females committing crimes that range from drug abuse to assaulting family members or intimate partners (Levintova, 2016; Morris and Perry, 2017).

David R. Frazier Photolibrary, Inc./Alamy Stock Photo

In 2016, Stanford University student Brock Turner was found guilty of raping an unconscious woman outside a fraternity party. Turner's father wrote the judge that his son was paying a "steep price" for "20 minutes of action," and should get probation. Rape carries a penalty of 2 to 14 years in state prison, but Turner was sentenced to six months in county jail and three years' probation. He was released after three months for good behavior (Golshan, 2016). Does this incident exemplify rape culture?

7-5b Women as Victims

Many of the serious crimes (murder and robbery) are committed by men against men. Women and girls, however, are almost always the victims of sexual assault, rape, intimate partner violence, stalking, sexual exploitation, sexual harassment, female infanticide, and other crimes that degrade and terrorize women.

Feminist scholars offer several explanations for women's victimization. First, **patriarchy**—a social system in which men control cultural, political, and economic structures—produces violence against women. In patriarchal societies, including the United States, because women have less access to power, they're at a disadvantage in creating and implementing laws that protect females (Price and Sokoloff, 2004; DeKeseredy, 2011). Such inequity diminishes women's control over their lives and increases their invisibility as victims.

Second, gender roles reinforce inequality: "Girls are rewarded for passivity and feminine behavior, whereas boys are rewarded for aggressiveness and masculine behavior" (Belknap, 2007: 243). As early as middle school, girls experience physical and sexual harassment in hallways, classrooms, and gym classes. Although the incidents are upsetting, girls often dismiss the boys' unwanted behavior as "only joking" (Espelage et al., 2016). In effect, then, both sexes internalize the belief that female victimization is normal. Because patriarchal societies rarely punish men who victimize women, or give them light sentences, many female victims feel trapped.

A third reason for women's abuse is living in a **rape culture**, an environment in which sexual violence is prevalent, pervasive, and perpetuated by the media and popular culture (Parrott and Parrott, 2015). Here are a few examples of the pervasiveness of rape and sexual assaults in U.S. culture:

- During their lifetimes, 3 percent of men and 17 percent of women have been victims of an attempted or completed rape (Rape Crisis Center, 2014);
- 27 percent of college women have experienced an attempted or completed rape (Lauerman, 2013);
- 12,214 active-duty military members reported sexual assaults during

patriarchy a social system in which men control cultural, political, and economic structures.

rape culture an environment in which sexual violence is prevalent, pervasive, and perpetuated by the media and popular culture.

2014 and 2015 (Department of Defense, 2016).

Steve Cole/Photodisc/Getty Images

The actual number of attacks is much higher because only 15 percent of service members and 33 percent of other victims report rapes and sexual assaults to police (Protect Our Defenders, 2016; Gramlich, 2017).

Across 77 countries, the number of rapes reported to the police ranges from fewer than 1 per 100,00 people in Japan, 37 in the United States, to 51 in England and Wales. Rape statistics vary considerably depending on the legal definition, recording practices, and other factors. Nonetheless, the variation across countries shows that sexual violence is preventable (United Nations Office on Drugs and Crime, 2017; see also Roden, 2017).

7-5c Critical Evaluation

Feminist sociologists have been at the forefront of raising awareness of crimes like date rape, intimate partner violence, and international sex trafficking. Some critics contend, however, that feminist explanations (1) don't show, specifically, *how* patriarchy victimizes women; (2) focus primarily on male but not female violence and crime, and (3) don't address the interdependent effects of gender, age, social class, and race/ethnicity (Friedrichs, 2009; Hunnicutt, 2009; Dao, 2013).

7-6 SYMBOLIC INTERACTION PERSPECTIVES ON DEVIANCE

For symbolic interactionists, deviance is socially constructed because it's in the eye of the beholder. For instance, only 47 percent of French people believe that extramarital affairs are immoral, compared with 84 percent of Americans and 94 percent of Palestinians and Turks ("Extramarital Affairs," 2014). Interactionists offer many explanations of deviance, but two of the best known are differential association theory and labeling theory.

differential association theory asserts that people learn deviance through interaction, especially with significant others.

labeling theory posits that society's reaction to behavior is a major factor in defining oneself or others as deviant.

7-6a Differential Association Theory

Differential association theory asserts that people learn deviance through interaction, especially with significant others like family members and friends. People become deviant, according to sociologist Edwin Sutherland, if they have more contact with significant others who violate laws than with those who are law-abiding. In effect, people learn techniques for committing criminal behavior, along with the values, motives, rationalizations, and attitudes that reinforce such behavior (Sutherland and Cressey, 1970). That is, deviance isn't inherent but learned through interaction with others.

Sutherland emphasized that differential association doesn't occur overnight. Instead, people are most likely to engage in crime if they're exposed to deviant values (1) early in life, (2) frequently, (3) over a long period of time, and (4) by significant others and reference groups (e.g., parents, siblings, close friends, and business associates).

Considerable research supports differential association theory. Almost 47 percent of state prisoners have a parent or other close relative who has also been incarcerated. Even before age 13, children who associate with peers who smoke, use drugs, or commit crimes are more likely to do so themselves because they learn these behaviors from their friends (Adamson, 2010; Roman et al., 2012; Huizinga et al., 2013). Thus, according to differential association theory, we're products of our socialization.

7-6b Labeling Theories

Have you ever been accused of something wrong that you didn't do? What about getting credit for something that you and others know you didn't deserve? In either case, did people start treating you differently? The reactions of others are the crux of **labeling theory**, which holds that society's reaction to behavior is a major factor in defining oneself or others as deviant. Some of the earliest sociological studies found that teenagers who misbehaved were tagged as delinquents. Such labeling changed the child's self-concept and resulted in more deviance and criminal behavior (Tannenbaum, 1938). During the 1950s and 1960s, two influential sociologists—Howard Becker and Edwin Lemert—extended labeling theory.

BECKER: DEVIANCE IS IN THE EYES OF THE BEHOLDER

According to Howard Becker (1963: 9), "The deviant is one to whom that label has successfully been applied;

deviant behavior is behavior that people so label" (emphasis in original). That is, it's not an act that determines deviance, but whether and how others react.

Some people are never caught or prosecuted for crimes they commit, and thus aren't labeled deviant. In other cases, people may be falsely accused (e.g., cheating on taxes) and stigmatized. In effect, then, deviance is in the eye of the beholder because societal reaction, rather than an act, labels people as law-abiding or deviant. Moreover, labeling can lead to secondary deviance.

LEMERT: PRIMARY AND SECONDARY DEVIANCE

Edwin Lemert (1951, 1967) developed labeling theory by differentiating between primary and secondary deviance. **Primary deviance** is the initial act of breaking a rule. Primary deviance can range from relatively minor offenses, like not attending a family member's funeral, to serious offenses, like rape and murder.

Even if people aren't guilty of primary deviance, labeling can lead to **secondary deviance**, rule-breaking behavior that people adopt in response to others' reactions. A teenager who's caught trying marijuana may be labeled a "druggie." If the individual is rejected by others, he or she may accept the label, associate with drug users, and become involved in a drug-using subculture. According to Lemert, a single deviant act will rarely result in secondary deviance. The more times a person is labeled, however, the higher the probability that she or he will accept the label and engage in deviant behavior.

iStock.com/Darren Mower

Do you think that this arrest is the result of primary or secondary deviance? Why?

There's considerable evidence that labeling affects people's lives. For example, those who have been jobless for a year or longer report stigmas in job searches: Employers assume that they're lazy, don't really want to work, or there's something wrong with them. Some companies have explicitly barred the long-term unemployed from certain job openings, telling them outright in job ads that they need not apply (see Chapter 11).

THE MEDICALIZATION OF DEVIANCE

In 1952, the American Psychiatric Association published the first edition of the *Diagnostic and Statistical Manual of Mental Disorders (DSM),* which describes mental health disorders. The number of disorders increased from 106 in 1952 to nearly 400 in 2013, and increased in size from "a spiral-bound pamphlet" to "a tome of nearly 1,000 pages" (Tavris, 2013: C5). Some of the new disorders include "binge eating" and "bereavement" (American Psychiatric Association, 2013).

The *DSM*'s ever-changing diagnoses and labels are an example of the **medicalization of deviance**, diagnosing and treating a violation of social norms as a medical disorder. The *DSM* has far-reaching effects, according to one of the past editors: "Anything you put in that book . . . has huge implications not only for psychiatry but for pharmaceutical marketing, research, for the legal system, for who's considered to be normal or not, for who's considered disabled. *And it has huge implications for stigma because the more disorders you put in, the more people get labels*" (Carey, 2010: 1, emphasis added; Chapter 14 discusses the costs of medicalizing deviance).

7-6c Critical Evaluation

Interaction theories help us understand deviance, but critics point to several weaknesses. Differential association theory doesn't explain (1) impulsive crimes triggered by rage or fear; (2) crimes committed by people who have grown up in law-abiding families; (3) why only a minority of children who grow up in disadvantaged communities join gangs and/or commit crimes; and (4) why only one sibling raised in the same family environment may commit juvenile or adult crime (Williams and McShane, 2004; Graif and Sampson, 2009; Breining et al., 2017).

Critics have faulted labeling theory for not explaining why (1) people are deviant before labeling occurs; (2) crime rates are higher in some parts of the country or at particular times of the year (as before holidays); and (3) labeling doesn't necessarily increase deviant behavior (Colvin and Pauly, 1983; Schurman-Kauflin, 2000; Benson, 2002). Conflict theorists, in particular, criticize symbolic interactionists for ignoring structural factors—like poverty and low-paid jobs—that create or reinforce deviance and crime (Currie, 1985).

primary deviance the initial act of breaking a rule.

secondary deviance rule-breaking behavior that people adopt in response to others' reactions.

medicalization of deviance diagnosing and treating a violation of social norms as a medical disorder.

Davis Barber/PhotoEdit

Does Labeling Deter Deviance? Not always. Despite federal laws, 45 states allow teachers accused of sexually abusing students to quit instead of notifying the police or education agencies. This practice, known as "passing the trash," allows sex predators to be re-hired for other teaching jobs. Administrators don't want to tarnish a school's reputation, to spend thousands in legal fees to fire tenured teachers or on potential lawsuits by the victims' families, or to tangle with teachers' unions that defend teachers accused of sexual abuse (Skinner, 2014; Reilly, 2016).

7-7 CONTROLLING DEVIANCE AND CRIME

Most of us conform to laws and cultural expectations. We usually conform, however, not because we're innately good, but because of **social control**, techniques and strategies that regulate people's behavior in society. The purpose of social control is to eliminate, or at least reduce, deviance.

7-7a Control Theory and Types of Social Control

Control theory proposes that deviant behavior decreases when people have strong social bonds with others. The bonds that strengthen our self-control are based on (1) *attachment* to other people (e.g., family, friends, and coworkers), (2) *commitment* to conformity (e.g., participating in community activities), (3) *involvement* in legitimate activities (e.g., getting an education, holding a job), and (4) *belief* in social norms and values (e.g., being moral, obeying laws) (Hirschi, 1969).

social control the techniques and strategies that regulate people's behavior in society.

control theory proposes that deviant behavior decreases when people have strong social bonds with others.

criminal justice system government agencies that are charged with enforcing laws, judging offenders, and changing criminal behavior.

Most conformity is due to the internalization of norms during the powerful socialization process. Because most of us care about the opinions of family members, friends, and teachers, we try to live up to their expectations (see Chapter 4). Thus, we conform because of *informal social controls* that we learn and internalize during childhood.

Formal social control, which also regulates social behavior, exists outside of the individual. For example, a college dean might threaten a student with suspension or expulsion, a business might install security cameras, or a police officer might arrest someone.

7-7b The Criminal Justice System and Social Control

Social institutions such as the family, education, and religion try to maintain *social* control over behavior; the criminal justice system has the *legal* power to control crime and punish offenders. The **criminal justice system** is made up of government agencies—including the police, courts, and prisons—that enforce laws, judge offenders, and try to decrease criminal behavior. The criminal justice system relies on three major strategies to control crime: prevention and intervention, punishment, and rehabilitation. Punishment, as you'll see, is the least effective approach.

PREVENTION AND INTERVENTION

Structural inequality—in employment, housing, education, and other areas—is the best predictor of violence and crime (Basken, 2016; Carpenter et al., 2016; Light and Ulmer, 2016). It costs U.S. taxpayers, annually, about $34,000 per inmate compared with $12,000 per K-12 public school pupil. A 10 percent increase in high school graduation rates decreases arrest rates by 9 percent. From 1979 to 2013, however, U.S. spending on prisons grew three times faster than on education ("The Right Choices," 2015; Kena et al., 2016; Stullich et al., 2016).

Crime prevention is much less expensive than punishment. Among adults, the average annual cost per outpatient for alcohol or drug abuse treatment is $4,700, compared with $24,000 for incarceration (National Institute on Drug Abuse, 2012). Nearly 28 percent of state and federal inmates are age 50 and older, and many have serious health problems. An older offender's annual medical costs can exceed $100,000 (Johnson and Beiser, 2013; Carson and Anderson, 2016).

Nationally, states with the most firearms and fewest restrictions have the highest gun-related violence, homicide, and suicide rates, but gun control laws are controversial. Many Americans (85 percent) support background checks, but only a majority (55 percent) wants stricter gun control laws. Advocating for harsher laws often surges after high-profile shooting incidents but then subsides again (Fleegler et al., 2013; Anestis and Anestis, 2015; Doherty, 2015; Newport, 2016; Parsons and Weigand, 2016).

Social Service Agencies and Community Outreach Programs. Many organizations—federal and state agencies, law enforcement professionals, social workers, and nonprofit groups—try to prevent and decrease crime (e.g., CrimeSolutions.gov), but some programs are more successful than others.

Compared with police-only interventions, when police and local social service workers respond together to an incident of intimate partner violence, victims experience significantly less violence immediately after the episode and months later. Why? The collaborative partnership provides victims with more protective strategies and resources (e.g., getting medical attention and protection orders, going someplace where a partner can't find them) (Messing et al., 2014).

Programs like "Scared Straight" are organized visits to prisons and face-to-face interaction with adult inmates to deter juvenile delinquency. Evaluations of scared-straight programs have concluded, however, that they may increase rather than deter adolescent deviance. The reasons are unclear, but may be due to the program participants' imitating the inmates' "brutal and vulgar" language and crimes. Still, the tours continue to be popular (Robinson and Slowikowski, 2011; Petrosino et al., 2013).

Police. The primary role of the police is to enforce society's laws, but can they prevent crimes? Police can deter some deviance by cruising high-risk areas in patrol cars or having more officers on foot patrols who return to the same street several times during a shift ("Policing Philadelphia . . . ," 2013). Concentrating on *hot spots,* areas of high criminal activity, reduces crime only temporarily, however, because criminals simply move to other parts of the city (U.S. Department of Justice, 1996; Braga, 2003).

Compared with 42 percent of White individuals and 31 percent of Latinos, just 14 percent of Black individuals have "a great deal" of confidence in their local police (Morin and Stepler, 2016). Negative views have spiked since 2015, after White police officers shot and killed unarmed Black men during routine law enforcement encounters in Louisiana, Missouri, Minnesota, and several other states. The deaths sparked protests in some cities, and violence and riots in Baltimore (see Chapter 16).

Minorities, particularly Black individuals, accuse police of discrimination, *racial profiling* (using race to stop and detain someone suspected of having committed a crime), and brutality. Many police officers (72 percent) say that the Black men's deaths are isolated incidents. Only 27 percent have ever fired their weapons while on duty, but they become aggressive because a "well-armed population" makes their jobs unpredictable and dangerous, and many suspects fight and resist arrest. Racial profiling is justified, according to some officers, because a high proportion of Black and Latino people commit crimes. An overwhelming 83 percent of Americans say they understand the risks that police face, but only 14 percent of officers agree, and believe that they're scapegoated for social problems ("Police Culture," 2015; Morin et al., 2017). Police have no control over macro-level factors—poverty, unemployment, limited educational opportunities, and neighborhood deterioration—that are associated with high crime rates.

Andrey Burmakin/Shutterstock.com

PUNISHMENT

Those who endorse a **crime control model** believe that crime rates increase when offenders don't fear apprehension or punishment. This model endorses a tough approach in sentencing, imprisonment, and capital punishment.

Sentencing. After someone has been found guilty of a crime or has pleaded guilty, a judge (and sometimes a jury) imposes a *sentence,* or penalty. A sentence can be a fine, *probation* (supervision instead of serving time in jail), incarceration, or the death penalty. Offenders who receive *parole* are released from prison before the end of their full sentence on condition that they check in regularly with an officer and obey the law.

Of the 6.8 million adults in the corrections system in 2015, 56 percent were on probation, 13 percent were parolees, and the rest were in a local jail or a state or federal prison. Since 2007, the number of parolees has increased slightly, while the number on probation and in jails and prisons has declined (Kaeble and Glaze, 2016; Kaeble and Bonczar, 2017).

Why? In 2016, only 14 percent of Americans said that the criminal justice system wasn't "tough enough," down from 65 percent in 2003, and 38 percent described drug crime sentences as "too tough" (McCarthy, 2016). It's not

crime control model proposes that crime rates increase when offenders don't fear apprehension or punishment.

Figure 7.3 The Number of U.S. Prisoners Has Nearly Tripled Since 1987

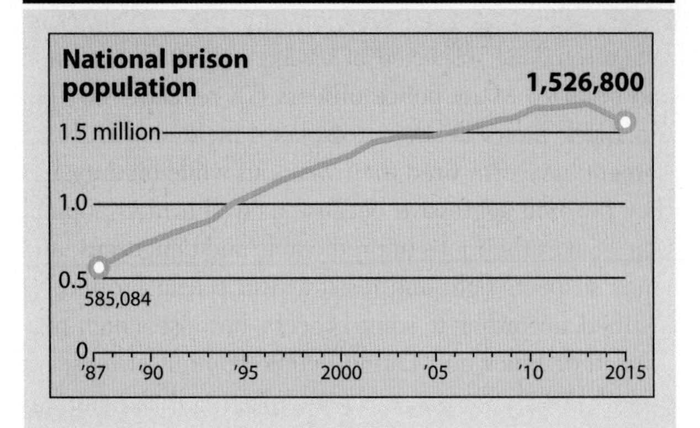

Sources: Based on Pew Center on the States, 2010, and Carson and Anderson, 2016.

clear whether (or how much) people's opinions affect the criminal justice system, but as budgets shrink, even states with reputations for being tough on crime are embracing more lenient policies. Some have begun to shorten the average number of years on probation and parole, which decreases supervision costs. Others are reducing the number of people sent to prison: It costs an average of $79 a day to keep an inmate in prison but only about $3.50 a day to monitor the same person on probation (Richburg, 2009; Porter, 2016).

Incarceration. The number of U.S. prisoners has surged since 1987 (*Figure 7.3*). Since 2008, and for the first time in history, 1 in every 100 American adults is in prison. The United States has less than 5 percent of the world's population, but more than 20 percent of the planet's prisoners, the majority of whom are incarcerated for nonviolent drug offenses. Besides the sheer number of inmates, the United States is the second highest (after Seychelles, Africa) in per capita inmates (698 per 100,000 population), and well ahead of Russia (445), Cuba (510), China (119), Germany (78), and India (33) (Walmsley, 2016).

Of all state and federal prisoners, 93 percent are men, but imprisonment rates vary by race/ethnicity. As *Figure 7.4* shows, incarceration rates are highest for Black individuals, both women

IT'S TEMPTING TO ASSUME THAT THE MORE PEOPLE BEHIND BARS, THE LOWER THE CRIME RATE.

and men. If current trends continue, 1 of every 3 Black males born in 2013 can expect to go to prison in his lifetime, as can 1 of every 6 Latino males—compared with 1 of every 17 White males. A major reason is arrest disparities. For example, White adolescents are 30 times more likely than Black youth to use cocaine and other "hard drugs," but Black adolescents are almost twice as likely to be arrested for drug offenses (The Sentencing Project, 2013; Rovner, 2016; Welty et al., 2016).

The imprisonment rate has been dropping since 2007 (Carson and Anderson, 2016). Arrest rates have declined, as you saw earlier, but prison counts have dropped primarily because of economic reasons. Driven by overcrowding and budget crises, many states are releasing low-level offenders before they serve their sentences. Releasing prisoners early has been controversial, but state officials maintain that their budgets can't afford the high incarceration costs (Goode, 2013; Kearney et al., 2014).

It's tempting to assume that the more people behind bars, the lower the crime rate. State spending on corrections surged from $17 billion in 1990 to $57 billion in 2015 (The Sentencing Project, 2017). However, *recidivism* (being

Figure 7.4 U.S. Prisoners by Race/Ethnicity and Sex, 2015

Imprisonment Rate per 100,000 U.S. Population

	Rate
MEN	
White	457
Black	2,613
Latino	1,043
Other*	929
WOMEN	
White	52
Black	103
Latina	63
Other*	90

0 500 1,000 1,500 2,000 2,500 3,000

*Includes Native Americans, Alaska Natives, Asians, Native Hawaiians, Pacific Islanders, and persons of two or more races.

Source: Based on Carson and Anderson, 2016, Appendix Table 4.

arrested for committing another offense after being released) has barely changed since 1999: Of those released from state prisons, 68 percent are arrested for a new crime within 3 years, and 77 percent within 5 years (Durose et al., 2014).

Capital Punishment. About half of Americans support *capital punishment* (the death penalty), down from 80 percent in 1994 (Oliphant, 2016). By mid-2015, more than two-thirds of the world's nations, including many developing countries, had abolished the death penalty or suspended executions. Thirty-one states have the death penalty, but the number of executions fell from 98 in 1999 to 10 in 2017 (Amnesty International, 2017; Death Penalty Information Center, 2017).

There are several reasons for fewer executions: (1) homicide rates have declined sharply since 1980; (2) the death penalty doesn't deter crime; (3) public support for the death penalty has waned; (4) hundreds of false convictions have been overturned; (5) Black males, especially if the victims are White, are substantially more likely to be executed than White males who murder Black victims; (6) most juries can now impose life sentences without the possibility of parole. Some also argue that the death penalty is a waste of money because inmates can spend 15 to 20 years appealing a sentence. In California, for example, it costs taxpayers $137 million annually for inmates who appeal a death sentence compared with $12 million for those serving life sentences (McPhate, 2016; Oliphant, 2016; Amnesty International USA, 2017; Death Penalty Information Center, 2017).

Jim West/Alamy Stock Photo

About 700,000 U.S. prisoners have jobs. Whether they work for the government or private businesses that have contracts with local correctional authorities, their earnings range from nothing to as little as 12 to 40 cents an hour (Vongkiatkajorn, 2016; "Prisons," 2017). Would paying inmates much higher wages decrease recidivism?

REHABILITATION

Rehabilitation, a third approach to controlling deviance, maintains that appropriate treatment can change offenders into productive, law-abiding citizens. Federal and state inmates produce more than $2 billion worth of products a year. They make everything from redwood canoes to specialty motorcycles and saddles. They also raise fish, milk cows and goats, teach dogs obedience, and manage vineyards at prison-run facilities (Alsever, 2014; "Prisons," 2017).

Are rehabilitation programs successful? Only if they provide employment after release. Other effective rehabilitation efforts include learning a trade while in prison, earning a high school diploma or college degree while in prison or after release, and receiving services that address several needs (e.g., housing, employment, *and* medical services) rather than just one (e.g., drug abuse counseling) (Cumberworth, 2010; Kelly, 2016).

STUDY TOOLS 7

READY TO STUDY? IN THE BOOK, YOU CAN:

- ☐ Check your understanding of what you've read with the Test Your Learning Questions provided on the Chapter Review Card at the back of the book.
- ☐ Tear out the Chapter Review Card for a handy summary of the chapter and key terms.

ONLINE AT CENGAGEBRAIN.COM WITHIN MINDTAP YOU CAN:

- ☐ Explore: Develop your sociological imagination by considering the experiences of others. Make critical decisions and evaluate the data that shape this social experience.
- ☐ Analyze: Critically examine your basic assumptions and compare your views on social phenomena to those of your classmates and other MindTap users. Assess your ability to draw connections between social data and theoretical concepts.
- ☐ Create: Produce a video demonstrating connections between your own life and larger sociological concepts.
- ☐ Collaborate: Join your classmates to create a capstone project.

8 Social Stratification: United States and Global

LEARNING OBJECTIVES

After studying this chapter, you will be able to...

8-1 Explain and illustrate social stratification systems and bases.

8-2 Describe the U.S. class structure and explain how and why social classes differ.

8-3 Describe poverty and explain why people are poor.

8-4 Compare the different types of social mobility, describe recent trends, and explain what factors affect mobility.

8-5 Describe global stratification, its variations and consequences, and the theoretical models that explain why inequality is universal.

8-6 Compare and evaluate the theoretical explanations of social stratification.

After finishing this chapter go to **PAGE 158** for **STUDY TOOLS**

In 2012, American multi-billionaire Larry Ellison—founder of Oracle, the world's biggest database company—bought the Hawaiian island of Lanai for $500 million. The same year, 17 percent of Hawaii's residents were living in poverty (*Hawai'i Free Press*, 2012; Kroll and Dolan, 2013). Many people in the United States and around the world are struggling to survive while others enjoy astonishing wealth. This chapter examines social stratification and why there are haves and have-nots.

WHAT DO YOU THINK?

Americans who are poor just aren't working hard enough.

1	2	3	4	5	6	7
strongly agree						strongly disagree

True or False?

HOW MUCH DO YOU KNOW ABOUT U.S. ECONOMIC EQUALITY?

1. A majority of Americans are middle class.
2. The most prestigious occupations have the highest incomes.
3. Americans age 65 and older have the highest poverty rates.
4. Immigration helps many native-born Americans move up the economic ladder.
5. Many Americans are poor, but poverty has decreased since the 1980s.
6. Among developed countries, the United States has one of the highest economic equality levels.

The answer for 4 is true; the rest are false. You'll see why as you read this chapter.

8-1 SOCIAL STRATIFICATION SYSTEMS AND BASES

Social stratification is a society's ranking of people based on their access to valued resources such as wealth, power, and prestige. All societies are stratified, but some more than others. Sociologists distinguish among stratification systems based on the extent to which they're open or closed. Before reading further, take the True/False quiz to see how much you know about U.S. economic inequality.

8-1a Closed and Open Stratification Systems

In a *closed stratification system*, movement from one social position to another is limited by ascribed statuses. An *open stratification system* allows movement up or down because people's achievements affect mobility. Closed stratification systems are considerably more fixed than open ones, but neither is completely open or closed.

Two closed stratification systems exist today: slavery and castes. In a **slavery system**, people own others as property and have almost total control over their lives. The slaves have been kidnapped, inherited, or given as gifts to pay a debt. They're bought and sold as commodities, sometimes multiple times. The United Nations banned all forms of slavery worldwide in 1948, but almost 36 million people are living as slaves across the globe. About 61 percent are in just five countries—India, China, Pakistan, Uzbekistan, and Russia—but modern slavery is also prevalent in parts of Africa, the Middle East, and Asia (Walk Free Foundation, 2014).

In a **caste system**, a second type of closed stratification system, people's positions are ascribed at birth and largely fixed. Their places in the hierarchy are primarily determined by inherited characteristics such as race, skin color, gender, family background, or nationality (see Chapter 5). People must marry

social stratification a society's ranking of people based on their access to valued resources such as wealth, power, and prestige.

slavery system people own others as property and have almost total control over their lives.

caste system people's positions are ascribed at birth and largely fixed.

within their caste; there are severe restrictions in their choice of occupation, residence, and social relationships; and people rarely move from one caste to another.

A good example is India, where a caste system has existed for more than 3,000 years. At the top were the Brahmins. On the bottom rung were the Dalits (formerly called "untouchables"), who performed the most menial and unpleasant jobs, including collecting human waste and cleaning streets. Fearing being "polluted," higher castes wouldn't interact with the Dalits (Mendelsohn and Vicziany, 1998).

India outlawed the caste system in 1949, but social distinctions are deeply entrenched, particularly in rural areas, and about 90 to 95 percent marry within their own castes ("Marriage in India," 2015). In 1950, India set up a quota system that reserves nearly half of government jobs and public college slots for members of the most disadvantaged castes. Some claim that the quota system is unfair because it benefits less qualified candidates. Others maintain that such complaints come primarily from groups who fear losing some of their deep-rooted privileges (Barstow and Raj, 2015; McCarthy, 2015).

In a **class system**—a relatively open stratification structure—people's positions are based on both birth and achievement. Because achieved characteristics (e.g., education, work skills, occupation) can change, people in open stratification systems can move from one social class to another. A **social class** is a group of people who have a similar standing or rank in a society based on wealth, education, power, prestige, and other valued resources. A large majority of Americans (60 percent) believe that hard work leads to success (Kohut, 2015), but is this really the case?

8-1b The Bases of Stratification

class system people's positions are based on both birth and achievement.

social class people who have a similar standing or rank in a society based on wealth, education, power, prestige, and other valued resources.

wealth economic assets that a person or family owns.

income the money a person receives, usually through wages or salaries, but can also include other earnings.

Although 40 percent of 18- to 34-year-olds still receive financial support from their parents, 28 percent expect to become millionaires in their lifetimes (Wile, 2015). Even if some become multimillionaires, and income is an important stratification factor, it's not the only one. Sociologists use a multidimensional approach that includes wealth, prestige, and power (Weber, 1946).

WEALTH

Wealth refers to the economic assets that a person or family owns. It includes money, property (e.g., real estate), stocks and bonds, retirement and savings accounts, personal possessions (e.g., cars and jewelry), and income. **Income** is the money a person receives, usually through wages or salaries, but it can also include rents, interest on savings accounts, stock dividends, royalties, or business proceeds. Income and wealth differ in several important ways:

- Wealth is *cumulative*. It increases over time, especially through investment, whereas income is usually spent on everyday expenses.
- Because wealth is accumulated over time, much of it can be *passed on to the next generation*. Inheritances and other monetary gifts offer opportunities to start a business, buy a first (or vacation) home, pursue college without debt, and increase assets (e.g., savings and stocks).
- Passing on wealth to the next generation *preserves privilege*. Wealthy people tend to have the greatest amount of *economic capital* (income and other monetary assets like property), *cultural capital* (advanced degrees and assets such as style of speech, table manners, and physical appearance), and *social capital* (networks comprised of influential people). Differential access to all three types of capital reinforces and reproduces the existing class structure and inequality (Bourdieu, 1984).

U.S. wealth and income inequality is staggering. As *Figure 8.1* shows, the top 1 percent of Americans own 37 percent of all wealth and 20 percent of all income—the highest levels ever recorded, and wealth

iStock.com/JeremyRichards

India's Dalits still perform unpleasant tasks such as burning corpses and removing garbage and human waste.

Figure 8.1 Who Has the Biggest Piece of Pie?

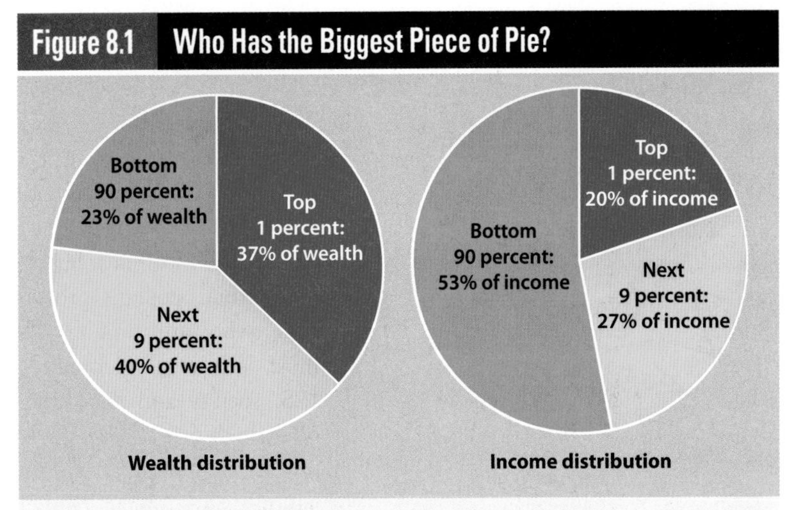

Sources: Based on Wolff, 2016, Table 2.

and income gaps have *increased*. In 2013, the wealthiest 10 percent of households held 77 of the country's wealth, up from 67 percent in 1989. The wealth of the bottom 90 percent fell by more than 3 percent; in 2013, this group owned only 23 percent of all wealth, about as much as in 1940 (Congressional Budget Office, 2016; Zucman, 2016).

Over the past three decades, the racial wealth divide has increased. Between 1983 and 2013, ultra-wealthy people in the top 0.001 percent, the Forbes 400, increased their wealth by 736 percent—from $700 million to $5.8 billion—while Black and Latino wealth rose only modestly (*Figure 8.2*). To put this extreme inequality in perspective, this page would have to be more than half a mile long to show the Forbes 400 wealth to scale (Asante-Muhammad et al., 2016).

Although racial wealth gaps have widened since the 1980s, Asians' wealth has increased significantly since 1989. Researchers predict that Asian wealth levels will soon surpass those of White wealth: Asian families already have the highest median household income, high educational levels, diversified assets, little or no credit card debt, and save money every year (Emmons and Noeth, 2015; see also Chapter 10).

Regardless of how the economy is doing, the rich keep getting richer. Since 1980, the average annual income of the bottom half of Americans (around $16,000) has barely changed, while that of the top 1 percent (nearly $1.4 million a year) grew 300 percent. In 2014, the average annual income of almost 235 million adults was only $64,600, compared with $1.3 million for the top 1 percent (2.4 million people) and $6 million for the top one-tenth of the top 1 percent (234,400 people). In early 2017, the United States had 536 billionaires (up from 413 in 2011) whose average net worth was $3.6 billion (Piketty et al., 2016; Saez, 2016; Kroll and Dolan, 2017).

PRESTIGE

A second basis of social stratification is **prestige**—respect or recognition attached to social positions. Prestige is based on many criteria, including wealth, family background, power, and accomplishments. Regarding accomplishments, for example, every college convocation acknowledges students who graduate *cum laude*, *magna cum laude*, and *summa cum laude*.

We typically evaluate people according to the kind of work they do. Higher prestige occupations include physicians, lawyers, pharmacists, college professors, architects, dentists, and teachers. Legislators, police officers, actors, librarians, realtors, and firefighters are examples of jobs at a medium prestige level. Occupations at the lower prestige level consist of janitors, carpenters, bartenders, garbage collectors, truck drivers, and food servers (Smith et al., 2011).

prestige respect or recognition attached to social positions.

Figure 8.2 The Color of Wealth

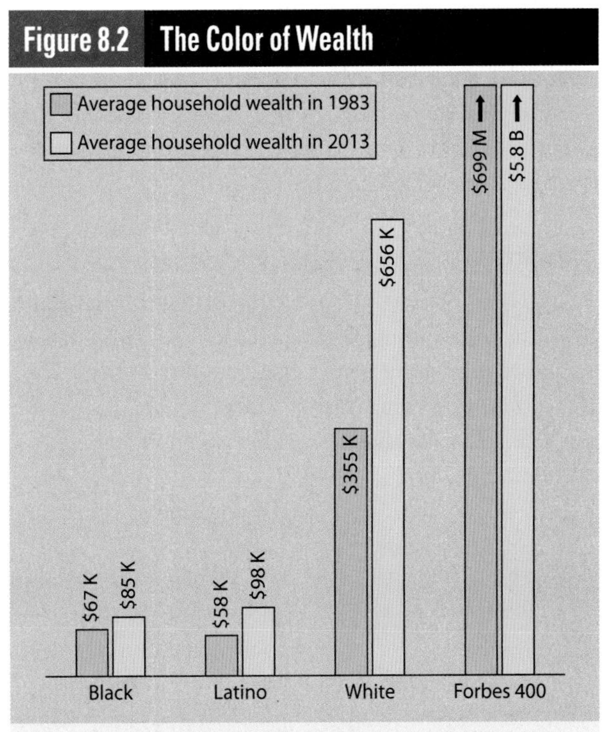

Note: In 2013, 95 percent of the Forbes 400 were white.

Source: Asante-Muhammed et al., 2016, 12.

The most prestigious occupations have several characteristics:

Lee Lorenz/The New Yorker Collection/The Cartoon Bank

- They *require more formal education* (college or postgraduate degrees) and/or extensive training. Physicians, for example, must fulfill internship and residency requirements after receiving a medical degree.
- They're primarily *non-manual and require more abstract thought.* All jobs have some physical activity, but an architect must use more imagination in designing a building than a carpenter, who usually performs very specific tasks.
- They *are paid more,* even though there are some exceptions. A realtor or a truck driver may earn more than a registered nurse, but registered nurses are likely to earn more over a lifetime because they have steady employment, health benefits, and retirement programs.
- They're *seen as socially more important.* An elementary school teacher may earn less than the school's janitor, but the teacher's job is more prestigious because teachers contribute more to a society's well-being.
- They *involve greater self-expression, autonomy, and freedom from supervision.* A dentist has considerably more freedom in performing her or his job than a dental hygienist. In effect, then, higher prestige occupations provide more privileges.

POWER

A third basis of social stratification is **power**—the ability to influence or control the behavior of others despite opposition. Wealthy people generally have more power than others in dominating top government and economic positions. Even if they don't hold a political office or a corporate position, the most powerful people make the nation's important decisions.

People who experience *status consistency* are about equal in terms of wealth, prestige, and power. A Supreme Court justice, for instance, is usually affluent, enjoys a

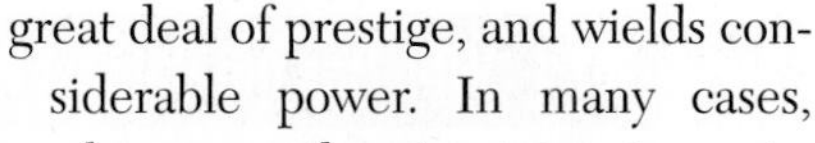

great deal of prestige, and wields considerable power. In many cases, however, there's *status inconsistency* (see Chapter 5) if a person ranks differently on stratification factors. Consider funeral directors. Their prestige is relatively low, but most have higher incomes than college professors, who are among the most educated people in U.S. society and have relatively high prestige.

We've looked at stratification systems and their bases, but how, specifically, do people in different social classes act? And how does social class affect our behavior?

power the ability to influence or control the behavior of others despite opposition.

socioeconomic status (SES) an overall ranking of a person's position in society based on income, education, and occupation.

8-2 SOCIAL CLASS IN AMERICA

Social class is a more important divider than race, ethnicity, gender, age, or other characteristics. To measure social class, sociologists typically use **socioeconomic status (SES)**—an overall ranking of a person's position in society based on income, education, and occupation. Some researchers ask people to identify the social classes in their communities (*reputational approach*), some ask people to place themselves in one of a number of classes (*subjective approach*), but most use SES indicators (*objective approach*).

In measuring social class, national surveys tend to use the subjective approach and closed-ended questions (see Chapter 2). Because the choices differ, so do respondents' self-identification. In a recent Gallup poll in which two of the five categories included the words "middle class," 62 percent of Americans identified themselves as upper-middle or middle class (Newport, 2017). In another national survey, however, where three of the five categories included the words "middle class," fully 87 percent said that they were upper-middle, middle, or lower-middle class (Doherty and Tyson, 2015).

Because there are different ways of measuring social class, sociologists don't always agree on the number of social classes. There's consensus, however, that there are four general classes: upper, middle, working, and lower. Except for the working class, many sociologists divide these classes further into more specific strata. There aren't rigid dividing lines, but *Figure 8.3* provides an overview of the U.S. social class structure. Besides income, education,

Figure 8.3 The American Class Structure

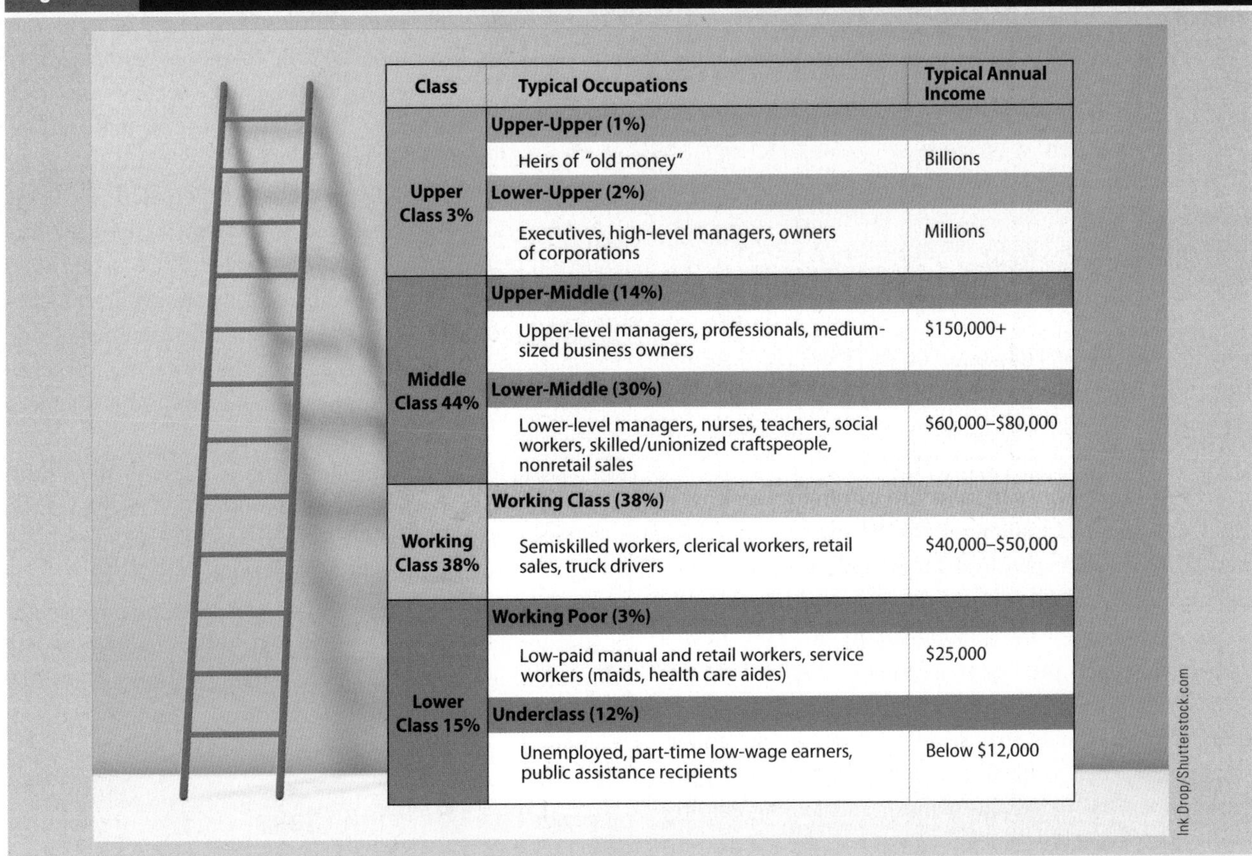

Class	Typical Occupations	Typical Annual Income
Upper Class 3%	**Upper-Upper (1%)**	
	Heirs of "old money"	Billions
	Lower-Upper (2%)	
	Executives, high-level managers, owners of corporations	Millions
Middle Class 44%	**Upper-Middle (14%)**	
	Upper-level managers, professionals, medium-sized business owners	$150,000+
	Lower-Middle (30%)	
	Lower-level managers, nurses, teachers, social workers, skilled/unionized craftspeople, nonretail sales	$60,000–$80,000
Working Class 38%	**Working Class (38%)**	
	Semiskilled workers, clerical workers, retail sales, truck drivers	$40,000–$50,000
Lower Class 15%	**Working Poor (3%)**	
	Low-paid manual and retail workers, service workers (maids, health care aides)	$25,000
	Underclass (12%)	
	Unemployed, part-time low-wage earners, public assistance recipients	Below $12,000

Ink Drop/Shutterstock.com

Sources: Based on U.S. Census Bureau, Current Population Survey..., 2013; Wysong et al., 2014; and BLS Reports, 2017.

and occupation, social classes also differ in values, power, prestige, social networks, and *lifestyles* (tastes, preferences, and ways of living).

8-2a The Upper Class

Two percent of Americans say they're upper class (Newport, 2017). Their self-identification doesn't necessarily reflect their actual SES, however, because the upper class comprises the *upper-upper class* and the *lower-upper class*.

THE UPPER-UPPER CLASS

Upper-upper-class people are the *old rich* who have been wealthy for generations. Because they value their privacy, upper-upper-class members rarely appear on the lists of wealthiest individuals published by *Forbes* or other sources.

An inherited fortune brings power. Upper-upper-class White males, in particular, shape the economic and political climate through a variety of mechanisms: Dominating the upper levels of business and finance, holding top political positions in the federal government, underwriting thousands of think tanks and research institutes that formulate national policies, and shaping public opinion through the mass media (Zweigenhaft and Domhoff, 2006; Hacker and Pierson, 2010).

THE LOWER-UPPER CLASS

The lower-upper class—which is much more diverse than the upper-upper class—is the *nouveau riche*, those with "new money." Some, like the Kennedys, amassed fortunes several generations ago. Others, like politician and businessman Jeb Bush, are augmenting their vast inheritances through a combination of speaking fees, corporate board memberships, investments, and consulting contracts that pay millions each year (O'Keefe et al., 2015). Many of the new rich—Oprah Winfrey, Beyoncé, and Mark Zuckerberg—worked for their income rather than inherited it.

Besides business entrepreneurs, the lower-upper class also includes high-level managers of international corporations, those who earn at least a million dollars a year, and some highly paid athletes and actors, but the lifestyles vary considerably. Some lower-upper class members live modestly, but many flaunt their new

Conspicuous consumption isn't limited to the super-rich. Many upper-middle-class people, in particular, use status symbols to show that they've made it by buying "almost rich" cars (like a Mercedes Benz that starts at $32,000), upscale kitchen appliances, designer handbags, expensive jewelry, and by taking lavish vacations.

Hemera Technologies/Getty Images

wealth. Sociologist Thorstein Veblen (1899/1953) coined the term *conspicuous consumption*, a lavish spending on goods and services to display one's social status and enhance one's prestige.

Most of the *nouveau riche* are extremely wealthy. If Bill Gates spent $1 million every single day, it would take him 218 years to spend all his money (Oxfam, 2014). Because they lack the "right" ancestry and have usually made their money by working for it, however, lower-uppers aren't accepted into "old money" circles that have strong feelings of in-group solidarity. Still, they engage in lifestyles and rituals that parallel those of the upper-upper class, such as having personal chefs, taking exotic vacations (often in private planes), and joining country clubs (Sherwood, 2010).

8-2b The Middle Class

For about 89 percent of Americans, being middle class means having a secure job and being able to save money for the future. This may help explain why—whether they earn $22,000 or $200,000 a year—people identify themselves as middle class (Brown, 2016; Grobart, 2016). There's no agreement on exactly where the middle class begins and ends, but there are some major differences between the *upper-middle class* and the *lower-middle class*.

THE UPPER-MIDDLE CLASS

Upper-middle-class members, although rich, live on earned income rather than accumulated or inherited wealth. The occupations of this group usually require a Ph.D. or advanced professional degree. People in this class include corporate executives and managers (but not those at the top), high government officials, owners of large businesses, physicians, and successful lawyers and stockbrokers. A higher level of education is associated with greater occupational prestige and autonomy, as well as job quality and security, but people in these occupations are three times more likely than those in the general population to work 50 or more hours per week (Dewan and Gebeloff, 2012).

THE LOWER-MIDDLE CLASS

The lower-middle class, more diverse than the upper-middle class, is composed of people in nonmanual occupations that require some training beyond high school, and professional occupations that require a college degree. Nonmanual jobs include office staff, low-level managers, owners of small businesses, medical and dental technicians, secretaries, police officers, and sales workers (e.g., insurance and real estate agents). Examples of professional occupations are nursing, social work, and teaching. Most families in the lower-middle class rely on two incomes to maintain a comfortable standard of living.

Unlike the upper-middle class, those in the lower-middle class have less autonomy and freedom from supervision, and there's little chance for advancement. Except for some retirement funds, most have only modest savings to cover emergencies. Many buy used or inexpensive late-model cars, eat out at middle-income restaurants, and take occasional vacations, but they rarely have the income to buy luxury products without going deeply into debt.

8-2c The Working Class

About 30 percent of Americans identify themselves as working class (Newport, 2016). This group includes skilled and semiskilled laborers (e.g., construction and assembly-line workers, truck drivers, auto mechanics, electricians). The semiskilled jobs typically require little training, are mechanized, and closely supervised. Most of the occupations are blue-collar, but some—clerks and retail sales workers—are white collar. The jobs don't require a college education, but offer little or no opportunity for advancement.

Working-class homeowners may experience foreclosure because of delinquent payments. Many use credit cards but then can barely pay the monthly minimum, don't have enough money to cover a $500 repair bill, and all of their savings may amount to only about four months of income. Debts become overwhelming when borrowers suffer setbacks like divorce, illness, or job loss (Bell, 2015; PEW Charitable Trusts, 2015).

8-2d The Lower Class

Only 8 percent of Americans identify themselves as lower class (Newport, 2016). People in this group are at the bottom of the economic ladder because they have little education, few occupational skills, work in minimum wage jobs, or are often unemployed. Most of the lower

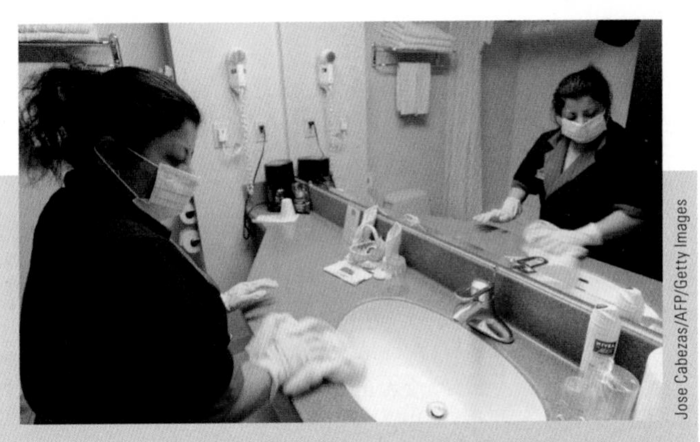

Jose Cabezas/AFP/Getty Images

Among America's working poor are hotel housekeepers. On average, those employed even at expensive hotels earn less than $19,000 a year working full time. The working poor do dirty and often demeaning work for next to nothing, barely scrape by, and live in constant fear of being fired (Wing and Schwartz, 2014).

class is poor, but sociologists distinguish between the *working poor* and the *underclass*.

THE WORKING POOR

Almost 9 million Americans are the **working poor**—people who work at least 27 weeks a year but whose wages fall below the official poverty level. Women are more likely than men to be among the working poor. Black and Latino individuals are more than twice as likely as Asian and White individuals to be in this group. Full-time workers are less likely than part-time workers to be working poor, but almost 4 percent of people usually employed full time are among the working poor, primarily because of low earnings (BLS Reports, 2017).

THE UNDERCLASS

The **underclass** are people who are persistently poor, residentially segregated, and relatively isolated from the rest of the population. Most are chronically unemployed or drift in and out of jobs. They may work erratically or part time, but their lack of skills, low education levels, and mental or physical disabilities make it difficult for them to find full-time jobs. Members of the underclass may have some earnings, but are often dependent on government programs, including Social Security, Medicaid, and veterans' benefits. Some also have income from criminal activities (Gilbert, 2011; see also Chapter 7).

8-2e How Social Class Affects Us

Social class, more than any other single variable, affects practically all aspects of our lives. Max Weber referred to the consequences of social stratification as *life chances*—opportunities to access social and economic resources that improve one's quality of life (e.g., food, housing, education). People's life chances vary across institutions. Regarding health, for example:

- Poor children are two times more likely than children in affluent families to have serious emotional or behavioral problems, and are more likely to be unhealthy in adulthood. Unhealthy adults earn less, spend less time in the labor force, and must often retire earlier (Reuben and Pastor, 2015; see also Chapter 14).

Glovatskiy/Shutterstock.com

Shkvarko/Shutterstock.com

- Living in a chronically disadvantaged neighborhood and experiencing violence increases anxiety and long-term stress, accelerating the onset of diseases (Harrell et al., 2014; Schanzenbach et al., 2016).
- Compared with the upper class, the lower class is four times more likely to be unhealthy, three times more likely to be unhappy, and twice as likely to be frequently stressed and depressed (Morin and Motel, 2012; Graham, 2015).
- The wealthiest Americans are now living 10 to 15 years longer than the poorest (Dickman et al., 2017).

8-3 POVERTY

There are two ways to define poverty: absolute and relative. **Absolute poverty** is not having enough money to afford the basic necessities of life, such as food, clothing, and shelter ("what I need"). **Relative poverty** is not having enough money to maintain an average standard of living ("what I want").

working poor people who work at least 27 weeks a year but whose wages fall below the official poverty level.

underclass people who are persistently poor and seldom employed, residentially segregated, and relatively isolated from the rest of the population.

absolute poverty not having enough money to afford the basic necessities of life.

relative poverty not having enough money to maintain an average standard of living.

Six Ways the Poor Pay More

FOOD: Not having a car to get to a supermarket chain or discount store means getting groceries at the corner store where staples are more expensive, produce is limited, and some products are stale.

TRANSPORTATION: About 45 percent of American households don't have access to public transportation. Old cars break down, can be expensive to fix, use a lot of gas, and can cost people a job.

DOING BUSINESS: Most neighborhood stores charge more for products because real estate and insurance are more expensive. Poor people also pay more for financial services. Banks are scarce and "check-cashing stores" charge up to 10 percent of a check's value to cash it.

CREDIT: Because poor people don't have credit, they must rely on "cash advance" stores if they can prove that they get a regular paycheck. If poor people qualify for such a loan, they pay an annual percentage rate of about 825 percent.

EVERYDAY HASSLES: The poor are hassled almost every day by bill collectors because of overdue payments, deal with Laundromat trips, carry groceries long distances, and contend with violence in high-crime areas.

WAITING: Poor people spend much of their time waiting—in food pantry lines, at blood banks to sell their blood, at bus stops to get to work, and for apartments in safer neighborhoods.

Sources: Brown, 2009; Bass and Campbell, 2013; Johnson, 2014 "Tackling Poverty," 2015.

8-3a The Poverty Line

The **poverty line** (also called the *poverty threshold*) is the minimal income level that the federal government considers necessary for basic subsistence. To determine the poverty line, the Department of Agriculture (DOA) estimates the annual cost of food that meets minimum nutritional guidelines and then multiplies this figure by three to cover the minimum cost of clothing, housing, health care, and other necessities. The poverty line, which in 2016 was $24,339 for a family of four (two adults and two children), is adjusted every year to include cost-of-living increases. If a family makes more than the poverty line, it's usually not eligible for public assistance (Semega et al., 2017).

Is the poverty line realistic? Some contend that the poverty line is too high. They argue, for example, that many people aren't as poor as they seem because the poverty threshold doesn't include the value of noncash benefits such as food stamps, medical services (primarily Medicare and Medicaid), public housing subsidies, and unreported income (Eberstadt, 2009; Haskins, 2015).

poverty line the minimal income level that the federal government considers necessary for basic subsistence (also called the *poverty threshold*).

Others claim that the poverty line is too low because it doesn't include child care and job transportation costs or the cost-of-living expenses, particularly housing, that vary considerably across states, regions, and urban-rural areas. According to some economists, a family of four needs almost $60,000 a year to cover basic necessities, a considerably higher amount than the official poverty line of just over $24,000. Some also argue that poverty estimates

Spencer Platt/Getty Images

One in three poor Americans live in the suburbs. Compared with central cities, many suburbs offer safer streets, better schools, and a greater number of low-paying jobs (e.g., retail sales, landscaping, restaurants). Rents are rising, however, and many of the suburban poor worry about the costs of maintaining a car to get to work (Kneebone and Berube, 2014; Kneebone, 2016).

exclude millions of Americans who live above the poverty line but rely on food banks, soup kitchens, and clothing thrift stores to survive (Fremstad, 2012; DeGraw, 2014).

8-3b Who Are the Poor?

In 2016, almost 13 percent of Americans (40.6 million people) were living in poverty, down from 15 percent in 2012. Between the ages of 25 and 60, 62 percent of Americans will experience at least one year of poverty (Rank and Hirschl, 2015; Semega et al., 2017). Both historically and currently, the poor share some of the following characteristics.

AGE

Children under age 18 make up only 23 percent of the U.S. population, but 33 percent (13.3 million) of the poor, up from a low of 16 percent in 2001. Americans aged 65 and older make up about 15 percent of the total population, but 9 percent (4.6 million) of the poor. The poverty rate of older Americans is at an all-time low, and lower than that of any other age group, because government programs, particularly Medicare and Medicaid, have generally kept up with the rate of inflation. In contrast, many programs for poor children have been reduced or eliminated since 1980 (Children's Defense Fund, 2014; Semega et al., 2017).

GENDER AND FAMILY STRUCTURE

Women's poverty rates are higher than men's (14 and 11 percent, respectively), but family structure is an important factor. In 2016, 5 percent of married-couple families, 13 percent of male-headed families, and 27 percent of female-headed families lived in poverty (Semega et al., 2017).

Feminization of poverty refers to the disproportionate number of the poor who are women. Because of increases in divorce, nonmarital childbearing, and low-paying jobs, single-mother families are four to five times more likely to be poor than married-couple families that have two wage earners, and more likely to be extremely poor over many years (Shriver, 2014; Irving and Loveless, 2015).

RACE AND ETHNICITY

In absolute numbers, there are more poor White people (17.3 million) than poor Latino (11.1 million), Black (9.2 million), or Asian people (1.9 million) (Semega et al., 2017). Proportionately, however, and as *Figure 8.4* shows, White and Asian Americans are less likely to be poor than other racial-ethnic groups.

Figure 8.4 Percentage of Americans Living in Poverty, by Race and Ethnicity, 2016

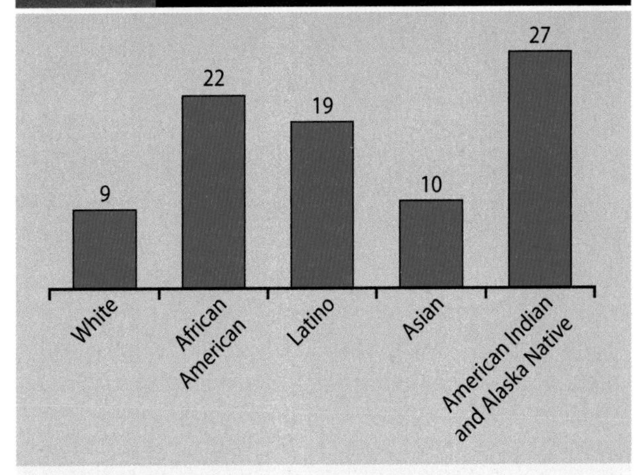

Sources: Based on "American Indian and Alaska Native...," 2016; and Semega et al., 2017, Table 3.

EDUCATION

Increasingly, education separates the poor from the nonpoor. In 2016, just 5 percent of college graduates were poor, compared with 13 percent of high school graduates and 25 percent without a high school diploma (Semega et al., 2017).

The more risk factors, the more likely that someone will experience poverty. For example, being a young, Black, unmarried female with young children and little education increases the odds of both poverty and extreme poverty (living below the bottom 10 percent of the income distribution) (Rank and Hirschl, 2015; Annie E. Casey Foundation, 2015).

8-3c Why Are People Poor?

Social scientists have a number of theories about poverty (see Cellini et al., 2008, for a nontechnical summary). Regarding two of the most common explanations, one blames the poor themselves, and the other emphasizes societal factors.

BLAMING THE POOR: INDIVIDUAL FAILINGS

More than a third of Americans believe that people are poor because they don't work hard enough. Republicans are more likely than Democrats (56 and 19 percent, respectively) to say that poverty is due to lack of effort rather than circumstances beyond people's control (Smith, 2017).

Such attitudes reflect a still influential *culture of poverty* perspective, which contends that people are poor because they're personally "deficient" or

feminization of poverty the disproportionate number of the poor who are women.

John Sturrock/Alamy Stock Photo

In 2016, almost 550,000 Americans were homeless on a given night, down by 15 percent since 2007. Most (68 percent) stayed in residential programs, 35 percent were people in families, and 22 percent were children under the age of 18. Unemployment, poverty, and a lack of affordable housing are the main reasons for homelessness (Henry et al., 2016).

"inadequate." If they worked harder, took initiative, were ambitious, and planned for the future instead of seeking immediate gratification, according to this perspective, poor people could succeed. Instead of being responsible and getting jobs, the poor rely on generous government handouts that erode an incentive to work (Lewis, 1966; Banfield, 1974; Murray, 1984, 2012). Thus, the most effective remedy to poverty is to change people's attitudes and behavior.

BLAMING SOCIETY: STRUCTURAL CHARACTERISTICS

In contrast to blaming the poor, most sociologists contend that macro-level structural factors create and sustain poverty. Technology, automation, industry relocation, and the growth of a low-wage service sector produce poverty conditions that are difficult for individuals to overcome. As the number of inner-city manufacturing jobs declined, neighborhood businesses collapsed, middle classes—both White and Black—moved to the suburbs, and the poor became more isolated and welfare-dependent (Oliver and Shapiro, 1995; Wilson, 1996; Conley, 1999). Thus, the most effective remedy to poverty is changing structural forces like unemployment and very low wages.

Which perspective is more accurate, blaming the poor or blaming society? Some people are poor because they're lazy and would rather get a handout than a job. Others "compound their misfortune by self-medicating or engaging in irresponsible, self-destructive behavior" (Kristof, 2016: SR1). Most people are poor, however, because of economic factors (low wages, job loss, lack of affordable housing, inability to afford health insurance), which, in turn, can result in acute health and employment problems.

social mobility movement from one social class to another.

intragenerational mobility movement up or down a social class over one's lifetime.

Temporary assistance for the nation's poorest families has fallen well below the poverty line in every state. In 16 states, a family of three receives less than $336 a month. Millions of poor Americans don't seek any help for food, shelter, or utilities because they don't know about the programs or have transportation and childcare difficulties in meeting some of the job-training requirements (Rodrigue and Reeves, 2015; Stanley et al., 2016).

Between 2014 and 2015, the poverty rate declined by 1.2 percent primarily because employers created more jobs, especially in the service sector, and raised minimum wages that attracted low-skill workers. For instance, a 32-year-old man, who had survived on odd jobs, lifted his family out of poverty when he became an assistant manager at a pizzeria with an annual salary of $40,000 and health benefits (Cohen, 2016).

On the other hand, and despite personal failings and deficiencies, millions of middle class Americans get government aid through programs for disabled individuals, veterans, college students, older people, and the unemployed. Wealthy farmers receive generous federal subsidies, and for every taxpayer dollar that helps poor families, six dollars benefit the rich who avoid paying taxes (Rampell, 2014; Eckhoff, 2015; Buchheit, 2015; Center on Budget and Policy Priorities, 2016).

8-4 SOCIAL MOBILITY

Social mobility is movement from one social class to another. Is the American dream of moving up the economic ladder a reality? About 62 percent of Americans, compared with 53 percent in 2012, believe that people can get ahead by working hard (Newport, 2016). Are they right?

8-4a Types of Social Mobility

Intragenerational mobility is movement up or down a social class over one's lifetime. If Ashley begins as a nurse's assistant, becomes a registered nurse, and then a physician's assistant (PA), she experiences intragenerational mobility. **Intergenerational mobility** is movement up or down a social class over two or more generations. If Ashley's parents were blue-collar workers, her upward movement to the middle class is an example of intergenerational mobility. Intragenerational and intergenerational mobility can be downward or upward.

8-4b Recent Trends in Social Mobility

Only 50 percent of Americans born in the 1980s, compared with 90 percent born in the 1940s, are economically better off than their parents. About 40 percent of children born in the lowest or highest fifth of the income distribution will still be there as adults. For those born in the bottom 10 percent in the 1980s, 30 percent are worse off than their parents, five times as many as for children born in the 1940s (Chetty, Grusky, et al., 2016; Reeves and Joo, 2016).

The middle class has been shrinking. Since 1971, the upper- and lower-income brackets have expanded, while the middle bracket has dropped from 61 to 50 percent (*Figure 8.5*). Because of such changes, some analysts describe the middle class as "eroding," "squeezed," and "sinking" (Galston, 2013; Porter, 2013; Trumbull, 2015).

Most social mobility moves are short because social classes are fairly rigid (Wysong et al., 2014). A child from a working-class family, for example, is more likely to move up to the lower-middle class than to jump to the lower-upper class. The same is true of downward mobility: Someone from the middle class is more likely to slide into the working class than to drop to the underclass. Upward intergenerational mobility has become more limited because the rungs of the economic ladder have grown further apart (Chetty et al., 2014). As you saw earlier, income inequality has increased, making it more difficult for even upper-middle class people to move up.

Across 24 middle- and high-income countries, the United States ranks only 16th in upward generational mobility and only slightly ahead of some developing countries, like Brazil and Chile (Corak, 2016). Countries with the greatest income inequality, such as the United States, are the most likely to have limited upward mobility (Greeley, 2013).

Figure 8.5 The Middle Class Has Been Shrinking

Percentage of adults in each income tier

Year	Lowest	Lower Middle	Middle	Upper Middle	Highest
2015	20%	9%	50%	12%	9%
2011	20%	9%	51%	12%	8%
2001	18%	9%	54%	11%	7%
1991	18%	9%	56%	12%	5%
1981	17%	9%	59%	12%	3%
1971	16%	9%	61%	10%	4%

Note: For 2015, the income for "lowest" was <$31,402; $31,402–$41,868 for "lower middle"; $41,869–$125,608 for "middle"; $125,609–$188,412 for "upper middle"; and >$188,412 for "upper."

Source: Based on Pew Research Center, 2015, "The American Middle Class is Losing Ground...," pp. 7, 16.

8-4c What Affects Social Mobility?

Upward mobility doesn't always reflect people's talents, intelligence, or hard work. Instead, much social mobility depends on structural, demographic, and family background factors, all of which are interrelated.

STRUCTURAL FACTORS

Macro-level variables, over which people have little or no control, affect social mobility in several ways. First, *changes in the economy* spur upward or downward mobility. During an economic boom, the number of jobs increases, and many people have an opportunity to move up. During recessions, such as the Great Recession, long-term unemployment leads to downward mobility (see Chapters 11 and 12).

Second, *government policies and programs* affect social mobility. Unlike the United States, countries that have promoted equality (e.g., Canada, Denmark, Finland, Norway) have the highest upward mobility rates. They have universal health care, which reduces the chance of people falling into poverty because of medical emergencies or poor health; provide affordable housing; and fund technical schools where young people learn a high-paying trade (Foroohar, 2011; Deparle, 2012).

Third, *immigration* fuels upward mobility. Because many recent immigrants take low-paying jobs, groups that are already in a country often move into higher-paying occupations (Haskins, 2008).

intergenerational mobility movement up or down a social class over two or more generations.

DEMOGRAPHIC FACTORS

Demographic factors also affect social mobility. Four of the most important are education, gender, race, and ethnicity, but place also has an effect.

Education. Higher education promotes upward mobility. Especially when the economy is slumping, people with a high school education or less often face long and frequent bouts of unemployment, must get by with temporary work, and may move down the socioeconomic ladder (Kochhar and Fry, 2015; Chetty et al., 2017).

Gender. Women's massive entry into the labor force since the 1980s has increased family income and many women's upward mobility. You'll see in later chapters that men's labor participation rates have been dropping, and women's educational attainment has risen faster than men's, but men's mobility is less affected than women's by a divorce, nonmarital children, or widowhood.

Race and ethnicity. Black and Latino middle classes have grown since the 1970s, but both groups still lag significantly behind White people and Asian Americans (see Chapters 10 and 11). Of Black children born into the bottom 20 percent of the income distribution, about half will still be there as adults, compared to less than one-quarter of White children. Economic, cultural, and social capital account for most of this mobility gap. Since the slavery era, White parents could accumulate skills and wealth that, with every generation, gave their children far better chances than Black children of moving up (Collins and Wanamaker, 2017; Matthew and Reeves, 2017)

AP Images/Kirsty Wigglesworth

J. K. Rowling worked as a teacher. After a divorce, she lived on public assistance, writing *Harry Potter and the Philosopher's Stone* during her daughter's naps. The book was published in 1997; by 2007, Rowling, a billionaire and one of the world's richest women, was considerably wealthier than the Queen of England.

Place. Where one lives can stimulate or dampen upward mobility. For example, the probability of a child born into the poorest fifth of the population in San Jose, California, making it to the top fifth is 13 percent compared with only 4 percent for a child born in Charlotte, North Carolina. These upward mobility disparities in different parts of the country are correlated with four factors: residential segregation (whether by income or race), the quality of schooling, how many children live with only one parent, and parents' social capital (Chetty et al., 2014; Berube and Holmes, 2015).

The longer that children live in better environments, the greater their chances of upward mobility. For example, children below age 13 whose parents move to lower-poverty neighborhoods are more likely to attend college, more likely to have substantially higher incomes

MCT/Tribune News Service/Getty Images

Children raised in poor families can be upwardly mobile. Sonia Sotomayor, whose parents were born in Puerto Rico, is the first Latina to serve on the U.S. Supreme Court. Her family first lived in a South Bronx tenement and later a working-class housing project. Sotomayor's father, an alcoholic, had a third-grade education, didn't speak English, and was a tool and die worker. He died when Sotomayor was nine years old. Her mother—a telephone operator and later a practical nurse—stressed the value of education. Sotomayor excelled in school, received a full scholarship to Princeton University, and put in long hours to catch up with her peers' academic knowledge.

as adults, and less likely to become single parents (Chetty and Hendren, 2016; Smeeding, 2017).

FAMILY BACKGROUND FACTORS

Family wealth affects social mobility. About 60 percent of all U.S. household wealth is inherited; most of it grows, spreads through families and dynasties, and endures for generations. Whether people inherit wealth or amass it themselves, children who grow up in high-income households can experience considerable upward mobility (Alvaredo et al., 2017; Roth, 2017).

In contrast, nearly half of U.S. adults don't have enough money to cover a $400 car repair or any other unexpected expense. Not having savings increases chronic stress. Constantly struggling to solve the crisis of the day makes it difficult to plan for the future, including saving for college (Larimore et al., 2016; Graham, 2016).

Only 7 of the 45 U.S. presidents came from the lower-middle class or below. Although Abraham Lincoln was born in a log cabin, his father was one of the wealthiest people in the community. President Donald Trump joined his father's thriving real estate company after college. He started his business with a $14 million loan from his father, received other monetary gifts from him over the years, and relied on his successful father's connections to build his own empire (Kessler, 2016). About a third of the students with low grades at Ivy League universities wouldn't be there if their parents weren't celebrities, well-known politicians, or others who donate millions to the school (see Chapters 11 and 13).

Our socialization affects what French sociologist Pierre Bourdieu (1984) called *habitus*—the habits of speech and lifestyle that determine where a person feels comfortable and knowledgeable. Upper-class parents cultivate flexibility, autonomy, and creativity because they expect their children to step into positions that require such characteristics. Poor and working-class parents stress obedience, honesty, and appearance—traits that many employers expect. High-income parents have social connections to jobs and admission to particular schools or colleges. Because of such resources, children of top earners are likely to grow up to be top earners themselves (Khan, 2012; Corak, 2013).

Many low SES households are resilient, but meager incomes limit upward mobility. Unlike high SES parents, low-income parents have less time, health, and job flexibility to read to their children every day, take them to museums, or attend school events. Children in high-earning families also benefit from high-quality schools, mentoring connections, enriching extracurricular activities, and skills for navigating bureaucracies that will help them succeed in schools and workplaces (Mitnik and Grusky, 2015; Putnam, 2015; Corak, 2016; see also Chapter 4).

8-5 GLOBAL STRATIFICATION

Global stratification refers to worldwide inequality patterns that result from differences in wealth, power, and prestige. All societies are stratified, but more than 75 percent of the world's population lives in countries where economic inequality has widened, increasing the possibility of social class tension and conflict (United Nations Development Program, 2013; World Economic Forum, 2013).

8-5a Wealth and Income Inequality

In 2016, global wealth reached $256 trillion, but how is it distributed? Among the world's richest,

- 1 percent own more wealth than the rest of the planet;
- 10 percent own 89 percent of all global wealth; and
- 8 people (6 of whom are Americans) own as much wealth as "half the human race" (Shorrocks et al., 2016; Hardoon, 2017).

There are enormous wealth disparities across regions (*Figure 8.6*), and the inequalities have swelled. The incomes of the top 1 percent have increased 60 percent since 1990, and the income growth of the ultra-rich 0.01 percent has been even greater. Even in more economically egalitarian countries like Norway and Sweden, the share of income going to the richest 1 percent increased by more than 50 percent between 1980 and 2012 (Oxfam, 2013; Fuentes-Nieva and Galasso, 2014).

Worldwide, income inequality is also pervasive. Across 131 countries, the richest 3 percent hold 20 percent of all income. The median per capita incomes in the wealthiest populations are more than 50 times those in the 10 poorest populations, all of which are in sub-Saharan Africa, but there's considerable economic stratification within countries. In oil-rich Nigeria, for instance, the richest 6 percent hold 40 percent of the country's total income (Phelps and Crabtree, 2013, 2014).

global stratification
worldwide inequality patterns that result from differences in wealth, power, and prestige.

Figure 8.6 Wealth per Adult, by Region and Selected Countries, 2016

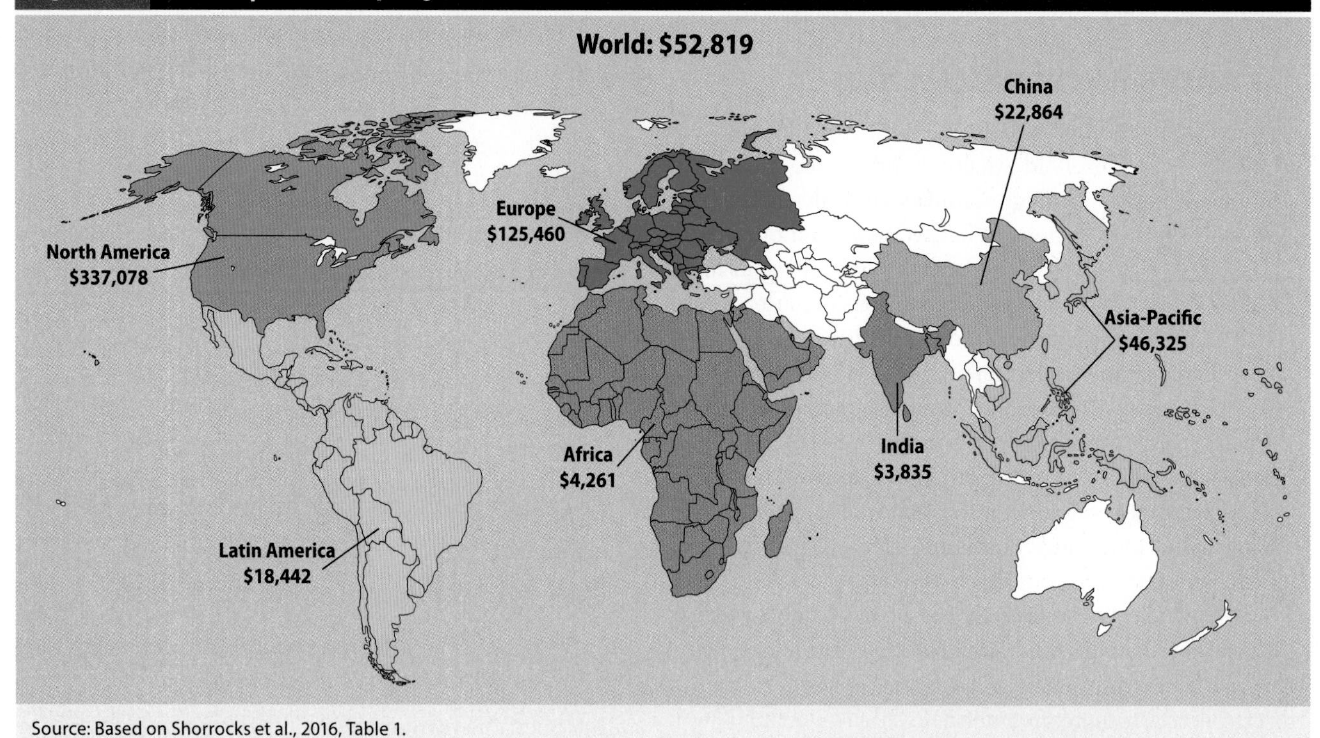

Source: Based on Shorrocks et al., 2016, Table 1.

Living in a wealthy country doesn't mean that people enjoy income equality. Among 35 high-income countries, the United States ranks 33rd in income equality—far behind other industrialized nations and just above some developing countries like Turkey and Mexico (OECD, 2016).

8-5b The Plight of Women and Children

Historically, currently, and across *all* nations, women and children experience the greatest inequality. An important measure of a country's health is its **infant mortality rate**, the number of babies under age 1 who die per 1,000 live births in a given year. The infant mortality rate ranges from a high of 97 in some low-income African nations to a low of 3 in affluent Western European countries (Kaneda and Bietsch, 2016).

Children's deaths have declined worldwide since 1990, but every year nearly 6 million poor children die before their fifth birthday; most of the deaths continue to be concentrated in sub-Saharan Africa Malnutrition—which makes children more vulnerable to severe diseases—is the underlying factor in 45 percent of all child deaths (World Health Organization, 2016).

infant mortality rate the number of babies under age 1 who die per 1,000 live births in a given year.

There's a strong association between high infant and child mortality rates and women's low education levels. Investing in women's education reduces unwanted and unplanned births; lowers infant, child, and maternal mortality; increases women's labor force participation; and can help lift a family out of poverty (World Bank and Collins, 2013; World Economic Forum, 2013).

Education doesn't guarantee women's greater economic equality, however. In both rich and poor countries, women are less likely than men to be employed because of cultural attitudes about women's roles and child rearing responsibilities. When women work outside the home, their average earnings are about half those of men: $11,000 and $20,000, respectively (World Economic Forum, 2016). Gender pay gaps vary across countries, but in all nations, women are paid less than men for similar work (see Chapters 9, 11, and 12).

8-5c Other Consequences of Global Stratification

Worldwide, 11 percent of the population (766 million people) live in *extreme poverty* (on less than $1.90 a day), but this is down from 35 percent (1.8 billion people) in 1990. This unprecedented decline is due, largely, to China's and India's economic growth, which has

El Greco/Shutterstock.com

benefited the rich and the poor. Despite such progress, extreme poverty rates have increased in sub-Saharan Africa, which now accounts for half of the world's extremely poor (World Bank, 2017).

Because of severe destitution, almost one in ten people worldwide suffer from chronic hunger and undernourishment. China and India have reduced their hunger rates, but in other developing regions (e.g., sub-Saharan Africa and Southern Asia, which includes Pakistan and Bangladesh), the number of undernourished people has increased. Some of the consequences include recurring illness, reduced work capacity, delayed physical and mental development, and stunted growth. For females, stunted growth can result in low birth weight babies and high child mortality rates (Martins, 2011; FAO, IFAD, and WFP, 2015).

People in poor countries report high levels of unhappiness, physical and mental health problems, and have little chance to escape poverty. Economic inequality has other costs: The world's richest 1 percent use their wealth to get political favors, including bypassing environmental laws; diminish other people's opportunities for upward mobility; and have a global network of tax havens worth almost $8 trillion within their own countries and offshore. Why should we care about tax havens? Taxing the income of the top 1 percent would (among other benefits) end extreme poverty, decrease political corruption, and ensure access to high-quality education and health care for all social classes (Helliwell et al., 2013; Lipton and Cresswell, 2016; Hardoon, 2017).

8-5d Why Is Inequality Universal?

Many theories try to explain why inequality is universal, but three of the most influential have been modernization theory, dependency theory, and world-system theory.

Modernization theory claims that low-income countries are poor because their leaders don't have the attitudes and values that lead to experimentation and using modern technology. Instead, policy makers adhere to traditional customs that isolate and prevent them from competing in a global economy. In effect, modernization theory blames poor nations for their poverty and other problems. After the key foundations of modernity and capitalism are in place, this perspective maintains, low-income countries will prosper.

Dependency theory contends that the main reason why low-income countries are poor is because they're pawns that high-income countries exploit and dominate. Rich nations wield an enormous amount of power by exporting jobs overseas, manipulating foreign aid, draining less powerful countries of their resources, penetrating other countries with multinational corporations, and coercing national governments to comply with corporate economic and political interests. In effect, according to dependency theorists, high-income countries benefit because the poor provide cheap labor and aren't powerful enough to protest even though they work in hazardous conditions and earn less than $1 a day.

iStock.com/jacus; Aly Song/REUTERS

We hear about the booming economies of India and China, but not everyone has benefited. A tent city in India houses many of the workers who earn about $1.30 a day building new office towers for the affluent. In China, a man in Shanghai begs as wealthier residents pass by.

More recently, *world-system theory*, similar to dependency theory, argues that "the economic realities of the world system help rich countries stay rich while poor countries stay poor" (Bradshaw and Wallace, 1996: 44). Countries like the United States that dominate the world economy control the economies of low-income countries because their workers depend on external markets for jobs.

High-income countries can extract raw materials (e.g., diamonds and oil) with little cost. They can also set prices, regardless of market values, for agricultural products that low-income countries export. Doing so forces many small farmers to abandon their fields because they can't pay for labor, fertilizer, and other costs (Carl, 2002; Alvaredo et al., 2013).

8-6 SOCIOLOGICAL EXPLANATIONS: WHY THERE ARE HAVES AND HAVE-NOTS

Why are societies stratified? *Table 8.1* summarizes the key points of the four theoretical perspectives. Let's begin by looking at a long-standing debate between functionalists and conflict theorists on why there are haves and have-nots.

8-6a Functionalist Perspectives: Stratification Benefits Society

Davis–Moore thesis the functionalist view that social stratification benefits a society.

Functionalists see stratification as both necessary and inevitable: Social class provides each individual a place in the social world,

Singer Lady Gaga had an estimated net worth of $275 million in 2016. Do her earnings reflect her contribution to society, especially compared with teachers, physicians, trash collectors, computer scientists, and others?

ChinellatoPhoto/Shutterstock.com

and motivates people to contribute to society. Without a system of unequal rewards, functionalists argue, many important jobs wouldn't be performed.

THE DAVIS–MOORE THESIS

Sociologists Kingsley Davis and Wilbert Moore (1945) developed one of the most influential functionalist perspectives on social stratification that persists today. The **Davis–Moore thesis**, as it's commonly called, asserts that social stratification benefits society. The key

Table 8.1 Sociological Explanations of Social Stratification

PERSPECTIVE	LEVEL OF ANALYSIS	KEY POINTS
Functionalist	Macro	• Fills social positions that are necessary for a society's survival • Motivates people to succeed and ensures that the most qualified people will fill the most important positions
Conflict	Macro	• Encourages workers' exploitation and promotes the interests of the rich and powerful • Ignores a wealth of talent among the poor
Feminist	Macro and micro	• Constructs numerous barriers in patriarchal societies that limit women's achieving wealth, status, and prestige • Requires most women, not men, to juggle domestic and employment responsibilities that impede upward mobility
Symbolic Interactionist	Micro	• Shapes stratification through socialization, everyday interaction, and group membership • Reflects social class identification through symbols, especially products that signify social status

arguments of the Davis–Moore thesis can be summarized as follows:

1. **Every society must fill a wide variety of positions and ensure that people accomplish important tasks.** Societies need teachers, doctors, farmers, trash collectors, plumbers, police officers, and so on.
2. **Some positions are more crucial than others for a society's survival.** Doctors, for example, provide more critical services to ensure a society's continuation than do lawyers, engineers, or bankers.
3. **The most qualified people must fill the most important positions.** Some jobs require more skill, training, or intelligence than others because they're more demanding, and it's more difficult to replace the workers. Pilots, for example, must have more years of training and aren't replaced as easily as flight attendants.
4. **Society must offer greater rewards to motivate the most qualified people to fill the most important positions.** People won't undergo many years of education or training unless they're rewarded by money, power, status, and/or prestige. If doctors and nurses earned the same salaries, there wouldn't be much incentive for people to spend so many years earning a medical degree.

According to the Davis–Moore thesis and other functionalist perspectives, then, stratification and inequality are necessary to motivate people to work hard and to succeed. For functionalists, social stratification is based on **meritocracy**—people's accomplishments. That is, people are rewarded for what they do and how well, rather than their ascribed status.

CRITICAL EVALUATION

Sociologist Melvin Tumin (1953) challenged the Davis–Moore thesis. First, he argued, societies don't always reward the positions that are the most important for the members' survival. If the multimillionaire professional athletes, actors, and pop musicians went on strike, many of us would probably barely notice. If, on the other hand, those with much lower income—whether they're doctors, garbage collectors, teachers, truck drivers, or mail carriers—refused to work, society would grind to a halt. Thus, according to Tumin, there's little association between earnings and the jobs that keep a society going.

Second, Tumin claimed, Davis and Moore overlook the many ways that stratification limits upward mobility. Where wealth is differentially distributed, access to education, especially higher education, depends on family wealth. As a result, large segments of the population are likely to be deprived of the chance to discover what their talents are, and society loses.

Third, Tumin criticized Davis and Moore for ignoring the critical role of inheritance. In upper classes, sons and daughters don't have to work because their inherited wealth guarantees a lifetime income and perpetuates privileges over generations.

Functionalist theories also don't explain why (1) upward social mobility is more limited in the United States than in other industrialized (and even some developing) countries; (2) so many college graduates can find only low-paying jobs; and (3) racial/ethnic income and wealth gaps persist across all social classes (Legatum Prosperity Index, 2013; McKernan et al., 2013).

8-6b Conflict Perspectives: Stratification Harms Society

Like Tumin, conflict theorists maintain that social stratification is dysfunctional because it hurts individuals and societies. Karl Marx's (1934) analysis of social class and inequality has had a lasting impact on modern sociology, especially conflict theory.

CAPITALISM BENEFITS THE RICH

Marx was aware that a diversity of classes can exist at any one time, but he predicted that capitalist societies would ultimately be reduced to two social classes: the capitalist class, or bourgeoisie, and the working class, or proletariat. The **bourgeoisie**, those who own and control capital and the means of production (e.g., factories, land, banks, and other sources of income), can amass wealth and power. The **proletariat**, workers who sell their labor for wages, earn barely enough to survive.

meritocracy a belief that social stratification is based on people's accomplishments.

bourgeoisie those who own and control capital and the means of production.

proletariat workers who sell their labor for wages.

Olivier Le Moal/Shutterstock.com

For conflict theorists, the economic struggles of the U.S. middle and working classes since the late 1970s weren't primarily the result of globalization and technological changes but, instead, a long series of government policies that overwhelmingly favored the rich. This policy of **corporate welfare** consists of an array of subsidies, tax breaks, and assistance that the government has created for businesses. For example:

- Taxpayers paid $7 trillion to bail out mismanaged financial institutions. The companies' executives still receive multimillion-dollar annual salaries and benefits (Ivry et al., 2015; Kiel and Nguyen, 2015; see also Chapter 11).
- Each household pays about $10,000 per year to subsidize corporations. Companies benefit enormously from the taxpayer-supported highways they use, law enforcement and judicial systems that protect their intellectual and physical property, public education that provides their workforce, and military personnel who safeguard corporate assets abroad. Other costs include corporations' unpaid taxes, oil producers' pollution and environmental disaster clean-ups, and taxpayer-funded research that results in patent-protected technology that reaps massive corporate profits (Buchheit, 2015, 2017).
- Among the most profitable 258 U.S. corporations, 18 paid no taxes between 2008 and 2015 (e.g., General Electric, Priceline.com), 24 paid zero taxes in four of the eight years, and 48 paid between 0 and 10 percent —well below the average employee's 30 percent tax rate (Frankel, 2017; Gardner et al., 2017).

Considerable data support conflict theorists' contention that growing inequality is dysfunctional for society. Among other social problems, ongoing economic inequity intensifies poverty, undermines people's trust in political and economic institutions, and, consequently, erodes national solidarity.

Whether their concern is driven by self-interest or a sense of justice, even powerful financiers and conservative economists have begun to worry about economic inequality. They fear that the increasing income and wealth gaps and low wages have made the United States less productive, have narrowed the consumer spending base, and have discouraged many people from entering or staying in the labor force. They also wonder if the huge gulf between the top 1 to 5 percent and the rest of society may lead to "oppressive taxes" on the super-rich, widespread social unrest, and even a revolution (Dumaine, 2015; Hightower, 2015).

corporate welfare subsidies, tax breaks, and assistance that the government has created for businesses.

gender stratification unequal access to wealth, power, status, prestige, and other valued resources because of one's sex.

CRITICAL EVALUATION

Conflict theories have their limitations. First, and despite Marx's prediction, even though corporate wealth has surged during the past 100 years while many people have become poorer, there have been some protests but no revolutions in capitalist countries. Americans aren't demanding greater economic equality, and many go into debt to purchase luxury items like TVs, jewelry, and pet supplies, and to replace expensive smartphones and other high-tech hardware every few years. Thus, according to critics, conflict theorists are exaggerating the existence and effects of economic inequality.

Second, some critics point out that conflict theorists overlook the fact that government programs have cut poverty by 40 percent since the 1960s. About 46 percent of American households pay no federal income taxes (and these tax breaks have nearly doubled since 1975) because they receive public assistance, get deductions for raising children under age 18, or don't report income. Moreover, the United States spends far more, per capita, than do many other wealthy countries on pensions, health care, education, unemployment, housing assistance, and similar benefits (Kirkegaard, 2015; "Inequality," 2017).

8-6c Feminist Perspectives: Stratification Benefits Primarily Men

For feminist scholars, functionalist and conflict theories are limited because they typically focus on men in describing and analyzing stratification and social class. As a result, women are largely invisible.

GENDER STRATIFICATION AND INEQUALITY

Gender stratification—unequal access to wealth, power, status, prestige, and other valued resources because of one's sex—contributes to systemic inequality. At the top rungs, only 13 percent of the world's billionaires are women, and 83 percent have inherited their fortunes. Only 16 percent of the top 1 percent of Americans are women. "The higher up you move in the income distribution," according to an influential economist, "the lower the proportion of women" (Piketty et al., 2016; Frank, 2017: BU3; Wealth-X, 2017).

In rich and poor countries, gender stratification leads to discrimination in the economy, politics, and access to medical services. Especially in low-income

Munir Uz Zaman/Getty Images

In 2013, more than 1,100 workers were killed and 2,000 were injured when an unsafe garment factory collapsed in Bangladesh. The country exports $20 billion in clothes annually to European and U.S. retailers, including Walmart. The $38-a-month minimum wage (for a six-day workweek) barely covers living expenses. More than 80 percent of the 4 million workers in Bangladesh's garment industry are women, mostly young and with little schooling (Al-Mahmood, 2013; Hammadi, 2013; Srivastava, 2013).

Gordana Sermek/Shutterstock.com

countries, almost 4 million girls and women are "missing" each year because of high death rates. About 20 percent are aborted because of a preference for sons, and more than a third die of childbirth and pregnancy complications. Besides excessive deaths, many women do much of the unpaid work (such as household chores and child rearing), earn less than men across all jobs, and hold only a minority of decision-making positions in public and private institutions (United Nations, 2015).

More than 100 countries ban women from certain jobs, accessing financial services, owning businesses, or conducting legal affairs. And a "global pandemic" of violence against females affects one in three worldwide, including young girls who become sex slaves (UN Women, 2014).

Feminist theorists contend that, in a patriarchal system, men dominate the stratification system: They control a disproportionate share of wealth, prestige, and power, and the feminization of poverty results in women's downward mobility. Women must often overcome economic inequities as well as juggle domestic and workplace responsibilities (see Chapters 11 and 12). The gender gaps in wealth, income, and household burdens make it harder for women to build a strong future for themselves and their families.

CRITICAL EVALUATION

Some critics point out that feminist theorists often focus only on poor women in showing how patriarchy affects stratification and social class (see Kendall, 2002). Another criticism is that many feminist scholars don't explain why so many women succeed despite patriarchal barriers. Third, feminist theories don't account for some striking cross-cultural variations. For example, women have greater wage equality and political power in some patriarchal and developing countries (e.g., Bolivia, Burundi, Cuba, Rwanda) than in many presumably more egalitarian industrialized nations (e.g., Australia, Canada, Denmark, United States) (World Economic Forum, 2016).

8-6d Symbolic Interaction Perspectives: People Create and Shape Stratification

Symbolic interactionists focus on how people create, change, and reproduce social classes. They address micro-level issues such as how people learn their social positions in everyday life and how such learning affects their attitudes, behavior, and lifestyles.

SOCIAL CONTEXTS SHAPE SOCIAL CLASS

People across social classes interact with and socialize their children differently. By age 2, children from higher-income homes know 30 percent more words than those from low-income homes. Early vocabularies increase reading comprehension and proficiency as early as kindergarten (Fernald et al., 2013).

Children as young as 4 years old are aware of and enact their social class. Among preschoolers, upper-middle-class children speak, interrupt, ask for help, and argue more often than do working-class children. Such behavior receives more adult attention and gives upper-middle-class children more opportunities to develop their language skills (Streib, 2011).

Social contexts also affect mobility. The social classes into which children are born affect their aspirations, the skills they value and can access, and their networks and resources. Doctors' children, for example, are more likely to become physicians themselves because they're exposed to medical discussions at home, are encouraged to pursue medicine as an occupation, and are embedded in social networks that provide medical school contacts (Jonsson et al., 2009).

You've seen that U.S. economic inequality is at its highest level since the 1930s, yet many Americans are relatively unconcerned. Fully 62 percent are satisfied with their opportunities to get ahead, and only 47 percent say that the rich–poor gap is a problem (DeSilver, 2014; Newport, 2016).

Why are so many people content instead of challenging policies that benefit primarily the top 1 percent? The average American knows very little about economic inequality (e.g., whether it's increasing or decreasing), where she or he is on the income spectrum, and believes (incorrectly) that there's more upward than downward mobility. Also, what people think they know is often wrong (Gimpelson and Treisman, 2015; Kraus and Tan, 2015). Thus, and because popular culture reinforces the idea that inequality is normal and inevitable, many people don't see the connection between structural inequality and their personal situation.

Many Americans aren't angry about rising inequality, moreover, because they embrace "mobility optimism"—a belief in the American Dream that anyone can succeed, and that they (or their children) will be rich someday (Manza and Brooks, 2016; see also Davidai and Gilovich, 2015). Political speeches and social media, particularly, fan beliefs that the American Dream is achievable.

CRITICAL EVALUATION

Symbolic interaction theories help us understand the everyday processes that underlie social stratification, but there are several weaknesses. First, interactionism, unlike the other perspectives, doesn't explain *why* stratification exists. Second, the theories don't explain why—despite the same family background, resources, and socialization—some siblings are considerably more upwardly mobile than their brothers and sisters. Third, conflict theorists, especially, fault symbolic interactionists for ignoring structural factors—like the economy, government policies, and educational institutions—that create and reinforce inequality.

STUDY TOOLS 8

READY TO STUDY? IN THE BOOK, YOU CAN:

- ☐ Check your understanding of what you've read with the Test Your Learning Questions provided on the Chapter Review Card at the back of the book.
- ☐ Tear out the Chapter Review Card for a handy summary of the chapter and key terms.

ONLINE AT CENGAGEBRAIN.COM WITHIN MINDTAP YOU CAN:

- ☐ Explore: Develop your sociological imagination by considering the experiences of others. Make critical decisions and evaluate the data that shape this social experience.
- ☐ Analyze: Critically examine your basic assumptions and compare your views on social phenomena to those of your classmates and other MindTap users. Assess your ability to draw connections between social data and theoretical concepts.
- ☐ Create: Produce a video demonstrating connections between your own life and larger sociological concepts.
- ☐ Collaborate: Join your classmates to create a capstone project.

9 Gender and Sexuality

Gary718/Shutterstock.com

LEARNING OBJECTIVES

After studying this chapter, you will be able to...

- 9-1 Differentiate between sex and gender and describe societal reactions to LGBTs.
- 9-2 Explain how gender stratification affects the family, education, workplace, and politics.
- 9-3 Describe contemporary sexual attitudes and practices, including sexual scripts and double standards.
- 9-4 Summarize abortion and same-sex marriage trends and explain why both issues are controversial.
- 9-5 Describe and illustrate gender and sexual inequality across cultures.
- 9-6 Compare and evaluate the theoretical explanations of gender and sexuality.

After finishing this chapter go to **PAGE 180** for **STUDY TOOLS**

Women pay more than men, and sometimes twice as much, for many things, including cars, mortgages, health care, high-end jeans from the same designer, and similar grooming products like moisturizers, deodorants, and even razors (Hill, 2015; Ngabirano, 2017). Why? This chapter examines how gender and sexuality affect our lives. First, however, take the True or False quiz to see how much you know about these topics.

WHAT DO YOU THINK?

Today, women have more employment advantages than men.

1	2	3	4	5	6	7
strongly agree						strongly disagree

True *or* False?

HOW MUCH DO YOU KNOW ABOUT GENDER AND SEXUALITY?

1. About 10 percent of the U.S. population is gay.
2. On average, Americans have sexual intercourse for the first time at about age 16.
3. Women make up about 20 percent of Congress.
4. Dating is dead on most college campuses.
5. U.S. abortion rates have declined since 1990.
6. Fathers now do almost as much child care and housework as mothers.

The answers to #3 and # 5 are true; the others are false. You'll see why as you read this chapter.

9-1 SEX, GENDER, AND CULTURE

Some differences between women and men are biological; others are social creations. Both biology and social factors shape a person's identity, but sex and gender aren't synonymous.

9-1a How Sex and Gender Differ

Many people use *sex* and *gender* interchangeably, but they're not the same. **Sex** refers to the biological characteristics with which we are born—chromosomes, anatomy, hormones, and other physical and physiological attributes. These attributes influence our behavior (e.g., shaving beards, wearing bras), but *don't determine* how we think or feel. Whether we see ourselves and others as feminine or masculine depends on gender, a more complex concept than sex.

Gender refers to learned attitudes and behaviors that characterize women and men. Gender is based on social and cultural expectations rather than on physical traits. Thus, most people are *born* either male or female, but we *learn* to be women or men because we internalize behavior patterns expected of each sex. In many societies, for example, women are expected to look young, thin, and attractive, and men are expected to amass as much wealth as possible.

9-1b Sex: Our Biological Component

Physical characteristics like breasts and beards indicate whether someone is a male or female, but sex isn't always clear-cut. Our cultural expectations dictate that we are female or male, but a number of people are "living on the boundaries of both sexes" (Lorber and Moore, 2007: 141). For example, **intersexual** individuals are people whose sex at birth isn't clearly either male or female. About 1 in 2,000 to 4,000 children born each year are classified as intersex because they're born with both male and female external genitals or an incomplete development of internal reproductive organs. Some parents seek surgery; others wait until a

sex the biological characteristics with which we are born.

gender learned attitudes and behaviors that characterize women and men.

intersexual a person whose sex at birth isn't clearly either male or female.

child is old enough to decide what to do (Bendavid, 2013).

SEXUAL IDENTITY AND SEXUAL ORIENTATION

Our **sexual identity** is an awareness of ourselves as male or female and how we express our sexual values, attitudes, and feelings. Our sexual identity incorporates a **sexual orientation**—an emotional or sexual attraction to sexual partners of the same sex, of the opposite sex, of both sexes, or neither sex:

- **Gay** Gay is a term referring to a person who is emotionally and sexually attracted to people of the same sex. While gay may be used to identify both men and women, gay men prefer to be called *gay* and gay women prefer to be called *lesbian(s)*. *Coming out* is a person's public announcement of a gay or lesbian sexual orientation
- **Heterosexuals**, often called *straight*, are attracted to people of the opposite sex.
- **Bisexuals**, sometimes called *bis*, are attracted to more than one gender.
- **Asexuals** lack any interest in or desire for sex.

Sexual orientation, like biological sex, isn't as clear-cut as many people believe. Alfred Kinsey (1948) and his associates' classic study found that most people weren't exclusively heterosexual or gay. Instead, they fell somewhere along a continuum in terms of sexual desire, attractions, feelings, fantasies, and experiences. Researchers have recently added *asexual* to Kinsey's classification (*Figure 9.1*).

sexual identity an awareness of ourselves as male or female and how we express our sexual values, attitudes, and feelings.

sexual orientation an emotional or sexual attraction to sexual partners of the same sex, of the opposite sex, of both sexes, or neither sex.

Gay those who are emotionally and sexually attracted to people of the same sex.

heterosexuals those who are sexually attracted to people of the opposite sex.

bisexuals those who are sexually attracted to more than one gender.

asexuals those who lack any interest in or desire for sex.

Most people's sexual identity corresponds with their biological sex, sexual attraction, and sexual behavior, but not always. Among Americans ages 15 to 44, for example, 86 percent *identify themselves* as straight, but 23 percent have had *same-sex experiences* (Moore, 2015).

HOW MANY AMERICANS ARE LGBT?

Since 2012, the share of adults identifying as lesbian, gay, bisexual, or transgender (LGBT) has increased from 3.5 percent to 4.1 percent. Women are more likely than men to identify as LGBT. Asians and Latinos are the most numerous and account for the largest increases since 2012 (*Table 9.1*).

Figure 9.1 Sexual Orientation Continuum

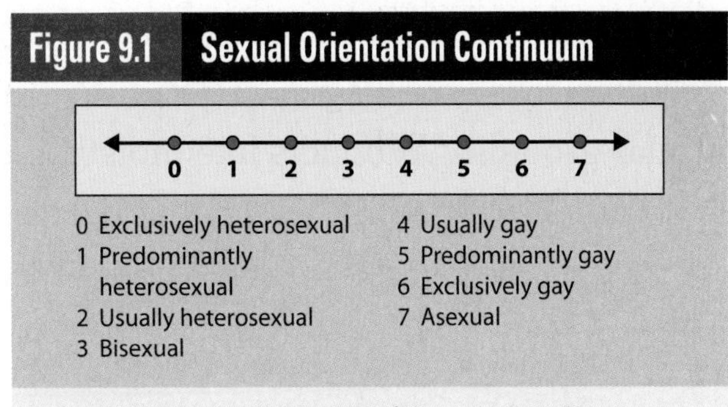

Sources: Kinsey et al., 1948, p. 638; and Kinsey Institute, 2011.

The LGBT population may have actually increased since 2012, but there are other explanations for the higher numbers. Because 63 percent of Americans (an all-time high) say that gay and lesbian relations are acceptable, people are more willing to identify as LGBT. Age is another factor: Millennials (people born between 1982 and 2004) are significantly more likely than older generations to reject traditional either/or categories (such as "man/woman" and "gay/straight"), to have LGBT friends, and to openly identify as LGBT (GLAAD, 2017; Jones, 2017; Steinmetz, 2017).

WHAT DETERMINES OUR SEXUAL ORIENTATION?

A Hong Kong billionaire offered $65 million to any man who succeeded in marrying his daughter after she eloped with her female partner. The offer attracted 20,000 suitors, but none were successful (Nichols, 2014). Like this father, 30 percent of Americans (but down from 56 percent in 1978) believe that being gay or lesbian is a "personal choice" or due to a person's upbringing (Jones, 2015).

Culture shapes people's sexual attitudes and behavior, but no one knows why we're straight, gay, bisexual, or asexual. Sexual orientation must have biological roots, according to some researchers, because gay and lesbian sexuality exists in all societies and, across cultures, the gay population is roughly the same—about 5 percent (Barash, 2012).

There's also growing scientific consensus that biological factors, particularly the early influence of sex hormones after conception and around childbirth, have a strong effect on sexual orientation (see LeVay, 2011, for a comprehensive summary of the studies). Other researchers speculate that a combination of genetic and cultural factors influence our sexual orientation (Slater, 2013).

Table 9.1	U.S. Adults Identifying as LGBT, by Selected Characteristics, 2012 and 2016							
			% BY GENDER AND RACE/ETHNICITY					
	% LGBT	Estimated Number of LGBT	Male	Female	White	Black	Latino	Asian
2012	3.5	8.3 million	3.4	3.5	3.2	4.4	4.3	3.5
2016	4.1	10.1 million	3.7	4.4	3.6	4.6	5.4	4.9

Source: Based on Gates, 2017.

9-1c Gender: Our Cultural Component

Gender doesn't occur naturally, but is socially constructed. This means that gender aspects may differ across time, cultures, and even groups within a society. Let's begin with gender identity.

GENDER IDENTITY

People develop a **gender identity**, a perception of themselves as either masculine or feminine, early in life. Many Mexican baby girls but not boys have pierced ears, for example, and hairstyles and clothing for American toddlers differ by sex. Gender identity, which typically corresponds to a person's biological sex, is part of our self-concept and usually remains relatively fixed throughout life.

Transgender is an umbrella term for people whose gender identity and behavior don't correspond with their birth sex. They comprise about 0.6 percent (1.4 million) of U.S. adults (Flores et al., 2016).

Because transgender is independent of sexual orientation, people may identify as heterosexual, gay, bisexual, or asexual. Facebook users can now choose their gender identity from more than 50 possibilities, but here are some of the most common transgender categories (American Psychological Association, 2014):

- *Transsexual* individuals are people who have permanently changed, or plan to change, their physical sex through medical intervention. Unlike transgender, transsexual is not an umbrella term to describe a person whose gender identity differs from their assigned sex. Transsexual is largely considered an outdated term and should only be used to describe someone who openly identifies as such.
- *Cross-dressers* wear clothing that's traditionally or stereotypically worn by another gender in their culture. People who cross-dress are usually comfortable with their assigned sex and don't wish to change it.
- *Genderqueer* are people who identify their gender as falling somewhere on a continuum between female and male, or a combination of gender identities and sexual orientations.

Gender expression is how a person communicates gender identity to others and includes

gender identity a perception of oneself as either masculine or feminine.

transgender people whose gender identity and behavior don't correspond with their birth sex.

gender expression how a person communicates gender identity to others.

PAUL J. RICHARDS/Getty Images

Olivier Douliery/Abaca Press/Washington/USA/Newscom

Christopher Beck, who received multiple military awards and decorations, retired from the elite U.S. Navy SEALs in 2011. A few years later, he became openly transgender, changing his name to Kristin Beck.

How Do Gender Roles Differ? While the world scrutinized Michelle Obama's gowns, dresses, shoes, jewelry, and hairdos, President Obama wore the same tuxedo and shoes during their eight years in the White House (Feldman, 2017).

Mike Theiler/Getty Images News/Getty Images

behavior, clothing, hairstyles, voice, or body characteristics. Cross-dressing, girls' frilly dresses, and men's business suits are all examples of gender expression. Even if a person's gender identity is constant, gender expression can vary from situation to situation and change over time. For example, between 2010 and 2014, the number of men's eyelid, facelift, and breast reduction cosmetic surgeries increased by 33 to 44 percent; half of American men now routinely use moisturizers, facial creams, or self-tanning lotions and sprays; and some of the National Basketball Association's "toughest players" have promoted products for Dove, La Mer, and other skin-care companies (Boyle, 2013; Holmes, 2013; American Society for Aesthetic Plastic Surgery, 2015).

GENDER ROLES, GENDER STEREOTYPES, AND SEXISM

Gender roles are the characteristics, attitudes, feelings, and behaviors that society expects of females and males. As you saw in Chapter 4, a major purpose of socialization, which begins at birth, is to teach people appropriate gender roles. As a result, we learn to become male or female through interactions with family members, teachers, friends, and the larger society.

gender roles the characteristics, attitudes, feelings, and behaviors that society expects of females and males.

gender stereotypes expectations about how people will look, act, think, and feel based on their sex.

sexism an attitude or behavior that discriminates against one sex, usually females, based on the assumed superiority of the other sex.

Americans are more likely now than in the past to pursue jobs and other activities based on their ability and interests rather than their sex. For the most part, however, our society still has fairly rigid gender roles and widespread **gender stereotypes**—expectations about how people will look, act, think, and feel based on their sex. We tend to associate stereotypically female characteristics with weakness and stereotypically male characteristics with strength. Consider, for example, how often we describe the same behavior differently for women and men:

- He's firm; she's stubborn.
- He's good with details; she's picky.
- He's honest; she's opinionated.
- He's raising good points; she's "bitching."
- He's experienced; she's "been around."
- He's enthusiastic; she's shouting.

Gender stereotypes fuel **sexism**, an attitude or behavior that discriminates against one sex, usually females, based on the assumed superiority of the other sex. In the late 1990s, and after receiving numerous rejections, a publisher finally accepted J. K. Rowling's manuscript of her Harry Potter book. Rowling followed the publisher's advice to sell the book under her initials, not her first name, Joanne. Even today, particularly for new science fiction and mystery authors, publishers instruct women to use male pseudonyms because "men prefer books written by men" (Cohen, 2012: D9).

A majority of women (63 percent), compared with 41 percent of men, believe that sexism makes it much harder for women to get ahead. Perhaps surprisingly, 62 percent of men aged 18 to 34, compared with 54 percent of those age 65 and older, say sexist obstacles that prevent women from succeeding "are now largely gone" (Fingerhut, 2016).

Men also experience sexism. Here's what one of my students wrote during an online discussion of gender roles:

Some parents live their dreams through their sons by forcing them to be in sports. I disagree with this but

Colin McConnell/Toronto Star/Getty Images

At an early age, sex-appropriate activities prepare girls and boys for future adult roles. As a result, there are few male ballet dancers and female auto mechanics.

want my [9-year-old] to be "all boy." He's the worst player on the basketball team at school and wanted to take dance lessons, including ballet. I assured him that this was not going to happen. I'm going to enroll him in soccer and see if he does better.

Is this mother suppressing her son's natural dancing talent? We'll never know because she, like many parents, expects her son to fulfill sexist gender roles that meet with society's approval.

9-1d Societal Reactions to LGBTs

People's attitudes toward LGBTs are mixed. There's been greater acceptance in some countries, but considerable repression in others.

GREATER ACCEPTANCE, BUT ...

Australian passports and birth certificates designate male, female, and transgender. In India, the 2011 national census for the first time offered three options: male, female, or a "third sex" that includes LGBTs. In Thailand, which has the world's biggest transgender population, an airline recruits "third sex" flight attendants. In the United States, in 2017 Oregon was the first state to allow residents to mark their sex as "not specified" on a driver's license.

In the United States, many jurisdictions, corporations, and small companies now extend more health care and other benefits to gay employees and their partners than to unmarried heterosexuals who live together. The U.S. Supreme Court and a growing number of states have legalized same-sex marriages, and large numbers of Americans support equal rights for LGBTs in the workplace and elsewhere (Von Drehle, 2014).

In 2012, the Army promoted the first openly gay female officer to brigadier general. In 2013, the Pentagon added benefits for same-sex partners, including services on U.S. military bases. Federal workers and Medicare recipients are now eligible for Sex Reassignment Surgery (Mach and Cornell, 2013; "Transgender Rights," 2015). And, since the mid-1990s, many LGBT characters have appeared in leading and supporting roles in popular TV programs (e.g., *Transparent, Modern Family, Gotham, This Is Us, Empire,* and *Game of Thrones*).

There's greater LGBT acceptance. However, Americans are about evenly divided on two issues: whether wedding-related businesses (like caterers and florists) should be required to serve same-sex couples and whether transgender people should be able to use public restrooms that correspond to their current gender identity rather than their birth sex (Masci, 2016; McCarthy, 2017).

Jonathan Daniel/Getty Images

In mid-2013, the Washington Wizards' Jason Collins appeared on the cover of *Sports Illustrated.* "I'm a 34-year-old NBA center, I'm Black, and I'm gay," he announced. Collins said that his teammates were supportive, but he retired a year later. In a recent survey, just 4 percent of LGBT adults described professional sports leagues as "friendly" toward LGBTs (Lipka, 2014).

... ALSO WIDESPREAD INTOLERANCE

According to one scholar, what makes gay people different from others is that "we are discriminated against, mistreated, [and] regarded as sick or perverted" (Halperin, 2012: B17). **Heterosexism**, a belief that heterosexuality is the only legitimate sexual orientation, pervades societal practices, laws, and institutions. For example, 13 percent of Americans—including judges, religious leaders, and politicians—want to undo recently achieved rights like gay marriage (McCarthy, 2016). Heterosexism can trigger **homophobia**, a fear and hatred of lesbians and gay men.

Homophobia often takes the form of *gay bashing:* threats, assaults, or acts of violence directed at LGBTs. Of the nearly 5,900 hate crimes reported to the police in 2015, 18 percent of the victims were LGBT, but much gay bashing isn't reported (see Chapter 7).

In high school, gay, lesbian, and bisexual students are three times more likely than straight students to be raped and skip school more often because they feel unsafe. At least a third have been bullied, and they're twice as likely as their heterosexual counterparts to be threatened or injured with a weapon while on school property. Among transgender people, 41 percent have attempted suicide sometime in their lives—nearly nine times the national average—due primarily to rejection by family and friends, and to harassment and violence at school, at work, and by police (Haas et al., 2014; Kann, Olsen et al., 2016).

Without warning, in mid-2017 President Trump

heterosexism belief that heterosexuality is the only legitimate sexual orientation.

homophobia a fear and hatred of lesbians and gay men.

tweeted that the government "will not accept or allow transgender individuals to serve in any capacity in the U.S. Military." If the president's ban is implemented, 18 countries will still allow openly transgender individuals to be members of their armed forces (LeBlanc, 2017).

9-2 CONTEMPORARY GENDER INEQUALITY

A recent study concluded that "it will take until 2085 for women to reach parity with men in leadership roles in government/politics, business, entrepreneurship, and nonprofit organizations" (Klos, 2013). You saw in Chapter 8 that there's still widespread gender stratification because of sex. *Gendered institutions* are social structures that enable and reinforce gender stratification. Let's begin with the family, remembering that institutions are interrelated (see Chapter 6).

9-2a Gender and Family Life

About 56 percent of married adults—with and without children—say that sharing household chores is "very important" to a successful marriage (Geiger, 2016). Men do more at home than they used to, but not as much as they say. Fathers spend more hours each week in paid work than do mothers, do less child care and housework, and have more leisure time (*Figure 9.2*), and many household chores are still gendered. On average, men are three times more likely to do home maintenance (e.g., repairing cars, lawn care); women are three times more likely to do the cooking, cleaning, and laundry ("American Time Use Survey ...," 2016).

Figure 9.2 How U.S. Parents Spend Their Time

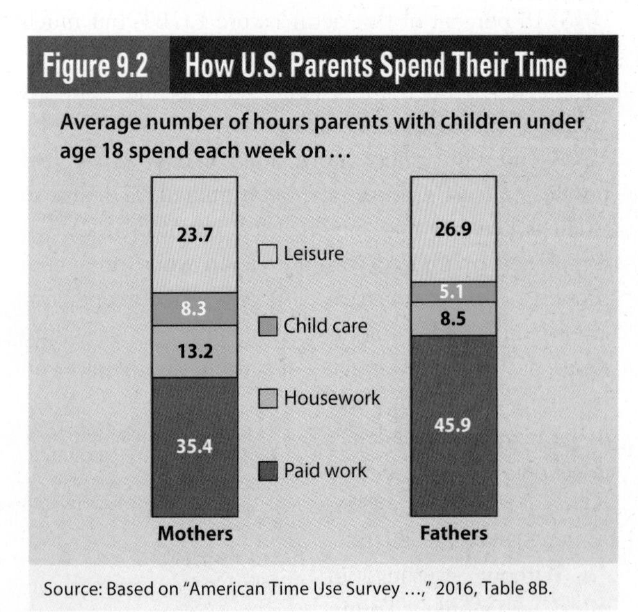

Source: Based on "American Time Use Survey ...," 2016, Table 8B.

For many women, there's nothing sexier than a man who does housework.

Gang Liu/Alamy Stock Photo

About 58 percent of both mothers and fathers say that parenting is "extremely important" to their identity, but parenting tasks are also gendered. Women do less housework than they used to, but devote twice as much time as men to child care. Fathers tend to do more of the enjoyable tasks (e.g., reading to children, playing with them, taking them to games), and are much more likely to "join in" with child care than to "take over" from mothers. In contrast, mothers do most of the daily, nonstop tasks like picking children up from school or day care and feeding, bathing, and putting them to bed. Compared with fathers, mothers' greater investments in children result in less happiness, more stress, and greater fatigue (Craig, 2015; Musick et al., 2016; Parker and Livingston, 2016).

Every year, the media feature and applaud stay-at-home dads, but their numbers are negligible. In 2016, 209,000 fathers (0.2 percent of all fathers) cared for children while their wives worked outside the home. A stay-at-home dad is usually a temporary role that's due to unemployment or health problems. In some cases, however, White, college-educated, upper-middle class men choose to be stay-at-home dads because their partners or wives have high incomes (Kane, 2015; "Father's Day ...," 2017).

9-2b Gender and Education

Despite substantial progress, there are gender differences at all educational levels. In public K–12 schools, as rank and pay increase, the number of women decreases. Among all full-time teachers, 76 percent at the elementary level are women; the number falls to 58 percent in high school. Among principals, the number of women drops from

Table 9.2 As Rank Increases, the Number of Female Faculty Decreases

RANK	PERCENTAGE OF FEMALE FACULTY MEMBERS
Instructor	57
Assistant Professor	50
Associate Professor	44
Professor	31

Note: Of the almost 791,400 full-time faculty in 2013, 45 percent were women.

Source: Based on Snyder et al., 2016, Table 315.20.

64 percent at elementary schools to 30 percent in high schools (Bitterman et al., 2013; Snyder et al., 2016).

Because women across all racial and ethnic groups are more likely than men to finish college, some observers have described this phenomenon as "the feminization of higher education." Even when women earn doctoral degrees in male-dominated STEM (science, technology, engineering, and math) fields, they're less likely than men to be hired (see Chapter 13). Once hired, women are less likely to be promoted. Since 2000, 45 percent of all Ph.D. degree recipients have been women (Snyder and Dillow, 2013), but as the academic rank increases, the number of female faculty decreases (*Table 9.2*). Such data contradict the description of higher education as feminized.

9-2c Gender and the Workplace

There has been progress toward greater workplace equality, but we still have a long way to go. In the United States (as around the world), many jobs are segregated by sex, there are ongoing gender pay gaps, and numerous women experience sexual harassment.

OCCUPATIONAL SEX SEGREGATION

Occupational sex segregation (sometimes called *occupational gender segregation*) is the process of channeling women and men into different types of jobs. As a result, a number of U.S. occupations are filled almost entirely by either women or men. Between 95 and 98 percent of all child care workers, secretaries, dental hygienists, and preschool and kindergarten teachers are women. Between 96 and 99 percent of all pilots, mechanics, plumbers, and firefighters are men. Women have made progress in a number of the higher-paying occupations, but 74 percent of chief executives, 80 percent of software developers, and 90 percent of engineers are men (Bureau of Labor Statistics, 2017).

Marmaduke St. John/Alamy Stock Photo

Of the nearly 3.1 million U.S. teachers at elementary and middle schools, only 22 percent are men. A major reason is low salaries compared with other occupations, but gender stereotypes are also a factor (Rich, 2014; Bureau of Labor Statistics, 2017).

The issue isn't women and men working in different spaces or locations, but that male-dominated occupations usually pay higher wages. And, as in education, women are much less likely than men to move up the occupational ladder (see Chapters 8 and 11).

THE GENDER PAY GAP

On average, full-time, year-round working women earn 80 to 81 cents for every dollar men earn (BLS News Release, 2017). To state this differently, *the average woman must work almost nine extra weeks every year to make the same wages as a man.*

This earnings difference between women and men is the **gender pay gap** (also called the *wage gap*, *pay gap*, and *gender wage gap*). Among year-round full-time workers, Asian women earn 91 percent as much as White men, White women earn 81 cents for every dollar a White man earns, and Latinas and Black women earn about 63 percent as much as White men. Across occupations, the gender pay gap ranges from 59 percent for financial advisors to zero for police patrol officers. Lower wages and salaries reduce women's savings, purchasing power, and quality of life, and they receive less income from Social Security and pensions after retirement (AAUW, 2017; BLS News Release, 2017; BLS Reports, 2017).

occupational sex segregation (sometimes called *occupational gender segregation*) the process of channeling women and men into different types of jobs.

gender pay gap the difference between men's and women's earnings (also called the *wage gap*, *pay gap*, and *gender wage gap*).

Figure 9.3 Gender Pay Gap, by Education, 2017

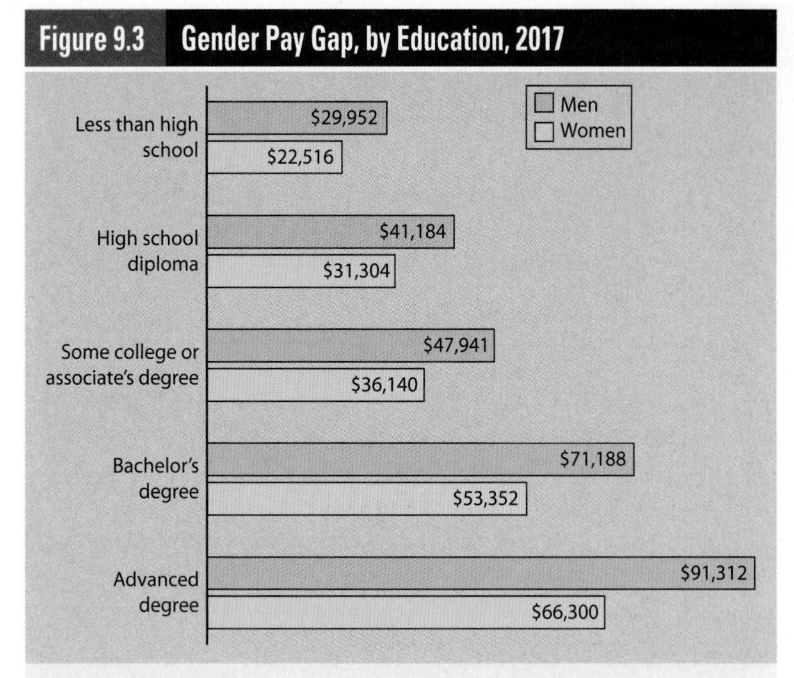

Note: These are median annual earnings of year-round full-time workers age 25 and older.

Source: Based on BLS News Release, "Usual Weekly Earnings...," 2017, Table 5.

Not only do women earn less than men at all educational levels, the higher the education level, the bigger the gender pay gap (*Figure 9.3*). Note that, as a group, women with advanced degrees earn less than men with a college degree.

Why is there a gender pay gap? Women tend to choose fields with lower earnings (e.g., health care and education), whereas men are more likely to major in higher paying fields (e.g., engineering and computer science). Women on average also work fewer hours than men, primarily to care for children or other family members. However, there's a pay gap after controlling for a number of variables—including occupation, hours worked, GPA, age, experience, and marital status. Between ages 25 and 45, the gender pay gap for college graduates, which begins close to zero, widens by 55 percentage points (AAUW, 2013; Blau and Kahn, 2016; Goldin et al., 2017).

A year after they graduate, women with Ph.D.s in science and engineering earn 31 percent less than do men. The pay gap disappears when women receive doctorates in better-paid fields (engineering versus mathematics or chemistry), *and* work in industry rather than government or higher education, *and* don't marry (to avoid moving because of a husband's job), *and* don't have children (Buffington et al., 2016; Goldin et al., 2017).

sexual harassment any unwanted sexual advance, request for sexual favors, or other conduct of a sexual nature that makes a person uncomfortable and interferes with her or his work.

SEXUAL HARASSMENT

Sexual harassment is any unwanted sexual advance, request for sexual favors, or other conduct of a sexual nature that makes a person uncomfortable and interferes with her or his work. It includes *verbal behavior* (e.g., pressure for dates and the threat of rape), *nonverbal behavior* (e.g., indecent gestures, display of sexually explicit posters, photos, or drawings), and *physical contact* (e.g., pinching, touching, or rape).

Sexual harassment occurs at all occupational levels. Female employees at the National Park Service have reported being groped, propositioned, verbally abused, and threatened with retaliation if they refused or reported incidents to supervisors. Female firefighters across the country have found their shampoo bottles filled with urine and semen on their bunks. Millions of female restaurant and home care workers and hotel room attendants are vulnerable to sexual harassment, and 60 percent of high-tech female workers in Silicon Valley have experienced unwanted sexual advances (Vassallo et al., 2015; Nguyen, 2016; Parker, 2016; Schrobsdorff, 2016).

About 25 percent of women have been sexually harassed at work, but 71 percent didn't file complaints: They feared being labeled as "difficult" or "too sensitive," and worried about retaliation and job security. Many employment contracts now have arbitration clauses that prevent workers from suing companies for sexual harassment (Ahn and Ruiz, 2015; Dias and Dockterman, 2016).

9-2d Gender and Politics

Unlike dozens of other countries, the United States has never had a woman serving as president or even vice president. In the U.S. Congress, 81 percent of the members are men. In other important elective offices (governor, mayor, state legislator), only a handful of the decision makers are women (*Table 9.3*). These numbers haven't changed much since the early 1990s.

Women's voting rates in the United States have been higher than men's since 1984. Why, in contrast, are there so few women in political office? There's a combination of reasons: (1) Women run for office at a far lower rate than

Table 9.3 U.S. Women in Elective Offices, 2017		
POLITICAL OFFICE	**TOTAL NUMBER OF OFFICE HOLDERS**	**PERCENTAGE WHO ARE WOMEN**
Senate	100	21
House of Representatives	435	19
Governor	50	10
State Legislator	7,383	25
Attorney General	50	14
Secretary of State	50	26
State Treasurer/ Chief Financial Officer	50	16
Mayor (100 largest cities)	100	20

Source: Based on Center for American Women and Politics, 2017.

men with similar credentials because women don't consider themselves qualified; (2) they have to do more than their male counterparts to prove themselves; (3) U.S. presidents appoint more men than women to important positions; and (4) there's a lingering sexism, among both men and women, that female politicians are both less feminine and compassionate than the average woman, and lack the leadership traits associated with male politicians (e.g., confident, assertive) (Schneider and Bos, 2014; Parker and Horowitz, 2015; see also Chapter 11).

9-3 SEXUALITY

In the movie *Annie Hall*, a therapist asks two lovers how often they have sex. The man rolls his eyes, and complains, "Hardly ever, maybe three times a week!" The woman exclaims, "Constantly, three times a week!" *Sexuality* is considerably more complex than just having sex, however, because it's a product of our sexual identity, sexual orientation and sexual scripts, and includes desire, expression, and behavior.

9-3a Contemporary Sexual Attitudes and Practices

Sex doesn't "just happen." It typically progresses through a series of stages such as approaching, flirting, touching, or asking directly for sex. Sexual attitudes and behavior can vary from situation to situation and change over time, including why we have sex.

WHY WE HAVE SEX

People have sex to reproduce and to experience physical pleasure, but there are other reasons. For example, almost a third of Americans aged 15 to 24 believe it's all right for unmarried 16-year-olds to have sexual intercourse "if they have strong affection for each other." Although the message is contradictory, parents reinforce the association between attraction and sex by telling teenagers "Don't have sex, but use condoms" (Mollborn, 2015; Daugherty and Copen, 2016).

Nationwide, 3 percent of male and 10 percent of female high school students have been physically forced to have unwanted sexual intercourse. Teenagers are also more likely to engage in sex at any early age if they use alcohol or other drugs or experience domestic violence (Kann, McManus et al., 2016). A study of nearly 2,000 college students identified 237 reasons for having sex that ranged from the physical (stress reduction) to the spiritual (to get closer to God) and from the altruistic (to make the other person feel good) to the spiteful (to retaliate against a partner who had cheated) (Meston and Buss, 2007).

SEXUALITY THROUGHOUT THE LIFE COURSE

Contrary to some stereotypes, adolescents aren't sexually promiscuous and older people aren't asexual. On average, Americans have sexual intercourse for the first time at about age 17, but don't marry until their mid-20s. Just 16 percent have had sexual intercourse by age 15, 30 percent by 16, 44 percent by 17, and almost 60 percent by age 18 (Guttmacher Institute, 2016).

Among teenagers aged 15 to 19, the percentage who ever had sexual intercourse declined from 51 percent in 1988 to 45 percent in 2013. By 2008, however, almost half of teens in this age group had had oral but not vaginal sex (Chandra et al., 2011; Martinez the Abma, 2015).

Adolescents who have oral sex prior to vaginal intercourse do so because it's "not really sex." Instead, they see it as a way to delay vaginal intercourse, to maintain one's virginity (especially among those who are religious), and to avoid the risk of pregnancy and STDs (Regnerus and Uecker, 2011; Copen et al., 2012).

By age 44, 93 percent of Americans have had vaginal intercourse, 87 percent have had oral sex, and 39 percent have had anal sex with an opposite-sex partner. Fewer than 4 percent of Americans identify as LGBT, but among

"Don't you have any sexual fantasies that don't involve me cleaning?"

people aged 18 to 44, 17 percent of women and 6 percent of men have had same-sex contact (Copen et al., 2016). Thus, as noted earlier, sexual identity, attraction, and behavior overlap.

A majority of adults ages 45 and older agree that a satisfying sexual relationship is important, but it's not their top priority. Marital sexual frequency may decrease because concerns about earning a living, making a home, and raising a family become more pressing than lovemaking. Others may be going through a divorce, dealing with unemployment, helping to raise grandchildren, or caring for aging parents—all of which sap people's sexual interest (ConsumerReports.org, 2009; Twenge et al., 2017).

As people age, they experience lower levels of sexual desire and some sexual activities, but a third of men and women age 70 and older report having sex at least twice a month. Poor health and inability to find a partner, rather than just advancing age, are more closely linked to declining sexual activity (Lee et al., 2015). Many couples in their seventies and eighties emphasize emotional intimacy and companionship, and are satisfied with kissing, cuddling, and caressing (Heiman et al., 2011; Lodge and Umberson, 2012).

9-3b Sexual Scripts and Double Standards

sexual script specifies the formal and informal norms for acceptable or unacceptable sexual behavior.

sexual double standard a code that permits greater sexual freedom for men than women.

We like to think that our sexual behavior is spontaneous, but all of us have internalized sexual scripts. A **sexual script** specifies the formal and informal norms for acceptable or unacceptable sexual behavior. Social scripts can change over time and across groups, but are highly gendered in two ways—women's increasing hypersexualization and a persistent sexual double standard.

THE "SEXY BABES" TREND

Sexualized social messages are reaching ever younger audiences, teaching or reinforcing the idea that girls and women should be valued for how they look rather than their personalities and abilities. For example, there are "bikini onesies" for infant girls, sexy lingerie for girls 3 months and older, and padded bras for 7- and 8-year-olds (that's right, for 7- and 8-year-olds!).

Many girls are obsessed about their looks, and from an early age, for a variety of reasons, including their mothers' role modeling. By age 9, girls start imitating the clothes, makeup, and behavior of mothers who dress and act in highly sexualized ways (Starr and Ferguson, 2012).

For many girls, constantly seeking "likes" and attention on social media is like being a contestant in a never-ending beauty pageant. The boom in selfie culture has increased girls' sexualization because validation is only a tap away, and "one of the easiest ways to get that validation is by looking hot. Sex sells, whether you're 13 or 35" (Sales, 2016: 26).

Media images also play a large role in girls' hypersexualization. Girls and boys see cheerleaders (with increasingly sexualized routines) on TV far more than they see female basketball players or other athletes. Women are now represented in more diverse TV roles—as doctors, lawyers, and criminal investigators—but they're often sexy ("hot"). Also, top female athletes regularly pose naked or semi-naked for men's magazines.

Who benefits from girls' and women's hypersexualization? Marketers who convince girls (and their parents) that being popular and "sexy" requires the right clothes, makeup, hair style, and accessories, create a young generation of shoppers and consumers who will increase business profits more than ever before (Lamb and Brown, 2007; Levin and Kilbourne, 2009).

THE SEXUAL DOUBLE STANDARD

Some believe that the **sexual double standard**—a code that permits greater sexual freedom for men than women—has faded. Others argue that it persists. Among U.S. adolescents, the higher the number of sexual partners, the greater the boy's popularity. In contrast, girls who have more than eight partners are far less popular than their less-experienced female peers. By age 44, many more men (21 percent) than women (8 percent) report having had at least 15 sex partners. And, over a lifetime, men are more likely than women to have sex outside of marriage (19 percent and 14 percent, respectively) (Kreager and Staff, 2009; Chandra et al., 2011; Drexler, 2012).

Another example of the sexual double standard is *hooking up*—which can mean anything from kissing to sexual intercourse. The prevalence of hooking up has increased only slightly since the late 1980s, but is now more common than dating at many high schools and colleges (Monto and Carey, 2014; Luff et al., 2016).

Hooking up has its advantages. Many men prefer hookups because they're inexpensive compared with dating. For women, hookups offer sex without becoming involved in time-consuming relationships that compete with schoolwork, dealing with boyfriends who become demanding or controlling, and experiencing breakups (Bogle, 2008; Rosin, 2012).

Hooking up also has disadvantages, especially for women, because it reinforces a sexual double standard. For example, men are more likely than women to perform sexual acts that a partner doesn't like; more than twice as many men as women experience an orgasm because the men typically don't satisfy a woman sexually; and women who hook up may get a reputation as "sluts" (England and Thomas, 2009; Armstrong et al., 2010, 2012). About half of women, compared with only 25 percent of men, have regretted having casual sex (Galperin et al., 2013).

Figure 9.4 U.S. Abortion Rates Have Decreased

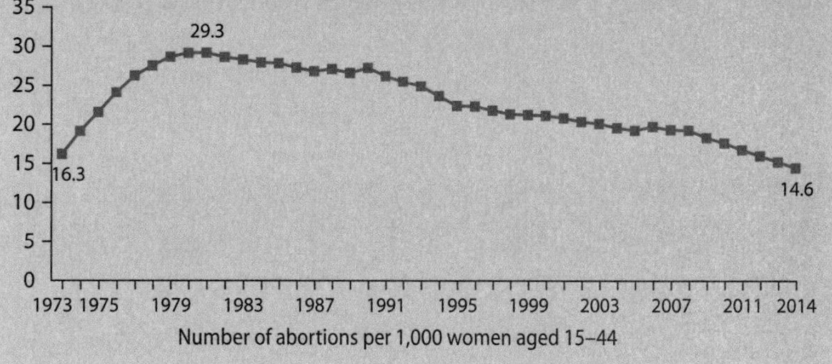

Source: Based on Guttmacher Institute, 2014, and Jones and Jerman, 2017.

CB2/ZOB/WENN.com/Newscom

Galia Slayen built a life-sized Barbie to show what she would look like if she were a real woman. (Slayen used a toy for the head because she wasn't able to create a proportional head) ("Life Size Barbie...," 2011).

9-4 SOME CURRENT SOCIAL ISSUES ABOUT SEXUALITY

Most Americans see sex as a private act, but others believe that the government should control some sexual behavior and decisions. People disagree about social policies on sex-related topics such as teenagers' birth control, prostitution, reproductive technologies (see Chapter 16), and teen pregnancy (see Chapter 12). Two of the most controversial and politically contested issues continue to be abortion and same-sex marriage.

9-4a Abortion

Abortion is the expulsion of an embryo or fetus from the uterus. It can occur naturally—in *spontaneous abortion* (miscarriage)—or be induced medically. Abortion was outlawed in the nineteenth century, but has been legal since the U.S. Supreme Court's *Roe v. Wade* ruling in 1973.

TRENDS

Every year, 40 percent of unintended pregnancies end in abortion. Over a lifetime, 33 percent of women have an abortion by age 45 (Guttmacher Institute, 2014). The *abortion rate*, or the number of abortions per 1,000 women ages 15 to 44, increased during the 1970s, then decreased, and has dropped to its lowest point since 1973 (*Figure 9.4*).

Why have abortion rates decreased? At least half of the states have restricted access to abortion, but most of the decline has been due to

abortion expulsion of an embryo or fetus from the uterus.

Figure 9.5 Who Has Abortions?

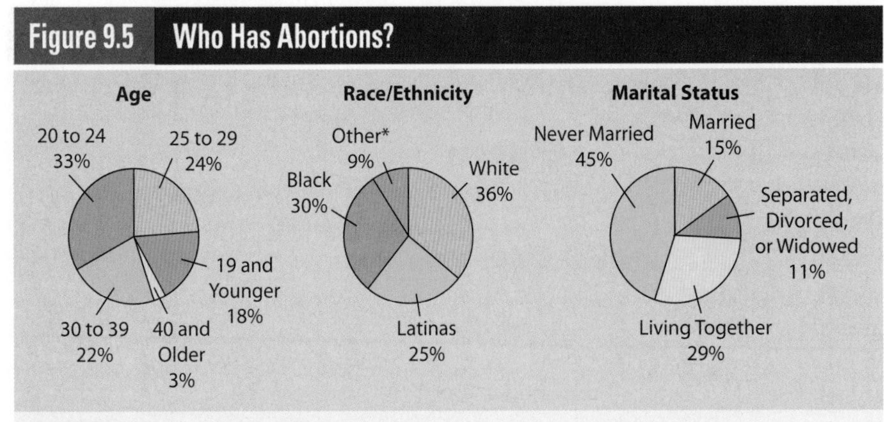

*Other refers to Asian/Pacific Islanders, American Indians, and Alaska Natives.
Sources: Based on Guttmacher Institute, 2013, 2014.

an overall drop in pregnancy rates, delaying childbearing, more effective usage of contraceptives, and greater access to emergency contraception that prevents pregnancy (Finer and Zolna, 2016; Dreweke, 2017).

Abortion is most common among women who are young (in their twenties), White, and never married (*Figure 9.5*). About 20 percent of women who get abortions have at least a college degree, but most are poor: 69 percent have incomes below or near the poverty level, and 74 percent are financially unable to support a baby. Low-income women are also much more likely than higher-income women to have experienced intimate partner violence that included being impregnated against their will. Abortion cuts across all religious groups, but 28 percent of women are Catholic and 37 percent are Protestant (including born-again/evangelical Christians) (Jones et al., 2013; Guttmacher Institute, 2014; Reeves and Venator, 2015).

WHY IS ABORTION CONTROVERSIAL?

Since its legalization, abortion has been one of the most persistently contentious issues in U.S. politics and culture. More Americans describe themselves as "pro-choice" (50 percent) than "pro-life" (44 percent), and 6 percent aren't sure. Both anti- and pro-abortion groups agree on some issues, such as requiring a patient's informed consent, but 28 percent want abortion to be illegal under all circumstances (Fingerhut, 2017).

Antiabortion groups believe that the embryo or fetus isn't just a mass of cells but a human being from the time of conception and, therefore, has a right to life. In contrast, abortion rights advocates point out that, at the moment of conception, the organism lacks a brain and other specifically and uniquely human attributes, such as consciousness and reasoning, and that a pregnant woman—not legislators—should decide whether or not to bear children.

Antiabortion groups maintain that abortion is immoral and endangers a woman's physical, mental, and emotional health. Whether abortion is immoral is a religious and philosophical question. On a physical level, a legal abortion in the first trimester (up to 12 weeks) is safer than driving a car, playing football, motorcycling, getting a penicillin shot, or continuing a pregnancy. There's also no evidence that having an abortion increases the risk of breast cancer or causes infertility (Sheppard, 2013; Pazol et al., 2014; Holloway, 2015).

What about mental and emotional health? Antiabortion activists argue that abortion leads to postabortion stress disorders, depression, and even suicide. National studies have consistently found that abortion poses no hazard to an adolescent or adult woman's mental health, doesn't increase emotional problems like depression or low self-esteem, and doesn't lead to drug or alcohol abuse or suicide. An unwanted pregnancy or being denied an abortion, not abortion, increases the risk of mental health problems (Academy of Medical Royal Colleges, 2011; Steinberg and Finer, 2011; Biggs et al., 2017).

A large majority (63 percent) of Americans want to keep abortion legal, but it's almost impossible for many women to get legal abortions in 89 percent of all U.S. counties. Since 2010, states have enacted 338 new abortion restrictions that cut public funding for low-income women, passed licensing requirements that closed abortion clinics, and limited access to medication abortion ("abortion pills") that legally ends a pregnancy in the first nine weeks. On the other hand, federal and state funding support thousands of "crisis pregnancy centers," usually next to abortion clinics, which distribute false medical information (e.g., abortion causes breast cancer, infertility, and suicide) and pressure women to continue an unwanted pregnancy (Daniels et al., 2016; Upadhyay, 2016; Gold and Nash, 2017; Guttmacher Institute, 2017).

Table 9.4 Why Do Americans Favor or Oppose Same-Sex Marriages?

What do you think? What other reasons can you add for each side of the debate?

Same-sex marriage should be legal because . . .	Same-sex marriage should be illegal because . . .
• Gay marriages strengthen families and long-term unions that already exist. Children are better off with parents who are legally married.	• Children need a mom and a dad, not two dads or two moms.
• There are no scientific studies showing that children raised by gay and lesbian parents are worse off than those raised by heterosexual parents.	• There are no scientific studies showing that children raised by gay and lesbian parents are better off than those raised by heterosexual parents.
• Every person should be able to marry someone that she or he loves.	• People can love each other without getting married.
• Gay marriages are good for the economy because they boost businesses such as restaurants, bakeries, hotels, airlines, and florists.	• What's good for the economy isn't necessarily good for society, especially its moral values and religious beliefs.

Sources: Bennett and Ellison, 2010; Olson, 2010; Sullivan, 2011; Sprigg, 2011; Whitehead, 2011; Bogage, 2015; Kaufman, 2015.

9-4b Same-Sex Marriage

Same-sex marriage (also called *gay marriage*) is a legally recognized marriage between two people of the same biological sex and/or gender identity. Although still controversial, same-sex marriage is becoming more acceptable in the United States and some other countries.

TRENDS

In 2015, the U.S. Supreme Court issued a landmark ruling (*Obergefell v. Hodges*) granting same-sex couples a constitutional right to marry. The 5-4 decision gave gay couples nationwide the same legal rights and benefits as heterosexual couples. With the Supreme Court's decision, the United States joined 21 other countries (so far) that allow same-sex marriage.

Prior to the ruling, gay marriage was illegal in 13 states. Some of these states' lawmakers urged their constituents to accept the new law. Others pressed their residents to "stand and fight by seeking a constitutional amendment banning gay marriage" (de Vogue and Diamond, 2015).

Opponents often invoke religion to defy gay marriage laws. For example, Roy Moore, chief justice of the Alabama Supreme Court, ordered the state's 68 probate judges to refuse to issue marriage licenses to same-sex couples. He defended his decision as "standing up for God" because "God ordained marriage as the union of one man and one woman." Alabama's judiciary suspended Judge Moore. The following year, a Democrat narrowly defeated Moore in an election for a U.S. Senate seat (Robertson, 2016; Cason, 2017). We'll examine same-sex marriages and families in Chapter 12, but why is gay marriage such a contentious issue?

WHY IS SAME-SEX MARRIAGE CONTROVERSIAL?

A large majority (64 percent) of Americans support same-sex marriage (up from 37 percent in 2006). Most of the opposition comes from people who are Republican, White, male, regularly attend religious services, live in the South, are 55 and older, and have conservative views on family issues (McCarthy, 2017; Masci et al., 2017).

Those who favor same-sex marriage argue that people should have the same rights regardless of sexual orientation. Those who oppose same-sex marriage contend that such unions are immoral, weaken traditional notions of marriage, and are contrary to religious beliefs. *Table 9.4* summarizes some of the major pro and con arguments in this ongoing debate.

9-5 GENDER AND SEXUALITY ACROSS CULTURES

There's considerable variation worldwide regarding gender inequality and sexual oppression. Such variations show that our behavior is learned, not innate.

same-sex marriage (also called *gay marriage*) a legally recognized marriage between two people of the same biological sex and/or gender identity.

9-5a Gender Inequality

Creatista/Shutterstock.com

A recent United Nations (2015) report concluded that women continue to face discrimination in access to work, economic assets, and participation in private and public decision making. They're also more likely than men to live in poverty, to be illiterate, and to experience violence.

In many countries, women's progress toward equality has been mixed. For example, Saudi Arabia, one of the wealthiest countries in the world, and which has some of the most educated women in the world (including STEM college and advanced degrees), ranks near the bottom in women's economic and political participation. In *all countries and regions*, the greatest gender gaps are in economic participation and political leadership (World Economic Forum, 2016).

ECONOMIC PARTICIPATION

Worldwide, 150 countries have at least one law that treats women and men differently, and 63 countries have five or more. The laws make it difficult for women to own property, open bank accounts, start businesses, and enter certain professions (World Bank, 2017).

Globally, about 75 percent of working-age men participate in the labor force, compared with 50 percent of working-age women, and women earn 24 percent less than men. In 85 percent of countries, women with advanced degrees have higher unemployment rates than men with similar levels of education (United Nations, 2015).

Countries that have closed education gaps and have high levels of women's economic participation—the Scandinavian countries, United States, Canada, New Zealand, and Australia—have strong economic growth (Worley, 2014). However, gender gaps still persist in senior positions, wages, and leadership. For example, Germany is Europe's No. 1 economy, but also has one of the largest pay gaps in the European Union. Of 191 executives on the management boards of Germany's 30 biggest companies, only 12 are women, a 20 percent decrease from a year before (de Pommereau, 2013; Webb, 2013).

POLITICAL LEADERSHIP

Worldwide, only 23 percent of national legislators are women. Rwanda has 64 percent, followed by nine countries where women hold 40 to 46 percent of the high-level political positions. The power is usually short-lived, however. Of 146 nations, only 56 (38 percent) have had a female head of government or state for at least one year in the past half-century. In 31 of these countries, women typically led for five years or less (Geiger and Kent, 2017; World Bank, 2017).

Of the 197 world leaders who are presidents or prime ministers, only 13 percent are women. Worldwide, women occupy only 22 percent of the positions in decision-making bodies. Of 193 countries, the United States ranks 101st in women's political leadership, well below many African, European, and Asian countries, and even below most of the Arab countries that many Westerners view as repressing women (Inter-Parliamentary Union, 2017).

9-5b Sexual Inequality

Globally, women have fewer rights and opportunities than men. There's been more acceptance of gay and lesbian sexuality in some countries, but heterosexism prevails.

VIOLENCE AGAINST FEMALES

Violence against women is a persistent problem. Worldwide, 35 percent of women have endured physical and/or sexual violence by an intimate partner or another male (World Health Organization, 2016). The rates are much higher in many countries. For example,

- In Afghanistan and some African countries, about 80 percent of girls—some as young as 8 years old—are forced into marriages; 87 percent of Afghan women have experienced physical, psychological, or sexual abuse (Peter, 2012; "Child Brides . . . ," 2014).
- In Pakistan, 90 percent of women undergo domestic violence in their lifetimes. As many as 5,000 females are victims of "honor killings" every year. An *honor killing* is the murder of a family member, almost always a female, who is considered to have shamed the family by being a rape victim or has been suspected of engaging in premarital or extramarital sex (Sahgal and Townsend, 2014; "Human Rights Violations," 2015).
- In the Democratic Republic of Congo, approximately 1,100 women are raped every day by soldiers, strangers, and intimate partners ("Human Rights Violations," 2015).
- In India, rape and gang rape are epidemic, but less than a quarter of reported crimes end in conviction. As many as 100,000 women a year are killed over

dowry disputes (the money or goods that a wife brings to her husband at marriage) (Harris, 2013; "Ending the Shame . . . ," 2013).

Female genital mutilation/cutting (FGM/C) is a partial or total removal of the female external genitalia. Most of the more than 200 million girls and women who have undergone FGM/C live in 29 African countries, Indonesia, and the Middle East. The mutilation occurs between 3 and 12 years old. The operator is typically an elderly village woman who uses a knife or other sharp object and doesn't administer an anesthetic. Countries justify FGM/C on the grounds that it controls a girl's sexual desires and preserves her virginity, a prerequisite for marriage. Although Nigeria, Egypt, and other countries have outlawed FGM/C, the practice remains widespread and widely accepted ("Female Genital Cutting," 2016; UNICEF, 2016).

VIOLENCE AGAINST MALES

You saw earlier that Americans are more accepting of gay and lesbian sexuality, and that more nations are legalizing gay marriage. In contrast, many countries in Asia, Africa, and the Middle East don't tolerate LGBTs. A vast majority of Africans (e.g., 98 percent in Ghana, 93 percent in Uganda, 88 percent in Kenya) say same-gender sexual behavior is unacceptable. Same-gender sexual behavior is illegal in 78 nations, including 34 of Africa's 54 countries. Gay men, particularly, may be legally tortured, stoned, imprisoned, or killed ("Deadly Intolerance," 2014; "Global Views on Morality," 2014; Pflanz, 2014; Baker, 2015).

Jean-Marc Bouju/Impact/HIP/The Image Works

Little girls like this one scream and writhe in pain during FGM/C (see text). Complications include hemorrhaging to death, a rupture that causes continual dribbling of urine or feces for the rest of the woman's life, severe pain during sexual intercourse, and death during childbirth if the baby can't emerge through the mutilated organs.

Russia's parliament recently banned LGBT relationships and forbade distributing material on gay rights. Russians are more accepting of extramarital affairs, gambling, and drinking alcohol (a major cause of men's death before age 55) than same-gender sexual behavior (Council for Global Equity, 2014; Poushter, 2014).

There are about 10 million transgender people in Asia and the Pacific. Governments in Bangladesh, India, Nepal, and Pakistan have recognized transgender people as a legal category that has rights, but many—often labelled "mentally ill" by the public—experience discrimination and violence. In China, the attackers are often the victim's relatives. In Fiji, 40 percent of *trans women* (male-to-female) have been raped. In Australia, 60 percent of *trans men* (female-to-male) suffer abuse from their partners ("Knife-edge Lives," 2016).

9-6 SOCIOLOGICAL EXPLANATIONS OF GENDER AND SEXUALITY

Gender and sexuality affect all people's lives, but why is there so much variation over time and across cultural groups? The four sociological perspectives answer this and other questions somewhat differently (*Table 9.5* summarizes these theories).

9-6a Functionalism

Functionalists view women and men as having distinct roles that ensure a family's and society's survival. These roles help society operate smoothly, and have an impact on the types of work that people do.

DIVISION OF GENDER ROLES AND HUMAN CAPITAL

Some of the most influential functionalist theories, developed during the 1950s, proposed that gender roles differ because women and men have distinct roles and responsibilities. A man (typically a husband and father) plays an *instrumental role* of economic provider; he's competitive and works hard. A woman (typically a wife and mother) plays an *expressive role*; she provides the emotional nurturance that sustains the family unit and supports the father/husband (Parsons and Bales, 1955; Betcher and Pollack, 1993).

Instrumental and expressive roles are complementary, and each person knows what's expected: If the house is clean, she's a "good wife"; if the bills are paid, he's a "good husband." The duties are specialized, but

Table 9.5 Sociological Explanations of Gender and Sexuality		
THEORETICAL PERSPECTIVE	**LEVEL OF ANALYSIS**	**KEY POINTS**
Functionalist	Macro	• Gender roles are complementary, equally important for a society's survival, and affect human capital. • Agreed-on sexual norms contribute to a society's order and stability.
Conflict	Macro	• Gender roles give men power to control women's lives. • Most societies regulate women's, but not men's, sexual behavior.
Feminist	Macro and micro	• Women's inequality reflects their historical and current domination by men, especially in the workplace. • Many men use violence—including sexual harassment, rape, and global sex trafficking—to control women's sexuality.
Symbolic Interactionist	Micro	• Gender is a social construction that emerges and is reinforced through everyday interactions. • The social construction of sexuality varies across cultures because of societal norms and values.

both roles are equally important in meeting a family's needs and ensuring a society's survival.

Such traditional gender roles help explain occupational sex segregation because people differ in the amount of human capital that they bring to the labor market. *Human capital* is the array of competencies—including education, job training, skills, and experience—that have economic value and increase productivity.

Arthur Greenberg/Alamy Stock Photo

From a functionalist perspective, what individuals earn is the result of the choices they make and, consequently, the human capital that they accumulate to meet labor market demands. Women diminish their human capital because they choose lower paying occupations (social work rather than computer science), as well as postpone or leave the workforce for childbearing and child care. When they return to work, women have lower earnings than men because, even in higher paying occupations, their human capital has deteriorated or become obsolete (Kemp, 1994).

WHY IS SEXUALITY IMPORTANT?

For functionalists, sexuality is critical for reproduction, but people should limit sex to marriage and forming families. Functionalists view sex outside of marriage as dysfunctional because most unmarried fathers don't support their children. The offspring often experience poverty and a variety of emotional, behavioral, and academic problems (Avellar and Smock, 2005).

You might be tempted to dismiss the functionalist view of limiting sex to marriage as outdated. Worldwide, however, sex outside of marriage is prohibited and *arranged marriages*—in which parents or relatives choose their children's future mates—are the norm. Most children agree to arranged marriages because of social custom and out of respect for their parents' wishes. The matches solidify relationships with other families and ensure that the woman's sexual behavior will be confined to her husband, avoiding any doubt about the offspring's parentage (see Benokraitis, 2015).

Some functionalists encourage marrying during one's mid-to-late twenties instead of delaying marriage. The benefits include enjoying more frequent sex, having an easier time getting pregnant, and being able to have more than one child than people who marry in their thirties or later (Wilcox, 2015).

CRITICAL EVALUATION

Critics fault functionalist gender role perspectives on three counts. First, even during the 1950s, White middle-class male sociologists ignored almost a third of the labor force that was composed of working-class, immigrant, and minority women who played *both* instrumental and expressive roles. Second, functionalists tend to overlook the fact that many people don't have a choice of playing only instrumental or expressive roles because most families rely on two incomes for economic survival. Third, the human capital model assumes that women have lower earnings than men because they "choose" lower paying occupations. As you saw earlier, however, there's a gender pay gap across *all* occupations, even those that require advanced degrees.

Functionalists tend to reject sexual relationships outside of marriage. Compared with married couples, for example, those who cohabit have poorer quality relationships and lower happiness levels. As you'll see in Chapter 12, however, marriage doesn't guarantee long or happy relationships.

Antigay discrimination, which is legal in 28 states, is dysfunctional. Companies don't attract young and talented LGBT workers, and forbidding transgender people to use a bathroom that corresponds to their gender identity has sparked considerable interpersonal and group conflict (Green, 2016; McCarthy, 2017).

9-6b Conflict Theory

For conflict theorists, gender inequality is built into the social structure. In both developing and industrialized countries, men control most of a society's resources and dominate women. Like functionalists, conflict theorists see sexuality as a key component of a society's organization, but they view sexuality as reflecting and perpetuating sexism and discrimination.

CAPITALISM AND GENDER INEQUALITY

Conflict theorists maintain that capitalism, not complementary roles, explains gender roles and men's social and economic advantages. Women's inequality is largely due to economic exploitation—both as underpaid workers in the labor force and unpaid domestic workers who care for children and aging family members. In effect, gender roles are profitable for business. Companies can require their male employees to work long hours or make numerous business trips and not worry about workers demanding payment for child care services that can cost up to $63,000 a year (Carey and Trap, 2014).

Women comprise 51 percent of the U.S. population, but Congress "looks much like the face of corporate America—overwhelmingly wealthy, white, and male" (Weathers, 2015). Women's underrepresentation in political institutions can help explain why occupational sex segregation, gender pay gaps, and sexual harassment persist.

IS GENDER INEQUALITY LINKED TO SEXUAL INEQUALITY?

From a conflict perspective, gender inequality gives men economic, political, and/or interpersonal power to control or dominate women's sexual lives. Most of the domestic violence and rape victims, in the United States and around the world, are women and girls. In workplace sexual harassment cases, the offender is typically a male supervisor. For example, 40 percent of women in the fast food industry have experienced sexual harassment by a manager, owner, or supervisor at least once a month. In prostitution and sex trafficking, almost all of the victims worldwide are poor women and girls (U.S. Department of State, 2016; Restaurant Opportunities Center United, 2017).

Particularly in the Middle East and some African countries, men dictate how women should dress and whether they can travel, work, receive health care, attend school, or start a business; dismiss women's charges of sexual assaults (including gang rapes); and blame girls for child rape because they're "seducing" older men (Neelakantan, 2006). Many women have internalized such sexism. For example, 25 to 38 percent of women in some African countries, Egypt, Palestine, and Indonesia believe that wives sometimes deserve beatings (e.g., for leaving the house without a husband's permission) (Kaneda and Bietsch, 2015; "The State of Arab Men," 2017).

CRITICAL EVALUATION

Critics point out several limitations of conflict theory. First, women aren't as powerless as some conflict theorists claim. Like men, women often barter to increase their economic and political power. Second, conflict theory often ignores women's exploitation of other women. In the United States, women comprise 19 percent of those involved in the sex trafficking industry. The thriving underground sex economy relies heavily on

A group of men recently surrounded a woman at a busy bus station in Nairobi, Kenya. They violently tore off her clothes because she was wearing a miniskirt, accused her of being a jezebel, and left her naked on the street. Bystanders watched. A week later, about 1,000 women (some dressed in miniskirts) protested the rising violence against girls and women, vowing to wear whatever clothes they want. Some men joined the protestors; others threw rocks or shouted lewd comments.

AP Images/Ben Curtis

nannies, secretaries, and escort services and brothels owned or operated by females (Dank et al., 2014).

Third, capitalism can discourage sexism. For years, Fox News and Bill O'Reilly, the station's popular political commentator, paid up to $13 million to settle several sexual harassment lawsuits. Fox fired O'Reilly only after more than 70 corporations dropped their ads—largely in response to social media campaigns—because the companies "couldn't afford to alienate women and their considerable purchasing power" (Abbey-Lambertz, 2017; Chira, 2017: B7).

9-6c Feminist Theories

Feminist scholars agree with conflict theorists that gender stratification benefits men and capitalism, but emphasize that women's subordination also includes their daily vulnerability to male violence (Katz, 2006). Feminist scholars, more than any other group of theorists, are especially concerned about men's controlling women's sexual lives.

LIVING IN A GENDERED WORLD

Men comprise 48 percent of the U.S. population age 18 and older, but they account for 98 percent of directors of the 700 top-grossing films since 2007, 97 percent of the televised news media's and ESPN's sports broadcasts, 96 percent of Fortune 500 CEOs, 83 percent of the largest private law firm partners, 81 percent of Congress, 73 percent of college/university presidents, 74 percent of federal and state judges, and 73 percent of generals and admirals (Cooky et al., 2015; Smith et al., 2015; Johnson, 2016; American Bar Association, 2017; Kidder et al., 2017; Merelli, 2017). In many of these sectors, women's representation has decreased or remained about the same for at least a decade.

Women often experience harsher sanctions than men for workplace offenses. In the financial services industry, for instance, fraud and forgery are punished by demotion or dismissal. Compared with female advisers, males engage in almost three times more fraud and forgery, and are more likely to be repeat offenders. Nonetheless, women are 20 percent more likely to be fired, and 30 percent less likely to find new jobs (Egan et al., 2017).

Sexism is prevalent in almost all workplaces and online environments. Social media has amplified feminist voices, but also silenced them. Once a writer is singled out by an anti-feminist men's group, she's deluged with hateful and threatening messages, both public and private. Males with the fewest skills and lowest status are the most likely to be threatened by and hostile toward an influx of talented women. Regardless of the reasons, some women no longer participate online because of the harassment (Goldberg, 2015; Kasumovic and Kuznekoff, 2015).

Feminist theorists emphasize that gender, race, and social class intersect to form a hierarchical stratification system that shapes people's experiences and behavior. Privileged women have less status than privileged men, but upper-class White men *and* women subordinate lower-class women *and* minority men (Andersen and Collins, 2010). Understanding the interconnections between gender, race/ethnicity, and social class provides a more comprehensive picture of living in a gendered world than does any single variable.

SEXUALITY, SOCIAL CONTROL, AND COMMERCIALIZING SEX

Men assert their power and control—across cultures and over time—through rape, intimate partner violence, sexual harassment, exploiting women through prostitution and pornography, forcing women to marry against their will, honor killings, FGM/C, and punishing women, but not men, for seeking a divorce or committing adultery (World Health Organization, 2013).

Sex is big business. Men's testosterone levels decline naturally with aging, but are also due to obesity and alcohol. Pharmaceutical companies "have seized on the decline in testosterone levels as pathological and applicable to every man." As a result, sales of testosterone-boosting drugs are surging even though they increase the risk of heart attacks (La Puma, 2014: A21; O'Connor, 2014).

In 2015, the Food and Drug Administration approved the drug flibanserin, marketed as Addyi, to increase premenopausal women's sexual desire. Some women's groups lobbied for the "female Viagra," but a recent study found that flibanserin resulted in "one-half additional satisfying sexual event per month." The medical researchers didn't define "one-half" of a "satisfying sexual event," but concluded that the drug significantly increased the risk of dizziness, sleepiness, nausea, and fatigue. They recommended an "integrative approach" that includes medical, psychiatric, psychological, and couple-relationship treatment (Jaspers et al., 2016: 457, 461).

Pornography, "sex drugs," and the earlier discussion of the "sexy babes" trend are just a few examples of the increasing *commercialization of sex*, making sexuality a commodity that can be sold for financial gain. Commercializing sex demeans both women and men, but the "products" are usually women.

CRITICAL EVALUATION

Feminist explanations are limited for several reasons. First, they pay little attention to parenting problems that many men face (try to find a diaper-changing station in the men's restroom, for example, or take time off from work to care for a sick child). Second, feminist analyses are inclusive, but this strength can also be a weakness. Regarding the gender pay gap, for example, should race, ethnicity, or social class be given priority in implementing change? Third, are feminist scholars overstating women's underrepresentation in some sectors? Since 2010, for instance, both cable and network shows have featured women in strong leading roles (e.g., *Girls, How to Get Away with Murder, Scandal, Madam Secretary, Nurse Jackie*) (O'Keefe, 2014; Blay, 2015). And, like conflict theorists, feminist scholars are sometimes accused of glossing over women's exploitation of others. In Iraq, for example, sex traffickers are often women who target the youngest girls because virgins bring the highest prices (Naili, 2011).

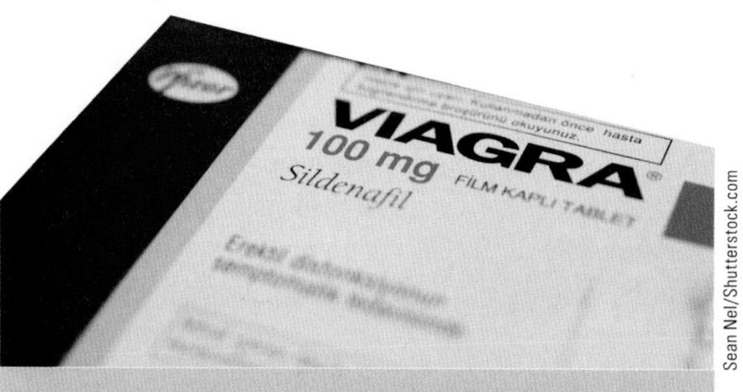

Sean Nel/Shutterstock.com

Many sexually healthy people in their 20s to 60s now take drugs like Viagra and Addyi. Are the drugs an example of commercializing sex?

9-6d Symbolic Interaction

Whereas functionalist, conflict, and some feminist theories are macro level, symbolic interactionists focus on the everyday processes that produce and reinforce gender roles. We "do" gender, sometimes consciously and sometimes unconsciously, by adjusting our behavior and our perceptions depending on the sex of the person with whom we're interacting (West and Zimmerman, 2009). Our sexual expression, similarly, isn't inborn but a product of socialization, and what families and other societal groups deem as appropriate and inappropriate behavior (Hubbard, 1990).

GENDER IS A SOCIAL CONSTRUCTION

For interactionists, gender is a social creation, and we learn gender roles through everyday social interaction. For example, when teachers tell girls and boys that both are equally capable in math and science, "the difference in performance essentially disappears" (Hill et al., 2010: 2). When, on the other hand, teachers discourage girls from pursuing math and science, girls fare worse than boys on exams, enroll in fewer advanced math and science courses in high school and college, and, consequently, choose lower-paying careers (Lavy and Sand, 2015). Thus, teachers' gender bias can affect occupational choices and earnings in adulthood.

Believing in gender differences can actually *produce* differences. Because men are self-confident about their worth, they're four times more likely than women to ask for a raise. When women do ask for a raise, they ask for 30 percent less than do men. Men in higher-level positions who aren't promoted often threaten to quit and are offered retention bonuses to stay. Women typically decide to work harder and try again next year (Lipman, 2015).

SEXUALITY IS ALSO A SOCIAL CONSTRUCTION

For interactionists, sexuality is also socially constructed. As with gender roles, we *learn* to be sexual and to express our sexuality differently over time and across groups because the people around us affect our attitudes and behavior. For example, students who attend conservative

Activists supporting the "kiss of love" campaign in New Delhi, India.

AP Images/Altaf Qadri

religious high schools are less likely than their public school counterparts to report same-sex attraction or to identify as LGBT in adolescence or young adulthood (Wilkinson and Pearson, 2013).

In many Middle Eastern countries, men have premarital sex but don't marry women who aren't virgins (Fleishman and Hassan, 2009). In the United States, as you saw earlier, men who have casual sex are "studs," whereas women are "sluts." Thus, sexual double standards are socially constructed.

In China, a university that planned to publicly shame students who engaged in "uncivilized behavior" (e.g., hugging or kissing in public) withdrew the policy after widespread protests. In much of India, people may be beaten for kissing on the street or even holding hands. To protest such moral policing by right-wing groups, two college students started a "Kiss of Love" campaign on Facebook. Despite arrests, hundreds of people joined the demonstration. Similar protests quickly spread to other cities (Ming, 2011; Bhardwaj, 2014).

CRITICAL EVALUATION

A common criticism is that interactionists ignore the social structures that create, maintain, or change gender roles and gender inequality. Many 18- to 32-year-olds plan to share earning and household/child care responsibilities equally with their future partners. An equal division of labor is unlikely, however, because current workplace and government policies don't support women's and men's balancing work and family life. Since 2001, more than 250,000 female soldiers have served as drivers, as pilots, and in other combat roles in Iraq and Afghanistan. It was only in 2015, however, that the Pentagon ended the formal ban on women in combat jobs, an important criterion for career advancement. Thus, people don't have as much ability to shape their lives as interactionists claim.

Interactionists emphasize that language, erotic images, and other symbols evoke sexual interest or desire, but they neglect the relationship between biological factors and sexual orientation. Thus, interactionism doesn't explain why siblings, even identical twins—who are socialized similarly—may have different sexual orientations. A third limitation is that interactionists don't explain why, historically and currently, women around the world are considerably more likely than men to be controlled and sexually exploited. Such analyses require macro-level analyses that examine religious, political, and economic institutions.

STUDY TOOLS 9

READY TO STUDY? IN THE BOOK, YOU CAN:

- ☐ Check your understanding of what you've read with the Test Your Learning Questions provided on the Chapter Review Card at the back of the book.
- ☐ Tear out the Chapter Review Card for a handy summary of the chapter and key terms.

ONLINE AT CENGAGEBRAIN.COM WITHIN MINDTAP YOU CAN:

- ☐ Explore: Develop your sociological imagination by considering the experiences of others. Make critical decisions and evaluate the data that shape this social experience.
- ☐ Analyze: Critically examine your basic assumptions and compare your views on social phenomena to those of your classmates and other MindTap users. Assess your ability to draw connections between social data and theoretical concepts.
- ☐ Create: Produce a video demonstrating connections between your own life and larger sociological concepts.
- ☐ Collaborate: Join your classmates to create a capstone project.

10 Race and Ethnicity

Henryk T. Kaiser/Getty Images

LEARNING OBJECTIVES

After studying this chapter, you will be able to...

10-1 Describe how and explain why U.S. racial and ethnic diversity has changed.

10-2 Define and give examples of race, ethnicity, and racial-ethnic group.

10-3 Show how immigration patterns have changed, and describe Americans' reactions to legal and undocumented immigrants.

10-4 Distinguish between dominant and minority groups, and describe the most common patterns of dominant-minority group relations.

10-5 Describe and illustrate the most common sources of racial-ethnic friction.

10-6 Compare the five major minority groups in terms of origins, social class, and achievements.

10-7 Compare and evaluate the theoretical explanations of race and ethnicity.

10-8 Describe how and explain why interracial and interethnic relationships are changing.

After you finish this chapter go to **PAGE 203** for **STUDY TOOLS**

In 2014, Coca-Cola aired an ad during the Super Bowl that portrayed U.S. ethnic diversity by featuring "America the Beautiful" sung in several languages. After the ad, Twitter lit up with criticism. Some of the commenters said that it's disrespectful to sing "America the Beautiful" in any language other than English, and many disparaged immigrants who don't learn English. Others pointed out that the song isn't our national anthem, perhaps it should've been sung in one of the Native American tongues that were here before Europeans arrived, and that we don't have an official national language (Sahgal, 2014).

WHAT DO YOU THINK?

The United States is a melting pot.

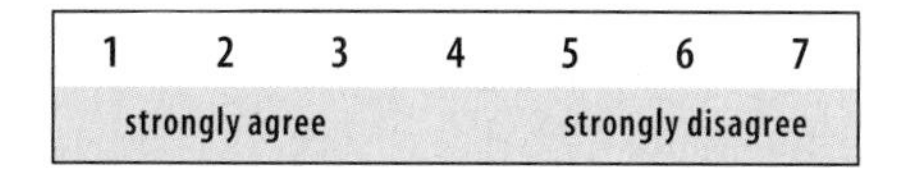

Such discourse illustrates our diversity and divisions. This chapter examines the impact of race and ethnicity on our lives, why racial-ethnic inequality is still widespread, and the growth of interracial and interethnic relationships. First, however, take the True or False quiz to see how much you know about these topics.

True or False?

HOW MUCH DO YOU KNOW ABOUT U.S. RACIAL AND ETHNIC GROUPS?

1. Race is determined biologically.
2. The United States has the most foreign-born residents of any country in the world.
3. People who are prejudiced also discriminate.
4. Among Asian Americans, the most economically successful are Asian Indians.
5. About 85 percent of Americans report being only one race.
6. Minority groups are small in number.

The answer to #4 is true; the others are false. You'll see why as you read this chapter.

10-1 U.S. RACIAL AND ETHNIC DIVERSITY

The United States is the most multicultural country in the world, a magnet that draws people from hundreds of nations and is home to millions of Americans who are bilingual or multilingual. Of the nearly 326 million U.S. population, 14 percent are foreign born (up from only 5 percent in 1965), and is expected to reach a historic 19 percent (78.2 million people) in 2065. America's multicultural umbrella includes at least 150 ethnic or racial groups and 350 languages (U.S. Census Bureau News, 2015; López and Radford, 2017).

By 2025, only 58 percent of the U.S. population is projected to be White—down from 86 percent in 1950 (*Figure 10.1*). By 2044, White people may make up less than half of the total population because Latinos and Asians are expected to double in size, and persons who identify themselves as two or more races will more than triple (Frey, 2014).

10-2 THE SOCIAL SIGNIFICANCE OF RACE AND ETHNICITY

All of us identify with some groups in terms of sex, age, social class, and other factors. Two of the most common and important sources of self-identification, as well as labeling by others, are race and ethnicity.

10-2a Race

A **racial group** refers to people who share visible physical characteristics, such as skin color and facial features, that members of

racial group people who share visible physical characteristics that members of a society consider socially important.

Figure 10.1 Racial and Ethnic Composition of the U.S. Population, 1950–2025

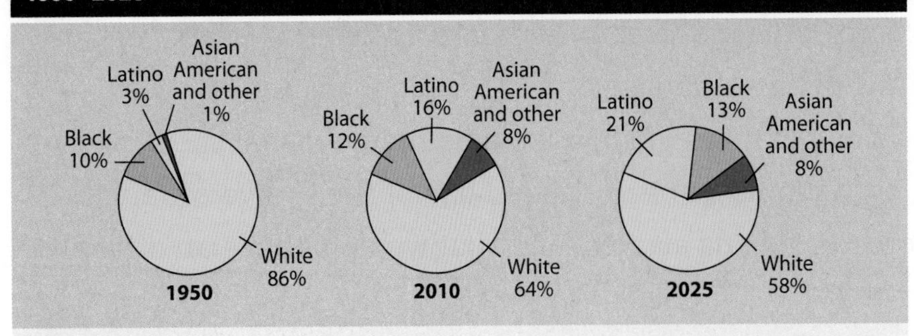

Note: "Asian American and other" includes American/Indian/Alaskan Native, Native Hawaiians and Pacific Islanders, some other race, and those who identify themselves with two or more races.

Sources: U.S. Census Bureau, 2008; Passel et al., 2011.

a society consider socially important. Contrary to the popular belief that race is determined biologically, it's a *social construction*, a societal invention that labels people based on physical appearance, social class, or other characteristics (see Daniel, 2014).

Possibly only six of the human body's estimated 35,000 genes determine the color of a person's skin. Because all human beings carry 99.9 percent of the same genetic material (DNA), the "racial" genes that make us look different are miniscule compared with the genes that make us similar (Graves, 2001; Pittz, 2005).

If our DNA is practically identical, why are we so obsessed with race? People react to the physical characteristics of others, and those reactions have consequences. Skin color, hair texture, and eye shape, for example, are easily observed and mark groups for unequal treatment. As long as we act on the basis of these characteristics, our life experiences will differ in access to jobs and other resources, how we treat people, and how they treat us.

10-2b Ethnicity

An **ethnic group** (from the Greek word *ethnos*, meaning "nation") refers to people who identify with a common national origin or cultural heritage. Cultural heritage includes language, geographic roots, food, customs, traditions, and religion. Ethnic groups in the United States include Puerto Ricans, Chinese, Serbs, Arabs, Swedes, Hungarians, Jews, and many others. Like race, ethnicity can be a basis for unequal treatment, as you'll see shortly.

ethnic group people who identify with a common national origin or cultural heritage.

racial-ethnic group people who have distinctive physical and cultural characteristics.

10-2c Racial-Ethnic Group

People who have distinctive physical and cultural characteristics are a **racial-ethnic group**. Some people use the terms *racial* and *ethnic* interchangeably, but remember that *race* refers to physical characteristics with which we're born, whereas *ethnicity* refers to cultural characteristics that we learn. The term *racial-ethnic* includes both physical and cultural traits.

Describing racial-ethnic groups has become more complex because the U.S. government allows people to identify themselves in terms of both race and ethnicity. In the 2000 and 2010 censuses, for example, Latinos could check off "Black" or "White" for race and "Cuban" for ethnic origin. Such choices generate dozens of racial-ethnic categories. People prefer some "labels" to others, and racial and ethnic self-identification varies.

10-2d What We Call Ourselves

In 1976, the U.S. government began to use the word *Hispanic* or *Latino* to categorize Americans who trace their roots to Spanish-speaking countries. About 70 percent say that it doesn't matter if they're referred to as *Latino* or *Hispanic*. If given a choice, however, 51 percent prefer to identify themselves by the family's

Jupiterimages/Stockbyte/Getty Images

There's much variation in skin color across and within groups. People of African descent have at least 35 different shades of skin tone (Taylor, 2003). So, can you determine someone's race simply by looking at her or him?

country of origin (e.g., Mexican, Cuban, Salvadoran) (Pew Hispanic Center, 2012; Jones, 2013).

Since 1900, the Census Bureau has used *Negro*, *Black* and *African American*. Currently, 65 percent of this group say it doesn't matter whether they're called *Black or African American*, and this trend hasn't changed much since 1991 (Jones, 2013). Many people, including African American scholars, use *Black* and *African American* interchangeably.

We see similar variations in the usage of *Native American* and *American Indian*. These groups prefer their tribal identities (e.g., Cherokee, Apache, Lumbi) to being lumped together under a single term. Among White Americans, in contrast, recent immigrants rarely refer to themselves by their country of origin (e.g., "Armenian" or "Armenian American"). Third and later generations may even be unsure of their ancestors' roots ("My dad's parents were Polish, but I don't know much about my mother's side").

JStaley401/Getty Images

10-3 OUR CHANGING IMMIGRATION MOSAIC

The United States, a nation of immigrants, has historically and currently both welcomed immigration and feared its consequences. What fuels public debate is the rising number of foreign-born people in the United States.

10-3a The Foreign-Born Population

The Census Bureau's definition of *foreign born* includes naturalized citizens; lawful permanent residents (immigrants); temporary migrants (e.g., foreign students); refugees; and unauthorized migrants (e.g., Mexicans crossing the U.S. border) ("Foreign Born," 2016). The United States has one of the highest foreign-born populations in the world, but there's been a significant shift in their country of origin. In 1900, almost 85 percent of the foreign born came from Europe compared with only 11 percent in 2015 (*Figure 10.2*). Today, the foreign born come primarily from Asia (mainly China and the Philippines) and Latin America (mainly Mexico) (López and Radford, 2017).

The United States admits more than 1 million immigrants every year—more than any other nation. A major change has been the rise of *undocumented* (also called *unauthorized* and *illegal*) *immigrants*—from 180,000 in the early 1980s to 11.1 million in 2015. They make up 26 percent of all foreign-born residents, and nearly 4 percent of the nation's population. Almost 53 percent are from Mexico, 21 percent from Central and Latin America, 13 percent from Asia, and 13 percent from other countries, including Canada, Europe, and Africa (Passel and Cohn, 2016). Most immigrants, legal and undocumented, come to the United States for the same reasons as in the past—religious and political freedom, economic and educational opportunities, and escape from wars and natural disasters (see Chapter 15).

Americans have usually opposed admitting large numbers of *refugees*, people who flee their country to escape war, persecution, or death. Since 1980, however, the United States has resettled about 3 million refugees, more than any other country. People from the Democratic Republic of Congo, Syria, Myanmar, Iraq, and Somalia accounted for 71 percent of all refugees admitted in 2016. Since 2002, 46 percent of all refugees have been Christian and 32 percent have been Muslim (Connor, 2016; Krogstad and Radford, 2017; Zong and Batalova, 2017).

Figure 10.2 Origins of U.S. Foreign-Born Population: 1900 and 2015

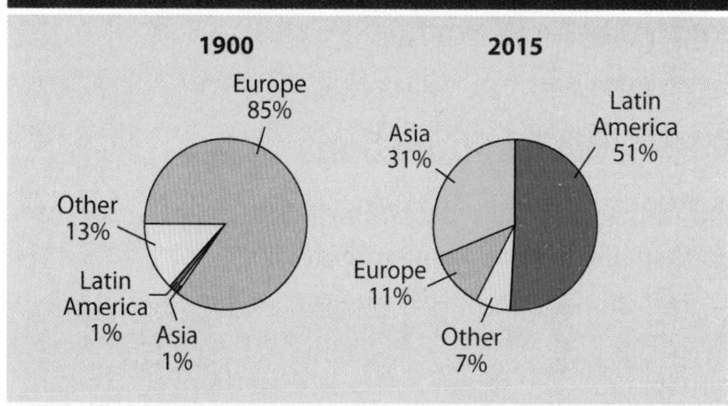

Note: Latin America includes the Caribbean, Central America (including Mexico), and South America.

Sources: Based on U.S. Department of Commerce, 1993, and U.S. Census Bureau, 2015 American Community Survey, Tables SO503–SO506. Accessed June 1, 2017 (factfinder.census.gov).

10-3b Americans' Reactions to Immigrants

American attitudes toward immigrants have grown more positive since the mid-1990s, but are still mixed: 63 percent say immigrants strengthen the country because of their hard work and talents; 27 percent believe that immigrants are a burden because "they take our jobs, housing, and health care." Americans are more divided about immigration levels: 49 percent are satisfied with current levels or believe they should be increased, but 49 percent want to reduce the stream of immigrants. On both issues—whether immigrants are more of a benefit or burden and immigration levels—Republicans, conservatives, older White people, and White working classes have more negative views than do Democrats, liberals, millennials, college-educated, and non-White people (Jones, 2016; Gates, 2017; López and Bialik, 2017).

Tim Boyle/Bloomberg/Getty Images

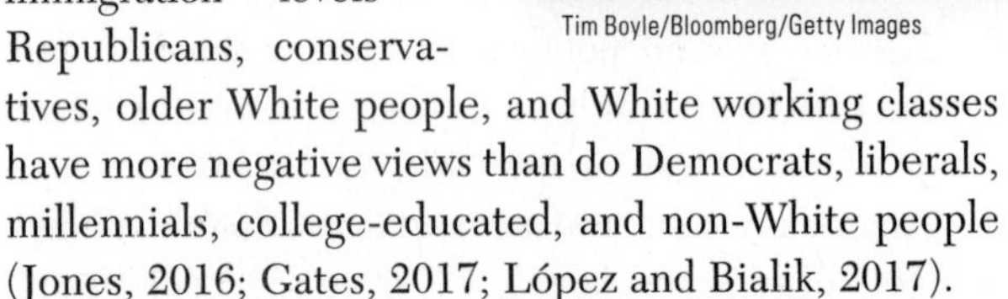

Immigrants have made significant contributions. Since the inception of the prestigious Nobel Prize in 1901, more than 100 of the 911 winners in medicine, chemistry, physics, and economics have been U.S. immigrants. Between 1880 and 2000, and although they earned considerably lower wages, immigrants generated more inventions and patents than their native-born counterparts. The inventions paved the way for long-term technological innovations in fields like computer science, engineering, and medicine. Foreign-born Americans have founded or co-founded about 25 percent of U.S. tech companies (including Intel, Yahoo, and Google) and more than 52 percent of the Silicon Valley start-ups between 1995 and 2005 (Najim, 2016; Agress, 2017; Akcigit et al., 2017).

10-3c Two Big Questions

Do immigrants take native-born Americans' jobs? And do they burden government budgets?

Of the nation's 161 million people in the labor force, 83 percent are U.S.-born, 15 percent are lawful immigrants, and 5 percent are unauthorized. Regardless of legal status, immigrants work in a variety of jobs and don't make up the majority of workers in any industry, but employment patterns differ markedly. Lawful immigrants are most likely to be in professional, management, or business and finance jobs (37 percent) or service jobs (22 percent). Unauthorized immigrants, by contrast, are most likely to be in service (32 percent) or construction jobs (16 percent) (López and Bialik, 2017).

About 45 percent of Americans believe the growing number of immigrants has hurt workers overall. Except for high school dropouts who have few skills that employers seek, there's little evidence that immigration has had a negative impact on native-born workers' wages or employment levels. Whether legal or undocumented, unskilled immigrants are more willing than their native-born counterparts to take grueling but low-paying jobs in service occupations (e.g., child care, maids, cooks), agriculture, and the seafood industry. "The State of American Jobs," 2016; Borjas, 2017; Gee et al., 2017).

Do immigrants burden government budgets? The answer is mixed. First-generation immigrants generally cost state and local governments about $57 billion a year, mainly because of health care and education services. The second generation, better educated and with higher wages and salaries, adds about $30 billion a year to government finances. The third generation pays almost $224 billion a year in taxes. Undocumented immigrants pay approximately $12 billion a year in state and local taxes—about 8 percent of their income compared with just 5 percent that the wealthiest 1 percent pays. Each immigrant, legal or undocumented, generates 1.2 local jobs because businesses need more people to meet the increased demand for goods and services (Hong and McLaren, 2015; Blau and Mackie, 2016; Gee et al., 2017).

On a number of measures, immigrants and their children—both legal and undocumented—outperform native-born Americans. Children of immigrants are more likely than their U.S.-born peers to live in a two-parent household; to have a parent with a secure job; to earn a college degree; to attend religious services regularly; and to be "prodigious job creators." They're also less likely than their native-born counterparts to be obese, disabled, or to suffer from chronic illnesses; to commit crimes; and to have nonmarital children (Ewing et al., 2015; Waters and Pineau, 2015; Landgrave and Nowrasteh, 2017; Livingston, 2016).

Design Pics Inc/Alamy Stock Photo

In Colorado, Iowa, and other states, refugees and immigrants from Asia, Africa, Mexico, and Central America fill $15-an-hour jobs that native-born workers spurn. According to a local resident who worked at a pork processing plant for nearly 40 years, "Even if pay were raised to $20 or $25 an hour, I don't think you could get White guys" (Cohen, 2017: A12; Etter and Singh, 2017).

10-4 DOMINANT AND MINORITY GROUPS

Race and ethnicity often make people feel like outsiders, or "the other." Otherness is being different or having characteristics that set you apart from the dominant group (Thorpe-Moscon and Pollack, 2014).

10-4a What Is a Dominant Group?

A **dominant group** is any physically or culturally distinctive group that has the most economic and political power, the greatest privileges, and the highest social status. As a result, it can treat other groups as subordinate. In most societies, for instance, men are a dominant group because they have more status, resources, and power than women (see Chapters 8 and 9).

Dominant groups aren't necessarily the largest in number. From the seventeenth century until 1994, about 10 percent of South Africans were White and had almost complete control of the Black population. Because of *apartheid*, a formal system of racial segregation, the Black residents couldn't vote, lost their property, and had minimal access to education and politics. Apartheid ended in 1994, but most Black South Africans are still a minority because White people "hold the best jobs, live in the most expensive homes, and control the bulk of the country's capital" (Murphy, 2004: A4).

10-4b What Is a Minority?

Sociologists describe Latinos, African Americans, Asian Americans, Middle Eastern Americans, and Native Americans as minorities. A **minority** is any group that may be treated differently and unequally because of their physical, cultural, or other characteristics. The characteristics include gender, age, sexual orientation, religion, ethnicity, or skin color. Minorities may be larger in number than a dominant group, but they have less power, privilege, and social status. For example, middle-class Black people are much more likely than their White counterparts to be called by debt collectors even though both groups have similar debt levels and repayment rates (Ruetschlin and Asante-Muhammad, 2013). *Table 10.1* offers other examples of everyday privileges associated with skin color.

10-4c Some Patterns of Dominant-Minority Group Relations

To understand some of the complexity of dominant-minority group relations, think of a continuum. At one end of the continuum is genocide; at the other end is pluralism (*Figure 10.3*).

GENOCIDE

Genocide is the systematic effort to kill all members of a particular ethnic, religious, political, racial, or national group. By 1710, the colonists in America had killed thousands of Native Americans, poisoned others, and promoted scalp bounties. In 1851, the governor of California officially called for the extermination of all Native Americans in the state (de las Casas, 1992;

dominant group a physically or culturally distinctive group that has the most economic and political power, the greatest privileges, and the highest social status.

minority people who may be treated differently and unequally because of their physical, cultural, or other characteristics.

genocide the systematic effort to kill all members of a particular ethnic, religious, political, racial, or national group.

Table 10.1 Am I Privileged?

Most White people don't feel privileged because they aren't wealthy. Nonetheless, they enjoy everyday benefits, and take them for granted, simply because they're members of the dominant group. Would you add other advantages of being White?

1. I can go shopping and feel fairly sure that I won't be followed or harassed by store detectives.
2. If a traffic cop pulls me over, I can be sure that I haven't been singled out because of my race.
3. I can be late to a meeting without having the lateness reflect on my race.
4. I can turn on the television and see people who look like me represented in positive ways and in a wide range of roles.
5. I live in a safe neighborhood with good schools.
6. I never think twice about calling the police when trouble occurs.
7. I can wear a hoodie and not worry about others thinking that I'm up to no good.
8. I don't get dirty looks if I listen to loud music at a gas station.

Sources: McIntosh, 1995; Independent Television Service, 2003; Williams, 2014.

Churchill, 1997). Worldwide, and between 1915 and 1995 alone, well over 74 million people have been victims of genocide in Africa, Cambodia, China, Eastern Europe, and Turkey (United Human Rights Council, 2004).

SEGREGATION

Segregation is the physical and social separation of dominant and minority groups. In 1954, the Supreme Court ruling in *Brown v. Board of Education* declared *de jure*, or legal, segregation unconstitutional. A variety of federal laws then prohibited racial segregation in public schools, as well as discrimination in employment, voting, and housing.

De facto, or informal, segregation has replaced *de jure* segregation in the United States and many other countries. Some *de facto* segregation may be voluntary, as when minorities prefer to live among their own racial or ethnic group. In most cases, however, *de facto* segregation is due to discrimination. In the case of equally qualified homeseekers, for instance, realtors sometimes show minorities fewer homes and apartments than they do White people, restricting their housing options, including access to neighborhoods with higher-performing schools, public playgrounds, and low crime rates (Turner et al., 2013; see also Chapter 15 on residential segregation).

segregation physical and social separation of dominant and minority groups.

acculturation the process of adopting the language, values, beliefs, and other characteristics of the host culture.

ACCULTURATION AND ASSIMILATION

Many minority group members blend into U.S. society through **acculturation**, the process of adopting the language, values, beliefs, and other characteristics of the host culture (e.g., learning English, celebrating Thanksgiving). Acculturation doesn't include intermarriage, but the newcomers merge into the host culture in

Figure 10.3 Continuum of Some Dominant-Minority Group Relations

Moviestore collection Ltd/Alamy Stock Photo

Based on true events, *Hidden Figures* depicts the lives of three Black women who helped pioneer space travel and desegregation at NASA.

most other ways. **Assimilation** involves conforming to the dominant group's culture, adopting its language and values, and intermarrying with that group. Journalists often use *assimilation* and *acculturation* interchangeably, but the former absorbs, rather than just changes, a group culturally.

PLURALISM

In **pluralism**, sometimes called *multiculturalism*, minority groups maintain many aspects of their original culture—including using their own language and marrying within their own racial or ethnic group—while living peacefully with the host culture. Pluralism is especially evident in urban areas with large racial and ethnic communities (e.g., "Little Italy," "Greek Town," "Little Korea," "Spanish Harlem"). U.S. minorities have numerous ethnic newspapers and radio stations, and the same constitutional rights (such as freedom of speech) as the dominant group. Nonetheless, people of various skin colors and cultures don't always experience the same social standing, and there can be considerable racial-ethnic friction.

10-5 SOME SOURCES OF RACIAL-ETHNIC FRICTION

During a recent trip to Switzerland, Oprah Winfrey asked a clerk at an expensive shop to show her a $37,000 purse. The clerk refused: "That one will cost too much, you won't be able to afford that." Winfrey didn't know for sure whether or not the snub was motivated by racism, but said that the incident was what "people with black or brown skin experience every day" (Goyette, 2013). Ahmed Mohamed, a 14-year-old Sudanese-American student in Texas, was arrested and suspended after teachers mistook his homemade digital clock for a bomb. Would people have reacted differently if Winfrey and Mohamed had been White? And were the reactions racist?

10-5a Racism

Racism refers to beliefs that one's own racial group is inherently superior to other groups. Using this definition, anyone can be racist if she or he believes that another group is inferior. It's a way of thinking about racial and ethnic differences that justifies and preserves the social, economic, and political interests of dominant groups.

Many White people believe that racism has been pretty much "solved." After all, we've elected a Black president twice, have

assimilation conforming to the dominant group's culture, adopting its language and values, and intermarrying with that group.

pluralism minority groups maintain many aspects of their original culture while living peacefully with the host culture.

racism beliefs that one's own racial group is inherently superior to other groups.

AP Images

The Washington Post/Getty Images

A number of colleges and professional teams have replaced their logos, nicknames, and mascots because many Native Americans have denounced them as racist. Some exceptions are the Cleveland Indians in baseball (left) and the Washington Redskins in football (right). Why do many people charge that the mascots and team names are demeaning? Why are the Minnesota Vikings and University of Notre Dame's Fighting Irish acceptable?

a Latina Supreme Court justice, and minorities now own many businesses. In fact, racism is widespread. A few recent examples—and all involving young White men— include the murder of nine African Americans in a Black church in South Carolina, and, at several universities, racist slurs and videos about lynching Black people (King, 2015; Schmidt, 2015). Racism fuels both prejudice and discrimination.

10-5b Prejudice

Prejudice is an *attitude* that prejudges people, usually in a negative way, who are different from "us" in race, ethnicity, religion, or other ways. If an employer assumes, for example, that White workers will be more productive than Black or Latino workers, she or he is prejudiced. Prejudice isn't one-sided because *anyone* can be prejudiced ("White people can't be trusted" or "Black women can't afford expensive purses"). Prejudice is most evident in stereotypes and scapegoating.

A **stereotype** is an oversimplified or exaggerated generalization about a group of people (see, for example, www.stuffwhitepeoplelike.com). Stereotypes can be positive ("Black people are great athletes") or negative ("Black people are violent"). Whether positive or negative, stereotypes distort reality. Some Black people are great athletes and some are violent, just like people in other groups. Whatever the intention, stereotypes lump people together and reinforce the belief that many traits are biological and fixed. Once established, stereotypes are difficult to change because people often dismiss any evidence to the contrary as an exception ("For an Asian, Jeremy Lin is a terrific basketball player").

Stereotypes can lead to a displacement of anger and aggression on **scapegoats**, individuals or groups whom people blame for their own problems or shortcomings ("They didn't hire me because the company wants Black people" or "I didn't get into that college because Asians are at the top of the list"). Minorities are easy targets because they typically differ in physical appearance and are usually too powerless to strike back.

At one time or another, almost all newcomers to the United States have been scapegoats. Especially in times of economic hardship, the most recent immigrants often become scapegoats ("Latinos are replacing Americans in construction jobs"). Prejudice, stereotypes, and scapegoating are attitudes, but they often lead to discrimination.

Pedalist/Shutterstock.com

prejudice an attitude that prejudges people, usually in a negative way.

stereotype an oversimplified or exaggerated generalization about a group of people.

scapegoats individuals or groups whom people blame for their own problems or shortcomings.

discrimination behavior that treats people unequally because of some characteristic.

individual discrimination unequal treatment on a one-to-one basis.

institutional discrimination unequal treatment because of a society's everyday laws, policies, practices, and customs.

10-5c Discrimination

Discrimination is *behavior* that treats people unequally because of some characteristic (e.g., race, ethnicity, gender, age, religion, sexual orientation). It encompasses all sorts of actions, ranging from social slights (e.g., not inviting minority coworkers to lunch) to rejection of job applications and racially motivated hate crimes. Discrimination can be subtle (e.g., not sitting next to someone) or blatant (e.g., racial slurs), and it occurs at individual and institutional levels.

Individual discrimination is unequal treatment on a one-to-one basis, usually by a dominant group member against someone in a minority group. Nationally, 60 percent of Black people, 52 percent of Latinos, and 30 percent of White people report experiencing discrimination because of their race or ethnicity. A recent study of Uber and Lyft drivers in Boston and Seattle found that Black people had to wait as much as 35 percent longer for some rides, and that drivers were more than twice as likely to cancel the rides of passengers with Black-sounding names than of passengers with White-sounding names (Ge et al., 2016; "On Views of Race...," 2016).

In **institutional discrimination** (also called *institutionalized discrimination*, *systemic discrimination*, and *structural discrimination*), minority group members experience unequal treatment because of a society's everyday laws, policies, practices, and customs. For example, sickle cell disease (SCD) affects three times as many Americans as cystic fibrosis (CF), but government spending on CF is four times higher than SCD. Why? About 90 percent of SCD patients are Black, whereas CF affects mostly White people. The CF community has more wealthy and powerful parents and advocates who pressure government to do research and provide treatment, and develop and fund private foundations, which, in turn, underwrite pharmaceutical companies to create drugs, even though a new drug only works for a handful of patients (Butler, 2015). Thus, although individuals may be unprejudiced, institutional policies and processes contribute to racially unequal access to health care.

Discrimination, both individual and institutional, also occurs *within* racial-ethnic groups. A study of New York City's Korean-owned nail salons found an "ethnic caste system." Among Asian manicurists, Koreans earned twice as much as their non-Korean counterparts. The latter frequently worked longer hours, were often forced into less desirable salons that had fewer customers and paltry tips, were told not to chat during their entire 12-hour shifts, and were assigned male customers that manicurists often dread because of the men's "thick toenails and hair-covered knuckles" (Nir, 2015).

10-5d Relationship Between Prejudice and Discrimination

Sociologist Robert Merton (1949) created a model showing how the relationship between prejudice and discrimination can vary. His model includes four types of people and their possible response patterns (*Table 10.2*).

Unprejudiced nondiscriminators aren't prejudiced and don't discriminate. They believe in the American creed of freedom and equality for all and cherish egalitarian values. They may not do much, however, individually or collectively, to change discrimination. In contrast, but equally consistent in attitude and action, are *prejudiced discriminators* who are both prejudiced and discriminate. They're willing to defy laws, such as not renting to minorities, because of their beliefs.

Unprejudiced discriminators aren't prejudiced but discriminate because it's expedient or in their own

"You look like this sketch of someone who's thinking about committing a crime."

David Sipress/The New Yorker Collection/The Cartoon Bank

self-interest to do so. If, for example, an insurance company charges higher automobile insurance premiums to people with low occupational and educational levels (often minorities), agents will implement these policies even though they themselves aren't prejudiced.

Prejudiced nondiscriminators are prejudiced, but don't discriminate. Despite their negative attitudes, they hire minorities and are civil in everyday interactions because they believe they must conform to antidiscrimination laws or situational norms. If, for example, most of their neighbors or coworkers don't discriminate, prejudiced nondiscriminators will go along with them.

Table 10.2 Relationship Between Prejudice and Discrimination

IS THE PERSON PREJUDICED?	DOES THE PERSON DISCRIMINATE? Yes	No
Yes	Prejudiced discriminator (e.g., a prejudiced person who attacks minority group members verbally or physically)	Prejudiced nondiscriminator (e.g., a prejudiced person who goes along with equal employment opportunity policies)
No	Unprejudiced discriminator (e.g., an unprejudiced person who joins a club that excludes minorities)	Unprejudiced nondiscriminator (e.g., an unprejudiced employer who hires minorities)

Source: Based on Merton, 1949.

10-6 MAJOR U.S. RACIAL AND ETHNIC GROUPS

Four states (California, Hawaii, New Mexico, and Texas) and the District of Columbia are now "majority-minority," meaning that minorities make up more than half of the population ("Millennials Outnumber...," 2015). Of the major racial-ethnic groups, some encounter more barriers than others, but all have numerous strengths that enhance U.S. society. *Figure 10.4* shows the racial-ethnic makeup of the U.S. population. Let's begin with White ethnic groups whose ancestors are from Europe.

10-6a White Americans: A Declining Majority

During the seventeenth century, English immigrants settled the first colonies in Massachusetts and Virginia. Other White Anglo-Saxon Protestants (WASPs), who included people from Wales and Scotland, quickly followed. Most of these groups spoke English. Some of the immigrants were affluent, but many were poor or had criminal backgrounds.

Figure 10.4 Major U.S. Racial and Ethnic Groups

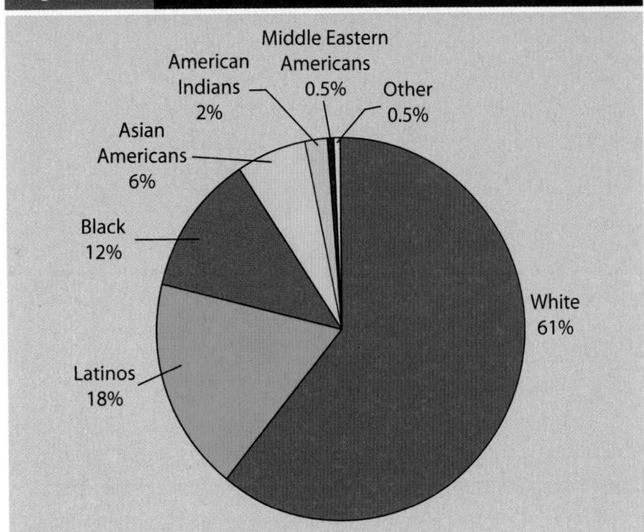

Note: The data are for a single race or ethnicity.

Source: Based on U.S. Census Bureau, 2015 American Community Survey, Table CP05. Accessed June 14, 2017 (factfinder.census.gov), and "American Indian and Alaska Native...," 2016.

DIVERSITY

About 53 percent of Americans (170 million people) identify their ancestry as European. The largest groups have ancestors from Germany, Ireland, England, Italy, Poland, France, and the Scandinavian countries (*Figure 10.5*).

CHARACTERISTICS AND CHANGES

WASPs generally looked down on later waves of immigrants from southern and eastern Europe. They viewed the newcomers as inferior, dirty, lazy, and uncivilized because they difered in language, religion, and customs. New England, which was 90 percent Protestant, was particularly hostile to Irish Catholics, characterizing them as irresponsible and shiftless (Feagin and Feagin, 2008).

All the later waves of European immigrants faced varying degrees of hardship in adjusting to the new land because the first English settlers had a great deal of power in shaping economic and educational institutions. In response to prejudice and discrimination, many of the immigrants founded churches, schools, and recreational activities that maintained their language and traditions (Myers, 2007).

Despite stereotypes, prejudice, and discrimination, European immigrants began to prosper within a few generations. They surmounted numerous obstacles and became influential in all sectors. Overall, they now fare much better financially than most other groups. This doesn't mean that all are rich. In fact, in absolute numbers, poor White people outnumber those of other racial-ethnic groups (see Chapter 8).

10-6b Latinos: The Largest Minority

Latinos are the youngest and largest racial or ethnic group, constituting 18 percent of the nation's population. The size and growth of the Latino population this century is due mainly to births in the United States, not recent immigration. On average, Latinas have three births each, one more than White, Black, and Asian or Pacific Islander women. Since the Great Recession in 2007, however, Latino immigration and birth rates have declined ("Hispanic Heritage Month...," 2016; Patten, 2016; Stepler and Lopez, 2016).

DIVERSITY

Worldwide, only Mexico has a larger Latino population (120 million) than the United States (57 million). More persons of Puerto Rican origin now live in the 50 states and the District of Columbia than in Puerto Rico (Cohn et al., 2014; "Hispanic Heritage Month...," 2016).

Some Latinos trace their roots to the Spanish and Mexican settlers who established homes and founded

Figure 10.5 European Americans by Origin, 2014

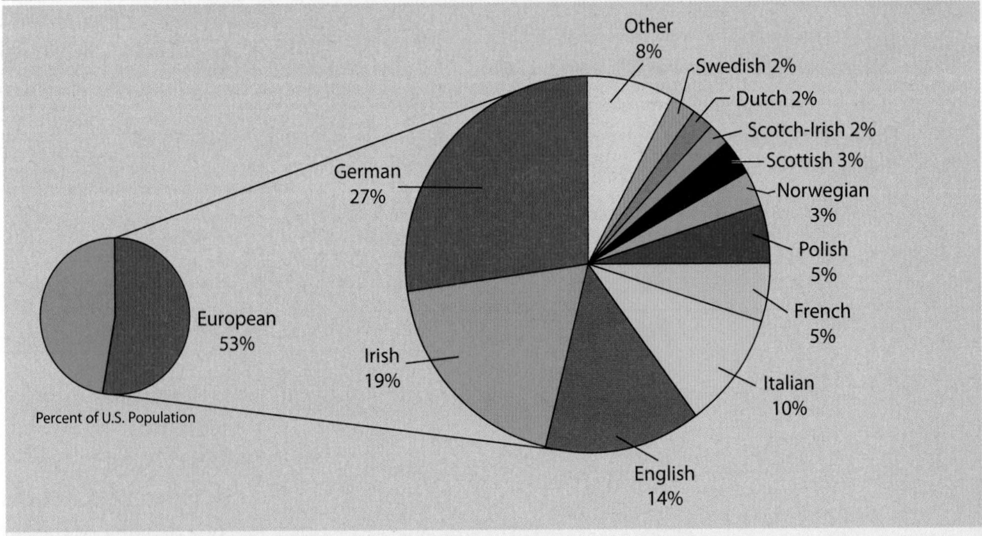

European 53%
Percent of U.S. Population
Other 8%
Swedish 2%
Dutch 2%
Scotch-Irish 2%
Scottish 3%
Norwegian 3%
Polish 5%
French 5%
Italian 10%
English 14%
Irish 19%
German 27%

Source: Based on U.S. Census Bureau, Selected Social Characteristics in the United States, 2015 American Community Survey, Table DP02. Accessed June 1, 2017 (factfinder.census.gov).

Jeffrey Macmillan for U.S. News & World Report

Dominican-born Alfredo Rodriguez is one of numerous successful Latino businessmen who are rebuilding neglected inner-city neighborhoods. In 1985, Rodriguez bought his first grocery store in Queens, New York, with the $25 a week his mother had been setting aside for him for a decade. In 2002, he purchased a supermarket in Newark, New Jersey, to meet the needs of local Latino shoppers. Five years later, his Xtra Supermarket had annual sales of $9 million (Rayasam, 2007).

cities in the Southwest before the arrival of the first English settlers on the East Coast. Others are recent immigrants or children of the immigrants who arrived in large numbers at the beginning of the twentieth century. Of the Latinos living in the United States, most are from Mexico (*Figure 10.6*), but Spanish-speaking people from different countries vary widely in their customs, cuisines, and cultural practices.

CHARACTERISTICS AND CHANGES

About 21 percent of Americans age 5 and older speak a language other than English at home. In this group, 73 percent speak Spanish. The number of U.S. residents who speak Spanish at home increased by 120 percent since 1990, whereas the number who speak Italian, French, German, and other Indo-European languages has declined (Ryan, 2013; "Hispanic Heritage Month…," 2016).

Latino median household income is lower than the national median, but higher than that of American Indian and Black households (*Figure 10.7*). About 40 percent of Latino families earn $50,000 a year or more, up considerably from only 7 percent in 1972, but 19 percent live below the poverty line (Semega et al., 2017).

Latinos have among the lowest education levels (*Figure 10.8*), but there's considerable variation across subgroups. Almost half of Venezuelans have a bachelor's degree or higher compared with only 8 to 9 percent of those from Guatemala, Mexico, and Salvador (Ogunwole et al., 2012).

As with other groups, Latinos' socioeconomic status reflects a number of interrelated factors, particularly education, occupation, English language proficiency, and recency of immigration. For example, 21 percent of second-generation Latinos have at least a college degree compared with 11 percent of foreign-born Latinos, and higher median household incomes—$48,400 and $34,600, respectively ("Second-Generation Americans," 2013).

Despite numerous economic, legal, and social barriers, 78 percent of second-generation Latinos, compared with 58 percent of all U.S. adults, believe that most people can get ahead if they're willing to work hard ("Second-Generation Americans," 2013).

Figure 10.6 Latinos by Origin, 2014

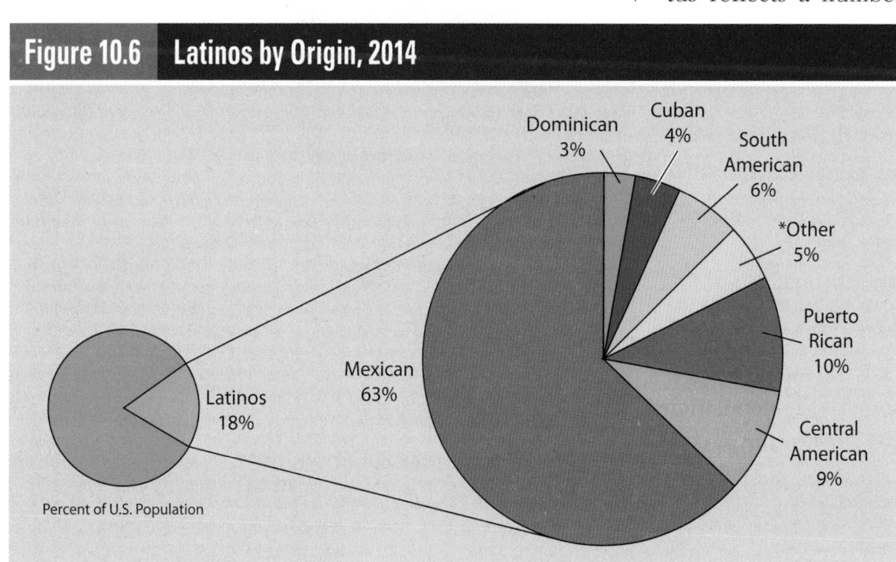

Note: Central American includes countries such as El Salvador, Honduras, and Guatemala; South American includes countries such as Argentina, Bolivia, and Venezuela.

*Includes people from Spain and those who didn't specify a country of origin.

Source: Based on U.S. Census Bureau, Hispanic or Latino Origin by Specific Origin, 2015 American Community Survey, Table B03001. Accessed June 15, 2017 (factfinder.census.gov).

Figure 10.7 U.S. Median Household Income, by Race and Ethnicity, 2016

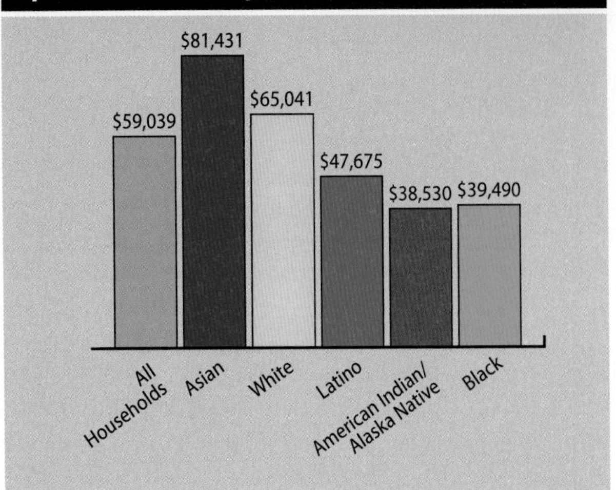

Sources: Based on "American Indian and Alaska Native...," 2016, and Semega et al., 2017, Table 1.

Such attitudes help explain why many Latinos are successful. Across 10 Latino groups, the second generation is working in better jobs with higher occupational status than their first-generation parents. Fully 72 percent expect their children will be better off financially than they themselves are right now. In 2014 alone, Latinos accounted for 40 percent of home ownership growth, the largest share among any racial or ethnic group. From 2007 to 2012, Latinos launched over 60 times more new small businesses than any other group. Since 2007, the number of Latina-owned businesses has grown by 137 percent, outpacing all other categories of minority women ("2012 Survey of Business Owners," 2015; Berenson, 2016; Lopez et al., 2016; Tran, 2016). Thus, despite Donald Trump's complaints that Latinos "are taking our money," they're driving much of America's economic growth.

10-6c African Americans: A Changing Minority

The 46 million African Americans are the second largest minority group, making up 12 percent of the population ("National African-American...," 2017). Over the past 40 years, Black people have made progress on many fronts, but racial gaps persist.

DIVERSITY

Most African Americans share a common characteristic: They're members of the only group ever brought to the United States involuntarily and legally enslaved. The term *African American* encompasses tremendous diversity, including native-born Americans with Black, White, American Indian, and/or Latino ancestors, as well as recent immigrants from Africa and elsewhere. Of the nearly 4 million foreign-born Black people, 50 percent are from the Caribbean, 36 percent from Africa, and 9 percent from a Central or South American country. The largest African-born populations are from Nigeria and Ethiopia (Anderson, 2015).

CHARACTERISTICS AND CHANGES

Black people make up 12 percent of the population but 22 percent of people living in poverty. The median family income of Black households is the lowest of all racial-ethnic groups (*Figure 10.7*). Between 2010 and 2016, the number of Black households with annual incomes of $50,000 or more rose from 36 to 41 percent (Semega et al., 2017).

A growing share of Black people are completing high school and college, but lag behind White and Asian people in getting a college or advanced degree (*Figure 10.8*). Since the mid-1990s, Black women have outpaced Black men (and Latinos) in educational attainment. In 2016, for example, 25 percent of Black women, compared with 22 percent of Black men, had completed four years of college or more. Black women have higher labor force participation rates than other minority or White women, and are far more likely to work after motherhood, but experience the highest gender wage gaps

Granger Historical Picture Archive/Alamy Stock Photo

Few people know that Madame C. J. Walker (1867–1919), who manufactured hair care products for Black women, was one of the first American female millionaires. Or that Dr. Charles R. Drew (1904–1950) was a renowned surgeon, teacher, and researcher. He founded two of the world's largest blood banks, saving untold lives during and since World War II.

Figure 10.8 Percentage with a Bachelor's Degree or Higher, by Race and Ethnicity, 2016

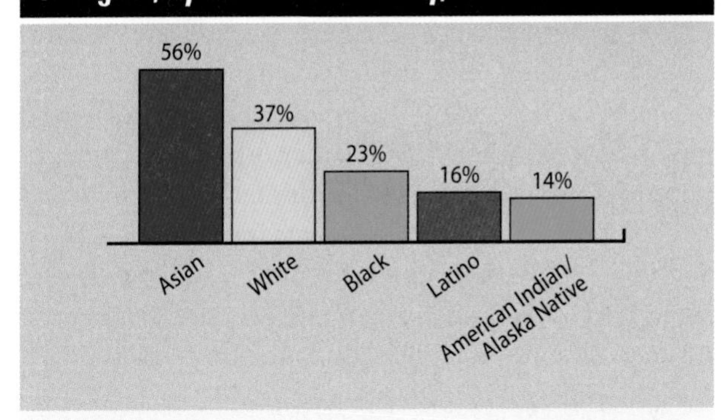

Sources: Based on U.S. Census Bureau, "Educational Attainment in the United States: 2016," Table 3, accessed June 2, 2017 (census.gov), and "American Indian and Alaska Native...," 2016.

(Holloway, 2016; "Educational Attainment...," 2017; see also Chapter 9).

The wage gap between college-educated Black and White Americans was larger in 2015 than in 1979, primarily because wage disparities build up over time. Some economists estimate that at least 30 percent of the Black–White wage gap is due to differential treatment in hiring, firing, and pay rather than differences in formal schooling or specific job skills (Fryer et al., 2011; Wilson and Rodgers, 2016).

Economic factors alone don't explain race inequality. For example, White mothers with less than a high school education have lower infant mortality rates than well-educated middle-class Black mothers. Affluent Black families, with annual incomes of more than $100,000, are four times more likely to live in poor neighborhoods than comparable White families. Some Black individuals prefer lower-income neighborhoods because of family ties, but most live in poorer areas because of segregation and White hostility in many suburbs (Eligon and Gebeloff, 2016; Matthew and Reeves, 2016).

Despite discrimination, many Black people are successful...The share of Black individuals in Congress is at an all-time high, increasing from 2 percent in 1971 to 9 percent in 2015. Also, the college completion gap between White and Black people has narrowed by 20 percent since 1976 ("2012 Survey of Business Owners," 2015; Bialik and Cilluffo, 2017).

10-6d Asian Americans: A Model Minority?

The nearly 23 million Asian Americans comprise almost 6 percent of the population. In 2012, Asians became the fastest-growing racial or ethnic group in the country; over 60 percent of the growth came from international migration. Hawaii is the only state with an Asian majority of 56 percent ("Asians Fastest-Growing...," 2013; "Asian-American and Pacific...," 2017).

DIVERSITY

Asian Americans encompass a broad swath of cultural groups. They come from at least 26 countries in East and Southeast Asia (e.g., China, Korea, Vietnam, Cambodia, the Philippines) and South Asia (especially India, Pakistan, and Sri Lanka), and speak at least 19 languages in the United States (Ryan, 2013; "Asian-American and Pacific...," 2017).

These diverse origins mean that there are huge differences in languages and dialects (and even alphabets), religions, cuisines, and customs. Chinese are the largest Asian American group, followed by Filipinos and Asian Indians (*Figure 10.9*). At least 83 percent of

Figure 10.9 Asian Americans by Origin, 2015

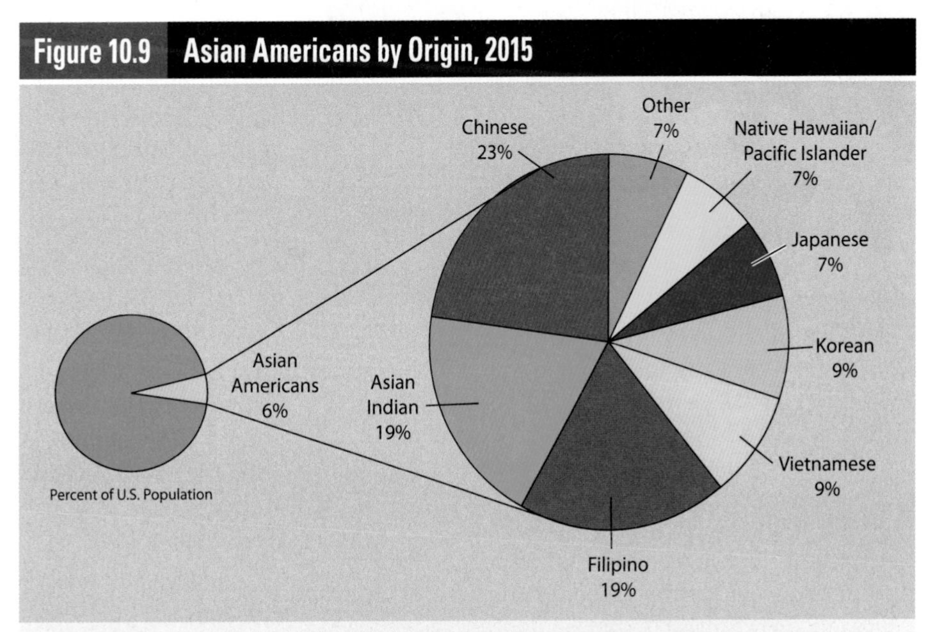

Note: "Other" includes people from at least 15 countries, including Laos, Cambodia, and Sri Lanka.

Source: Based on U.S. Census Bureau, 2015 American Community Survey, Tables B02018 and B02019. Accessed June 15, 2017 (factfinder.census. gov).

Asian Americans trace their roots to only six countries—China, India, Japan, Korea, the Philippines, and Vietnam (Taylor et al., 2012).

CHARACTERISTICS AND CHANGES

Compared with White people and other racial-ethnic groups, many Asian Americans are doing well. They have the highest education levels (*Figure 10.8*), and almost 22 percent, compared with 12 percent of the general population, have a graduate or professional degree. Because of their high education levels, almost half of Asian Americans are in highly skilled and well-paying occupations such as information technology, science, engineering, and medicine ("Asian-American and Pacific...," 2017). Consequently, they have the highest median household income (*Figure 10.7*).

Because of their educational and economic success, Asian Americans are often hailed as a "model minority." Such labels are misleading, however, because there's considerable variation across subgroups. For example, more than 70 percent of people from India and Taiwan have a bachelor's degree or higher compared with only 12 to 14 percent of Hmong, Laotians, and Cambodians (Ogunwole et al., 2012).

Many Asian college students succeed, and despite discrimination, because they work harder than most of their peers. Stereotypes of Asian Americans as hard-working and high-achieving can enhance performance: Teachers and guidance counselors place the students in advanced classes, the honors track, and give them extra help with homework and college applications. On the other hand, the model minority stereotype pigeonholes Asian students into specific careers and doesn't give them the chance "to see themselves as something other than doctors, engineers, or accountants," assumes that all minorities can overcome institutionalized discrimination, and holds Asian Americans to unrealistically high standards (Kay et al., 2013; Leung, 2013; Lee and Zhou, 2015).

Applauding Asian educational attainment obscures other barriers. Compared with White people, for example, Asians are nearly twice as likely to hold PhDs, law degrees, MBAs, or MDs, yet earn 5 percent less. Among technology companies—including those in Silicon Valley that hire a disproportionately high number of Asian professionals—Asians are vastly underrepresented in upper management: "Whites are the only group whose proportion increases as they rise through the ranks" (Chin, 2016: 70; Hilger, 2016).

Alex Wong/Getty Images News/Getty Images

Asian Indian students have won the Scripps National Spelling Bee 10 years in a row, and all but four of the last 18 years. In 2015, 13-year-old Vanya Shivashankar and 14-year-old Gokul Venkatachalam were co-champions. The winning words were *scherenschnitte* and *nuntak*. (Am I the only one who had to look them up?) Asian Indians' extraordinary performance has been attributed to factors like hard work, sacrifice, pursuing a championship over many years, and growing up in households where one or both parents are well-educated professionals who set high academic goals (Heim, 2015).

10-6e Native Americans: A Growing Nation

Native Americans used to be called the "vanishing Americans" but have "staged a surprising comeback" due to higher birth rates, a longer life expectancy, and better health services (Snipp, 1996: 4). The 6.6 million Native Americans and Alaska Natives (AIANs) make up almost 2 percent of the population, and are expected to increase to nearly 3 percent by 2060 ("American Indian and Alaska Native...," 2016).

DIVERSITY

Like Asian Americans, Native Americans and Alaska Natives are a heterogeneous group. Of the 566 federally recognized tribes, 8 have more than 100,000 members. The Cherokee, with almost 820,000 members, is the largest, followed by the Navajo and Choctaw (Norris et al., 2012).

Tribes speak 150 native languages, although many are quickly vanishing. For example, only about 25 people nationwide speak Comanche. Tribes also vary widely in their religious beliefs and cultural practices. Thus, a Comanche-Kiowa educator cautions, "Lumping all Indians together is a mistake. Tribes ... are sovereign nations and are as different from another tribe as Italians are from Swedes" (Pewewardy, 1998: 71; Mangan, 2013; U.S. Census Bureau News, 2015).

CHARACTERISTICS AND CHANGES

AIANs are a unique minority group because they're not immigrants and have been in what is now the United States longer than any other group. They have experienced centuries of subjugation, exploitation, and political exclusion (e.g., not having the right to vote until 1924). Some tribes are still trying to reclaim billions of dollars that the federal government has squandered or mismanaged (Wilkinson, 2006; Volz, 2012; Kindy, 2013).

Many AIANs are better off today than they were a decade ago, but long-term institutional discrimination has been difficult to shake. For example, 27 percent live below the poverty line compared with 13 percent of the general population. The median household income of AIANs is slightly higher than that of Black households, but lower than that of other racial-ethnic groups (*Figure 10.7*).

AIAN educational levels have increased, but only 14 percent have a bachelor's or advanced degree or higher compared with 36 percent of the general population. One of four civilian-employed AIANs works in a management or professional occupation. Even when AIANs are similar to White people in age, gender, education level, marital status, state of residence, and other factors, their odds of being employed are 31 percent lower than those of White people. Such data suggest that AIANs are experiencing discrimination in the labor market (Austin, 2013; "American Indian and Alaska Native...," 2016).

Despite numerous obstacles, AIANs have made considerable economic progress by insisting on self-determination and the rights of tribes to run their own affairs. For example, 244 tribes now generate about 43 percent of all U.S. casino gaming revenue. The Justice Department allows tribes to grow or sell marijuana on their lands, even in states that ban the practice. Some tribes, "mindful of the painful legacy of alcohol abuse in their communities," are opposed to selling or using marijuana on their territory. Others see marijuana sales as a huge source of future income, similar to cigarette sales and casino gambling (Cano, 2015; "2013 Indian Gaming Industry Report," 2015).

10-6f Middle Eastern Americans: An Emerging Minority

The Middle East is "one of the most diverse and complex combinations of geographic, historical, religious, linguistic, and even racial places on Earth" (Sharifzadeh, 1997: 442). It encompasses about 30 countries, including Armenia, Turkey, Israel, Iran, Afghanistan, Pakistan, and 22 Arab nations (e.g., Algeria, Iraq, Kuwait, Saudi Arabia, and the United Arab Emirates).

DIVERSITY

Of the 61 million people in the United States who speak a language other than English at home, about 5 percent speak Middle Eastern languages such as Armenian, Arabic, Hebrew, Persian, or Urdu.

Because of the continued immigration from the Middle East and North Africa, Arabic is the fastest

iStock.com/dcdebs

AP Images/Jeff Robbins

Mohegan Sun in southern Connecticut is one of the largest casinos in the United States. It has spent some of its profits on college scholarships, a $15 million senior center, and health insurance for tribal members. In contrast, some of the poorest tribes, such as the Navajo and Hopi, who have rejected gaming for religious reasons, have many members who live in poverty without kitchen facilities (like stoves or refrigerators) or indoor plumbing.

Steve Jobs—the late co-founder, chairman, and CEO of Apple Inc.—was born to unmarried university students. His father was a Syrian-born Muslim, and his mother was a U.S.-born Catholic of Swiss descent. The father put the baby up for adoption because his girlfriend's family wouldn't allow her to marry an Arab ("Steve Jobs . . .," 2011).

growing and seventh most commonly spoken non-English language in the United States (Brown, 2016). There are at least 1.9 million Americans of Arab descent, accounting for about 0.5 percent of the population (Brown, 2016). Those who identify themselves as Arab Americans come from many different countries (*Figure 10.10*). As in the case of Asian American families, Middle Eastern families make up a heterogeneous population that is a "multicultural, multiracial, and multiethnic mosaic" (Abudabbeh, 1996: 333).

CHARACTERISTICS AND CHANGES

Middle Eastern Americans tend to be better educated and wealthier than other Americans. About 46 percent have a college or advanced degree compared with 36 percent of the general population. In 2010, the median income of Arab American households was almost $57,000 compared with $52,000 for all U.S. households. As in other groups, however, there are wide variations. Lebanese have higher median family incomes ($67,300) than Iraqis ($32,000) (Asi and Beaulieu, 2013; Motel and Patten, 2013).

Not all Middle Eastern Americans are successful. Lebanese and Syrians have the lowest poverty rates, 11 percent, compared with more than 26 percent of Iraqis, and 23 percent of all foreign-born Middle Easterners (Arab American Institute Foundation, 2012; Motel and Patten, 2013).

Michigan has the country's highest concentration of residents with Middle Eastern, particularly Arabic, roots. As the auto industry recovers, companies are hiring immigrants and refugees from Syria, Iraq, Somalia, Yemen, and Myanmar. According to a manufacturing recruiter: "They work really hard, and that's what companies are looking for" (Green, 2016: 18).

10-7 SOCIOLOGICAL EXPLANATIONS OF RACIAL-ETHNIC INEQUALITY

Four major sociological theories help us understand racial-ethnic relations. (*Table 10.3* summarizes the key points of each perspective.)

Figure 10.10 Arab Americans by Origin, 2010

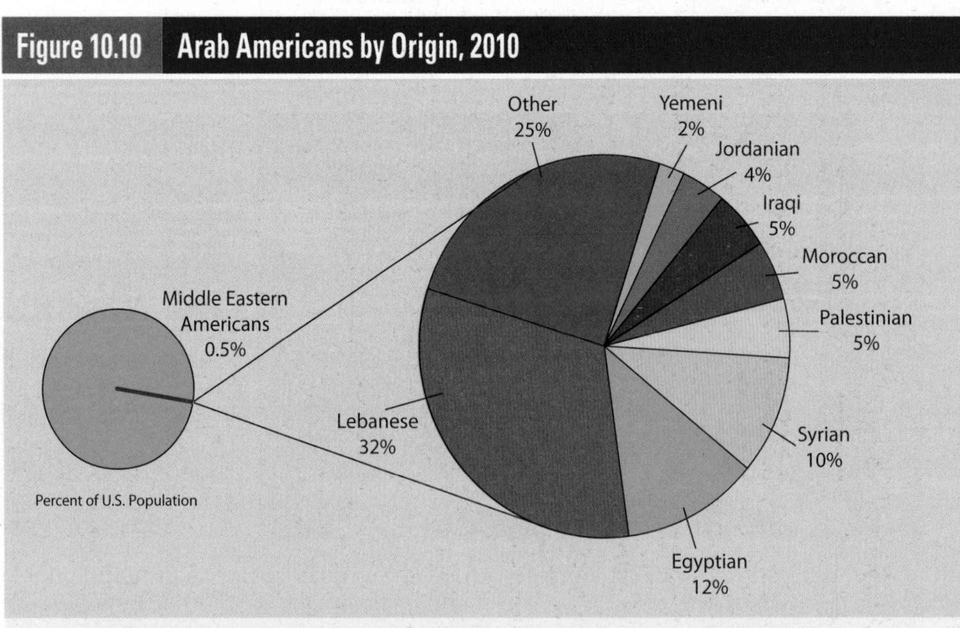

Note: "Other" includes those from the Middle East and North Africa.

Source: Based on Asi and Beaulieu, 2013, Table 1.

Table 10.3	Sociological Explanations of Racial-Ethnic Inequality	
THEORETICAL PERSPECTIVE	LEVEL OF ANALYSIS	KEY POINTS
Functionalist	Macro	Immigration provides needed workers; acculturation and assimilation increase social solidarity; racial-ethnic inequality can be dysfunctional, but benefits dominant groups.
Conflict	Macro	There's ongoing strife between dominant and minority groups; powerful groups maintain their advantages primarily through economic exploitation; race is a more important factor than social class in perpetuating racial-ethnic inequality.
Feminist	Macro and micro	Minority women suffer from the combined effects of racism and sexism; gendered racism occurs within and across racial-ethnic groups.
Symbolic Interactionist	Micro	Because race and ethnicity are socially constructed, social interaction can increase or reduce racial and ethnic hostility; antagonistic attitudes toward minorities, which are learned, can be lessened through cooperative interracial and interethnic contacts.

10-7a Functionalism

People who criticize immigrants for not becoming Americanized quickly enough reflect a functionalist view of racial-ethnic relations. That is, if a society is to work harmoniously, newcomers must adopt the dominant group's values, goals, and particularly language. Doing so increases a society's cultural solidarity.

STABILITY AND COHESION

Immigration is functional for the host nation if it gains needed workers. Highly educated and skilled immigrants fill important positions in medicine, science, and business. Many employers also rely on immigrants to work in fields, orchards, and vineyards at low wages; others are actively recruiting immigrants for decent-paying but "dirty" jobs at meat and fish factories that native-born Americans avoid (Newkirk and Douban, 2012; Davey, 2014).

Functionalists view racial-ethnic inequality as dysfunctional, but attribute much of the inequity to individual failings and a lack of acculturation. There are incentives for not acculturating, however. For example, hundreds of Puerto Ricans get federal disability benefits, and regardless of work experience or level of education, because they're considered less employable if they can't speak English. Office of the Inspector General, 2015).

CRITICAL EVALUATION

Functionalist explanations are limited in several ways. First, acculturation and assimilation increase social solidarity, but can also have negative outcomes. For example, second-generation immigrants are more likely than their foreign-born peers to join gangs, commit crimes, and experience obesity and other health problems (Bersani, 2014; Akbulut-Yuksel and Kugler, 2016; see also Chapter 14).

Second, by focusing on order and stability, functionalists ignore racial-ethnic inequalities that often spawn tension and discord. Instead of implementing an immigration policy, President Trump tried to prevent refugees and people from seven majority-Muslim countries from entering the United States. His action polarized many Americans—particularly the 64 percent who think the nation's growing racial and ethnic diversity makes the country a better place ("In First Month…," 2017).

Third, functionalism doesn't explain why, despite acculturation and assimilation, minorities experience exclusion. For instance, 61 percent of Black people and 45 percent of White people say that U.S. race relations are "generally bad," and 43 percent of Black people believe that America will never make the changes needed for Black people to achieve equal rights with White people (Geiger, 2016; Bialik and Cilluffo, 2017). Functionalists acknowledge that inequality is dysfunctional, but this isn't their major focus. Thus, they seem to accept discrimination as inevitable (Chasin, 2004).

10-7b Conflict Theory

Conflict theorists see ongoing strife between dominant and minority groups. Dominant groups try to protect their power and privilege, whereas subordinate groups struggle to gain a larger share of societal resources.

ECONOMIC AND SOCIAL CLASS INEQUALITY

For conflict theorists, capitalism creates and sustains racial-ethnic inequality. According to a classic explanation, there's a "split labor market." Jobs in the *primary labor market*, held primarily by White workers, provide better wages, health and pension benefits, and some measure of job security. In contrast, workers in the *secondary labor*

market (e.g., fast-food employees) are largely minorities and easily replaced. Their wages are low, there are few fringe benefits, and working conditions are generally poor (Doeringer and Piore, 1971; Bonacich, 1972).

Such economic stratification pits minorities against each other and low-income White people. Because these groups compete with each other instead of uniting against exploitation, capitalists don't have to worry about increasing wages or providing safer work environments.

Social class doesn't always protect minorities from economic inequality. Because of residential segregation, for instance, middle-income Black and Latino households are much more likely than White ones to live in poor neighborhoods, exposing children to inadequate schools and more crime. At leading technology companies like facebook and Google, Black and Latino people still make up only 2 to 3 percent of college-educated officials and managers (Andrews, 2015; Kang, 2015; Reardon et al., 2015).

CRITICAL EVALUATION

Conflict theories have several drawbacks. First, discrimination isn't always as conscious and deliberate as some conflict theorists claim. To increase diversity, Google and Facebook have for some time recruited at historically Black colleges, trained employees about unconscious bias, and rewarded recruiters and managers for hiring Black, Latino, and female engineers (McGirt, 2017; Zarya, 2017).

Second, conflict theories are better at explaining racial-ethnic competition than cooperation. Forty-two percent of Americans (up from only 13 percent in 2010) worry a "great deal" about U.S. race relations. This surge is likely due to high-profile police shootings of unarmed Black men and some of President Trump's racist comments about Mexicans and Middle Eastern refugees (Swift, 2017). Despite such divisive rhetoric, Americans generally work together and cooperate to improve race relations, especially at the local level. Finally, economic inequality reinforces racial-ethnic inequality, but racism existed hundreds of years before the rise of capitalism.

gendered racism the overlapping and cumulative effects of inequality due to racism *and* sexism.

10-7c Feminist Theories

Walk through almost any hotel, large discount store, nursing home, or fast-food restaurant in the United States. You'll notice two things: Most of the low-paid employees are women, and predominantly minority women. For feminist scholars, such segregation of minority women is due to gendered racism.

GENDERED RACISM

Gendered racism refers to the overlapping and cumulative effects of inequality due to racism *and* sexism. Many White women encounter discrimination on a daily basis (see Chapter 9). Minority women, however, are also members of a racial-ethnic group, bringing them a double dose of inequality. If social class is included, some minority women experience *triple oppression*. Many affluent women, in particular, have no qualms about exploiting recent immigrants, especially Latinas, who perform demanding housework at very low wages (Hondagneu-Sotelo, 2001).

In 2015, Viola Davis became the first Black woman to win an Emmy for best actress in a drama series (*How to Get Away with Murder*). In her acceptance speech, Davis said, "The only thing that separates women of color from anyone else is opportunity."

Gendered racism also occurs *within* racial-ethnic groups. According to a Black male sociologist, scholars rarely discuss Black male privilege, which is characterized by having advantages over Black women. Examples include being promoted more often and getting higher pay than their Black female counterparts who are equally skilled and educated (National Public Radio, 2010).

Minority women are eager to reach high-level positions: 48 percent, compared with 37 percent of White women, aspire to be a top executive. Minority women make up one-third of the workforce, but less than 4 percent of executives and only 0.4 percent of CEOs at the largest companies. Some label minority women's roadblocks a "concrete ceiling." In contrast to a "glass ceiling" that White women might shatter, concrete ceilings are more difficult to break through (Piazza, 2016; Calacal, 2017; see also Chapter 6 on glass ceilings).

AP Images/Dave Martin

After Hurricane Katrina in 2005, the caption on a photo of a young Black man wading through water described him as "looting."

Chris Graythen/Getty Images news/Getty Images

A caption described a White man and a light-skinned woman as "finding" goods at a flooded local grocery store.

For symbolic interactionists, images shape our perceptions of racial and ethnic groups.

CRITICAL EVALUATION

Feminist perspectives can be faulted on two counts. First, because all of us have internalized institutional discrimination, minority group members are also guilty of reinforcing gendered racism in schools, workplaces, and elsewhere. Black girls as young as 15 have to cope with Black men's sexual harassment, stereotypes that Black girls and women are sexually promiscuous, and many Black men's preference for women with lighter complexions and facial features that are closer to European standards of beauty (Friedman, 2011; Thomas et al., 2011). Second, feminist scholars seldom explore women's contribution to gendered racism (e.g., affluent Black women and Latinas who exploit domestic and farm workers).

10-7d Symbolic Interaction

According to symbolic interactionists, we learn attitudes, norms, and values throughout the life course. Because race and ethnicity are constructed socially by thoughts and conversations, labeling, selective perception, and social contact can have powerful effects on everyday intergroup relations.

LABELING, SELECTIVE PERCEPTION, AND THE CONTACT HYPOTHESIS

Labeling can increase racial tension and conflict, because people attach meaning to symbols and act according to their subjective interpretation of the symbols. Students on some college campuses have demanded the removal of paintings and Civil War statues and the renaming of buildings that, according to students, honor slave owners, segregationists, and white supremacists. The protests antagonize many alumni whose generous donations support academic and sports programs, scholarships, and libraries ("Universities," 2015).

Selective perception also affects racial attitudes and behavior. When White people get proof of being privileged, they may insist that *other* people, not they themselves, benefit personally from privilege (Phillips and Lowery, 2015). People who see what they want and ignore any evidence that challenges their perceptions are unlikely to support policies to reduce racial inequality.

Negative images create and reinforce racial and ethnic stereotypes, but people can decrease labeling and selective perception. The **contact hypothesis** posits that the more people get to know minority group members personally, the less likely they are to be prejudiced against that group. Such contacts are most effective when dominant and minority group

contact hypothesis posits that the more people get to know members of a minority group personally, the less likely they are to be prejudiced against that group.

members have approximately the same ability and status (e.g., both are coworkers or bosses), when they share common goals (e.g., work on a project), when they cooperate rather than compete, and if an authority figure supports intergroup interaction (e.g., an employer requires White supervisors to mentor minority workers) (Allport, 1954; Kalev et al., 2006; Carrell et al., 2015).

CRITICAL EVALUATION

Interactionism is limited for three reasons. First, it's not clear why labeling, selective perception, and racial bias are more common among some people than others, especially when they're similar on a number of variables such as social class, gender, age, religion, race, and ethnicity (Dovidio, 2009).

Second, symbolic interaction tells us little about the social structures that create and maintain racial-ethnic inequality. For instance, people who aren't prejudiced can foster discrimination by simply going along with the inequitable policies that have been institutionalized in education, the workplace, and other settings (i.e., the unprejudiced discriminators described earlier).

A third criticism is that contact between and among racial-ethnic groups doesn't necessarily increase acceptance and understanding. For instance, darker- and lighter-skinned Black people report more everyday discrimination even from Black people than those with a medium tone (Monk, 2015; Jones et al., 2016).

10-8 INTERRACIAL AND INTERETHNIC RELATIONSHIPS

President Obama, the son of a Black father from Kenya and a White mother from Kansas, identified himself as "Black" on the 2010 census questionnaire, although he could have checked off both Black and White or "other." The president is just one of the growing number of Americans who are biracial or multiracial.

10-8a Growing Multiracial Diversity

The 2000 U.S. Census for the first time allowed people to mark more than one race, which generated about 126 categories and combinations. In 2010, 97 percent of Americans reported being only one race, but almost 3 percent (9 million people) self-identified as being two or more races, up from 2.4 percent in 2000. Every state saw its multiple-race population jump by at least 8 percent, but Native Hawaiians/Pacific Islanders were the most likely to identify themselves as belonging to two or more races (*Figure 10.11*).

AP Images/ Kevin Cederstrom

White supremacist Craig Cobb made headlines worldwide when he tried, unsuccessfully, to establish a white-only town in North Dakota. Shortly after that, Cobb's DNA test showed that he was 14 percent Black. How awkward, a journalist observed ("DNA Reveals...," 2013: 8).

A recent survey found that 17 percent of U.S. adults could be considered multiracial based on how people describe themselves as well as their parents' and grandparents' racial backgrounds. The percentage may be even higher because 61 percent of adults with a mixed racial background don't consider themselves "multiracial" (Parker et al., 2015; Patten, 2016).

10-8b Interracial Dating and Marriage

Laws against **miscegenation**, marriage or sexual relations between a man and a woman of different races,

Figure 10.11 Percentage of Americans Who Identified Themselves as Being Two or More Races, 2010

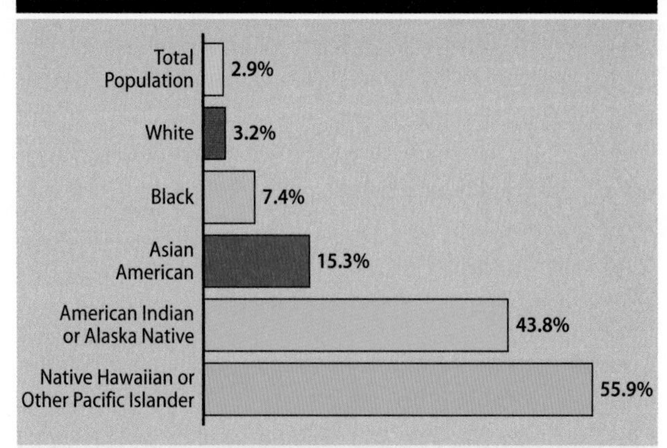

Note: Of the almost 309 million U.S. population in 2010, only 0.3 percent said they were three or more races.

Source: Based on Jones and Bullock, 2012, Figure 10.

existed in America as early as 1661. It wasn't until 1967 that the U.S. Supreme Court's *Loving v. Virginia* decision overturned antimiscegenation laws nationally.

Racial-ethnic intermarriages have increased slowly—from only 0.7 percent in 1970 to 10 percent (11 million people) in 2015. In that year, 17 percent of newlyweds had a spouse of a different race or ethnicity. Latinos and Asians have higher intermarriage rates than Black and White people, but recent overall increases have been fueled by rising intermarriage rates among Black and White newlyweds (*Figure 10.12*).

The increase in intermarriage is due to many interrelated factors—both micro and macro level—that include everyday contact and changing attitudes. In 2000, 31 percent of Americans said they would oppose an intermarriage in their family. That share fell to 10 percent in 2016. Only 12 percent of White people, 9 percent of Black people, and 3 percent of Latinos say they would oppose a close relative marrying someone of a different race or ethnicity (Livingston and Brown, 2017).

Proximity and education also affect intermarriage rates. We tend to date and marry people we see on a regular basis. The higher the educational level, the greater the potential for intermarriage, because educated minority group members often attend integrated colleges, and their workplaces and neighborhoods are more racially mixed than in the past (Chen and Takeuchi, 2011; Wang, 2012).

People often marry outside of their racial-ethnic groups because of a shortage of potential spouses within their own group. Because the Arab American population is so small, 80 percent of U.S.-born Arabs have non-Arab spouses. Acculturation also affects intermarriage rates. Intermarriage is more common among second-generation immigrants (15 percent) than those in the first generation (8 percent). By the third generation, 31 percent of Latinos and Asian Americans have a spouse of a different race or ethnicity (Kulczycki and Lobo, 2002; "Second-Generation Americans," 2013).

Figure 10.12 U.S. Intermarriage Rates Vary by Race and Ethnicity

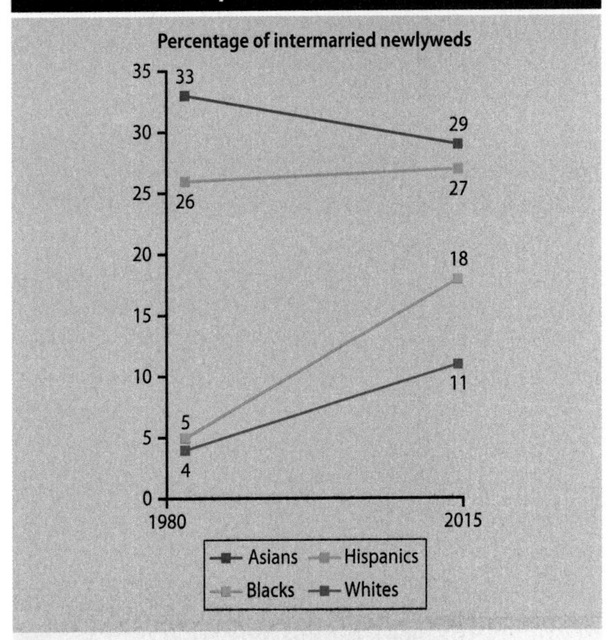

Source: Based on Livingston and Brown, 2017: 11.

The White majority is shrinking as multiracial marriages surge. According to demographer William Frey (2015: 191), the rise in intermarriages and multiracial children shows "an unmistakable trend toward a softening of racial boundaries" that should lead to greater equality in education, politics, the economy, and other institutions.

miscegenation marriage or sexual relations between a man and a woman of different races.

STUDY TOOLS 10

READY TO STUDY? IN THE BOOK, YOU CAN:

- ☐ Check your understanding of what you've read with the Test Your Learning Questions provided on the Chapter Review Card at the back of the book.
- ☐ Tear out the Chapter Review Card for a handy summary of the chapter and key terms.

ONLINE AT CENGAGEBRAIN.COM WITHIN MINDTAP YOU CAN:

- ☐ Explore: Develop your sociological imagination by considering the experiences of others. Make critical decisions and evaluate the data that shape this social experience.
- ☐ Analyze: Critically examine your basic assumptions and compare your views on social phenomena to those of your classmates and other MindTap users. Assess your ability to draw connections between social data and theoretical concepts.
- ☐ Create: Produce a video demonstrating connections between your own life and larger sociological concepts.
- ☐ Collaborate: Join your classmates to create a capstone project.

11 The Economy and Politics

Andrew Lichtenstein/Corbis News/Getty Images

LEARNING OBJECTIVES

After studying this chapter, you will be able to...

11-1 Compare and illustrate the different types of global economic systems.

11-2 Describe corporations and explain how corporate and political power are interwoven.

11-3 Explain how and why macro-level variables have changed the U.S. economy and jobs.

11-4 Compare, illustrate, and evaluate the theoretical explanations of work and the economy.

11-5 Compare and illustrate the different types of global political systems.

11-6 Explain how power and authority differ, and the different types of authority.

11-7 Describe the U.S. political system, and explain who votes, who doesn't, and why.

11-8 Compare, illustrate, and evaluate the theoretical explanations of politics and power.

After finishing this chapter go to **PAGE 228** for **STUDY TOOLS**

In mid-2017, 58 percent of Americans said that economic conditions were good (up from only 28 percent in 2014). The same percentage also said that they didn't trust political leaders. Financial worries have decreased since the Great Recession, but dissatisfaction with government has increased (Jones, 2016; Stokes, 2017). In this chapter, we look at how the economy and government affect people's everyday lives.

WHAT DO YOU THINK?

The country would be better off if there were more women in political office.

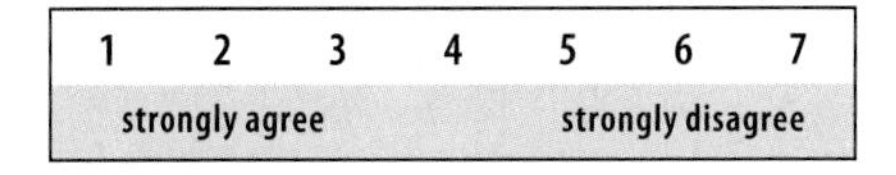

1	2	3	4	5	6	7
strongly agree						strongly disagree

The **economy** determines how a society produces, distributes, and consumes goods and services. In **politics** individuals and groups acquire and exercise power and authority and make decisions. Societies worldwide differ in the kinds of economic and political systems they develop because of factors such as globalization and technology. Let's begin by looking at global economic systems.

11-1 GLOBAL ECONOMIC SYSTEMS

The two major economic systems around the world are capitalism and socialism. A handful of countries endorse communism, but in actual practice, economies are usually some mixture of both capitalism and socialism.

11-1a Capitalism

Ideally, *capitalism*, an economic system based on the private ownership of property and the means of production, has four essential characteristics (Smith, 1776/1937; Heilbroner and Thurow, 1998):

- **Private ownership of the means of production.** Property (e.g., real estate, banks, utilities) belongs to individuals or organizations rather than the state or the community.
- **Market competition.** Owners of production compete in deciding what goods and services to produce and setting prices that offer consumers the greatest value in price and quality.
- **Profit.** Selling something for more than it costs to produce generates an accumulation of wealth for individuals and companies.
- **Investment.** By investing profits, capitalists can increase their own wealth. Workers, too, can save and invest their money.

In reality, capitalism doesn't function ideally because of abuses, greed, and worker exploitation. It usually results in monopolies and oligopolies rather than a free market that encourages competition. Federal laws prohibit the establishment of a **monopoly**, the domination of a particular market or industry by one person or company. With little or no competition, a company would be able to dictate prices and lower the quality of goods and services.

An **oligopoly**, which is legal, is the domination of a market by a few large producers or suppliers. In the mid-1980s, 50 companies controlled 90 percent of all U.S. media. Today, it's six companies—including CBS, Time Warner, and the Walt Disney Corporation. There were hundreds of airlines in the mid-1980s, but four giant airlines (United, Delta, American, and Southwest) now control 85 percent of the U.S. market. CEOs maintain that merging companies increases efficiency and safety. Critics argue that oligopolies raise prices, stifle competition, and reduce consumer choices. Companies that dominate industries make huge profits, but workers' wages and salaries tend to decline (Yglesias, 2013; Koenig and Mayerowitz, 2015; Coy, 2016). Moreover, entrepreneurs have little chance of breaking into an industry dominated by an oligopoly.

11-1b Socialism

Socialism is an economic system based on the public ownership of

economy determines how a society produces, distributes, and consumes goods and services.

politics individuals and groups acquire and exercise power and authority and make decisions.

monopoly domination of a particular market or industry by one person or company.

oligopoly domination of a market by a few large producers or suppliers.

socialism an economic system based on the public ownership of the production of goods and services.

the production of goods and services. Ideally, socialism is the opposite of capitalism and has the following characteristics:

- **Collective ownership of property.** The community, rather than the individual, owns property. The state owns utilities, factories, land, and equipment, but distributes them equally among all members of a society.
- **Cooperation.** Working together and providing social services to all people are more important than competition.
- **No profit motive.** Private profits that are fueled by greed and exploitation of workers are forbidden.
- **Collective goals.** The state is responsible for all economic planning and programs. People are discouraged from accumulating individual profits and investments, and are expected to work for the greater good.

There have been many socialist governments during the last 150 years, but none has reflected pure socialism. Competition and individual profits are officially forbidden, but government officials, top athletes, and high-ranking party members enjoy more freedom, larger apartments, higher incomes, and greater access to education and other resources than the rank and file.

11-1c Communism

Communism is a political and economic system in which property is communally owned and all people are considered equal. Whereas socialism allows some free market economy, and people receive resources according to how hard they work, communism demands that all production be owned by the public, and that people receive resources according to their need.

Karl Marx (1867/1967) believed that socialist states would evolve into communist societies. This hasn't happened and most formerly communist countries have collapsed. Today, only China, Cuba, Laos, North Korea, and Vietnam practice some form of communism, but have also incorporated capitalism. In contrast to its principles, communism has failed, according to historians, because of widespread corruption and mismanagement, oppression, imprisonment or execution of people who questioned coercive policies, economic inefficiency that resulted in shortages of food and other goods, and few rewards for working hard (Brown, 2010; Johnson, 2010; Pollick, 2014).

Walt Disney Company is an example of an oligopoly. Among other holdings, it owns publishing companies, web portals, 19 major television and cable stations, at least 60 radio stations, numerous magazines, and 14 theme parks and resorts around the world. All of these outlets promote and sell Disney products, increasing the corporation's profits.

11-1d Mixed Economies

Welfare capitalism (also called *state capitalism*) is an economic system that combines private ownership of property, market competition, and a government's regulation of many programs and services. Many countries, including the United States, have some form of welfare capitalism. Most industry is private, but the government owns or operates some of the largest industries and services (e.g., education, transportation, postal services) that benefit the entire population (Esping-Andersen, 1990; Hacker, 2002).

Crony capitalism is an economy based on close relationships between business people and the government. In this system, the government gives wealthy people and large corporations preferential treatment through special tax breaks (called *tax credits* and *deductions*), direct

communism a political and economic system in which property is communally owned and all people are considered equal.

welfare capitalism (also called *state capitalism*) an economic system that combines private ownership of property, market competition, and a government's regulation of many programs and services.

Nir Elias/REUTERS

China is an example of a country with a mixed economy: It espouses communism, practices socialism, and has endorsed many aspects of capitalism. China's economic boom hasn't benefited all of its citizens. Many urban centers are thriving—offering those in upper and middle classes high-rise apartments, stores, and restaurants. In contrast, millions of Chinese, including many older people, like the one pictured here, survive by scouring trash bins for plastic bottles to recycle.

payment or loans (called *subsidies*, not welfare), and provides grants that rarely require competition in an open market. From 2000 to 2012, for example, all but one of the Fortune 100 companies received almost $1.3 trillion from the federal government in tax credits, contracts, grants, and various subsidies (Andrzejewski, 2014; see also the discussion of corporate welfare in Chapter 8).

Large companies drive both capitalism and mixed economies. Corporations, in particular, wield enormous influence both at home and abroad.

11-2 CORPORATIONS AND THE ECONOMY

In 2005, Hamdi Ulukaya, a Turkish immigrant, founded Chobani, a yogurt company that's now valued at about $5 billion. In 2016, Ulukaya told his 2,000 full-time employees that they'd receive shares worth up to 10 percent of the company if it goes public or is sold. Many workers would become millionaires (Strom, 2016). Unlike Ulukaya, most corporate heads have luxurious lifestyles, eliminate employees' pensions, retire as billionaires, and pass on their massive wealth to their heirs (see Chapter 8). Corporate and political power are interwoven, but let's begin by looking at some of the characteristics of corporations.

11-2a Corporations

A **corporation** is an organization that has legal rights, privileges, and liabilities apart from those of its members. Until the 1890s, there were only a few U.S. corporations—in textiles, railroads, and the oil and steel industries. Today, there are almost 6 million, most created for profit, but a mere 0.5 percent bring in 90 percent of all corporate income (U.S. Census Bureau, 2012).

Whereas many industrialized nations have closed corporate tax loopholes, the United States has expanded them. As a result, many corporations exercise more power than do governments, and have amassed enormous wealth. A number of U.S. corporations use legal tax loopholes to avoid paying tens of billions of dollars every year on overseas income; 29 companies alone have more cash than the U.S. Treasury Department; and the profits of the Fortune 500 corporations have soared (Durden, 2011; Niquette and Rubin, 2014; see also Chapter 8).

11-2b Conglomerates

A **conglomerate** is a corporation that owns a collection of companies in different industries. Conglomerates emerged during the 1960s and grow by acquiring companies through mergers. Mergers might increase the value of shareholders' stock, but typically make chief executives "truly, titanically, stupefyingly rich" (Morgenson, 2004: C1). An example of a conglomerate is Kraft Foods, which owns companies that produce snacks, beverages, pet foods, a variety of groceries and convenience foods, and has ties with other corporations such as Starbucks (see www.kraft.com/brands).

Conglomerates diversify business risk by participating in a number of different markets, and can ward off emerging competition. Conglomerates waned after the 1980s, but are making a comeback. Google, for example, has set up companies to create driverless cars, extend human lifespans, and develop online education ("From Alpha to Omega," 2015).

11-2c Interlocking Directorates

In an **interlocking directorate**, the same people serve on the boards of

corporation an organization that has legal rights, privileges, and liabilities apart from those of its members.

conglomerate a corporation that owns a collection of companies in different industries.

interlocking directorate the same people serve on the boards of directors of several companies or corporations.

Roy Delgado/CartoonStock.com

directors of several companies or corporations. An estimated 15 to 20 percent of corporate directors sit on two or more corporate boards. Some interlocking directorates are especially powerful because they include past U.S. presidents and past members of Congress who provide connections in Washington, D.C. Interlocking directors develop similar economic perspectives; those who sit on two or more corporate boards are especially likely to receive appointments to influential government advisory committees (Domhoff, 2013). Thus, a very small group of people has the power to shape the national agenda.

11-2d Transnational Corporations and Conglomerates

Interlocking directorates have become more influential than ever because of the proliferation of transnational corporations. A **transnational corporation** (also called a *multinational corporation* or an *international corporation*) is a large company that's based in one country but operates across international boundaries. By moving production plants abroad, large U.S. corporations can avoid trade tariffs, bypass environmental regulations, and pay low wages. On the positive side, many corporate giants (e.g., Johnson & Johnson, Microsoft, Apple) invest heavily in local and global education, medical research, and developing economic opportunities (Preston, 2016; Gray, 2017).

The most powerful are **transnational conglomerates** (also called *multinational conglomerates*), corporations that own a collection of different companies in various industries in a number of countries. General Electric is the world's biggest transnational conglomerate. It owns hundreds of companies in the United States, and has subsidiaries in more than 170 countries ("GE Fact Sheet," 2017).

transnational corporation (also called a *multinational corporation* or an *international corporation*) a large company that's based in one country but operates across international boundaries.

transnational conglomerate (also called a *multinational conglomerate*) a corporation that owns a collection of different companies in various industries in a number of countries.

work a physical or mental activity that produces goods or services.

deindustrialization social and economic change that reduces industrial activity, especially manufacturing.

11-3 WORK IN U.S. SOCIETY TODAY

Economic problems vary by social class, but much financial hardship has to do with changes in **work**, a physical or mental activity that produces goods or services. Even those who work hard have lost economic ground because of macro-level variables such as deindustrialization, globalization, offshoring, and weakened labor unions.

11-3a Deindustrialization and Globalization

Many Americans have been casualties of **deindustrialization**, a process of social and economic change that reduces industrial activity, especially manufacturing. The number of manufacturing jobs plummeted from a peak of almost 20 million in 1979 to 11 million in 2010. Manufacturing added more than 800,000 new jobs after 2010, but of its 77 industries, only 19 are projected to add jobs by 2022—primarily in motor vehicle parts manufacturing, and animal slaughter and processing (Torpey, 2014).

Avatar_023/Shutterstock.com

Many U.S. factories today need only a few skilled workers to manage the high-tech machines that have replaced people.

A major reason for deindustrialization is that, beginning in the early 1960s, employers easily replaced workers with the lowest skill levels, usually those on assembly lines, with automation. Machines and robots have also replaced many blue collar jobs that provided a comfortable standard of living in the past. Robots are commonplace in manufacturing jobs, but employers report a shortage of workers with science, technology, engineering, or math degrees who can program robots, repair sophisticated hardware, and design software that runs a factory's operations (Gregory, 2017; Manyika et al., 2017).

The manufacturing industry is also facing a major shortage of skilled workers like welders, ironworkers, and brick masons. An estimated two million such jobs will go unfilled over the next decade because employers can't find enough people with the required skills, and despite above-average wages. Entry-level welders, for example, can move quickly from $17 to $30 an hour (Cohen, 2015; Payne and Somerville, 2015).

Deindustrialization was hastened by **globalization**, the growth and spread of investment, trade, production, communication, and new technology around the world. One example of globalization is a car that's assembled in the United States with practically all of its parts manufactured and produced in Germany, Japan, South Korea, or developing countries.

Proponents argue that globalization creates millions of jobs and brings affordable goods and services (e.g., cell phones) to billions of people around the world. Critics contend that globalization has reduced the number of jobs in Western economies. They also maintain that globalization benefits primarily the world's most powerful transnational corporations: It increases their profits by expanding their worldwide base of consumers, exploits poor people in developing countries, creates greater income inequality between workers and the global elite, and gives corporations unprecedented political power (Schaeffer, 2003; Porter, 2014).

11-3b Offshoring and Labor Unions

Offshoring refers to sending work or jobs to another country to cut a company's costs at home. Sometimes called *international outsourcing* or *offshore outsourcing*, the transfer of manufacturing jobs overseas has been going on since at least the 1970s.

iStock.com/P_Wei

Between 2001 and 2011, U.S. companies moved more than 2.7 million jobs to China, 77 percent of them in manufacturing and primarily blue-collar (Scott, 2012).

Figure 11.1 Median Annual Earnings of an Accountant in the United States and India

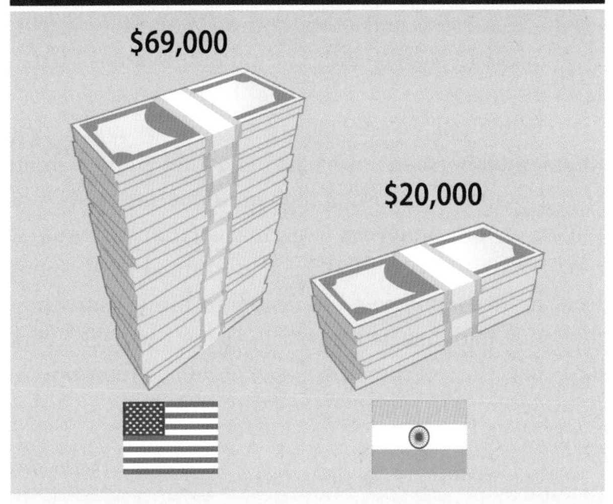

Sources: Based on Labor Force Statistics, Current Population Survey, 2017, Table 39, "Median Weekly Earnings of Full-Time Wage and Salary Workers by Detailed Occupation and Sex" (bls.gov/cps); Krumwiede, 2017.

During the same period, U.S. firms also offshored 28 percent of high-level, well-paid information technology (IT) jobs, including those in accounting, computer science, and engineering (National Science Board, 2012). Most of the offshored jobs go to India and China, but many have also moved to Canada, Hungary, Mexico, Poland, Russia, Egypt, Venezuela, Vietnam, and South Africa.

Companies can get accounting services in India that cost much less than in the United States (*Figure 11.1*). In Mexico, the average General Motors worker earns wages and benefits that cost less than $4 an hour compared with $55 an hour in the United States. Because of such large wage differences, American consumers can purchase goods and services at low prices, and the majority doesn't care where products are made (Black, 2010; Guarino, 2013; Newport et al., 2014).

Deindustrialization, globalization, and offshoring have weakened *labor unions*, organized groups that seek to improve wages, benefits, and working conditions. Union membership has dropped sharply—from 35 percent of the workforce in the mid-1950s to 11 percent in 2016. Five

globalization the growth and spread of investment, trade, production, communication, and new technology around the world.

offshoring sending work or jobs to another country to cut a company's costs at home.

states prohibit unionization; 27 states have significantly restricted collective bargaining rights; and anti-union groups have helped defeat efforts to unionize autoworkers in Tennessee and other states (Whitesides, 2014; Bookman and Neeley, 2017; "Union Members—2016," 2017).

Do we still need unions? Opponents argue that union members are overpaid, and that unions drain state resources because of high pensions and salaries of public sector employees (e.g., teachers, nurses, sanitation workers, police). Critics also contend that unions have limited employers' flexibility in hiring and firing decisions, and that ever-increasing labor and health care costs have forced some employers to move their operations overseas to remain competitive (McKinnon, 2011; Schlesinger, 2011).

Proponents argue that states have experienced budget deficits because of the housing crisis and a recession that was due to Wall Street greed and not overpaid union members. Others point out that union members have made numerous concessions like decreasing their wages and benefits. Most important, historically, unions have benefited almost all workers by insisting on paid holidays and vacations, greater workplace safety, and overtime pay (Stepan-Norris, 2015).

Source: *fightfor15.org*

11-3c How Americans' Work Has Changed

The U.S. economy started adding more jobs in 2012, but millions of Americans still have only low-paying and part-time jobs. If these strategies fail, they find themselves among the unemployed.

LOW-WAGE JOBS

About half of U.S. jobs pay less than $15 an hour. After taxes, the wages of a full-time year-round worker who earns $15 an hour is only about $26,000 a year, and many jobs pay much less. The federal minimum wage, which rose from $6.55 to $7.25 an hour in 2009, increased the wages of less than 4 percent of the workforce. The District of Columbia and 29 states now have minimum wages that exceed the federal minimum wage. If the wage had been raised since 1968 at the same rate as the average worker's growth in productivity, the federal minimum wage would be nearly $18.50 an hour (Cooper, 2015; DeSilver, 2017; Hoxie, 2017).

Almost 3 percent (2.2 million) of workers 18 and older are at or below the federal minimum wage. Another 30 percent (20.6 million) are "near-minimum" workers who earn less than $10.10 an hour (DeSilver, 2017). Thus, most jobs don't pay enough to allow workers to save money or get ahead. In 2016, Walmart, the nation's largest low-wage employer, increased the hourly wage to $10 an hour. If Walmart paid its workers $12 an hour and passed every penny of the costs to consumers, the average customer would pay just 46 cents more per shopping trip, or about $12 a year (Cooper, 2015; Montlake, 2016). Thus, many Americans are being paid less for doing more.

Except for registered nurses, whose median annual salary is about $69,000, eight of the occupations expected to have the most jobs added through 2024 are low paying, under $24,000 a year. These occupations include home health aides and personal care aides who help older and disabled people. Other fast-growing but low-paying jobs—from about $22,000 to $25,000 a year—include retail salespersons, cooks, stock clerks, janitors, housekeeping cleaners, and fast-food workers ("Employment Projections," 2017).

PART-TIME, TEMPORARY, AND GIG WORK

Of the almost 27 million part-timers (those who work less than 35 hours a week), 20 percent are part time involuntarily, because they can't find suitable full-time employment or employers have reduced their hours. The "alternative workforce"—which includes working for temporary help agencies, as freelancers, or as independent contractors—rose from 11 percent in 2005 to 16 percent in 2015 (15 million and nearly 24 million, respectively) (Katz and Krueger, 2016; "Economic News Release," 2017).

Rapid technological advances have spurred a *gig economy* (also called a sharing, on-demand, platform, collaborative, or digital economy). Although there's no official definition, *gig work* is a project or task for which an individual, often through a digital marketplace, receives payment (e.g., Uber, Airbnb, TaskRabbit, Postmates, Etsy). Because definitions of gig work vary, prevalence rates range from 1 to 33 percent of the workforce (Katz and Krueger, 2016; Torpey and Hogan, 2016; Morgan, 2017).

The benefits of alternative and gig work include job flexibility and autonomy, few formal education or experience requirements, turning a hobby or pastime into a source of income, and supplementing a full-time job. On the other hand, people generally earn less than in traditional occupations, must cover their own health care costs, and may face unpredictable hours and financial instability (Smith, 2016; Hicks, 2017; Morgan, 2017).

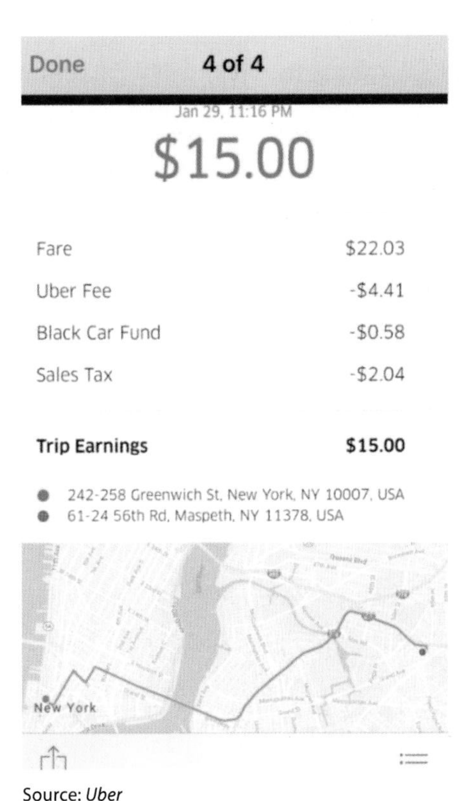

Source: *Uber*

UNEMPLOYMENT AND DISCOURAGED WORKERS

The U.S. unemployment rate surged between 2008 and 2010, but has dropped since then (*Figure 11.2*). Despite the drop, only 63 percent of Americans age 16 and older have a job or are looking for one—the lowest share of the population participating in the labor force since 1978. The share of men aged 25 to 54, the prime of working life, has been declining since 1965. As a result, almost 12 percent (7 million men) aren't employed or looking for work. Large numbers are Black, veterans, or ex-prisoners; have a high school diploma or less; are on painkillers; or receive disability benefits (Eberstadt, 2016; Krueger, 2016; White House Council of Economic Advisers, 2016).

Figure 11.2 U.S. Unemployment Rate, 2000 to 2017

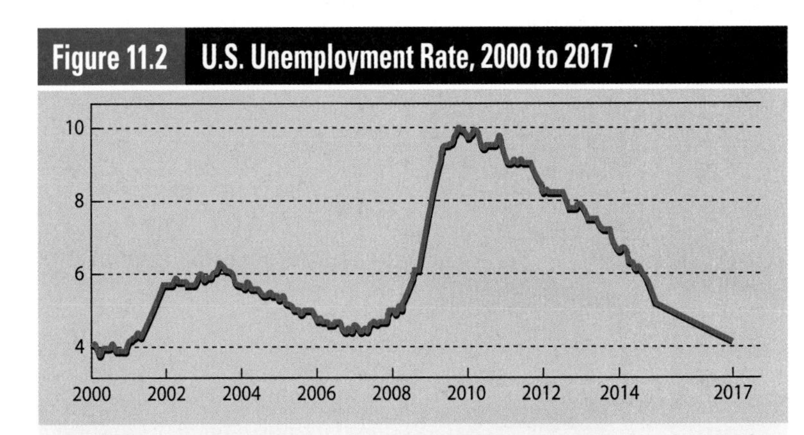

Source: Bureau of Labor Statistics, 2017, "Unemployment Rate." Accessed June 8, 2017 (data.bls.gov).

About 24 percent of the jobless are *long-term unemployed*, people who have searched for but not found a job for 27 weeks or longer. The longer that people are unemployed, the harder it is to find a job: Only 11 percent find steady, full-time employment a year later. Many of the long-term unemployed become **discouraged workers**, people who stop looking for work because they believe that job hunting is futile. About 5 percent of Americans are discouraged workers who aren't included in official unemployment rates (Center for Economic and Policy Research, 2015; BLS News Release, 2017).

JOB SATISFACTION AND PRODUCTIVITY

A whopping 69 percent of Americans report being "not engaged" or "actively disengaged" in their jobs, meaning that they're uninvolved, unhappy that their needs aren't being met, and put in as little time as possible (Harter and Adkins, 2017). Service workers are less engaged than managers and executives (28 percent and 40 percent, respectively), but 60 percent of the latter aren't engaged or are actively disengaged. Across generations, people born after 1980 are the least engaged, primarily because they believe that their jobs don't allow them to use their talents and strengths (Adkins, 2015).

In most European countries, the typical workweek is 35 hours—comparable with the U.S. definition of part-time work. Americans work nearly 25 percent more hours than Europeans, which amounts to an additional 258 hours per year. Because taxes are much higher in Europe than in the United States and Europeans get generous pensions, they have less incentive to work longer hours. Despite longer workweeks, the United States ranks fifth in productivity compared with 34 other countries. Luxembourg, the most productive country, has an average workweek of just 29 hours. Luxembourg's workplace and government policies encourage employees to make time for family and

discouraged workers people who stop looking for work because they believe that job hunting is futile.

Jeff Zelevansky/The New York Times/Redux

In good times and in bad, Black men have the highest unemployment rates. And compared with their White counterparts, Black workers are unemployed longer after their unemployment insurance benefits end. Pictured here are laid-off workers examining listings at a job fair.

Table 11.1 Vacation Days, on Average, that Workers Receive and Use

	Receives	Uses
France	30	30
Spain	30	26
Italy	30	25
Germany	30	28
Denmark, Norway, and Sweden	25	25
Japan	20	10
Ireland	21	21
United States	15	12
South Korea	15	8

Source: Based on Expedia's Vacation Deprivation Study, 2016.

other commitments, only 3 percent work 50 hours or more per week, and all employees have a minimum of five weeks of paid vacation (Bick et al., 2017; OECD, 2017).

The United States is the only industrialized nation in the world that doesn't legally guarantee workers a paid vacation; 25 percent of Americans have no paid vacation *and* no paid holidays. In contrast, the typical worker, especially in Western Europe, has at least 6 weeks of paid vacation, regardless of job seniority or the number of years worked (Ray et al., 2013; Greenfield, 2015). U.S. workers have the least vacation days and don't use all of them (*Table 11.1*). Workers in France receive and take 30 days, but 90 percent complain that they're "vacation deprived" ("Vacation Habits…," 2013).

In 2014, 41 percent of Americans didn't take any vacation days. The rest shrank their vacations to a few days or long weekends because they couldn't afford a vacation, had too much to do on the job, wanted to impress their boss with their dedication, or feared being replaced while away. Employees who forfeit paid time off don't get more raises or bonuses and report higher stress than those who take all of their vacation time, but employers save almost $53 billion a year when people work for free (U.S. Travel Association, 2014; Dickey, 2015; Mudallal, 2015).

Laura Gangi Pond/Shutterstock.com

11-3d Women and Minorities in the Workplace

One of the most dramatic changes in the United States during the twentieth century was the increase of women in the labor force (*Table 11.2*). Many factors have contributed to the surge in women's employment, especially since the 1970s, including the growth in the number of college-educated women (who, consequently, had more job opportunities), an increase in the number of working single mothers, and the higher costs of homeownership that require two incomes.

Largely because of their higher educational attainment, 29 percent of women in two-income marriages bring home the bigger paycheck, up from

Table 11.2 Women and Men in the Labor Force, 1890–2016

Percentage of Men and Women in the Labor Force			Women as a Percentage of All Workers
Year	Men	Women	
1890	84	18	17
1900	86	20	18
1920	85	23	20
1940	83	28	25
1960	84	38	33
1980	78	52	42
1990	76	58	45
2016	66	54	47

Source: Based on Labor Force Statistics, Current Population Survey, Table 3, "Employment Status," 2016. Accessed May 1, 2017 (bls.gov/cps).

only 4 percent in 1970 ("Wives Who Earn More . . . ," 2014). As a group, however, women have lower earnings than men in both the highest and lowest paying occupations, and the wage gaps are greater in high-income jobs (*Figure 11.3*).

There are earnings disparities across racial-ethnic groups, but the differences are especially striking by sex. As you examine *Figure 11.4*, note two general characteristics. First, earnings increase—across all racial-ethnic groups and for both sexes—as people go up the occupational ladder. But across all occupations, men have higher earnings than women of the same racial-ethnic group. At the bottom of the occupational ladder are Black women and Latinas, with the latter faring worse than any of the other groups. Thus, both gender *and* race-ethnicity affect earnings. (See Hegewisch et al., 2015, for a detailed analysis of the gender wage gap, by race and ethnicity, since 1955.)

Figure 11.3 Women Earn Less Than Men Whether They're CEOs or Cooks

These are five of the highest and lowest paid occupations of full-time, year-round U.S. workers in 2016. How might you explain why the earnings differ by sex, especially in the highest paid jobs?

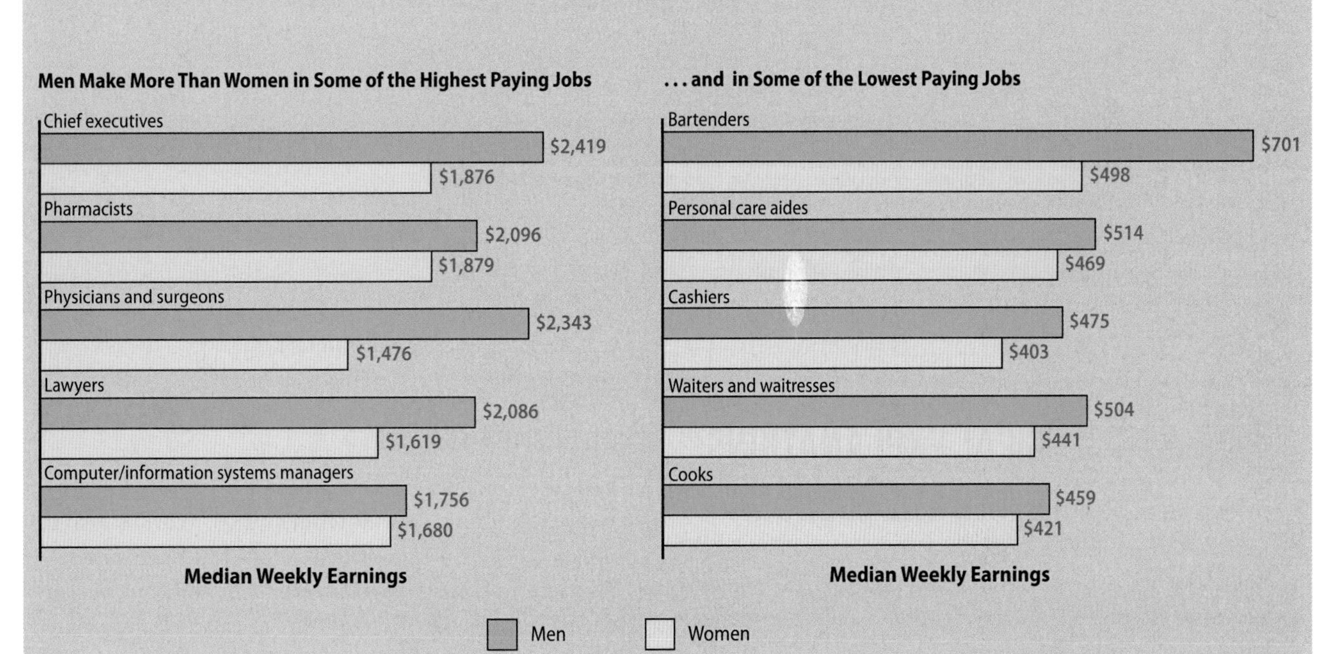

Note: Some of the differences between men's and women's median weekly earnings may seem small, but multiply each number by 52 weeks. Thus, in annual earnings, male physicians and surgeons average more than $121,836 compared with only $76,752 for females.

Source: Based on Labor Force Statistics, Current Population Survey, 2017, Table 39, "Median Weekly Earnings of Full-Time Wage and Salary Workers by Detailed Occupation and Sex." Accessed July 1, 2017 (bls.gov/cps).

Figure 11.4 Median Weekly Earnings of Full-Time Workers by Occupation, Sex, Race, and Ethnicity

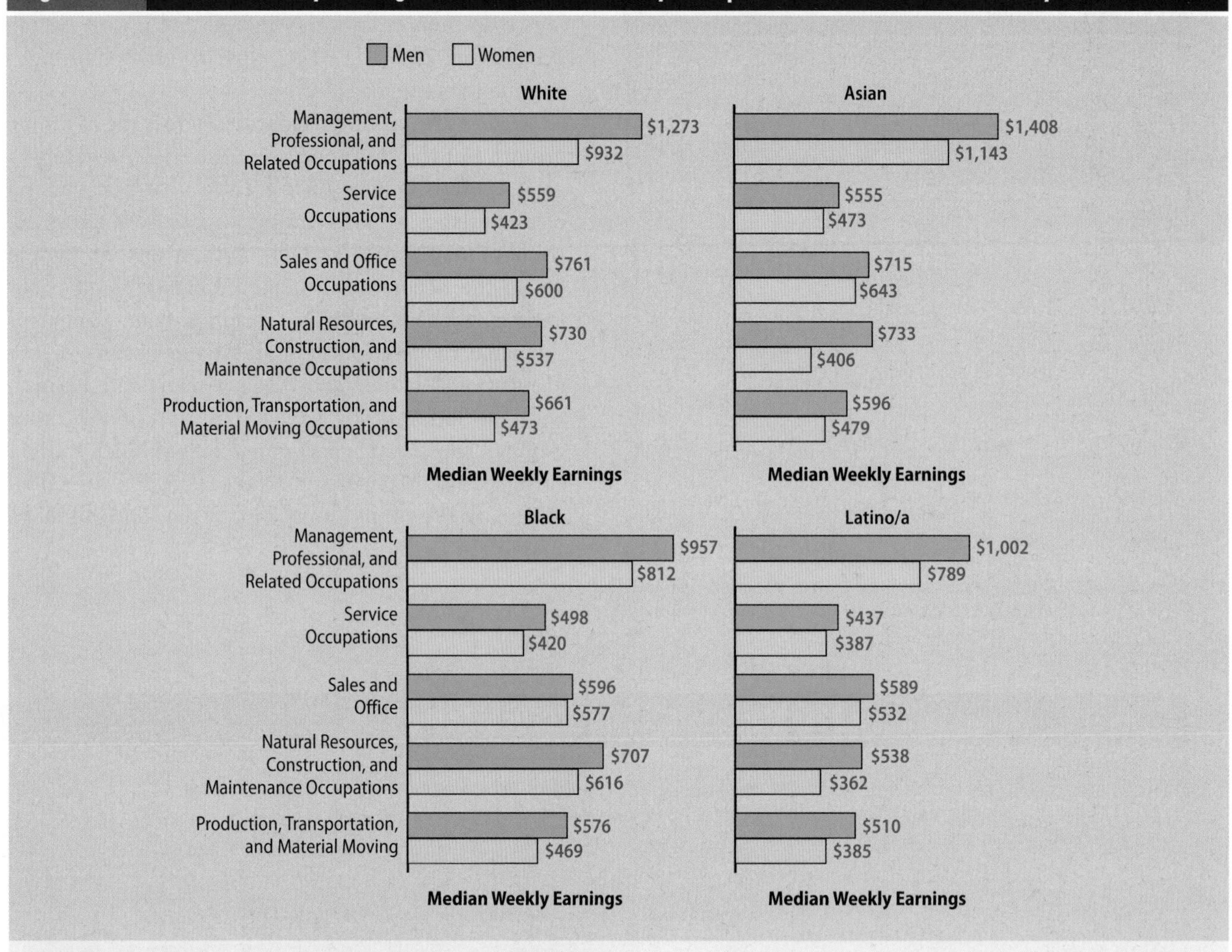

Note: Examples of the occupations include the following:
Managerial and Professional—Executives, managers, public administrators
Service—Private household workers, police, firefighters, people in food and health services, janitors
Sales and Office—Supervisors, sales representatives, office administrators
Natural Resources, Construction, and Maintenance—Farmers, fishing industries, construction workers, auto mechanics, machine repairers
Production, Transportation, and Material Moving—Truck drivers, assembly-line workers, equipment cleaners

Source: Based on Bureau of Labor Statistics, *The Editor's Desk*, 2011. Accessed April 15, 2014 (bls.gov/opub/ted/2011/ted_20110914.htm).

11-4 SOCIOLOGICAL EXPLANATIONS OF WORK AND THE ECONOMY

Sociological theories offer different insights on the economy and its impact on society. *Table 11.3* summarizes each perspective's key points.

11-4a Functionalism

Functionalists typically emphasize the benefits of work and the economy. They also see capitalism as bringing prosperity to society as a whole.

SOME KEY ISSUES

For functionalists, work is necessary for a society's survival and defining its members' roles. Work is also important because it connects people to each other: As jobs become more specialized, a small group of workers is responsible for getting the work done, and coworkers can get to know one another (Parsons, 1954, 1960; Merton, 1968). Such networks increase workplace solidarity and enhance a sense of belonging to a group where people listen to each other's ideas, interact, and "share a vision for the work [they] do together" (Gardner, 2008: 16).

Table 11.3	Sociological Explanations of Work and the Economy	
Theoretical Perspective	**Level of Analysis**	**Key Points**
Functionalist	Macro	Capitalism benefits society; work provides an income, structures people's lives, and gives them a sense of accomplishment.
Conflict	Macro	Capitalism enables the rich to exploit other groups; most jobs are low-paying, monotonous, and alienating; productivity isn't always rewarded.
Feminist	Macro and micro	Gender roles structure women's and men's work experiences differently and inequitably.
Symbolic Interactionist	Micro	How people define and experience work in their everyday lives affects their workplace behavior and relationships with coworkers and employers.

Besides providing income, work has social meaning. For many Americans, jobs offer a sense of accomplishment and feeling valued; stability, order, and a daily rhythm; and a network of interesting social and professional contacts (Katzenbach, 2003; Brooks, 2007). Functionalists also maintain that wage inequities spur people to persevere and to set higher goals. Low-paying jobs, for example, can motivate people to work harder or obtain further education to move up the economic ladder (see Chapter 8).

CRITICAL EVALUATION

Critics point out that work is a significant source of stress for 69 percent of Americans and often leads to myriad health problems (American Psychological Association, 2014). Many people have unstimulating and low-paid jobs that require repetitive and routine tasks, instead of providing a connection to a product or a group (see the discussion of McDonaldization in Chapter 6). Even in higher paying jobs, only a third of physicians and nurses report being engaged in their work (Gallup, 2013).

Functionalists also gloss over many U.S. corporations' disinterest in workers' well-being. In the private sector, 30 percent of full-timers and 74 percent of part-timers don't have paid sick days (Williams and Gault, 2014). Even when business improves, many employers convert full-time jobs to part-time or temporary positions, and cut health benefits (Schultz, 2011).

11-4b Conflict Theory

Whereas functionalists emphasize the economy's benefits, conflict theorists argue that capitalism creates social problems. They contend, for example, that globalization has led to job insecurity, and that a handful of transnational conglomerates have enormous power.

SOME KEY ISSUES

For conflict theorists, low wages alienate employees rather than motivate them to work harder, as functionalists claim. Many people are stuck in low-paying, monotonous jobs and don't always get promotions regardless of their skills, attitudes, perseverance, or productivity (Clements, 2012).

Conflict theorists also argue that capitalism enables the rich and powerful to exploit other groups, resulting in huge wealth disparities. In most industrialized countries, CEOs earn 10 to 25 times more than the average worker. In some of the largest U.S. companies, CEOs earn 1,000 times more than their employees. Between 1978 and 2014, CEO pay increased by 937 percent compared with only 10 percent for workers. Because executive compensation boards are "packed with insiders who parcel

AP Images/Connecticut Post, Ned Gerard

For functionalists, the economy provides order and stability. Many unemployed men—including recent Asian and Latino immigrants—join the military because it offers steady work, college education benefits, and an opportunity for career advancement.

ARE CONFLICT THEORISTS TOO QUICK TO BLAME CAPITALISM FOR ECONOMIC PROBLEMS?

out rewards to their friends," it's not unusual for CEOs to be paid at least $60 million a year. And, unlike other industrialized countries, the U.S. government doesn't cap corporate pay (Bivens and Mishel, 2015; Krantz, 2015; "Executive Pay," 2016: 18; Weaver, 2016).

Instead of higher wages, annual raises, and health benefits, more businesses are rewarding employees with perks (e.g., free gym memberships, $100 gift cards, pet health insurance). Working-class families' eroding job opportunities and wages have led to personal problems such as marital discord and mental health problems. As economic problems accumulate, "deaths of despair"—death by drugs, alcohol, or suicide—increase, especially among middle-aged White men with low education levels (Mui, 2015; Case and Deaton, 2017; see also Chapter 14).

ivector/Shutterstock.com

In 2017, the median CEO compensation was nearly $16 million. In contrast, the median annual wages for full-time workers was only $44,980—a CEO-to-worker pay ratio of 356 to 1 (Economic News Release, 2017; Marcec, 2017).

CRITICAL EVALUATION

Critics fault conflict theorists for emphasizing economic constraints rather than choices. For example, almost a third of households making more than $75,000 a year live paycheck-to-paycheck because they enjoy "lifestyle purchases" like dining out and entertainment ("SunTrust: Many Higher Income . . . ," 2015).

Also, according to functionalists, conflict theorists underestimate people's ability to get ahead. Many young people are thronging schools that teach high-tech skills in welding and other trades, fast-food workers are unionizing to raise the industry's minimum wage to $15.00 an hour, and 63 percent of small-business owners find qualified new employees through word-of-mouth referrals (Berfield, 2013; Jacobe, 2013; Phillips, 2014). Thus, many Americans are using several strategies to succeed economically.

11-4c Feminist Theories

Feminist scholars agree with conflict theorists that there's widespread workplace inequality. They emphasize, however, that gender is a critical factor in explaining the inequity.

SOME KEY ISSUES

At all income levels and in all occupations, women, especially Latinas and Black women, earn less than men (*Figure 11.4*). There's a 40 percent gender wage gap—after accounting for race, ethnicity, education, occupation, workplace experience, and other factors—that's presumably due to gender discrimination (Blau and Kahn, 2016). Both historically and currently, as you saw in Chapters 6 and 9, women have lower earnings than men because of occupational sex segregation, glass ceilings, and glass escalators.

Employment rates increased after the recession, but the gains for women have been largely in low-paying jobs, particularly waitressing, in-home health care, food preparation, and housekeeping. From 2009 to 2012, 60 percent of the increase in women's employment, versus 20 percent for men, was in jobs that pay less than $10.10 an hour. Women make up less than half of the labor force but hold more than two-thirds of all low-wage jobs (e.g., personal and home health aides, fast-food workers, lowest-paid retail positions). These female-dominated low-wage jobs, which will be the fastest-growing until 2024, will leave millions of women and families struggling to make ends meet (National Women's Law Center, 2014; Robbins and Vogtman, 2016).

Many women's earnings suffer from a **motherhood penalty** (also called *motherhood wage penalty* or *mommy penalty*), a pay gap between women who are and aren't mothers. On average, women without children make 90 cents to a man's dollar, married mothers make 73 cents to a man's dollar, and single mothers make only 60 cents to a man's dollar (Rowe-Finkbeiner, 2012; Entmacher et al., 2014).

After declining for several decades, the share of stay-at-home mothers rose from 23 percent in 1999 to 29 percent in 2012. There are many reasons for this increase, but low wages and motherhood penalties are at the top of the list. Many mothers in two-income households have dropped out of the labor force because child care costs rose more than 70 percent from 1985 to 2011 (Cohn et al., 2014).

CRITICAL EVALUATION

Critics note that men don't always dominate women in the workplace. There are many situations where female supervisors have power and authority over women and men, and top female managers aren't always concerned about gender wage gaps nor support qualified women's promotions (Huffman, 2013). Second, many feminist scholars maintain that capitalism exploits women by crowding them into lower-paying occupations. However, there's considerable economic gender inequality in socialist, communist, and mixed economies in both industrialized and developing nations (Cudd and Holstrom, 2010).

Some people also question whether the economy reinforces sex discrimination or simply reflects cultural sexism. For example, 51 percent of Americans, including women, say that children are better off if the mother is at home and doesn't work; only 8 percent say the same about stay-at-home dads (Cohn et al., 2014). Thus, patriarchal values may be as important as discrimination in explaining many women's economic inequality.

11-4d Symbolic Interaction

Symbolic interactionists rely on micro-level approaches to explain the day-to-day meaning of work. They're especially interested in how work shapes people's self-identity.

SOME KEY ISSUES

Symbolic interaction has provided numerous insights on how people define and experience work. Regardless of income level, 63 percent of U.S. workers who are engaged in their jobs, and even 42 percent of those who aren't engaged, say they would continue to work in their current job if they won a $10 million lottery. All of them might quit, of course, but such responses show that work provides many people with "a source of identity, purpose, and satisfaction that money alone may not replace" (Harter and Agrawal, 2013).

iStock.com/CatLane

Would you work if you won a $10 million lottery? Why or why not?

Interactionists also study how people are socialized into their jobs and the informal rules that shape behavior. In medicine, surgeons teach their interns and residents that it's normal to make some mistakes, but that carelessness and continued errors are unprofessional (Bosk, 1979). Assembly-line workers also control coworkers who overproduce or underproduce by punishing or rewarding them (see Chapter 6).

CRITICAL EVALUATION

The most common criticism is that symbolic interaction, although providing in-depth analyses, sacrifices scope. Studies of small work groups can tell us much about people's interaction. We don't know, however, if the findings are applicable to large companies or different parts of the country because the research is based on small and nonrepresentative samples.

Another limitation is that interactionism neglects macro-level social forces that affect people's work and choices. Consider government: When it allows monopolies, consumers must pay high prices for low-quality goods and services because there are no options. Low wages and discrimination, which most politicians ignore, alienate workers and increase personal and family stress. Thus, structural obstacles can thwart individual choices and efforts.

11-5 GLOBAL POLITICAL SYSTEMS

Every society has a **government**, a formal organization that has the authority to make and enforce laws. Governments are expected to maintain order, provide social services, regulate the economy, establish educational systems, create armed forces to discourage (real or imagined) attacks by other countries, and ensure their residents' safety.

motherhood penalty (also called *motherhood wage penalty* or *mommy penalty*) a pay gap between women who are and aren't mothers.

government a formal organization that has the authority to make and enforce laws.

Figure 11.5 Political Freedom Around the World, 2017

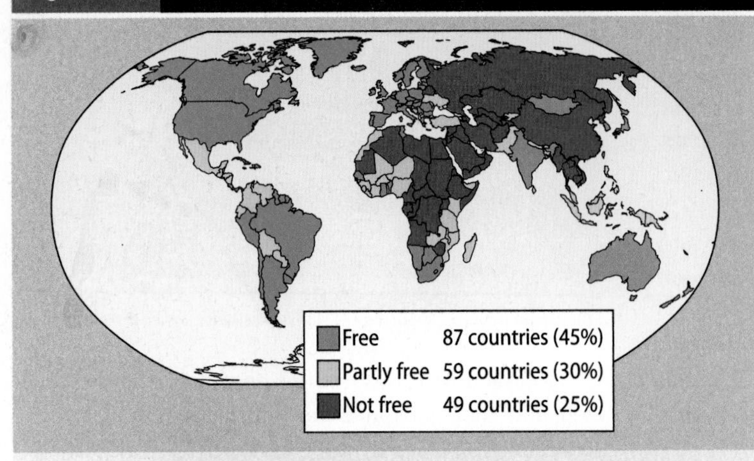

This map shows the degree of political freedom and civil liberties in 195 countries around the world. *Political freedom* includes free and fair elections, competitive parties, and no discrimination against minorities. *Civil liberties* include freedom of speech and the press, to practice one's religion, and to discuss political issues without fear of physical violence or intimidation by the government.

Source: Freedom House, 2015; Puddington and Roylance, 2017.

Worldwide, governments vary from democratic to totalitarian, but there are also authoritarian governments and monarchies. Let's begin by looking at democracy.

11-5a Democracy

A **democracy** is a political system in which, ideally, citizens have control over the state and its actions. Democracies are based on several principles:

- Individuals participate in decisions, and select leaders who are responsive to the majority of the people's wishes.
- Suffrage (the right to vote) is universal, and elections are frequent, free, fair, and secret.
- The government recognizes individual rights, such as freedom of speech (including dissent), press, and assembly, and the right to organize political parties whose members compete for public office.
- The "rule of law" requires everyone to obey the law and to be held accountable if they violate it.

Worldwide, 36 percent of the world's population (2.7 billion people) live in countries with repressive governments that deny basic political and civil rights. People in sub-Saharan and North Africa, the Middle East, China, Eurasia, and Russia are the least likely to have political freedom (*Figure 11.5*). Despite some countries' attempts to establish democracies, democracy has declined around the world for 11 consecutive years (Puddington and Roylance, 2017).

democracy a political system in which, ideally, citizens have control over the state and its actions.

totalitarianism the government controls almost every aspect of people's lives.

11-5b Totalitarianism and Dictatorships

At the opposite end of the continuum from democracy is **totalitarianism**, a political system in which the government controls almost every aspect of people's lives. Totalitarianism has several distinctive characteristics (Taylor, 1993; Tormey, 1995; Arendt, 2004):

- A pervasive ideology that legitimizes state control and instructs people how to act in their public and private lives.
- A single political party controlled by one person, a *dictator*—a supreme, sometimes idolized leader—who stays in office indefinitely.
- A system of terror that relies on secret police and the military to intimidate people into conformity and to punish dissenters.
- Total control by the government over other institutions, including the military, education, family, religion, economy, media, and all cultural activities, including the arts and sports.

Timofeev Sergey/Shutterstock.com

Some political analysts believe that Russia is moving from authoritarianism to totalitarianism. Vladimir Putin, Russia's prime minister/president since 1999, censors much of the Internet content, crushes gay and other civil rights activists, represses his regime's critics, and is resurrecting the KGB—a secret police force and spy network involved in all aspects of people's everyday lives (Ryzhkov, 2015; Soldatov, 2016).

In 1995, at age 3, Oyo Nyimba Kabamba Iguru became the world's youngest monarch when he was crowned as king of the Ugandan Kingdom of Toro. At age 18, he officially took control of the kingdom.

Peter Busomoke/Getty Images

Contemporary examples of the most repressive totalitarian governments include Central African Republic, North Korea, Saudi Arabia, Somalia, Sudan, Syria, Iran, and Uzbekistan. Within these countries and territories, state control over daily life is pervasive and wide-ranging, independent organizations and political opposition are banned or suppressed, and those who criticize the government are imprisoned, tortured, or killed (Puddington and Roylance, 2017).

11-5c Authoritarianism and Monarchies

Many nations have some version of democracy or totalitarianism. However, a number of countries are characterized by **authoritarianism**, a political system in which the state controls the lives of citizens but permits some degree of individual freedom. Authoritarian governments, like Russia, may permit elections, often controlled, but prohibit free speech, punish those who publicly question state policies, and imprison people without giving them a trial.

In a **monarchy**, the oldest type of authoritarian regime, power is based on heredity and passes from generation to generation. A member of a royal family, usually a king or queen, reigns over a kingdom. A monarch's power and authority are legitimized by tradition and religion. There are more than 40 monarchies around the world. In some countries—especially in the Middle East and parts of Africa—monarchs have absolute control. Those in Norway, Denmark, Belgium, Britain, Japan, and many other countries have little political power because they're limited by democratic constitutions and serve primarily ceremonial roles (Glauber, 2011).

We see, then, that political systems vary worldwide. Despite the variation, political leaders wield considerable power and authority.

11-6 POLITICS, POWER, AND AUTHORITY

Power and authority are important concepts in sociological analyses of politics. The use and misuse of power and authority affect the quality of our everyday lives.

11-6a Power

Power is the ability of a person or group to influence others, even if they resist. For example, a government can quash a peaceful protest or, conversely, protect people's right to protest. *Legitimate power* comes from having a role, position, or title that people accept as legal and appropriate (e.g., Congress has the right to pass laws, the police have the right to enforce laws). *Coercive power* relies on force or the threat of force to impose one's will on others (e.g., a dictator who imprisons or executes protestors).

Power is relational. Lawmakers affect their constituents' lives because of the laws they pass on taxes, the environment, and other issues. Voters can protest, refuse to follow laws, and vote politicians out of office.

Whether legitimate or coercive, power—especially political power—is about controlling others. Because people may revolt against sheer force, many governments depend on authority to establish order, shape people's attitudes, and control their behavior.

11-6b Authority

Authority, the legitimate use of power, has three characteristics. First, people *consent* to authority because they believe that their obedience is for the greater good (e.g., following traffic rules). Second, people see the authority as *legitimate*—valid, justifiable, and necessary (e.g., paying taxes to support public education, increase national security, and remove trash and snow). Third, people accept authority because it's *institutionalized*, accepted by a large number of people, in organizations (e.g., police departments and government agencies).

Max Weber (1925/1978) described three ideal types of legitimate authority: traditional, charismatic, and rational-legal (*Table 11.4*). Remember that ideal types

authoritarianism the state controls the lives of citizens but permits some degree of individual freedom.

monarchy power is based on heredity and passes from generation to generation.

power the ability of a person or group to influence others, even if they resist.

authority the legitimate use of power.

Table 11.4 Weber's Three Types of Authority

Type of Authority	Description	Source of Power	Examples
Traditional	Power is based on customs, traditions, and/or religious beliefs.	Personal	Medieval kings and queens, emperors, tribal chiefs
Charismatic	Power is based on exceptional personal abilities or a calling.	Personal	Adolf Hitler, Gandhi, Martin Luther King, Jr.
Rational-legal	Power is based on the rules and laws that are inherent in an elected or appointed office.	Formal	U.S. presidents, members of Congress, state officials, police, judges

are models that describe the basic characteristics of any phenomenon (see Chapters 1 and 5). In reality, these types of authority often overlap.

TRADITIONAL AUTHORITY

Traditional authority is power based on customs that justify the ruler's position. The source of power is personal because the ruler inherits authority on the basis of long-standing customs, traditions, or religious beliefs. Traditional authority is most common in nonindustrialized societies where power resides in kinship groups, tribes, and clans. In many African, Middle Eastern, and Asian countries, power is passed down among men within a family line (Tétreault, 2001; see also Chapter 8).

ChinaFotoPress/Getty Images

First Lady Michelle Obama is an example of a charismatic political leader. She worked to end childhoood obesity, helped military families, and visited other countries as a goodwill ambassador.

CHARISMATIC AUTHORITY

Charismatic authority is power based on exceptional individual abilities and characteristics that inspire devotion, trust, and obedience. Like traditional authority, charismatic authority is personal and reflects extraordinary deeds or even a belief that a leader has been chosen by God, but the leaders don't pass their power down to their offspring.

Charismatic leaders can inspire loyalty and passion whether they're heroes or tyrants. Examples of the latter include historical figures Adolf Hitler, Napoleon, Ayatollah Khomeini, and Fidel Castro. These and other dictators have been spellbinding orators who radiated magnetism, dynamism, and tremendous self-confidence, and promised to improve a nation's future (Taylor, 1993).

traditional authority power based on customs that justify the ruler's position.

charismatic authority power based on exceptional individual abilities and characteristics that inspire devotion, trust, and obedience.

rational-legal authority power based on the belief that laws and appointed or elected political leaders are legitimate.

RATIONAL-LEGAL AUTHORITY

Rational-legal authority is power based on the belief that laws and appointed or elected political leaders are legitimate. Unlike traditional and charismatic authority, rational-legal authority comes from rules and regulations that pertain to an office rather than to a person. For example, anyone running for mayor must have specific qualifications (e.g., U.S. citizenship). When a new mayor (or governor or other politician) is elected, the rules don't change because power is vested in the office rather than the person currently holding the office.

MIXED AUTHORITY FORMS

People may enjoy more than one type of authority. Some historians believe that presidents Abraham Lincoln, Theodore Roosevelt, John F. Kennedy, and Ronald Reagan enjoyed charismatic appeal beyond their rational-legal authority. In some countries, including Japan and England, emperors, kings, and queens have traditional authority and perform symbolic state functions (e.g., attending ceremonies). Both of these countries also have parliaments that exercise legal-rational authority in determining laws and policies.

11-7 POLITICS AND POWER IN U.S. SOCIETY

Politics plays a critical role in our everyday lives. Three important components of the political system are political parties, special-interest groups, and voters.

11-7a Political Parties

A **political party** is an organization that tries to influence and control government by recruiting, nominating, and electing its members to public office. Political parties are avenues for citizens to shape public policy at the local, state, and national levels.

FUNCTIONS OF U.S. POLITICAL PARTIES

Political parties engage in a range of activities—from mailing flyers and calling voters to drafting laws if a candidate is elected. Parties perform a number of vital political functions, especially recruiting candidates for public office, organizing elections, and if elected, running the government.

Ideally, that's how political parties *should* work. In reality, parties perform some of these functions more effectively than others. For example, parties typically concentrate on winning elections rather than passing laws that benefit the general population once candidates are in office.

THE TWO-PARTY SYSTEM

Many democracies around the world have a number of major political parties: 3 in Canada, 5 in Germany, 9 in Italy and Holland, 34 in Israel, and more than 1,000 in India. Thus, compared with many other countries, the two-party system of Democrats and Republicans in the United States is uncommon. Political parties base their activities on an *ideology*, a set of ideas that constitute a person's or group's beliefs, goals, expectations, and actions. *Table 11.5* gives some examples of major ideological differences between Democrats and Republicans.

Until the late 1980s, about equal percentages of Americans identified themselves as either Republicans or Democrats. Because of dissatisfaction with both parties, however, 28 percent identify themselves as Republicans, 29 percent as Democrats, and 39 percent as independent, down from a record high of 43 percent in 2014. A large majority (60 percent) of Americans say that we need a third major party that would do a better job of representing the people (McCarthy, 2015; Jones, 2017). Because there's no national Independent Party,

political party an organization that tries to influence and control government by recruiting, nominating, and electing its members to public office.

Table 11.5 How Do Democrats and Republicans Differ?

Issue	Many Democrats Believe That...	Many Republicans Believe That...
Economy	The government should raise taxes on the wealthy and corporations, and expand programs for the poor.	The government should lower taxes on the wealthy and corporations to encourage investment and economic growth; poor people are too dependent on the government.
Abortion	Women should have the right to choose abortion in practically all cases; the government should pay for abortions for low-income women.	The unborn should be protected in all cases; there should be no public money for abortions; *Roe v. Wade* should be overturned.
Women's Rights	Significant obstacles still make it harder for women to get ahead than men; women should have paid child care and family leave, and equal pay.	The obstacles that made it harder for women to get ahead are largely gone; there are as many or more women than men in powerful positions; women have equal or more financial stability than men.
Education	The government should strengthen public schools by raising salaries for teachers and decreasing classroom sizes. It shouldn't use tax dollars to help students attend private schools.	The government should increase state and local control of schools. It should use tax dollars to fund students to attend private schools of their choice if their local school is underperforming.
Immigration	Immigrants help the economy by creating small businesses, taking jobs Americans don't want, and accepting lower wages that keep the costs of goods and services down; immigrants have enriched U.S. culture through food, music, and the arts.	Immigrants hurt the economy by driving down wages and decreasing the incomes of Americans who work in the lowest-paying jobs, increasing crime rates, and paying few (if any) taxes.

Sources: Doherty et al., 2014; Wilke and Newport, 2014; Dugan, 2015; Fingerhut, 2016, 2017; McCarthy, 2017; PerryUndem, 2017; Swift, 2017.

In 5 to 4 decisions in 2010 and 2014, the Supreme Court overturned previous limits on federal campaign donations. Now, super PACs can contribute as much as they want.

however, independents usually vote for Democrats, Republicans, or don't vote at all.

Especially at the national level, political candidates routinely promise that, if elected, they'll reform campaign spending, curb lobbying, and stop wasting taxpayer money. Instead, special-interest groups play an important role in political outcomes.

11-7b Special-Interest Groups

A **special-interest group** (also called an *interest group*) is made up of people who seek or receive benefits or special treatment. Whereas political parties include diverse individuals, special-interest groups are usually made up of people who are similar in social class and political objectives. There are thousands of U.S. special-interest groups that include everything from agribusiness to unions. Special-interest groups use many tactics to influence the government, but the most effective are campaign contributions and lobbying.

CAMPAIGN CONTRIBUTIONS

Eighty-four percent of Americans agree that money has too much influence in our political campaigns, but it's often the most important factor in winning elections. Money doesn't guarantee a victory, but candidates with little financing usually have little public visibility. Among all presidential and congressional candidates, the spending doubled from $3.1 billion in the 2000 election to nearly $6.5 billion in the 2016 election (Montanaro et al., 2014; Kamp, 2016; Center for Responsive Politics, 2017).

special-interest group (also called an *interest group*) a group of people that seeks or receives benefits or special treatment.

political action committee (PAC) a special-interest group that raises money to elect one or more candidates to public office.

lobbyist someone hired by a special-interest group to influence legislation on the group's behalf.

Contributions come from individuals and national, state, and local party committees, but the most powerful are **political action committees (PACs)**, special-interest groups that raise money to elect one or more candidates to public office. PACs can contribute directly to candidates but have donation limits. Super PACs can't contribute directly to political candidates, but may raise and spend unlimited funds to advocate for or against candidates.

PACs have mushroomed—from 608 in 1974 to almost 15,700 in 2014 (35 percent represent corporate interests). During 2016 alone, 2,389 super PACs collected nearly $3 billion (Center for Responsive Politics, 2017). Some PACs, like the International Brotherhood of Electrical Workers, support primarily Democrats, whereas the National Rifle Association contributes only to Republicans. A number of PACs, however, contribute to both parties to cover their bases regardless of who's elected. PACs and super PACs provide access to politicians and obligate lawmakers to pass legislation that a PAC wants.

LOBBYISTS

A **lobbyist** is someone hired by a special-interest group to influence legislation on the group's behalf. At the federal level, the number of lobbyists rose from 10,400 in 1998 to almost 11,200 in 2016, an average of 21 lobbyists per congressional member. Some of the best-paid lobbyists include former members of Congress, former congressional aides, and lawmakers' relatives (Drutman and Furnas, 2014; "Lobbying Database," 2017).

In 2016, lobbyists spent $3.2 billion to send members of Congress and their family members on expensive vacations, and, most importantly, made generous campaign contributions. In return, corporations and other organizations receive favorable legislation, such as Congress passing lax gun and banking laws, approving corporate mergers, and not investigating monopolies (Schweizer, 2013; "Top Spenders," 2017).

Lobbying also generates *earmarks* (known commonly as "pork")—funding requests by Congress to provide federal money to companies, projects, groups, and organizations in their district—that don't require external competitive bidding. After considerable public pressure, Congress banned earmarks in 2010, but pork spending continues. In 2016, taxpayers paid $5.1 billion, a nearly 90 percent increase since 2014, for pork barrel projects. Examples include $1 billion for a destroyer that the Pentagon didn't ask for, $164 million to fund education programs that the Obama administration didn't request, and $10 million to provide farmers with electricity that they don't need (Citizens Against Government Waste, 2016).

During the 2016 presidential campaign, Donald Trump repeatedly promised to "drain the swamp" of special-interest groups and lobbyists. Shortly after taking office, however, he packed his administration with corporate executives, corporate lawyers, and lobbyists: "Auto industry lobbyists are setting transportation policy . . . Wall Street is in control of financial policy and regulatory agencies, and corporate defense lawyers staff the key positions in the Justice Department" (Zibel, 2017).

11-7c Who Votes, Who Doesn't, and Why

In 2016, 62 percent of Americans voted in the presidential election, about the same as in 2012. These were larger than usual percentages, but lower than the numbers of those who voted in presidential elections during the 1950s and 1960s (File and Crissey, 2010; File, 2017).

Of 110 countries around the world that hold democratic presidential elections, 69 have higher voter turnout rates than the United States; in 17 countries, at least 80 percent of the eligible population votes (Friedman, 2012; International Institute for Democracy and Electoral Assistance, 2014).

Why are U.S. voting rates so low? Who votes, who doesn't, and why reflect demographic characteristics, attitudes about politics, and situational and structural factors.

DEMOGRAPHIC FACTORS

Many demographic factors (e.g., marital status, geographic region, religion) affect registration and voting. The most important, however, are sex, age, social class, and race and ethnicity.

Sex. In every presidential election since 1996, women—and across all racial/ethnic groups—have voted at higher rates than men. In the 2016 election, 58 percent of women and 54 percent of men voted. The biggest gender voting gap was among Black people—Black women voted at higher rates than Black men by 9 percentage points. The reasons for this difference are unclear, but may reflect higher turnouts by women who believe that social services, abortion rights, and health care coverage are threatened (Jones, 2012; U.S. Census Bureau, 2017).

Age. Historically, and across all elections, voting rates increase with age. In 2016, voting rates were higher among older than younger people (*Table 11.6A*). A key factor is registration. Only 55 percent of young people (ages 18–24) were registered to vote compared with 81 percent of those age 65 and older (U.S. Census Bureau, 2017).

Young people have lower registration rates because they're more geographically mobile than older people and are less likely to reregister after a move. Moreover, they often feel uninformed about politics, believe that elections aren't relevant in their lives, and may be preoccupied with major life events such as going to college and finding jobs (Schachter, 2004; Child Trends Data Bank, 2013). In effect, young people's low voting rates means that they have little effect on political processes.

Social Class. Social class has a significant impact on voting behavior. At each successive level of educational attainment, the voting rate increases, and those with advanced degrees are the most likely to vote (*Table 11.6B*). People with higher educational levels are usually more informed about and interested in the political process and more likely to believe that their vote counts (File and Crissey, 2010; "The Politics of Financial Insecurity," 2015).

Voting rates also increase with income levels (*Table 11.6C*). People with higher incomes are more likely to be employed and to have assets (e.g., houses, stock) and, therefore, are more likely to vote to protect or increase their

Table 11.6 Voter Rates in the 2016 Presidential Election, by Selected Characteristics

A. Age	
18–24	39%
25–44	49%
45–64	62%
65–74	70%
75 and older	66%
B. Education	
High school, no diploma	29%
High school graduate	47%
Some college or associate degree	61%
Bachelor's degree	69%
Advanced degree	74%
C. Annual Family Income	
Under $20,000	38%
$20,000 to $29,999	43%
$30,000 to $39,999	50%
$40,000 to $49,999	56%
$50,000 to $74,999	62%
$75,000 to $99,999	67%
$100,000 to $149,999	73%
$150,000 and over	77%

Source: Based on U.S. Census Bureau, 2017, Tables 5 and 7.

How Would *You* Describe Congress?

In 2010, the Pew Research Center asked Americans to use one word that described Congress. The responses were put into a "word cloud." The larger the word, the more times it was used. What one word would *you* use to describe Congress?

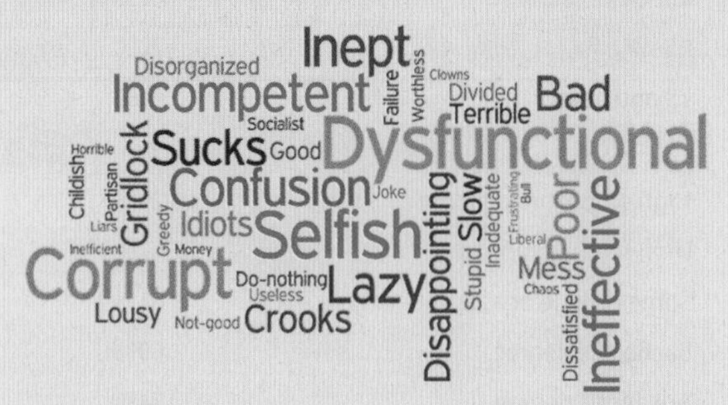

Word cloud graphic, created using http://wordle.net, from "Congress in a Wordle", Mar. 22, 2010, The Pew Research Center For the People & the Press, a project of the Pew Research Center.

resources. In contrast, low-income people usually have few assets and may be too disillusioned with the political system to vote ("Who Votes . . .," 2006; File and Crissey, 2010).

Race and Ethnicity. Across racial-ethnic groups, there are more native-born or naturalized White people who are eligible to vote. They're also the most likely to register and to vote. Minorities, especially Black people, played a pivotal role in Barack Obama's 2008 and 2012 presidential wins. In 2016, in contrast, 64 percent of White people versus 53 percent of minorities voted. Among minority groups, 60 percent of Black people voted; it was the lowest Black turnout since 2000 (Frey, 2017).

In the past and currently, Black voting rates have been higher than those of Latinos and Asians, but their turnout fell between 2012 and 2016 (*Figure 11.6*). About 75 percent of Asians and Latinos were born in the United States, but both groups had much lower voter registration rates (about 57 percent) than Black people (70 percent) and White people (74 percent) (U.S. Census Bureau, 2017).

The racial-ethnic voting gaps may close in the future as immigrants become eligible for citizenship, and as Latinos and Asians become more familiar with registration and voting practices. Low Asian voter turnout is puzzling, however: They have the highest education and income levels (see Chapters 8 and 10) but, even among the college-educated, voter turnout has lagged behind White and Black individuals (Krogstad, 2014; Frey, 2015).

ATTITUDES

Millions of Americans have a low opinion of the government and Congress: Only 20 percent trust the government to do what's right; 25 percent say that the government is the most important problem facing the United States; 75 percent (compared with 66 percent in 2009) believe there's widespread corruption; and trust in political leaders has plummeted—from 63 percent in 2004 to 42 percent in 2016. A day before the election, 61 percent of Americans rated Donald Trump as "totally unfavorable" and 52 percent said the same about Hillary Clinton. They were the most negatively reviewed presidential candidates since 1956, "and probably ever" (Clifton, 2016; Jones, 2016; Saad, 2016; "Public Trust . . .," 2017; Reinhart, 2017).

Such distrust and dissatisfaction helps explain why many people don't vote: They believe that their lives won't improve regardless of who's elected. On the other hand, few incumbents are voted out of office election after election. Why? Only 17 percent of registered voters say that members of Congress should be re-elected, but almost half believe that *their* congressional representatives are doing a good job (Dugan and Hoffman, 2014). Perhaps, then, "the government we have is the government we deserve" (Cillizza, 2014).

Figure 11.6 Voting Rates in Presidential Elections, by Race and Ethnicity: 1996–2016

Sources: Based on File, 2017, Figure 2, and Frey, 2017.

SITUATIONAL AND STRUCTURAL FACTORS

A quarter of registered voters didn't cast a ballot in the 2016 presidential election because they didn't like the candidates or campaign issues, but situational and structural factors also affect elections. Some of the high turnout rates in other countries are partly due to *compulsory voting*: The government imposes fines for not voting, may cut off public assistance benefits, or requires evidence of voting to get a passport or a driver's license. Such sanctions send the message that voting isn't a privilege but a civic responsibility. In many European and other countries, voters are registered automatically when they pay taxes or receive public services, elections are held on weekends, and people vote electronically (Seward, 2012; Samuelson, 2016; U.S. Census Bureau, 2017).

Instead of making voting easier, 34 states (particularly those controlled by Republican legislatures) have passed or are considering laws that will make voter registration more difficult. Some of the new rules require government-issued photo identification (e.g., a driver's license or passport) and a birth certificate or proof of citizenship. Some states have reduced early voting, limited the time polls are open, done away with same-day voter registration, and moved polling locations outside of areas with large racial-ethnic populations. Proponents contend that the new laws will crack down on voter fraud; opponents argue that the new regulations will stifle voter turnout, especially among students, the poor, and minorities, who tend to vote for Democrats. Of the more than 1 billion ballots cast from 2000 through 2014, there were only 31 documented incidents of voter fraud (Whiteaker et al., 2014; Rios, 2016; Underhill, 2017).

Because many Americans don't vote, who has the most power? And do government leaders represent the average citizen? Such questions have generated considerable debate among sociologists and other social scientists.

SOCIOLOGICAL PERSPECTIVES ON POLITICS AND POWER

According to symbolic interaction, power is socially constructed through interactions with others; people learn to be loyal to a political system—whether it's a democracy or a monarchy—and to show respect for its symbols and leaders. For the most part, however, sociologists rely on functionalist, conflict, and feminist theories to explain political power. *Table 11.7* summarizes these perspectives.

11-8a Functionalism: A Pluralist Model

For functionalists, the people rule through **pluralism**, a political system in which power is distributed among a variety of competing groups in a society (Riesman, 1953; Polsby, 1959; Dahl, 1961).

SOME KEY ISSUES

For functionalists, the government is an important institution that creates and enforces laws, regulates elections, and protects people's civil liberties. In the pluralist model, individuals have little direct power over political decision making but can influence government policies

pluralism a political system in which power is distributed among a variety of competing groups in a society.

Table 11.7 Sociological Explanations of Political Power

	Functionalism: A Pluralist Model	Conflict Theory: A Power Elite Model	Feminist Theories: A Patriarchal Model
Who has political power?	The people	Rich upper-class people—especially those at top levels in business, government, and the military	White men in Western countries; most men in traditional societies
How is power distributed?	Very broadly	Very narrowly	Very narrowly
What is the source of political power?	Citizens' participation	Wealthy people in government, business corporations, the military, and the media	Being White, male, and very rich
Does one group dominate politics?	No	Yes	Yes
Do political leaders represent the average person?	Yes, the leaders speak for a majority of the people.	No, the leaders are most concerned with keeping or increasing their personal wealth and power.	No, the leaders are rarely women who have decision-making power.

Figure 11.7 Pluralist and Power Elite Perspectives of Political Power

PLURALIST MODEL

Power is dispersed among multiple groups that influence the government. For example,

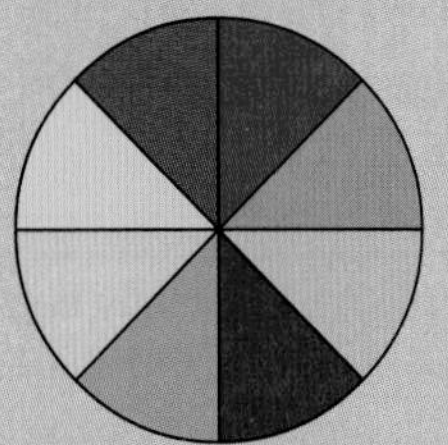

- government employees
- victims' rights groups
- labor unions
- banks and other financial institutions
- realtors and home builders
- teachers
- environmental groups
- women's rights groups

POWER ELITE MODEL

Power is concentrated in a very small group of people who make all the key decisions. For example,

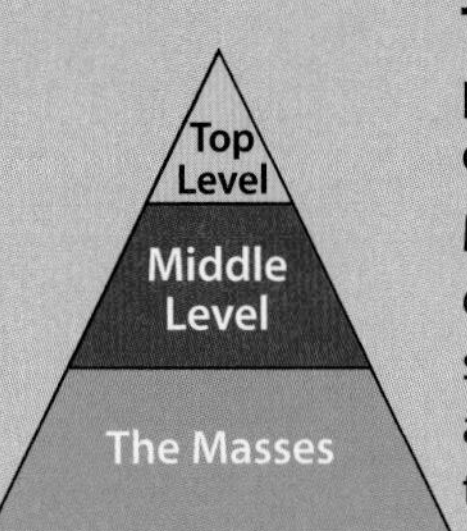

Top level (1 percent)—CEOs of large corporations, high-ranking lawmakers in the executive branch, and top military leaders

Middle level (8 percent)—most members of Congress, lobbyists, entrepreneurs of small businesses, leaders of labor unions and other interest groups (in law, education, medicine, etc.), and influential media commentators

The masses (91 percent)—people who are unorganized and exploited and either don't know or don't care about what's going on in government

through special-interest groups—unions, professional organizations, and so on (*Figure 11.7*). The various groups rarely join ranks because they concentrate on single issues such as health care, pollution, or education. This focus on different issues fragments groups, but also results in a broad representation of interests and a distribution of power.

Because there are a number of single-issue groups, functionalists maintain, there are multiple leaderships (Dye and Ziegler, 2003). As a result, many leaders, not just a few, can shape decisions that represent numerous groups and issues. Pluralists note that people also have power outside of interest groups: They can vote, run for office, contact lawmakers, and put specific issues on a ballot. Therefore, there are continuous checks and balances as individuals and groups vie for power and try to influence laws and policies.

Pluralists believe that groups resolve differences by bargaining and compromising. As this process unfolds, the government is a neutral referee: It reduces outright hostility, helps groups achieve their goals to at least some degree, and ensures that no particular group has absolute authority. Thus, group competition benefits society as a whole.

CRITICAL EVALUATION

Does pluralism work as democratically as functionalists maintain? Critics argue that interest groups have unequal resources. The poor and disadvantaged rarely have the skills and educational backgrounds to organize or promote their interests. In contrast, wealthy individuals and organizations can influence government through lobbyists, political contributions, and personal connections.

power elite a small group of influential people who make the nation's major political decisions.

Critics also maintain that pluralists aren't realistic about the emergence of multiple leaderships because the affluent control the government. For example:

- President Trump's Cabinet, the wealthiest in U.S. history, is made up of multi-millionaires and billionaires. More than half of the members of Congress are millionaires or multi-millionaires, and five are worth $100 million or more (Choma, 2015; Buchanan et al., 2017).
- All nine Supreme Court Justices are millionaires or billionaires (Levinthal et al., 2017).
- Cabinet memberships and Cabinet-level positions are dominated by business executives who have been abolishing or watering down pollution and banking regulations that conflict with corporate interests (Flesher and Biesecker, 2017; Walsh, 2017).

Because the wealthy control the political system, according to critics, the government isn't a neutral referee that benefits the general population.

11-8b Conflict Theory: A Power Elite Model

Conflict theorists contend that the United States is ruled not by pluralism but a **power elite**, a small group of influential people who make the nation's major political decisions. Sociologist C. Wright Mills (1956) coined the term *power elite* to describe a pyramid of power that he believed characterized American democracy.

Christos Georghiou/Shutterstock.com

KEY CHARACTERISTICS

According to Mills, the power elite is made up of three small but dominant groups of people at the top level who run the country: political leaders, corporate heads, and military chiefs (*Figure 11.7*). Practically all of the members of these groups are White, Anglo-Saxon, Protestant men who form an inner circle of power. Those at the middle power level include members of Congress, lobbyists, influential media commentators, and others. The bottom level, and the largest and least powerful group, the masses, is composed of everyone else.

The power elite tolerates the masses—including their elections and laws—but in the end simply does what it wants. In 2013, almost 80 percent of Americans wanted limits on the amount of money that congressional candidates could spend on their campaigns (Saad, 2013). The following year, the Supreme Court did just the opposite by striking down laws that limited contributions to political candidates, parties, and PACs.

Like Mills, contemporary conflict theorists maintain that the United States isn't a democracy because of the close ties and "revolving door" between the power elite in politics, business, and the military. For example, fewer than 31,390 people—less than 0.01 percent of the nation's population—contributed more than $1.6 billion to political campaigns in 2012, and nobody was elected to Congress without the economic elite's money. In return, politicians pass favorable legislation such as $83 billion in taxpayer-financed subsidies to the country's banks, award earmarks and lucrative military and other contracts, and may later serve as corporate lobbyists (Gilson, 2014; Kristof, 2014; Winship, 2015).

CRITICAL EVALUATION

Functionalists accuse conflict theorists of exaggerating the power elite's influence. They point out, for example, that "the masses" aren't just puppets but vote, mobilize, support particular interest groups, and protest current political and corporate policies.

Also, according to critics, the power elite perspective assumes, incorrectly, that the ruling class acts as a unified force in protecting its interests. Because the Democratic and Republican parties and their top officials endorse very different agendas, the power elite are rarely unified.

Some also question whether elites are always as self-serving and self-interested as conflict theorists contend. For example, past administrations have reduced smoking, pollution, and unsafe automobiles despite considerable lobbying and resistance by wealthy and influential corporations.

11-8c Feminist Theories: A Patriarchal Model

In a recent national poll, 75 percent of women and men said that both sexes are equally good political leaders (Associated Press and NORC, 2016). Why, then, are there so few female officeholders?

SOME KEY ISSUES

Feminist theorists maintain that women are generally shut out of the most important political positions because, in a patriarchal society, men rule. A "gender gap in political ambition" begins early in the socialization process. Parents often encourage their sons, not their daughters, to think about politics as a career and to join political clubs in college. In adulthood, family, friends, and party leaders are more likely to urge men than women to promote themselves and to run for a political office. Thus, despite comparable educational and occupational backgrounds, men are almost 60 percent more likely than women to view themselves as "very qualified" to run for office (Lawless and Fox, 2013).

When women run for political office, they're almost as likely to be elected as men, but must overcome stereotypes and double standards. According to 40 percent of Americans, women who seek the highest political offices are held to higher standards than men. Some voters believe that mothers with young children who run for office won't be able to balance work and family life. In contrast, men routinely include their children in political advertising to project an image of being a father and family man, a traditional gender role. Voters are also more likely to expect female than male candidates to be likeable, honest, and ethical (Hayes and Lawless, 2016; Barbara Lee Family Foundation, 2017; Ditmar, 2017).

During the 2016 presidential campaign, some political observers rebuked both Democrat Hillary Clinton and Republican Carly Fiorina for appearing too stern and not smiling enough. No one criticized Donald

Shortly after the 2016 election, Vice President Mike Pence convened an influential group of Republicans to discuss, among other things, removing Obamacare insurance companies' requirements covering maternity, newborn, and pregnancy care (Filipovic, 2017). *Where are the women!*

The White House

Trump's or Bernie Sanders' smiles or scowls. Talk show commentators (particularly Fox News) and male journalists accused Clinton of being "shrill," "angry," and "grating." Sanders, who often shouted, was described as "enthusiastic" and "tough." Except for a handful of female journalists, few news correspondents or talk show hosts objected to Trump's continuous lewd, degrading, and sexist comments about women or his multiple sexual assaults (Christina, 2016; Holloway, 2016; Ditmar, 2017).

CRITICAL EVALUATION

Critics fault feminist explanations for several reasons. First, there are powerful women on some of the most influential congressional committees. Women have attained top leadership posts—president, prime minister, or its equivalent—in more than 70 countries in Europe, Latin America, and the Asia-Pacific. Angela Merkel in Germany and Theresa May in Britain run two of Europe's most powerful, and patriarchal, nations (Kiefer, 2015; Bennhold and Gladstone, 2016).

Second, a larger number of female officeholders doesn't guarantee greater gender equality. For example, some Republican women in Congress have opposed equal pay for equal work laws because (they maintain) well-qualified women are already paid the same as men, and because such legislation would increase the number of civil lawsuits and hurt businesses (Lowery, 2014; McDonough, 2014).

One might also question whether patriarchy is the root of political power differences between women and men. Some of the world's most patriarchal societies (e.g., in parts of Africa, Latin America, and the Middle East) have more women in high-ranking political positions than does the United States, which professes gender equality. Women have the fewest rights in societies where male political leaders are religious zealots who don't believe in gender equality (see Jones, 2014). Thus, religion may be more important than patriarchy in quashing women's political leadership.

STUDY TOOLS 11

READY TO STUDY? IN THE BOOK, YOU CAN:

- ☐ Check your understanding of what you've read with the Test Your Learning Questions provided on the Chapter Review Card at the back of the book.
- ☐ Tear out the Chapter Review Card for a handy summary of the chapter and key terms.

ONLINE AT CENGAGEBRAIN.COM WITHIN MINDTAP YOU CAN:

- ☐ Explore: Develop your sociological imagination by considering the experiences of others. Make critical decisions and evaluate the data that shape this social experience.
- ☐ Analyze: Critically examine your basic assumptions and compare your views on social phenomena to those of your classmates and other MindTap users. Assess your ability to draw connections between social data and theoretical concepts.
- ☐ Create: Produce a video demonstrating connections between your own life and larger sociological concepts.
- ☐ Collaborate: Join your classmates to create a capstone project.

12 Families and Aging

Masterfile

LEARNING OBJECTIVES

After studying this chapter, you will be able to…

12-1 Describe how families are similar and different in the United States and worldwide.

12-2 Describe how and explain why U.S. families are changing.

12-3 Describe, illustrate, and explain why intimate partner violence, child maltreatment, and elder abuse occur.

12-4 Describe, illustrate, and explain how the U.S. older population is changing, and its impact on our society.

12-5 Compare and evaluate the theoretical explanations of families and aging.

After finishing this chapter go to **PAGE 252** for **STUDY TOOLS**

American families vary greatly in structure, dynamics, and racial-ethnic diversity. This chapter examines the ways that families and aging are changing, both in the United States and globally. Before reading further, take the True or False quiz to see how much you already know about these topics.

WHAT DO YOU THINK?

I would undergo medical treatments to live to age 120 or longer.

1	2	3	4	5	6	7
strongly agree						strongly disagree

True or False?

HOW MUCH DO YOU KNOW ABOUT CONTEMPORARY U.S. FAMILIES AND AGING?

1. Nonmarital teenage births have increased over the past 20 years.
2. College-educated women are less likely than those with a high school diploma to live with a man outside of marriage.
3. Divorce rates have increased since the 1990s.
4. The most common type of family is married couples with children.
5. Baby boomers (those born between 1946 and 1964) are the fastest growing segment of the population.
6. Asian Americans are more likely than Latinos to live in a home with three or more generations.

The answers to #2 and #6 are true; the others are false. You'll see why as you read this chapter.

12-1 WHAT IS A FAMILY?

Ask five of your friends to define *family*. Their definitions will probably differ not only from each other's but also from yours. For our purposes, a **family** is an intimate group consisting of two or more people who (1) have a committed relationship, (2) care for one another and any children, and (3) share activities and close emotional ties. This definition includes households (e.g., foster families, same-sex couples) whose members aren't related by birth, marriage, or adoption.

Contemporary households are complex; family structures vary across cultures and have changed over time. In some societies, a family includes uncles, aunts, and other relatives. In others, only parents and their children are viewed as a family.

12-1a How Families Are Similar Worldwide

The family, a social institution, exists in some form in all societies. Worldwide, families are similar in fulfilling some functions that have persisted over time.

FAMILY FUNCTIONS

Families vary considerably in the United States and globally but fulfill five important functions that ensure a society's survival (Parsons and Bales, 1955):

- **Sexual activity.** Every society has norms regarding who may engage in sexual relations, with whom, and under what circumstances. U.S. laws ban sexual intercourse with someone younger than 18 (or 16 in some states), but several states allow 12- and 13-year-old girls and 14-year-old boys to marry with parental permission, and some countries permit marriage with girls as young as 8 (McClendon and Sandstrom, 2016; Sandstrom and Theodorou, 2016; see also Chapter 9).

One of the oldest rules that regulate sexual behavior is the **incest taboo**, cultural norms and laws that forbid sexual intercourse between close blood relatives (e.g., brother and sister, father and daughter, uncle and niece).

family an intimate group consisting of two or more people who (1) have a committed relationship, (2) care for one another and any children, and (3) share activities and close emotional ties.

incest taboo cultural norms and laws that forbid sexual intercourse between close blood relatives.

- **Procreation and socialization.** Procreation is an essential family function because it replenishes a country's population. Through socialization, children acquire language; absorb the accumulated knowledge, attitudes, beliefs, and values of their culture; and learn the social and interpersonal skills they need to function effectively in society (see Chapters 3 and 4).
- **Economic security.** Families provide food, shelter, clothing, and other material resources for their members.
- **Emotional support.** Families supply the nurturance, love, and emotional sustenance that people need to be happy, healthy, and secure. Our friends may come and go, but our family is usually our emotional anchor.
- **Social class placement.** Social class affects all aspects of family life. Initially, our social position is based on our parents' social class, but we can move up or down the social hierarchy in adulthood (see Chapter 8).

Some sociologists include *recreation* as a basic U.S. family function. This function isn't critical for survival, but since the 1950s many parents have spent much more time with their children on leisure activities (e.g., visiting amusements parks, playing video games together) that strengthen interpersonal bonds.

MARRIAGE

Marriage, a socially approved mating relationship that people expect to be stable and enduring, is also universal. Countries vary in their specific norms and laws dictating who can marry whom and at what age, but marriage everywhere is an important rite of passage that marks adulthood and its related responsibilities, especially providing for a family.

ENDOGAMY AND EXOGAMY

All societies have formal or informal rules about the "right" marriage partner. **Endogamy** (often used interchangeably with *homogamy*) is a cultural practice of marrying within one's group. The partners are similar in religion (e.g., Catholics marrying Catholics), race or ethnicity (e.g., Black people marrying Black people), social class (e.g., college-educated marrying college-educated), and/or age (e.g., young people marrying young people). Across the Middle East and Africa, marrying a first or second cousin is not only common but desirable. In Egypt, for example, 40 percent of the population marry a family member. The benefits include knowing a lot about one's relatives and ensuring that property stays in the family, but the practice increases the chance of genetic diseases (such as cystic fibrosis and thalassemia, a blood disorder) in their children ("Consanguineous Marriage," 2016).

Exogamy (often used interchangeably with *heterogamy*) is a cultural practice of marrying outside one's group, such as not marrying one's relatives. In the United States, 25 states prohibit marriage between first cousins, but violations are rarely prosecuted (National Conference of State Legislatures, 2014). Even when there are no formal laws, cultural norms and values (as well as social pressure) usually limit our marital partner choices.

marriage a socially approved mating relationship.

endogamy (often used interchangeably with *homogamy*) cultural practice of marrying within one's group.

exogamy (often used interchangeably with *heterogamy*) cultural practice of marrying outside one's group.

nuclear family a family form composed of married parents and their biological or adopted children.

12-1b How Families Differ Worldwide

Families also differ around the world. Some variations affect the family's structure, whereas others regulate where people reside and who has the most household power and authority.

NUCLEAR AND EXTENDED FAMILIES

In Western societies, the typical family form is the **nuclear family** composed of married parents and their biological or adopted children. In much of the world, however,

"Empty-nesters. They're hoping to sell before the flock tries to move back in."

age fotostock/Alamy Stock Photo

In China's Himalayas, the Mosuo are a matriarchal society. A family consists of a woman, her children, and the daughters' offspring. An adult man will join a lover for the night and then return to his mother's or grandmother's house in the morning. Children resulting from these unions belong to the female and take her surname, she and her relatives raise them, and daughters are preferred to sons. The Mosuo population is decreasing, however, as more young people marry outside the group or move to cities for work (Qin, 2015).

the most common family form is the **extended family**, composed of parents, children, and other kin (e.g., uncles and aunts, nieces and nephews, cousins, grandparents).

As the number of single-parent families increases in industrialized countries, extended families are more common. By helping out with household tasks and child care, other adult members make it easier for a single parent to work outside the home.

Many Americans assume that the nuclear family is the most common arrangement, but such families have declined in number. In 2014, just 14 percent of children lived with a stay-at-home mother and a working father—down from 50 percent in 1960 (Livingston, 2015).

RESIDENCE PATTERNS

In a **patrilocal residence pattern**, newly married couples live with the husband's family. In a **matrilocal residence pattern**, they live with the wife's family. In a **neolocal residence pattern**, the couple sets up their own residence.

Around the world, the most common residence pattern is patrilocal. In industrialized societies, married couples are typically neolocal. Since the early 1990s, however, the tendency for young married adults to live with the parents of either the wife or the husband—or sometimes with the grandparents of one of the partners—has increased. Such "doubled-up" U.S. households have always existed, but escalated during the 2007–2009 recession for economic reasons (Parker, 2012).

One result is a **boomerang generation**, young adults who move back into their parents' home after living independently for a while or never leave home in the first place. Seven years after the 2007–2009 recession, for example, 33 percent of 18- to 34-year-olds lived with their parents. Of those, one in four didn't work or go to school (Vespa, 2017).

You saw in Chapter 4 that many young adults don't feel a need to set up their own homes because, among other reasons, the stigma of living with parents has faded. Some observers predict that boomerangers will become more numerous. Many "helicopter-parented, trophy-saturated, and self-centered" millennials, who were raised by coddling baby boomers, are catering to their own children even more. They want to be their kids' friends rather than authority figures. Instead of helicopter parenting, millennials are "drone parenting"—they still hover, but are now "following and responding to their kids rather than directing and scheduling them" (Steinmetz, 2015: 38, 41). Parents try to launch their children into the adult world, but like boomerangs, some keep coming back.

AUTHORITY AND POWER

Residence patterns often reflect who has authority and power in the family. In a **matriarchal family system**, the oldest females (usually grandmothers and mothers) control cultural, political, and economic resources and, consequently, have power over males. Some Native American tribes were matriarchal, and in some African countries, the oldest women have considerable authority and influence. For the most part, however, matriarchal societies are rare.

Worldwide, a more typical pattern is a **patriarchal family system**, in which the oldest males

extended family a family form composed of parents, children, and other kin.

patrilocal residence pattern newly married couples live with the husband's family.

matrilocal residence pattern newly married couples live with the wife's family.

neolocal residence pattern a newly married couple sets up its own residence.

boomerang generation young adults who move back into their parents' home or never leave it in the first place.

matriarchal family system the oldest females control cultural, political, and economic resources and, consequently, have power over males.

patriarchal family system the oldest males control cultural, political, and economic resources and, consequently, have power over females.

(grandfathers, fathers, and uncles) control cultural, political, and economic resources and, consequently, have power over females. In some patriarchal societies, women have few rights within or outside the family; they may not be permitted to work outside the home or attend college. In other patriarchal societies, women may have considerable decision-making power in the home but few legal or political rights, such as getting a divorce or running for political office (see Chapter 9).

In an **egalitarian family system**, both partners share power and authority fairly equally. Many Americans think they have egalitarian families, but our families tend to be patriarchal. For example, employed women shoulder almost twice as much housework and child care as men, and are more likely than men to provide caregiving to aging family members (see Chapter 9).

COURTSHIP AND MATE SELECTION

Sociologists often describe U.S. dating as a **marriage market**, a courtship process in which prospective spouses compare the assets and liabilities of eligible partners, and choose the best available mate. Marriage markets don't sound very romantic, but open dating fulfills several important functions: recreation and companionship; a socially acceptable way of pursuing love; opportunities for sexual intimacy and experimentation; and finding a spouse (Benokraitis, 2015).

Many societies, in contrast, discourage open dating and have **arranged marriages** in which parents or relatives choose the children's spouses. An arranged marriage is a family rather than an individual decision that increases solidarity between families and preserves endogamy. Children may have veto power, but they believe that if partners are compatible, love will result. In some of India's urban areas, arranged marriages also rely on nontraditional methods (e.g., online dating services) to find prospective spouses (Cullen and Masters, 2008; see also Chapter 9).

egalitarian family both partners share power and authority fairly equally.

marriage market prospective spouses compare the assets and liabilities of eligible partners and choose the best available mate.

arranged marriage parents or relatives choose the children's spouses.

monogamy one person is married exclusively to another person.

serial monogamy individuals marry several people, but one at a time.

polygamy a man or woman has two or more spouses.

MONOGAMY AND POLYGAMY

In **monogamy**, one person is married exclusively to another person. Where divorce and remarriage rates are high, as in the United States, people engage in **serial monogamy**. That is, they marry several people but one at a time—they marry, divorce, remarry, divorce, and so on.

Polygamy, in which a man or woman has two or more spouses, is subdivided into *polygyny*—one man married to two or more women—and *polyandry*—one woman is married to two or more men. Nearly 1,000 cultures around the world allow some form of polygamy, either officially or unofficially (Epstein, 2008).

Although rare, there are pockets of polyandry in some remote and isolated parts of India, and among the Pimbwe in western Tanzania, Africa. Polyandry serves several functions: The family is more likely to survive in harsh environments if there's more than one husband to provide food, and if one husband dies, the others care for the widow (Borgerhoff Mulder, 2009; Polgreen, 2010).

In contrast to polyandry, polygyny is common in many societies, especially in some regions of Africa, South America, and the Middle East. Men benefit from polygyny in several ways: They can have many legal sexual partners, more chances to have male heirs, more income if some of the wives are employed, and high social status because they can support multiple wives and children (Al-Jassem, 2011; Nossiter, 2011).

Western and industrialized societies forbid polygamy, but there are pockets of isolated polygynous groups in the United States, Europe, and Canada. The Church of Jesus Christ of Latter-Day Saints (Mormons) banned polygamy in 1890 and excommunicates members who follow such beliefs. Still, males of the Fundamentalist Church of Jesus Christ of Latter Day Saints (FLDS), a polygynous sect that broke away off from the mainstream Mormon Church more than a century ago, head

Adnan Abidi/Reuters

India's Ziona Chana has the world's largest polygynous family—39 wives, 94 children, and 33 grandchildren (so far). They all live together in a 100-room mansion (Sykes, 2015).

an estimated 300,000 families in Texas, Arizona, Utah, and Canada. Wives who have escaped from these groups have reported forced marriage between men in their 60s and girls as young as 10 years old, sexual abuse, and incest (Janofsky, 2003).

12-2 HOW U.S. FAMILIES ARE CHANGING

American families have changed considerably since the 1950s. Some of the most important changes are related to marriage, divorce, cohabitation, nonmarital childbearing, and single-parent families.

12-2a Marriage and Divorce

The United States has one of the highest marriage and divorce rates in the world. By age 65, 95 percent of Americans have been married at least once; over a lifetime, about half of first marriages end in divorce (Vespa et al., 2013). Despite the high divorce rate, most people aren't disillusioned about marriage. Indeed, nearly 85 percent who divorce remarry, half of them within 4 years, and 8 percent have been married three times or more (Kreider and Ellis, 2011; Livingston, 2014).

TRENDS

U.S. marriage and divorce rates rose steadily during the twentieth century, but have declined since 1990 (*Figure 12.1*). In 1960, only 9 percent of U.S. adults age 25 and older had never been married. By 2016, 52 percent had never been married—an historic high. Marriage rates are expected to decrease further because the number of newlyweds has been falling since 2012 (Wang and Parker, 2014).

In 2004, Massachusetts became the first state to legalize same-sex marriage. In 2017, 10 percent of LGBT adults were married to a same-sex spouse, and 13 percent were married to an opposite-sex partner. About half of people who self-identify as LGBT are bisexual, helping to explain the high proportion who are married to opposite-sex partners. Men (11 percent) are more likely than women (9 percent) to be married to a same-sex partner. The marriage rates are highest among males age 50 and older, but 32 percent of this age group have never married, compared with 11 percent of their non-LGBT counterparts (Jones, 2017; see also Chapter 9). Thus, despite the Supreme Court's legalization of same-sex marriage in 2015, LGBT marriages haven't surged. Like heterosexuals, LGBTs may be postponing marriage or decide to not marry.

In 1970, California was the first state to pass a *no-fault divorce* law; neither partner needs to prove guilt or wrongdoing (e.g., adultery, desertion). Today, in all states, couples can simply give "irreconcilable differences" or "incompatibility" as a valid reason for divorce. As these laws changed, marital dissolutions became quick and cheap, and divorce rates rose to historically high levels, particularly in the 1980s and 1990s. Divorce rates are high, but *lower* today than they were between 1980 and 2009 (*Figure 12.1*). In 2017, 73 percent of Americans said that divorce was morally acceptable, compared with only 53 percent in 1954 (Dugan, 2017). Thus, and despite greater public acceptance, the divorce rate has fallen to its lowest point in decades.

Figure 12.1 U.S. Marriage and Divorce Rates, 1870–2014

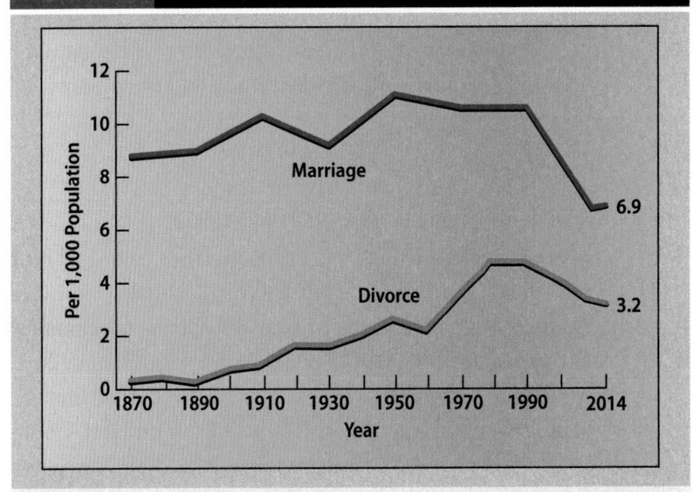

Sources: Based on Plateris 1973, Table 1; and CDC/NCHS National Vital Statistics System, 2017.

WHY ARE MARRIAGE AND DIVORCE RATES FALLING?

There are several macro-level reasons for declining marriage and divorce rates. First, U.S. values have been changing. In 2010, 39 percent of Americans said that marriage is becoming obsolete, up from 28 percent in 1978 (Cohn et al., 2011; Fry, 2012). Just 45 percent of millennials say that marriage is an important step in becoming an adult, and only 44 percent of Americans believe that having children is a "very important" reason to marry. Only 31 percent think that premarital sex is immoral, open marriage markets provide many opportunities for nonmarital sex, and nonmarital births are now socially acceptable (Cohn, 2013; Jones, 2017; Vespa, 2017). Thus, the traditional reasons for marriage have waned.

Second, the economy affects marriage and divorce rates. Economic depressions, recessions, and unemployment tend to delay marriage, especially for men. Because of the recent recession, 20 percent of 18- to 34-year-olds have postponed marriage. Moreover, when incomes plummet and people are insecure about their jobs, unhappy married couples tend to stay together: They can't afford to divorce and risk the possibility of not being able to maintain separate households (Cohn, 2012; Martin et al., 2014).

Demographic variables also affect marriage and divorce rates. The median age at first marriage is 30 for men and almost 28 for women, compared with 23 for men and 21 for women in 1970 ("Marital Status," 2017).

There are now almost as many single as married Americans (128 million and 129 million, respectively) and, for the first time, single women outnumber married women. Marriage has become an option, rather than a goal, primarily because women are free to pursue a higher education, to have a career, and to cohabit. Thus, many women have become choosier about whether and when to marry (Traister, 2016; "Unmarried and Single . . .," 2017).

About 64 percent of Americans with college degrees are married, compared with 47 percent of those with a high school diploma or less, who are more likely to cohabit than marry. The education-marriage relationship holds even at higher levels. Among women in their early 40s, for example, 80 percent with a Ph.D. or professional degree are married, compared to 63 percent with a bachelor's degree. The highly educated tend to marry other highly educated people, and are almost twice as likely as people with less education to have marriages that last at least 20 years. In effect, then, there's a growing "marriage gap" between the most and least educated (Wang, 2015; Reeves et al., 2016). Such data also challenge stereotypes of highly educated women as sad and lonely "old maids" who lavish attention on a brood of cats.

Americans with a bachelor's or graduate degree also have lower divorce rates than those without college degrees. By age 46, 30 percent of people with a college degree or higher are divorced compared with 59 percent with less than a high school diploma (Aughinbaugh et al., 2013). College graduates have lower divorce rates not because they're smarter but because going to college postpones marriage. As a result, better-educated couples are often more mature and capable of dealing with personal crises. They also have higher incomes and health care benefits, both of which lessen marital stress over financial problems.

cohabitation two unrelated and unmarried people live together and are in a sexual relationship.

Victoria Roberts/The New Yorker Collection/The Cartoon Bank

"It's National We're History Month."

There are also *micro-level (individual) reasons* for falling marriage and divorce rates. About 84 percent of unmarried Americans say that love is a "very important" reason to marry. As a result, a third are still waiting for their "ideal mate" or "one true love" (Wang and Parker, 2014).

The most common micro-level reasons for divorce include infidelity, communication and financial problems, substance and spousal abuse, premarital doubts, continuous disagreements about how to raise and discipline children, and expecting to change a partner after marriage (see Benokraitis, 2015, for a discussion of these studies). Because many people are delaying marriage, they're usually more mature in handling the challenges of married life, which decreases the likelihood of divorce.

12-2b Cohabitation

Cohabitation is an arrangement in which two unrelated and unmarried people live together and are in a sexual relationship (shacking up, in plain English). Because it's based on emotional rather than legal ties, "cohabitation is a distinct family form, neither singlehood nor marriage" (Brown, 2005: 33).

TRENDS

Married couples comprise 49 percent of all households, a sharp decline from 78 percent in 1950. The decline is due to falling marriage rates and rising cohabitation. The number of adults in heterosexual cohabiting relationships surged from 430,000 in 1960 to almost 18 million in 2016. This number climbs by another 860,000 if we include same-sex cohabiters. Despite the high numbers, only 7 to 9 percent of the population is cohabiting in any given year ("Characteristics of Same-Sex Households," 2017; Stepler, 2017).

Figure 12.2 Cohabitation, by Age Groups, 2007 and 2016

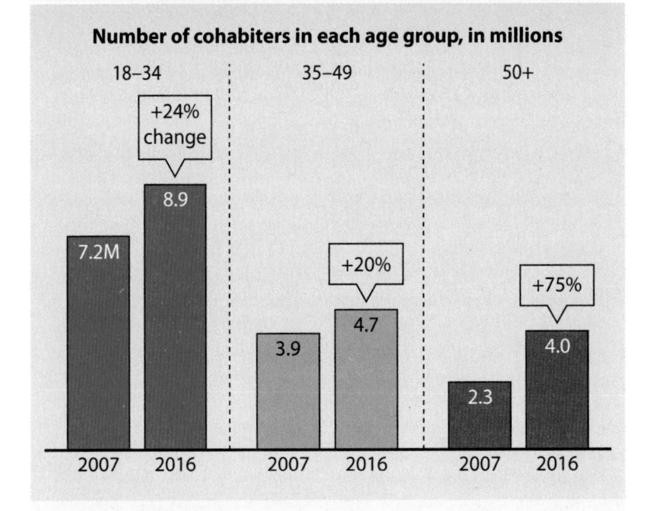

Source: Stepler, 2017.

About half of cohabiters are younger than 35. Since 2007, however, the number of cohabiters aged 50 and older has increased faster than other groups (*Figure 12.2*). Because of high divorce rates and a growing share of people who have never been married in this age group, more people are available for cohabitation (Stepler, 2017).

By age 44, 65 percent of women—compared with 33 percent in 1987—have cohabited. Cohabitation is a common experience at all education levels, but the likelihood of cohabiting decreases as women's educational levels increase (e.g., 76 percent of women with less than high school vs. 58 percent of women with a bachelor's or advanced degree). On average, college-educated women cohabit for the shortest period (17 months) and are more likely than those without college degrees to transition to marriage (Copen, 2013; Vanorman and Scommegna, 2016).

Black women are less likely to have ever cohabited (59 percent) than White women (67 percent) or Latinas (64 percent). Among all cohabiting adults, significantly more Asians (46 percent) live with a partner of a different race or ethnicity than Latinos (24 percent), Black people (20 percent), or White people (12 percent) (Vanorman and Scommegna, 2016; Livingston, 2017). The high rate of Asian interracial cohabitation, like intermarriage, may be due to this group's small population size and shortage of available partners within their own group (see Chapter 10).

WHY HAS COHABITATION INCREASED?

On a micro level, some people drift gradually into *dating cohabitation*, when a couple that spends a great deal of time together decides to move in together. Dating cohabitation is essentially an alternative to singlehood because the decision may be based on a combination of reasons (e.g., convenience, finances, companionship, and sexual accessibility), but there's no long-term commitment. In this type of cohabitation, and especially among young adults, there's considerable *serial cohabitation*, living with different sexual partners over time. Even if there's an unplanned pregnancy, the man, especially, may decide to move on to another cohabiting arrangement (Manning and Smock, 2005; Wartik, 2005).

For many people, premarital cohabitation is a step between dating and marriage. In *premarital cohabitation*, the couple lives together before getting married. They may or may not be engaged but plan to marry. Such "almost-married" cohabitation may be especially attractive to partners who wonder if they can deal successfully with problems that arise from differences in personalities, interests, finances, ethnicity, religion, or other issues.

On a macro level, and across all ethnic groups, 3 out of 4 people who cohabit say that they're delaying marriage because "Everything's there except money." Many low-income women don't want to marry because they believe that their live-in partners will be poor providers, unemployed, unfaithful, irresponsible fathers, or immature even though "he's the love of my life" (Xie et al., 2003; Lichter et al., 2006).

Those with high income and education levels have little to gain from cohabitation. Unlike people in lower socioeconomic groups, they're more likely to have jobs, to afford their own housing, and, consequently, to have

Do you think that women or men benefit more from cohabitation? Why?

Table 12.1 Some Benefits and Costs of Cohabitation

Benefits	Costs
• Couples can pool their resources instead of paying for separate housing, utilities, and so on. They can also have the emotional security of an intimate relationship but maintain their independence by spending time with their friends and family members separately (McRae, 1999; Fry and Cohn, 2011).	• U.S. laws don't specify a cohabitant's rights and responsibilities. For example, there's no automatic inheritance if a partner dies without a will, and it's more difficult to collect child support from a cohabiting partner than an ex-spouse (Silverman, 2003; Grall, 2013).
• Couples who postpone marriage have a lower likelihood of divorce because being older is one of the best predictors of a stable marriage (Copen et al., 2012).	• Compared with married couples, cohabitants have a poorer quality of relationship and lower levels of happiness and satisfaction (Sassler et al., 2012; Wiik et al., 2012).
• Couples find out how much they really care about each other when they have to cope with unpleasant realities (e.g., a partner who doesn't pay bills or rarely showers).	• Cohabitation dilutes intergenerational ties. Compared with their married peers, the longer people live together, the less likely they are to give or receive help from their parents, and to be involved in extended family activities (Eggebeen, 2005).
• Children in cohabiting households reap economic advantages by living with two adult earners instead of a single mother (Lundberg et al., 2016).	• Because cohabiting parents are more than twice as likely as married parents to break up, children's academic, emotional, behavioral, and financial problems often increase (Fomby and Estacion, 2011; Rackin and Gibson-Davis, 2012).

more options in living independently even though they're involved romantically (Fry and Cohn, 2011; Sassler and Miller, 2011).

Does cohabitation lead to better marriages? Recent studies show that women who cohabit are no more likely to divorce than those who didn't cohabit if there's a commitment (definite plans to marry) and the cohabitation is short. However, marital success also depends on the cohabitors' age, socioeconomic status, commitment, and attitudes toward marriage (Reinhold, 2010; Manning and Cohen, 2012). Cohabitation, like any other relationship, has benefits and costs (*Table 12.1*).

12-2c Nonmarital Childbearing

More than 9 in 10 Americans aged 45 or older either have children (86 percent) or wish they had (7 percent). Such positive attitudes toward childbearing are about the same as in 1990, but since then U.S. birth rates have dropped by 11 percent (Martin et al., 2013; Newport and Wilke, 2013).

A notable exception is nonmarital childbearing, which has increased significantly since 1970. Nearly half of American men ages 15 to 44 report that at least one of their children was born outside of marriage, and 31 percent say that all of their children were born outside of marriage (Livingston and Parker, 2011).

TRENDS

In 1950, only 3 percent of all U.S. births were to unmarried women. In 2016, there were more than 1.6 million, accounting for 40 percent of all births. Thus, 4 in 10 American babies are now born outside of marriage, a new record (Hamilton et al., 2017).

Births to unmarried women vary widely across racial-ethnic groups. White women have more nonmarital babies than do other groups. Proportionately, however, nonmarital birth rates are highest for Black women and lowest for Asian American women (*Figure 12.3*).

As in marriage and cohabitation, nonmarital births vary by social class. Among college graduates, only 13 percent of recent births were outside of marriage compared with 57 percent for high school dropouts (*Table 12.2A*).

Figure 12.3 Births to Unmarried Mothers, by Race and Ethnicity, 2016

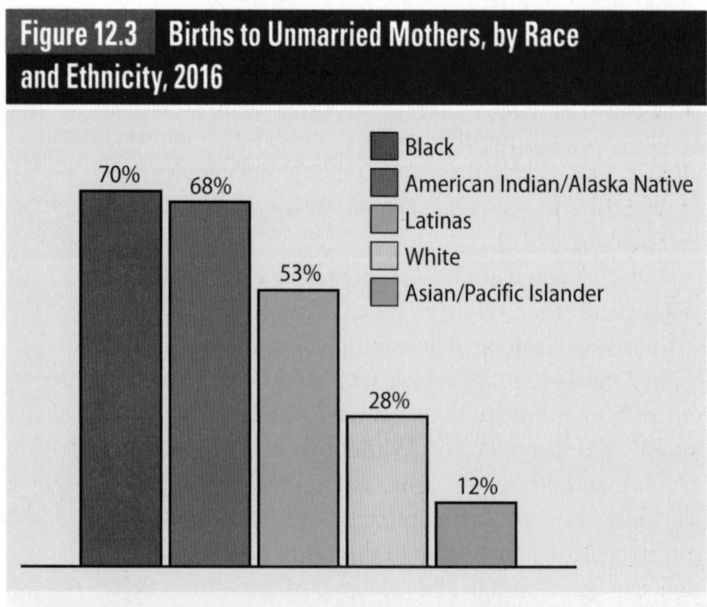

Source: Based on Hamilton et al., 2017, Table 3.

Table 12.2 Births to Unmarried Mothers, by Education and Household Income

A. Educational Attainment	
Less than high school	57%
High school graduate	52%
Some college	40%
Bachelor's degree	13%
Graduate or professional degree	7%
B. Annual Household Income	
Under $10,000	77%
$10,000 to $24,999	62%
$25,000 to $49,999	45%
$50,000 to $99,999	30%
$100,000 to $199,999	20%
$200,000 and above	13%

Source: Based on U.S. Census Bureau, 2015 American Community Survey 1-Year Estimates, Table S1301, and "Fertility of Women in the United States: 2016," Current Population Survey, Table 7. Accessed July 10, 2017 (factfinder.census.gov and census.gov/hhes/fertility/data).

The percentage of nonmarital births decreases as household income increases (*Table 12.2B*). Consequently, the poorest and least-educated women are the most likely to have babies outside of marriage, decreasing the children's chances of moving up the social class ladder (Sawhill and Venator, 2014).

Unmarried teenage births peaked in the early 1990s, have declined steadily since then, and are now at their lowest level since 1968. Teenagers account for 13 percent of all nonmarital births compared with 78 percent for women ages 20 to 34. Thus, the number of births to unmarried women ages 20 to 34 is six times greater than for teenagers. Still, 89 percent of teenage births are nonmarital, compared with 52 percent for women in their 20s, and 22 percent for women in their 30s. Moreover, 20 percent of unmarried teenagers have a repeat birth: 86 percent have a second child, and 15 percent have 3 to 6 children (Gavin et al., 2013; Martin et al., 2017).

WHY HAS NONMARITAL CHILDBEARING INCREASED?

On a *micro (individual) level,* many teens and young adults use contraception inconsistently or incorrectly because they think they or their partners are sterile, they believe that not taking the pill a few times "doesn't really

THE NUMBER OF BIRTHS TO UNMARRIED WOMEN AGES 20 TO 34 IS SIX TIMES GREATER THAN FOR TEENAGERS.

matter," or the males don't want to use condoms. Others think that unplanned pregnancies are predetermined or controlled by outside forces like God or fate ("It was meant to happen" or "It's God's plan") (Frohwirth et al., 2013; Lindberg et al., 2016).

Unintended pregnancies have decreased by 18 percent since 2008. Nonetheless, 81 percent of pregnancies among never married non-cohabiters, 75 percent of teen pregnancies, and 56 percent of pregnancies among cohabiting women are unintended. Even if the pregnancies are unwanted, women may not get abortions because of personal or religious beliefs (Curtin et al., 2014; Finer and Zolna, 2016).

Demographic variables also affect nonmarital childbearing. Social class is a key factor in nonmarital births (*Table 12.2*). Educated women are more likely than less-educated women to postpone parenthood until marriage, to be more aware of family planning, and to have more decision-making power in their relationships, such as insisting that men use condoms (Livingston and Cohn, 2013).

Among college-educated women, 41 percent of Black women, compared with 22 percent of Latinas and only 9 percent of White women, choose single motherhood. Nonmarital childbearing is a "rational choice" for Black women who don't want to marry less educated Black men or across racial lines, believe that they may never marry, or that marriage will come too late for them to bear children (Keels, 2014).

Macro-level variables increase nonmarital birth rates in several ways. Until the early 1970s, nonmarital births were rare and kept secret, young people were forced into "shotgun" marriages if the girl was pregnant, and young women—especially White women—were pressured to give their babies up for adoption. Now, almost 75 percent of Americans aged 15 to 44 believe that "it's okay for an unmarried female to have and raise a child." Because attitudes have changed, nonmarital childbearing has become normal. In contrast, only 2 percent of children in Japan and South Korea are born outside of marriage because unmarried mothers are stigmatized (Daugherty and Copen, 2016; "Marriage in Japan," 2016; Chamie, 2017).

A recent study found that the reality TV show *16 and Pregnant* reduced U.S. teen births by almost 6 percent in the 18 months following its release (Kearney and Levine, 2014). Should middle schools require adolescents to watch such programs?

Fewer teens, particularly in nonmetropolitan areas, are receiving formal sex education in school than they did a decade ago. Among all 15- to 17-year-olds, 58 percent didn't get information about birth control and 88 percent weren't taught about STDs and HIV/AIDS before they had sex for the first time (Lindberg et al., 2016).

Government policies also affect nonmarital birth rates. Publicly funded family planning services that offer counseling and contraceptives, particularly for teenagers and young adults, reduce unintended nonmarital pregnancies and births. Many states, however, fund only sex-abstinence education, and can now deny federal money to organizations that include contraception services. In 2013, Texas closed nearly half of the state's abortion clinics. Having to travel 200 or more miles for medical services reduced abortion rates by 47 percent. Reducing access to family planning clinics and abortion services can increase nonmarital births, especially among teens (Cunningham et al., 2017; Superville, 2017; see also Chapter 9).

In what some researchers describe as a pattern of "negative assimilation," 41 percent of second-generation Latinas and Asian mothers are unmarried, compared with 23 percent of recent immigrant women ("Second-Generation Americans," 2013). Thus, many second-generation women are espousing U.S. values that births outside of marriage are acceptable.

12-2d Single-Parent Households

Most children spend the majority of their childhood living with two parents, but single-parent families are increasingly common. Women head most one-parent families.

TRENDS

In 2016, 69 percent of children under age 18 lived with married parents, down from 77 percent in 1980. The number of one-parent families, on the other hand, has nearly doubled—from 16 percent in 1975 to 31 percent in 2015. Compared with other families with children under 18, mother-only families

- make up nearly 25 percent of all families and 78 percent of one-parent families;
- comprise 54 percent of Black, 29 percent of Latino, 19 percent of White, and 12 percent of Asian families;
- make up 28 percent of poor families; and
- are nearly twice as likely as father-only families to be living in poverty, even when the mother works full-time and year-round (Women's Bureau, 2016).

Asian American children are the most likely to grow up in two-parent homes (*Figure 12.4*). Until 1980, married couple families were the norm in Black families. Since then, Black children have been more likely than children in other racial-ethnic groups to grow up with

Figure 12.4 Where U.S. Children Live, by Race and Ethnicity, 2016

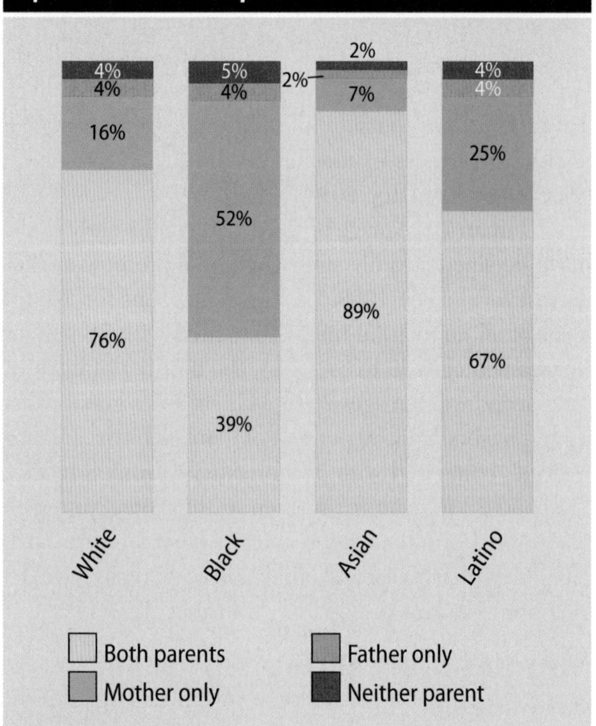

Note: "Both parents" includes married and cohabiting couples.

Source: Based on U.S. Census Bureau, Current Population Survey, 2016 Annual Social and Economic Supplement, Table C3, April 6, 2017. Accessed July 11, 2017 (census.gov).

only one parent, usually the mother. Among Latinos, 67 percent of children live in two-parent families, down from 78 percent in 1970 (Lugaila, 1998; *Figure 12.4*).

Nationally, of the 690,000 same-sex households, 18 percent are raising children under age 18. About 200,000 of these children have married or cohabiting same-sex parents; more than a million LGBT adults who aren't in a couple are raising approximately 2 million children. Even though same-sex individuals and couples are less likely than different-sex couples to be raising children, they're three times more likely to adopt or foster a child (Gates, 2014, 2015). Thus, large numbers of lesbians and gay men—single, partnered, or married—are parents.

WHY HAVE FEMALE-HEADED HOUSEHOLDS INCREASED?

On a *micro level*, having sex at an early age, and not using contraception at all, incorrectly, or sporadically increases the likelihood of unintended or unwanted pregnancies and mother-only families, especially among teenagers. Among unmarried parents, 63 percent have multiple partners. In these relationships, most of the fathers invest little time and money in their children because the fathers are young, have low education levels, or are incarcerated. The fathers don't live with their children, know little about them, and move from one sexual relationship to another (Scommegna, 2011; Dodson and Luttrell, 2011).

Mass media routinely feature the "single mother by choice" (SMC). Whether in movies, in newspapers, or on TV shows, SMCs are typically White, well-educated, and successful women in their 30s and 40s who decide to have and raise a baby without a partner. According to some sociologists, however, SMCs are "a relatively rare phenomenon," more common in the media than in reality, because they make up less than 5 percent of women, most of whom are older and well-educated (Hayford and Guzzo, 2015; Braff, 2016).

Demographic variables, especially social class, also affect the rise of female-headed households. College-educated people, as you've seen, tend to postpone marriage, marry rather than cohabit, and delay parenthood—factors that decrease the likelihood of divorce, one-parent households, and, consequently, children's experiencing poverty (Redd et al., 2011).

Not all female-headed households are poor, of course. Among mother-only families, 17 percent of the women have at least a college degree, and 38 percent own their own homes (Vespa et al., 2013). Instead of waiting for "Mr. Right," some women with economic resources are deciding to raise children on their own.

Karen Struthers/Shutterstock.com

Is the absent Black father a stereotype? A recent national study of White, Black, and Latino men found that, among fathers who don't live with their children, Black fathers were the *most* likely to parent their children from birth to age 18. The parenting included, and on a daily basis, feeding, bathing, dressing, playing with and reading to children, taking them to and from activities, and helping them with or checking homework (Jones and Mosher, 2013).

In many Black and Latino families, adult males (e.g., sons, brothers, uncles, grandfathers) often provide emotional and financial support to female kin who head households (Sudarkasa, 2007). Many black families also welcome **fictive kin**, nonrelatives who they accept as part of the family. Fictive kin have strong bonds with biological family members and provide important services (e.g., caring for children when young mothers are employed or negligent) (Billingsley, 1992; Dilworth-Anderson et al., 1993). A variation of fictive kin involves single mothers—many of whom are White, unmarried, and college educated—who turn to one another for companionship and child care help (Bazelon, 2009).

At the *macro level*, values and views about single parenthood have shifted. The share of Americans who view the growing trend of single mothers as a "big problem" decreased from 71 percent in 2007 to 64 percent in 2013 (Wang et al., 2013).

The economy also affects the number of female-headed households. As in the case of cohabitation and nonmarital childbearing, low-income women often drift into parenthood with low-income or unemployed men. Men may embrace parenthood and resolve to be good fathers, but already weak bonds with the child's mother can deteriorate further. Seeking financial security and stability, women gain little or nothing by marrying the babies' fathers or other low-income men (Edin and Nelson, 2013).

fictive kin nonrelatives who are accepted as part of a family.

12-3 FAMILY CONFLICT AND VIOLENCE

Conflict is a normal part of family life, but violence is *not* normal. Over a lifetime, we're much more likely to be assaulted or killed by a family member or current/former spouse, boyfriend, or girlfriend than by a stranger (Truman and Langton, 2014).

12-3a Intimate Partner Violence

Intimate partner violence (IPV) is abuse that occurs between people in a close relationship. The term *intimate partner* refers to current and former spouses, couples who live together, and current and former boyfriends or girlfriends. IPV, which ranges from a single episode to ongoing abuse, includes three prevalent types of behavior:

- *Physical abuse* is threatening, trying to hurt, or hurting someone using physical force (e.g., throwing objects, pushing, grabbing, slapping, biting, choking, beating).
- *Sexual abuse* is threatening or forcing a partner to take part in a sex act without her or his consent (e.g., unwanted anal, oral, or vaginal sex).
- *Psychological abuse* (also called *emotional abuse*) is aggressive behavior that threatens, humiliates, manipulates, or controls another person (e.g., criticism, rejection, isolation, name-calling, bullying, stalking).

IPV, which is pervasive in U.S. society (*Figure 12.5*), begins early in life. Among 12- to 18-year-olds in a recent or current dating relationship, 69 percent have been victims of IPV and 63 percent have perpetrated abuse. Among adults, 36 percent of females and 29 percent of males have been IPV victims at some time in their lives, and nearly half have experienced psychological abuse (Black et al., 2011; Copen et al., 2013; Taylor and Mumford, 2016).

Of the nearly 13,500 homicides in 2015, 30 percent involved IPV; 80 percent of the victims were females. When victims survive assaults, 40 percent of females compared with 5 percent of males suffer a serious physical injury (e.g., gunshot or knife wounds, internal injuries, broken bones) (Catalano, 2013; Federal Bureau of Investigation, 2016). Women are also much more likely than men to experience IPV over a lifetime regardless of age, race or ethnicity, or social class (Black et al., 2011; Truman and Morgan, 2014).

There are nearly 1.3 million intimate partner victimizations every year, but this number is conservative.

intimate partner violence (IPV) abuse that occurs between people in a close relationship.

Figure 12.5 Intimate Partner Violence, Sexual Violence, and Stalking in Lifetime

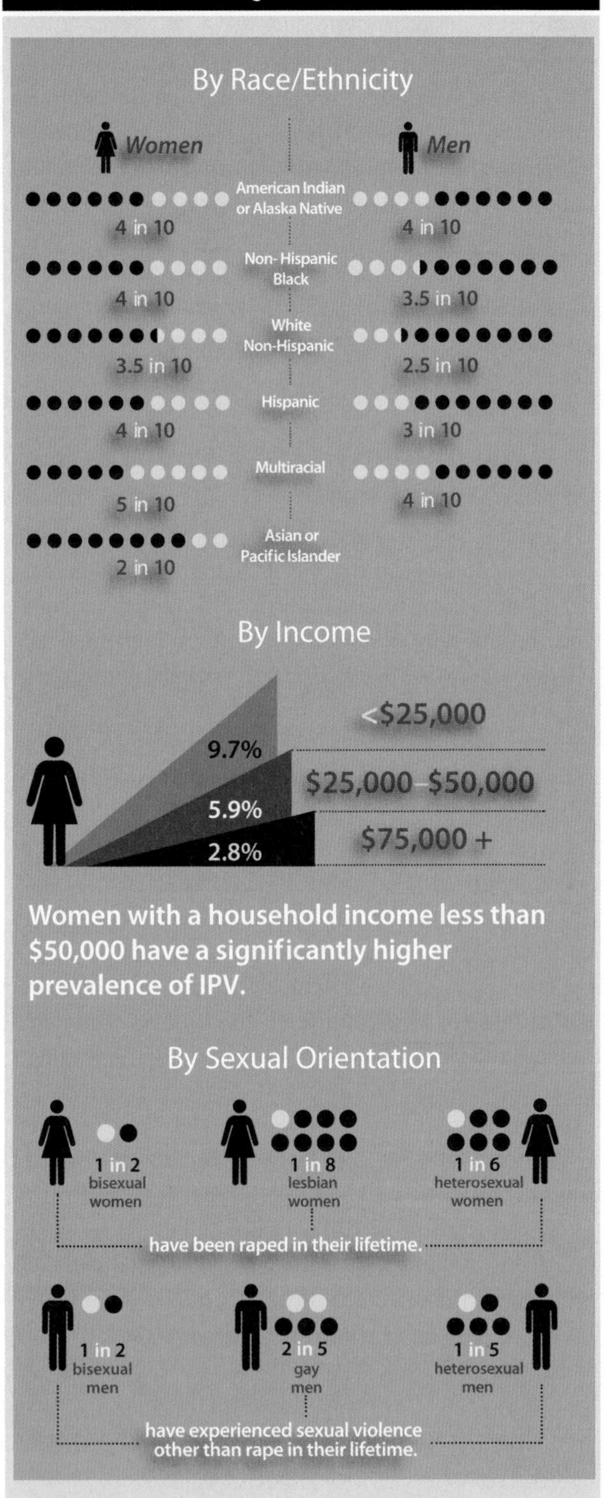

Source: Centers for Disease Control and Prevention, Infographic, 2014. Accessed October 8, 2015 (cdc.gov/violenceprevention/nisvs/infographic.html).

An estimated 44 percent of victims don't report IPV to the police because they're ashamed, believe that no one can help, or fear reprisal (Reaves, 2017).

WHY DOES INTIMATE PARTNER VIOLENCE OCCUR?

The reasons for IPV are due to interrelated individual, demographic, and societal factors. On a *micro level*, violence escalates if one or both partners use alcohol or other drugs, if they have more children than they can afford (which intensifies financial problems), or if either partner has been raised in a violent household, has low self-esteem, and a controlling personality. The most common disagreements that can escalate into abuse and violence have four sources: gender role expectations (who does what housework); money (saving and spending); children (especially discipline); and infidelity, both personally and online (Spivak et al., 2014; Benokraitis, 2015).

Regarding *demographic variables*, IPV often begins at a young age: 22 percent of females and 15 percent of males are between 11 and 17 years old, and 47 percent of females and 39 percent of males are between 18 and 24 years old (Black et al., 2011). IPV cuts across all social classes, but women living in households with an annual income of less than $7,500 are nearly four times more likely to be abused than those living in households with an annual income of $75,000 or more (*Figure 12.5*).

Macro-level variables, particularly unemployment and poverty, increase the likelihood of financial stress and violence. The absence of legal or social sanctions against IPV and a scarcity of shelters discourage victims from trying to escape abuse (Matjasko et al., 2013). A study of girls aged 11 to 16 who had experienced sexual assault concluded that girls (and later women) often didn't report the incidents because such violence had been "normalized in their communities." The girls were ashamed and feared retribution, but also believed that men "can't help it," perceived everyday harassment and abuse as "normal male behavior," and assumed that male authority figures (e.g., police officers, judges) would accuse the girls of overreacting (Hlavka, 2014).

12-3b Child Maltreatment

Child maltreatment (also called *child abuse*) includes a broad range of behaviors that can result in serious harm, including physical and sexual abuse, neglect, and emotional mistreatment. The victims often experience several types of maltreatment.

Figure 12.6 Types of Child Maltreatment, 2015

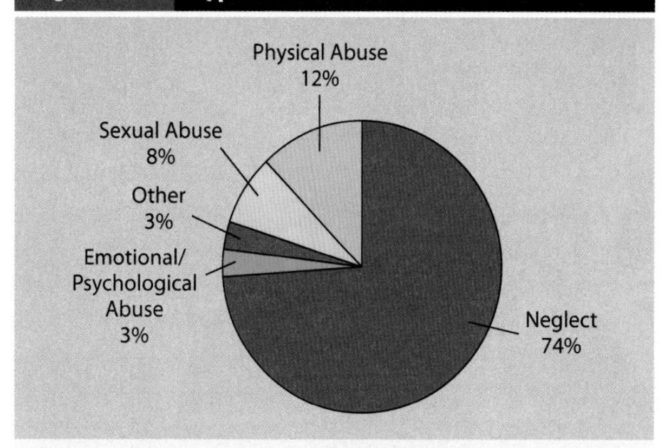

Note: "Neglect" includes medical neglect (almost 2 percent of these cases). "Other" includes categories that some states report, such as a parent's drug/alcohol abuse.

Source: Based on U.S. Department of Health and Human Services, 2017, Table 3-11.

TRENDS

U.S. child maltreatment rates fluctuate, but have decreased since 2011. In 2015, nearly 684,000 children (9.2 per 1,000 children) experienced child maltreatment, and 1,670 died of abuse and neglect (U.S. Department of Health and Human Services, 2017). Because only a fraction of the cases are reported, however, federal agencies assume that millions of children experience abuse and neglect every day.

Neglect is the most common type of abuse (*Figure 12.6*). The most vulnerable children, those younger than 3 years old, account for 28 percent of all child victims; another 19 percent are ages 3 to 5. Victimization rates per 1,000 children vary by race and ethnicity: 15 for African Americans, 14 for Native American/Alaska Natives, 10 for multiracial, 9 for Pacific Islanders, 8 (each) for White and Latino individuals, and under 2 for Asians (U.S. Department of Health and Human Services, 2017).

Almost 81 percent of the perpetrators are one or both parents; in 37 percent of the cases, the abusers are mothers alone. An additional 7 percent of assailants are relatives, and 4 percent are the parents' intimate partners, usually boyfriends (U.S. Department of Health and Human Services, 2013).

WHY DOES CHILD MALTREATMENT OCCUR?

On a *micro level*, child mistreatment is most common in households where there is parental substance abuse or mental illness,

child maltreatment (also called *child abuse*) a broad range of behaviors that can result in serious emotional or physical harm.

Anne Marie Fox/Lions Gate/Everett Collection

The movie *Precious* (2009) received many awards, but some critics contended that the film stereotyped Black female teenagers—especially those in low-income households—as obese, illiterate, and living with abusive mothers (Amusa, 2010). If you've seen the movie, do you agree with such criticism or not? Why?

adult or sibling violence, and one or more children have emotional, developmental, or physical disabilities (Finkelhor et al., 2011). *Demographic variables* that increase the likelihood of child abuse include living with a young parent, living in one-parent or stepparent households—particularly those with low socioeconomic status—and experiencing parental conflict before, during, and after a hostile divorce (Truman and Smith, 2012; White and Lauritsen, 2012). On a *macro level*, economic hardship, unemployment, and poverty increase the likelihood of stress that leads to child abuse, including infant deaths due to severe head traumas (Eckenrode et al., 2014).

Almost 26 percent of all children live in homes where parents or other adults are violent (Hamby et al., 2011). Whether children are targets of abuse or see it, violent childhood experiences are associated with lifelong developmental problems, including depression, delinquency, suicide, alcoholism, low academic achievement, unemployment, and medical problems in adulthood (Hibbard et al., 2012; Sacks et al., 2014).

12-3c Elder Abuse and Neglect

Elder abuse (also called *elder mistreatment*) is any knowing, intentional, or negligent act by a caregiver or other person that causes harm to people age 65 or older. This term includes physical, psychological, and sexual abuse; neglect; isolation from family and friends; deprivation of basic necessities such as food and heat; not providing needed medications; and financial exploitation.

elder abuse (sometimes called *elder mistreatment*) any knowing, intentional, or negligent act by a caregiver or other person that causes harm to people age 65 or older.

TRENDS

A national study of people ages 60 and older found that almost 12 percent had experienced at least one of the following types of mistreatment: emotional (4.6 percent), physical (1.6 percent), sexual (0.6 percent), financial (5.2 percent), and neglect (6 percent) (Acierno et al., 2010). The actual rates are probably much higher because only an estimated 20 percent of all elder abuse and neglect is reported (Administration on Aging, 2013).

Who are the victims? About 83 percent are White; the average age is 76; 76 percent are women; 84 percent live in their own homes; 86 percent have a chronic disease or other health condition; 57 percent are married or cohabiting; 53 percent haven't graduated from high school; 50 percent suffer from dementia, Alzheimer's, or other mental illness; 46 percent feel socially isolated; and the average combined household income is less than $35,000 a year (Acierno et al., 2010; Jackson and Hafemeister, 2011). These data suggest that the most likely victims of elder abuse are those who are the most vulnerable physically, mentally, socially, and financially.

Who are the perpetrators? Most are adult children, spouses or cohabiting partners, or other family members. Less than a third are acquaintances, neighbors, or nonfamily service providers. The average age of the abuser is 45; 77 percent are White; 61 percent are males; 82 percent have a high school diploma or less; 50 percent abuse alcohol and/or other drugs; 46 percent have a criminal record; 42 percent are financially dependent on the elder; 37 percent live with the older person; 29 percent are chronically unemployed; and 25 percent have mental health problems (Jackson and Hafemeister, 2011).

WHY DO ELDER ABUSE AND NEGLECT OCCUR?

On a *micro level*, abuse of alcohol and other drugs is more than twice as likely among family caregivers who abuse elders as among those who don't. Both victims and offenders often report a childhood history of witnessing or experiencing family violence, poor family relationships in the past and currently, and communication problems. In addition, older people with cognitive impairment—due to dementia (deteriorated mental condition) after a stroke, or the onset of Alzheimer's disease—are abused at higher rates than those without such disabilities (Heisler, 2012; National Center on Elder Abuse, 2012).

On a *macro level*, a shared residence is a major risk factor for elder mistreatment because the caregiver(s) may depend on the older person for housing, whereas the elder is dependent on the caregiver(s) for physical help. These situations compound the likelihood of everyday tensions and conflict. Financial stress may also increase the risk of abuse. Unlike low-income families, those in the middle class aren't eligible for admission to public facilities, yet few can pay for the in-home nursing care, high-quality nursing homes, and other services that upper-class families can afford (Acierno et al., 2010; Jackson and Hafemeister, 2011).

12-4 OUR AGING SOCIETY

One of our friends, 55, was shocked and insulted when he ordered a meal at a fast-food restaurant, and the young worker called out "one senior meal!" The *Baltimore Sun* (2012) recently reported on the death of an "elderly" man in a house fire. He was 62.

AP Images/Itsuo Inouye

Between the ages of 65 and 91, avid mountaineer Hulda Crooks scaled Mount Whitney (the highest mountain in the continental United States) 23 times. She died at the age of 101 in 1997.

12-4a When Is "Old"?

What images come to mind when you hear the word *old*? In a recent national survey, people in their 40s said that a person is old at age 63, those in their 70s said age 75, and a 90-year-old woman said that a woman isn't old "until she hits 95" (*AARP Magazine*, 2014: 40). So, how old is "old"?

It depends on whom you ask because "old" is a social construction. In some African nations, where people rarely live past 50 because of diseases and civil wars, 40 is old. In industrialized societies, where the average person lives to age 78, 40 is considered young. Still, regardless of how we feel physically, society usually defines old in chronological age. In the United States, for instance, people are deemed old at age 65, 66, or 67 because they can retire and become eligible for Medicare and Social Security benefits.

Many older Americans are vigorous and productive, but others experience physical and mental limitations as they age. There are significant differences between the *young-old* (65–74 years old), the *old-old* (75–84 years old), and the *oldest-old* (85 years and older) in health, living independently, and working. Generally, for example, a 75-year-old is much less likely than an 85-year-old to need caretaking.

12-4b Life Expectancy

Life expectancy is the average expected number of years of life remaining at a given age. American children born in 2010 have a life expectancy of almost 79 years (compared with 71 in 1970 and 47 in 1900). Despite our nation's enormous wealth, the United States ranks just 42nd globally in life expectancy (Schwartz, 2017).

Life expectancy varies by sex, social class, and race-ethnicity. Historically, currently, and across all racial-ethnic groups, women live longer than men, and Asians have the longest lifespans (*Figure 12.7*).

The higher rates of cigarette smoking, heavy drinking, gun use, employment in hazardous occupations, and risk-taking in recreation and driving are responsible for many males' shorter lifespans. The life expectancy gender gap has narrowed since the 1990s, however, because there's been an increase in women's smoking, use of alcohol and other drugs, obesity (which increases the risk of hypertension and heart disease), and stresses due to multiple roles (e.g., juggling employment while caring for children and older family members) ("Catching Up," 2013; see also Chapter 14).

Generally, higher socioeconomic levels increase life expectancy. People with a college degree and higher are more likely than less educated adults to be employed, to be financially secure, and to have employment-related health insurance. They're also more likely to exercise, not smoke, drink alcohol in moderation, and maintain a healthy body weight.

Asian Americans die from the same diseases as other groups, but do so later in life. They're the

life expectancy the average expected number of years of life remaining at a given age.

Figure 12.7 U.S. Life Expectancy at Birth, by Sex and Race-Ethnicity, 2015

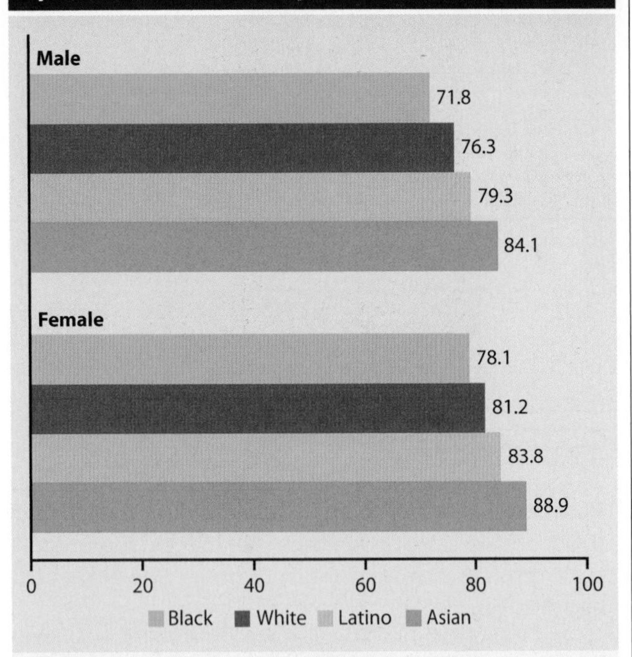

Source: Based on National Center for Health Statistics, 2017, Figure 6; USA Life Expectancy, 2016.

highest-income and best-educated minority in the United States (see Chapter 10). About 97 percent live in or near a major city with access to high-quality health care and have strong family support networks during illness. They're also more likely than other racial-ethnic groups to cook healthful traditional meals, smoke less, and maintain a healthy body weight (Acciai et al., 2015). There are no definitive reasons for Asians' longevity, but these factors prolong life.

12-4c How Our Graying Nation Is Changing

The 65 and older population is booming, and becoming more ethnically and racially diverse. For example,

- 49 million Americans (15 percent of the total population) are 65 or older, a dramatic increase from just 4 percent in 1900, and will comprise 21 percent of U.S. residents by 2030.
- One of the fastest-growing groups is the oldest-old, whose numbers increased from 100,000 in 1900 to 6.3 million in 2015. By 2030, this group will comprise almost 3 percent of the population.
- By 2030, 28 percent of the 65-and-older population will be minority, up from 20 percent in 2010; 23 percent of the 85-and-older population will be minority, up from 15 percent in 2010 (Colby and Ortman, 2015; "The Nation's Older Population...," 2017; "Older Americans Month...," 2017).

baby boomers people born between 1946 and 1964.

old-age dependency ratio (also called the *elderly support ratio*) the number of working age (18 to 64) adults for every person aged 65 and older.

Our aging population has four important (and interrelated) implications. First, as you'll see shortly, health care and caregiving needs, services, and costs will surge. Second, the number of older Americans is increasing, whereas the proportion of young people is falling (*Figure 12.8*). As a result, many adult children—even in their 60s—will be caring for aging parents, grandparents, and other older relatives for more years than in the past.

Third, disability rates among older Americans have increased, particularly among **baby boomers** (people born between 1946 and 1964) and those 10 years younger than baby boomers. The higher disability rates are due to better medical diagnoses, but baby boomers are more likely than the previous generation to be obese and to have diabetes and high blood pressure (Scommegna, 2013). Thus, the boomer generation might be sicker and for more years than their predecessors who had lower life expectancies.

Fourth, as the number of older people increases, so does the **old-age dependency ratio** (also called the *elderly support ratio*)—the number of working age (18 to 64) adults for every person aged 65 and older. The ratio is an approximation because some people work into their late 60s and 70s, but it's a useful measure of the burden of workers who support the older population. The old-age dependency ratio has dropped from 14 workers per

Figure 12.8 The Young and the Old in the United States, 1900–2030

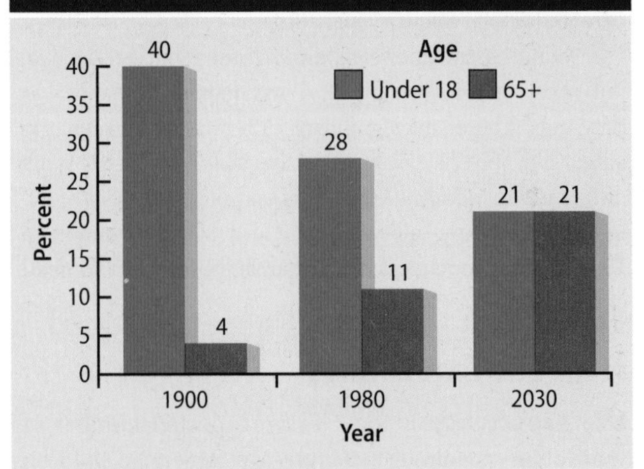

Sources: Based on U.S. Senate Special Committee on Aging et al., 1991: 9, and Colby and Ortman, 2015, Figure 4.

older person in 1900. By 2020, only three working adults will be supporting each older person, and the ratio could drop to two by 2060 (Ortman et al., 2014; Mather et al., 2015). Thus, many people will have to pay much higher federal and state taxes to support our ever-growing older population.

12-4d Some Current Aging Issues

The *sandwich generation* is composed of midlife people who care for aging parents, are raising a child under age 18, or supporting an adult child. About 71 percent who do so are ages 40 to 59 (Parker and Patten, 2013). In an aging society, the sandwich generation and older people experience many changes, including the rise of multigenerational households, work and retirement options, right-to-die issues, and competition for scarce resources.

MULTIGENERATIONAL HOUSEHOLDS

The share of the U.S. population living in multigenerational households rose from 12 to 16 percent between 1980 and 2010. Asians (28 percent) are more likely to live in multigenerational households than Black (26 percent), Latino (25 percent), and White (14 percent) households.

Why are several generations living under one roof? Graying boomers may move in with their adult children and grandchildren to avoid poverty. More grandparents are raising children because drug-addicted parents neglect them, abandon them, or die from an overdose. Adult children may move back home with their parents (often with children and girlfriends/boyfriends in tow) because of unemployment, low wages, or divorce. Extended family living provides intergenerational support, including child care for employed parents. Cultural values, particularly among recent immigrants, reflect long-standing practices of caring for aging parents and grandparents in one's own home (Ellis, 2013; Fry and Passel, 2014; Wiltz, 2016).

The effects of multigenerational households are mixed. Grandparents who live with their adult children and grandchildren may provide crucial economic and child care support. On the other hand, if the multigenerational household is struggling financially, if the grandparents are also raising their own children, and if there's mother–grandmother conflict, grandparents experience emotional problems, including unhappiness, stress, worry, and anger (Deaton and Stone, 2013).

WORK AND RETIREMENT

Many older Americans are postponing retirement. In 2017, almost 19 percent of people aged 65 and older were working—up from 14 percent in 2000 (Kromer and Howard, 2013; Bureau of Labor Statistics, 2017). The labor force participation rates have increased for both men and women, even for people aged 70 or over (*Figure 12.9*).

Figure 12.9 The U.S. Labor Force Is Growing Older

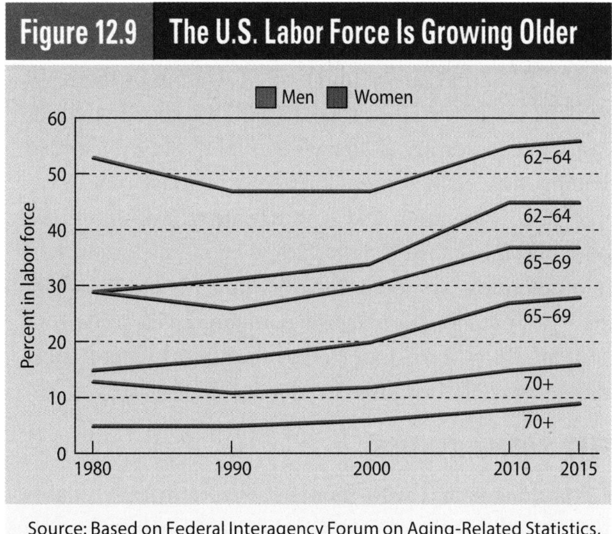

Source: Based on Federal Interagency Forum on Aging-Related Statistics, 2016, Table 12.

Most work out of necessity. A consequence of longer life expectancy is that people need to finance more years of retirement, but are they doing so? Almost a third of employed adults with $10,000 or more in investments give financial assistance to an adult child, a parent, or both (Saad, 2017). The contributions increase family stability and strengthen family ties, but also deplete retirement savings.

Since the early 1990s, many large companies (especially in the steel, airline, and auto industries) have greatly reduced or eliminated employee pension plans, but only about half of private sector workers participate in a 401(k) or similar retirement savings program. Among those with savings, the median account value is just $148,00 for households aged 65 and older. In 2015, the average Social Security payment for people aged 65 and older was only $1,342 a month, but 60 percent rely on Social Security for half or more of their family income ("Who Pays the Bill?" 2013; Center on Budget and Policy Priorities, 2016; Federal Interagency Forum on Aging-Related Statistics, 2016).

Social class is a key factor in living well after retirement. People in higher social classes have more savings, pensions, and property (including homes with paid-off mortgages); higher Social Security income because they had higher lifetime earnings; and 401(k)s that have increased in value over the years. Because of such assets, 64 percent of nonretirees with annual household incomes of $75,000 or more expect to live comfortably in retirement, compared with only 32 percent with annual incomes below $30,000 (Norman, 2016).

For older women, particularly, working is often a necessity rather than a choice. Did you notice, in *Figure 12.8*, that larger numbers of women in their 60s and 70s are in the labor force? You'll see shortly that divorce rates after age 50 have been increasing. Divorced women, especially if they don't get the house or a favorable legal settlement, may have to enter (or re-enter) the labor force because they have few economic resources. In other cases, women in their 60s and 70s are staying in the workforce longer because, compared with prior generations, they have more education, work experience, and access to better jobs (Goldin and Katz, 2016).

THE RIGHT TO DIE

Living longer and with more years of chronic diseases and disabilities raises questions about later-life choices. Both historically and currently, one of the most controversial issues involves end-of-life decisions. Oregon's Death with Dignity Act, which took effect in 1997, authorized lethal prescriptions when two doctors agreed that a patient would die within six months and freely chose *physician-assisted suicide* (also called *physician-aid-in-dying*). Modeling Oregon's law, six states and the District of Columbia have passed similar legislation.

In a recent Gallup poll, 73 percent of Americans said that doctors should be allowed to "end the patient's life by some painless means" (Wood and McCarthy, 2017). Advocates argue that terminally ill people who face a long and painful death and huge medical costs should have the legal right to die with dignity on their own terms instead of lingering in a nursing home, hospital, or hospice. Opponents maintain that actively ending a life, regardless of a person's frailty or suffering, is a moral violation, and that patients might be pushed to die early for the caregiver's convenience (Eckholm, 2014).

Ken Tannenbaum/Shutterstock.com

Medical care for patients age 65 and older in the last year of life accounts for about 25 percent of annual Medicare costs (Cubanski et al., 2016). Is this a good investment of our resources?

COMPETITION FOR SCARCE RESOURCES

When Congress passed the Social Security Act in 1935, life expectancy was about 62 years compared with almost 79 today. In the future, the growing older population will put a significant strain on the nation's health care services and retirement income programs.

The older population is about half as large as the 18 and younger population. Nonetheless, the federal government spends nearly $6 on older people for every $1 it spends on children (Hahn et al., 2014). The older people get, the higher the medical costs. For example,

- Among Medicare recipients, the average annual health care costs are almost $17,000 for someone aged 65 to 74, compared with almost $26,000 for people aged 85 and over (Federal Interagency Forum on Aging-Related Statistics, 2016).
- A study of 1.8 million Medicare recipients who died in 2008 found that nearly 20 percent had surgery in the last month of life, and nearly 10 percent had surgery in the last week of life. Among those undergoing end-of-life surgery, almost 60 percent were 80 and older (Kwok et al., 2011).
- About 60 percent of prostate cancer is diagnosed in men aged 65 or older. It costs $93,000 per patient to treat advanced prostate cancer; the treatment prolongs life by an average of four months (Beil, 2012).

Many older people maintain that they deserve all the medical benefits they can get because they paid taxes over many years. In fact, an average two-earner couple that retired in 2015 will draw about $212,000 more from Medicare and Social Security than they paid in taxes to support these programs (Steuerle and Quakenbush, 2013).

Younger generations can't count on federally financed health care and retirement benefits in the future. Medicare funds are supposed to last until 2026, and Social Security through 2033 (The Board of Trustees..., 2013), but there are no guarantees. As baby boomers age, the costs of these programs will mushroom. Because boomers have higher educational levels than previous generations, they're likely to demand more and more expensive health care services (see Chapter 14).

Despite rising health care costs, 20 percent of Americans want to live to age 100—about 21 years longer than the current average U.S. life expectancy. Moreover, 38 percent support medical treatments that would allow them to live to at least age 120 (Lipka and Stencel, 2013; Lugo et al., 2013). Who'll pay for such radical life extensions?

12-5 SOCIOLOGICAL EXPLANATIONS OF FAMILY AND AGING

The four sociological perspectives are useful in understanding families and aging. *Table 12.3* summarizes the key points of these theories.

12-5a Functionalism

We began this chapter by looking at the vital functions that families perform—such as procreation and economic security—that promote societal stability and individual well-being. Functionalists recognize that families differ in structure (e.g., nuclear vs. extended), but believe that their similar functions ensure a society's continuity.

STABILITY AND ACTIVITY

For functionalists, marriage, followed by procreation, is critical in fostering social order and cohesion. Parenthood is a crucial social role. One of its tasks, socialization, is essential to maintaining any culture. Problems arise, for instance, when parents can't or don't provide their children with the necessary financial and emotional support or when they divorce.

Kinship ties involve responsibilities, but they also help keep families together, especially in time of trouble. Extended family members offer financial assistance, emotional support, and help to care for young children and aging adults. Nearly 1.5 million employed grandparents are responsible for most of the basic care of grandchildren who live with them ("National Grandparents Day...," 2017).

Retirement benefits society because it provides younger people jobs. Even in retirement, however, **activity theory** proposes that many older people remain engaged in numerous roles and activities, including work. Moreover, those who are active adjust better to aging and are more satisfied with their lives (see Atchley and arusch, 2004, for a summary of some of this research).

CRITICAL EVALUATION

Functionalism has several weaknesses. First, it's questionable whether some family functions are as universal or necessary as functionalists claim. Procreation often occurs outside of marriage and divorce rates have tripled among people aged 65 and older since 1990. Also, the government has assumed some of the family's functions, including caring for some children (as in foster homes) and the aged (through Medicare, for example) (Lundberg and Pollak, 2007; Stepler, 2017).

Second, is activity theory as representative of older people as some functionalists claim? People may continue to work even in their 80s, but usually not by choice. They can't afford to retire, even though they're in poor health and unhappy in their low-income jobs. As health deteriorates, many older people become less active and more isolated (Lee, 2009).

activity theory proposes that many older people remain engaged in numerous roles and activities, including work.

Table 12.3 Sociological Perspectives on Families and Aging

Theoretical Perspective	Level of Analysis	Key Points
Functionalist	Macro	Families are important in maintaining societal stability and meeting family members' needs. Older people who are active and engaged are more satisfied with life.
Conflict	Macro	Families promote social inequality because of social class differences. Many corporations view older workers as disposable.
Feminist	Macro and micro	Families both mirror and perpetuate patriarchy and gender inequality. Women have an unequal burden in caring for children as well as older family members and relatives.
Symbolic Interactionist	Micro	Families construct their everyday lives through interaction and subjective interpretations of family roles. Many older family members adapt to aging and often maintain previous activities.

Third, 10 percent of the huge wave of baby boomers say that they'll never retire, primarily because they've saved very little and carry too much debt (Harter and Agrawal, 2014). This means that there isn't always an orderly progression to retirement and opening up job slots to the younger generation.

12-5b Conflict Theory

Conflict theorists agree that families serve important functions, but point out that some groups benefit more than others. Families are sources of social inequality that mirror the larger society, and the inequities persist from birth to old age (Cruikshank, 2009).

INEQUALITY, SOCIAL CLASS, AND POWER

For conflict theorists, families in high-income brackets have the greatest share of capital, including wealth that they can pass down to the next generation. An inheritance reduces the likelihood that all families can compete for resources such as education, decent housing, and health care (see Chapter 8).

Economic inequality affects all aspects of family life. Child abuse and neglect are highest in U.S. counties with the greatest gaps between rich and poor. Moreover, the more distance between these social classes, the less likely that the poor will receive needed services and support that reduce child maltreatment rates (Eckenrode et al., 2014).

Unequal access to resources continues into old age. Most employers insist that they value older workers' loyalty, work ethic, reliability, and experience, but many are less likely to hire or retain older people because they're usually more expensive than younger workers. Because many large companies have cut their pension plans, numerous older workers must work long after they expected to retire (Kromer and Howard, 2013; Winerip, 2013).

CRITICAL EVALUATION

Conflict theory is limited for several reasons. First, conflict theorists tend to overlook the fact that many families, especially those in the middle and upper classes, fund public assistance programs such as Medicaid (a social health care program for low-income individuals and families). They also pay billions each year to cover the costs of unintended pregnancies and the resulting infant care expenses for poor women (Sonfield and Kost, 2013).

Second, conflict theory links family inequality to capitalism and social class, but there's also considerable family inequality in countries that aren't capitalist (see Chapters 8 and 11). Thus, social class affects but doesn't determine whether a child will succeed economically.

Michael Hall/Taxi/Getty Images

The average 65-year-old can now expect to live for another 20 years, half of them free of disability. For functionalists, older adults are happiest when they stay active and maintain social relationships.

Third, many older people don't have to struggle for resources. As you saw earlier, the U.S. government's spending on children has declined but increased for people age 65 and older. Older people, regardless of social class, have generous health care coverage through Medicare or Medicaid. They vote in large numbers, have influential lobbyists in Congress, and flood lawmakers with hundreds of thousands of letters and email messages when there's a threat of cutting Medicare or Social Security payments (Johnson, 2011; see also Chapter 11).

12-5c Feminist Theories

Feminist scholars agree with conflict theorists that there's considerable inequality between low-income and wealthy families in accessing necessary resources. However, feminist theorists emphasize the inequality of gender roles in families, especially in patriarchal societies (including the United States).

GENDER ROLES AND PATRIARCHY

For feminist scholars, families both mirror and perpetuate patriarchy and gender inequality. In most countries, males pass laws about property and inheritance rights, marriage and divorce, and many other regulations that give men authority over women. The United States is similar to many other nations in men (particularly lawmakers and Catholic Church officials) controlling women's decisions about reproduction and access to abortion that, in turn, increases unwanted pregnancies and nonmarital birth rates.

Male aggression against women and children is common in patriarchal societies where men hold most of the power, status, and privilege. Females, on the other

Maggie Steber/National Geographic Creative

In most Latino communities, the *quinceañera* (pronounced "keen-say-ah-NYAIR-ah") is an important coming-of-age ritual that celebrates a girl's entrance into adulthood on her fifteenth birthday. It's an elaborate and dignified religious and social event that reinforces strong ties with family, relatives, and friends.

hand, are marginalized and socialized to accept male domination. In many cases, legal, political, and religious institutions don't take violence against women and children seriously (Lindsey, 2005). Recently, for instance, a judge in Florida "sentenced" a man charged with domestic battery to take his wife, the victim, to Red Lobster for dinner and then bowling (McEwan, 2012). Thus, some judges still view domestic violence charges as frivolous.

Out of 190 nations, the United States, Papua New Guinea, and Swaziland are the only countries that have no national paid parental leave policy. As a result, caregivers, who are predominantly women, must often leave their jobs or work only part-time to care for children and aging parents. Working part-time or stepping out of the workforce reduces women's earnings, career advancement, retirement income, and their families' living standards. Among those who take time off from work following the birth or adoption of a child, women are nearly twice as likely as men (25 percent and 13 percent, respectively) to say that doing so had a negative impact on their job or career (Fry, 2017; Horowitz, 2017; Rossin-Slater, 2017).

CRITICAL EVALUATION

Feminist explanations of families and aging have several limitations. First, according to critics, feminist scholars overstate women's domination by men. For example, a record 40 percent of all households with children under the age of 18 include mothers who are either the sole or primary source of family income, up from just 11 percent in 1960 (Wang et al., 2013).

A second criticism is that feminist theories tend to gloss over data which show that intimate partner violence is often mutual. Several dozen studies have found that women initiate from 30 to 73 percent of violent incidents (Straus, 2011, 2014). You'll recall that mothers mistreat their children, and almost 40 percent of the perpetrators of elder abuse are women (Jackson and Hafemeister, 2011).

Third, the poorest older adults are minority women, but social class and marital status are important factors. Older married women and those in higher socioeconomic levels are less likely to experience poverty in later life than women who are single (never married, divorced, or widowed) and from lower socioeconomic levels (Hokayem and Heggeness, 2014; Hunter, 2014). Thus, social class may be more significant than patriarchy in shaping women's later life outcomes.

12-5d Symbolic Interaction

For symbolic interactionists, people create subjective meanings of what a family is and does. Thus, people learn, through interaction with others, how to act as a parent, a grandparent, a teenager, a stepchild, and so on throughout the life course.

LEARNING FAMILY AND AGING ROLES

Throughout the socialization process, family members establish trust and develop emotional bonds. Between 90 and 97 percent of Americans believe that it's important for *both* mothers and fathers to provide discipline and emotional support, and to teach values ("The New American Father," 2013).

Interactionists often use exchange theory to explain mate selection and family roles. The fundamental premise of **exchange theory** is that people seek through their social interactions to maximize their rewards and minimize their costs. In mate selection, people trade their resources (e.g., wealth, good looks, youth, and/or status) for more, better, or different assets. People may stay in unhappy marriages and

exchange theory people seek through their social interactions to maximize their rewards and minimize their costs.

other intimate relationships because the rewards seem equal to the costs. Many women tolerate abuse because they fear loneliness or losing the economic benefits that a man provides (Khaw and Hardesty, 2009; see also *Table 12.1* on the benefits and costs of cohabitation).

A well-known psychologist who interviewed more than 200 couples over a 20-year period found that the difference between lasting marriages and those that split up was a "magic ratio" of 5 to 1. That is, if partners have five positive interactions for every negative one, the marriage is likely to be stable over time (Gottman, 1994). Thus, we can improve our family relationships by learning to interact in positive ways.

Stereotypes about older people are deeply rooted in U.S. society. About 84 percent of Americans age 60 and older encounter **ageism**, discrimination against older people, including insulting jokes, disrespect, and patronizing behavior (Roscigno, 2010).

Our language is full of ageist words and phrases that stereotype and disparage older people (e.g., *biddy, old bat, old fart, old fogey, geezer, over the hill, doddering*). Because it's legal to ask about age (but not sex or race), age discrimination is a major barrier in getting a job. If employers assume that older workers are less productive, "technophobic," or inflexible, older applicants—even those applying for low-skilled jobs—are half as likely as people aged 29 to 31 to get a callback. Ageism, especially in hiring practices, helps explain why people try to look younger. In 2016, for example, people aged 40 to 54 had the most plastic surgery (American Society of Plastic Surgeons, 2017; EEOC, 2017; Newmark et al., 2017; Rosenblatt, 2017).

ageism discrimination against older people.

continuity theory older adults can substitute satisfying new roles for those they've lost.

Continuity theory posits that older adults usually maintain the same activities, behaviors, social roles, personalities, and past relationships. The theory also proposes that older people adapt to changes by substituting satisfying new roles for those they've lost (Atchley and Barusch, 2004). For example, a retired music teacher can join a local chorus or orchestra. Thus, developing new roles may lessen some of the emotional distress due to ageism.

CRITICAL EVALUATION

A common criticism is that symbolic interaction, a micro-level perspective, doesn't address macro-level constraints. For example, U.S. nonmarital teen births have fallen primarily because of structural factors (e.g., greater access to family planning services, expanded educational opportunities for disadvantaged young women) rather than teen attitudes about childbearing (Kearney and Levine, 2014).

Second, exchange theory is limited because people don't always calculate the potential costs and rewards of every decision. In the case of women who care for older family members, genuine love, concern, and a sense of obligation override cost-benefit decisions, even when the person receiving care is abusive (Jackson and Hafemeister, 2013).

Third, continuity theory doesn't distinguish normal aging from disease and disability. Older people who experience chronic illness may be too sick to continue their activities and relationships. Continuity theory also overlooks structural obstacles that discourage older people from pursuing current or new roles. Poverty, for example, can affect physical and mental health which, in turn, weakens social networks and limits housing, medical, and recreational options.

STUDY TOOLS 12

READY TO STUDY? IN THE BOOK, YOU CAN:

- ☐ Check your understanding of what you've read with the Test Your Learning Questions provided on the Chapter Review Card at the back of the book.
- ☐ Tear out the Chapter Review Card for a handy summary of the chapter and key terms.

ONLINE AT CENGAGEBRAIN.COM WITHIN MINDTAP YOU CAN:

- ☐ Explore: Develop your sociological imagination by considering the experiences of others. Make critical decisions and evaluate the data that shape this social experience.
- ☐ Analyze: Critically examine your basic assumptions and compare your views on social phenomena to those of your classmates and other MindTap users. Assess your ability to draw connections between social data and theoretical concepts.
- ☐ Create: Produce a video demonstrating connections between your own life and larger sociological concepts.
- ☐ Collaborate: Join your classmates to create a capstone project.

13 Education and Religion

FatCamera/Getty Images

LEARNING OBJECTIVES

After studying this chapter, you will be able to…

13-1 Differentiate between education and schooling.

13-2 Compare and evaluate the sociological perspectives on education.

13-3 Identify, describe, and illustrate some of the contemporary issues in U.S. education.

13-4 Explain how religion, religiosity, and spirituality differ.

13-5 Describe and illustrate religious organization and some of the major world religions.

13-6 Explain how and why U.S. religion is diverse, and whether or not secularization is increasing.

13-7 Compare and evaluate the sociological perspectives on religion.

After finishing this chapter go to **PAGE 281** for **STUDY TOOLS**

Education and religion, two important social institutions, teach values, shape attitudes, maintain traditions, bring people together, control behavior, and grapple with social change. The second half of this chapter examines U.S. and global religions. Let's begin with education.

WHAT DO YOU THINK?

I'm optimistic about getting a good job after graduating from college.

1	2	3	4	5	6	7
strongly agree						strongly disagree

13-1 EDUCATION AND SOCIETY

Education transmits attitudes, knowledge, beliefs, values, norms, and skills to a society's members. Education can be formal or informal and can occur in a variety of settings. **Schooling**, a narrower term, is formal training and instruction provided in a classroom setting.

U.S. education and schooling have undergone four significant changes since the beginning of the twentieth century: Universal education has expanded, community colleges have flourished, public higher education has burgeoned, and student diversity has increased as more women and racial-ethnic groups enrolled at colleges and universities. As a result of these and other changes, a record number of Americans have completed high school (92 percent) and obtained a bachelor's or higher degree (36 percent) (McFarland et al., 2017).

13-2 SOCIOLOGICAL PERSPECTIVES ON EDUCATION

Sociologists agree that education and schooling are important, but offer different insights into their purpose and outcomes. *Table 13.1* summarizes the four major perspectives.

13-2a Functionalism: What Are the Benefits of Education?

For functionalists, education contributes to society's stability, solidarity, and well-being, and provides people with an opportunity for upward mobility. In their analyses, functionalists distinguish between manifest and latent functions.

MANIFEST FUNCTIONS OF EDUCATION

Some of the functions of education are *manifest*; that is, they're open, intended, and visible. For example:

- Schools are *socialization agencies* that teach children how to get along with others and prepare them for adult economic roles (Durkheim, 1898/1956; Parsons, 1959).
- Education *transmits knowledge and culture*. Schools teach skills like reading, writing, and counting; they also instill cultural values that encourage competition, achievement, and democracy (see Chapter 3).
- Similar values increase *cultural integration*, the social bonds that people have with each other and with the community at large, and *societal cohesion*.
- Education promotes *cultural innovation*. Faculty at research universities receive billions of dollars every year to develop computer technology, treatments for diseases, and programs to address social problems.
- Education *benefits taxpayers* because more highly educated people tend to pay higher taxes, are less likely to rely on public assistance programs, and lead healthier lifestyles, reducing state and federal health care costs (Baum et al., 2013).

Many Americans agree with functionalists that a college degree is "very important": 74 percent say that a college degree leads to a better quality of life, and in a survey of incoming first-year college students, 88 percent said that the most important reason for going to college is to get a good job (Schneider, 2013; Calderon and Sorenson, 2014; Jones, 2016). Among college graduates, 74 percent said their education helped them grow intellectually (Parker et al., 2011).

Besides expanding a person's intellectual horizons, education increases earnings. On average, college graduates earn 60 to

education transmits attitudes, knowledge, beliefs, values, norms, and skills.

schooling formal training and instruction provided in a classroom setting.

Table 13.1 Sociological Explanations of Education

Theoretical Perspective	Level of Analysis	Key Points
Functionalist	Macro	Contributes to society's stability, solidarity, and cohesion and provides opportunities for upward mobility
Conflict	Macro	Reproduces and reinforces inequality and maintains a rigid social class structure
Feminist	Macro and micro	Produces inequality based on gender
Symbolic Interactionist	Micro	Teaches roles and values through everyday face-to-face interaction and behavior

80 percent more per year than high school graduates. Even academically marginal students who manage to get a four-year degree have earnings that are as high, on average, as those of their peers with higher grades (Oreopoulos and Petronijevic, 2013).

As peoples' educational attainment rises, their earnings increase (*Figure 13.1*). Over a lifetime, education affects earnings five times more than other demographic factors like sex, race, ethnicity, and age (Schramm et al., 2013; Daly and Bengali, 2014). By age 64, people with a bachelor's degree earn six times more than those with only a high school diploma (Selingo, 2013).

Regardless of race, ethnicity, family background, and marital status, people with a bachelor's degree live longer, report better physical and emotional health, and are better able to handle stress because of their economic and social position. They're also more likely than non-college graduates to hold jobs that offer a greater sense of accomplishment, more autonomy, more opportunities for creativity, and more social interaction with coworkers (Rheault and McGeeney, 2011; Oreopoulos and Petronijevic, 2013).

What people study affects their financial payoff, but lifetime earnings vary by occupation, even with the same major field of study. Someone with a degree in the social sciences who works in management can earn $3.4 million versus $1.9 million in education (Julian, 2012; Carnevale and Cheah, 2013). Thus, the combination of what people study in college *and* the careers they pursue after graduation can make a big difference in lifetime earnings.

Figure 13.1 Education Pays Off

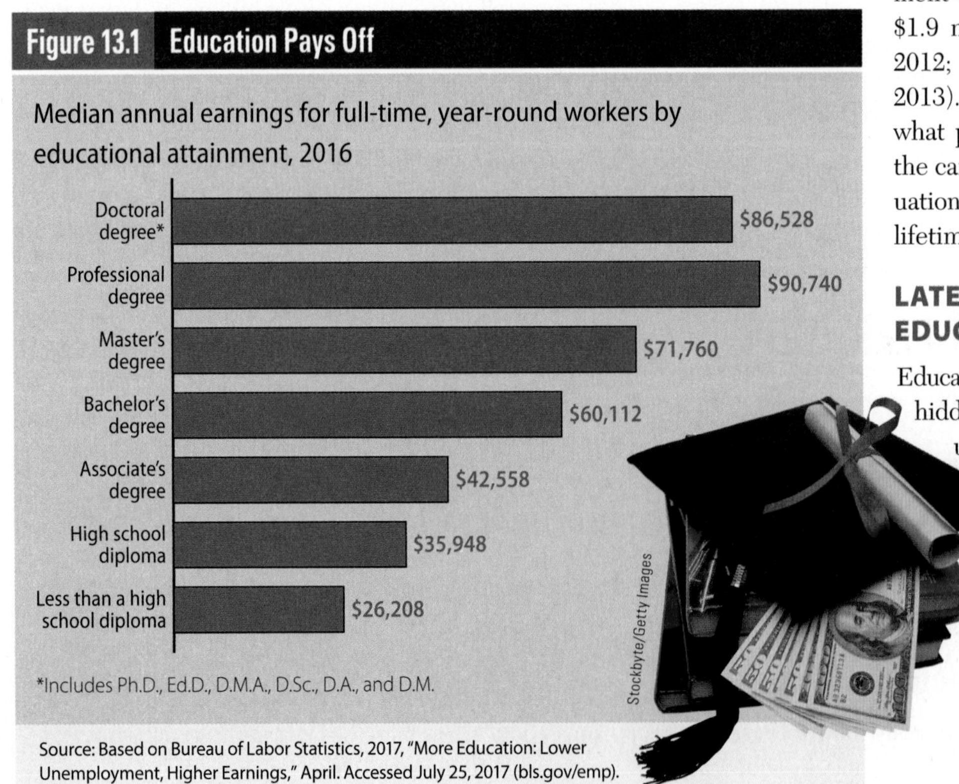

Source: Based on Bureau of Labor Statistics, 2017, "More Education: Lower Unemployment, Higher Earnings," April. Accessed July 25, 2017 (bls.gov/emp).

LATENT FUNCTIONS OF EDUCATION

Education also has *latent functions*—hidden, unstated, and sometimes unintended consequences. For example:

- Schools *provide child care*, particularly after-school programs, for the growing number of single-parent and two-income families.

- High schools and colleges are *matchmaking institutions* that bring together unmarried people.
- Education *decreases job competition*; the more time that young adults spend in school, the longer the jobs of older workers are safe.
- Educational institutions *create social networks* that can lead to jobs or business opportunities.
- Education is *good for business.* Thousands of companies offer services that tutor and test students and produce textbooks and related materials.

CRITICAL EVALUATION

Do functionalists exaggerate education's benefits? About 35 percent of the world's billionaires don't have a bachelor's degree, and some even dropped out of high school; 18 members of the 115th Congress have only a high school diploma (Wealth-X and UBS, 2014; Manning, 2017). Moreover, some of the highest paying jobs (e.g., radiation therapists, dental hygienists, commercial pilots, registered nurses) that will be in high demand in the coming decades require only a two-year degree or vocational training (Ogunro, 2012; Reich, 2015).

Also, according to some critics, functionalists tend to gloss over education's dysfunctions, including rising college costs and student loan debt, topics we'll consider shortly. Conflict theorists, especially, argue that educational institutions produce and reproduce inequality.

13-2b Conflict Theory: Does Education Perpetuate Social Inequality?

From preschool to graduate school, conflict theorists maintain, education creates and perpetuates social inequality based on social class, race, and ethnicity. Schools are also gatekeepers that control and maintain the status quo.

BananaStock/Jupiter Images

Figure 13.2 Attainment of Bachelor's Degree or Higher, by Race and Ethnicity, 1970 and 2016

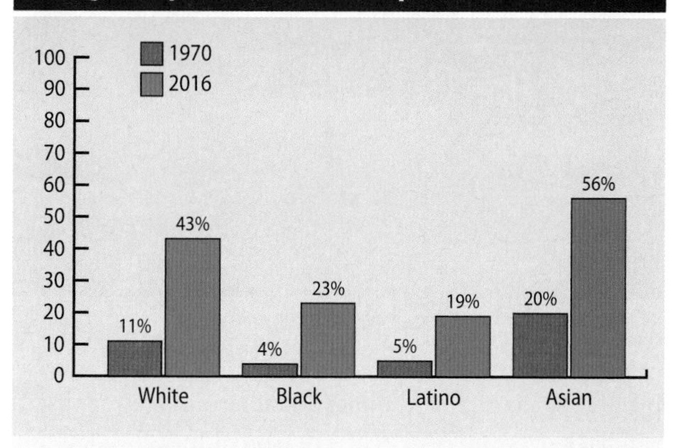

Note: 1970 data aren't available for American Indians/Alaska Natives and Pacific Islanders. In 2016, 10 percent and 20 percent, respectively, had a bachelor's or higher degree.

Sources: Based on U.S. Census Bureau, 2012, Table 229; McFarland et al., 2017, Figure 4.

SOCIAL CLASS AND ACHIEVEMENT GAPS

Countries like the United States, with high income and wealth inequality levels, also have large achievement gaps (Chmielewski and Reardon, 2016; see also Chapter 8). An **achievement gap** is a persistent and significant disparity in academic performance between different groups of students (e.g., males and females, White students and minorities, higher-income and lower-income students). The most common measures of achievement gaps are standardized test scores, GPAs, high school dropout rates, and college enrollment and completion rates.

College enrollments have surged since 1990, especially among minority groups, but who graduates? Asian Americans have made the largest gains (*Figure 13.2*), but there's considerable variation across subgroups. Of the many Asian subgroups, for example, 72 percent of Asian Indians have at least a bachelor's degree compared with only 14 percent of Cambodians, Hmong, and Laotians (Ramakrishnan and Ahmad, 2014). Such variations, as with other minority subgroups, are due to many factors, including English language proficiency, but the best predictor of educational attainment is social class.

Fifty years ago, the largest achievement gap was between Black and White people; today it's between social classes. The gap between children from high- and low-income families has grown by about 40 percent since the 1960s (Reardon, 2011), but a mix of race/ethnicity and social class variables

achievement gap a persistent and significant disparity in academic performance between different groups of students.

Blend Images/Ariel Skelley/Getty Images

Low- and middle-income students who receive free help from tax professionals to fill out the complicated federal financial aid forms are substantially more likely than students who receive only financial information to submit a college application, to enroll the following fall, and to be awarded financial aid (Bettinger et al., 2012).

produces accumulated educational advantages or disadvantages over the course of a person's life

CUMULATIVE REASONS FOR ACHIEVEMENT GAPS

When the 2007–2009 recession began, a number of schools experienced federal, state, and local budget cuts. Five years after the recession, class sizes at many public schools increased, but staff positions—including teachers, reading and math specialists, and guidance counselors—continued to decrease. Wealthier communities, in contrast, increased property taxes to accommodate higher enrollments (Rich, 2013).

Many of the nation's schools now have a high level of "double segregation" because students are increasingly separated not only by race but also by income. Nationally, 40 percent of Black students attend a high-poverty public school compared with only 6 percent of White students. Educational problems linked to poverty *and* racial segregation include less-experienced and less-qualified teachers, high teacher turnover, inadequate facilities (e.g., computers, science labs), outdated textbooks, high dropout rates, and school buildings that are falling apart (Reich, 2014; Carnoy and García, 2017).

Poverty and inferior schooling are significant academic roadblocks for low-income students. They're less likely to attend high schools that offer high-level courses in mathematics and science. This can mean the difference between passing or failing required courses during the first few years of college. Even when top students from low-income public high schools enter selective colleges, they often struggle to maintain passing grades. ("Selective" schools are those that admit a relatively low percentage of applicants.) Many college youths from low- and middle-income families must also work part-time or full-time and, consequently, have more stress and less time for academic work (Ross et al., 2012; Black et al., 2014).

hidden curriculum school practices that transmit nonacademic knowledge, values, attitudes, norms, and beliefs.

credentialism an emphasis on certificates or degrees to show that people have certain skills, educational attainment levels, or job qualifications.

EDUCATION AND SOCIAL CONTROL

Functionalists see education as an avenue for upward mobility. Conflict theorists maintain that education restricts upward mobility because of a hidden curriculum, credentialism, and privilege.

Hidden Curriculum. Every school has a formal curriculum that includes reading, writing, and learning other skills. Schools also have a **hidden curriculum**, practices that transmit nonacademic knowledge, values, attitudes, norms, and beliefs that legitimate "economic inequality and the staffing of unequal work roles" (Bowles and Gintis, 1977: 108).

Schools in low-income and working-class neighborhoods tend to stress obedience, following directions, and punctuality so that students can fill low-paid jobs (e.g., restaurants, nursing homes, hospitals) that require these characteristics (Kozol, 2005). Schools in middle-class neighborhoods emphasize proper behavior and appearance, cooperation, conforming to rules, and deference to authority because many of these students will go to college and work in bureaucracies that require such attributes (Hedges, 2011). In contrast, selective schools encourage leadership, creativity, independence, and people skills—all prized traits in elite circles (Persell and Cookson, 1985; see also Chapter 8). In effect, then, the hidden curriculum reproduces the existing class structure and provides workers for jobs and occupations in the stratification hierarchy.

Credentialism. Have you noticed that many faculty, doctors', lawyers', and dentists' offices are usually wallpapered with framed degrees? Such tangible symbols of achievement reflect **credentialism**, an emphasis on certificates or degrees to show that people have certain skills, educational attainment levels, or job qualifications.

Functionalists maintain that credentialism rewards people for their accomplishments, sorts out those who are the most qualified for jobs, and stimulates upward social mobility. Conflict theorists contend, however, that for many positions, people can gain skills on the job with a few weeks of training or succeed because of ability or other factors.

Because of a large supply of high school graduates, employers can demand higher levels of education even though some jobs (e.g., retail sales, law enforcement) don't require a college degree for competent performance, a process called *credential inflation*. As more people obtain a college degree, its value diminishes, and students from low-income families, who are the least likely to have access to a college education, fall further behind (Bollag, 2007).

Largely because of rising college costs and student loan debt, postsecondary certificates have become increasingly popular. A *certificate* is a credential showing that a person has completed educational courses in a specific field (e.g., information systems security, business and office management). Certificates take anywhere from a few months to two years to complete, match skills employers want with a job candidate's qualifications, can be "stacked" as students move in and out of college or a shifting job market, are high paying in some fields, and can be stepping stones to further education, including a college or advanced degree (Blumenstyk, 2015; Mangan, 2015).

Universal Pictures/Fotos International/Getty Images

How important is graduating from a prestigious college or university? Nationally, 84 percent of business leaders say that the amount of knowledge a job candidate has in a particular field is "very important." Only 9 percent say that where a person attended school is very important (Calderon and Sidhu, 2014). Steven Spielberg, the well-known movie producer and director, graduated from California State University at Long Beach after being rejected by the more prestigious University of Southern California and University of California at Los Angeles film schools.

Bob Daemmrich/Alamy Stock Photo

Privilege. According to one observer, "We're the only rich nation to spend less educating poor kids than we do educating kids from wealthy families" (Reich, 2014). Many colleges claim that they're committed to admitting talented low-income students, but award more financial aid and merit-based scholarships to students from high-income families. Among students with similar high school GPAs and SAT scores, the proportion of high-income students receiving scholarships from colleges, the federal government, or the states has increased since 1995, but the proportion for low-income students has fallen (Burd, 2013; Carey, 2013; Mettler, 2014).

Among the nation's 342 selective four-year public colleges, almost a third of all students come from wealthy households compared with just 8 percent from low-income families. A mere 4 percent of all colleges and universities (138 institutions) hold 75 percent of all postsecondary endowment wealth. Instead of supporting low-income students, "many super wealthy colleges are playgrounds for the children of the wealthiest in our country and the world" (Nichols and Santos, 2016: 1; Halikias and Reeves, 2017).

Another privileged group is *legacies*, the children of alumni who have "reserved seats" regardless of their accomplishments or ability. For example, President George W. Bush—who had mediocre high school grades and standardized test scores—was admitted as a legacy at Yale University, which his father and grandfather had attended (Golden, 2006). Legacies also include students with inferior academic records whose parents, including celebrities, make million-dollar donations. Among selective colleges and universities, almost 75 percent use legacies, which account for up to 30 percent of some universities' student body (Massey, 2007; Hurwitz, 2011).

CRITICAL EVALUATION

Conflict theories have several weaknesses. First, a hidden curriculum doesn't necessarily determine job placement because many students aren't passive recipients of educational systems. Below the bachelor's degree level, for example, nearly 25 million U.S. adults have acquired professional certificates and licenses in wide-ranging fields. These and other "alternative credentials" have labor market value that increases employment and earnings (Ewert and Kominski, 2014). Conflict theorists may denounce credentialism and credential inflation, but employers often view credentials as useful initial screening tools that signal job candidates' discipline and drive.

Second, do conflict theorists overstate the importance of social class achievement gaps? A study tracked more than 5,200 Asian American and White students from kindergarten through high school and found that the former, regardless of social class, performed better than their White counterparts. The researchers attributed the Asian American students' greater success to cultural beliefs that achievement is learned rather than innate, parental pressures to succeed, and, especially among recent immigrants, ethnic community resources such as private tutoring and vital information about navigating the education system (Hsin and Xie 2014; see also Chapter 10).

13-2c Feminist Theories: Is There a Gender Gap in Education?

Since 1982, women have been graduating from college at higher rates than men. Some view this trend as an indicator of greater gender equality; others are alarmed about the supposed "male crisis" in and "feminization" of higher education (Mullen, 2012; see also Chapter 9). Has women's progress come at the cost of men? Or are there still gender gaps, particularly in higher education?

WHO'S GETTING DEGREES

Equal numbers of girls and boys are high school graduates. Women earn more associate's, bachelor's, and master's degrees than men, but their percentage of professional and doctoral degrees drops considerably (*Figure 13.3*).

Across all racial-ethnic groups, students from affluent families are the most likely to earn a bachelor's degree, but men are more likely to do so than women (Mullen, 2012). Among people with a bachelor's or an advanced degree, Asian males (59 percent) have higher rates than Asian females (53 percent), White males and females are about the same (37 percent), and Latinos have slightly higher rates than Latinas (16 percent and 15 percent, respectively). Only Black women (25 percent) have

Figure 13.3 Educational Attainment, by Sex, 2016

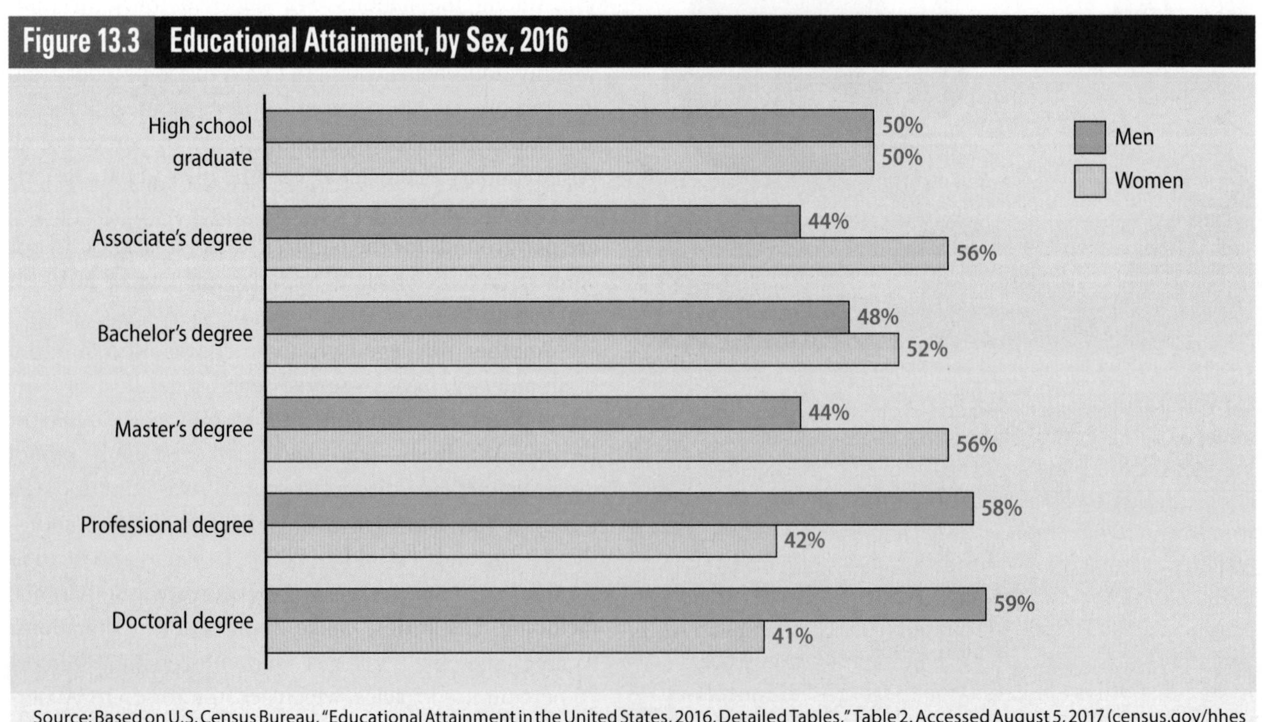

Source: Based on U.S. Census Bureau, "Educational Attainment in the United States, 2016, Detailed Tables," Table 2. Accessed August 5, 2017 (census.gov/hhes/socdemo/education).

higher rates than their male counterparts (22 percent) ("Educational Attainment…," 2017). Such data contradict the description of higher education as feminized, but why do women achieve more degrees than men up to the master's level?

In both high school and college, women spend more time than men studying, earn better grades, hold more leadership positions, and are more involved in student clubs and community volunteer work. In high school, boys are twice as likely as girls to describe school as a "waste of time." This may be one reason why women are less likely to drop out of high school or college. Regardless of the reasons for women's success, many colleges have been giving men preferential treatment in admissions to avoid large gender imbalances in their student bodies (Gewertz, 2009; Kahlenberg, 2010; "Gender, Education, and Work," 2015).

MAJORS AND CAREERS

A major gender gap is women's underrepresentation in the high-paying fields of science, technology, engineering, and mathematics (STEM). In elementary, middle, and high school, girls and boys take math and science courses in roughly equal numbers, and about the same numbers leave high school planning to pursue STEM majors in college. After a few years, however, men outnumber women in nearly every STEM field, and in some—mathematics, engineering, and computer science—women earn only 10 to 20 percent of the bachelor's degrees. Their numbers decline further at the graduate level and yet again in the workplace (Landivar, 2013; Snyder and Dillow, 2013; Munoz-Boudet and Ravenga, 2017).

Why is there a gradual attrition? In college, females are initially as persistent as men in a STEM major and earn higher grades. However, they're less satisfied than men with the core courses and more likely to doubt their ability to succeed in a male-dominated discipline. As a result, women's self-confidence falters and they change majors (Shapiro and Williams, 2012; Jagacinski, 2013).

The exit from a STEM major is also associated with having few female faculty role models, and some science professors' beliefs that female students won't benefit from mentoring because they're less competent than men. "Chilly climates"—uncomfortable work environments for women—are common in male-dominated jobs, including STEM occupations. Chilly climates decrease women's job satisfaction and increase the likelihood of leaving a job (Moss-Racusin et al., 2012; Kahn and Ginter, 2017).

A recent survey of 27 prominent U.S. universities (including all but one of the Ivy League schools) found that 23 percent of undergraduate women and 5 percent of undergraduate men were victims of non-consensual sexual contact—ranging from penetration to sexual touching. Even in the most serious assaults that included penetration, 72 percent of victims didn't report the assaults because they were "…embarrassed, ashamed or thought it would be too emotionally difficult" or "…didn't think anything would be done about it" (Cantor et al., 2015).

CRITICAL EVALUATION

A common criticism is that feminist scholars address women's education barriers but not their choices. Socialization, gender stereotypes, and teachers' expectations affect our behavior (see Chapters 4 and 9), but it's still unclear why many women choose fields of study that they know are on the lower end of the pay scale (e.g., health, education).

Some fault feminist theorists for being more interested in women's than men's educational attainment gaps. There's also the question of why feminist scholars devote little attention to issues such as why boys are more disruptive in elementary school. The disruptions lead to suspensions and decrease the chance of attending college by at least 16 percent for each suspension (Bertrand and Pan, 2011).

13-2d Symbolic Interaction: How Do Social Contexts Affect Education?

None of us is born a student or a teacher. Instead, these roles, like others, are socially constructed (see Chapters 3–5). For symbolic interactionists, education is an active *process* that involves students, teachers, peers, and parents and which includes tracking, labeling, and student engagement.

TRACKING

Beginning in kindergarten, practically all schools sort students by aptitude. Such sorting results in **tracking**, assigning students to specific educational programs and classes on the basis of test scores, previous grades, or perceived ability.

Some educators believe that tracking is beneficial because students learn better in groups with others like themselves, and it allows teachers to develop curricula for students with similar ability. Many interactionists maintain, however, that tracking creates and reinforces inequality.

- High-track students take classes that involve critical thinking, problem solving, and creativity that high-status occupations require. Low-track students take classes that are limited to simple skills (e.g., punctuality and conformity) that usually characterize lower status jobs.
- High-track students have more homework, better quality instruction, and more enthusiastic teachers. One result is that high-track students are more likely to see themselves as "bright," whereas low-track students see themselves as "dumb" or "slow."
- The effects of tracking are usually cumulative and lasting. Teachers tend to have low expectations for low-track students, who therefore fall further behind every year in reading, mathematics, and interaction skills (Oakes, 1985; Hanushek and Woessman, 2005).

Middle school and high school become even more stratified as high-track students are sorted into gifted, honors, and advanced courses. In college, students continue to be tracked and sorted into honors programs and accelerated undergraduate courses.

LABELING

Tracking often leads to labeling, a serious problem because "there's a widespread culture of disbelief in the learning capacities of many of our children, especially children of color and the economically disadvantaged" (Howard, 2003: 83). Labeling, in turn, can result in a *self-fulfilling prophecy*. That is, students live up or down to teachers' expectations and evaluations that are influenced by a student's social class, skin color, hygiene, accent, and test scores (see Chapter 5).

tracking assigning students to specific educational programs and classes on the basis of test scores, previous grades, or perceived ability.

Implicit bias unconscious prejudices or stereotypes that affect our attitudes, actions, and decisions.

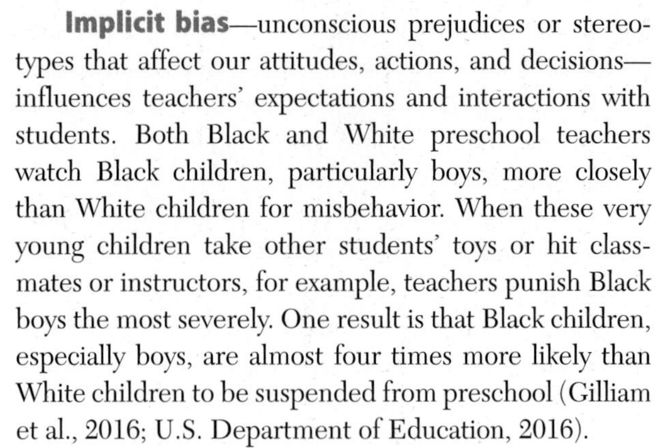

Implicit bias—unconscious prejudices or stereotypes that affect our attitudes, actions, and decisions—influences teachers' expectations and interactions with students. Both Black and White preschool teachers watch Black children, particularly boys, more closely than White children for misbehavior. When these very young children take other students' toys or hit classmates or instructors, for example, teachers punish Black boys the most severely. One result is that Black children, especially boys, are almost four times more likely than White children to be suspended from preschool (Gilliam et al., 2016; U.S. Department of Education, 2016).

Implicit bias and labeling also affect girls' education. Among seventh- to twelfth-grade students, for example, 46 percent of females, compared with 62 percent of males, are confident that they can learn computer science. Only 4 to 10 percent of parents and teachers think that girls are interested in computer science. In contrast, 64 percent say that boys are interested in computer science and more likely to succeed at it than girls. Viewing computer science as a male field helps explain why the share of women earning a bachelor's degree in computer science slipped from 37 percent in the mid-80s to barely 18 percent by 2008. Women's drop in computing majors is due to many factors, including chilly classroom and work environments (U.S. Department of Education, 2012; Keating and English, 2015; Sherman, 2015). Teachers' and parents' implicit biases, however unintentional, shape students' choices and outcomes.

STUDENT ENGAGEMENT

Many schools assess performance not only through tests but also through *student engagement*, how involved students are in their own learning. A team of researchers who studied more than 2,500 elementary school classrooms concluded that the typical U.S. child has only a 1 in 14 chance of being in a school that encourages her or his engagement. Because of standardized tests, teachers in public elementary schools spend most of the day on basic reading and math drills and little time on problem solving, reasoning, science, and social studies (Pianta et al., 2007).

iStock.com/Mevans

By the time they reach middle school, minority students—particularly Black and Latino students—believe that teachers discipline them more harshly than their White peers for similar behavior (e.g., not having a hall pass, disobedience). Students' perceptions that teachers are biased often lead to distrust, defiance, disengagement, underperformance, acting out, and suspensions or

Dean Drobot/ShutterStock.com

Large shares of full-time undergraduates—46 percent at community colleges and 42 percent at 4-year colleges—are employed (McFarland et al., 2017). Is working while in college a necessity or a preference? And, how does balancing work and academic obligations affect student engagement?

expulsions. Teachers can counteract such self-fulfilling prophecies by sending the message that their students are capable, valued, and respected (Yeager et al., 2017).

U.S. high school students aren't as engaged in their education as they could be: 26 percent admit that they usually don't do their homework. The students who are most likely to be disengaged are from low-income families, Black or Latino, don't live with both biological parents, attend financially strapped urban public high schools that are typically overcrowded and understaffed, and have fewer resources such as computers and even textbooks (Planty et al., 2008; U.S. Department of Education, 2016).

About 20 percent of first-year college students have difficulty learning course material and getting help with coursework. Students who are the most likely to be engaged in their coursework have mentors, participate in faculty projects, have internships, and are involved in several extracurricular activities (Busteed, 2015; National Survey of Student Engagement, 2016). Such opportunities encourage engagement and decrease attrition.

On average, even full-time college students study only 14 hours a week compared with 24 hours a week in 1961 (Babcock and Marks, 2011). This is well below the 24 to 30 hours faculty members say students should be spending on class preparation if they're taking three courses.

According to one college instructor, "Education is the only business in which the clients want the least for their money" (Perlmutter, 2001: B1). However, some analysts also blame professors for students' not studying. Many faculty have watered down their required readings, say nothing when students don't prepare for classes or do assignments, and often give easy exams to avoid complaints about low grades. Only 55 percent of first-year college students and 61 percent of seniors say their courses "challenged them to do their best work" (National Survey of Student Engagement, 2013: 9). When faculty dilute their course requirements and tests, some students don't study because they're bored by courses that don't stimulate them (Benton, 2011; Glenn, 2011).

CRITICAL EVALUATION

One weakness is that interactionists presume that once a person (or group) is labelled, a self-fulfilling prophecy will follow. People can overcome labelling effects by changing a situation. In middle school, for example, placing higher-achieving minority children into "gifted" tracks and programs significantly boosts their academic performance—even in the poorest neighborhoods of large urban school districts—because students have high-quality teachers and textbooks, as well as positive peer pressure to learn and succeed (Card and Giuliano, 2016).

Another limitation is that interactionists, because of their micro-level analysis, overlook the macro-level structural constraints that affect education. Educational success varies greatly depending on factors such as the availability of high-quality preschools and kindergartens, neighborhood resources (like property taxes) devoted to education, and well-funded and well-staffed high schools. Even more importantly (as you'll see shortly), politicians—with little or no experience in education—sometimes establish policies that perpetuate racial/ethnic and social class inequality.

13-3 SOME CURRENT ISSUES IN U.S. EDUCATION

In an open-ended question about "the most important problem facing this country today," only 5 percent of Americans named education (Riffkin, 2014). A dysfunctional government and weak economy ranked the highest, but the U.S. education system has problems at all levels, some more controversial than others.

13-3a Elementary and Middle Schools

Advanced economies, including ours, rely not on physical labor but on "cognitive labor" that requires formal analytical abilities, written communications, and specific technical knowledge—skills that people acquire and cultivate beginning in preschool (Autor, 2014). Three of the ongoing issues regarding elementary and middle schools are low test scores, achievement gaps, and effective teaching.

LOW TEST SCORES AND ACHIEVEMENT GAPS

Children who read proficiently by the end of the third grade are more likely to do well in other subjects, including mathematics, and to graduate from high school. Since 1990, eighth-grade reading and math scores have increased, but there are racial and ethnic gaps. On a scale of 0 to 500, Asian/Pacific Islander students have the highest scores in both subjects, and Black students the lowest, but both Black and Latino students have narrowed achievement gaps (*Figure 13.4*).

Social class explains some of the variations. In 2015, 48 percent of 3- to 5-year-olds whose parents had either a graduate or professional degree, compared with 29 percent of those whose parents had only a high school diploma, were enrolled in kindergarten, preschool, and nursery school programs that provided enriching educational experiences. Nearly 80 percent of fourth graders in low-income families are below proficiency in reading compared with 49 percent of higher-income children (National Center for Education Statistics, 2013; McFarland et al., 2017).

Social class alone doesn't explain student performance, however. Homework also affects test scores. Many elementary and middle school children aren't getting or doing their homework, but parents and the public often blame teachers for the students' poor academic performance (Loveless, 2014).

EFFECTIVE TEACHING

A multi-year project concluded that the most accurate way to evaluate elementary school teachers is to use a three-pronged approach that includes student test scores, multiple classroom observations, and student evaluations (Cantrell and Kane, 2013), but 40 states use only student test scores to evaluate teachers. Much research shows that test scores are invalid and unreliable measures of teacher effectiveness. Other factors affect test scores, including a teacher's quality of undergraduate education, years of experience, being assigned to low-track or high-track classrooms, whether or not students study, and school funding (Kalogrides et al., 2013; Amrein-Beardsley, 2014; Polikoff and Porter, 2014).

EDUCATION AND POLITICS

Many teachers believe that politics threaten public education. As this book goes to press, the Trump administration has rolled back consumer protections for student loan borrowers, slashed federal education funding, and expanded private school *vouchers*. (Vouchers allow parents to use taxpayer money to pay for all or some of a child's private school fees.) The administration has also proposed cutting federal funding for teacher training, after-school programs, and work-study programs for college students, as well as implementing affirmative action policies for White students.

13-3b High Schools

High schools, like elementary and middle schools, must grapple with low achievement scores and effective teaching. In addition, the SAT is becoming an increasingly controversial issue.

INTERNATIONAL COMPARISON

Among 71 countries, U.S. 15-year-olds rank 38th in math and 24th in both science and reading. U.S. students are

Figure 13.4 Average Reading and Mathematics Scores of U.S. 8th-Grade Students, by Race and Ethnicity

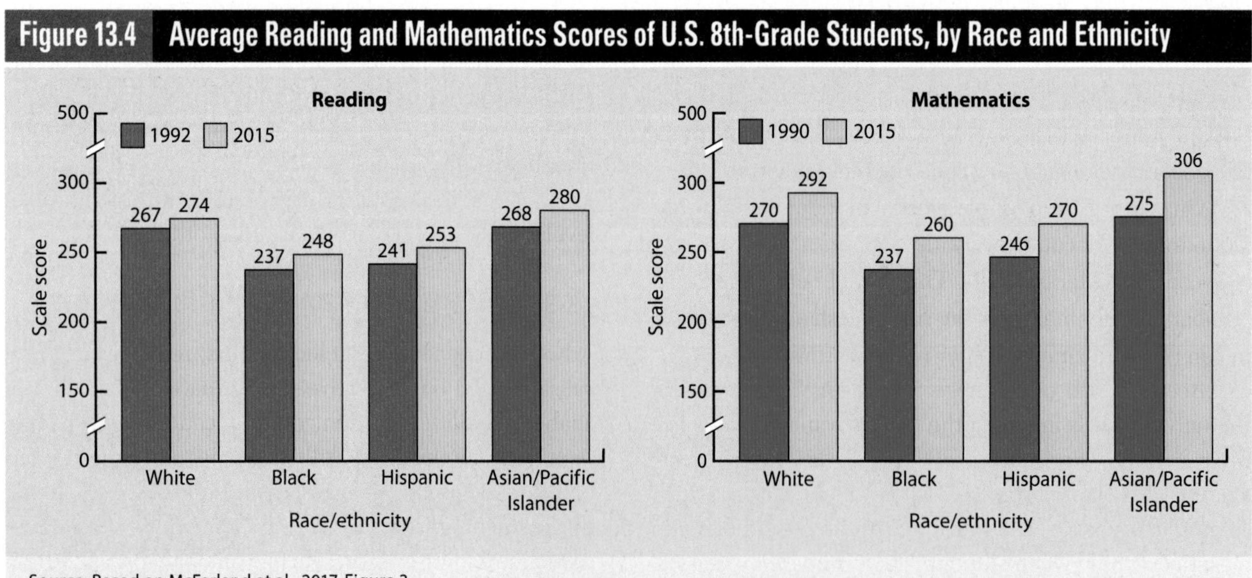

Source: Based on McFarland et al., 2017, Figure 3.

About 6 percent of Americans ages 16 to 24 are high school dropouts, down from 11 percent in 2000. The dropout rate, which is slightly higher for males than for females, is highest among foreign-born Latinos and youth whose families are in the bottom 25 percent of all family incomes (Snyder et al., 2016; McFarland et al., 2017).

Hola Images/Getty Images

roughly in the middle in reading and science, but math performance, which is below average, has declined since 2009. In Singapore, which consistently has the highest achievement, average high school students are roughly three years ahead of their American peers in math. Compared with 35 other wealthy industrial nations, the United States ranks only 30th in math and 19th in science ("Education," 2016; Sparks, 2016; DeSilver, 2017).

The United States is one of the wealthiest countries in the world; spends more per student from the ages of 6 to 15 than 30 other industrialized countries; ranks above average in the share of high-SES high school students; and has fewer non-English speaking students than other Western countries, including Canada (Carnoy and Rothstein, 2013; OECD, 2013). Why, then, are U.S. students doing so poorly? A major difference between the United States and other developed nations is teacher preparation and quality.

TEACHER PREPARATION AND QUALITY

American teachers spend, on average, 20 to 50 percent more hours teaching in class than do their Australian, Canadian, and Finnish counterparts, all of whose students outrank Americans in international tests (Lawrence, 2013). A combination of structural factors helps explain why, despite their individual efforts, many U.S. teachers aren't as effective as they could be.

First, the top-performing countries accept only the top applicants for education programs. Finland, Singapore, and South Korea recruit the top 5 to 30 percent of high school graduates, but accept only 10 percent who have high standardized test scores in science, math, and reading (Auguste et al., 2010; Sawchuck, 2012).

Several recent reports described U.S. teacher preparation programs as "an industry of mediocrity" that has low or no academic entry standards, easy coursework, and gives out many easy A's (Greenberg et al., 2013; Putnam et al., 2014). In top-performing countries, 100 percent of teachers are in the top third of their college graduating classes. In the United States, 47 percent of kindergarten through twelfth-grade teachers come from the bottom third. "In other words, we hire lots of our lowest performers to teach, and then we scream when our kids don't excel" (Cloud, 2010: 48).

Second, compared with teachers in top-performing countries, a higher number of U.S. teachers don't have a bachelor's degree or higher in the subjects they teach. For example, 30 percent of public high school teachers who teach math and 54 percent who teach chemistry didn't major in these subjects (Hill et al., 2015).

Third, the top-performing countries offer teachers competitive salaries. In South Korea and Singapore, teachers on average earn more than lawyers and engineers. In these countries, Finland, and many others, teachers are highly regarded and enjoy the same prestige as physicians and other high-status professions and have lifelong careers, resulting in very low teacher attrition. In contrast, 40 to 50 percent of new U.S. teachers leave the profession within their first 5 years on the job. High turnover rates are especially harmful to students in low-performing schools: Replacing a teacher—especially a good one—is costly, disruptive, and demoralizing for both students and parents (Ronfeldt et al., 2011; "Schools in Finland," 2016; Startz, 2016).

Fourth, high-performing countries endorse egalitarian, student-directed policies and learning environments. Unlike the United States, they (1) enroll most children in high-quality preschools; (2) direct more resources to the most disadvantaged pupils; (3) delay steering children into academic or vocational programs until they're 15 or 16 years old, which decreases SES gaps due to tracking; (4) have small class sizes (one teacher for every 12 high school students in Estonia, one of the top-performing countries); (5) apply rigorous, consistent standards across all classrooms; and (6) give principals (not politicians) considerable autonomy over resources, curricula, and other school policies ("Education," 2016; OECD, 2016; Ripley, 2016).

THE GROWING CONTROVERSY OVER SATs, ACTs, AND AP COURSES

In 2013, about 3.5 million high school students took either the SAT or the ACT, standardized college entrance examinations. The administrators/owners of both exams say that the scores indicate where students are falling

Alexander Raths/Shutterstock.com

Finnish high school students routinely score at or near the top on international reading, science, and math tests. Why?

behind in college readiness, measure students' capability to do college-level work, and predict college success (College Board, 2013; ACT, Inc., 2014).

A growing number of critics, including college officials, contend that the tests are little more than gatekeeping tools: They exclude lower socioeconomic students from higher education, don't predict college success, and the test scores reflect accumulated life advantages and disadvantages rather than the ability to do college-level work. As a result, 850 four-year colleges don't require applicants to submit SAT or ACT scores (Botstein, 2014; Fraire, 2014; Hiss and Franks, 2014; Reeves, 2017).

Advanced Placement (AP) offers high school students college-level curricula and exams. About 78 percent of degree-granting colleges and universities give course credit to students who get high scores on the exams (Berrett, 2014). Thirty percent of high school students take at least one AP exam; 27 percent are low income, and only 48 percent of this group score high enough to get college credit (College Board, 2014).

Like the ACT and SAT, AP is becoming more controversial. First, AP courses are rarely as demanding as college courses and don't offer the same content. Second, to increase a high school's prestige, superintendents and principals may pressure teachers to offer AP courses and students to take them. Both groups may be unprepared to do so, and failure can make students and teachers feel inferior. Finally, only about 45 percent of the nation's public high schools offer AP courses, and those that do so are typically in high-income and predominantly White neighborhoods (Simon, 2013; Berrett, 2014).

meritocracy a system that rewards people because of their individual accomplishments.

13-3c Colleges and Universities

Public postsecondary institutions often feel embattled by numerous constituencies and external pressures. Many colleges and universities are grappling with affirmative action issues, grade inflation and cheating, low graduation rates, and rising student loan debt.

AFFIRMATIVE ACTION

In higher education, *affirmative action* refers to admission policies that provide equal access for groups, particularly women and minorities, that have been historically excluded or underrepresented. The enrollment rates of these groups has increased steadily since the 1970s, but using race and ethnicity in college admission decisions has been, and continues to be, a hotly debated topic (see "The Affirmative Action Debate in Higher Education").

In 2006, Michigan voters approved an amendment to the state's constitution to ban using race as a factor in deciding who's admitted to the state's public universities. In 2014, the U.S. Supreme Court (*Schuette v. Coalition to Defend Affirmation Action*) ruled that Michigan's ban was constitutional. The court's decision legitimated similar laws in seven other states (Schmidt, 2014). Affirmative action opponents praised the decision for reinforcing **meritocracy**, a system that rewards people because of their individual accomplishments. We say things like "work hard and you'll get ahead" and "pull yourself up by your bootstraps," but is that what happens? As you've seen, the best predictor of who goes to college isn't ability but income.

GRADE INFLATION AND CHEATING

Grade inflation is widespread across all education levels, and some children start cheating on tests as early as the third grade. The consequences of grade inflation and cheating in postsecondary education are more serious, however, because someone who cheated her or his way through college or graduate school may be your dentist, lawyer, doctor, or tax accountant.

Grade Inflation. Has an A replaced the C of the 1970s? In 1971, only 26 percent of first-year college students expected to earn at least a B average in college, compared with 52 percent in 1995, and a whopping 70 percent in 2010 (Pryor, 2011), and even though students are studying less than in the past.

Many faculty give high grades because it decreases student complaints, involves less time and thought in grading exams and papers, and reduces the chances of students' challenging a grade. Some faculty believe that they can get favorable course evaluations from students

> **WOULD YOU WANT A MILITARY OFFICER, A FINANCIAL PLANNER, A DENTIST, OR A DOCTOR WHO CHEATED HER OR HIS WAY THROUGH COLLEGE OR GRADUATE SCHOOL?**

by handing out high grades, and others accept students' view of high grades as a reward for simply showing up in class. Inflating grades decreases student attrition and satisfies administrators, especially when state legislators base funding on graduation rates (Bartlett and Wasley, 2008; Gillespie, 2014).

Cheating. Despite grade inflation, cheating is common in college and graduate school. Many studies have uncovered cheating at the Air Force Academy and in many graduate programs, including business administration and dentistry.

Policy analyst David Callahan (2004) contends that dishonesty is prevalent because we live in a "cheating culture," which rationalizes deceit (e.g., "everybody's doing it"), and rewards cutting corners to make money and to succeed. As competition increases and more people become afraid of falling behind, even usually honest people don't feel guilty for cheating at tax time, at work, or in higher education. For many college students, cheating "isn't a big deal," perhaps because they see it everywhere—on Wall Street, by sports stars, and by celebrities. Political (and particularly White House) dishonesty is so common, *Time* magazine offers a weekly list of "Facts vs. Alternative Facts," meaning lies (see Chapter 7 on widespread corporate, white collar, and consumer retail fraud).

More than two-thirds of college students admit having cheated at least once on tests and assignments: "To them, higher education is just another transaction, less about learning than about obtaining a credential" (Wolverton, 2016: A42). Students cheat because they can. At least 41 percent of faculty don't take disciplinary action against students they know are cheating (Coren, 2011). Students are rarely caught and, if caught, seldom punished. Despite widespread grade inflation and cheating, millions of students don't graduate from college.

GRADUATION RATES

The United States has one of the lowest college graduation rates in the developed world. Only 59 percent of full-time students at four-year colleges complete a bachelor's degree within 6 years, and 29 percent at 2-year colleges earn a certificate or associate's degree within 3 years (McFarland et al., 2017).

Once on campus, 40 percent must take one or more remedial courses. Dropout rates are higher for males than females, part-time than full-time students, and

The Affirmative Action Debate in Higher Education

Supporters argue that affirmative action programs...

- give disadvantaged and underrepresented students a needed boost in overcoming past and current discrimination
- have doubled or tripled the number of minority applications, resulting in student populations that are more representative of the surrounding community
- have increased minorities' upward social and economic mobility
- promote diversity and encourage people to work together effectively in a multicultural society

Opponents argue that affirmative action programs...

- are discriminatory because they're based on race rather than academic achievement
- benefit primarily middle- and upper-class minorities
- are condescending because they imply that minorities can't succeed without preferential treatment
- admit students who often can't live up to an institution's academic standards, resulting in high failure rates

What else would you add to either side of the debate?

Sources: Messerli, 2012; Krishnamurthy and Edlin, 2014.

Keith Brofsky/Getty Images

students from lower-income families. At four-year colleges, the more selective the institution, the higher the graduation rate: 35 percent at colleges with open admission policies, 61 percent at colleges that accept 50 percent or more of applicants, and 85 percent at colleges that accept only 25 percent of applicants (Snyder et al., 2016).

Higher graduation rates at more selective institutions don't necessarily mean that the students are more intelligent than those at open admission colleges. Instead, selective colleges require higher GPAs and standardized test scores, both of which screen out students who may be unmotivated or unprepared for college. Selective colleges also have higher tuition and fees that, in turn, generate more financial aid, on-campus resources and support services, and enriching extra-curricular activities—all of which increase the likelihood of earning a degree.

Social class and family relationships also affect college completion. Students whose family income is in the top 25 percent are nearly five times more likely to attain a bachelor's degree than those from the lowest 25 percent (58 percent vs. 12 percent). Fully three-quarters of undergraduates are managing some combination of family responsibilities, jobs, and commuting to class. Working while in college, living at home, and experiencing family-related stresses (e.g., parents' divorce, job loss, illness) decrease the odds of graduating from college (Princiotta et al., 2014; Wilbur and Roscigno, 2016; Cahalan et al., 2017). Students also drop out because they get sick, accept attractive job offers, or decide that college isn't for them, but cost is a major reason for attrition.

STUDENT LOAN DEBT

Student loan debt, which has almost tripled since 2004 to more than $1.4 trillion, is now second only to mortgage debt. About 70 percent of students who graduated from public and private nonprofit four-year colleges in 2015 had debt, which ranged from $3,000 to $53,000 and averaged $30,100 (up from $9,500 in 1993). Those at for-profit colleges (e.g., University of Phoenix) owed much more—almost $40,000 on average (Institute for College Access & Success, 2016; Student Loan Report, 2017).

The rising student loan debt depends on factors like an institution's endowment, in-state and out-of-state tuition prices, whether the college is public or private, and the growing number and costs of full-time administrators (*Chronicle of Higher Education*, 2014). However, a major reason for the rise is that, since 2000, public spending on higher education has dropped 30 percent even though enrollment at public colleges has jumped 34 percent (Kena et al., 2014). As state funding has decreased, tuition has increased (*Figure 13.5*), and so has borrowing to cover tuition, fees, and, for many students, the costs for room and board.

13-4 RELIGION AND SOCIETY

Like education, religion has a profound effect on society. Some form of religion exists in all societies and cultures. Throughout history, religion has been a central part of the human experience, affecting the family, economy, and other social institutions. Why is religion important to many people? And does it always benefit society? The four theoretical perspectives offer different insights in answering these and other questions.

Figure 13.5 Why College Costs More Than It Used To

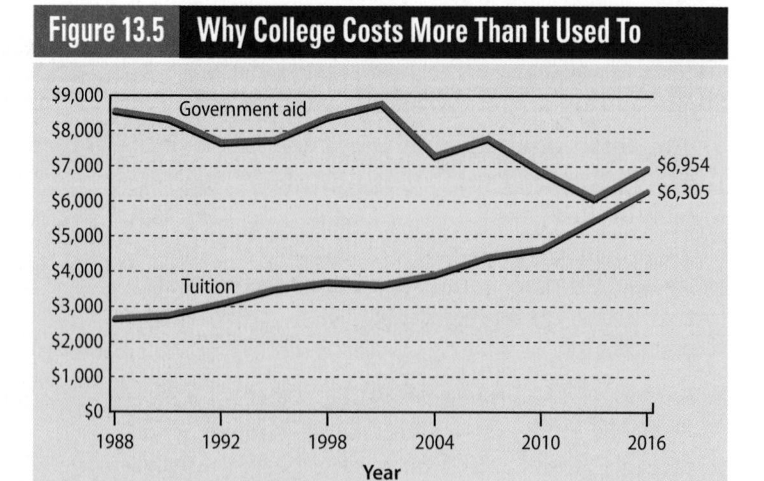

Note: These are public higher-education revenues per student, in 2016 dollars.
Source: Based on SHEEO, 2017, Figure 4.

Mike Peters/Cartoonist Group

13-4a What Is Religion?

Religion is a social institution that involves shared beliefs, values, and practices related to the supernatural. It unites believers into a community, but customs and practices differ across cultures and groups. For example, being Catholic requires confessing one's sins to a priest, whereas silent confessions directly to God are more common among Protestants, Jews, Muslims, and other religious groups.

Émile Durkheim (1961) distinguished between the sacred and the profane. **Sacred** refers to anything that people see as awe-inspiring, supernatural, holy, and not part of the physical world. In contrast, **profane** refers to the ordinary and everyday elements of life that aren't related to religion. Sociologists differentiate religion from religiosity and spirituality.

13-4b Religion, Religiosity, and Spirituality

Religion is a belief system, but religious expression can vary. When sociologists examine **religiosity**, how people demonstrate their religious beliefs, they find that religion and religiosity differ. For example, 53 percent of Americans say that religion is "very important" in their lives, but only 36 percent attend worship services once a week or more (Newport, 2016).

Spirituality is a personal quest to feel connected to a reality greater than oneself. About 37 percent of Americans describe themselves as "spiritual but not religious." Within this group, however, 44 percent pray at least once a day, and 55 percent are "absolutely certain" that there's a God (Lugo et al., 2012). Thus, religious people are spiritual, but spiritual people aren't necessarily religious.

RELIGIOUS ORGANIZATION AND MAJOR WORLD RELIGIONS

People manifest their religious beliefs most commonly through organized groups, including cults, sects, denominations, churches, and ecclesia. The major world religions differ in their membership and beliefs.

13-5a Cults (New Religious Movements)

A **cult** is a religious group devoted to beliefs and practices that are outside of those accepted in mainstream society. Many sociologists use **new religious movement (NRM)** rather than *cult* because the media have used the latter term in derogatory ways to describe any unfamiliar, new, or seemingly bizarre religious group (Roberts, 2004).

NRMs usually organize around a **charismatic leader** (like Jesus) whom followers see as having exceptional or superhuman powers and qualities (see Chapter 11). Some NRMs have become established religions. The early Christians were a renegade group that broke away from Judaism, and Islam, the world's second largest religion, began as a cult around Muhammad.

religion a social institution that involves shared beliefs, values, and practices related to the supernatural.

sacred anything that people see as awe-inspiring, supernatural, holy, and not part of the natural world.

profane the ordinary and everyday elements of life that aren't related to religion.

religiosity how people demonstrate their religious beliefs.

cult a religious group devoted to beliefs and practices that are outside of those accepted in mainstream society.

new religious movement (NRM) term used instead of *cult* by most sociologists.

charismatic leader someone that followers see as having exceptional or superhuman powers and qualities.

Joyfull/Shutterstock.com

Hare Krishna is the popular name for the International Society of Krishna Consciousness (ISKCON), a new religious movement based in Hinduism. Established in the United States in 1965, its practices include an austere and simple life, vegetarianism, abstinence from drugs and alcohol, chanting, and evangelism.

Most contemporary cults are fragmentary, loosely organized, and temporary, but others (e.g., the Church of Scientology) have developed into lasting and highly bureaucratic international organizations.

13-5b Sects

A **sect** is a religious group that has broken away from an established religion. People who begin sects are usually dissatisfied members who believe that the parent religion has become too secular and has abandoned key original doctrines. Like cults, some sects are small and disappear after a time, whereas others become established and persist. Examples of sects that have persisted include the Amish, the Jewish Hassidim, Jehovah's Witnesses, Quakers, and Seventh-Day Adventists (Bainbridge, 1997). Some sects develop into denominations.

sect a religious group that has broken away from an established religion.

denomination a subgroup within a religion that shares its name and traditions and is generally on good terms with the main group.

church a large established religious group that has strong ties to mainstream society.

ecclesia (also called a *state religion*) an official religious organization that claims everyone in society as its members.

13-5c Denominations

A **denomination** is a subgroup within a religion that shares its name and traditions and is generally on good terms with the main group. Denominations can form slowly or develop rapidly, depending on factors like geography, immigration, and a country's birth rate. Some scholars describe a denomination as somewhere between a sect and a church. Like sects, denominations have a professional ministry. Unlike sects, denominations view other religious groups as valid and don't make claims that only they possess the truth (Hamilton, 2001).

Denominations typically accommodate themselves to the larger society instead of trying to dominate or change it. As a result, people may belong to the same denomination as did their grandparents or great-grandparents. Denominations exist in all religions, including Christianity, Judaism, and Islam. In the United States, the many Protestant denominations include Episcopalians, Baptists, Lutherans, Methodists, and Evangelicals.

13-5d Churches

A **church** is a large established religious group that has strong ties to mainstream society. Because leadership is attached to an office rather than a specific leader, new generations of believers replace previous ones, and members follow tradition or authority rather than a charismatic leader. As in a denomination, people are usually born into a church, but may later decide to leave it.

Yienkeat/Shutterstock.com

Churches (e.g., Roman Catholic Church, Greek Orthodox Church) are typically bureaucratically organized, have formal worship services and trained clergy, and often maintain some degree of control over political or educational institutions (see Chapter 11). Because churches are an integral part of the social order, they often become dependent on, rather than critical of, the ruling classes (Hamilton, 2001).

Radomir/Shutterstock.com

13-5e Ecclesiae

An **ecclesia** (also called a *state religion*) is an official religious organization that claims everyone in society as its members. Membership is automatic at birth, the religion is part of the country's cultural identity, and there's a close association between religious and state officials. Examples include the Anglican Church in England, the Lutheran Church in Norway and Sweden, and Islam in Iran, Iraq, Afghanistan, and other Middle East countries.

An ecclesia has some of the same characteristics as a church, such as a formal structure and professionally trained and designated clergy. A major difference is an ecclesia's power. In some societies, particularly in the Middle East, ecclesiae have immense authority because there's little or no separation between church and state. In others (e.g., England, Sweden, Norway), ecclesiae have little or no influence on government laws and regulations or most cultural practices.

13-5f Some Major World Religions

Worldwide, the largest religious group is Christians, followed by Muslims. If the world's population is represented as an imaginary village of 100 people, it has about:

- 32 Christians
- 23 Muslims
- 16 Unaffiliated (people who may or may not believe in God but don't identify with any particular religious group)
- 15 Hindus
- 7 Buddhists
- 6 Folk religionists (includes followers of African traditional religions, Chinese folk traditions, Native American religions, and Australian aboriginal religions)
- 1 Other (includes Jews, Baha'is, Sikhs, Jains, Shintoists, and many others) (based on Pew Research Center, 2015)

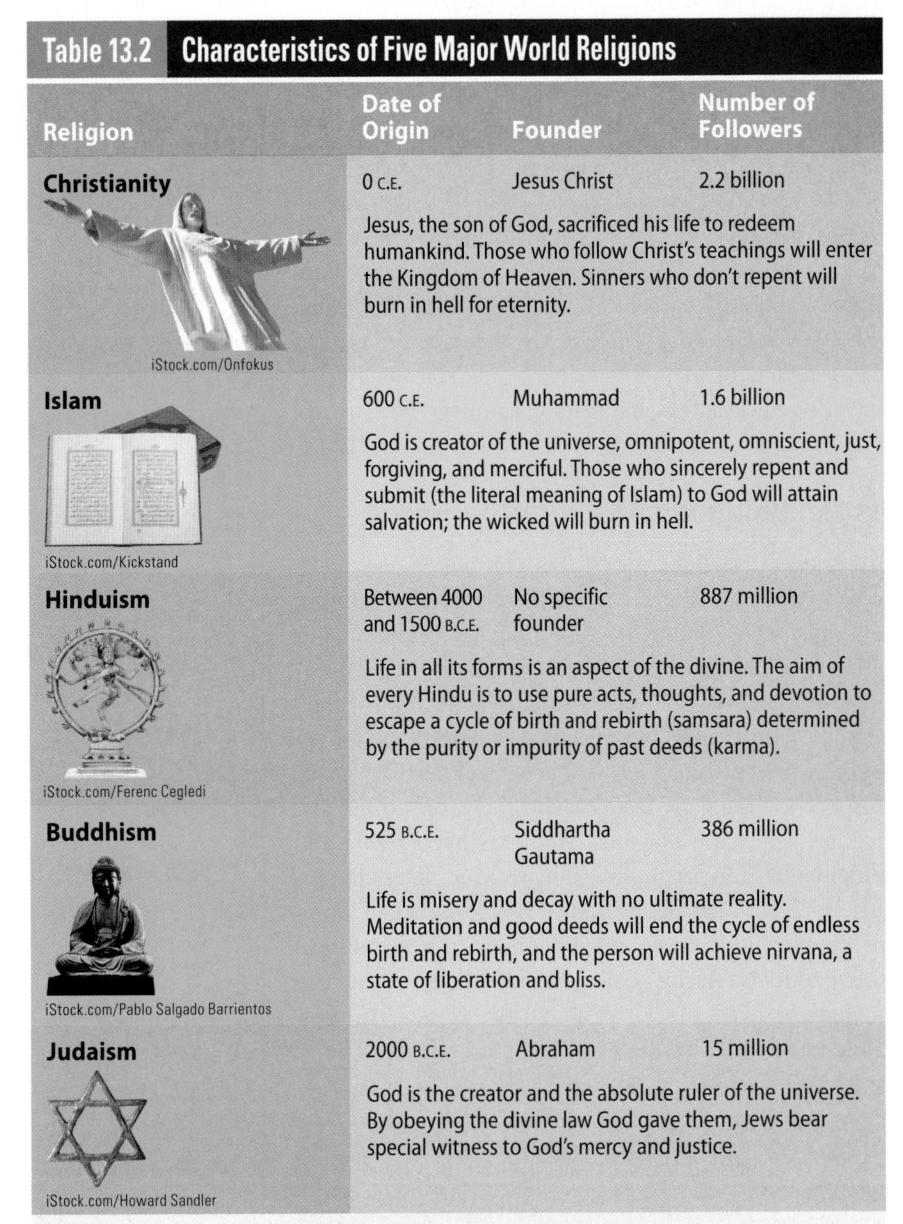

Table 13.2 Characteristics of Five Major World Religions

Religion	Date of Origin	Founder	Number of Followers
Christianity (iStock.com/Onfokus)	0 C.E.	Jesus Christ	2.2 billion
	Jesus, the son of God, sacrificed his life to redeem humankind. Those who follow Christ's teachings will enter the Kingdom of Heaven. Sinners who don't repent will burn in hell for eternity.		
Islam (iStock.com/Kickstand)	600 C.E.	Muhammad	1.6 billion
	God is creator of the universe, omnipotent, omniscient, just, forgiving, and merciful. Those who sincerely repent and submit (the literal meaning of Islam) to God will attain salvation; the wicked will burn in hell.		
Hinduism (iStock.com/Ferenc Cegledi)	Between 4000 and 1500 B.C.E.	No specific founder	887 million
	Life in all its forms is an aspect of the divine. The aim of every Hindu is to use pure acts, thoughts, and devotion to escape a cycle of birth and rebirth (samsara) determined by the purity or impurity of past deeds (karma).		
Buddhism (iStock.com/Pablo Salgado Barrientos)	525 B.C.E.	Siddhartha Gautama	386 million
	Life is misery and decay with no ultimate reality. Meditation and good deeds will end the cycle of endless birth and rebirth, and the person will achieve nirvana, a state of liberation and bliss.		
Judaism (iStock.com/Howard Sandler)	2000 B.C.E.	Abraham	15 million
	God is the creator and the absolute ruler of the universe. By obeying the divine law God gave them, Jews bear special witness to God's mercy and justice.		

Note: C.E. (Common Era) is the nondenominational abbreviation for A.D. (Anno Domini, Latin for "In the year of our Lord") and B.C.E. (Before the Common Era) is the nondenominational abbreviation for B.C. (Before Christ).

Sources: Based on a number of sources including the Center for the Study of Global Christianity, 2007; "Religions of the World...," 2007; and "Global Christianity...," 2011.

Note that no religious group comes close to being a global majority, that the third largest group is religiously unaffiliated, and that non-Christians outnumber Christians 2 to 1. By 2050, however, Muslims are expected to surpass Christians as the world's largest religious group and will make up nearly one-third of the world's population. The main reasons for Islam's growth is that Muslims have more children than members of the other major religious groups. Also, because Muslims have the youngest median age (20 versus 30 for non-Muslims), a larger share will soon be having children, accelerating the growth of the Muslim population (Lipka, 2017).

Five religious groups in particular have had a worldwide impact on economic, political, and social issues. *Table 13.2* provides a brief overview of these groups.

13-6 RELIGION IN THE UNITED STATES

Among U.S. adults, 92 percent believe in God, 3 percent don't, and 5 percent don't know or don't care (Lipka, 2016). For sociologists, religiosity is a better measure of being religious than simply asking people whether they believe in God (or a universal spirit) and which religion they follow. Religiosity includes a number of variables, but the most common are religious belief, affiliation, and participation.

13-6a Religious Belief

Some 53 percent of Americans say that religion is very important in their lives, down from 70 percent in 1965 (Newport, 2016). Not surprisingly, religion isn't important for *agnostics* (people who say that it's impossible to know whether there's a God), *atheists* (those who believe that there's no God), or others who are skeptics. Nonetheless, about 25 percent of Americans, including some atheists and agnostics, embrace the tenets of some Eastern religions or elements of New Age spirituality that include reincarnation, meditation, astrology, and the evil eye (casting curses and evil spells) (Lugo et al., 2012).

13-6b Religious Affiliation

Americans' religious identity is shifting. Since 2007, the share of Christians has declined, while the number of adults who say they have no religious preference or affiliation has grown to 23 percent (*Table 13.3*). A vast majority (78 percent) of the unaffiliated, often called "nones," were raised in a specific religion but no longer believe the teachings, dislike formal religion, have lost respect for religious leaders who focus on power and money, see many religious people as hypocritical and judgmental, and are more willing to admit not being religious than in the past. "Nones" include a broad swath of people but tend to be millennials, males, political independents, Asians, college graduates, or in higher-income brackets (Taylor et al., 2014; Lipka, 2016; Pew Research Center, 2016).

Table 13.3 Religious Affiliation in the United States, 2014

		Percentage Change Since 2007
CHRISTIAN	**71%**	**−8%**
Protestant	47	−5
Catholic	21	−3
Mormon	2	–
OTHER FAITH	**6**	**+2**
UNAFFILIATED	**23**	**+7**
Atheist	3	+2
Agnostic	4	+2
Nothing in particular	16	+4

Source: Based on Pew Research Center, 2015, 40.

Another recent trend is the declining number of Americans who identify with a specific Protestant denomination (e.g., Baptist, Lutheran, Methodist); only 30 percent did so in 2016, compared with 50 percent in 2000. Some are "nones," but 42 percent of Americans—especially those who were raised as Catholics or Protestants—have switched religions because of moving to another location, intermarrying, divorcing, or feeling more welcome at a new place of worship (Pew Research Center, 2015, 2016; Newport, 2017).

13-6c Religious Participation

Turning to the third measure of religiosity, religious participation, among Americans who say that religion is important to them, 30 percent seldom or never attend religious services. Thus, many people are more likely to believe in a religion than to practice it by attending services regularly. Mormons (75 percent), Protestants (53 percent), Muslims (50 percent), and Catholics (45 percent) have higher attendance rates at religious services than do Jews or other non-Christians (19 percent) (Newport, 2014; Sandstrom and Alper, 2016).

13-6d Some Characteristics of Religious Participants

Americans differ in their beliefs and affiliations. Religious participation also varies by sex, age, race and ethnicity, and social class.

SEX

American women are generally more religious than men. They're more likely than men to say that religion is "very important" in their lives (60 percent vs. 47 percent), to pray every day (64 percent vs. 47 percent), and to attend religious services at least once a week (40 percent vs. 32 percent), and they are less likely to be "nones" (19 percent vs. 27 percent) ("The Gender Gap…," 2016).

The religious gender gap is due to a mix of factors. It may be that women are expected to be more pious and spiritual because, especially as nurturers, they transmit religious values to their children (see Chapter 9). Because of lower labor force participation rates, women have more time for religious activities, where they may forge friendships. Women give higher priority to religion because it offers a sense of self-identity and well-being, especially for women who experience poverty, poor health, or domestic violence ("The Gender Gap…," 2016). Also, women, on average, live longer than men, and religiosity increases with age.

AGE

Generally, Americans age 65 and older are more likely than younger people to describe themselves as religious, to say that religion is very important in their lives, to pray, and to attend services at least weekly. These age-related differences may reflect several factors: Older Americans grew up decades ago when church attendance was higher, they seek spiritual comfort as elderly friends and relatives die, they want to lessen a sense of isolation or loneliness, and they're preparing for death (Taylor et al., 2009).

Among adults under age 33, 36 percent are "nones," compared with just 11 percent who are 72 and older. Young adults are also much more likely to be unaffiliated than earlier generations were at a similar stage in their lives. Among other reasons, many baby boomers taught their children to be independent, think for themselves, and "find their own moral compass." As a result, perhaps, millennials are less trusting of major institutions such as religion, the economy, and government (Masci, 2016).

iStock.com/Shelly_Au

RACE AND ETHNICITY

Black individuals are more religious than any other U.S. racial or ethnic group. They're the most likely to report a religious affiliation, to belong to a church, to pray every day, and to attend religious services every week. The vast majority (71 percent) are Protestant, primarily Baptist and Evangelical; 5 percent are Catholic; and 18 percent are "nones" (Pew Research Center, 2015).

Latinos' religious landscape has been shifting. Of the largest affiliations, 48 percent are Catholic (down from 67 percent in 2010), 19 percent are evangelical Protestant, and 20 percent are "nones" (up from 10 percent in 2010). Some Catholic and evangelical Protestant churches have been especially successful in attracting recent Latino immigrants because the churches offer services in Spanish, the ceremonies are expressive rather than formal, and the clergy are more responsive to their members' social and economic needs ("Select-a-Faith," 2014; Pew Research Center, 2017).

Asian Americans may be the most diverse religious group in America. Nationally, 17 percent are Catholic, 16 percent are Hindu, 16 percent are mainline or evangelical Protestant, 6 percent are Buddhist, and 6 percent are Muslim. A large share, 31 percent, are "nones" (Pew Research Center, 2015).

Figure 13.6 Education and Religion

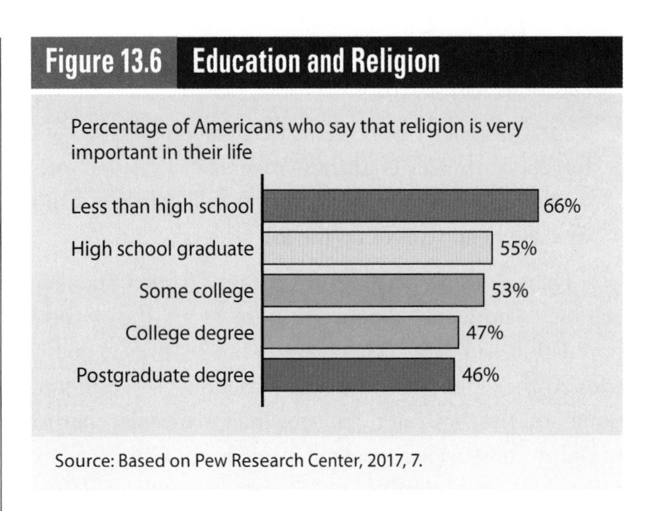

Source: Based on Pew Research Center, 2017, 7.

SOCIAL CLASS

As education increases, the importance of religion decreases (*Figure 13.6*). Generally, college graduates are less likely than people with less schooling to believe in God, to pray, and to attend religious services. They're also more likely (11 percent) than adults with a high school education (4 percent) to describe themselves as atheists or agnostics. Among the "nones," 30 percent of atheists and agnostics have an annual family income of $100,000 or more, compared with 17 percent of Christians (Pew Research Center, 2017).

13-6e Secularization: Is Religion Declining or Thriving?

Industrialized nations have been experiencing **secularization**, a process in which religion loses its social and cultural influence. Is secularization endangering religion? Or is religion flourishing?

IS SECULARIZATION INCREASING?

You've seen that the number of "nones" is growing, attendance at religious services has dropped, and that fewer Americans say that religion is "very important" in their lives. There's other evidence that secularization is rising in the United States and globally. For example:

- 41 percent of Americans have a "great deal" of confidence in the church and organized religion, down from 68 percent in 1975 (Newport, 2017).
- 70 percent of Americans attended religious services on Christmas Eve or Day when they were children; only 54 percent do so in adulthood (Cooperman et al., 2013).
- 68 percent of people worldwide describe themselves as religious,

secularization religion loses its social and cultural influence.

down from 77 percent in 2005 (WIN-Gallup International, 2013).

- The European Court of Justice, which passes laws for 28 European countries, recently ruled that private employers can ban Muslim female workers from wearing head scarves on the job (Bilefsky, 2017).

There's also evidence of greater "forced secularization" worldwide. About 77 percent of the world's population—up from 58 percent in 2007—lives in countries with high government restrictions on religion. Some of the laws and policies ban particular faiths, prohibit conversions, or limit preaching. Others allow religion-related conflict or terrorism, mob violence against religious minorities, and harassment over religious attire (Grim and Cooperman, 2014; Henne, 2015).

IS RELIGION THRIVING?

Many sociologists contend that the prevalence of secularization has been greatly exaggerated. For example:

- 62 percent of Americans say that it's important for a U.S. president to have strong religious beliefs (Smith, 2017).
- 72 percent endorse Christmas displays on government property (Pew Research Center, 2014).
- Nearly all states allow exemptions if vaccinations conflict with parents' religious beliefs; so far, 34 states and the District of Columbia don't prosecute parents for child abuse or manslaughter even if denying medical treatment for religious reasons results in a child's death or lifelong disability (Sandstrom, 2015, 2016).

fundamentalism belief in the literal meaning of a sacred text.

Pascal Deloche/Corbis Documentary/Getty Images

- The U.S. Constitution never mentions God or the divine. In contrast, all 50 state constitutions use "God," "Creator," "divine," "Lord," "Supreme Being," and similar religious references at least once (Sandstrom, 2017).

Fundamentalism, a belief in the literal meaning of a sacred text (e.g., Christian Bible, Muslim Qur'an, Jewish Torah), is widespread in practically all countries except Europe. About 25 percent of Americans believe that the Bible should be taken literally because it's the word of God (Saad, 2017).

The U.S. Constitution established a separation of church and state. Over the years, that separation has become fuzzier. Some southern states have passed laws that allow public schools and government buildings to

AP Images/Harry Cabluck

Brendan Smialowski/Getty Images

In 2005, the U.S. Supreme Court ruled that a 6-foot-high monument containing the Ten Commandments outside the Texas state capitol was constitutional, even though it's on government property. In 2014, the court ruled that prayer in civic life, including legislative bodies, is not only constitutional, but can use explicit Christian or other religious language. Do such decisions illustrate greater secularization or religiosity?

"WE ESTABLISH NO RELIGION IN THIS COUNTRY. WE COMMAND NO WORSHIP. WE MANDATE NO BELIEF, NOR WILL WE EVER. CHURCH AND STATE ARE AND MUST REMAIN SEPARATE." RONALD REAGAN, REPUBLICAN, 40TH U.S. PRESIDENT (1911–2004).

display traditional Christmas scenes and symbols; many public elementary and high schools allow prayer before class, permit students to leave classes for religious instruction, support religious student clubs that faculty sponsor, and offer after-school evangelical programs. A growing number of public colleges and universities have approved building privately funded "religious dorms" (Bruinius, 2013; Campo-Flores, 2013; Lawrence, 2013).

One of the most divisive issues has been over teaching evolution in public school science classes. For scientists, it's "a fact" that human beings (and other creatures) evolved from earlier animals through a process known as "natural selection" over several million years (National Science Board, 2014). In contrast, *creationism*, based on a fundamentalist interpretation of the Bible, argues that God created humans in their present form about 10,000 years ago. Forty-two percent of Americans espouse creationism, and another 31 percent believe in evolution, but under God's guidance. Only 19 percent believe God has nothing to do with evolution (Newport, 2014).

Despite the constitutional separation of church and state, since 2006 local governments have passed 67 anti-evolution education bills that teach creationism rather than evolution in science classes. In early 2017, eight states proposed laws to protect teachers who present creationism as a scientific theory (Matzke, 2016; Embury-Dennis, 2017).

The government has blurred the distinction between church and state in other ways. In 2014, the U.S. Supreme Court ruled that for-profit companies don't have to pay for their employees' contraceptives—as the Affordable Care Act (Obamacare) mandated—if doing so conflicted with the owners' religious beliefs. Betsy DeVos, the U.S. Secretary of Education, is an ardent supporter of using taxpayer-funded vouchers for private religious schools and has a lifelong dedication to "building God's kingdom through education" (Rizga, 2017).

The Johnson Amendment, enacted in 1954, prohibits tax-exempt institutions like churches from endorsing or opposing political candidates. Even though 90 percent of the nation's Evangelical board members and nearly 100 religious organizations have urged Congress to maintain the Johnson Amendment, President Trump has pledged to "get rid of and totally destroy" the law (Beckwith, 2017; Goodstein and Shear, 2017).

Another indicator that religion is booming in the United States and elsewhere is the prevalence of **civil religion** (sometimes called *secular religion*), integrating religious beliefs into secular life. The Pledge of Allegiance was written in 1892; the phrase "under God" was added in 1954. Other examples of civil religion include the phrase "In God We Trust" on U.S. currency, prayer in public schools, a legally mandated National Day of Prayer held on the first Thursday of May, and many local jurisdictions' closing public schools, public libraries, and government offices on Good Friday, a Christian holy day.

13-7 SOCIOLOGICAL PERSPECTIVES ON RELIGION

"Religion is a social glue"

Religion, cohesiveness of culture caused by religion

nany ... ion? ... um- ... on.

per- ... s to ... e of ... let's

of important functions at individual, community, and societal levels. All of the

civil religion (sometimes called *secular religion*) integrating religious beliefs into secular life.

iStock.com/Paul Maguire

following, according to functionalists, contribute to a society's survival, stability, and solidarity:

- *Belonging and identity.* Communal worship and rituals increase social contacts, develop a sense of acceptance and identity, and reinforce people's feeling of belonging to a group. During Christmas, 86 percent of Americans, including "nones," attend holiday activities with family or friends (Cooperman et al., 2013).
- *Purpose and emotional comfort.* Religion provides meaning in life and offers hope for the future. Religiosity is highest among the world's poorest countries because religion helps people cope with daily struggles to survive (Crabtree, 2010; WIN-Gallup International, 2013).
- *Well-being.* Very religious Americans report being healthier and experiencing less depression and worry than those who aren't religious (Newport et al., 2012; Newport and Himelfarb, 2013).
- *Social service.* Religious groups raise millions of dollars for the victims of natural disasters, distribute food and clothing, find shelter for displaced families, and help the poor and sick.
- *Social control.* Because many religious people fear punishment in the afterlife, they try to follow societal rules. Doing so suppresses deviant and antisocial behavior and promotes societal cohesion (Durkheim, 1961; Shariff and Aknin, 2014).

RELIGION PROMOTES SOCIAL CHANGE

Religion can also spearhead social change. Mohandas Gandhi (1869–1948), a spiritual leader in India, worked for his country's independence from Great Britain through nonviolence and peaceful negotiations. In the United States, religious leaders, especially the Reverend Martin Luther King, Jr., were at the forefront of the civil rights movement during the late 1960s.

In 2011, after decades of debate, the Presbyterian Church, one of the nation's largest Protestant denominations, approved ordaining people with same-sex partners as ministers, elders, and deacons. Numerous Lutheran churches have also welcomed gay pastors. In 2014, and despite many of its congregations' threats of leaving, the Presbyterian Church voted to allow same-sex marriages. Other religious groups (e.g., United Church of Christ, Episcopal Church, Quakers, Reform and Conservative Judaism) have followed (Goodstein, 2014). Pope Francis created a tribunal to prosecute bishops who protect priests suspected of sexual child abuse. He has made it faster, easier, and cheaper for Roman Catholics to obtain marriage annulments and insists that priests welcome (rather than reject) divorced Catholics and their children.

Max Weber (1920/1958) asserted that religion sparks economic development. His study of Calvinism, a Christian sect that arose in Europe during the sixteenth century, led to his coining the term *Protestant ethic*, a belief that hard work, diligence, self-denial, frugality, and economic success were signs of God's favor, and that prosperity indicated divine blessing. According to Weber, the harder the early Calvinists worked and saved, the more likely they were to accumulate money, to become successful, and to drive the growth of capitalism.

Was Weber right about the relationship between the Protestant ethic and the rise of capitalism? The data are mixed. Some studies show that people are diligent not because of religious beliefs but simply because amassing savings provides resources when disaster strikes (as when droughts wipe out crops). On the other hand, a study of 59 industrialized and developing countries found that religious beliefs that encourage hard work and thrift spur economic growth (Cohen, 2002; Barro and McCleary, 2003).

Table 13.4 Sociological Explanations of Religion

Theoretical Perspective	Level of Analysis	Key Points
Functionalist	Macro	Religion benefits society by providing a sense of belonging, identity, meaning, emotional comfort, and social control over deviant behavior.
Conflict	Macro	Religion promotes and legitimates social inequality, condones strife and violence between groups, and justifies oppression of poor people.
Feminist	Macro and micro	Religion subordinates women, excludes them from decision-making positions, and legitimizes patriarchal control of society.
Symbolic Interactionist	Micro	Religion provides meaning and sustenance in everyday life through symbols, rituals, and beliefs, and binds people together in a physical and spiritual community.

John Gress/REUTERS

Megachurches are Christian congregations that have a regular weekly attendance of more than 2,000 and tend to be evangelical. They represent only 0.5 percent of all U.S. churches, but their number has almost tripled (to more than 1,600) since 2000 (Bird and Thumma, 2011). From a functionalist perspective, why do you think megachurches are so popular?

IS RELIGION DYSFUNCTIONAL?

Functionalists emphasize its benefits, but also recognize that religion can be dysfunctional when it harms individuals, communities, and societies. Religious intolerance can spark conflict between groups (e.g., vandalizing churches, mosques, and synagogues) and attacks on religious minorities. The United States has a long history, beginning with the Pilgrims in 1620, of people discriminating against other religious groups even though they themselves sought religious freedom (Davis, 2010).

The Roman Catholic Church and the Vatican protected predator priests for at least five decades (United Nations Committee on the Rights of the Child, 2014). Overall, religion meets important psychological needs, but strong group identity may breed narrowmindedness, ethnocentrism, and bigotry toward nonmembers.

CRITICAL EVALUATION

One weakness is that functionalists, by emphasizing religion's benefits, imply that religion is indispensable to leading a good life. People who aren't religious do many good things for society, and some religious people commit heinous acts. Primarily because of priests' sexual abuse of young children, Americans' rating of the clergy's honesty and ethics fell from 67 percent in the mid-1980s to 47 percent in 2013 (Swift, 2013).

A second limitation is functionalists' exaggerating religion's positive effect on physical and emotional health. Jews, who are one of the least religious U.S. groups, and "nones," including atheists, report higher well-being rates than do religious groups (Newport et al., 2012; Leurent et al., 2014).

A third, and major, criticism is that functionalists gloss over numerous dysfunctional aspects of religion that generate wars, terrorism, and genocide. Conflict theory addresses these issues.

13-7b Conflict Theory: Religion Promotes Social Inequality

From a conflict perspective, religion promotes and reinforces social inequality. Throughout history and currently, religion has created discord and divisiveness within groups, between groups, and in the larger society.

"THE OPIUM OF THE PEOPLE"

Much conflict theory reflects the work of Karl Marx (1845/1972), who described religion as "the sigh of the oppressed creature" and "the opium of the people" because it encouraged passivity and acceptance of social and economic inequality. Marx viewed religion as a form of **false consciousness**, an acceptance of a system of beliefs that prevents people from protesting oppression. Contemporary conflict theorists don't view religion as an opiate, but they agree with Marx that religion often justifies violence and promotes social inequality.

RELIGION JUSTIFIES INTOLERANCE AND VIOLENCE

For thousands of years, many governments and religious leaders have condoned or perpetrated widespread violence in the name of religion. Religion promotes conflict when religious groups differentiate between "we" and "they" ("We're right and they're wrong," "Our God is 'the *real* God'"). Such self-righteousness condones aggression, oppression, and brutality within and across societies. For example:

- The "eternal war" between Sunnis and Shiites, two major Islamic sects, dates back more than 1,300 years,

false consciousness an acceptance of a system of beliefs that prevents people from protesting oppression.

and there are long-standing conflicts and wars between Muslims and Jews in Israel and Palestine (Crowley, 2014; "Religion: The New Strife," 2016).

- In several of Asia's Buddhist-majority nations, monks have incited bigotry and violence—mostly against Muslims (Beech, 2013).
- In 2014 alone, religion-related terrorism and wars displaced almost 34 million people from their homes (Henne and Kishi, 2016).
- Among 198 countries, 53 percent experienced widespread government harassment of religious groups in 2015, up from 43 percent in 2014 ("Global Restrictions...," 2017).

"What goes on down there in the name of religion is turning me into an atheist."

David Sipress/The New Yorker Collection/The Cartoon Bank

RELIGION PROMOTES SOCIAL INEQUALITY

Conflict theorists see religion as a tool dominant groups use to control society, protect their own interests, and derail social change. Roman Catholic bishops, for example, have opposed comprehensive health care because it might fund abortions, have publicly chastised pro-choice Catholic politicians, and have campaigned against same-sex marriage—all of which fuel social inequality and conflict (Doyle, 2011).

Private evangelical schools, colleges, and universities receive massive state and federal financial aid that diverts tax revenues from public schools and colleges. Many of the schools' textbooks describe gay and lesbian individuals and abortion rights supporters as "evil," and are hostile toward other religions, including nonevangelical Protestants, Jews, and Catholics (Berkowitz, 2011; Tabachnick, 2011).

In some southern states, conservatives maintain that "religious freedom" should allow businesses to refuse services to gay and transgender people (Bruinius, 2016). Doing so denies people basic civil liberties, seals LGBTs off from society, and creates resentment.

CRITICAL EVALUATION

Functionalists may overemphasize consensus and harmony, but conflict theorists, according to some critics, often ignore religion's role in challenging exclusionary policies. For example, more than 800 churches and synagogues, while also fighting deportation battles in courts, have mobilized to provide sanctuary for law-abiding undocumented immigrants and their U.S.-born children (Rodgers, 2017).

A second criticism is that conflict theorists attach too much credence to Marx's concept of false consciousness. Instead of passively submitting to religious oppression, many people revolt, complain, or resist. Many have sued employers, sometimes successfully, who denied requests to not work on holy days, and whose secular dress codes don't allow head scarves, turbans, or long, shaggy beards (Trottman, 2013). Whether religious discrimination is real or imagined, many Americans protest, demonstrate, and seek legal remedies.

13-7c Feminist Theories: Religion Subordinates and Excludes Women

Feminist scholars agree with conflict theorists that religion can foster violence and inequality. They go further, however, by criticizing organized religions as sexist, patriarchal, and shutting women out of leadership positions.

SEXISM, PATRIARCHY, AND SUBORDINATION

"Religion is probably the most important force in continuing the oppression of women worldwide," a feminist blogger writes, and "religiosity and sexism go hand in hand" (Marcotte, 2014). From a feminist perspective, most religions are patriarchal: They emphasize men's experiences and a male point of view and see women as subordinate to men. According to the apostle Paul, "Wives should submit to their husbands in everything" (*Ephesians* 5: 24). The idea that Eve was created out of Adam's rib is often used to justify men's domination of women. Many religious teachings and institutions continue to propagate such beliefs (see Paludi and Ellens, 2016).

In Orthodox Judaism, a man's daily prayers include this line: "Blessed art thou, O Lord, our God, King of the Universe, that I was not born a woman." The Qur'an tells Muslims that men are in charge of women and that women should obey men. Almost all contemporary religions worship a male deity, and none of the major world religions treat women and men equally (Gross, 1996; Jeffreys, 2011).

Feminist scholars offer alternatives to patriarchal interpretations of Scripture. Some contend that Jesus was a feminist. For example, he defended women against men, encouraged women's intellectual pursuits when it wasn't the norm, and there are numerous passages in the Bible about women spreading Jesus' teachings (Gross, 1996; Ferguson, 2014).

Muslim feminists, similarly, note that men have interpreted sacred Islamic texts to ensure male dominance and control. Women were among some of Muhammad's earliest converts, and the Qur'an has numerous passages that establish women's equal rights in inheritance and family roles (Menissi, 1991, 1996). Thus, women's subordination in many Islamic societies isn't due to religious tenets but to men's interpretations of Scripture to maintain their power and privilege (Smith, 1994).

In the Middle East and North Africa, 35 percent of the countries have religious police who, among other things, enforce strict segregation of the sexes and punish women who are perceived as behaving or dressing improperly (Theodorou and Henne, 2014). Thus, it's women's fault if men have lustful thoughts or assault women.

EXCLUSION OF WOMEN FROM LEADERSHIP POSITIONS

In 2014, women earned 33 percent of theology degrees, a dramatic increase from only 2 percent in 1970. Still, women make up only 21 percent of the nation's clergy, lead only 11 percent of congregations, and earn 27 percent less than their male counterparts (Chaves and Eagle, 2015; Emmert, 2015; Snyder et al., 2016).

Roman Catholicism, Orthodox Judaism, and the Mormon Church don't allow women to be ordained because "women should serve and not lead." U.S. Catholic women contribute $6 billion a year during Sunday Masses, but "the presence of women anywhere within the institutional power structure is virtually nil" (Miller, 2010: 39; Gibson, 2012). Mormon women can be missionaries and provide "welfare and compassionate service," but are excluded from the priesthood and the all-male central leadership that governs the church (Kantor and Goodstein, 2014).

Some Protestant denominations justify women's exclusion from leadership positions based on biblical passages such as "I permit no woman to teach or have authority over men; she is to keep silent" (I *Timothy* 2: 11–12). Many Protestant groups—including Southern Baptists and evangelical born-again Christians—interpret this and similar passages to mean that women should never, under any circumstances, instruct men, within or outside of religious institutions. Even in liberal Protestant congregations, female clergy tend to be

Jim Urquhart/REUTERS

Emmanuel Dunand/Getty Images

In 2014, the Mormon Church excommunicated Kate Kelly (left), leader of the Ordain Women group. According to some U.S. Catholic bishops, the ideas of Sister Simone Campbell (right), and other nuns who advocate ordaining women, border on heresy. Both women are human rights lawyers.

relegated to specialized ministries with responsibilities for music, youth, or Bible studies (Banerjee, 2006; Richie, 2013).

CRITICAL EVALUATION

Feminist perspectives on religion are limited in several ways. First, some feminist Muslim scholars have criticized Western feminists for misreading Islamic and other sacred scriptures, and reducing practically all discussions of gender to the *hijab* (a veil or scarf that Muslim women wear) instead of focusing on justice for both women and men in marriage, employment, and other areas (Fakhraie, 2009; Daneshpour, 2015).

Second, religious women aren't as oppressed as many feminist scholars claim. For instance, an increasing number of Muslim women exercise authority within Islamic institutions as preachers and interpreters of religious text. Even in the most conservative religions, women don't blindly submit to religious dogma. Instead, they challenge existing doctrines, stop volunteering at church, don't attend weekly services, and establish their own parishes (Padgett, 2010; Miller, 2012).

13-7d Symbolic Interaction: Religion Is Socially Constructed

For symbolic interactionists, religion isn't innate but socially constructed. As a result, people can learn and interpret the same religion differently across cultures and over time. Symbols, rituals, and beliefs are three of the most common means of learning and internalizing religion.

SYMBOLS

A *symbol* is anything that stands for or represents something else to which people attach meaning (see Chapter 3). Many religious symbols are objects (a cross, a steeple, a Bible), but also include behaviors (kneeling or bowing one's head), words ("Holy Father," "Allah," "the Prophet"), and physical appearance (e.g., wearing head scarves, skull caps, turbans, clerical collars).

Religious symbols, like all symbols, are shorthand communication tools. Some interactionists define a religion as "a system of symbols" because it's a community that's unified by its symbols (Berger and Luckmann, 1966; Geertz, 1966). Religious symbols can also be divisive (e.g., a Ten Commandments monument on government property that alienates non-Christians and nonbelievers).

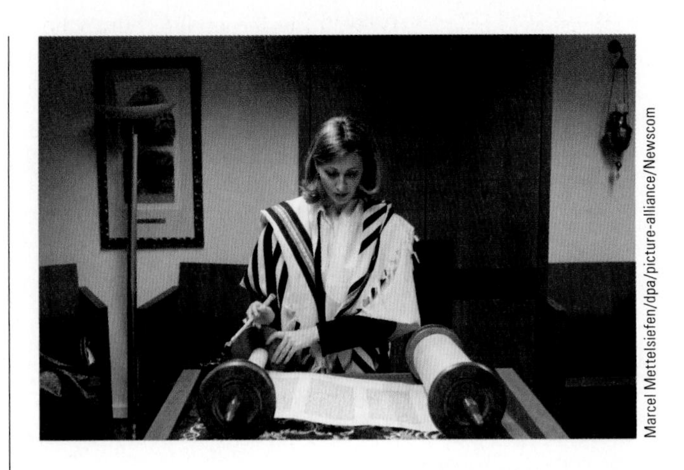

Marcel Mettelsiefen/dpa/picture-alliance/Newscom

RITUALS

A *ritual* (sometimes called a *rite*) is a formal and repeated behavior that unites people (see Chapter 3). Religious rituals, like secular ones, strengthen a participant's self-identity (Reiss, 2004). Religious rites of passage—*bat mitzvah* for girls and *bar mitzvah* for boys in the Jewish community, and first communion and confirmation for Catholic children—reinforce the individual's sense of belonging to a particular religious group. A group's rituals symbolize its spiritual beliefs and include a wide range of practices—praying, chanting, fasting, singing, dancing, and offering sacrifices.

All religions have rituals that mark significant life events like birth and marriage (see Chapters 3, 5, and 12). Death rituals are probably the most elaborate and sacred worldwide. They vary across religious groups and socie-ties, but all of them comfort the living and show respect for the dead.

BELIEFS

Rituals and symbols come from *beliefs*, convictions about what people think is true. Religious beliefs can be passive (believing in God but never attending services) or active (participating in rituals and ceremonies). Beliefs bind people together into a spiritual community.

One of the strongest beliefs worldwide is that prayer is important. Islam requires prayer five times a day. In the United States, 55 percent of U.S. adults say that they pray every day, and for a variety of reasons, including feeling close to God, as well as requesting better health, more money, and cures for sick pets. Even 20 percent of the "nones" pray every day (Wicker, 2009; Lipka, 2016). Prayer offers psychological and spiritual benefits such as comfort and a sense of unity among those who pray together, but depending on God can also diminish people's motivation to actively shape their own lives (Schieman, 2010).

CRITICAL EVALUATION

A common criticism is that interactionists' focus on micro-level behavior ignores the ways that religion promotes social inequality at the macro level. Conflict theorists and feminist scholars, especially, maintain that people often use religion to justify violence and women's subordination.

Some critics also wonder if interactionists paint too rosy a picture of religion even on a micro level because people's beliefs can wreak considerable havoc. The Islamic zealots who targeted the World Trade Center on 9/11 didn't see themselves as "crazed terrorists" but as "true believers": They were carrying out "God's will" by imposing their religious beliefs on others or destroying their "religious enemies" (Juergensmeyer, 2003).

STUDY TOOLS 13

READY TO STUDY? IN THE BOOK, YOU CAN:

- ☐ Check your understanding of what you've read with the Test Your Learning Questions provided on the Chapter Review Card at the back of the book.
- ☐ Tear out the Chapter Review Card for a handy summary of the chapter and key terms.

ONLINE AT CENGAGEBRAIN.COM WITHIN MINDTAP YOU CAN:

- ☐ Explore: Develop your sociological imagination by considering the experiences of others. Make critical decisions and evaluate the data that shape this social experience.
- ☐ Analyze: Critically examine your basic assumptions and compare your views on social phenomena to those of your classmates and other MindTap users. Assess your ability to draw connections between social data and theoretical concepts.
- ☐ Create: Produce a video demonstrating connections between your own life and larger sociological concepts.
- ☐ Collaborate: Join your classmates to create a capstone project.

14 Health and Medicine

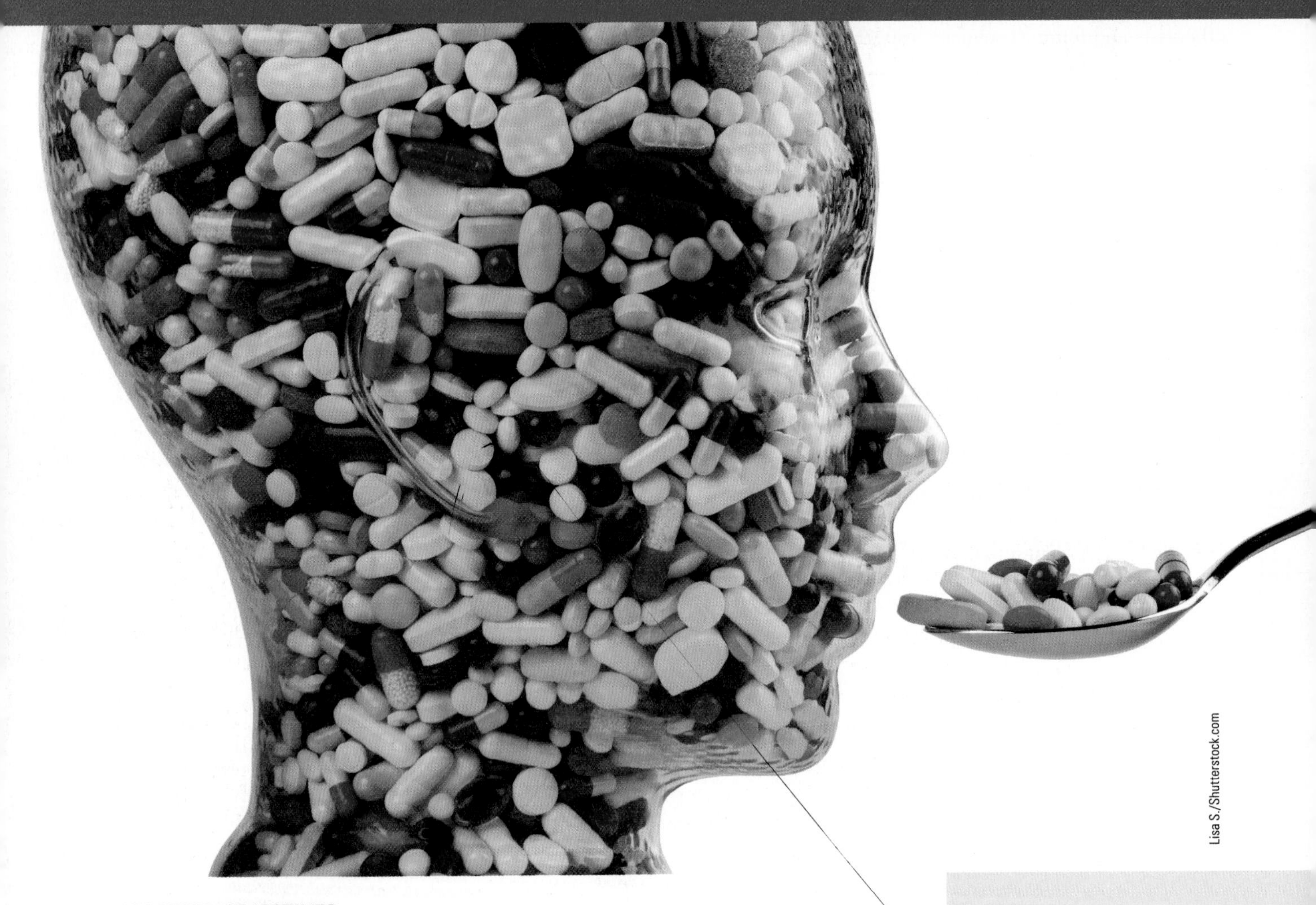

Lisa S./Shutterstock.com

LEARNING OBJECTIVES

After studying this chapter, you will be able to...

14-1 Describe social epidemiology and explain why there are global health gaps.

14-2 Explain how and why environmental, demographic, and lifestyle factors affect U.S. health and illness. Illustrate your explanation with specific examples.

14-3 Compare U.S. with other countries' health care systems in terms of access, costs, and outcomes.

14-4 Compare and evaluate the theoretical explanations of health and medicine.

After finishing this chapter go to **PAGE 301** for **STUDY TOOLS**

Most of us would probably agree with American poet Ralph Waldo Emerson, who wrote, in 1860, “Health is the first wealth.” **Health** is the state of physical, mental, and social well-being. We can usually determine physical health by using objective measures like weight and blood pressure. Mental and social well-being are more difficult to gauge because they rely on subjective definitions that change over time.

WHAT DO YOU THINK?

The government should be responsible for making sure that all Americans have health care.

1	2	3	4	5	6	7
strongly agree				strongly disagree		

14-1 GLOBAL HEALTH AND ILLNESS

How does health differ around the world? And why? Social epidemiology addresses both questions.

14-1a Social Epidemiology

All people experience *disease*, a disorder that impairs a person's normal physical and/or mental condition. **Social epidemiology** examines how societal factors affect the distribution of disease within a population. Why, for example, do people live much longer in some countries than in others?

In answering this and other questions, epidemiologists look at two factors. One is *incidence*, the number of new cases of a health problem that occur in a given population during a given time period (e.g., in 2015, there were 1.5 million new diabetes cases among Americans aged 18 and older). The other measure is *prevalence*, the total number of cases (extent) of an illness or health problem within a population or at a particular point in time (e.g., in 2015, 10 percent of the U.S. population had diabetes) (*National Diabetes Statistics Report,* 2017). Epidemiological studies show large health differences across countries.

14-1b Global Health Disparities

The World Bank classifies countries into four broad groups based on gross national income (GNI) per capita. *Figure 14.1* shows the number and location of these countries and compares them on life expectancy, infant mortality, and per capita health expenditures. These three variables are among the most frequently used to measure health status. They indicate a population's living standards, people's average socioeconomic status, and the quality and financial resources that a country devotes to **health care**, the prevention, treatment, and management of illness.

The 80 high-income countries make up only 13 percent of the world's population, but have the greatest access to clean water, sanitation, food, and health services. In contrast, more than a third of the world's population lacks basic sanitation, almost half are at risk of dying of malaria, and 44 percent of people in low- and lower-middle-income countries can't afford medicine to treat infections and diseases (World Health Organization, 2015).

High income populations live longer, have low infant mortality rates, and spend more on health (*Figure 14.1*), but experience “diseases of wealth” (e.g., diabetes, heart disease, various cancers). Compared with 16 other high-income countries, the United States has a large and growing “health disadvantage” and ranks last or near-last in nine key health areas including drug-related deaths, chronic lung and heart disease, obesity, and the prevalence of HIV and AIDS (Woolf and Aron, 2013).

14-2 HEALTH AND ILLNESS IN THE UNITED STATES

Why do Americans have a health disadvantage, including high disability rates? Compared with other high-income nations, non-medical social determinants contribute to large health disparities in the United States.

health the state of physical, mental, and social well-being.

social epidemiology examines how societal factors affect the distribution of disease within a population.

health care the prevention, management, and treatment of illness.

Figure 14.1 Worldwide Health Gaps

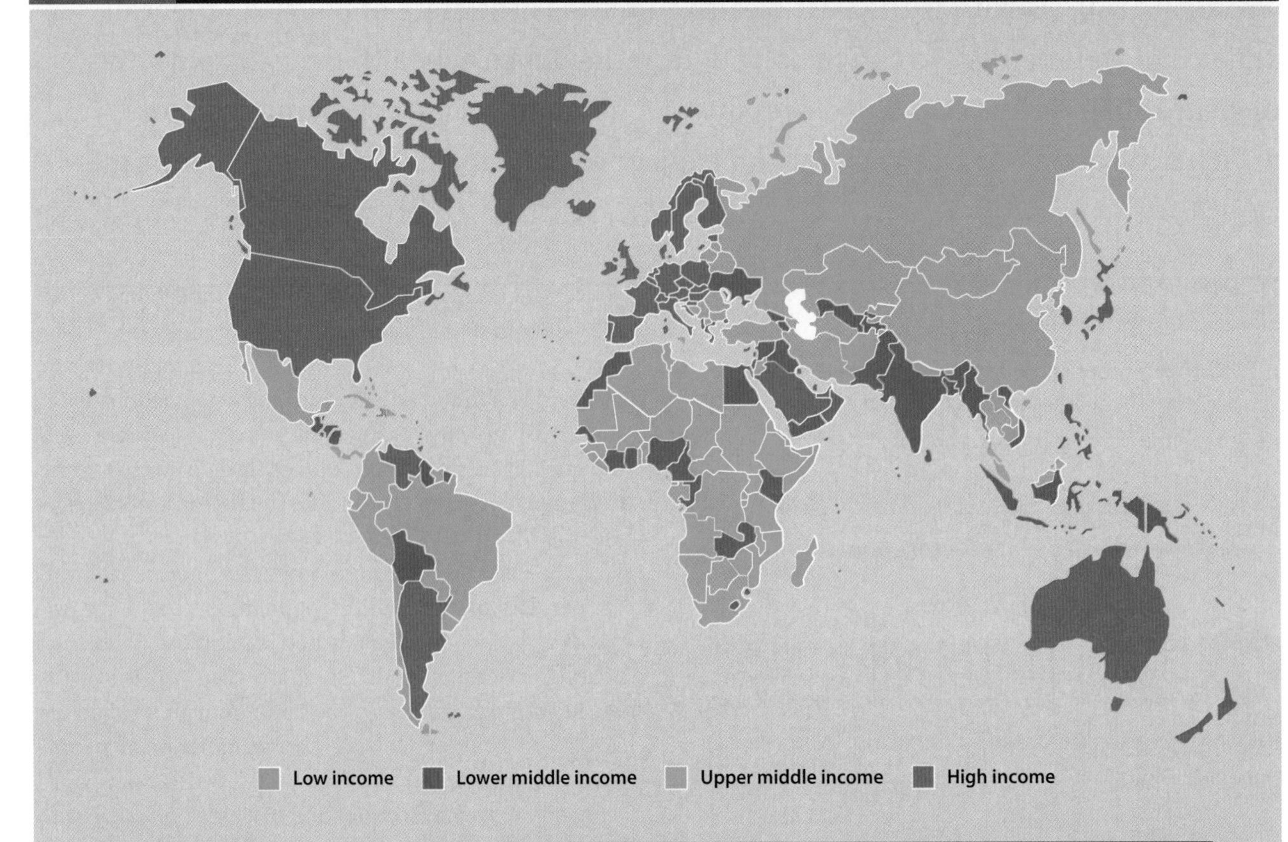

Income Group and Number of Countries	GNI	Life Expectancy at Birth	Infant Deaths per 1,000 Live Births	Per Person Total Expenditure on Health*
Low income (31)	$1,045 or less	59	53	$83
Lower middle income (51)	$1,046–$4,125	67	44	$235
Upper middle income (53)	$4,126–$12,735	74	16	$766
High income (80)	$12,736 or more	79	5	$4,516

*These figures include private, employer, and government expenditures in U.S. dollars.

Sources: Based on World Bank, 2015; World Health Organization, 2015, Tables 1 and 7.

14-2a Understanding Disability

A **disability** is any physical or mental impairment, temporary or permanent, that limits a person's ability to perform a basic life activity. These activities include walking; doing errands alone; bathing, dressing, or feeding oneself; and working.

Almost a third of Americans aged 18 and over have a disability. Age is a major predictor of disability, affecting 58 percent of adults age 65 and older compared with 25 percent of those ages 18 to 64. Disability also increases with age—from 26 percent for people aged 65 to 74 to 73 percent for those age 85 and older. Besides age, disability rates are higher for women (36 percent) than men (27 percent), for people living in poverty (41 percent versus 23 percent who aren't poor), and are highest among Native Americans (41 percent) (National Center for Health Statistics, 2017).

disability any physical or mental impairment that limits a person's ability to perform a basic life activity.

The number of Americans experiencing one or more disabilities rose from 61 million in 1997 to almost 77 million in 2015 (National Center for Health Statistics, 2017). Why the increase? First, as you'll see shortly, behavior previously considered normal has been medicalized. Second, medical advances help many people survive diseases, car accidents, and wars, but they may

Christophe Calais/Corbis Historical/Getty Images

Advances in battlefield medicine and body armor mean than many soldiers are surviving injuries, but they have shattered bones, brain damage, and amputated limbs, and may suffer from posttraumatic stress disorder (PTSD) or other mental health problems. Among post–9/11 veterans, 34 percent have a service-connected disability (U.S. Department of Veterans Affairs, 2017).

have lifelong health problems. Third, as the proportion of the population age 65 and older increases, so does the proportion living with disabilities.

14-2b Social Determinants of Health and Illness

No single variable explains well-being. About 65 percent of cancers are the result of "bad luck" because stem cells make mistakes copying healthy DNA. This means that 35 percent of cancers are preventable (Tomasetti and Vogelstein, 2015). Whether it's cancer or other diseases, environmental, demographic, and lifestyle factors—all of which are interrelated—have a significant impact on health and illness.

ENVIRONMENTAL FACTORS

Many environmental hazards affect our bodies. For example:

- There's a strong association between early-life exposure to air pollution and autism, schizophrenia, and learning disabilities (Roberts et al., 2013; Allen et al., 2014).
- Thousands of children in Flint, Michigan—a predominantly poor, Black city—are expected to suffer irreversible developmental problems because of lead contamination in the water supply (Milman and Glenza, 2016).
- Many toxic chemicals in food and everyday products (e.g., detergents, shampoos, shaving products, and makeup) have been linked to asthma, some types of cancer, birth defects, and children's impaired brain development (Blake, 2014; Grandjean and Landrigan, 2014; Attina et al., 2016).
- Norovirus—a very contagious virus—is the leading cause of disease outbreaks from contaminated food. About 21 million Americans get sick from norovirus each year, 71,000 are hospitalized, and about 800 die. Outbreaks on cruise ships account for only 1 percent of all norovirus; 70 percent of cases are due to food workers who are infected, don't wash their hands, or go to work when sick (Hall et al., 2014).

Access to health care can prolong life, but an estimated 250,000 Americans die each year because of preventable medical errors (e.g., doctors operating on the wrong patient or body part, patient infections). Medication errors—misdiagnosing an illness, prescribing the wrong drugs, or giving patients drugs that interact dangerously—injure about 1.3 million Americans and kill about 7,000 every year. Moreover, thousands of patients a year leave operating rooms with surgical sponges in their bodies that cause infections, lifetime digestive problems, removal of intestines, and even death (Eisler, 2013; Laliberte, 2016; Makary and Daniel, 2016).

A growing problem is antibiotic resistance, which kills an estimated 23,000 Americans each year. Bacteria have become more resistant because of the overuse of antibiotics: People take them for illnesses for which antibiotics aren't effective, and farmers feed them to animals to promote growth. No new antibiotics have been discovered since 1987, largely because pharmaceutical companies reap much higher profits by developing very expensive and specialized medicines for small populations with chronic conditions, including multiple sclerosis, cancer, HIV, and hepatitis C ("The Drugs Don't Work," 2014: 54; Tozzi, 2014).

DEMOGRAPHIC FACTORS

Religion, family size, marital status, urban-rural residence, and other variables affect health. Four of the most important demographic factors are age, gender, social class, and race and ethnicity.

Age. Not surprisingly, age is the single best predictor of illness and death. Death rates drop sharply shortly after birth, begin to rise at about age 45, and escalate, particularly after age 78.

From birth through elementary school, children have health problems over which they have no control.

Nearly 1 in 100 U.S. babies are born with *fetal alcohol spectrum disorders*, a range of permanent birth defects caused by a mother drinking alcohol during pregnancy. The disorders include intellectual infirmities, speech and language delays, poor social skills, mental retardation, and physical abnormalities like congenital heart defects (National Organization on Fetal Alcohol Syndrome, 2012; Weinhold, 2012).

Many adolescents have health problems because of their own choices, like smoking, abusing alcohol and other substances, and texting while driving. Such risky behavior may be due to peer pressure, to the ongoing development of some parts of the teen brain, and to increasing independence as adolescents break away from parental supervision (see Chapter 4).

Health problems emerge and increase during one's late thirties because of genes, lifestyle choices (which we'll address shortly), and because physical decline is normal and inevitable as we age. No matter how well tuned we keep our bodies, the parts start wearing down: Reflexes slow, hearing and eyesight dim, and stamina and muscle strength decrease.

After age 65, chronic rather than acute diseases comprise the majority of health problems, including disability. **Chronic diseases** (e.g., asthma and high blood pressure) are long-term or lifelong illnesses that develop gradually or are present from birth. In contrast, **acute diseases** (e.g., chicken pox) are illnesses that strike suddenly and often disappear rapidly but can cause incapacitation and sometimes death. Chronic diseases increase as people age (*Figure 14.2*). *Dementia*, a loss of mental abilities, most commonly occurs during one's 70s. The most debilitating form of dementia is **Alzheimer's disease**, a progressive, degenerative disorder that attacks the brain and impairs memory, thinking, and behavior.

Figure 14.2 Chronic Illness Increases with Age

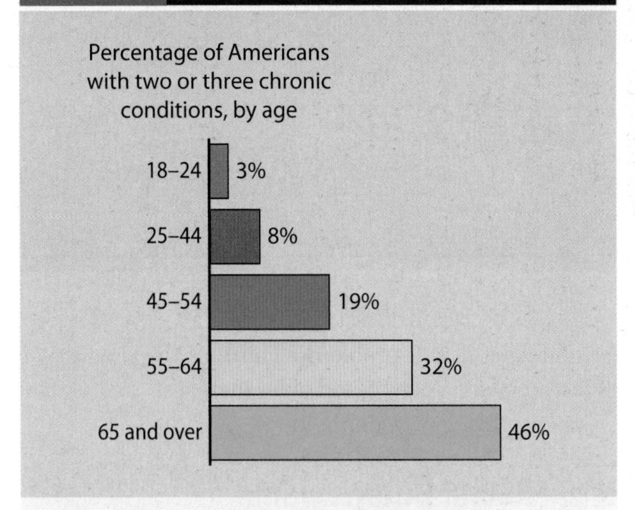

Source: Based on National Center for Health Statistics, 2017, Table 9.

Gender. The gender gap in life expectancy has decreased since 1975, but women, on average, still live longer than men (see Chapter 12). Men of all racial-ethnic groups are two to three times more likely than women to die in motor vehicle crashes, to be victims of homicide, to smoke and drink alcohol, to abuse drugs, and to work in dangerous occupations (e.g., construction, law enforcement). Women are also more likely than men to have a regular physician, to see a doctor when something's wrong, and to keep medical appointments (Levine, 2014; National Center for Health Statistics, 2017).

Depression is the single most common mental health problem among adolescents, begins at about

chronic diseases long-term or lifelong illnesses that develop gradually or are present from birth.

acute diseases illnesses that strike suddenly and often disappear rapidly.

Alzheimer's disease a progressive, degenerative disorder that attacks the brain and impairs memory, thinking, and behavior.

PhotoPlus Magazine/Future Publishing/Getty Images

Thomas Trutschel/Photothek/Getty Images

UniversalImagesGroup/Getty Images

Three reasons why women live longer than men.

age 12, and affects 30 percent of U.S. high school students. Girls (20 percent) are more likely than boys (9 percent) to experience depression, but the reasons are similar: substance abuse, bullying (either as a victim or perpetrator), sexual or physical abuse, parental divorce, and a family history of mental disorders. In adulthood, and across all ages, women are more likely than men to be taking antidepressant medications (Murphey et al., 2013; Federal Interagency Forum on Child and Family Statistics, 2017).

Depression may lead to suicide. Among high school students, females are twice as likely as males to attempt suicide. Across all ages, however, males are four times more likely than females to actually kill themselves, and the highest rates are among White men aged 75 and older (Curtin et al., 2016).

Social Class. Social class has a strong effect on health. The wealthiest Americans live 10 to 15 years longer than the poorest. As early as age 45, the lower the family income, the greater the likelihood that adults will experience two or more chronic diseases. Living in poor neighborhoods increases the likelihood of stress (which affects diet, smoking, and alcohol/drug usage), a sedentary lifestyle because of limited recreational facilities, and less access to health care (Beckles and Truman, 2013; Hummer and Hernandez, 2013; Dickman and Woolhandler, 2017).

Policymakers have tried to eliminate *food deserts*, geographic areas where access to fresh, healthy, and affordable food is limited or nonexistent because grocery stores are too far away. Social class disparities persist even when consumers have access to the same food, however. Supermarkets, compared with convenience stores, offer more healthy foods. Because low-income workers might make long commutes or have several jobs to support their families, they are more likely to choose convenience than to cook healthy meals from scratch ("Dietary Inequality," 2016; Zagorsky and Smith, 2017).

People in lower social classes are also more likely to have dangerous jobs. In 2015, construction and electrical workers, truck drivers, and miners accounted for more than half of the nearly 5,000 job-related deaths. Almost 3 million experienced workplace injuries or illnesses caused, among other things, by falls, overexertion, and accidents using equipment (BLS News Release, 2016; "Employer-Reported . . . ," 2016).

Race and Ethnicity. A good measure of a population's health is its *infant mortality rate*, the number of babies under age 1 who die per 1,000 live births in a given year. Infants born to Black women are more than

Figure 14.3 U.S. Infant Mortality Rates by Mother's Race and Ethnicity, 2015

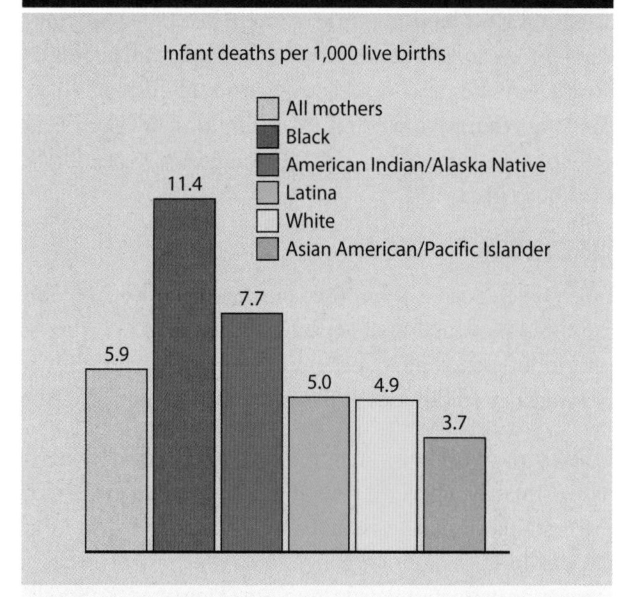

Source: Based on National Center for Health Statistics, 2017, Table 11 and Figure 7.

twice as likely to die before age 1 as those born to White, Asian, and Hispanic women (*Figure 14.3*). The higher Black infant mortality rate may be due to a combination of factors: More babies born before 24 weeks who don't survive, less access to quality health care, and Black mothers' greater likelihood of having some conditions (high blood pressure, diabetes, obesity) that are associated with negative birth outcomes (Chen et al., 2014; MacDorman et al., 2014).

Even when infants survive and thrive, race and ethnicity affect lifespan. You saw in Chapter 12 that Black individuals have lower life expectancy rates than Asian, Latino, and White individuals. The death rate for Black people has declined about 25 percent since 1999. Compared with White people, however, Black individuals ages 18 to 49 are twice as likely to die from heart disease and, between ages 35 to 64, are 50 percent more likely to have high blood pressure and diabetes ("African American Health," 2017).

Black people are twice as likely as White people to experience the death of two or more family members by age 30, and 90 percent more likely to experience four or more deaths by age 65 (Umberson et al., 2017). Poverty, unemployment, and family members' deaths contribute to poor health and lower life expectancy.

Despite generally low education and income levels, Latinos have higher life expectancy rates than Black or White individuals. One reason may be that only the healthiest individuals migrate to the United States. Another may be that, compared with other racial and ethnic groups,

Latinos are the least likely to engage in harmful behavior or to die from drug abuse. The longer Latinos live in the United States, however, the higher their rates of heart disease, high blood pressure, some types of cancer, and diabetes. Acculturation may result in improved access to health care, but also to adopting harmful behaviors (e.g., smoking, excessive alcohol consumption), less physical activity, and unhealthy diets (Dominguez et al., 2015; Riosmena et al., 2015).

LIFESTYLE FACTORS

Lifestyle choices can improve or impair our health. The top three preventable lifestyle health hazards, in order of priority, are smoking, obesity, and substance abuse. Sexually transmitted diseases also cause infections and illness.

Smoking. Worldwide and in the United States, tobacco use, primarily cigarette smoking, is the leading cause of preventable disease, disability, and death. U.S. smoking has declined, is higher among men than women, and decreases as education levels increase (*Figure 14.4*). Smoking rates among people living below the poverty line are nearly twice as high as among those above the poverty line (Jamal et al., 2016).

There are now more former than current smokers, but tobacco use is responsible for about 1 in 5 deaths annually, and, on average, smokers die about 10 years earlier than nonsmokers. Because tobacco harms nearly every human organ, smoking is linked to cancer, heart disease, stroke, Type 2 diabetes, macular degeneration, erectile dysfunction, lung diseases (including emphysema and bronchitis), and birth defects (U.S. Department of Health and Human Services, 2014).

Figure 14.4 Cigarette Smoking Has Decreased, but Varies by Gender and Education Level

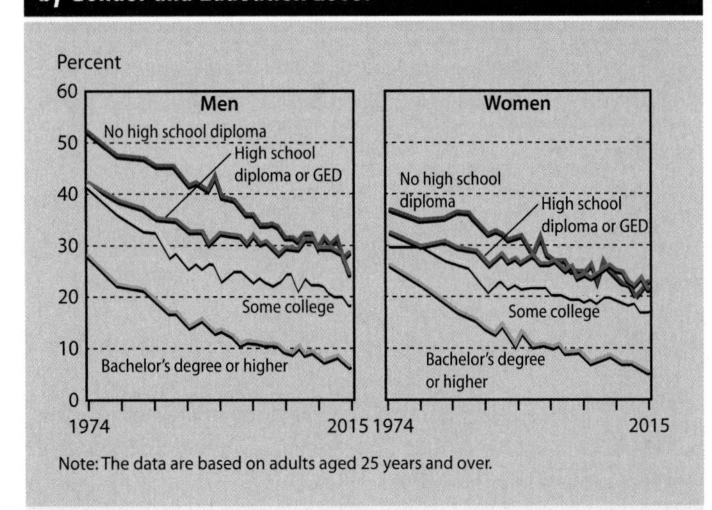

Note: The data are based on adults aged 25 years and over.

Source: National Center for Health Statistics, 2017, Figure 10.

Diego Cervo/Shutterstock.com

From 2013 to 2016, the number of middle and high school students using electronic cigarettes almost tripled, rising from 780,000 to 2.2 million students (Jamal et al., 2017). To attract adolescents, the e-cigarette industry uses celebrities, cartoon characters, and "kid-friendly" flavors like chocolate, gummy bears, and mint candy (Zimmerman, 2014).

Prevention is difficult because the tobacco industry spends almost $10 billion a year to market its products; the number of smoking scenes among the top-grossing children's movies rose from 564 in 2010 to 809 in 2016, a 43 percent increase; and 69 percent of middle and high school students are exposed to electronic cigarette ads in retail stores, on the Internet, in magazines or newspapers, on TV, or in the movies (Singh et al., 2016; Tynan et al., 2017). During 2016, states collected almost $26 billion from tobacco taxes and legal settlements, but spent less than 2 percent of the money on tobacco control programs (Jamal et al., 2016).

Public health experts are debating whether electronic cigarettes, or e-cigarettes, are safer than tobacco cigarettes. E-cigarettes deliver nicotine in a liquid smoke-free vapor (often called "vaping"), but don't contain the same carcinogens and other toxic chemicals as regular cigarettes. According to the U.S. Surgeon General, because e-cigarettes contain nicotine, they can harm the developing adolescent brain and lead to addiction. Among adolescents who have never smoked, for example, those who try e-cigarettes are likely to start smoking conventional cigarettes within a year. Some researchers maintain, however, that it's too soon to know if vaping leads to smoking traditional cigarettes, because both usages have been declining among middle and high school students (U.S. Department of Health and Human Services, 2016; Hines et al., 2017; Kozlowski and Warner, 2017).

Obesity. Obesity is the second leading and preventable cause of disease, disability, and death. As you saw in Chapter 6, childhood and adult obesity has increased since the early 1970s. Preschoolers who are overweight or obese are 5 times more likely than normal-weight children to be overweight or obese as adults, increasing their risk for heart disease, high blood pressure, strokes, diabetes, osteoporosis, and several types of cancer. Nonetheless, almost 79 percent of parents of obese preschoolers describe their children as "about the right weight" (Cunningham et al., 2014; Duncan et al., 2015).

In 1990, not a single state had an obesity rate above 15 percent. Now *all* states have obesity rates above 20 percent and 22 states have obesity rates equal to or greater than 30 percent ("Obesity Prevalence Maps," 2015). Some researchers predict that, at the current rate, 51 percent of American adults will be obese in 2030 (*Figure 14.5*), and 9 percent of that group will be severely obese (Finkelstein et al., 2012).

Figure 14.5 U.S. Adults Keep Getting Fatter

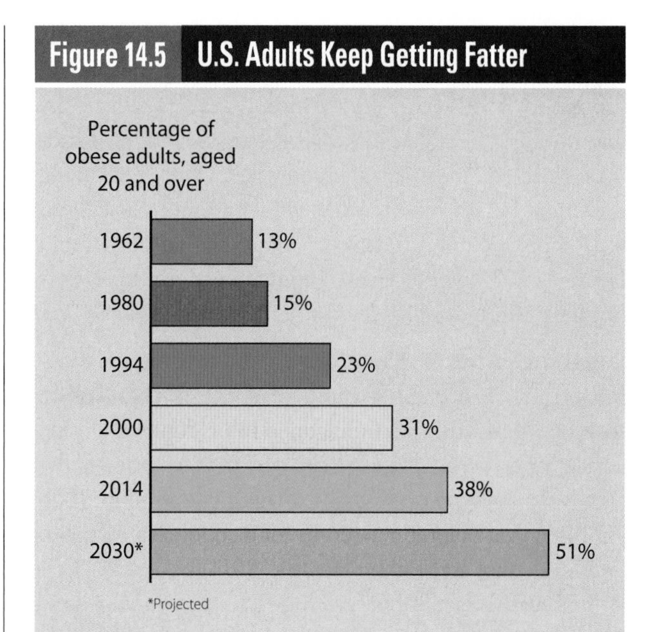

Source: Based on Finkelstein et al., 2012, and National Center for Health Statistics, 2015, Table 64.

Our choices affect weight. Unhealthy eating habits start before age one, when parents introduce babies to French fries, soft drinks, potato chips, cookies, processed juice, and other "junk food." By 23 months, toddlers may be consuming *the equivalent of 9.2 teaspoons of sugar a day* (Cha, 2016).

Adults are getting fatter because of poor eating habits, large food portions, being sedentary, and eating frequently at fast-food restaurants, but structural factors also affect obesity rates. Healthy food is less expensive than during the 1950s, more widely available than ever before, and easy to prepare. However, a smaller share of income now buys many more calories, including sugar-sweetened beverages and prepared food, that are high in unhealthy carbohydrates or fat. Thus, since 1970, the average American's consumption of calories has risen by about 20 percent, fueling an "obesity epidemic" (Sturm and An, 2014).

How Much Sugar Are You Really Eating?

The American Heart Association recommends limiting our daily dose of added sugar to 24 grams (about 6 teaspoons) for women and 36 grams (about 9 teaspoons) for men.

Low-fat diets rarely reduce weight because nearly 60 percent of Americans' calories come from "ultra-processed" foods (e.g., soft drinks, pizza, frozen meals) that are high in added sugar (Steele et al., 2016). Products with added sugar include those with packaging statements like "made with whole grain," "excellent source of calcium," "fat-free," "good source of Vitamin D," or "100% juice." A small container of Greek yogurt can have three or more teaspoons of added sugar.

Substance Abuse. The nation's third leading lifestyle-related and preventable cause of death is **substance abuse**, a harmful overindulgence in or dependence on a drug or other chemical. Nearly 85,000 Americans die each year because of excessive drinking, and more than 4 million visit emergency rooms for alcohol-related problems ("Alcohol and Public Health," 2014).

Excessive alcohol use includes *binge drinking*, having 4 or more drinks on a single occasion for women or 5 or more drinks on a single occasion for men, generally within about 2 hours. *Heavy drinking* involves consuming 15 or more drinks per week for men, and 8 or more drinks for women. Some 25 percent of Americans are binge drinkers and 7 percent are heavy alcohol users (National Center for Health Statistics, 2017).

Excessive alcohol use results in myriad short- and long-term health risks that include injuries, violence, risky sexual behaviors, miscarriage and stillbirth, physical and mental birth defects, alcohol poisoning, unemployment, psychiatric problems, heart disease, several types of cancer, and liver disease. *Table 14.1* summarizes some of the consequences of excessive college drinking for students, their families, and the larger community.

Table 14.1 Some Consequences of Abusive College Drinking

Researchers estimate that, each year, among college students aged 18 to 24...	
1,825	Die from alcohol-related unintentional injuries, including motor vehicle crashes
599,000	Are unintentionally injured while drunk
696,000	Are assaulted by another student who's been drinking
97,000	Are victims of alcohol-related sexual assault or date rape
400,000	Have unprotected sex while intoxicated, and more than 100,000 were too drunk to remember if they consented to having sex
3.4 million	Drive while drunk
25%	Have academic problems because of their drinking, including missing classes, doing poorly on exams or papers, earning lower grades overall, and flunking out of college

Sources: Based on National Institute on Alcohol Abuse and Alcoholism, 2013, 2014.

Illicit (illegal) *drugs* include marijuana/hashish, cocaine and crack, heroin, hallucinogens (e.g., LSD, PCP), inhalants, and any prescription-type psychotherapeutic drug, including stimulants and sedatives, used nonmedically. More than 10 percent of Americans (about 34 million people) age 12 and older use illegal drugs (*Figure 14.6*). Illicit drug use is more common among men than women and among people aged 16 to 25, and least common among Asian Americans (National Center for Health Statistics, 2017).

Some states have legalized marijuana, but it's the most commonly used illicit drug (*Figure 14.6*). In the late 1990s, doctors started prescribing more and stronger opioids to treat arthritis, back problems, diabetes, and other chronic pain.

Opioids include illegal drugs (e.g., heroin), synthetic drugs (e.g., fentanyl), and legally prescribed pain relievers (e.g., oxycodone, morphine). As opioid-driven drug overdose deaths surged (*Figure 14.7*), health practitioners blamed opioids for "our nation's deadliest drug epidemic ever." According to a national drug commission, the death toll for drug overdoses is "equal to Sept. 11 every three weeks" (Goodnough, 2017: A14).

A third of Americans who have taken prescription opioids admit being addicted, but defend the drug for dramatically relieving pain and allowing them to walk, work, and pursue other activities. Some manufacturers, however, can't find enough workers—even for jobs that pay $25 an

substance abuse a harmful overindulgence in or dependence on a drug or other chemical.

Figure 14.6 Illicit Drug Use Among Americans Aged 12 and Older, 2015

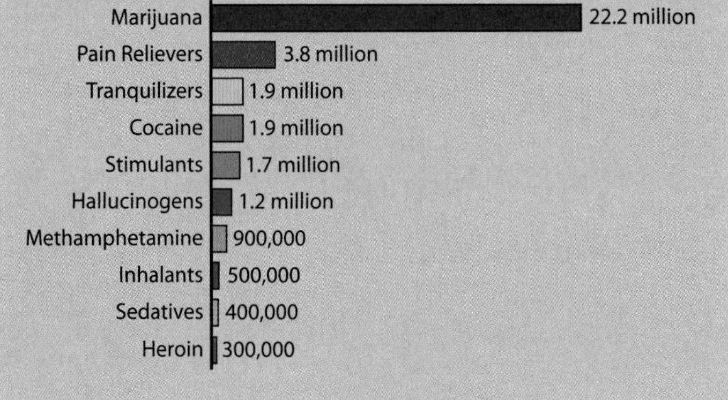

Source: Based on Center for Behavioral Statistics and Quality, 2016, Figure 1.

Figure 14.7 The Number of U.S. Drug Overdoses Has Quadrupled Since 2000

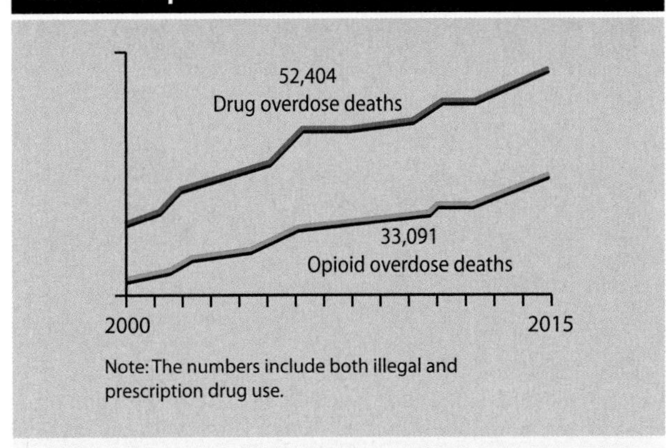

Source: Based on material from National Institute on Drug Abuse, 2017.

hour and offer full benefits—because 25 to 50 percent of the applicants fail drug tests (Clement and Bernstein, 2016; Schwartz, 2017).

Sexually Transmitted Diseases. There are nearly 20 million new sexually transmitted disease (STD) cases in the United States every year. People infected with STDs are two to five times more likely than uninfected individuals to contract HIV, the virus that causes AIDS. About 1.1 million Americans are living with HIV, including 15 percent who aren't aware of their infection. Men who have sex with men (MSM)—whether straight, gay, or bisexual—represent only 2 percent of the population but 67 percent of all new HIV infections. Of those diagnosed with HIV, 9 percent are drug injectors, 26 percent are age 55 and older, 45 percent are Black, and 25 percent are women. Overall, 86 percent of women contract HIV through heterosexual sex. Most of the women aren't aware that their partners are MSM, inject drugs, or have recently tested positive for HIV (Spiller et al., 2015; Centers for Disease Control and Prevention, 2016; "HIV in the United States...," 2017).

We've looked at some of the reasons why some people are healthier than others. Who gets health care? And who pays for it?

14-3 HEALTH CARE: UNITED STATES AND GLOBAL

Health care encompasses a number of components. One of the most important is **medicine**, a social institution that deals with illness, injury, and other health problems (see also Chapter 6 on institutions).

14-3a U.S. Health Care Coverage

In 2010, Congress passed the Patient Protection and Affordable Care Act, also called the Affordable Care Act (ACA) or "Obamacare." The ACA's goal was to give more Americans under age 65 affordable, quality health insurance, and to reduce the growth in health care spending.

Has the ACA worked? The uninsured decreased from 18 percent (42 million people) in 2013, before the ACA took effect, to 10 percent (28 million) in 2017. The ACA generated the greatest gains for people living below or just above the poverty level, Latinos, and foreign-born noncitizens—the groups least likely to have health insurance (Barnett and Vornovitsky, 2016; "Health Insurance Coverage," 2017; National Center for Health Statistics, 2017).

Who are the 28 million uninsured? People who live in states where Medicaid doesn't cover the "nearly poor," are undocumented immigrants, haven't enrolled in Medicaid, or have incomes above the federal poverty level but can't afford the programs offered by employers or private insurers. The largest group of the uninsured—about 46 percent—doesn't have coverage for a variety of reasons: Paying out-of-pocket as needed is less expensive than health insurance, they rely on free or low-cost clinics, go without care or medicine, and/or use emergency rooms (Robert Wood Johnson Foundation, 2015; Kaiser Family Foundation, 2016).

How do politics affect health policies? Repealing Obamacare has been the rallying cry of the Republican Party ever since the law passed. During the 2016 presidential election, Donald Trump described Obamacare as "disastrous" and vowed—repeatedly and to thunderous applause—to "immediately repeal and replace Obamacare."

Nationally, however, 35 percent of Americans, many of whom voted for Trump and were recipients of the ACA, didn't know that the ACA and Obamacare are the same (Dropp and Nyhan, 2017). Congress hasn't repealed or replaced Obamacare as this book goes to press. *Table 14.2* presents some of the pros and cons of the ACA.

14-3b Who Pays for Health Care?

About 67 percent of Americans have private health insurance, through an employer or union, that they buy directly from an insurance company or through someone outside the household. Of the insured, employer-based insurance covers 56 percent of the population; public health insurance funded by taxpayers (Medicare, Medicaid, military) covers another 37 percent (Barnett

medicine a social institution that deals with illness, injury, and other health problems.

Table 14.2 Some Obamacare Pros and Cons

Pros	Cons
• Millions of uninsured Americans now have health insurance. Young adults can stay under their parents' health plans until age 26.	• People who don't have health insurance (through an employer, Medicare, Medicaid, the military, or a private company) risk paying a penalty.
• Covers pre-existing conditions (e.g., diabetes, heart disease, cancer, HIV/AIDS), and there are no time limits on care.	• Healthy people pay for patients with pre-existing conditions. The costliest five percent of these patients account for nearly half of all health care spending.
• Lowers overall health care costs in the long run, because preventive services (e.g., screening for cancer or other diseases) have no out-of-pocket payments.	• Tens of millions of low- and middle-income people with private insurers earn too much to qualify for Obamacare or Medicaid, but struggle to pay medical bills.
• Offers access to and comparisons of health insurance options through exchanges.	• Some major insurers have dropped out of exchanges, especially in rural counties, leaving consumers with few choices and rising premium costs.

Sources: Kim and Roland, 2015; Cox and Levitt, 2017; Infinit Healthcare, 2017; Tozzi, 2017; Tracer and Recht, 2017.

and Vornovitsky, 2016; see also Chapter 12 on Medicare and Medicaid).

Since 1999, workers at both large and small firms have been paying higher premiums, deductibles, copayments, and other out-of-pocket costs, whereas wages, including those of middle-class families, have stagnated.

Most people pay a variety of state and federal taxes to support programs like Medicare and Medicaid. Medicare pays almost 60 percent of all health care costs for people age 65 and over, regardless of income (Federal Interagency Forum on Aging-Related Statistics, 2016). Thus, even billionaires are eligible for Medicare. Medicaid, another government insurance program, provides medical care, regardless of age, for people living below the poverty level (see Chapter 12).

14-3c Why Is U.S. Health Care So Expensive?

The United States spends more on health care than any other nation in the world—per person, nationally, and as a percentage of the gross domestic product. We spend more on health care than 34 other high-income countries, but rank last among the top 11 in quality of care, access to care, efficiency, and health outcomes (Davis et al., 2014; OECD, 2014).

By 2022, health care expenses will consume $1 of every $5 in the economy (*Table 14.3*). Why, then, are many Americans worse off than their counterparts in other high-income countries? One reason is that from a third to almost half of U.S. health care spending is wasteful or fraudulent (Couffinhal and Socha-Dietrich, 2017; U.S. Department of Justice, 2017).

In Canada, Germany, Great Britain, France, Japan, Sweden, and other high-income countries, the government picks up most of the health care bill. Patients and health care workers don't have to submit bills to several insurance providers, resulting in considerable administrative savings. The government, nonprofit organizations, or large groups (cities and industries) have considerable power in keeping down drug costs and setting prices for health care providers and services. There are also private hospitals, but most are nonprofit (OECD, 2014).

In the United States, "hospitals are the most expensive part of the world's most expensive health system." Even nonprofit hospitals are businesses: Up to 26 percent

Table 14.3 The Increasing Cost of U.S. Health Care, 1980–2022

	1980	1990	2000	2010	2022
Average cost per person	$1,100	$2,864	$4,878	$8,402	$14,664
National health expenditure	$256 billion	$724 billion	$1.4 trillion	$2.6 trillion	$5 trillion
Percent of gross domestic product (GDP)	9%	13%	14%	18%	20%

Notes: The "average cost per person" includes medical care, supplies, drugs, and health insurance. All numbers are in current dollars; those for 2022 are projected.

Sources: Based on Centers for Medicare & Medicaid Services, 2010, Table 1, and 2014, Table 1.

of their revenues are profits, executives can earn almost $6 million a year, and they sell as many services as possible at the highest price (Brill, 2013; "Prescription for Change," 2013).

Unlike other countries where the government sets a national price for each drug, the United States lets pharmaceutical companies charge whatever they want. Congressional lawmakers—heavily lobbied by the pharmaceutical industry, the American Hospital Association, and other groups—have forbidden Medicare, the nation's largest medical insurer and the world's largest buyer of prescription drugs, to negotiate drug prices or the cost of health products. Thus, for example, the price of EpiPen, a lifesaving injection for severe allergic reactions, jumped more than 400 percent—from $94 to $609—after a pharmaceutical company acquired the product. A drug that cures Hepatitis C, a liver disease, costs just $4 a pill in India, compared with $1,000 in the United States (Agorist, 2016; Atwater, 2016).

Medical technology also increases health care costs. U.S. doctors perform 71 percent more CT scans than do doctors in Germany. Technology can save lives, but U.S. doctors, compared with those in many other countries, are more likely to adopt new and expensive machines that are highly profitable for both physicians and hospitals. Technology helps us live longer, but also increases a patient's, insurer's, and nation's medical costs (Brill, 2015; Frakt, 2017). *Figure 14.8* provides more examples of why U.S. health care costs are so high. Because of the high costs, each year up to 750,000 Americans engage in *medical tourism*, traveling to another country for medical care ("Medical Tourism," 2015).

14-4 SOCIOLOGICAL PERSPECTIVES ON HEALTH AND MEDICINE

Sociologists focus on different aspects of health, health care, and medicine. *Table 14.4* summarizes these perspectives.

14-4a Functionalism: Good Health and Medicine Benefit Society

For functionalists, good health and medicine are critical for a society's survival and stability. Thus, countries try to develop a medical care system that, ideally, benefits the entire population regardless of age, sex, race, ethnicity, social class, or other characteristics.

HEALTH CARE AND OTHER INSTITUTIONS

Health care systems are connected to many institutions. Despite ongoing debates about the ACA and the government's requiring people to buy health insurance, millions of individuals and families are now eligible for government assistance to pay for coverage, and can get free preventive health services (e.g., diabetes tests, colonoscopies, routine vaccinations).

Figure 14.8 Should You Travel Abroad for Medical Care?

Would you be better off living outside the United States when it comes to health care costs? Except for Argentina, all of these countries have higher life expectancy rates than the United States.

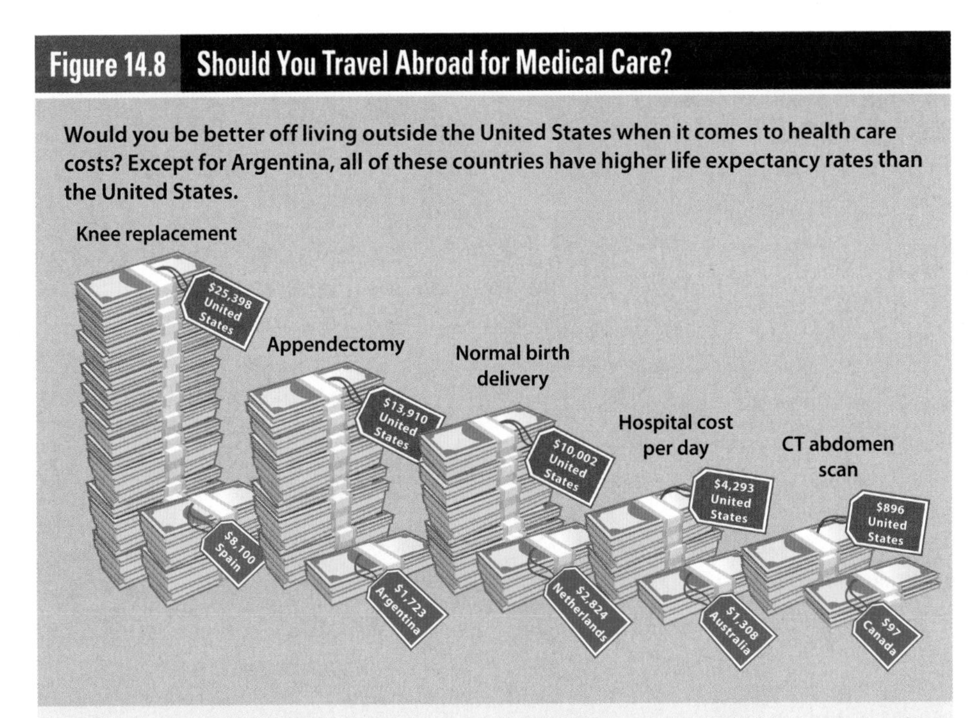

Source: Based on International Federation of Health Plans, 2014. See also Howard, 2014, for a comparison of surgery costs in the United States and other countries.

The government is deeply involved in health care in other ways. It funds much of the scientific research at universities as well as federal agencies like the Centers for Disease Control and Prevention and the National Institutes of Health, both of which deal with prevention, treatment, and health care policy. At the national and state levels, numerous government agencies are responsible for passing and enforcing regulations regarding new drugs, medical procedures, and access to medical care.

Health care affects the economy because it's the nation's largest employer. "Demand for health services

Table 14.4 Sociological Perspectives on Health and Medicine

Perspective	Level of Analysis	Key Points
Functionalism	Macro	• Health and medicine are critical in ensuring a society's survival and are closely linked to other institutions. • Illness is dysfunctional because it prevents people from performing expected roles. • Sick people are expected to seek professional help and get well.
Conflict	Macro	• There are gross inequities in the health care system. • The medical establishment is a powerful social control agent. • A drive for profit ignores people's health needs.
Feminist	Macro and micro	• Women are less likely than men to receive high-quality health care. • Gender stratification in medicine and the health care industry reduces women's earnings. • Men control women's health.
Symbolic Interaction	Micro	• Illness and disease are socially constructed. • Labeling people as ill increases their likelihood of being stigmatized. • Medicalization has increased the power of medical associations, parents, and mental health advocates and the profits of pharmaceutical companies.

will rise as Obamacare expands insurance and Americans grow older, fatter, and sicker" ("The Health Paradox," 2013: 27). Our huge health care spending threatens funding for other public programs, including education and transportation. Functionalists would point out, however, that the trillions spent on health care have generated millions of jobs at hospitals, drug companies, insurance companies, nursing homes, and technology firms (see Chapter 11).

Besides generating jobs, health care systems shape corporate policies which, in turn, benefit families. Recently, for instance, the pharmacy chain CVS banned sales of all tobacco products, Target is promoting employees' wellness and adding more organic and natural food to grocery aisles, and some restaurants (e.g., McDonald's, Wendy's, Red Lobster) have taken sodas off their children's menus and reduced sodium and fat in kids' meals (D'Innocenzio, 2015; Martin, 2015).

THE SICK ROLE

Talcott Parsons (1951), an influential sociologist, introduced the concept of the **sick role**, a social role that excuses people from normal obligations because of illness. In Parsons' model, sick people aren't responsible for their condition and, therefore, have legitimate reasons for not performing their usual social roles ("I missed the exam because I had the flu").

sick role a social role that excuses people from normal obligations because of illness.

Parsons emphasized that the sick role is temporary, and that individuals must seek medical help to hasten their recovery. Otherwise, people will view them as hypochondriacs, slouchers, and malingerers who aren't living up to their responsibilities. Thus, from a functionalist perspective, the sick role is legitimate if it's short-lived, but dysfunctional if people feign illness to shirk their duties long term in the family, the workplace, or other groups.

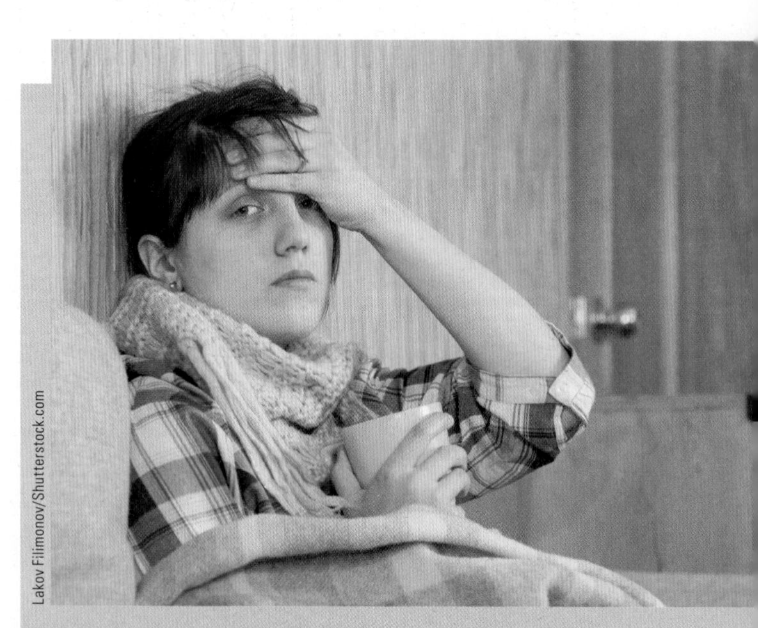

Lakov Filimonov/Shutterstock.com

How sick is too sick to miss work or class? Are menstrual cramps enough? What about a severe cold? A hangover? And when does being sick become dysfunctional—after a day, a week, a month?

THE PHYSICIAN AS GATEKEEPER

From a functionalist perspective, physicians play a key gatekeeping role in limiting the sick role. They verify a person's condition as sick and provide an excuse for temporarily not performing necessary roles. They also designate a patient as "recovered" and ready to meet societal role expectations again. Doctors' specialized knowledge gives them considerable authority in defining health and illness that's unmatched by other health care providers, such as nurses and pharmacists.

CRITICAL EVALUATION

Functionalist theories are limited for several reasons. First, health care policies can be divisive. For example, 52 percent of Americans say that the government should pay for health care, but 45 percent disagree (McCarthy, 2016). Moreover, most of the job growth in the health care sector has been in low-paying occupations (hospital orderlies, nursing home assistants, lab technicians) (Ross et al., 2014). Thus, many people aren't benefiting from the booming health care industry.

Second, and despite Obamacare, 25 percent of Americans say that either they or members of their family don't receive needed medical care because they can't afford the out-of-pocket costs. About 80 percent of low-wage workers don't receive any paid sick leave. When workers have paid sick days, 69 percent don't take them, even when they're ill, because they worry about falling behind or having coworkers take on extra work (Henderson, 2016; Frankel, 2017). Thus, many people don't or can't assume even a short-lived sick role.

Third, because physicians are powerful gatekeepers, many prosper by prolonging sick roles. The more times that doctors see patients or visit them at the hospital, and the more tests that doctors order, the higher their salaries. Conflict theorists, especially, maintain that much of the health care industry is dysfunctional because access to health care varies considerably, and only the wealthy don't have to worry about receiving and paying for the best available medical care.

14-4b Conflict Theory: Health Care and Medicine Don't Benefit Everyone

For conflict theorists, medicine and the health care industry benefit some groups much more than others. In contrast to functionalists, conflict theorists argue that the medical system reinforces social inequality, exerts social control to maintain the status quo, and is often driven by a profit motive rather than a concern for people's well-being.

SOCIAL CONTROL OF MEDICINE AND HEALTH CARE

From a conflict perspective, those at the top of the medical hierarchy have considerable control. Physicians, for instance, have almost absolute power in diagnosing an illness, providing treatment, and deciding on medical procedures.

The health care industry can maintain the status quo because of the **medical-industrial complex**, a network of business enterprises that influences medicine and health care. The medical-industrial complex includes many groups—doctors, nurses, hospitals, lawyers, nursing homes and hospices, insurance companies, drug manufacturers, accountants, banks, and real estate and construction businesses.

Who benefits from the medical-industrial complex? All of those involved, especially pharma (also called Big Pharma), the pharmaceutical drug industry, and hospitals. The annual profit margins of some prominent drug manufacturers, like Amgen, are 43 percent, compared with 3 to 6 percent for Ford and General Motors (Love, 2017).

Big hospitals have been buying up previously independent cancer clinics and doctors' offices. Doing so gives them the power to raise prices for private insurers and government programs, particularly Medicare. As a result, they can charge higher rates, sometimes three times as much, for cancer drugs and outpatient visits and services (Bach, 2015; Bai and Anderson, 2015).

The percentage of Americans taking three or more prescription drugs increased from 11 percent in 1998 to 23 percent in 2014. The prices for dozens of drugs that treat everything from blood pressure to multiple sclerosis have doubled since 2007, and the prices of some generic drugs have soared by 1,000 percent or more. Drug prices keep rising, despite widespread complaints, because Big Pharma can charge whatever it wants (Jaret, 2015; Love, 2017; National Center for Health Statistics, 2017).

The United States and New Zealand are the only developed countries that allow direct-to-consumer prescription drug marketing ("Ask your doctor about…"). The ads, which began in the 1980s, increase a drug's sales. People who see drug advertising are more likely to ask their doctors for a particular brand-name medication. Doctors may feel pressure to accommodate their patients' requests, which, in turn, increases a drug's price and sales (Timmermans and Oh, 2010; Niederdeppe et al., 2013).

medical-industrial complex a network of business enterprises that influences medicine and health care.

End-of-life care is a billion dollar industry. Compared with their nonprofit counterparts, for-profit hospices spend less on nursing per patient, provide fewer services, and push out patients whose care becomes expensive. Medicare provides about 90 percent of funding for hospices and nursing homes, but rarely punishes either for violating health and safety rules (Hallman and Shifflett, 2014; Whoriskey and Keating, 2014).

HEALTH, PROFIT, AND WASTE

From a conflict perspective, health care is big business. The median household income in 2015 was $56,515—what the average health care CEO made in less than a day. Much of a CEO's earnings (nearly $19 million in 2015) are based on increasing stock prices rather than improving patient care. That means selling more drugs, raising prices above inflation, performing more tests and procedures, and getting more people into the hospital (Herman, 2017).

Large hospitals are crowded with "senior VPs, VPs of this, that, and the other" whose salaries are two to three times higher than those of surgeons. A Wisconsin surgeon discovered that a brief outpatient appendectomy he had performed for a fee of $1,700 generated over $12,000 in hospital bills. Thus, "the biggest bucks are earned not through the delivery of care, but from overseeing the business of medicine" (Healthcare-NOW! 2014; Rosenthal, 2014: SR4).

The medical-industrial complex profits in other ways. Some popular diet and other supplements contain dangerous chemicals that pose health risks. A number of the FDA's top officials are former lobbyists or industry leaders. In other cases, sugar and food industries fund doctors who publish articles in prominent medical journals that play down the link between sugar and heart disease (O'Connor, 2015; Kearns et al., 2016).

At least 30 percent of federally funded health care is wasteful. "Improper" payments (e.g., to ineligible beneficiaries, to dead people, and for unallowable services) totaled $144 billion in 2016, a 250 percent increase since 2003. Between 10 and 20 percent of all surgeries are unnecessary. Doctors often use aggressive and expensive treatment in the last month of life for three-quarters of patients younger than 65 with terminal cancer. Patients and families may insist on such treatment, not realizing that it'll bring "nothing but emotional and physical misery," because physicians, nurses, and other professionals aren't trained in end-of-life conversations (Chen et al., 2016; "Uncle Sam Overpays," 2017).

Other wasteful spending includes the expensive services that have the same health outcomes as lower-priced alternatives, preventable hospital injuries and readmissions, unnecessary tests and treatments to guard against liability in malpractice lawsuits, high prices for drugs, and prescriptions for drugs that work no better than a placebo. Huge agencies like Medicare and Medicaid are particularly vulnerable to waste. The private contractors employed to pay the government's millions of health care bills are under pressure to process claims as quickly and inexpensively as possible. "The cheapest way to process a claim is to pay it without question" (Rosenberg, 2016; Office of Inspector General, 2017; "Uncle Sam Overpays," 2017).

CRITICAL EVALUATION

The most common criticism is that conflict theorists often overlook the contributions of health care systems. Without them, people would suffer more, die at a younger age, and have a lower quality of life.

Second, doctors, medical scientists, and some medical associations (e.g., American Academy of Family Physicians) have been among the most vocal critics of unneeded surgery, tests, and procedures. Examples include cardiac stress tests and recommending vitamins and other supplements that don't benefit health. Others have urged ending expensive end-of-life care that simply prolongs the dying process. Most recently, nearly 120 of the nation's leading oncologists have publicly protested drug companies' "out of control greed" which is bankrupting cancer patients (Guallar et al., 2013; Agnvall, 2014; Tefferi et al., 2015).

Finally, most people aren't simply victims of a malicious medical-industrial complex. Many make unhealthy lifestyle choices (unprotected sex, drug abuse), and complain rather than change their behavior when employers impose health insurance penalties for smoking

Is Being Fashionable Hazardous to Women's Health?

Squeezing into skinny jeans and tight pants can cause nerve compression, numbness, and digestive problems. Narrow-toed shoes and stilettos wreak havoc: blisters, bunions, corns, calluses, hammertoes, nerve damage, back and neck pain, painful inflammation, stress fractures, and ankle sprains. Still, according to a podiatrist, "Women will wear their high-heeled shoes until their feet are bloody stumps" (Ianzito, 2013: 3; Gleiber, 2014).

Many cosmetics contain chemicals that can trigger skin problems (e.g., itching, rashes, acne). Others—contaminated with bacteria, yeasts, or molds—can lead to a range of problems from simple rashes to serious infections that can cause swelling and breathing difficulties. A recent study that tested 32 commonly sold lipsticks and lip glosses found that they contained lead, cadmium, chromium, aluminum, and five other metals—some at potentially toxic levels that are linked to stomach tumors. Some hair products contain hazardous ingredients like formaldehyde and mercury that can result in losing large clumps of hair (Liu et al., 2013; Lipton and Abrams, 2016).

Except for some hair color additives, the FDA doesn't require companies to test products for safety or list toxic ingredients. Japan, Canada, and the European Union have banned more than 500 cosmetic products containing toxic ingredients that are still sold in the United States (Rano and Houlihan, 2012). Skin Deep (ewg.org/skindeep) ranks the safety of a range of cosmetic products.

Maron/Age Fotostock

and obesity. Others join forces with drug companies to lobby the FDA to decrease current restrictions on potent opioids (Meier and Lipton, 2013; Brody, 2015).

14-4c Feminist Theories: Health Care and Medicine Benefit Men More Than Women

Both feminist and conflict theorists emphasize the connection between health and inequality. Feminist scholars go further by addressing the health costs of being a woman, gender stratification in medicine and health care, and men's control over women's choices.

THE HEALTH COSTS OF BEING A WOMAN

Women and men face different health care issues. For example, women are more likely than men to

- get inaccurate results from treadmill stress tests to detect heart problems, because the scoring system is based on "only middle-aged men" (Cremer et al., 2017);
- have chronic pain, be prescribed pain relievers, receive higher doses, use opioids for longer periods, become addicted, and, early in a pregnancy, risk birth defects and preterm births (Ailes et al., 2015; American Society of Addiction Medicine, 2016);
- be unpaid family caregivers, even if they're ill, disabled, or old (Hooyman, 2016); and
- suffer a fatal heart attack, because they often receive less aggressive treatment than men (Sagon, 2017).

For decades, heart disease "has been studied in men, for men, and by men." Even though heart disease is the leading cause of women's death, and kills more women than men, only 35 percent of participants in heart-related studies are women. Such research gaps "leave women's health to chance" (Johnson et al., 2014; Merz, 2015: A16).

During their reproductive years, 33 percent of U.S. women have a cesarean section (C-section), up from only 5 percent in 1970. The main reason for the increase is hospital administrators' and maternity clinicians' fear of lawsuits if a vaginal birth has a negative outcome (Morris, 2014). Because C-sections are major surgical procedures, the risks include infection, blood clots, and injury to other organs, particularly the bladder.

Table 14.5 Women Earn Less Than Men in Health Care Occupations, 2016

	Men	Women	Percentage of Women in This Occupation
All health care practitioners	**$69,264**	**$53,872**	**75**
Pharmacists	$108,992	$95,628	60
Physicians and surgeons	$121,836	$76,752	38
Physical therapists	$70,096	$67,912	65
Registered nurses	$65,572	$59,436	89
Clinical laboratory technologists and technicians	$55,692	$43,368	67
Emergency medical technicians and paramedics	$42,692	$33,800	34

Note: The data are median annual earnings of full-time workers.

Source: Based on Bureau of Labor Statistics, 2017, Table 39.

U.S. maternal mortality rates have increased. American women are now three to four times as likely to die from pregnancy-related complications as their counterparts in Britain, the Czech Republic, Germany, Greece, Finland, Italy, or Japan, and Black women's maternal mortality rates are more than three times higher than those of White women. The upturn is probably due to increasingly common health conditions (e.g., obesity, hypertension, diabetes) that make delivery more dangerous ("Maternal Mortality," 2015; "Pregnancy Mortality Surveillance System," 2017).

GENDER STRATIFICATION IN HEALTH CARE OCCUPATIONS

In health care jobs, men consistently earn more than women (*Table 14.5*). Among physical therapists, for example, males earn almost $71,000 a year compared with $68,000 for females even when both have the same level of education, years of experience, and work responsibilities (see also Chapters 8 and 11).

Feminist scholars attribute much of the gender wage gap to male doctors' gatekeeping. Because many registered nurses, pharmacists, and physical therapists (occupations that have more women than men) now receive doctoral degrees, they want to use the honorific title of "doctor." Doing so would win more respect from patients, help women land top administrative jobs that pay more, provide more autonomy in treating patients, and bring higher fees from health insurers. Physicians oppose the idea for several reasons: They want to maintain their prestige in the health care industry; they treat patients first, whereas nurses and others play only secondary roles; and they have considerably more education and training to diagnose and treat illness and disease (Harris, 2011).

MEN'S CONTROL OF WOMEN'S HEALTH

Feminist scholars maintain that men control many aspects of women's health. Catholic bishops, all of whom are men, condemn contraceptives as sinful, even though 98 percent of U.S. Catholic women have used birth control (Jones and Dreweke, 2011).

In 2014, the Supreme Court ruled, 5-4 (*Burwell v. Hobby Lobby Stores*), that privately held for-profit corporations don't have to pay for their employees' contraceptive coverage, as the ACA requires, if doing so conflicts with the owners' religious beliefs. In effect, a religious objection can trump a federal law, and businesses can shift insurance costs to taxpayers or private insurers. According to some Johns Hopkins doctors, "The Supreme Court decision has started us down a dangerous path on which the religious beliefs of a third party enter the examination room and interfere with the doctor-patient relationship" (Singal et al., 2014: 17; see also Rosenfeld, 2014). All of the justices who ruled against women's contraceptive coverage were males.

In many states, conservative lawmakers have passed laws to restrict or block women's ability to get abortions (see Chapter 9). They aren't passing similar laws to limit the availability of Viagra and other erectile dysfunction medications for men because "Viagra is a wonderful drug" (Beadle, 2012). The Hobby Lobby chain doesn't have to pay for IUDs and emergency contraceptive pills, but covers Viagra and other erectile dysfunction drugs. The White House recently announced that it's cutting a $214 million teen pregnancy prevention program that the Obama administration funded. President Trump's chief of staff, who oversees adolescent health, maintains that abstinence is the most effective approach in preventing unwanted teen pregnancy (Belluck, 2017).

CRITICAL EVALUATION

Feminist theories are limited in several ways. First, they sometimes gloss over the fact that social class, rather than gender, has a big effect on people's health and receiving health care services (Montez and Zajacova, 2013).

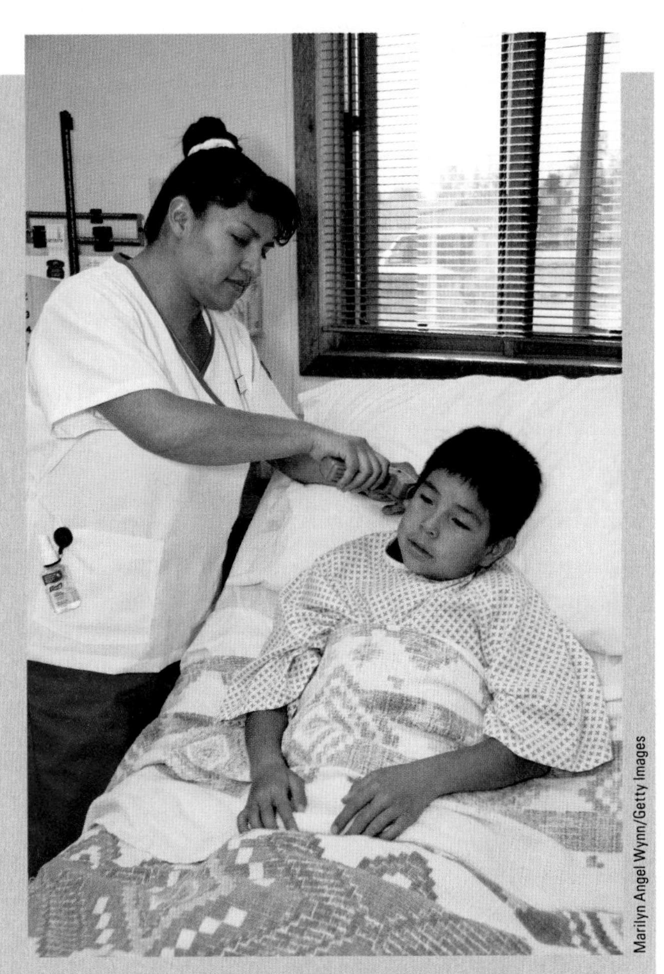

Marilyn Angel Wynn/Getty Images

Nurse practitioners (NPs), 92 percent of whom are women, complete a master's or doctoral program that includes diagnosing and treating patients, managing acute and chronic illnesses, and prescribing medications. Because of a projected shortage of doctors, NPs are pressing 34 states to relax laws requiring doctors to oversee their work. Many doctors' groups, led by the AMA, are fighting the NPs, arguing that less supervision would jeopardize patient safety. Several national studies have found no evidence that NPs provide lower-quality care than physicians (Beck, 2013; Pettypiece, 2013). How might you explain the AMA's resistance from a feminist perspective?

Second, men also experience gender bias in health care. They suffer one-third of all hip fractures but, because osteoporosis is considered a "woman's disease," there's little research on men, and guidelines about testing older men's bone density are vague (Sagon, 2017).

Third, and like conflict theorists, feminist scholars rarely address lifestyle choices that affect people's health. Among adult women, for example, 14 percent smoke, 67 percent are overweight or obese, and at least 8 percent use illicit drugs (National Center for Health Statistics, 2017). Finally, women aren't always passive but stand up to men's domination of health laws. Recently, for example, a Republican-controlled Senate failed to repeal Obamacare because two Republican women voted "no," and despite "intense pressure" from Republican male leaders (Leonhardt, 2017: A23).

14-4d Symbolic Interaction: The Social Construction of Health and Illness

Symbolic interactionists focus on how we define and construct views about health, illness, and medicine, and then implement these definitions in everyday life. Social constructions include labeling, stigmatizing behavior, and medicalizing attitudes and behaviors as normal or sick.

THE SOCIAL CONSTRUCTION OF ILLNESS

Medical models assume that illness is anything that deviates from normal biological functioning, and that an illness has specific features that a doctor can recognize. In contrast, interactionists view illness and medicine as social constructions that can change over time. In 1956, the AMA declared that alcoholism is a treatable disease that's due to a genetic predisposition rather than a person's lacking moral character or self-discipline. And, in 2013, the AMA officially designated obesity as a disease—rather than a lifestyle habit—that requires research funding and medical intervention, including surgery and drugs (Newberry, 2013; Pollack, 2013).

With or without medical intervention, individuals construct and manage their illnesses differently. Some people's social worlds shrink when they become immersed in the day-to-day aspects of managing a chronic illness like rheumatoid arthritis: They become increasingly cut off from everyday routines if they're unable to work, spend less time with family and friends, or can't move about freely. Others create new self-identities by describing themselves as survivors (of breast cancer, for instance), exchange information about treatment options, and participate in local or national fundraising events (Conrad and Barker, 2010).

Social class also affects social constructions of illness. As educational level rises, people report poorer health, presumably because they're more aware of medical knowledge and evaluate their own physical well-being more critically (Schnittker, 2009).

THE MEDICALIZATION OF HEALTH

In 1952, the American Psychiatric Association (APA) published the first edition of the *Diagnostic and Statistical Manual of Mental Disorders* (DSM), which has become the global "bible of mental illness." The DSM relies on subjective definitions and criteria, but is influential in labeling (or unlabeling) mental illness. The number of disorders increased from 106 in 1952 to nearly 400 in 2013. Some of the new mental illnesses include "binge eating" (frequent overeating), "bereavement" (mourning the loss of a loved one), "caffeine intoxication" (drinking too much coffee), "disruptive mood dysregulation disorder" (children's temper tantrums), and "hoarding" (accumulating too much stuff) (American Psychiatric Association, 2013).

Anne Kitzman/Shutterstock.com

The DSM's ever-changing diagnoses and labels are an example of **medicalization**, a process that defines a nonmedical condition or behavior as an illness, disorder, or disease that requires medical treatment. The APA once classified "homosexuality" as a psychological disorder, but dropped sexual orientation from its roster of mental illnesses in 1973. The DSM affects patients, doctors, insurers, pharmaceutical companies, and taxpayers because psychiatrists and other physicians use it to bill insurance companies. DSM-based diagnoses also determine whether people get special services at school, qualify for disability benefits, are stigmatized, and even whether they can adopt children.

Medicalization is a lucrative business. Pharmaceutical corporations have reaped enormous profits because everyday normal anxieties, discomforts, and stresses (frustration with traffic, boredom with routine housekeeping chores, feeling sad or personally insecure) can be "fixed" by popping a Prozac or other pill (Herzberg, 2009). The more behaviors that the DSM defines as mental illness, the more likely psychiatrists are to increase their number of patients. Mental rights advocates and parents also benefit from medicalization, as when children who are diagnosed with attention-deficit hyperactivity disorder (ADHD) get insurance coverage or special treatment, including more attention from teachers and health specialists (Conrad and Barker, 2010; Rochman, 2012).

According to critics, the DSM encourages misdiagnosis, overdiagnosis, the medicalization of normal behavior, and the prescription of a large number of unnecessary drugs, especially in the case of ADHD. In 2014, more than 10,000 2- to 3-year-olds (that's right, 2- to 3-year-olds!), particularly those on Medicaid, were being medicated for ADHD. Between ages 2 and 5, 44 percent of children diagnosed with ADHD take medications (Schwarz, 2014; Danielson et al., 2017).

medicalization a process that defines a nonmedical condition or behavior as an illness, disorder, or disease that requires medical treatment.

Attitudes vary by country, but many European parents, teachers, and doctors are reluctant to use ADHD medication for what they consider routine childhood behavioral problems. Even though some Europeans believe that ADHD is a disease, "parents are loath to get their child labeled" (Kelley, 2013: 26). Thus, America's high ADHD rates and drug sales illustrate cultural differences in the medicalization of health and illness.

LABELING AND STIGMA

From an interactionist perspective, conditions and behaviors that are diagnosed and labeled as illness or disease change over time, differ among groups, and vary across countries. Medicalization and labeling stigmatize some illnesses and diseases more than others. Male impotence, which commonly increases as men age and was stigmatized, is now called "erectile dysfunction," and treated with drugs, regardless of age, to enhance a man's sexual experience. Ads and commercials ("Cialis is ready when you are") have flooded magazines, newspapers, and online sites.

Women's sexuality is also being medicalized. The FDA recently approved the sale of flibanserin to increase postmenopausal women's sex drive. In clinical trials, only 10 percent of women said that the pill increased their sexual desire. Flibanserin has dangerous side effects, like fainting, because it interacts with other medications and alcohol. Women's sex drives fluctuate throughout life, diminish due to bad relationships or stress, and there's nothing wrong with low sexual desire (Nagoski, 2015; Weisman, 2015; see also Chapter 9). Pharmaceutical companies, however, can reap huge profits by convincing women that their sexuality is abnormal.

Labeling and stigma can also prevent seeking medical help. Because of fatigue, feeling overwhelmed, and sleep deprivation, almost 30 percent of resident physicians (doctors still being trained) have either severe symptoms or a diagnosis of depression. Few seek medical help, however, because admitting such problems carries a stigma that might affect obtaining a medical license (Mata et al., 2015; Hill, 2017).

CRITICAL EVALUATION

Symbolic interaction theories are limited for several reasons. The most common criticism is that they don't address structural factors such as government policies and practices that can ensure better health for everyone. For example, Black people experience much poorer health than White people because of racial differences in access to health care, neighborhood poverty, and other macro-level factors. Public policies could reduce such disparities, but haven't done so (Johnson, 2017).

Second, millions of people successfully resist medicalization. Many deaf people have ignored recommendations to get cochlear implants (devices that can increase hearing). They don't see deafness as a medical disability but as a social reality that helps them form a community and identify with other deaf people (Conrad and Barker, 2010). A "fat pride" movement argues that obesity's health risks are exaggerated and focuses, instead, on changing discrimination against overweight people (see Chapter 16).

Third, interactionists emphasize that health and illness are social constructions, but there are many serious health problems whether or not they're medicalized. People die from unknown diseases, for example. Others suffer from "contested" illnesses, including chronic fatigue syndrome, that many physicians question because the symptoms are difficult to diagnose and treat (Institute of Medicine, 2015).

Tom Williams/Getty Images

How do authority figures affect people's health perceptions and behavior? Dr. Mehmet Oz is a highly respected and published cardiothoracic surgeon, and host of the popular *The Dr. Oz Show*. Products he endorses on his show "are almost guaranteed to fly off the shelves" (Weathers, 2014). A study of 40 randomly selected episodes of *The Dr. Oz Show* found that scientific data supported only 46 percent of Dr. Oz's recommendations. The researchers cautioned consumers to be skeptical about any information on television medical talk shows because only "up to one half of recommendations are based on believable or somewhat believable evidence" (Korownyk et al., 2014: 4).

STUDY TOOLS 14

READY TO STUDY? IN THE BOOK, YOU CAN:

- ☐ Check your understanding of what you've read with the Test Your Learning Questions provided on the Chapter Review Card at the back of the book.
- ☐ Tear out the Chapter Review Card for a handy summary of the chapter and key terms.

ONLINE AT CENGAGEBRAIN.COM WITHIN MINDTAP YOU CAN:

- ☐ Explore: Develop your sociological imagination by considering the experiences of others. Make critical decisions and evaluate the data that shape this social experience.
- ☐ Analyze: Critically examine your basic assumptions and compare your views on social phenomena to those of your classmates and other MindTap users. Assess your ability to draw connections between social data and theoretical concepts.
- ☐ Create: Produce a video demonstrating connections between your own life and larger sociological concepts.
- ☐ Collaborate: Join your classmates to create a capstone project.

15 Population, Urbanization, and the Environment

US Army Photo/Alamy Stock Photo

LEARNING OUTCOMES

After studying this chapter, you will be able to…

15-1 Explain how and why populations change, and evaluate the population growth theories.

15-2 Describe how and explain why global and U.S. cities are changing, describe the consequences of urbanization, and compare and evaluate the theoretical explanations of urbanization.

15-3 Describe and illustrate the major environmental issues, and discuss whether sustainable development is achievable.

After finishing this chapter go to **PAGE 325** for **STUDY TOOLS**

In mid-2017, the world hit a population milestone of nearly 7.6 billion people, and is projected to increase to 9.8 billion by 2050. Our planet has a record number of inhabitants, and many are living longer than ever before. This chapter examines how such changes affect the world's population, urbanization, and environment. Let's begin with population.

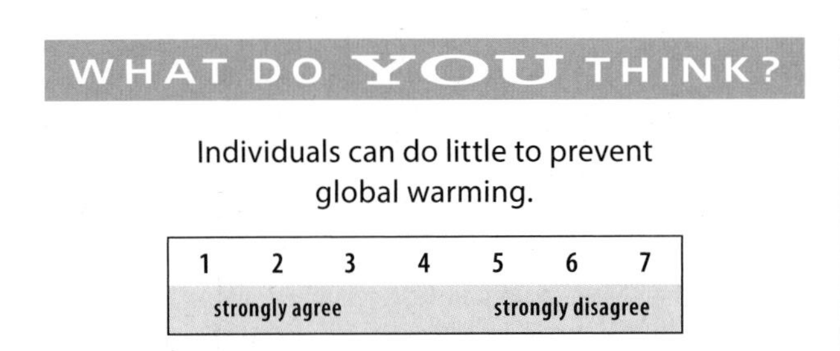

15-1 POPULATION CHANGES

Global population has grown rapidly since 1800, and more than tripled since 1900 (*Figure 15.1*). By 2100, the world population will climb to 11.2 billion, and nearly all of the growth will occur in six developing African countries (United Nations Population Division, 2017).

Figure 15.1 World Population Growth Through History

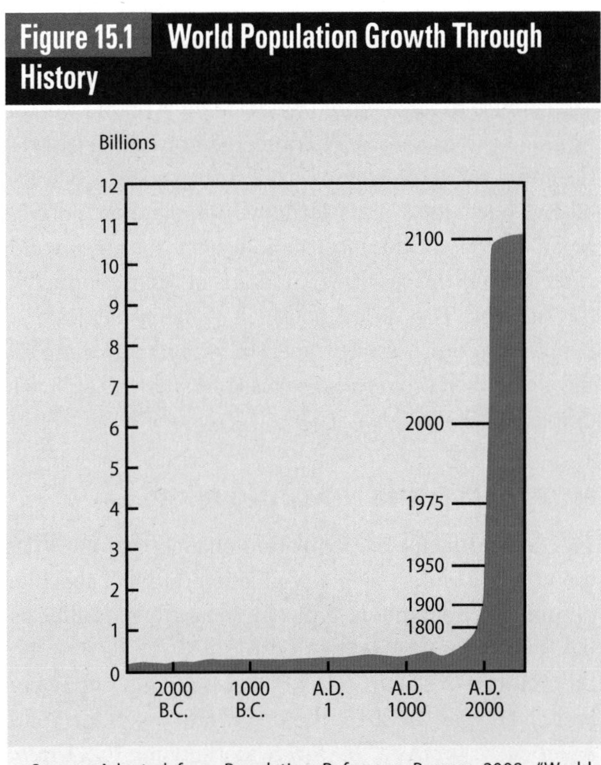

Source: Adapted from Population Reference Bureau, 2008, "World Population Projections to 2100," accessed at www.prb.org.

The world adds about 400,000 people each day, but population changes vary across countries. By 2050, India, with nearly 1.7 billion people, will surpass China as the world's largest country, Nigeria will replace the United States as the third largest country, and three African nations will be among the most populous (*Table 15.1*).

Information about these and other changes comes from **demography**, the scientific study of human populations. A **population** is a group of people who share a geographic territory. A territory can be as small as a town or as vast as the planet, depending on a researcher's focus. Demographers analyze populations in terms of size, composition, distribution, and change. They also study personally relevant topics, including your probability of getting married or divorced, the kind of job you'll probably have, how many times you'll move, and how long you'll probably live (McFalls, 2007).

15-1a Why Populations Change

Populations are never static. Their growth or decline involves three key factors: fertility (how many people are born), mortality (how many die), and migration (how many move from one area to another).

FERTILITY: ADDING NEW PEOPLE

The study of population changes begins with **fertility**, the number of babies born during a specified period in a particular society. Demographers use several fertility measures depending on the level of specificity needed. One of the most commonly used measures is the **crude birth rate** (also called the *birth rate*), the number of live births for every 1,000 people in a population in a given year. "Crude" implies that the rate is a general measure of a society's childbearing. It's based on the total

demography the scientific study of human populations.

population people who share a geographic territory.

fertility the number of babies born during a specified period in a particular society.

crude birth rate (also called the *birth rate*) the number of live births per 1,000 people in a population in a given year.

Table 15.1 World's Ten Most Populous Countries, 2017 and 2050

2017			2050		
	Country	Population (In Millions)		Country	Population (In Millions)
1	China	1,410	1	India	1,659
2	India	1,339	2	China	1,364
3	United States	324	3	Nigeria	411
4	Indonesia	264	4	United States	390
5	Brazil	209	5	Indonesia	322
6	Pakistan	197	6	Pakistan	307
7	Nigeria	191	7	Brazil	233
8	Bangladesh	165	8	Bangladesh	202
9	Russia	144	9	Democratic Republic of Congo	197
10	Mexico	129	10	Ethiopia	191

Source: Based on United Nations Population Division, 2017, Table S.3.

population rather than more specific measures (e.g., a woman's age or marital status). In 2016, the crude birth rate was 20 worldwide, 36 for Africa, 12 for the United States, and 11 for Europe (Kaneda and Bietsch, 2016).

Another standard measure, the **total fertility rate (TFR)**, is the average number of children born to a woman during her lifetime. In the early 1900s, U.S. women had an average TFR of 3.5 compared with 2.5 in 2016. Worldwide, TFRs are much higher in the least developed nations (4.3) than in developed nations (1.7), ranging from a low of under 1.2 in some Asian and European countries to 6.0 and higher in 9 African nations (Kaneda and Bietsch, 2016). The U.S. population has increased primarily because of high TFRs among foreign-born women. In 2014, there were 58.3 births per 1,000 U.S.–born women ages 15 to 44, compared with 84.2 births for foreign-born mothers (Bialik, 2017).

total fertility rate (TFR) the average number of children born to a woman during her lifetime.

mortality the number of deaths in a population during a specified period.

crude death rate (also called the *death rate*) the number of deaths per 1,000 people in a population in a given year.

To maintain a stable population, a woman must have an average of two children, the *replacement rate* for herself and her partner. TFRs above 2.1 indicate that a country's population is increasing and getting younger. Rates below 2.1 mean that a country's population is decreasing and getting older. The older a country's population, the more difficult it is to support people who are living into their eighties and longer and to provide resources for the young (United Nations Population Division, 2017; see also Chapter 12 on the old-age dependency ratio).

In nearly all countries, fertility tends to fall as people grow richer and women are better educated. Some nations, particularly those in East Asia, worry that their "ultra low" TFRs of 1.3 or less aren't at replacement levels. In Tokyo, Seoul, Bangkok, and other Asian cities, female college graduates outnumber males and female labor force participation is high. According to cultural norms, however, women should give up work after the birth of a child. And, employed or not, women do at least three more hours of housework a day than men. Consequently, many women aren't marrying or are postponing marriage, reducing their likelihood of ever having children. Because nonmarital births are taboo and rare in Asia, TFRs have plummeted ("Asia's New Family Values," 2015).

Italy and Japan have among the lowest fertility rates in the world (1.4 and 1.5, respectively). Italy's government has promoted a "Fertility Day" campaign to encourage couples to have more babies. Many working women would like to have more children, but don't do so because the government offers limited child care services. Because of low birth rates and restrictive immigration policies, Japan has one of the most rapidly aging societies worldwide. One of the results is hundreds of "ghost houses" in the suburbs. The vacant houses have been inherited by people who don't want or need them, but they can't sell the abandoned properties because of a shortage of buyers (Soble, 2015; Pianigiani, 2016).

MORTALITY: SUBTRACTING PEOPLE

The second factor in population change is **mortality**, the number of deaths in a population during a specified period. Demographers typically measure mortality using the **crude death rate** (also called the *death rate*), the number of deaths per 1,000 people in a population in a given year. In 2016, the death rate was 6 for South America, 7 for Asia, 8 worldwide and for the United States, 11 for Europe, but 14 or higher for 8 African nations (Kaneda and Bietsch, 2016).

AP Images/John Heilprin

Agencja Fotograficzna Caro/Alamy Stock Photo

Africa's Niger and South Sudan have the world's highest TFRs, 7.6 and 6.7, respectively. The reasons for the high fertility include poverty and women's limited access to education and birth control. In contrast, some European towns are facing "slow death" due to low fertility and immigration rates. Germany is trying to boost its shrinking TFR (1.5 compared with 2.5 in the 1960s) by offering families tax breaks and free day care for all children age 1 and older (Rossi and Jucca, 2014; Kaneda and Bietsch, 2016).

Developed countries, compared with most developing countries, have lower death rates, but also large proportions of people age 65 and older who are no longer employed but require costly medical services. Thus, a better measure of a population's health is the *infant mortality rate*, the number of deaths of infants younger than 1 year per 1,000 live births. Generally, as the standard of living improves—access to clean water, adequate sanitation, and medical care—the infant mortality rate decreases. In 2016, the infant mortality rate was 5 in developed countries, 59 in the least developed countries, and 85 or higher in six African nations (Kaneda and Bietsch, 2016).

Life expectancy is the average number of years that people who were born at about the same time can expect to live. Globally, life expectancy is 70 for males and 74 for females. Again, however, there are considerable variations across countries—from a high of 84 in Hong Kong and Japan to a low of 44 in several African nations. The United States, with a life expectancy of 79, ranks below at least 25 other industrialized countries and only slightly higher than less developed countries such as Cuba and Uruguay (Kaneda and Bietsch, 2016). In almost every society, people with a higher socioeconomic status live longer and healthier lives because their occupations are physically safe, and their resources include clean drinking water, sanitation, and medical services.

MIGRATION: ADDING AND SUBTRACTING PEOPLE

The third demographic factor in population change is **migration**, the movement of people into or out of a specific geographic area. There are two types of migration: international and internal. *International migration*, movement to another country, includes *emigrants* (people who move out of a country) and *immigrants* (people who move into a country).

Migrants are people who choose to leave their country to seek a better life elsewhere. *Refugees* (sometimes called *displaced persons*) are people who are forced to flee their country—often with no warning—to escape war, persecution, torture, or death. An international migrant can have several of these characteristics. For example, my family members and I were refugees who fled Lithuania (by horse and wagon) from Communism during World War II, emigrated to displaced persons camps in Germany, and then immigrated to the United States.

International Migration. The number of international migrants grew from about 79 million in 1960 to nearly 250 million in 2015 (a 200 percent increase), and they make up more than 3 percent of the world's population today. The largest flows have moved from one developing country to another (as from Indonesia to Saudi Arabia) or from a developing to an industrialized country (as from Mexico to the United States). Also, most emigrants move to a neighboring country (from Mexico to the United States, for example, rather than from Mexico to Canada). The United States has 20 percent of the world's migrants (nearly 47 million), more than any other country in the world. Proportionately, however, only 14 percent of the U.S. population is foreign born, compared with 75 percent in several Persian Gulf countries

migration the movement of people into or out of a specific geographic area.

(e.g., United Arab Emirates, Kuwait), 28 percent in Australia, and 22 percent in Canada (Connor, 2016).

There are currently more refugees worldwide than at any time in recorded history. At the end of 2016, nearly 66 million people had been forcibly displaced from their homes because of conflict, violence, or human rights violations. More than half of all current global refugees come from just three countries—Syria, Afghanistan, and South Sudan (United Nations High Commissioner for Refugees, 2017). The most common reason for other international migration is economic, but also includes fleeing bad schools, high crime rates, and natural disasters. Thus, the reasons for international migration today are similar to those of European immigrants who came to the United States during the twentieth century.

Internal Migration. International migration has increased, but 92 percent of adults worldwide who change residences do so because of *internal migration*, movement within a country. Over a five-year period, 24 percent of U.S. adults move within the country, similar to rates reported in other advanced economies, including New Zealand, Finland, and Norway (Esipova et al., 2013). In parts of rural Spain, people have migrated to cities for jobs, education, and access to public transit and health care. As a result, abandoned villages are for sale. One village, where the asking price is about $230,000, includes 100 acres with half a dozen houses and two sprawling farms with room for 70 cattle. The biggest houses "have hardwood floors and five bedrooms overlooking an orchard with peaches, figs, walnuts, apples, and pears . . . and a little river full of trout" (Frayer, 2015).

Between 2015 and 2016, 11 percent of Americans moved to a different residence. Why do people move? The most common reasons (42 percent) are housing-related (e.g., "Wanted new or better home/apartment," eviction), 28 percent are family-related (e.g., change in marital status), and 20 percent are job-related (e.g., new job, job transfer, to look for work) ("Americans Moving . . . ," 2016).

Social class and natural disasters also affect migration. By 2015, there were more Puerto Ricans living stateside than on the island, primarily because many low-income residents sought better jobs and schools in the United States. After Hurricane Maria devastated much of the island in 2017, between 10 and 14 percent of Puerto Rico's 3.4 million people, including many young professionals, moved to the mainland or planned to do so (Dorell, 2017; Luscombe, 2017).

sex ratio the proportion of men to women in a population.

So far, more than 12 million Syrians (more than half of the population) have fled the civil war that began in 2011 or have been internally displaced from their homes (United Nations High Commissioner for Refugees, 2017). "Many have been on the run for a year or more... moving from village to village, up to as many as 20 times, before they finally made it across an international border" (Bengali, 2014: 10). The death of Syrian refugee Aylan Kurdi, 3, sparked an outburst of compassion worldwide.

15-1b Population Composition and Structure

Demographers study age and gender to understand a population's composition and structure. Two of the most common measures are *sex ratios* and *population pyramids*.

SEX RATIOS

A **sex ratio** is the proportion of men to women in a population. A sex ratio of 100 means that there are equal numbers of men and women; a ratio of 95 means that there are 95 men for every 100 women (fewer males than females). Worldwide, without human intervention, 103 to 107 boys are born for every 100 girls. In the United States, the sex ratio is 105 at birth, but skewed in many countries: 115 in China, 112 in Armenia and India, and 111 in Vietnam (*World Factbook*, 2017; World Population Review, 2017).

These sex ratio imbalances are attributed primarily to *female infanticide* (sometimes called *gendercide*)—the intentional killing of baby girls. In many Asian countries, including China and India, there's a preference for boys because males are expected to carry on the family name, care for elderly parents, inherit property, and play a central role in family rituals. Consequently, and particularly in rural areas, hundreds of thousands of female infants die every year because of neglect, abandonment, and starvation. Others are aborted after ultrasound scanners reveal the child's sex ("Sex Selection," 2017).

China is a good example of some of the unintended negative consequences of lopsided sex ratios. In 1979, China passed a one-child policy to combat poverty and overpopulation. The policy succeeded in limiting population growth because of at least 600 million abortions and sterilizations. The country's TFR fell from 6 in the late 1960s to 1.5 by 2010, but well below the 2.1 replacement rate to maintain a constant population (Eberstadt, 2013; "Family Planning...," 2014).

Because of the unbalanced sex ratio for the last 30 years, by 2020 China will have about 35 million more young men than women. That means a large number of men won't be able to find wives—a problem that has already increased the illegal trafficking of women from poorer neighboring countries, including Cambodia, Myanmar, and Vietnam—and there'll be millions of aging bachelors (Tsai, 2012; "The Marriage Squeeze...," 2015).

To reduce the burden of elder care, the government eased the policy in 2013 and in 2015 by allowing married couples to have two children. The urban, educated middle class—the group that China's leaders want to see increase its family size—hasn't shown much interest in doing so because of high costs of living and expenses in raising a second child. China's working-age population started declining in 2015, resulting in fewer workers to support a growing older population (Buckley, 2015; Walsh, 2015).

POPULATION PYRAMIDS

A **population pyramid** is a graphic depiction of a population's age and sex distribution at a given point in time. As *Figure 15.2* shows, Mexico is a young country: Much of its population is under age 45 (which also means that many women are in their childbearing years), and there are relatively few people 65 years and older. In contrast, Italy is an old country, and the United States is somewhere in the middle.

The shape of the pyramid (a triangle for Mexico, a rectangle for the United States, and a diamond for Italy) has future implications for young and old countries. Italy has a relatively small number of women ages 15 to 44 (in their reproductive years) and a bulge of people ages 45 to 79. This suggests that there'll be a scarcity of workers to support an aging population and a greater need for social services for older people than for children and adolescents (Kochhar and Oates, 2014; see also Chapter 12). Thus, population pyramids give us a snapshot of a country's demographic profile and indicate some of the problems that countries are likely to face in the future.

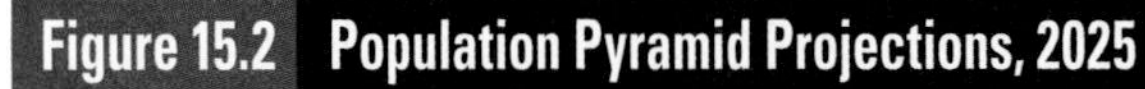

Figure 15.2 Population Pyramid Projections, 2025

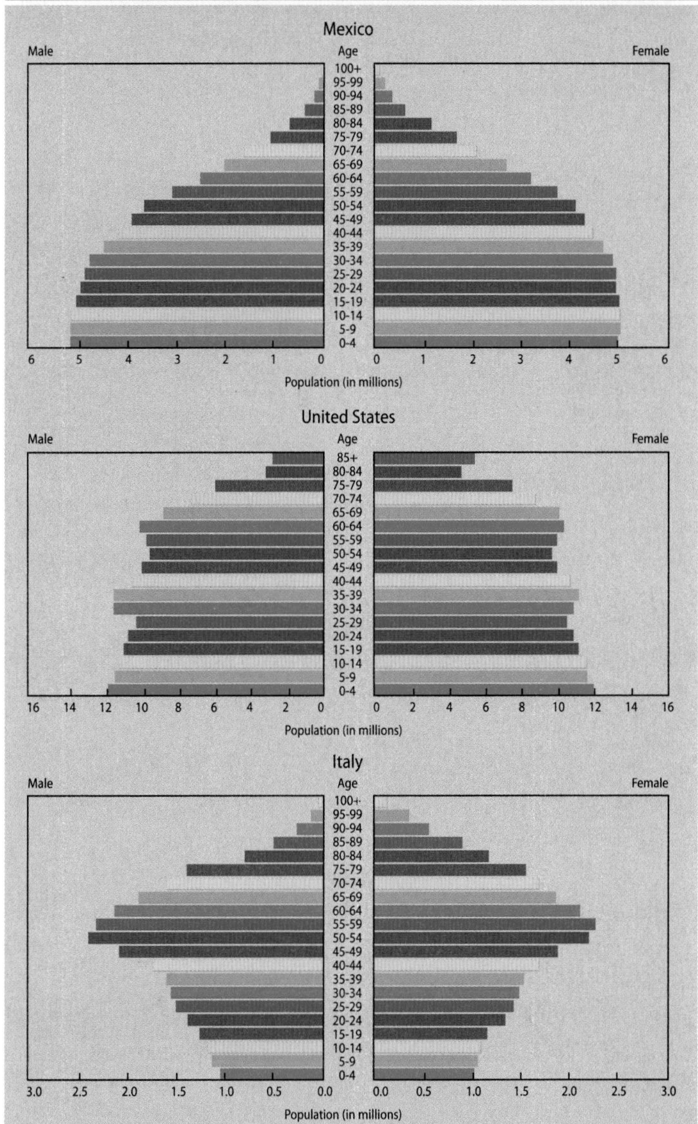

Source: U.S. Census Bureau, International Database, www.census.gov/ipc/www/idb/pyramids.html. Accessed January 20, 2007.

15-1c Population Growth: A Ticking Bomb?

Some of the world's largest countries, many of them in the developing world, will grow even more by 2050 (*Table 15.1* on page 304). So, has population growth gotten out of hand? There are many views on this question, but two of the most influential have been Malthusian theory (which argues that the world can't sustain its unprecedented population surge), and demographic transition theory (which maintains that population growth is slowing).

population pyramid a graphic depiction of a population's age and sex distribution at a given point in time.

MALTHUSIAN THEORY

For many demographers, population growth is a ticking bomb. They subscribe to **Malthusian theory**, which maintains that the population is growing faster than the food supply needed to sustain it. This theory is named after Thomas Malthus (1766–1834), an English economist, clergyman, and college professor who argued that humans are multiplying faster than the earth's ability to produce sufficient food.

Thomas Malthus

Hulton Archive/Stringer/Getty images

According to Malthus (1798/1965), population grows at a *geometric rate* (2, 4, 8, and so on), whereas the food supply grows at an *arithmetic rate* (1, 2, 3, 4, and so on). That is, two parents can have 4 children and 16 grandchildren within 50 years. The available number of acres of land, farm animals, and other sources of food can increase in that time period, but certainly not quadruple. In effect, then, the food supply will not keep up with population growth. Because there are millions of parents, the results could be catastrophic, with masses of people living in poverty or dying of starvation.

Malthus posited that two types of checks affect population size. *Positive checks* (famine, disease, war) limit reproduction, raise the death rate, and lower the overall population. *Preventive checks* (contraception, postponing marriage, abortion, and extramarital sex) also limit reproduction by reducing birth rates and, consequently, ensure a higher standard of living for all (Malthus, 1872/1991).

Malthusian theory has had a lasting influence. *Neo-Malthusians* (or New Malthusians) agree that the population is exploding beyond food supplies. The world's population reached its first billion in 1800. In the 200 years that followed, the world added another 5 billion people (*Figure 15.1* on page 303). As a result of this growth, according to some influential neo-Malthusians, the earth has become a "dying planet"—a world with insufficient food and a rapidly expanding population that pollutes the environment (Ehrlich, 1971; Ehrlich and Ehrlich, 2008).

Malthusian theory the population is growing faster than the food supply needed to sustain it.

demographic transition theory population growth is kept in check and stabilizes as countries experience economic and technological development.

An estimated one in nine people in the world suffer from chronic hunger that prevents them from working and having a normal life. We already produce enough food for 14 billion people, far more than we'll ever need to feed the projected world population of nearly 10 billion in 2050. Hunger is due to poverty and inequality, not scarcity: People can't afford to buy food or the land to grow it. Internal wars also restrict distribution, fueling crises in the countries that need food the most (Druker, 2015; World Food Programme, 2017).

DEMOGRAPHIC TRANSITION THEORY

Some demographers are more optimistic than neo-Malthusians. **Demographic transition theory** maintains that population growth is kept in check and stabilizes as countries experience economic and technological development, which, in turn, affects birth and death rates. According to this theory, population growth changes as societies undergo industrialization, modernization, technological progress, and urbanization. During these processes, a nation goes through four stages (*Figure 15.3*), from high birth and death rates to low birth

Figure 15.3 The Classical Demographic Transition Model

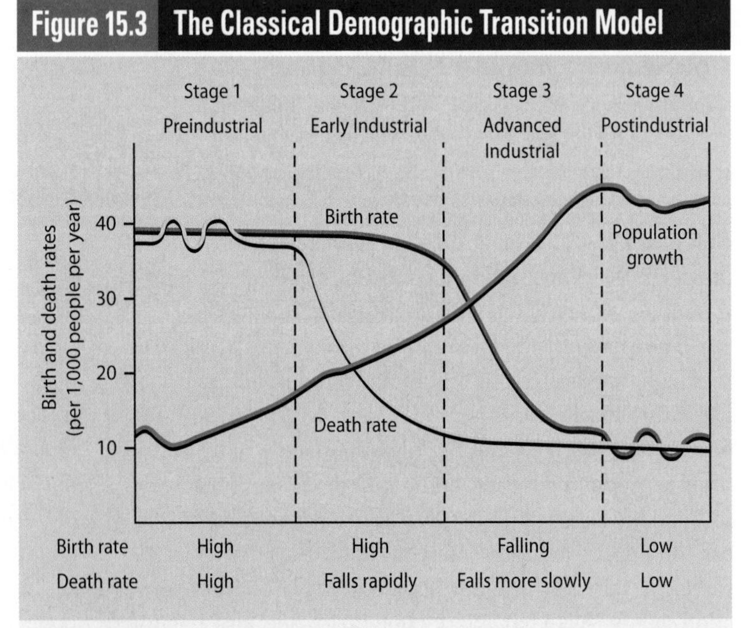

Source: U.S. Census Bureau, International Database, www.census.gov/ipc/www/idb/pyramids.html (accessed January 20, 2007); Max Roser, ourworldindata.org (accessed August 19, 2017).

Sean Gallup/Getty Images

A number of countries, particularly in Europe and parts of Asia, have below-replacement fertility rates. For some policy analysts, the recent influx of refugees and migrants is an opportunity to shore up labor forces with young workers whose taxes will support graying and dwindling native-born populations (OECD/European Union, 2015; Portes, 2015).

and death rates. Put simply, the greater the human development, the lower the fertility and death rates.

- **Stage 1: Preindustrial society.** In this initial stage, there's little population growth. The birth rate is high because people rarely use birth control: They want as many children as possible to provide unpaid agricultural labor and support parents in old age, but a high death rate offsets the high birth rate. No country exists in this current stage of the demographic transition.
- **Stage 2: Early industrial society.** There's significant population growth because the birth rate is higher than the death rate. The birth rate may even increase over what it was in Stage 1 because mothers and their children enjoy improved health care. Couples may still have large numbers of children because they fear that many of them will die, but the death rate declines because of better sanitation, better nutrition, and medical advances (e.g., immunizations and antibiotics). Most of the world's poorest countries—including sub-Saharan Africa, Afghanistan, and Yemen—are currently in Stage 2.
- **Stage 3: Advanced industrial society.** As the infant mortality rate declines, parents have fewer children. Effective birth control reduces family size. The decrease in child care responsibilities, in turn, enables women to work outside the home. Brazil, Mexico, and some Asian countries are currently in Stage 3.
- **Stage 4: Postindustrial society.** In this stage, the demographic transition is complete, and the society has low birth and death rates. Women tend to be well educated and to have full-time jobs or careers. If there's little immigration, the population may even decrease because the birth rate is low. This is the case today in Canada, Japan, Singapore, Hong Kong, Australia, New Zealand, the United States, and Europe.

CRITICAL EVALUATION

The dire predictions of Malthus and his successors that global population growth would lead to worldwide famine, disease, and poverty haven't come true. Still, 1.2 billion people (22 percent of the world's population) live in abject poverty, subsisting on less than $1.25 a day (see Chapter 8).

Despite neo-Malthusians' fears, global fertility is half of what it was in 1972. The population of some industrialized countries is declining because people aren't having enough babies to replace themselves. These countries are experiencing **zero population growth (ZPG)**, a stable population level that occurs when each woman has no more than two children.

Fearing that there won't be enough young workers to pay for social security and the rising cost of health care for aging populations, some low-birth nations with fertility rates below ZPG are paying women to have more children. Russia gives mothers with one child $12,500 for each additional baby; Japan has expanded its day care facilities and offers families a monthly allowance of $145 per child younger than 15; Germany and France have liberal parental leave; South Korea turns off its office lights once a month so people have more time to go home and multiply and offers other incentives, including cash gifts, for staff who produce more than two babies; China is encouraging married couples to have two children; and Singapore, besides giving couples almost $5,000 for having a child, lets families with babies go to the front of the queue for high quality, affordable public housing (Hales, 2014; Nechepurenko, 2014; "Pro-natalism," 2015).

Some neo-Malthusians maintain, however, that it's irresponsible for *any* country to encourage higher fertility rates. They worry about the consequences of adding 3 billion more inhabitants to the planet in less than 50 years, especially for many developing countries "with desperate economic outlooks" (Sachs, 2005; Shorto, 2008).

One result of population growth is urban growth. Cities attract people because of jobs and cultural activities, but urbanization has also created numerous problems.

zero population growth (ZPG) a stable population level when each woman has no more than two children.

North Wind/North Wind Picture Archives

During the Industrial Revolution, many cities, like this one in Hamburg, Germany, constructed canals that provided an inexpensive means of transportation and distributing goods.

15-2 URBANIZATION

If you've flown over the United States, you've probably noticed that people tend to cluster in and around cities. After sunset, some areas glow with lights, whereas others are engulfed in darkness. The average person, in the United States and worldwide, is more likely to live in a city than a rural area, and this trend is rising.

A *city* is a geographic area where a large number of people live relatively permanently and make a living primarily through nonagricultural activities. **Urbanization**, which increases the size of cities, is the movement of people from rural to urban areas. Most of this discussion focuses on U.S. cities, but let's begin with a brief look at global urbanization.

15-2a Urbanization: A Global View

In 2008, for the first time in history, a majority of the world's population lived in urban areas. By 2050, 66 percent of the world's population is projected to be urban (United Nations, 2014).

ORIGIN AND GROWTH OF CITIES

Cities are one of the most striking features of modern life, but they've existed for centuries. About 7,000 years ago, for example, people built small cities in the Middle East and Latin America to protect themselves from attackers and to increase trade. By 1800, 56 cities in Western Europe had a population of 40,000 or more (Chandler and Fox, 1974; De Long and Shleifer, 1992).

urbanization people's movement from rural to urban areas.

megacities metropolitan areas with at least 10 million inhabitants.

Before the Industrial Revolution, which began in the late eighteenth century, urban settlements in Europe, India, and China developed largely because people figured out how to use natural resources (e.g., mining coal and transporting water for irrigation and consumption). The Industrial Revolution spurred ever-increasing numbers of people to move to cities in search of jobs, schooling, and improved living conditions. As a result, the urban population surged—from 3 percent of the world's population in 1800 to 14 percent in 1900 (Sjoberg, 1960; Mumford, 1961). As industrialization advanced, urbanization increased.

WORLD URBANIZATION TRENDS

Between 1920 and 2014, the world's urban population increased from 270 million to 3.9 billion, and is expected to rise to 6.3 billion by 2050. Africa and Asia are home to nearly 90 percent of the world's rural population, but both are urbanizing faster than other regions (*Table 15.2*). Between 2014 and 2050, just three countries—India, China, and Nigeria—are expected to account for 37 percent of the world's urban population growth (United Nations, 2014).

Many of the world's largest cities are **megacities**, metropolitan areas with at least 10 million inhabitants. In 1950, the world had only two megacities: Tokyo (11.3 million) and New York-Newark (12.3 million). By 2030, there'll be 41 megacities. Only two of them will be in the United States—New York-Newark

Table 15.2 Urbanization Around the World

Percentage of People Living in Urban Areas	1950	2014	2050 (Projected)
World	**29**	**54**	**66**
Africa	14	40	56
Asia	18	48	64
Latin America and the Caribbean	41	80	86
North America	64	81	87
Europe	51	73	82
Oceania	62	71	74

Source: Based on United Nations Department of Economic and Social Affairs, 2012, Table 2, and United Nations, 2014, Table 1.

Qilai Shen/Bloomberg/Getty Images

China's Beijing, with nearly 21 million people, is the world's 11th largest megacity. Rentable bikes are popular, but bike lanes are often blocked by parked cars, can end without warning, and have a "menagerie of moving objects," including "delivery guys on motorcycles, tourists meandering in rickshaws, and construction crews hauling bricks in rusted carts" (Demographia, 2017; Larson, 2017: 50).

(19.9 million), and Los Angeles-Long Beach-Santa Ana (13.3 million)—and both will be much smaller than Tokyo (37.2 million), Delhi (36.1 million), Shanghai (30.8 million), and other Asian cities. Thus, the number of megacities is growing, and they're much bigger than in the past (United Nations, 2014).

Should the explosive growth of cities and megacities concern us? Generally, cities provide jobs, offer better health services, and have more educational opportunities, but not everyone benefits from such advantages. The urban poor are often crowded into slums where children are less likely to be enrolled in school, sanitation is inadequate, and the economic gap between the haves and have-nots is widening (Laneri, 2011).

15-2b Urbanization in the United States

Like many other countries, the United States is becoming more urban. How, specifically, has the urban landscape changed? And what are some of the consequences?

HOW URBAN AMERICA IS CHANGING

During the Industrial Revolution, millions of Americans in agricultural areas migrated to cities to find jobs. As a result, between 1900 and 2000, the urban population surged from 39 to 79 percent. Rural areas cover 97 percent of the nation's land, but contain only 19 percent of the population (about 60 million people), down from 55 percent in 1910 (Riche, 2000; U.S. Census Bureau, 2016).

Suburbanization. As cities grew more crowded, dirtier, and noisier, urban growth sparked *suburbanization*, people's moving to communities just outside a city. In 1920, about 15 percent of Americans lived in the suburbs compared with more than half today (Palen, 2014).

During the 1950s, two-thirds of urban dwellers moved to suburbs. The federal government, fearing a return to the economic depression of the 1930s, underwrote the construction of much new housing in the suburbs. The general public obtained low-interest mortgages, veterans were offered the added incentive of being able to purchase a home with a $1 down payment, and massive highway construction programs enabled commuting by car. As a result, suburbs mushroomed (Rothman, 1978).

Suburban life is appealing. Some of the benefits include more privacy and space, one's own yard and garages, better schools, safer streets, and, if there are jobs nearby, shorter commutes. Suburban elites have sprawling "McMansions," considerable acreage, and gated communities.

Especially since 1980, however, much of the social and physical separation between cities and suburbs has blurred. More Latino, Asian, and Black individuals now live in the suburbs than in the city, representing up to 35 percent of the suburban population in more than a third of the nation's largest metropolitan areas. One in three poor Americans now live in suburbs and crime rates, although lower than in cities, have increased (Kneebone and Berube, 2013; Frey, 2014).

Edge Cities and Exurbs. Originally, most suburbs were bedroom communities for commuters with jobs in the city. Over the last few decades, suburbanization has generated **edge cities**, business centers that are within or close to suburban residential areas and include offices, schools, shopping, entertainment, malls, hotels, and medical facilities.

People have also created **exurbs**, small, usually prosperous communities beyond a city's suburbs. About 18 percent of Americans live in exurbs, and counties containing far-flung exurbs are growing faster than many urban counties. The average exurbanite is White, middle or upper-middle class, married with children, and a "super commuter" (who travels two or more hours a day for work) (Berube et al., 2006; Badger, 2015; Frey, 2015).

edge cities business centers that are within or close to suburban residential areas.

exurbs small, usually prosperous communities beyond a city's suburbs.

Patrick Lienin/AGE Fotostock

Is this your dream house? Compared with the average American home of around 2,500 square feet, the typical "tiny house" is between 100 and 400 square feet. Living smaller is becoming popular because of environmental concerns and much lower financial costs. Prices range from $4,000 to $200,000 depending, like a traditional house, on location, cost of the land, building materials, and so on. Some residents and local governments refuse to allow tiny houses, fearing they'll drive down property values (Willett, 2015; Beitsch, 2016; Sullivan, 2017).

Metropolitan Statistical Areas. Together, suburbs, edge cities, and exurbs form **metropolitan statistical areas** (MSAs), also called *metro areas*, that consist of a central city of at least 50,000 people and the urban areas linked to it. The vast majority of Americans (84 percent) live in 366 MSAs. Since 2015, 10 of the 15 fastest-growing metro areas have been in the South, with four of the top five in Texas (e.g., Frisco and McKinney in the Dallas–Fort Worth–Arlington MSA) (U.S. Census Bureau Newsroom, 2017).

Since 2011, and for the first time in nearly a hundred years, metro areas—particularly those with more than 1 million people—have grown faster than suburbs (Frey, 2014). Two groups—young professionals and baby boomers—have driven the trend in metro living. Because many young professionals are postponing marriage and parenthood (see Chapter 12), they don't have to worry about the quality of schools. Moreover, many work in cities, want to reduce commute times, and can't afford the down payment for a house in the suburbs. A number of baby boomers, particularly those who are retiring, are moving to metro areas for a variety of reasons (e.g., better public transportation, greater access to cultural events, social services for older people, no longer being healthy enough to maintain a suburban house) (Kneebone and Berube, 2013; Westcott, 2014).

metropolitan statistical area (MSA, also called *metro area*) a central city of at least 50,000 people and urban areas linked to it.

urban sprawl rapid, unplanned, and uncontrolled spread of development into regions adjacent to cities.

SOME CONSEQUENCES OF URBANIZATION

Cities offer many benefits, including a vast array of culturally diverse restaurants, shops, and activities, but urbanization also creates problems. Some of the drawbacks include urban sprawl, increased traffic congestion, a scarcity of affordable housing, and residential segregation.

Urban Sprawl. Urban sprawl—the rapid, unplanned, and uncontrolled spread of development into regions adjacent to cities—is widespread. According to some estimates, urbanization in the Southeast will increase by up to 190 percent by 2060. Development on that scale will result in losing 15 percent of agricultural land, 12 percent of grasslands, and 10 percent of forests (Terando et al., 2014).

Urban sprawl has created rapid *job sprawl*, which occurs when companies move jobs from metropolitan areas to suburbs. The more distant the suburb, the less likely low-income people are to hear about employment opportunities through informal networks, to afford houses in these areas, and to have transportation to the jobs (Raphael and Stoll, 2010).

Some large metro areas (e.g., Washington, D.C., New York, Boston, Atlanta, Miami, and Denver) are implementing "walkable urbanism" to curb urban sprawl. That is, the developments are neighborhoods where everyday destinations (e.g., work, school, restaurants) are concentrated and within walking distances. For the most part, however, most urban development is built around driving rather than walking (Leinberger and Lynch, 2014).

Traffic Congestion. Generally, the only way to get around in urban sprawl areas is by automobile. This means that most suburban households face the costs of buying, fueling, insuring, and maintaining multiple cars. The U.S. population has increased by 23 percent over the last 25 years, but total highway miles have increased by only 5 percent, resulting in greater traffic congestion within and outside cities. More than 10 million Americans (8 percent of all people who don't work at home) now travel 60 minutes or longer one way, a proportion that has increased by 95 percent since 1990. Traffic snarls and long commutes increase air pollution and stress, waste fuel and time, and decrease the time that people have for family and leisure activities (McKenzie, 2013; Jones, 2017).

Lack of Affordable Housing. In some cases, the poor are pushed out by **gentrification**, a process in which upper-middle-class and affluent people buy and renovate houses and stores in downtown urban neighborhoods. Governments in many older cities encourage gentrification to increase dwindling populations, to revitalize urban areas, and to augment tax revenues. Gentrification can benefit an entire community because it breaks up concentrated poverty pockets, the new residents may demand improvements in schools and crime control, and an influx of retail stores can generate new jobs. Rent increases, however, have displaced many low-income residents and small businesses. Also, some longtime Black residents have complained that aggressive policing tactics, like stopping and frisking people on the street, increase when wealthier people move into a neighborhood (Buntin, 2015; Mendoza, 2016).

Gentrification improves old city neighborhoods and increases property values, but also displaces low-income residents.

Residential Segregation. Since 1970, racial residential segregation has decreased. Metropolitan areas, especially, are becoming racially more diverse (*Figure 15.4*). Among metropolitan areas with a population of 500,000 or more, the least segregated are in the South and West, and the most segregated are mainly in the Northeast and Midwest. There may be a "continued easing of the color line" because racial segregation has waned even in long-standing "hypersegregated" cities like Chicago and New York. The nation's 20 most multiethnic metropolitan regions are now "global neighborhoods" that have substantial numbers of White, Black, Latino, and Asian individuals (Iceland et al., 2013: 119; Frey, 2015).

Figure 15.4 U.S. Metropolitan Areas Are Getting Racially More Diverse

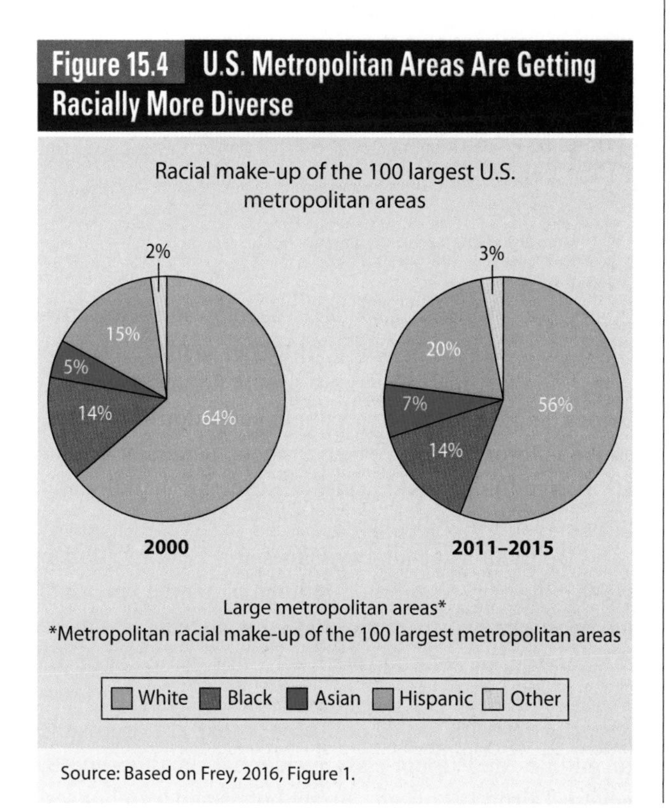

Source: Based on Frey, 2016, Figure 1.

Whereas racial residential segregation has decreased, social class residential segregation is more pronounced than in the past, and the most residentially segregated are low-income Black and Latino individuals. Since 2000, for example, the number of Black and Latino people living in high-poverty ghettos and slums has nearly doubled, rising from 7 to almost 14 million people (Bischoff and Reardon, 2013; Jargowsky, 2015).

Residential segregation along income lines has increased primarily because of the long-term rise in economic inequality (see Chapters 8 and 11). Some wealthy neighborhoods have even fought off state laws that mandate constructing affordable housing for low- and middle-income families. According to the high-income homeowners, building less expensive houses would raise crime rates, lower property values, and "There's plenty of affordable housing *in neighboring communities*" (McCabe, 2014: 39, emphasis added).

15-2c Sociological Explanations of Urbanization

How and why do cities change? In answering this and other questions, functionalists underscore urban development, conflict

gentrification upper-middle-class and affluent people buy and renovate houses and stores in downtown urban neighborhoods.

Table 15.3 Sociological Explanations of Urbanization

Perspective	Level of Analysis	Key Points
Functionalist	Macro	Cities serve many important social and economic functions, but urbanization can also be dysfunctional.
Conflict	Macro	Driven by greed and profit, large corporations, banks, developers, and other capitalist groups shape cities' growth or decline.
Feminist	Macro and micro	Whether they live in cities or suburbs, women generally experience fewer choices and more constraints than men.
Symbolic Interactionist	Micro	City residents differ in their types of interaction, lifestyles, and perceptions of urban life.

theorists emphasize the impact of capitalism and big business, feminist scholars focus on women's safety and space, and symbolic interactionists examine the quality of city life (*Table 15.3* summarizes these perspectives).

Figure 15.5 Four Models of City Growth and Change

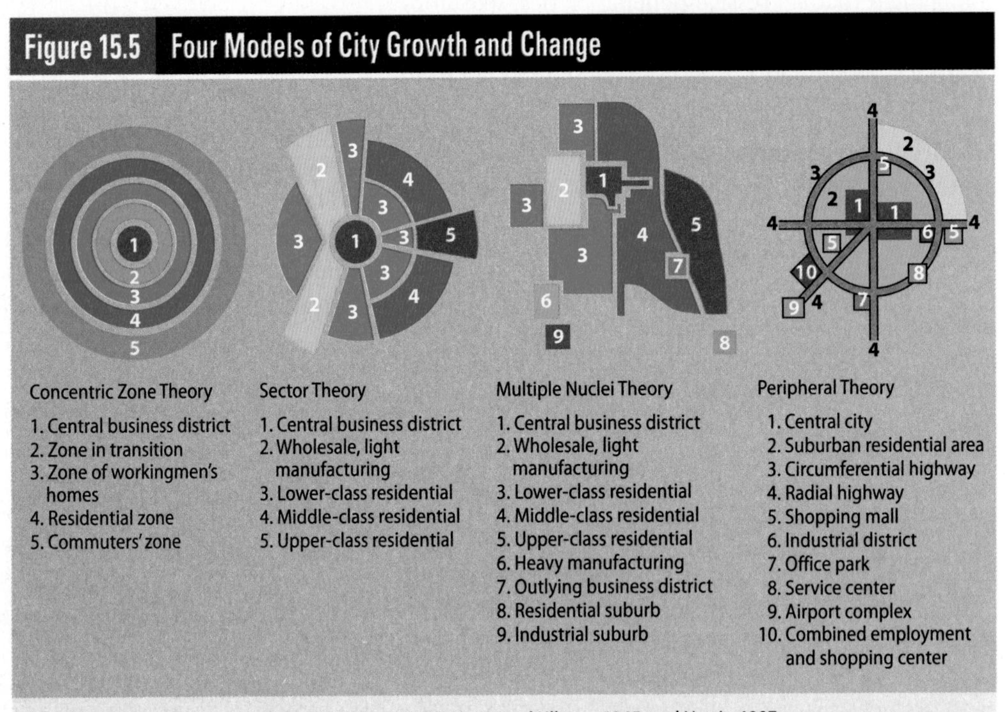

Source: Based on Park and Burgress, 1921; Hoyt, 1939; Harris and Ullman, 1945; and Harris, 1997.

FUNCTIONALISM: HOW AND WHY CITIES CHANGE

Urban ecology studies the relationships between people and urban environments. Over the years, social scientists have revised urban ecology theories (*Figure 15.5*).

Sociologists Robert Park and Ernest Burgess (1921) proposed *concentric zone theory* to explain the distribution of social groups within urban areas. According to this model, a city grows outward from a central point in a series of rings. The innermost ring, the central business district, is surrounded by a zone of transition, which contains industry and poor-quality housing. The third and fourth rings have housing for the working and middle classes. The outermost ring is occupied by people who live in the suburbs and commute daily to work in the central business district.

urban ecology studies the relationships between people and urban environments.

Economist Homer Hoyt's (1939) *sector theory* refined concentric zone theory. He proposed that cities develop in sectors (instead of rings) that radiate from the central business district depending on various economic and social activities. Some sectors are predominantly industrial, some contain stores and offices, and others, generally farther away from the central business district, are middle- and upper-class residential areas.

Geographers Chauncey Harris and Edward Ullman (1945) developed another influential model, *multiple nuclei theory*, which proposed that a city contains more than one center around which activities revolve. For example, a "minicenter" often includes an outlying business district with stores and offices that are accessible to middle- and upper-class residential neighborhoods, whereas airports typically attract hotels and warehouses.

As cities grew after World War II, these models no longer described urban spaces. Thus, Harris (1997) proposed a *peripheral* theory of urban growth, which emphasized the development of suburbs around a city but away from its center. As suburbs and edge cities burgeon, highways that link the city's central business district to outlying areas and beltways that loop around the city provide relatively easy access to airports, the downtown, and surrounding areas.

Functionalists also examine urban dysfunctions, including overcrowding, poverty, deviant behavior, and environmental destruction. From 2000 to 2010, for instance, the number of vacant housing units—many of them in older industrial cities like Baltimore and Detroit—increased by 4.5 million, or 44 percent. Abandoned buildings can increase crime rates and a neighborhood's social disorganization (Williams, 2013).

audioscience/Shutterstock.com

Slums are one of the most dysfunctional by-products of urbanization. Megacities—particularly those in India, Egypt, Pakistan, Kenya, and Mexico—have the world's largest slums.

CONFLICT THEORY: THE IMPACT OF CAPITALISM AND BIG BUSINESS

For functionalists, people's choices shape urban changes. In contrast, **new urban sociology**, a perspective heavily influenced by conflict theory, views urban changes as largely the result of decisions made by powerful capitalists and high-income groups. That is, economic and political factors favorable to the rich, not ordinary citizens, determine urban growth or decline. When a local government wants to rejuvenate parts of the inner city, it typically offers tax breaks, changes zoning laws, and allows real estate, construction, and banking industries to seek profits with little regard for the needs of low-income households or the homeless (Macionis and Parrillo, 2007).

Baltimore, like many other cities, subsidizes private development. Taxpayers have financed billions of dollars in city projects for multinational real estate and banking corporations. The profits are channeled to CEOs and stockholders, while city residents continue paying for (inadequate) schools, parks, libraries, and other public buildings and spaces (Kreitner, 2016).

For conflict theorists, urban space is a commodity that's bought and sold for profit. It's not the average American, they argue, but bankers, developers, politicians, and influential businesspeople who determine how to use urban space. Because increasing a property's value is a higher priority than community needs, poor and low-income people are crowded into dilapidated neighborhoods (Logan and Molotch, 1987; Gottdiener and Hutchison, 2000).

Some analysts are especially critical of the tech sector for being "bad urbanists." To illustrate, in 2011 Twitter received generous tax incentives to move its new headquarters into one of San Francisco's poorest neighborhoods. City officials expected that the company's presence would revitalize the community. Instead, Twitter employees ate all their meals in the dining area, rarely leaving the building to shop because small businesses couldn't afford the area's skyrocketing rents. The neighborhood's evictions increased 38 percent by 2013 because longtime residents couldn't pay the soaring rent increases (Arieff, 2013; Butler, 2014; Steinmetz, 2014).

FEMINIST THEORIES: GENDER, POVERTY, AND SAFETY

Feminist theories emphasize gender-related urban constraints. Women's exclusion from urban planning results in "a male perspective and is in men's interest" (WUNRN, 2017). For example, many women fear the city, especially urban public spaces like streets, parks, and public transportation (Domosh and Seager, 2001; Norman, 2015). They see these places as risky for their physical safety, despite the fact that most violence against women occurs at home. A few cities provide public transportation (e.g., minivans that operate seven nights a week) to prevent crimes against women, usually minority women, who must travel to work after 8:00 p.m. and return home before dawn. For the most part, however, such services are rare (Hayden, 2002).

People living below the poverty level are predominantly female, Black, single parents, and urban

new urban sociology views urban changes as largely the result of decisions made by powerful capitalists and high-income groups.

(Proctor et al., 2016). Because of a lack of affordable housing, many women and their children may be forced to live under the same roof with an aggressor. If public transportation isn't available or adequate, low-income women are cut off from job opportunities. Low-income mothers are expected to work. However, low-wage jobs, coupled with a lack of affordable child care, create role conflicts, undermine mental health, and increase the likelihood of intimate partner violence (Jacobs et al., 2016; WUNRN, 2017).

SYMBOLIC INTERACTION THEORY: HOW PEOPLE EXPERIENCE CITY LIFE

Symbolic interactionists are most interested in the impact of urban life on city residents. In a classic essay, sociologist Louis Wirth (1938: 14) described the city as a place where "our physical contacts are close, but our social contacts are distant."

Wirth defined the city as a large, dense, and socially and culturally diverse area. These characteristics produce *urbanism*, a way of life that differs from that of rural dwellers. Wirth saw urbanites as more tolerant than residents of small towns or rural areas of a variety of lifestyles, religious practices, and attitudes.

He also believed that urbanism has negative consequences, including alienation, friction because of physical congestion, impersonal relationships, and a disintegration of kinship and friendship ties. Some studies have supported Wirth's theory of urbanism (see Guterman, 1969), but others have challenged his views. In a recent national study, people living in large metropolitan areas scored higher than those living in small towns and rural areas on physical and emotional health, access to basic necessities, and being satisfied with life (Witters, 2010).

CRITICAL EVALUATION

Functionalists tend to overlook urbanization's negative political and economic impact, especially when profit and greed guide urban planning. Conflict theory seems to assume that residents are helpless victims as developers and corporations raze low-income houses. In fact, environmental groups have had considerable success in pushing through legislation to maintain and even increase open public spaces and build energy-saving homes in low-income neighborhoods (Moore, 2008).

Feminist sociologists have made important contributions through studies of the everyday lives of low-income women, especially in central cities (see Chapter 11), but urbanization has received much less attention. Urbanites are more diverse than some symbolic interactionists claim. People living in cities aren't necessarily more self-centered or isolated than those in small towns or rural areas. Instead, many have close family bonds, friends, and satisfying relationships with coworkers (Wilson, 1993; Kotkin, 2016). Also, symbolic interaction doesn't show how political, educational, religious, and economic factors shape people's experiences of city life (Hutter, 2007).

ecosystem a community of living and non-living organisms that share a physical environment.

You've seen that the world's population is growing rapidly and becoming more urbanized. Both population growth and urbanization are taxing the planet's limited resources.

15-3 THE ENVIRONMENT

Consider the following:

- Each year, Americans throw away 31 to 40 percent of their food—50 percent more than in the 1970s—that ends up in landfills (Gunders, 2012; Neff et al., 2015).
- The average American throws away 5 pounds of solid waste every day. As waste decomposes in landfills, it gives off many gases, including methane, which contributes to global warming (Powell et al., 2015).
- Up to 3.5 million Americans get sick every year from polluted beach water that causes a wide range of diseases, including ear-nose-eye infections, hepatitis, skin rashes, and respiratory illnesses (Dorfman and Haren, 2014).
- The United States uses 1.2 billion pounds of pesticides a year, but only 0.01 percent reaches the intended target. The other 99.99 percent contaminates the food, air, and water (Hsu et al., 2014).

Such environmental problems threaten our **ecosystem**, a community of living and non-living organisms that share a physical environment. Plants, animals, and humans depend on each other for survival. Because the ecosystem is interconnected worldwide, what happens in one country affects others. Water, air pollution, and climate change are three interrelated factors that are endangering the global ecosystem.

15-3a Water

An expanding world population, extreme weather patterns, and industrial pollution are jeopardizing already limited water supplies. More than 768 million people worldwide don't have clean water, and 1.8 billion drink contaminated water. Contaminated water, inadequate

Danshutter/Shutterstock.com

Among developing countries, 25 percent of girls aren't in school compared with 17 percent of boys. One reason for this difference is that girls are more likely than boys to be responsible for collecting the family's water, making it difficult for them to attend school ("Global WASH Fast Facts," 2017).

availability of water, and lack of access to sanitation together contribute to 88 percent of deaths from diarrheal diseases. In some developing countries, families spend up to 25 percent of their income to purchase water, and many women and children spend up to 6 hours a day carrying it home (World Health Organization and UNICEF, 2014; WWAP, 2014).

AVAILABILITY AND CONSUMPTION

Worldwide, water scarcity is our biggest problem. Just 3 percent of the earth's water is fresh, and two-thirds of it is locked up in the ground, glaciers, and ice caps. That leaves about 1 percent of the earth's water for the world's more than 7 billion people. The world's demand for water has tripled over the last half century. Water scarcity, which affects at least 40 percent of people around the world, is projected to increase. More than half of Earth's 37 largest aquifers (underground layers of rock that store water) are being depleted "at alarming rates" to keep pace with demands from agriculture, growing populations, and industries like mining (WWAP, 2014; Richey et al., 2015; United Nations, 2015).

Some refer to water as "blue gold" because it's becoming one of the earth's most precious and scarce commodities. Water shortages—rather than oil or

How Much Water It Takes to Make...

In 1950, Americans used 150 billion gallons of water every day compared with 400 billion gallons today. It takes water to make everything. For example, it takes:

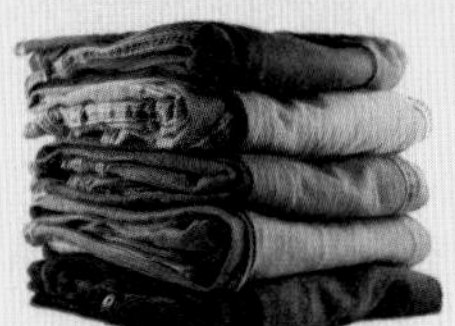
Jacob Kearns/Shutterstock.com

2,600 gallons to make a pair of blue jeans

Stocksnapper/Shutterstock.com

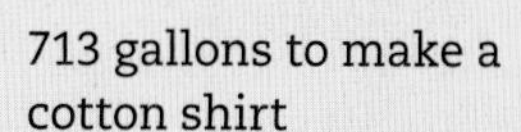
713 gallons to make a cotton shirt

Christopher Elwell/Shutterstock.com

634 gallons to make an average hamburger

magicoven/Shutterstock.com

53 gallons to produce a cup of to-go latte coffee

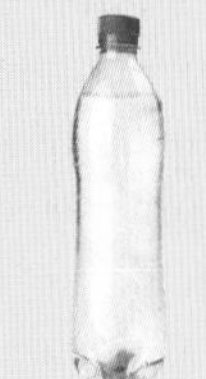
Evgeny Karandaev/Shutterstock.com

1.5 gallons to produce an average 18 ounces of bottled water

Sources: Based on Connell, 2011; Postel, 2012.

diamonds—are behind conflicts and even wars in a number of countries. According to a past Secretary-General of the United Nations, "Too often, where we need water, we find guns . . ." (World Water Assessment Program, 2009: 20; Reynolds, 2012).

Industrialized nations not only have greater access to clean water than the developing world, they use more and pay less for it. The average person in the United States uses about 163 gallons of water per day compared with 40 gallons in Germany, 30 in Denmark, 23 in China, and less than 3 in Mozambique. On average, Americans pay only $.48 per gallon compared with $1.65 a gallon in Denmark. In the developing world, people typically pay five times as much as Europeans (Maxwell, 2012). Thus, in many countries, clean water is a luxury rather than a basic human right.

THREATS TO WATER SUPPLIES

Precipitation (in the form of rain, snow, sleet, or hail) is the ecosystem's main source of water. Clean water has been depleted for many reasons, including pollution, privatization, and waste.

Pollution. Every year, more than 860 billion gallons of sewage, pesticides, fertilizers, automotive chemicals, and trash spoil the country's freshwater. Industrial sites and farm fertilizers are two of the major pollution sources. As a result, 20 percent of the nation's lakes and 55 percent of its rivers and streams can't provide drinking water or support healthy aquatic life (EPA, 2013, 2014).

For almost two years, the residents of Flint, Michigan—a predominantly low-income, Black city—complained of burning skin, hair loss, seizures, strange red splotches on their hands and faces, and anemic children. The mayor and city officials repeatedly told residents the water was fine, until an environmental engineering professor confirmed that Flint's water supply was contaminated with lead. In mid-2016, an independent panel concluded that "the Flint water crisis is a clear case of environmental injustice" (Flint Water Advisory Task Force, 2016: 55).

Environmental racism, also called *environmental injustice*, refers to the exposure of poor people, especially minorities, to environmental hazards. Because of environmental policies and practices, poor neighborhoods are routinely the site of toxic waste, sewerage treatment plants, landfills, illegal dumps, incinerators, polluting power plants, and air and water pollution. These and other hazards increase the likelihood of illnesses and diseases, including cancer, asthma, and children's stunted brain development (Krajicek, 2016).

environmental racism (also called *environmental injustice*) exposure of poor people, especially minorities, to environmental hazards.

Few corporate polluters are fined, and many are exempt from major environmental laws, including the 1974 Safe Water Drinking Act. Consider hydraulic fracturing, or *fracking*, a process that injects water, sand, and chemicals at high pressure to extract gas and oil from rock that lies deep underground. Many U.S. lawmakers—some of them with deep ties to industry that finances political campaigns—endorse fracking to increase energy supplies (Blake, 2014).

Fracking has risks because gases can escape into drinking water. Some of the 750 chemicals injected into the ground contaminate water, increasing the risks of cancer, infertility, and other health problems (Kassotis et al., 2014; EPA, 2015).

Fracking has also increased the number and magnitude of earth tremors and earthquakes from Colorado to the Atlantic coast. In Oklahoma, the number of earthquakes measuring 3 or higher on the Richter scale jumped from 1 in 2007 to 907 in 2015. No place in the world has ever experienced earthquakes at such a rate in such a short time. Cities and counties nationwide have passed 430 measures to control or ban fracking, but companies typically prevail despite opposition (Shauk and Olson, 2014; Skoumal et al., 2015; Philips, 2016).

Privatization. Water is a big business because of *privatization*, transferring some or all of the assets or operations of public systems into private hands. Perrier, Evian, Coca-Cola, PepsiCo—and particularly the French giants Vivendi and Suez—have been buying the rights to extract water in the United States and other countries

J Pat Carter/Getty Images News/Getty Images

Oklahoma's state officials have only recently acknowledged the connection between fracking and earthquakes (Sanburn, 2016).

at will from aquifers, then bottling and selling it around the world.

In 2014, bottled water companies spent more than $84 million on advertising to compete with each other and to convince consumers that bottled water is healthier and safer than soda and tap water. In fact, tap water is safer. FDA regulations allow bottles to be contaminated with small amounts of bacteria, including *E. coli*. And, unlike tap water, which is always moving, bacteria increase after bottling because microbes "can attach to the plastic, munch on organic material in the water, and multiply" (Lohan, 2015; Matanoski, 2017: 21).

Bottling water is lucrative for corporations, but there are many environmental drawbacks. It depletes local water supplies, whether the water comes from municipal sources (40 percent) or local springs (60 percent). Moreover, about 86 percent of the empty plastic bottles in the United States clog landfills instead of being recycled (Food and Water Watch, 2013).

iStock.com/gmutlu

For-profit companies own approximately 10 percent of U.S. community water systems. Proponents maintain that, compared with government, private businesses improve water infrastructure, decrease local mismanagement (as in Flint), operate more efficiently, and spend billions of dollars fixing problems. Opponents contend that, on average, privatized water is 58 percent more expensive than the average public water utility charges, and that a business is more likely than the government to place a lien on consumers' property if they fall behind in paying water and sewer bills (Food & Water Watch, 2016; Ivory et al., 2016; Millsap, 2016).

Waste. Most water problems are due to human mismanagement and waste, not nature. Of all available water worldwide, agriculture consumes about 70 percent, industry uses 20 percent, and 10 percent is residential (WWAP, 2014). In agriculture, many irrigation systems are inefficient, farmers and agribusiness (large agriculture companies) often grow water-hungry crops like cotton and sugarcane in arid areas, and pesticide and chemical fertilizer runoff from fields pollutes streams, rivers, and lakes.

A significant water pipe bursts, on average, every 2 minutes somewhere in the country, losing 7 billion gallons of water a day. Most of the nation's pipes, especially in cities along the eastern seaboard, are nearly 200 years old, and some are made of wood. A water pipe that leaks or bursts wastes huge amounts of drinking water, damages streets and homes, and seeps dangerous pollutants into drinking water. Despite such hazards, replacing a city's substandard pipes has decreased. From 1890 to 1920, for instance, Chicago replaced aging water and sewage pipes at a rate of about 75 miles a year. That number has declined steadily over the years, dropping to 30 miles a year in 2003 (Duhigg, 2010; Pearlstine, 2013).

15-3b Air Pollution, Global Warming, and Climate Change

Global warming is a serious environmental problem. Let's begin with air pollution, the major cause of climate change and global warming.

AIR POLLUTION: SOME SOURCES AND CAUSES

Worldwide, 92 percent of people breathe unhealthy outdoor air. About three million die each year from cardiovascular and respiratory diseases linked to outdoor air pollution (WHO, 2016).

There are many reasons for air pollution, but four are the most common. First, a major source of air pollution is the burning of *fossil fuels*, substances obtained from the earth, including coal, petroleum, and natural gas. The exhaust gases of cars, trucks, and buses contain poisons—sulfur dioxide, nitrogen oxide, carbon dioxide (CO_2), and carbon monoxide. Power plants that produce electricity by burning coal or oil account for nearly 40 percent of U.S. carbon dioxide emissions.

Second, manufacturing plants that produce consumer goods pour pollutants into the air. Formaldehyde-based vapors that can lead to cancer and respiratory problems are emitted by many household and personal care products: pressed wood (often used in furniture), plastic grocery bags, waxed paper, latex paints, detergents, nail polish, cosmetics, and shampoos ("Formaldehyde," 2012; see also Chapter 14).

Third, winds blow contaminants in the air across borders and oceans. Air pollution originating in Europe has been tracked to Asia, the Arctic, and even rural areas in the western United States where there's little industry or automobile traffic (Cooper et al., 2010; Zhang et al., 2017).

A fourth reason for air pollution is that government policies and enforcement vary from one administration to another. The Bush administration blocked the efforts of 18 states to cut emissions from cars and trucks because

Raj K Raj/Hindustan Times/Getty Images

Delhi, India, is the world's most polluted city, but air quality in Pakistan's urban areas is, on average, worse (Smith, 2017).

it believed that tougher regulations would hurt the U.S. economy ("No Action on Greenhouse Gases," 2008).

President Obama promised to decrease air pollution, but environmental groups accused his administration of being too soft on the fossil-fuel industry by not limiting their carbon dioxide emissions. The rules also excluded Native American reservations—which have some of the most polluting coal-fired power plants in the country—from meeting state goals to reduce carbon dioxide emissions by 2030 (Eilperin and Bernstein, 2014; Magill, 2014).

U.S. presidents of both political parties have attended U.N.–sponsored global climate change conventions since 1995, supporting programs to reduce global pollution. In 2017, however, President Trump pulled out of the Paris talks to reduce planet-warming emissions. He has also removed dozens of Obama-era environmental rules that would have closed hundreds of coal-fired power plants and replaced them with wind and solar farms, and would have limited oil and gas drilling in America's Arctic and Atlantic waters (Davenport and Rubin, 2017; Rushe et al., 2017).

global warming increase in the average temperature of earth's atmosphere.

greenhouse effect heating of earth's atmosphere because of the presence of certain atmospheric gases.

climate change a change in overall temperatures and weather conditions over time.

GLOBAL WARMING AND THE GREENHOUSE EFFECT

Global warming is an increase in the average temperature of earth's atmosphere. The warming is due to several factors. Some are natural, such as changes in solar radiation, the earth's orbit, and the frequency or intensity of volcanic activity. At least 95 percent of global warming is due to human activities, however, and not natural climate swings (Intergovernmental Panel on Climate Change, 2014).

Global warming begins with the **greenhouse effect**, the heating of earth's atmosphere because of the presence of certain atmospheric gases. When the sun's heat enters the atmosphere, some of it is absorbed by earth's surface, and some of it is reflected back to space. Greenhouse gases in the atmosphere trap some of this heat. Heat is necessary to support life, but when greenhouse gases increase, earth becomes warmer than it would be otherwise, endangering public health and the welfare of current and future generations (EPA, 2009).

Air pollutants, primarily CO_2, ignite the greenhouse effect. Between 2000 and 2010, greenhouse-gas emissions grew at 2.2 percent a year—almost twice as fast as in the previous 30 years. The 20 largest global economies (e.g., China, United States, European Union, India) account for 82 percent of global CO_2 emissions that come from burning fossil fuels, primarily from coal-fired power plants. Some of the countries are shifting to natural gas and other energy sources, but coal-fired plants produce 46 percent of their CO_2 emissions (Intergovernmental Panel on Climate Change, 2014; Olivier et al., 2016).

China emits almost twice as much CO_2 as the next-biggest polluter, the United States. Since 2014, however, China's CO_2 emissions have dropped by nearly 6 percent, meeting its Paris climate agreement to get 15 percent of its energy from clean sources. The United States, although one of the world's biggest polluters, is going in the opposite direction as President Trump pursues policies to bring back coal and to cut government spending on clean energy development (McKenna, 2017).

SOME EFFECTS OF CLIMATE CHANGE

Climate change is a change in overall temperatures and weather conditions over time. Global warming probably began thousands of years ago when humans started changing the planet by cutting down and burning forests to grow food (Fischman, 2009). Such *deforestation*, clearing massive amounts of trees, affects global climate changes because forests recycle carbon dioxide into oxygen.

Regardless of when it began, climate change is "unequivocal," has "moved firmly into the present," and is having a profound impact on every ecosystem worldwide. Since record-keeping began in 1880, 2014 was the hottest year on earth, and 9 of the 10 warmest years have

occurred since 2000. Between 1750 and 2005, seven countries produced 63 percent of all global warming. The United States, accounting for 20 percent of all climate-changing pollution, was the leading offender, followed by China and Russia (8 percent each). Scientists predict that extreme weather, especially heat waves, will produce heavier rainfall in some regions; more floods, landslides, and uncontained wildfires; stronger hurricanes and tornadoes; and more intense droughts around the world (Matthews et al., 2014; NOAA National Centers . . . , 2015).

Many changes have already occurred because of climate change. For example:

- Higher CO_2 levels are turning oceans more acidic, resulting in some shellfish (e.g., clams and crabs) dying or not developing (Spotts, 2009).
- Global weather- and climate-related disasters rose from 743 between 1971 and 1980 to 3,496 between 2001 and 2010. Climate change will lead to an increase in big storms (like Hurricanes Harvey and Irma) that cause floods, landslides, and other natural disasters (World Meteorological Organization, 2014; Prein et al., 2017).
- In 2016, earth reached its highest temperature on record. Excessive heat in the United States claims more lives each year than floods, lightning, tornadoes, and hurricanes combined (National Oceanic and Atmospheric Administration, 2014; Blunden and Arndt, 2017).
- The world's largest ice sheets in Antarctica and western Greenland are melting by an average of 3 feet a year. The Arctic Ocean could be largely free of sea ice by the late 2030s, only two decades from now. The melting glaciers could contribute to sea levels rising 3 to 6 feet by 2100, flooding up to 70 percent of the populations in Florida, Georgia, South Carolina, and Louisiana (Hauer et al., 2016; AMAP, 2017; Blunden and Arndt, 2017).
- Some Alaskan native villages are facing disaster. Climate change has thinned the Arctic ice so much that it's become too dangerous to hunt whales, a traditional means of sustenance, or even to live in the state's coastal communities. As the ice melts, the towns are in "imminent danger" from erosion and being washed to sea during storms (Mooney, 2015).

Such data show that our planet is ailing. Can people slow some of earth's devastation through better environmental policies and practices, particularly by endorsing sustainable development?

15-3c Is Sustainable Development Possible?

Sustainable development refers to economic activities that don't threaten the environment. There are reasons to be both pessimistic and optimistic about achieving sustainable development.

REASONS TO BE PESSIMISTIC

Which country is the greenest? Not the United States. For example:

- A study of environmental performance ranked 180 countries on factors like water quality, air pollution, and protecting the ecosystem. Iceland and the Scandinavian countries were the top four. The United States ranked 26th, and below some developing countries like Slovakia and Croatia (Hsu et al., 2016).
- In a study of 128 nations, the United States ranked only 18th in environmental quality (Porter et al., 2017).

sustainable development economic activities that don't threaten the environment.

Steffen Thalemann/Getty Images

Fluffy and ultra-plush toilet paper is made by chopping down and grinding up the pulp of trees that are decades or even a century old. U.S. corporations have introduced "earth-friendly" toilet paper, but "customers are unwavering in their desire for the softest paper possible" (Fahrenthold, 2009: A1).

- Among the world's 23 top energy-consuming countries, the United States ranks only 8th in energy-efficiency policies and programs, and below developing countries like China (Kallakuri et al., 2016).

Why does the United States rank so low in environmental performance? There are many reasons, but three important and interrelated factors are profit, politics, and personal choices.

Profit. Does profit—capitalism's basic drive—override environment hazards? Exxon (now ExxonMobil), the world's largest oil company, started conducting cutting-edge climate research in the mid-1950s. When the internal studies showed that rising CO_2 levels were likely to cause global warming, the top executives cut much of the climate research budget, began to organize against air pollution regulations, and funded climate denial research and activities (Banerjee et al., 2015).

Even well-intentioned businesses dismantle eco-friendly ventures if they don't bring quick profits. For example, both Chevron Corporation and BP (British Petroleum) pulled back from clean energy projects because the profits from non-renewable oil and gas sales were much higher (Upton, 2014).

Some environmentalists are especially dismayed by *greenwashers*, companies and other organizations that pollute the planet while presenting an environmentally responsible public image. In 2000, for instance, BP rebranded itself "Beyond Petroleum" and changed its logo to a green-and-yellow sunburst, but the majority of its business produces fossil fuels (Elgin, 2014).

Politics. You saw earlier that environmental policies reflect a president's views. So far, the current administration has halted a study of the public risks of coal mining by the National Academies of Sciences, Engineering, and Medicine; closed an Energy Department office that works with other countries to develop clean energy technology; and eliminated a rule to prevent coal mining companies from dumping toxic debris into nearly streams. President Trump has also appointed a chief of the EPA who contends that nature, not human activities (e.g., burning fossil fuels), causes global warming, and who dismissed at least five academic scientists from the EPA that will be replaced by chemical and fossil fuel company applicants. Despite such changes, some of President Trump's supporters have criticized him for not going far enough in rolling back environment regulations (Biesecker, 2017; Davenport, 2017; Freking and Daly, 2017; Friedman and Plumer, 2017; Goodell, 2017; Plumer, 2017).

These and other political pressures, policies, and practices help explain America's low environmental rankings globally. The United States is "one of the very few large energy-consuming economies that doesn't have national energy reduction targets," across all sectors, to lower greenhouse gas emissions (Kallakuri et al., 2016).

A recent national survey of 1,500 middle and high school science teachers found that 30 percent—because of personal beliefs or pressure from parents or school administrators—taught that global warming is due to natural, not human, causes. Also, 62 percent didn't know that more than 95 percent of climate scientists attribute recent global warming to human activities. A conservative think tank known for attacking climate science has sent 200,000 copies of a glossy book to public school teachers (and some college professors) that describes global warming as "another fake crisis" created by Democrats (Plutzer et al., 2016; Stager, 2017). Thus, politics also fuels ignorance of and misinformation about environment issues.

Personal Choices. About 75 percent of Americans say that "the country should do whatever it takes to protect the environment." However, we recycle or

Fotos593/Shutterstock.com

Almost 15 million tons of plastics enter oceans each year (Geyer et al., 2017). Because plastic trash doesn't degrade, it kills oceanic marine life and endangers ecosystems.

compost only 35 percent of municipal waste, even though 73 percent of the population has curbside recycling. Also, people want cleaner energy and endorse wind farms, but "not in my backyard" (NIMBY) (EPA, 2016; Sustainable Packaging Coalition, 2016; Anderson, 2017; Hemphill and Perry, 2017). Thus, are Americans eco-friendly only when it's convenient?

ARE AMERICANS ECO-FRIENDLY ONLY WHEN IT'S CONVENIENT?

REASONS TO BE OPTIMISTIC

There's been considerable progress since 1970, when the United States celebrated its first Earth Day. Let's look briefly at government, business, and individual contributions to the environment.

Government. Primarily because of federal rules and regulations, the United States ranks second among 23 of the world's top energy-consuming countries in how efficiently buildings use energy. It has the most mandatory appliances and equipment standards, and stringent energy codes for new residential and commercial buildings (Kallakuri et al., 2016).

Most cars no longer burn leaded gasoline, ozone-destroying chlorofluorocarbons (CFCs) have been generally phased out, and total emissions of the six major air pollutants declined by 54 percent during the same period as the U.S. population increased by 47 percent. The U.S. recycling rate increased from less than 10 percent in 1980 to over 35 percent in 2014. As a result, landfill trash decreased from 89 to 53 percent during this interval (Sperry, 2008; EPA, 2016).

Current efficiency standards for appliances, lighting, and other equipment that were implemented during the early 1980s will save the United States the equivalent of two years of energy use ($1.1 trillion by 2035) and slash greenhouse gas emissions. Also, because of serious droughts, local governments in California and some Texas cities are recycling wastewater, including "toilet-to-tap" drinking water (Lowenberger et al., 2012; Erbentraut, 2016).

Whereas the Trump administration is pushing for more exploitation of oil, gas, and coal, business and

Elena Elisseeva/Shutterstock.com

AP Images/Brandon Tabiolo

Many individuals and businesses are contributing to sustainable development. Hawaii has some of the nation's greenest cities, admired for their excellent air quality and extensive use of rooftop solar panels. KFC, a restaurant chain, is introducing edible cups at Seattle's Best Coffee, part of Starbucks, to decrease the environmental impact of packaging (Strom, 2015).

political leaders in seven Republican-dominated states have embraced clean energy sources. In fact, 69 percent of the nation's wind power comes from states that elected Trump. Tapping the wind and sun is an economic strategy that "creates manufacturing jobs, puts steady money in the hands of farmers who host wind turbines, and lures big employers who want renewable power" (Gillis and Popovich, 2017: A22).

Some government-level progress has been mixed. For example, the EPA has imposed stiffer penalties on automakers that overstate their fuel efficiency, but allows the use of 82 chemicals, including pesticides that Europe restricts or bans (Hakim, 2015; Kessler, 2015).

Businesses. Many companies have found that being green is good for their profits, image, and environment, and are now more eco-friendly. For example:

- Each pair of Levi's new Waste<Less jeans is composed of eight recycled plastic bottles (Berfield, 2012).
- Ford, Honda, and Toyota use recycled materials—jeans, sweaters, plastic bottles, and sugar cane by-products—for dashboards, tires, seat covers, and cushions (Fleming, 2014).
- Several companies, including Best Buy and Sprint, have already met their 2020 goal of cutting energy consumption by 20 percent (Martin, 2016).
- Perdue Farms, the country's fourth-largest poultry producer, no longer uses growth-producing antibiotics that pose health risks to people. The chickens grow just as fast and efficiently as in the past, and getting rid of the drugs hasn't affected the company's profitability (Philpott, 2016).
- Among others, Ikea, Nike, and H&M (a clothing retail company) use sustainable cotton grown by farmers who use less water and fewer pesticides (Kessenides, 2017).
- The founder of Patagonia, a popular outdoor clothing and gear store, has run TV commercials urging viewers to protest the Trump administration's selling of public land to corporations (Flasphaler, 2017).

Economic self-interest rather than a concern for the environment sometimes drives such eco-friendly practices. Some corporations are eager to reduce their reliance on water-intensive products like cotton and nuts. Regardless of motives, environmentally friendly policies promote sustainability.

Figure 15.6 Are Stricter Environmental Laws Worth the Cost?

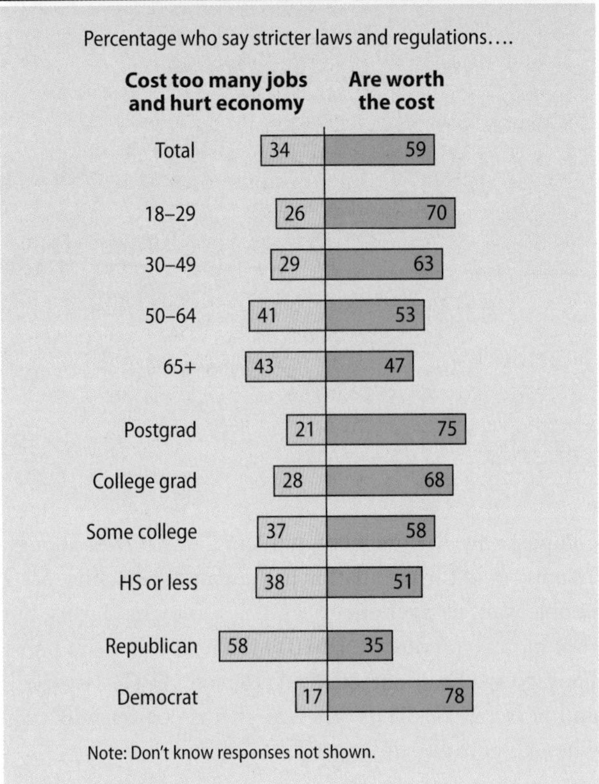

Percentage who say stricter laws and regulations....

	Cost too many jobs and hurt economy	Are worth the cost
Total	34	59
18–29	26	70
30–49	29	63
50–64	41	53
65+	43	47
Postgrad	21	75
College grad	28	68
Some college	37	58
HS or less	38	51
Republican	58	35
Democrat	17	78

Note: Don't know responses not shown.

Source: Based on Bialik, 2016.

Individuals. A majority of Americans (59 percent) say stricter environmental laws and regulations are worth the expense, compared with 34 percent who believe such regulations cost too many jobs and hurt the economy. Perceptions differ quite a bit by age, education level, and political affiliation (*Figure 15.6*). Americans are also divided, especially along party lines, over increasing offshore drilling, nuclear power plants, fracking, and coal mining. The only unity is that the vast majority support expanding solar panel and wind turbine farms (89 percent and 83 percent, respectively) (Funk and Kennedy, 2016).

People's environmental beliefs and behavior aren't always consistent. About 75 percent of Americans say they're concerned with helping the environment as they go about their lives, but only 15 percent always bring their own shopping bags, and only 12 percent always choose cleaning products based on whether the ingredients would help or hurt the environment (Funk and Kennedy, 2016). Once again, then, are many of us eco-friendly only when it's convenient?

STUDY TOOLS 15

READY TO STUDY? IN THE BOOK, YOU CAN:

- ☐ Check your understanding of what you've read with the Test Your Learning Questions provided on the Chapter Review Card at the back of the book.
- ☐ Tear out the Chapter Review Card for a handy summary of the chapter and key terms.

ONLINE AT CENGAGEBRAIN.COM WITHIN MINDTAP YOU CAN:

- ☐ Explore: Develop your sociological imagination by considering the experiences of others. Make critical decisions and evaluate the data that shape this social experience.
- ☐ Analyze: Critically examine your basic assumptions and compare your views on social phenomena to those of your classmates and other MindTap users. Assess your ability to draw connections between social data and theoretical concepts.
- ☐ Create: Produce a video demonstrating connections between your own life and larger sociological concepts.
- ☐ Collaborate: Join your classmates to create a capstone project.

16 Social Change: Collective Behavior, Social Movements, and Technology

Chet Strange/Getty Images News/Getty Images

LEARNING OBJECTIVES

After studying this chapter, you will be able to...

16-1 Compare, illustrate, and evaluate the major types of collective behavior, their theories, and functions.

16-2 Compare, illustrate, and evaluate the major types of social movements, their theories, and functions.

16-3 Describe and illustrate recent technological advances, their benefits, costs, and ethical controversies.

After finishing this chapter go to **PAGE 345** for **STUDY TOOLS**

This chapter examines **social change**, the transformations of societies and social institutions over time. Some collective behavior is short-lived with few long-term societal changes; others, particularly social movements, can have lasting effects. Technology has also played a critical role in sparking social change. Let's begin with collective behavior.

WHAT DO YOU THINK?

I sometimes feel overwhelmed in keeping up with texting, Twitter, Facebook, and other social media.

1	2	3	4	5	6	7
strongly agree						strongly disagree

COLLECTIVE BEHAVIOR

Do you consume energy drinks? Have a tattoo? Look for the label "natural" on food packages? Have you ever joined a club or sports team? Texted? Taken a selfie? If so, you've engaged in collective behavior.

16-1a What Is Collective Behavior?

Collective behavior is the spontaneous and unstructured behavior of a large number of people. Collective behavior encompasses a wide range of actions, including riots, fads, fashion, panic, rumors, and responses to disasters.

Collective behavior has two important characteristics. First, it's an act rather than a state of mind. You may *feel* panic when a tornado threatens your town, but you don't engage in collective behavior until you actually *leave* your home and head for a safer location.

Second, collective behavior varies in its degree of spontaneity and structure. Panic, the least structured form of collective behavior, is typically short-lived. Fads (e.g., diets, "ugly" Christmas sweaters) are more structured. They may last several years, are planned, and may be expensive. Other forms of collective behavior, including pro- and antiabortion groups, become highly institutionalized social movements that include a staff, budget, and lobbying.

16-1b Why Collective Behavior Occurs

Sociologists have proposed many explanations of collective behavior. Four of the most influential are contagion, convergence, emergent norm, and structural strain theory. Each has strengths and weaknesses.

CONTAGION THEORY

Contagion theory proposes that individuals act emotionally and irrationally due to a crowd's almost hypnotic influence. This perspective is rooted in the work of French scholar Gustave Le Bon (1841–1931), who asserted that people in crowds, particularly mobs and riots, change radically. The anonymity of the crowd, a feeling of power, and an infectious "mob mind" embolden people to abandon personal responsibility and to engage in antisocial, and often violent, behavior (Le Bon, 1896/1968).

American sociologists later refined Le Bon's theory. They emphasized "milling," a process of people moving about, talking to one another, and becoming increasingly excited. Milling produces a "circular reaction" in which the participants become more intense and more unified in attacking a target (Blumer, 1969). Because people feel anonymous in crowds, they abandon self-control and adopt behavior that they'd reject in other settings (Brown and Goldin, 1973).

CONVERGENCE THEORY

Contagion theory asserts that crowds are highly suggestible and out of control. In contrast, *convergence theory* proposes that crowds consist of like-minded people who deliberately assemble in a place to pursue a common goal. Thus, collective behavior occurs when people who share similar values, beliefs, attitudes, emotions, and goals come together in, or converge on, a certain location. Instead of the crowd affecting individuals, as contagion theory claims, convergence theory argues that individuals influence a crowd. Examples of convergence theory include marches, rallies, and acts of civil disobedience that oppose a government's domestic or foreign policies.

social change transformations of societies and social institutions over time.

collective behavior spontaneous and unstructured behavior of a large number of people.

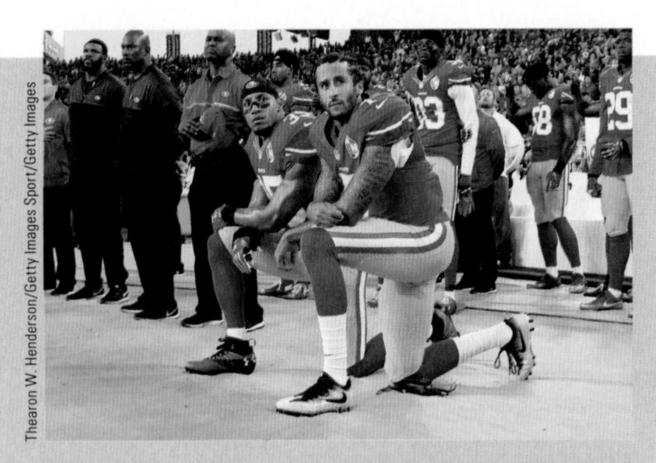
Thearon W. Henderson/Getty Images Sport/Getty Images

In 2016, NFL quarterback Colin Kaepernick (front) and some of his teammates started "taking a knee" instead of standing during the national anthem, to protest racism and police violence. Players across the NFL and at some high school and college football games did the same to show solidarity. Which collective behavior theory, if any, explains these protests? Why?

EMERGENT NORM THEORY

Emergent norm theory posits that crowd members establish rational behavioral norms in response to a precipitating crisis. When Hurricane Harvey deluged Houston and surrounding towns, for example, people used their own boats, Jet Skis, and even air mattresses to rescue strangers from roofs and partly submerged cars, and volunteered at shelters. In effect, then, people aren't spontaneously "infected" with the emotions of others, as contagion theory postulates. Instead, crowd members develop norms as a situation unfolds (Turner and Killian, 1987).

STRUCTURAL STRAIN THEORY

Structural strain theory (also called *value-added theory*) proposes that collective behavior occurs if six conditions are present (Smelser, 1962). These conditions are "value-added" in the sense that each condition leads to the next one, ending in an episode of collective behavior:

1. *Structural conduciveness* refers to the social conditions that foster collective behavior. When channels for expressing a grievance either aren't available or fail, like-minded people may resort to protests.
2. *Structural strain* occurs when societal problems (e.g., discrimination, war) make people angry, frustrated, or interfere with their everyday lives.
3. In the *growth and spread of a generalized belief* stage, people begin to see a situation as a widespread problem instead of their own fault, attribute the problem to a person or group, and believe that something should be done.
4. *Precipitating factors* are incidents or events that trigger action.
5. During the *mobilization for action* stage, leaders emerge.
6. In the *social control* stage, opposing groups prevent, interrupt, or repress those advocating social change. Government officials, the police, community and business leaders, courts, the mass media, and other social control agents—all of whom benefit from the status quo—may ridicule or quash the emerging collective behavior.

CRITICAL EVALUATION

Over time, scholars found little evidence for contagion theory's claim that crowds are typically irrational or out of control. Crowds can sway emotions, but can also be effective in meeting their objectives, and even mobs can be focused and rational (Couch, 1968; McPhail, 1991).

Convergence theories are useful in describing crowds that are deliberate and planned, but have several limitations. They don't explain why some people who share the same attitudes join a protest, rally, or demonstration while others don't, and don't account for changes in crowd behavior, no matter how well planned (Berk, 1974; McPhail, 1991).

Emergent norm theory helps explain why many crowds are orderly and focused rather than mob-like. Nonetheless, the theory doesn't explain which norms emerge, why, and how they differ in crowds. In the case of large crowds—those numbering in the thousands, for instance—how are emergent norms disseminated so quickly and accepted by the participants? It's also not clear why some participants conform to emergent norms while others ignore them and act irrationally (e.g., looting stores and vandalizing property) (Brown and Goldin, 1973; Berk, 1974).

Structural strain theory helps to predict when and where collective behavior might occur, and offers insights on why, at every stage, collective behavior may either fade or escalate. The model doesn't explain all forms of collective behavior, however. In the case of fads and rumors, as you'll see shortly, all six stages don't necessarily occur. A second limitation is that the sequence of stages isn't necessarily the same as Smelser outlined. The theory also doesn't say how much structural strain must exist to spark collective behavior (Rule, 1988; Locher, 2002).

A day after the CDC confirmed the first U.S. Ebola case in Texas, and emphasized that the risk of infection was low, there were hundreds of social media rumors that the disease had spread to Iowa and other states. Why did the rumors proliferate? Almost 67 percent of Americans believed that Ebola spreads easily (it doesn't), 40 percent expected a major outbreak in the United States (highly unlikely), and 26 percent were personally terrified that they or someone in their immediate family would get the Ebola virus (also highly unlikely) (Harvard School of Public Health, 2014).

Center for Disease Control

16-1c Types of Collective Behavior

There are many forms of collective behavior; some are more fleeting or harmful than others (Turner and Killian, 1987). Let's begin with rumors, one of the most common types of collective behavior.

RUMORS

There were widespread rumors that on January 1, 2000, a glitch (Y2K) in operating systems would cause computers around the world to crash, leading to global power outages, banks losing all of their customers' accounts, and even airplanes falling from the skies. None of this occurred.

A **rumor** is unfounded information that people spread quickly. Through modern communication technology, a rumor can spread to millions of people within seconds, and incite riots, panic, or widespread anxiety. Because of the Y2K rumor, thousands of people built underground shelters, and millions stocked up on bottled water, canned food, batteries, and medical supplies.

Most rumors (that rock stars Elvis Presley and John Lennon are alive, for example) are harmless. Others can be damaging. After a woman claimed that she found part of a human finger in her cup of Wendy's beef chili, the restaurant's business dropped by half nationally, and rumors warning people to stop eating fast food altogether spread over the Internet (Richtel and Barrionuevo, 2005). The woman admitted, ultimately, that she had planted the finger to try to get a lucrative settlement. Nonetheless, some customers are still leery about eating at fast food restaurants.

Rumors are typically false, so why do so many people believe them? First, rumors often deal with an important subject about which—especially during uncertain economic times or natural disasters—people are anxious, insecure, or stressed. This makes people especially suggestible. In Hurricane Katrina's aftermath, the media reported numerous rumors of carjacking, murders, thefts, and rapes, the overwhelming majority of which subsequently proved to be false.

Second, there's often little factual information to counter a rumor, or people distrust the sources of information. During Y2K, the people who stockpiled groceries and so forth didn't believe computer scientists or federal officials who assured the general public that there wouldn't be a calamity. Third, rumors offer entertainment, diversion, and drama in our otherwise mundane daily lives (Campion-Vincent, 2005; Heath, 2005). Gossip and urban legends are two of the most common types of rumor.

Gossip. **Gossip** is the act of spreading rumors, often negative, about other people's personal lives. Someone once said that "no one gossips about other people's virtues." Because of its tendency to be derogatory, gossip makes us feel superior ("Did you know that Margie just had breast implants? Isn't she pathetic?"). Gossip is also interesting, entertaining, and exposes hypocrisy (Epstein, 2011).

Sometimes, individuals gossip to control other people's behavior and to reinforce a community's moral standards. Comments about someone's drug abuse or marital infidelity strengthen norms of what's deviant or unacceptable, and bonds us with like-minded people (see Chapter 4). In other cases, people gossip because they resent or envy someone's physical appearance or accomplishments. Besides hurting personal

rumor unfounded information that people spread quickly.

gossip rumors, often negative, about other people's personal lives.

and workplace relationships, gossip often creates fear, resentment, and stress—all of which can increase the likelihood of illness (Sunstein, 2009; Drexler, 2014).

Urban Legends. Another form of rumor is **urban legends** (also called *contemporary legends* and *modern legends*), stories—funny, horrifying, or just odd—that supposedly happened somewhere. Some of the most common and enduring urban legends, but with updated variations, deal with food contamination, like the finger at Wendy's. Others have targeted politicians. We still hear, for example, that President Obama wasn't born in the United States, even after he produced his Hawaii birth certificate. (Snopes.com is a useful website that debunks urban legends and Internet rumors.)

Health-related urban legends also persist. In 1998, medical researchers showed conclusively that there's no relationship between autism and vaccinations for measles, mumps, and rubella. Despite the scientific evidence, during the 2015 Republican presidential debates, Donald Trump vigorously asserted that childhood vaccines can cause autism. Two other presidential candidates—Dr. Ben Carson, a retired neurosurgeon, and Senator Paul Rand, an ophthalmologist—didn't disagree with Trump. Doctors, health organizations, and autism advocacy groups quickly criticized and tried to correct the misinformation, but such "ongoing antivaccine noise by celebrities" helps explain mumps and measles outbreaks in the United States and Europe (Sifferlin, 2014: 7; Welch, 2015; see also Chapters 2, 11, and 14).

urban legends (also called *contemporary legends* and *modern legends*) rumors about stories that supposedly happened somewhere.

panic collective flight, often irrational, from a real or perceived danger.

mass hysteria an intense, fearful, and anxious reaction to a real or imagined threat by large numbers of people.

Why do urban legends persist much longer than gossip? First, they reflect contemporary anxieties and fears—about contaminated food, unscrupulous companies, and corrupt and unresponsive governments. Second, urban legends are cautionary tales that warn us to watch out in a dangerous world. There have been tales that sunscreens cause blindness, and that women have died sniffing perfume samples sent to them in the mail. Third, we tend to believe urban legends because we hear them from people we trust, especially family members, coworkers, and friends. Fourth, rejecting misinformation and outright lies is hard work because doing so requires rethinking already-held beliefs. Finally, urban legends—like the one about alligators living in New York City's sewer system or that AriZona tea contains human urine—are fun to tell and "too beguiling to fade away" (Brunvand, 2001; Ellis, 2005; Emery, 2015).

PANIC AND MASS HYSTERIA

In 2003, an indoor fireworks display to kick off a heavy metal concert in West Warwick, Rhode Island, set off a fire that killed 100 people and injured 200 others. As thick black smoke poured through the audience, hundreds of patrons stampeded for the front door (even though there were three other exits), trampling and crushing those who had fallen beneath them.

Most of the deaths in West Warwick weren't due to the fire but to **panic**—a collective flight, often irrational, from a real or perceived danger. The danger seems so overwhelming that people desperately jam an escape route, jump from high buildings, leap from a sinking ship, or sell off their stock. Fear drives panic: "Each person's concern is with his [or her] own safety and personal security, whether the danger is physical, psychological, social, or financial" (Lang and Lang, 1961: 83).

Panic is similar to **mass hysteria**—an intense, fearful, and anxious reaction to a real or imagined threat by large numbers of people. In mid-2009, the

iStock.com/sx70; iStock.com/Canon_Bob

STR/AFP/Getty Images

Every year, more than 2 million Muslims make a pilgrimage to Mecca (in Saudi Arabia) to participate in the Hajj, an important religious ritual. In 2015, more than 2,400 of the pilgrims were crushed to death during a stampede. The suggested causes included heat (110 degrees Fahrenheit), people rushing to complete the rituals and pushing against each other in opposite directions, confusion among the many first-timers, and panic as people fell and were trampled (Gladstone, 2015; Yan, 2015).

World Health Organization issued an alert that H1N1 (the "swine flu") would become a worldwide epidemic and result in numerous deaths. Millions of Americans stood in line for hours, sometimes overnight, because the vaccine was initially in short supply, but the swine flu never materialized on the predicted scale. Unlike panic, which usually subsides quickly, mass hysteria may last longer because warnings—especially about health and food—reinforce our general fears and anxieties about life's dangers.

FASHION

As the"Betty Crocker Makeover" on the next page shows, General Mills has been modernizing "Betty" since the 1930s. Doing so may reinforce people's cynical views about corporate manipulation. On the other hand, updating Betty's image and fashion may help explain why General Mills has successfully attracted new generations of consumers over the decades, and why Betty Crocker products are now sold worldwide.

Fashion is a popular way of dressing during a particular time or among a particular group of people. Over the years, Black women's hairstyles have changed—from Afros in the late 1960s, to straightened hair during the 1980s, to braids, cornrows, dreadlocks, hair extensions, and coloring more recently. All reflect gender politics, racial solidarity, generational differences, identity, and changing images of beauty (Banks, 2000; Desmond-Harris, 2009).

People who want to be fashionable buy clothes with prominent labels, but what's fashionable changes from year to year. Thus, U.S. adults regularly wear only about 20 percent of their wardrobe (Smith, 2013). Fashion also includes periodic changes in the popularity of furniture, music, language usage, books, automobiles, sports, recreational activities, the names parents give their children, and even the dogs that people own.

Why do fashions, particularly clothes, change fairly quickly? One reason is that designers, manufacturers, and retailers must continuously generate demand for new products and services to maintain a profit. Retailers have created a number of "shopping holidays" to move merchandise, including clothes (e.g., Amazon's "Prime Day," Kohl's senior discount Wednesdays, and many large department stores' "Black Friday" sales the day after Thanksgiving). Second, many people keep up with fashion because they don't want to seem different, out-of-date, or dowdy. Third, shopping for new clothes and other products decreases the boredom of everyday routine. Also, clothes and other merchandise are status symbols that signal being an insider: "Others...will admire me for...making stylish choices" (Best, 2006: 85–86; for classic analyses of fashion and collective behavior, see Veblen, 1899/1953; Packard, 1959; and Bourdieu, 1984).

FADS AND CRAZES

A **fad** is a trend that's popular for a short time. Fads include *products* (e.g., bean bag chairs, hoverboards, pet rocks), *activities* (e.g., twerking, step aerobics, grapefruit diets), and *popular personalities and television characters* (e.g., the Lone Ranger during the 1950s, the Kardashians more recently).

A *craze*, a type of fad, is a temporary, widespread, and obsessive enthusiasm for a particular activity or object. Recent examples include fidget spinners, Hatchimals, and Powerball lotteries that pay millions of dollars. Crazes, like fads, are usually harmless, but there are exceptions. For instance, two men who were following Pokémon Go cartoon creatures on their smartphones suffered injuries when they walked off a cliff in California (Hernandez, 2016).

Why do fads and crazes pop up? A major reason is profit. Because children and adolescents

fashion a popular way of dressing during a particular time or among a particular group of people.

fad a trend that's popular for a short time.

BETTY CROCKER MAKEOVER

1936

1955

1965

1968

1972

1980

1986

1996

Source: General Mills, Inc.

In 1921, General Mills created Betty Crocker, a fictitious woman, to answer thousands of questions about baking that came in from consumers every year. As fashions and hairstyles changed, so did Betty Crocker's image. The original image of a stern, gray-haired older woman has changed over the years so that, by 1996, she had a darker complexion and wore casual attire. Can you think of other brands that have changed their image over the years to keep up with changing trends?

are especially likely to adopt fads, manufacturers create numerous products and activities that they hope will catch on. The products include toys, sportswear, and new cereals. The hottest fads are usually the must-have Christmas toys that children plead for every year. A few months later, the toy may be thrown away or stuffed in the back of a closet.

Some people dismiss fads as "ridiculous" or "silly," but they serve several functions. In a mass society, where people often feel anonymous, a fad can develop strong in-group feelings and a sense of belonging, especially among people who have similar interests and attitudes. Fads can also be fun, promise to resolve a nagging problem (e.g., being overweight), and help us keep up with technological changes (Marx and McAdam, 1994; Best, 2006).

Most fads are soon forgotten, but some become established. Pez candy dispensers, which originated in 1952 and cost 49 cents, are still inexpensive (under $2.00), are sold in more than 60 countries, and have been continuously updated to include popular television characters like the Simpsons (Paul, 2002). Other fads, like streaking (running around nude in public places), reemerge from time to time ("Streaking," 2005).

disaster an unexpected event that causes widespread damage, destruction, distress, and loss.

public a group of people, not necessarily in direct contact with each other, who are interested in a particular issue.

DISASTERS

A **disaster** is an unexpected event that causes widespread damage, destruction, distress, and loss. Some disasters are due to *social causes* like war, genocide, terrorist attacks, and civil strife. Some are due to *technological causes*, including oil spills, nuclear accidents, burst dams, building collapses. Others—like floods, landslides, earthquakes, hurricanes, and tsunamis—have environmental origins (Marx and McAdam, 1994).

Disasters often inspire organized behavior rather than chaos. Instead of panicking, most people are rational, cooperative, and altruistic. They often care for family members instead of fleeing, and thousands of volunteers offer financial, medical, and other help.

PUBLICS, PUBLIC OPINION, AND PROPAGANDA

A **public** is a group of people, not necessarily in direct contact with each other, who are interested in a particular issue. A public is different than the general public, which consists of everyone in a society.

There are as many publics as there are issues—abortion, gun control, pollution, health care, and same-sex marriage, to name just a few. Even within one organization or institution, there may be several publics that are concerned about entirely different issues. At a college, students may be most concerned about the cost of tuition, faculty may spend much time discussing instructional technology, and maintenance employees may be most interested in wages.

Rachel Weill/FoodPix/Jupiter Images

The interaction within a public is often indirect rather than face-to-face—as through the mass media, social media, blogs, or professional journals. Because publics aren't organized groups with memberships, they're often transitory. Publics expand or contract as people lose or develop interest in an issue. A public may surge during a highly publicized and controversial incident, like removing a patient's life support, but then evaporate quickly. In other cases, publics organize and become enduring social movements (a topic we'll examine shortly).

Some publics express themselves through **public opinion**, widespread attitudes on a particular issue. Public opinion (1) is a verbalization rather than an action; (2) is about a matter that concerns many people; and (3) involves a controversial issue (Turner and Killian, 1987). Like publics, public opinions wax and wane over time. People's interest in crime and education drops, for example, when they're more concerned about pressing issues such as jobs and income.

Public opinion can be swayed by **propaganda**, spreading information (or misinformation) to influence people's attitudes or behavior. Propaganda isn't a type of collective behavior, but affects it in several important ways. First, it can create attitudes that will arouse collective outbursts (e.g., strikes or riots). Second, propaganda can attempt or succeed in preventing collective outbursts, as when corporations convince employees that job losses are due to a weak economy rather than to offshoring (see Chapter 11). Third, propaganda tries to gain followers for a cause, whatever it might be (Smelser, 1962). Propaganda is institutionalized in advertising, political campaign literature, and government policies. It's conveyed in many ways: the mass media, social media, political speeches, religious groups, rumor, and symbols (e.g., flags, bumper stickers).

Library of Congress Prints and Photographs Division [LC-USZC4-3859]

One of the best-known examples of propaganda in the United States is this Uncle Sam poster, designed in 1917, and used to recruit soldiers for World Wars I and II. Opponents of the Vietnam War used it as an antiwar poster during the 1960s and 1970s.

CROWDS

Much collective behavior is scattered geographically, but crowds are concentrated in a limited physical space. A **crowd** is a temporary gathering of people who share a common interest or participate in a particular event. Whether it's a few dozen people or millions, crowds come together for a specific reason, such as a religious leader's death, a concert, or a riot.

Crowds differ in their motives, interests, and emotional level:

- A *casual crowd* is a loose collection of people who have little in common except for being in the same place at the same time and participating in a common activity or event. There's little if any interaction, the gathering is temporary, and there's little emotion. Examples include people watching

public opinion widespread attitudes on a particular issue.

propaganda spreading information (or misinformation) to influence people's attitudes or behavior.

crowd a temporary gathering of people who share a common interest or participate in a particular event.

iStock.com/Tanuki Photography

"Public mourning" includes strangers coming together to show their respect to the dead. Less than a day after Apple's co-founder Steve Jobs' death in 2011, fans across the globe created memorials in Japan and other countries. Is this kind of collective behavior an example of a conventional, expressive, or acting crowd? Why?

a street performer, spectators at the scene of a fire, and shoppers at a busy mall.

- A *conventional crowd* is a group of people that assembles for a specific purpose and follows established norms. Unlike casual crowds, conventional crowds are structured, their members may interact, and they conform to rules that are appropriate for the situation. Examples include people attending religious services, funerals, graduation ceremonies, and parades.
- An *expressive crowd* is a group of people who show strong emotions toward some object or event. The feelings—which can range from joy to grief—pour out freely as the crowd reacts to a stimulus. Examples include attendees at religious revivals, revelers during Mardi Gras, and enthusiastic fans at a football game.
- An *acting crowd* is a group of people who have intense emotions and a single-minded purpose. The event may be planned, but acting crowds can also be spontaneous. Examples include people fleeing a burning building, soccer fans storming a field, and college students having a water balloon fight.
- A *protest crowd* is a group of people who assemble to achieve a specific goal. Protest crowds demonstrate their support of or opposition to an idea or event. Most demonstrations—like antiwar protests, boycotts, and labor strikes—are usually peaceful. Peaceful protesters can become aggressive, however, resulting in destruction and violence (Blumer, 1946; McPhail and Wohlstein, 1983).

mob a highly emotional and disorderly crowd that uses force, the threat of force, or violence against a specific target.

riot a violent crowd that directs its hostility at a wide and shifting range of targets.

One type of crowd can easily change into another. A conventional crowd at a nightclub can turn into an acting crowd if a fire erupts; people panic and flee for safety. Any of the five types of crowds—from casual to protest—can become a mob or a riot.

MOBS

A **mob** is a highly emotional and disorderly crowd that uses force, the threat of force, or violence against a specific target. The target can be a person, group, or property. Mobs often arise when people are demanding radical societal changes, like removing a corrupt government official. In other cases, especially when authority breaks down, people who take advantage of a situation may engage in mob behavior.

A week after the Charlottesville, Virginia clashes, an angry mob—chanting "No KKK, no fascist U.S.A.!"—pulled down a statue honoring Confederate soldiers at a court house in Durham, North Carolina. The crowd stomped and spat on the statue until the police arrived. After attacking, a mob tends to dissolve quickly.

RIOTS

Compared with mobs, riots usually last longer. A **riot** is a violent crowd that directs its hostility at a wide and shifting range of targets. Unlike mobs, which usually have a specific target, rioters unpredictably attack whomever or whatever gets in their way. Most riots arise out of long-standing anger, frustration, or dissatisfaction that may have smoldered for years or even decades. Some of these long-term tensions are due to discrimination, poverty, unemployment, economic deprivation, or other unaddressed grievances.

There are numerous protests in the United States every year, but race riots have been the most violent and destructive. More than 150 U.S. cities experienced race riots after the assassination of Martin Luther King Jr. in 1968. In 1992, riots broke out in 11 cities after four White police officers were acquitted of beating Rodney King, a Black motorist, in Los Angeles. The violence resulted in deaths and considerable property damage because of fires and looting.

In 2014 and 2015, riots erupted in Ferguson, Missouri, and Baltimore, Maryland, after the controversial deaths of Black men at the hands of the police. As in the past, looting and/or burning local businesses accompanied both of the riots. A few years later, white supremacists marched into Charlottesville, Virginia, to protest the planned removal of a statue of Robert E. Lee, a Confederate general, from a city park. Many carried Confederate flags, gave Nazi salutes, and displayed swastikas and KKK symbols. Counterprotesters, including some with Black Lives Matter signs, confronted the marchers, violence ensued, and a car that barreled into a group of anti-supremacist demonstrators killed a young woman.

Riots are usually expressions of deep-seated hostility, but this isn't always the case. In the 2010 Championship Series, riots broke out after the Los Angeles Lakers beat the Boston Celtics. Such "celebration riots" are due to extreme enthusiasm and excitement rather than anger or frustration, and can lead to "an orgy of gleeful destruction" (Locher, 2002: 95).

Much collective behavior, like a mob or riot, is spontaneous and short-lived. In contrast, because social movements are usually structured and enduring, they create or suppress social changes.

Table 16.1 Five Types of Social Movements

Movement	Goal	Examples
Alternative	Change some people in a specific way	Alcoholics Anonymous, transcendental meditation
Redemptive	Change some people, but completely	Jehovah's Witnesses, born-again Christians
Reformative	Change everyone, but in specific ways	Gay rights advocates, Mothers Against Drunk Driving (MADD)
Resistance	Preserve status quo by blocking or undoing change	Antiabortion groups, white supremacists
Revolutionary	Change everyone completely	Right-wing militia groups, Communism, ISIS in the Middle East

16-2 SOCIAL MOVEMENTS

There are hundreds of social movements in the United States alone. Why are they so prevalent? And why do they matter?

16-2a What Is a Social Movement?

A **social movement** is a large and organized group of people who want to promote or resist a particular social change. "Social movements are as American as apple pie. The abolition of slavery, women's right to vote, unions, open admissions to public colleges and student aid, and Head Start are all changes in our society that were won through social movements" (Ewen, 1998: 81–82).

Unlike other forms of collective behavior, social movements are goal-oriented, deliberate, structured, and can have a lasting impact on a society. And in contrast to many other forms of collective behavior (like crowds, mobs, and riots), the people who make up a social movement are dispersed over time and space, and usually have little face-to-face interaction (Turner and Killian, 1987; Lofland, 1996).

Some U.S. social movements, like white supremacists, are relatively small. Others are large and have subgroups that appeal to different segments of the population. The U.S. environmental movement has at least 50 smaller groups, including Earth First!, Greenpeace, the National Audubon Society, the Union of Concerned Scientists, and the Wilderness Society.

16-2b Types of Social Movements

Sociologists generally classify social movements according to their goals (changing some aspect of society or resisting change) and the amount of change they seek (limited or widespread). Some social movements are perceived as more threatening than others because they challenge the existing social order (*Table 16.1*).

Alternative social movements focus on changing some people's attitudes or behavior in a specific way. They typically emphasize spirituality, self-improvement, or physical well-being. These movements are the least threatening to the status quo because they seek limited change and only for some people. For instance, millions of non-Asian Americans, influenced by Asian religions, have embraced yoga, meditation, and healing practices like acupuncture (Cadge and Bender, 2004).

Redemptive social movements (also called *religious* or *expressive movements*) propose a dramatic change, but only for some people. They're usually based on spiritual or supernatural beliefs, promising some form of salvation or rebirth. Examples include any religious movements that

social movement a large and organized group of people who want to promote or resist a particular social change.

actively seek converts (e.g., Jehovah's Witnesses, some Christian evangelical groups) (see Chapter 13).

Reformative social movements want to change everyone, but only in a particular way. These movements, the most common type in U.S. society, don't want to replace the existing economic, political, or social class arrangements. Examples include groups that champion the rights of disabled individuals, gay people, crime victims, fat people, and animals.

Resistance social movements (also called *reactionary movements*) try to preserve the status quo by blocking change or undoing change that has already occurred. Resistance movements are often called *countermovements* because they usually form immediately after an earlier movement has created change. For example, antiabortion groups that arose in the United States shortly after the Supreme Court decision in *Roe v. Wade* (1973), which legalized abortion, seek to reverse that decision.

Revolutionary social movements want to completely destroy the existing social order and replace it with a new one. These movements range from utopian groups that withdraw from society and try to create their own to terrorists who use violence and intimidation. Examples of the latter include ISIS in the Middle East and Boko Haram in Africa.

Black Lives Matter, a civil rights movement, emerged after a White man was acquitted of killing an unarmed Black teenager in a Florida suburb in 2013. The group's chapters regularly protest racial profiling, police brutality, and racial inequality in the criminal justice system.

16-2c Why Social Movements Emerge

A social movement is "an answer either to a threat or a hope" (Touraine, 2002: 89), but not everyone who feels threatened or hopeful joins a social movement. Why not? Let's look at four explanations, beginning with the oldest.

MASS SOCIETY THEORY

Early on, sociologists believed that the people who formed social movements felt powerless, insignificant, and isolated in modern mass societies, which are impersonal, industrialized, and highly bureaucratized. Thus, according to *mass society theory*, social movements offer a sense of belonging to people who feel alienated and disconnected from others (Kornhauser, 1959).

CRITICAL EVALUATION

Mass society theory may explain why some people form extreme political movements like Fascism and Nazism, but subsequent research has shown that movement organizers are typically not isolated but well-integrated into their families and communities. Also, historically, many political activists in the United States, including those behind the civil rights and women's rights movements during the late 1960s, weren't powerless but came from relatively privileged backgrounds (McAdam and Paulsen, 1994).

relative deprivation a gap between what people have and what they think they should have compared with others in a society.

RELATIVE DEPRIVATION THEORY

Relative deprivation theory is broader than mass society theory. **Relative deprivation** is a gap between what people have and what they think they should have compared with others in a society. What people think they should have includes money, social status, power, or privilege.

Relative deprivation theorists note two other elements. First, people often feel that they *deserve* better than they have ("I've worked hard all my life"). Second, they believe that they *can't attain their goals through conventional channels* ("I've contacted my senators, but they ignore me"). Thus, shared beliefs combined with unfulfilled expectations can trigger change-oriented social movements (Davies, 1979; Morrison, 1971).

Relative deprivation also includes people who want to protect their declining power and privilege. For example, the white supremacist/neo-Nazi movement attracts people, particularly low-income White people, who believe that minorities and immigrants are threatening the economic and political status that White people enjoyed in the past.

CRITICAL EVALUATION

Relative deprivation theory helps explain why some social movements emerge. Critics point out, however, that there's a certain degree of relative deprivation in all societies, but people don't always react by forming or joining social movements. Relative deprivation theory also doesn't explain why some people join movements even though they don't see themselves as deprived, and don't expect to gain anything personally if the movement succeeds (Gurney and Tierney, 1982; Johnson and Klandermans, 1995).

RESOURCE MOBILIZATION THEORY

It takes more than feeling alienated (mass society theory) or disadvantaged (relative deprivation theory) to sustain a social movement. Instead, according to *resource mobilization theory*, a social movement will succeed if it can put together (or mobilize) an organization and leadership dedicated to advancing its cause (Oberschall, 1995; Gamson, 1990). Other important resources include money, devoted volunteers, paid staff, access to the media, effective communication systems, special technical or legal knowledge and skills, equipment, physical space, alliances with like-minded groups, and lobbyists who influence legislation (see Chapter 11).

CRITICAL EVALUATION

A key contribution of resource mobilization theory is its emphasis on structural factors (like organization and leadership) in explaining why some social movements thrive whereas others shrivel. A major criticism, however, is that resource mobilization theory largely ignores the role of relative deprivation in a social movement's formation. If there aren't large numbers of dissatisfied people to initiate a movement, even plentiful resources won't be able to sustain it (Klandermans, 1984; Buechler, 2000).

NEW SOCIAL MOVEMENTS THEORY

New social movements theory, which became prominent during the 1970s, emphasizes the linkages between culture, politics, and ideology. Unlike the earlier perspectives, new social movements theory proposes that many recent movements (like those that work for peace and environmental protection) promote the rights and welfare of *all* people rather than specific groups in particular countries (Laraña et al., 1994; Melucci, 1995). Thus, new social movements theory is especially interested in "the struggle to liberate the voices of the dispossessed" (Schehr, 1997: 6).

IF THERE AREN'T LARGE NUMBERS OF DISSATISFIED PEOPLE TO INITIATE A MOVEMENT, EVEN PLENTIFUL RESOURCES WON'T BE ABLE TO SUSTAIN IT.

For these theorists, recent social movements differ from older ones in several ways. They attract a disproportionate number of people who are well-educated and relatively affluent, who represent a wide array of professions (e.g., educators, scientists, actors, businesspeople, political leaders), and who share a broad goal—improving the quality of life for all people around the world. Moreover, new social movements pursue goals or advance values that may not personally benefit its members, such as eradicating diseases in developing countries (Obach, 2004).

CRITICAL EVALUATION

Unlike earlier perspectives, new social movements theory helps us understand collective behavior that crosses international boundaries. According to some critics, however, neither this perspective nor the groups that it examines are novel. Some social movements (like feminism and environmentalism) have been around for a long time and still focus on the same basic issues, like women's second-class citizenship and population growth. In addition, some scholars point out that educated middle-class or wealthy activists were as common in the old social movements as in more recent ones (Rose, 1997; Buechler, 2000; Sutton, 2000).

Critics also contend that new social movements theory often overstates people's altruistic motivations. Many people who join environmental groups do so for reasons referred to as NIMBY (not in my backyard). That is, they're concerned about their own community's environment, but show little interest in ecological threats to people elsewhere.

These four theories, despite their limitations, offer insights about social movements because "no

Figure 16.1 Typical Stages of a Social Movement

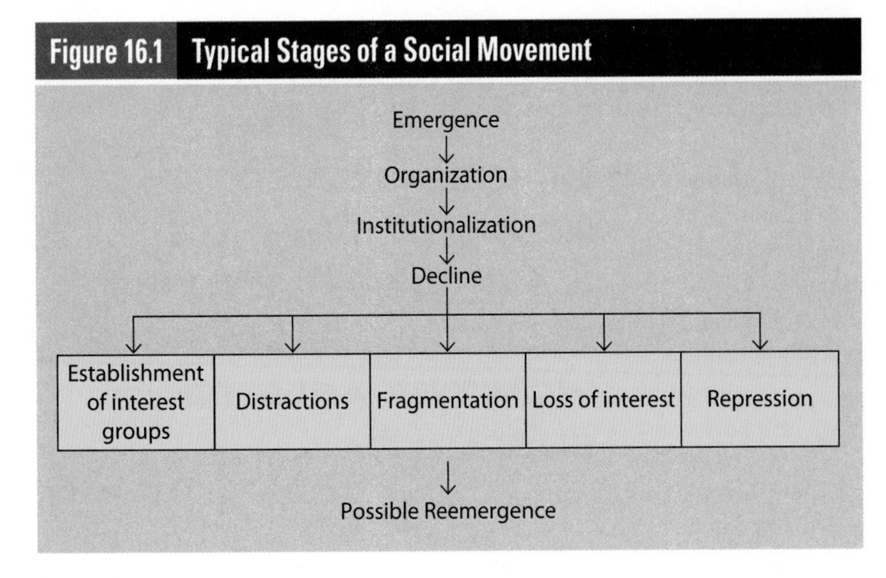

single theory is sufficient to explain the complexities of any social movement" (Blanchard, 1994: 8). The next question is why some social movements flourish and others collapse.

16-2d The Stages of Social Movements

Most social movements are short-lived. Some never really get off the ground; others meet their goals and disband. Social movements generally go through four stages: emergence, organization, institutionalization, and decline (*Figure 16.1*) (Tilly, 1978).

EMERGENCE

During *emergence*, the first stage of a social movement, a number of people are distressed about some condition and want to change it. One or more individuals, serving as agitators or prophets, emerge as leaders. They verbalize the feelings of the discontented, crystallize the issues, and push for action. If leaders don't get much support, the movement may die. If, on the other hand, the discontent resonates among growing numbers of people, public awareness increases, and the movement attracts like-minded people.

ORGANIZATION

Once people's consciousness has been raised, the second stage is *organization*. The most active members form alliances, seek media coverage, develop strategies and tactics, recruit members, and acquire the necessary resources. The organization establishes a division of labor in which leaders make policy decisions and followers perform necessary tasks like preparing mass mailings, developing websites, and responding to phone calls and emails. At this stage, the movement may develop chapters at local, regional, national, and international levels (see Obach, 2015).

INSTITUTIONALIZATION

As a movement grows, it becomes *institutionalized* and more bureaucratic: The number of staff positions increases, members draw up bylaws, the organization may hire outsiders (writers, attorneys, and lobbyists) to handle some of the necessary tasks, and the leaders may spend more of their time on speaking tours, in media interviews, and at national or international meetings. As the social movement grows and becomes more bureaucratic and self-sufficient, the original leaders may move on to better paying and more influential positions in government or the private sector.

DECLINE

Almost all social movements end sooner or later. ("Why Is the NRA a Successful Social Movement?" describes one exception.) This *decline*, the last stage, may take several forms:

- If a social movement is successful, it can become an *interest group* and a part of society's fabric. A small antismoking movement that began in the mid-1970s

AP Images/Damian Dovarganes

Food movements have burgeoned worldwide, especially in urban centers. Many U.S. cities have institutionalized food carts and trucks by passing or revising regulations that support food vending. Doing so encourages neighborhood entrepreneurs to use available city spaces, attracts "foodies," and offers residents and tourists a variety of ethnic, organic, and gourmet "culinary experiences" (Hanser and Hyde, 2014: 47).

Why Is the NRA a Successful Social Movement?

An overwhelming majority of Americans want more gun control, such as preventing mentally ill people from buying guns (89 percent), performing background checks of buyers for private sales and gun show sales (84 percent), and creating a federal database to track gun sales (71 percent).

A brief look at the stages of the National Rifle Association (NRA) helps explain why, despite public consensus to the contrary, this social movement has successfully increased gun rights. For example, 15 states, so far, allow people to carry firearms in plain view in public spaces, including bars, schools, and churches.

1. *Emergence* The NRA was founded in 1871 by two former Union veterans to improve their troops' shooting skills. The NRA built a rifle range on Long Island, New York, and organized rifle competitions and clubs. The leaders attracted people—soldiers, hunters, and sportsmen—who shared the common goal of becoming better marksmen.
2. *Organization* In 1871, NRA leaders formed a board of directors to elect a president, elected other officers (vice president, treasurer, and so on), moved the headquarters to Washington, D.C., established a committee to lobby for laws that befitted the NRA, and created a legislative affairs division to update members on upcoming bills. In effect, then, the NRA set up a division of labor, developed strategies, and formed alliances with political decision makers.
3. *Institutionalization* After the passage of the Gun Control Act of 1968, the NRA hired well-paid lobbyists, funded political action committees (see Chapter 11), supported gun-rights candidates, and opposed gun-control candidates. It developed multilayered executive staff positions, elected or paid celebrity spokespersons (e.g., Chuck Norris, Charlton Heston), and formed official chapters in every state.

   The NRA also implemented a "savvy business plan" that offers members—for only $25 a year—official membership ID cards, discounts "on everything from rental cars to hearing aids," free items (rosewood hunting knives, heavy-duty duffel bags), a free subscription to one of their many magazines, and insurance for them and their guns. According to an NRA official, its millions of members "give us small amounts of money [that] add up to an annual budget of more than $300 million" ("Gun Laws: A Shot and a Beer," 2014). Thus, the NRA embedded its fundamental values and objectives into culturally acceptable activities.

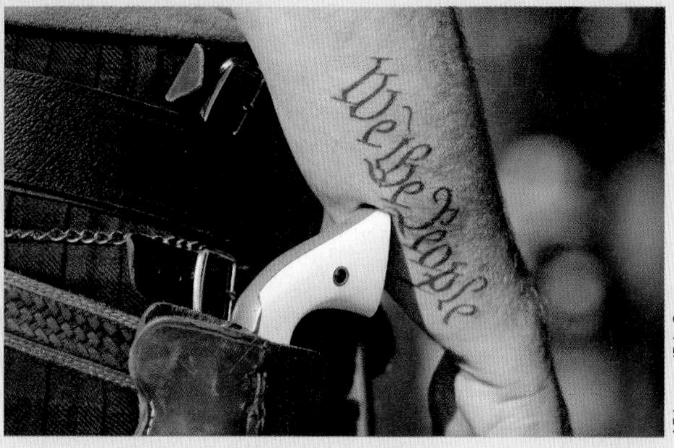

AP Images/Eric Gay

4. *Decline* The NRA isn't facing imminent decline. It offers shooting, training, and gun safety programs for many diverse groups (e.g., women, children, hunters, competitive shooters, law enforcement), which broadens the NRA's member base and loyalty. It has a network of over 15,000 affiliated businesses, associations, and clubs that generate significant revenue and help lobby against gun control.

   Perhaps most importantly, to many Americans, guns represent freedom and independence. As a result, 21 percent of gun-owners, compared with only 12 percent of non-gun-owners, contact public officials at least once a year, urging more lenient gun laws.

Sources: Davidson, 1998; Rodengen, 2002; Melzer, 2009; Winkler, 2011; Bellini, 2012; BBC News, 2016; Law Center to Prevent Gun Violence, 2017; Parker et al., 2017; and undated NRA online sites.

now has the enthusiastic support of numerous prestigious organizations, including the American Cancer Society, the American Heart Association, the American Medical Association, the World Health Organization, and governing bodies within and outside the United States (Wolfson, 2001).

- Those involved in a social movement may become *distracted* because the group loses sight of its original goals, their enthusiasm diminishes, or both. In the mid-1960s, when Ralph Nader initially condemned the automobile industry for car safety defects, he gained a large following. As Nader and his consumer rights groups expanded their focus to include environmental issues and corporate crimes, many of the initial followers lost interest.

- A social movement may experience *fragmentation* because the participants disagree about goals, strategies, or tactics. The environmental movement encompasses numerous groups that focus on different issues—air quality, marine life, land use, and global warming. Participants may also drift away because of time constraints, strategy disagreements, health problems, or similar reasons (see Blee, 2015).
- Social movements may decline because of *repression*. Many autocratic governments quash dissent. A government can crush an emerging social movement by arresting protestors and imprisoning or even executing leaders (see Chapter 11).

Social movements can wane for a combination of reasons. In 2011, the Occupy Wall Street (OWS) movement protested socioeconomic inequality, corporate greed, and corporate power over the U.S. government. The group's slogan, "We are the 99%," referred to the income inequality between the top 1 percent and the rest of the population. Despite similar demonstrations in 70 major U.S. cities and over 600 communities, and 900 cities worldwide, the movement faded away. Why? OWS raised awareness of the injustice of this nation's inequality, but ultimately failed because it lacked leadership, clear objectives, solutions, organization, and a long-term commitment (Madrick, 2013; Sandbu, 2013).

A social movement that declines can sometimes experience a resurgence. For example, there have been several waves of the women's rights movement in the United States since the mid-nineteenth century. The first wave ensured women's right to vote in 1920; the second wave expanded employment and educational rights during the 1960s and 1970s; and the third wave, during the 1990s, focused on economic and other inequalities experienced by women of different social classes, sexual orientations, and nationalities. Despite considerable backlash over the past hundred years, American feminism has appealed to diverse groups, agitated in a multitude of spheres ranging from athletics to religion, and has used a variety of strategies to press for social change (Cobble et al., 2014).

16-2e Why Social Movements Matter

On an *individual level*, many of us enjoy a variety of rights as workers, consumers, voters, and victims—rights that we owe to highly dedicated people who were determined to change inequitable laws and customs.

On an *institutional level*, social movements can change general practices. Shopping for healthy food is much easier today than it was before the 1990s, when veggie burgers, tofu, and nutrition labeling were practically nonexistent. Now, mainstream grocery stores have large organic food sections, a variety of fruits and vegetables, and breads and cereals made with whole grains, nuts, and less salt, sugar, and chemical additives. The "farm-to-table" movement, with its emphasis on locally grown produce and meat, has benefited people's health and regional economies. In effect, then, vegetarian and consumer groups have changed the way many farms operate and have loosened corporate control of food products (O'Connor, 2015; Kowitt, 2015).

On a *societal level*, social movements have had a major impact in the United States and globally. Most of the world's great religions began as protest movements (see Chapter 13). Also, democratic governments in the United States and other countries grew out of the activities of revolutionary groups that sought greater political and economic freedom (Giugni et al., 1999; della Porta and Diani, 1999).

You've seen that social movements can generate or resist change. Technology is another important source of societal change.

16-3 TECHNOLOGY AND SOCIAL CHANGE

This brief "history" of how to cure an earache has circulated on the Internet:

- 2000 B.C.—Here, eat this root.
- 1000 A.D.—That root is heathen, say this prayer.
- 1850 A.D.—That prayer is superstition, drink this potion.
- 1940 A.D.—That potion is snake oil, swallow this pill.
- 1985 A.D.—That pill is ineffective, take this antibiotic.
- 2000 A.D.—That antibiotic is artificial. Here, eat this root.

As this anecdote suggests, despite technological progress over the centuries, some of the old remedies are

One Teenager + Technology = Social Change

In 2012, Sarah Kavanagh—a high school sophomore in Hattiesburg, Mississippi—started a Change.org petition to get PepsiCo to remove brominated vegetable oil (BVO) from its Gatorade products. BVO contains a chemical compound that keeps sports and citrus-flavored drinks more uniform by preventing oily ingredients from separating. It's also used as a flame retardant on upholstered furniture and children's products. Sarah gathered more than 200,000 signatures, and PepsiCo eliminated BVO in early 2013. She then created another successful petition that resulted in Coca-Cola's removing BVO from all its beverages by the end of 2014 (Strom, 2014; Velasco, 2014).

JAMES EDWARD BATES/The New York Times/Redux

enjoying renewed popularity (peoplespharmacy.com, for instance, offers home treatments for everything from arthritis to whooping cough). Still, technological advances have brought enormous social changes, including benefits, costs, and ethical concerns.

16-3a Some Recent Technological Advances

Technology, the application of scientific knowledge for practical purposes, is vital to human life: "For good or ill, [technologies] are woven inextricably into the fabric of our lives, from birth to death, at home, in school, in paid work" (MacKenzie and Wajcman, 1999: 3; see Chapter 3 for early technological inventions).

In the next 50 years, technology is likely to change our lives more dramatically than ever before. Some companies are working on "smart" pills to improve mental ability and restore brains that have been impaired by disease or injury. Our houses may be built with sensors that automatically test for anthrax, environmental contaminants, allergens, and radioactivity, and with devices that can defend against chemical and biological agents (Rubin, 2004; Murphy, 2005). Whether such predictions will become reality is anyone's guess, but seven companies—including General Motors, Google, Tesla, and Nissan—plan to sell driverless cars by 2020. In the meantime, technological advances are changing our lives. Let's look briefly at a few of the most influential.

COMPUTER TECHNOLOGY

In 1887, English mathematician Charles Babbage designed the first programmable computer. Since then, computers have gone through seven generations of evolution. *Artificial intelligence (AI)*, a branch of computer science, is developing theories and computer systems to perform tasks that normally require human intelligence (e.g., speech recognition, decision making, and understanding). AI advances have spurred the development of *robots*, machines that are programmed to perform humanlike functions. Industrial robots, known as "collaborative robots" or "cobots," already work alongside people in doing tasks like stacking heavy packages. For $2,000, you can now buy a laundry-folding robot. By 2025, according to some experts, AI and robots will affect nearly every aspect of daily life (Smith et al., 2014; Salkin, 2017).

Robots have performed repetitive, dirty, or dangerous human tasks, including toxic waste cleanup, mining, minefield sweeping, underwater and space exploration, and locating bombs. Using the most up-to-date information, Watson, IBM's supercomputer, can answer thousands of questions that doctors and nurses might ask about, including the symptoms and treatment of most diseases. It can also answer lawyers' questions on bankruptcy and other topics, complete with citations from legislation or case law. In time, robots may be

technology the application of scientific knowledge for practical purposes.

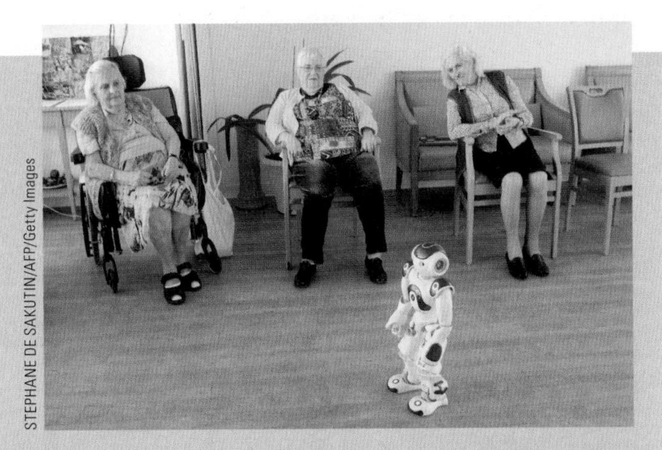

STEPHANE DE SAKUTIN/AFP/Getty Images

Some European nursing homes are using Zora, a robot, to teach their residents exercises and to chat with children in their classrooms. One woman, who hadn't spoken in four months, surprised everyone when she blurted out "I'm well" after Zora asked how she was doing, and then carried on a brief conversation (McFarland, 2015).

able to discuss stock market investment strategies, give you advice about a personal problem, read a book to you in any desired language or in a voice of either sex, cook and serve your meals, and be emotionally savvy companions who cheer you up when you're unhappy ("Artificial Intelligence," 2015; Kharif, 2015).

BIOTECHNOLOGY

Biotechnology is a broad term that applies to all practical uses of living organisms. It covers anything from the use of microorganisms (e.g., yeast) to ferment beer to *genetic engineering*, sophisticated techniques that can change the makeup of cells and move genes across species to produce new organisms.

In medicine, biotechnology includes genetic testing/screening that can determine a child's biological parents, a person's ancestry, inherited disorders, and the chance of developing or passing on a genetic disease. Pig heart valves are already transplanted into patients. It may soon be possible to transplant pig livers, hearts, and other organs, saving some of the 22 people a day who die waiting for an organ transplant. Scientists also predict that it won't be too long before amputees will be able to control their prosthetics using their minds ("Biomedical Engineering," 2017; Kolata, 2017).

NANOTECHNOLOGY

Another promising innovation is *nanotechnology*, the ability to build objects one atom or molecule at a time. The key characteristic of these objects is tiny size. Nanotechnology is based on structures measured in nanometers, a unit of measurement equal to 1 billionth of a meter, or 1/80,000th the width of a human hair.

Scientists are also working on nanoparticles that hunt down and kill tumor cells, and nanosensors—microscopic robots with legs, propellers, and cameras—that would live in the bloodstream and send messages to smartphones about signs of infection, an impending heart attack, or other issues that are early warnings for disease or death. Swallowing little pills that contain nanosensors may become a routine part of doctors' office visits in the future (Cha, 2014; "Hunting as a Pack," 2014).

These and other technological innovations promise longer and healthier lives in the future, but what about the present? What are some of the current benefits and costs of technological changes?

16-3b Some Benefits and Costs of Recent Technologies

The ever-accelerating pace of change means that we're using technologies that didn't exist just a decade ago. The average American embraces modern technology, but is also wary of some advancements. For example, a majority doesn't want some medical enhancements that could improve people's cognitive or physical capacities (*Table 16.2*).

A computer scientist has noted that every technology has a "dark side" because practically every benefit also has a cost (Lohr, 2011: A1). Here are a few contemporary examples.

COLLABORATIVE CONSUMPTION

- *Benefits*: Technology has spurred "collaborative consumption" (also called a "sharing economy" and "access economy") that enables people to earn and save money by sharing goods and services. Airbnb matches budget travelers and people with a spare room or other lodging for rent, RelayRides

Table 16.2 Americans Are Divided About Some Medical Advances

Percentage who say they wouldn't want these enhancements for themselves or their children	
Gene editing to greatly reduce a baby's risk of disease	50
Synthetic blood for much improved physical abilities	63
Brain chip implant for much improved cognitive abilities	66

Source: Based on Funk et al., 2016.

SS MIRZA/AFP/Getty Images

Are restaurant jobs at risk?

allows people to rent your car, and taxi-like services (e.g., Lyft, Uber, Sidecar) bring drivers and passengers together.

- *Costs*: The sharing economy has created conflict: Airbnb hosts have sometimes ignored zoning laws, don't always pay city or state taxes, and neighbors have complained about noise and a steady stream of strangers. Taxicab companies have protested that ridesharing has decreased their business because the drivers have avoided all regulations, inspections, fees, and insurance requirements (Goodale, 2015; Stein, 2015).

3-D PRINTING

- *Benefits*: 3-D printing—a process for making a physical object from a three-dimensional digital model—is revolutionizing manufacturing. Because the process can reduce manufacturing costs by 25 percent or more, the CEO of an aluminum corporation describes 3-D printing as "the beginning of... a second Industrial Revolution" (Geier, 2014: 76). Among other things, 3-D printing produces medical implants (such as knee replacements), synthetic heart valves, prosthetics that are easy to use, models of buildings, artificial human bones, and even cars that are safe and fuel efficient. Desktop 3-D printers produce a wide range of objects like toys, clocks, ceramics, and wrenches.
- *Costs*: As with any new technology, 3-D printers have negative impacts: They consume 50 to 100 times more electrical energy than traditional molds, increase usage of plastics that aren't recycled, and emit toxic particles when used in the home that can pose health risks. Moreover, people can print weapons, including plastic guns, that metal detectors or x-ray scanners can't spot (Gilpin, 2014).

HEALTH AND MEDICINE

- *Benefits*: People who use health-related mobile gadgets and apps can monitor their blood pressure, calories, and physical fitness. The number of robotic procedures increases by 30 percent each year. Robotic surgery reduces physician fatigue because they don't have to stand over the patient for hours, and a robot's "hands" can reach into tight spots and move in ways that human hands can't. In addition, minimally invasive robotic surgeries usually result in less blood loss and faster recoveries because there's a smaller incision to heal.
- *Costs*: Health-related apps encourage "iPochondria" among people who are anxious about their health which, in turn, increases doctors' workloads and insurance costs. Robotic surgery is expensive, the outcomes are generally similar to traditional operations, there are no national training standards for robotic surgery, one company dominates the market, and adverse events inceased 34 percent between 2011 and 2012 alone (Langreth, 2013; "M-Health...," 2014).

MONITORING SYSTEMS

- *Benefits*: New technology allows older people to live at home rather than in an institutional setting. Sensors can track people's medications and falls. Adult children can monitor an aging parent or relative remotely via cellular connections, seeing if they go outside or get home safely after short drives.
- *Costs*: Medicaid pays for some aging-at-home technology, but Medicare doesn't. Even if families can afford the technology, some older people feel intimidated by even one-button devices and don't use them (Abrahms, 2014; Tsukayama, 2014).

JOBS AND PRIVACY ISSUES

Some policy analysts fear that technology will produce more jobs for robots than for humans, even in occupations like teaching, sales, nursing, and stock trading. Driverless cars could displace millions of truck drivers, bus drivers, and others who drive for a living. Others contend that computers will never replace jobs that require "people skills" like teamwork, creativity, managing diverse employees, leadership, decision-making, and face-to-face interaction with clients or customers (Colvin, 2014; Arntz et al., 2016).

There's considerably more consensus that almost every technological advance reduces privacy and increases data breaches. In 2016 alone, hackers obtained personal

William Haefeli/The New Yorker Collection/The Cartoon Bank

"Hey! Elbows off the table."

data from more than one billion Yahoo accounts. During the presidential election, the Russian government infiltrated the Democratic National Committee's computer network, hoping to undermine Hillary Clinton's campaign. The hackers also targeted voting equipment in at least 21 states, and created several hundred fake Facebook and Twitter accounts to spread anti-Clinton messages during the election. Government auditors and other investigators have warned the White House for several decades about cybersecurity weaknesses, but the warnings have usually been ignored (Davis, 2015; Yeung, 2016; Perlroth et al., 2017).

Much software, including smartphone apps and games like Pokémon Go, have lengthy, legalese-filled "terms of service" agreements. Users rarely read these documents, but clicking on "I agree" gives a company "perpetual, irrevocable, transferable, worldwide, royalty-free license to use, copy, modify ... publicly display, publicly perform, and distribute your User Content" (Ziccarelli, 2016: 15).

Many "data mining" companies routinely collect information as people click from site to site. Much of this web tracking is done anonymously, but a new crop of "snooper" sites makes it easier than ever before for anyone with Internet access to assemble and sell personal information, including your name, Social Security number, address, what you buy, whom you love, and which sites you've visited.

Google compiles enough data to build comprehensive portfolios of most users—who they are, where they go, and what they do. To create a new stream of profit, Facebook sells marketers detailed information about its users. Marketers, in turn, can flood millions of websites and mobile apps with targeted ads. At many department stores, cameras have become so sophisticated that companies can analyze what shoppers are looking at, and even their mood (Efrati, 2013; "Marketing in the Digital Age," 2015; Komando, 2016).

Only 14 percent of Americans are "very confident" that government agencies or businesses will keep the data they collect private and secure. However, millions provide information—including gender, age, and income—for as little as a "$1 off" online coupon or a $5.00 to $10.00 prepaid gift card, and agree to be tracked over GPS, Wi-Fi, and cellular networks. The companies then sell the data to storeowners, online retailers, and app developers (Clifford and Hardy, 2013; Olmstead and Smith, 2017).

16-3c Some Ethical Issues

DNA testing has given millions of people information about their genetic predispositions for diseases like cystic fibrosis, cancer, and Huntington's disease (an incurable neurological disorder). Having such information helps doctors and patients make better health care decisions, but the expanding use of genetic testing has an unforeseen consequence: More people who don't show any symptoms are being told they have genes for potentially fatal diseases that may develop "next month, next year, when you are 60 years old, or never" (Marcus, 2013: D1).

Those affected become "patients-in-waiting" who undergo continuous screening and worry about their condition. Asymptomatic patients with an inherited risk for a genetic heart muscle disease may have a defibrillator implanted, which involves regular maintenance, possible equipment failures, infections, and other complications. Some doctors believe that such preventive measures for patients-in-waiting are unethical because they're unnecessary, risky, and people live in limbo not knowing whether or when a disease will develop (Marcus, 2013).

In a recent breakthrough, an international team of scientists reported successfully editing the DNA in human embryos. The technique could potentially prevent a long list of inheritable diseases, such as cystic fibrosis, heart disorders, Huntington's disease, and breast and ovarian cancer. The findings, however, "set off alarm bells among critics around the world" who fear that such research could lead to "designer babies." Wealthy parents, some worry, could use the technology to produce smarter, taller, healthier, and stronger offspring that would have a competitive edge. Designer babies aren't technically possible, but some geneticists are denouncing any experiments that involve genetically modified human embryos as unethical and irresponsible (Ma et al., 2017; Stein, 2017).

Another ethical dilemma is that biotechnological advances, like other technological developments, are most readily available to higher-income people. Screening expectant mothers for fetal abnormalities like Down syndrome is becoming more standard, but not all insurers cover the $1,200 to $2,700 costs of more sophisticated screening tests for high-risk women (Lewis, 2014). Insurance also rarely covers pediatric prosthetics, such as up to $40,000 for a hand that'll have to be replaced about every two years as a child grows (Cohn, 2014). Thus, technological advances are fraught with both promise and pitfalls.

STUDY TOOLS 16

READY TO STUDY? IN THE BOOK, YOU CAN:

- ☐ Check your understanding of what you've read with the Test Your Learning Questions provided on the Chapter Review Card at the back of the book.
- ☐ Tear out the Chapter Review Card for a handy summary of the chapter and key terms.

ONLINE AT CENGAGEBRAIN.COM WITHIN MINDTAP YOU CAN:

- ☐ Explore: Develop your sociological imagination by considering the experiences of others. Make critical decisions and evaluate the data that shape this social experience.
- ☐ Analyze: Critically examine your basic assumptions and compare your views on social phenomena to those of your classmates and other MindTap users. Assess your ability to draw connections between social data and theoretical concepts.
- ☐ Create: Produce a video demonstrating connections between your own life and larger sociological concepts.
- ☐ Collaborate: Join your classmates to create a capstone project.

REFERENCES

The references that are new to this edition are printed in red.

A

"2012 Survey of Business Owners." 2015. American Factfinder, Table SB1200CSA01. Accessed October 3, 2015 (factfinder.census.gov).

"2013 Indian Gaming Industry Report." 2015. 500 Nations. Accessed October 4, 2015 (500nations.com).

"75%—A Nation of Flag Wavers." 2011. Pew Research Center. Accessed July 15, 2011 (www.pewresearch.org).

AAA Foundation for Traffic Safety. 2015. "2014 Traffic Safety Culture Index." January. Accessed July 21, 2015 (aaafoundation.org).

Aaron, Daniel G., and Michael B. Siegel. 2017. "Sponsorship of National Health Organizations by Two Major Soda Companies." *American Journal of Preventive Medicine* 52 (January): 20–30.

AARP Magazine. 2014. "You're Old, I'm Not." February 3, 40–42.

AAUW. 2013. *The Simple Truth About the Gender Pay Gap, Fall 2013 Edition*. Washington, DC: American Association of University Women.

AAUW. 2017. "The Simple Truth About the Gender Pay Gap." Spring. Accessed May 20, 2017 (aauw.org).

Abbey-Lambertz, Kate. 2017. "How Activists Destroyed Bill O'Reilly's Reputation with Advertisers." *Huffington Post*, March 11. Accessed March 13, 2017 (huffingtonpost.com).

Abelson, Reed. 2016. "Survey Finds Workers Pay a Bigger Share for Health Care." *New York Times*, September 15, B2.

Abrahams, Jessica. 2015. "Are Men Natural Born Criminals? The Prison Numbers Don't Lie." *The Telegraph*, January 13. Accessed April 22, 2017 (telegraph.co.uk).

Abrahms, Sally. 2014. "Is This the End of the Nursing Home?" *AARP Bulletin*, March, 20, 22.

Abrams, Rachel. 2016. "Barbie Now Comes in Tall, Short and Curvy." *New York Times*, January 29, B2, B9.

Abudabbeh, Nuha. 1996. "Arab Families." Pp. 333–346 in *Ethnicity and Family Therapy*, 2nd edition, edited by Monica McGoldrick, Joe Giordano, and John K. Pearce. New York: Guilford Press.

Academy of Medical Royal Colleges. 2011. "Induced Abortion and Mental Health: A Systematic Review of the Mental Health Outcomes of Induced Abortion, Including Their Prevalence and Associated Factors." National Collaborating Centre for Mental Health, December. Accessed February 9, 2013 (www.aomrc.org.uk).

Acciai, Francesco, Aggie J. Noah, and Glenn Firebaugh. 2015. "Pinpointing the Sources of the Asian Mortality Advantage in the USA." *Journal of Epidemiology and Community Health* 69 (October): 1006–1011.

Acierno, Ron, Melba Hernandez, Amanda B. Amstadter, Heidi S. Resnick, Kenneth Steve, Wendy Muzzy, and Dean G. Kilpatrick. 2010. "Prevalence and Correlates of Emotional, Physical, Sexual, and Financial Abuse and Potential Neglect in the United States: The National Elder Mistreatment Study." *American Journal of Public Health* 100 (February): 292–297.

ACT, Inc. 2014. "Improving College and Career Readiness and Success for Everyone." Accessed June 21, 2014 (www.act.org).

Adams, Bert N., and R. A. Sydie. 2001. *Sociological Theory*. Thousand Oaks, CA: Pine Forge Press.

Adamson, David M. 2010. "The Influence of Personal, Family, and School Factors on Early Adolescent Substance Use." Rand Health. Accessed August 20, 2011 (www.rand.org).

Adkins, Amy. 2015. "Majority of U.S. Employees Not Engaged Despite Gains in 2014." Gallup, January 28. Accessed October 7, 2015 (gallup.com).

Adkins, Amy. 2015. "Only 35% of U.S. Managers Are Engaged in Their Jobs." Gallup, April 2. Accessed August 22, 2015 (gallup.com).

Administration on Aging. 2013. "Protect Seniors in the Year of Elder Abuse Prevention." Accessed November 30, 2014 (www.aoa.gov).

"Adultery in New England: Love Free or Die." 2014. *The Economist*, April 19, 24.

"African American Health." 2017. Centers for Disease Control and Prevention, May. Accessed August 15, 2017 (cdc.gov).

Agence France-Presse. 2013. "Bill Gates' 'Hand in Pocket' Draws Criticism in South Korea." April 23. Accessed November 10, 2013 (www.rawstory.com).

Agnvall, Elizabeth. 2014. "10 Tests to Avoid." *AARP Bulletin*, March, 12, 14, 16.

Agorist, Matt. 2016. "A Pill That Cures Hepatitis Costs Just $4, but If You Live in America It's $1,000." AlterNet, February 22. Accessed August 15, 2017 (alternet.org).

Agress, Lynne. 2017. "Immigrants: Our Best and Brightest." *Baltimore Sun*, May 2, 11.

Aguila, Raul. 2014. "What We Hate About Holiday Travel." *Consumer Reports*, November, 7.

Aguilar, Leslie, and Linda Stokes. 1995. *Multicultural Customer Service: Providing Outstanding Service Across Cultures*. New York: McGraw Hill.

Ahn, Lauren, and Michelle Ruiz. 2015. "Survey: 1 in 3 Women Has Been Sexually Harassed at Work." Cosmopolitan, February 16. Accessed May 22, 2017 (cosmopolitan.com).

Ailes, Elizabeth C., et al. 2015. "Opioid Prescription Claims among Women of Reproductive Age—United States, 2008–2012." *MMWR* 64 (2): 37–41.

Akbulut-Yuksel, Mevlude, and Adriana D. Kugler. 2016. "Intergenerational Persistence of Health in the U.S.: Do Immigrants Get Healthier as They Assimilate?" National Bureau of Economic Research, February. Accessed June 20, 2017 (nber.org).

Akcigit, Ufuk, John Grigsby, and Tom Nicholas. 2017. "Immigration and the Rise of American Ingenuity." National Bureau of Economic Research, February. Accessed June 20, 2017 (nber.org).

Akechi, Hironori, Atsushi Senju, Helen Uibo, Yukiko Kikuchi, Toshikazu Hasegawa, and Jari K. Hietanen. 2013. "Attention to Eye Contact in the West and East: Autonomic Responses and Evaluative Ratings." *PLoS One* 8 (3): e59312. Accessed December 5, 2013 (www.plosone.org).

Akers, Ronald L. 1997. *Criminological Theories: Introduction and Evaluation*, 2nd edition. Los Angeles: Roxbury.

"Alcohol and Public Health." 2014. Centers for Disease Control and Prevention, March 14. Accessed July 30, 2014 (www.cdc.gov).

Al-Jassem, Diana. 2011. "Women in Polygamous Marriages Suffering Psychological Torture." ArabNews, March 8. Accessed September 12, 2012 (www.arabnews.com).

Allen, Joshua, et al. 2014. "Early Postnatal Exposure to Ultrafine Particulate Matter Air Pollution: Persistent Ventriculomegaly, Neurochemical Disruption, and Glial Activation Preferentially in Male Mice." *Environmental Health Perspectives*, June 5, advance publication. Accessed July 30, 2014 (www.ehponline.org).

Allen, Kathy Grannis. 2015. "Retailers Estimate Holiday Return Fraud Will Cost $2.2 Billion in 2015." National Retail Federation, December 17. Accessed April 14, 2017 (nrf.com).

Allport, Gordon W. 1954. *The Nature of Prejudice*. Reading, MA: Addison-Wesley.

Al-Mahmood, Syed Z. 2013. "Bangladesh to Raise Workers' Pay." *Wall Street Journal*, May 13, B4.

Alsever, Jennifer. 2014. "Prison Labor's New Frontier." *Fortune*, June 16, 60.

Altintas, Evrim. 2016. "The Widening Education Gap in Developmental Child Care Activities in the United States, 1965–2013." *Journal of Marriage and Family* 78 (February): 26–42.

Alter, Charlotte. 2016. "Seeing Sexism from Both Sides: What Trans Men Experience." *Time*, June 27, 24.

Altonji, Joseph G., Sarah Cattan, and Iain Ware. 2010. "Identifying Sibling Influence on Teenage Substance Use." National Bureau of Economic Research, October. Accessed July 24, 2011 (www.nber.org).

Alvaredo, Facundo, Anthony B. Atkinson, Thomas Piketty, and Emmanuel Saez. 2013. "The Top 1 Percent in International and Historical Perspective." *Journal of Economic Perspectives* 27 (Summer): 3–20.

Alvaredo, Facundo, Bertrand Garbinti, and Thomas Piketty. 2017. "On the Share of Inheritance in Aggregate Wealth: Europe and the USA, 1900–2010." *Economica* 84 (April): 239–260.

AMAP. 2017. "Snow, Water, Ice and Permafrost in the Arctic: Summary for Policy-makers." Accessed August 20, 2017 (amap.no).

American Academy of Pediatrics. 2016. "American Academy of Pediatrics Announces New Recommendations for Children's Media Use." October 21. Accessed February 28, 2017 (aap.org).

American Association of Suicidology. 2009. "Elderly Suicide Fact Sheet." June 23. Accessed December 17, 2009 (www.suicidology.org).

American Bar Association. 2017. "A Current Glance at Women in the Law: January 2017." Accessed May 25, 2017 (americanbar.org/women).

"American Indian and Alaska Native Heritage Month: November 2016." 2016. U.S. Census Bureau News, November 2. Accessed November 22, 2016 (census.gov).

American Psychiatric Association. 2013. *Diagnostic and Statistical Manual of Mental Disorders*, 5th edition. Arlington, VA: American Psychiatric Association.

American Psychological Association. 2014. "About Transgender People, Gender Identity, and Gender Expression." Accessed March 2, 2015 (apa.org).

American Psychological Association. 2014. "Are Teens Adopting Adults' Stress Habits?" February 11. Accessed April 12, 2014 (www.stressinamerica.org).

American Society for Aesthetic Plastic Surgery. 2015. "Cosmetic Surgery National Data Bank Statistics: 2014." Accessed August 18, 2015 (surgery.org).

American Society of Addiction Medicine. 2016. "Opioid Addiction: 2016 Facts & Figures." Accessed August 15, 2017 (asam.org).

American Society of Plastic Surgeons. 2017. "2016 Plastic Surgery Statistics Report." Accessed July 20, 2017 (plasticsurgery.org).

American Sociological Association. 1999. *Code of Ethics and Policies and Procedures of the ASA Committee on Professional Ethics.* Accessed January 10, 2010 (www.asanet.org).

"American Times Use Survey—2015 Results." 2016. Bureau of Labor Statistics News Release, June 24. Accessed May 20, 2017 (bls.gov/tus).

"Americans Moving at Historically Low Rates, Census Bureau Reports." 2016. Census Bureau News, November 16. Accessed August 20, 2017 (census.gov).

Amnesty International. 2017. "Amnesty International Global Report: Death Sentences and Executions 2016." Accessed April 28, 2017 (amnesty.org).

Amnesty International USA. 2017. "Death Penalty Cost." Accessed April 20, 2017 (amnestyinternationalusa.org).

Amrein-Beardsley, Audrey. 2014. "Recommended Readings on VAMs." VAMboozled. Accessed June 21, 2014 (www.vamboozled.com).

Amusa, Malena. 2010. "'Precious' Pushes Past Controversy to Oscar Night." Women's eNews, March 5. Accessed March 8, 2010 (www.womensenew.org).

Andersen, Margaret L., and Patricia Hill Collins. 2010. "Why Race, Class, and Gender Still Matter." Pp. 1–16 in *Race, Class, and Gender: An Anthology,"* 7th edition, edited by Margaret L. Andersen and Patricia Hill Collins. Belmont, CA: Wadsworth.

Anderson, James F., and Laronistine Dyson. 2002. *Criminological Theories: Understanding Crime in America.* Lanham, MD: University Press of America.

Anderson, Monica. 2015. "6 Facts About Americans and Their Smartphones." Pew Research Center, April 1. Accessed August 14, 2015 (pewresearch.org).

Anderson, Monica. 2015. "A Rising Share of the U.S. Black Population Is Foreign Born." Pew Research Center, April 9. Accessed October 4, 2015 (pewresearch.org).

Anderson, Monica. 2016. "Parents, Teens and Digital Monitoring." Pew Research Center, January 7. Accessed March 14, 2017 (pewresearch.org).

Anderson, Monica. 2017. "For Earth Day, Here's How Americans View Environmental Issues." Pew Research Center, April 20. Accessed August 25, 2017 (pewresearch.org).

Anderson, Monica, and Andrew Perrin. 2017. "Tech Adoption Climbs Among Older Adults." Pew Research Center, May 17. Accessed June 14, 2017 (pewresearch.org).

Andrews, Edmund L. 2015. "Stanford Study Finds Blacks and Hispanics Typically Need Higher Incomes Than Whites to Live in Affluent Neighborhoods." Stanford Report, June 25. Accessed October 4, 2015 (news.stanford.edu).

Andrzejewski, Adam. 2014. "The Federal Transfer Report." Open the Books, March. Accessed April 12, 2014 (www.openthebooks.com).

Anestis, Michael D., and Joye C. Anestis. 2015. "Suicide Rates and State Laws Regulating Access and Exposure to Handguns." *American Journal of Public Health* 105 (October): 2049–2058.

Angwin, Julia, Terry Parris Jr., and Surya Mattu. 2016. "Facebook Doesn't Tell Users Everything It Really Knows About Them." AlterNet, March 4. Accessed December 29, 2016 (alternet.org).

Annie E. Casey Foundation. 2014. "Race for Results: Building a Path to Opportunity for All Children." Accessed June 21, 2014 (www.aecf.org).

Aptekar, Sofya. 2016. "Celebrating New Citizens, Defining the Nation." *Contexts*, 15 (Spring): 46–51.

Arab American Institute Foundation. 2012. "Quick Facts About Arab Americans." Accessed March 22, 2014 (www.aaiusa.org).

Arendt, Hannah. 2004. *The Origins of Totalitarianism.* New York: Schocken.

Arieff, Allison. 2013. "What Tech Hasn't Learned from Urban Planning." *New York Times*, December 13. Accessed September 19, 2014 (www.nytimes.com).

Ariès, Phillippe. 1962. *Centuries of Childhood.* New York: Vintage.

Armstrong, Elizabeth A., Laura Hamilton, and Paula England. 2010. "Is Hooking Up Bad for Young Women?" *Contexts* 9 (Summer): 22–27.

Armstrong, Elizabeth A., Paula England, and Alison C. K. Fogarty. 2012. "Accounting for Women's Orgasm and Sexual Enjoyment in College Hookups and Relationships." *American Sociological Review* 77 (June): 435–462.

Arntz, Melanie, Terry Gregory, and Ulrich Zierahn. 2016. "The Risk of Automation for Jobs in OECD Countries: A Comparative Analysis." OECD Publishing, Paris. Accessed September 2, 2017 (oecd-library.org).

"Artificial Intelligence." 2015. *The Economist*, October 3, 81–82.

ASA Research Department. 2013. "Recruitment and Retention of Sociology Majors." *ASA Footnotes* 41 (January): 1, 4.

Asante-Muhammad, Dedrick, Chuck Collins, Josh Hoxie, and Emanuel Nieves. 2016. "The Ever-Growing Gap: Without Change, African-American and Latino Families Won't Match White Wealth for Centuries." CFED and Institute for Policy Studies, August. Accessed May 13, 2017 (ips-dc.org).

Asch, Solomon. 1952. *Social Psychology.* Englewood Cliffs, NJ: Prentice-Hall.

"The Ashley Madison Hack." 2015. *The Economist,* August 22, 14.

Asi, Maryam, and Daniel Beaulieu. 2013. "Arab Households in the United States: 2006–2010." U.S. Census Bureau, May. Accessed March 22, 2014 (www.census.gov).

"Asia's New Family Values." 2015. *The Economist*, August 22, 36.

"Asian-American and Pacific Islander Heritage Month: May 2017." 2017. U.S. Census Bureau, March 14. Accessed June 20, 2017 (census.gov).

"Asians Fastest-Growing Race or Ethnic Group in 2012, Census Bureau Reports." 2013. Newsroom, U.S. Census Bureau, June 13. Accessed March 22, 2014 (www.census.gov).

Associated Press and NORC. 2016. "Hillary Clinton's Candidacy and the State of Gender Discrimination in the United States." August. Accessed June 26, 2017 (apnorc.org).

Association of Certified Fraud Examiners. 2016. "Report to the Nations on Occupational Fraud and Abuse: 2016 Global Fraud Study." Accessed April 2, 2016 (acfe.com).

AT&T Newsroom. 2015. "Smartphone Use While Driving Grows Beyond Texting to Social Media, Web Surfing, Selfies, Video Chatting." May 19. Accessed July 21, 2015 (att.com).

Atchley, Robert C., and Amanda S. Barusch. 2004. *Social Forces and Aging: An Introduction to Social Gerontology*, 10th edition. Belmont, CA: Wadsworth.

Attina, Teresa M., et al. 2016. "Exposure to Endocrine-Disrupting Chemicals in the USA: A Population-Based Disease Burden and Cost Analysis." *The Lancet: Diabetes & Endocrinology* 4 (December): 996–1003.

Attinasi, John J. 1994. "Racism, Language Variety, and Urban U.S. Minorities: Issues in Bilingualism and Bidialectalism." Pp. 319–347 in *Race,* edited by Steven Gregory and Roger Sanjek. New Brunswick, NJ: Rutgers University Press.

Atwater, Peter. 2016. "The Wild EpiPen Price Hike Points to a Looming Pharmaceutical Crisis." *Time,* September 12–19, 32.

Aughinbaugh, Alison, Omar Robles, and Hugette Sun. 2013. "Marriage and Divorce: Patterns by Gender, Race, and Educational Attainment." *Monthly Labor Review* (October): 1–19.

Auguste, Byron, Paul Kihn, and Matt Miller. 2010. "Closing the Talent Gap: Attracting and Retaining Top-Third Graduates to Careers in Teaching." McKinsey & Company, September. Accessed March 20, 2012 (www.mckinsey.com).

Aunola, Kaisa, and Jari-Erik Nurmi. 2005. "The Role of Parenting Styles in Children's Problem Behavior." *Child Development* 76 (November/December): 1144–1159.

Auster, Ellen R., and Ajnesh Prasad. 2016. "Why Do Women Still Not Make It to the Top? Dominant Organizational Ideologies and Biases by Promotion Committees Limit Opportunities to Destination Positions." *Sex Roles* 75 (5/6): 177–196.

Austin, Algernon. 2013. "Native Americans and Jobs." Economic Policy Institute, December 17. Accessed March 22, 2014 (www.epi.org).

Autor, David H. 2014. "Skills, Education, and the Rise of Earnings Inequality Among the 'Other 99 Percent.'" *Science* 344 (May): 843–851.

Avellar, Sarah, and Pamela Smock. 2005. "The Economic Consequences of the Dissolution of Cohabiting Unions." *Journal of Marriage and Family* 67 (May): 315–327.

Azofeifa, Alejandro, Margaret E. Mattson, and Rob Lyerla. 2015. "Driving Under the Influence of Alcohol, Marijuana, and Alcohol and Marijuana Combined Among Persons Aged 16–25 Years—United States, 2002–2014." *Morbidity and Mortality Weekly Report* 64 (48): 1325–1329.

B

Babbie, Earl. 2013. *Social Research Counts*. Belmont, CA: Cengage.

Babcock, Philip, and Mindy Marks. 2011. "The Falling Time Cost of College: Evidence from Half a Century of Time Use Data." *The Review of Economics and Statistics* 83 (May): 468–478.

Bach, Peter B. 2015. "Seeking a Cure for Drug-Price Insanity." *Fortune*, September 15, 63–64.

Bachrach, Deborah, and Jonah Frohlich. 2016. "Retail Clinics Drive New Health Care Utilization and That Is a Good Thing." Health Affairs Blog, May 20. Accessed April 3, 2017 (healthaffairs.org).

Badger, Emily. 2015. "New Census Data: Americans Are Returning to the Far-Flung Suburbs." *Washington Post*, March 26. Accessed August 20, 2017 (washingtonpost.com).

Bagri, Neha Thirani, and Nida Najar. 2015. "Chicken: It's What's for Dinner in an Indian State, Even Among Caged Carnivores." *New York Times*, March 30, A4.

Bahadur, Nina. 2014. "'Misconception' Reveals the Dark, Misleading World of Crisis Pregnancy Centers." *Huffington Post*, September 18. Accessed September 29, 2015 (huffingtonpost .com).

Bai, Ge, and Gerard F. Anderson. 2015. "Extreme Markup: The Fifty US Hospitals with the Highest Charge-to-Cost Ratios." *Health Affairs* 34 (6): 922–928.

Bainbridge, William S. 1997. *The Sociology of Religious Movements.* New York: Routledge.

Baker, Aryn. 2015. "Out in Africa." *Time*, August 17, 34–39.

Bales, Robert F. 1950. *Interaction Process Analysis*. Reading, MA: Addison-Wesley.

Baltimore Sun. 2012. "Elderly Man Dies in Fire at Single-Family Home." December 9, 5.

Bandura, Albert, and Richard H. Walters. 1963. *Social Learning and Personality Development.* New York: Holt, Rinehart & Winston.

Bandy, Tawana. 2012. "What Works for Male Children and Adolescents: Lessons from Experimental Evaluations of Programs and Interventions." Child Trends Research Brief, August. Accessed September 22, 2012 (www .childtrends.org).

Banerjee, Neela. 2006. "Clergywomen Find Hard Path to Bigger Pulpit." *New York Times*, August 26, A1, A12.

Banerjee, Neela, Lisa Song, and David Hasemyer. 2015. "Exxon: The Road Not Taken." Inside Climate News, September 16. Accessed August 20, 2017 (insideclimatenews.org).

Banfield, Edward C. 1974. *The Unheavenly City Revisited.* Boston: Little, Brown.

Banks, Ingrid. 2000. *Hair Matters: Beauty, Power, and Black Women's Consciousness.* New York: New York University Press.

Barash, David P. 2012. "The Evolutionary Mystery of Homosexuality." *Chronicle Review*, November 21, B4–B5.

Barbara Lee Foundation. 2017. "Modern Family: How Women Candidates Can Talk About Politics, Parenting, and Their Personal Lives." Accessed June 26, 2017 (barbaraleefoundation.org).

Barnes, Robert. 2011. "Limits on Video Games Rejected." *Washington Post*, June 28, A1.

Barnett, Jessica C., and Marina S. Vornovitsky. 2016. "Health Insurance Coverage in the United States: 2015." U.S. Census Bureau, Current Population Reports, September. Accessed August 15, 2017 (census.gov).

Barreto, Michelle, Micheal K. Ryan, and Manuela T. Schmitt, eds. 2009. *The Glass Ceiling in the 21st Century: Understanding Barriers to Gender Equality*. Washington, DC: American Psychological Association.

Barro, Josh. 2015. "How Unpopular Is Trump's Muslim Ban? Depends on How You Ask." *New York Times*, December 16, A21.

Barro, Robert J., and Rachel M. McCleary. 2003. "Religion and Economic Growth Across Countries." *American Sociological Review* 68 (October): 760–781.

Barry, Ellen. 2013. "Policing Village Moral Codes as Women Stream to India's Cities." *New York Times*, October 20, A6.

Barry, Ellen, and Mansi Choksi. 2013. "Gang Rape in India, Routine and Invisible." *New York Times*, October 27, A1.

Barstow, David, and Suhasini Raj. 2015. "Caste Quotas in India Come Under Attack." *New York Times*, August 31, A1, A3.

Barthel, Michael, Amy Mitchell, and Jesse Holcomb. 2016. "Many Americans Believe Fake News Is Sowing Confusion." Pew Research Center, December 15. Accessed December 17, 2016 (pewresearch.org).

Bartlett, Thomas, and Paula Wasley. 2008. "Just Say 'A': Grade Inflation Undergoes Reality Check." *Chronicle of Higher Education*, September 5. Accessed June 21, 2014 (www .chronicle.com).

Bartlett, Tom. 2011. "Caffeine Is Definitely Good/Bad for You." *Chronicle of Higher Education*, January 20. Accessed January 25, 2011 (www .chronicle.com/blogs).

Barton, Allen H. 1980. "A Diagnosis of Bureaucratic Maladies." Pp. 27–36 in *Making Bureaucracies Work*, edited by Carol H. Weiss and Allen H. Barton. Beverly Hills, CA: Sage.

Basken, Paul. 2016. "Data Could Help Scholars Persuade, If Only They Were Willing to Use It." *Chronicle of Higher Education*, January 29, A23–A24.

Basken, Paul. 2016. "To Curtail Violence, Researchers Say, Reduce Economic Inequality." *Chronicle of Higher Education*, September 2, A34.

Bass, Frank, and Dakin Campbell. 2013. "Poor Neighborhoods See Branches Disappear." *Bloomberg Businessweek*, May 13–19, 48–49.

Bates, Nancy, and Theresa J. DeMaio. 2013. "Measuring Same-Sex Relationships." *Contexts* 12 (Winter): 66–69.

Baum, Sandy, Jennifer Ma, and Kathleen Payea. 2013. "Education Pays 2013: The Benefits of Higher Education for Individuals and Society." College Board. Accessed June 21, 2014 (www.collegeboard.org).

Baumrind, Diana. 1968. "Authoritarian versus Authoritative Parental Control." *Adolescence* 3 (11): 255–272.

Baumrind, Diana. 1989. "Rearing Competent Children." Pp. 349–378 in *Child Development Today and Tomorrow*, edited by William Damon. San Francisco: Jossey-Bass.

Bax, Pauline. 2013. "In Ghana, Death Has Become Big Business." *Bloomburg Businessweek*, August 29–September 1, 23–26.

Bazelon, Emily. 2009. "2 Kids + 0 Husbands = Family." *New York Times Magazine*, February 1, 30.

BBC News. 2016. "How Has the US Gun Lobby Been So Successful?" January 27. Accessed September 2, 2017 (bbc.com).

Beadle, Amanda Peterson. 2012. "Minnesota Senator Thinks Abortion Pill Is Wrong, While Viagra Is a 'Wonderful Drug.'" Alternet, May 3. Accessed May 24 (www .alternet.org).

Beard, Henry, and Christopher Cerf. 2015. *Spinglish: The Definitive Dictionary of Deliberately Deceptive Language*. New York: Blue Rider Press.

Beck, Melinda. 2013. "Nurse Practitioners Seek Right to Treat Patients on Their Own." *Wall Street Journal*, August 15, A3.

Becker, Howard S. 1963. *Outsiders: Studies in the Sociology of Deviance.* New York: Free Press.

Becker, Howard, and Ruth H. Useem. 1942. "Sociological Analysis of the Dyad." *American Sociological Review* 7 (February): 3–26.

Becker, Sam. 2015. "Places Where People Tip (and Don't Tip) Around the World." Money & Career Cheat Sheet, September 4. Accessed February 14, 2017 (cheatsheet.com).

Beckles, Gloria L., and Benedict I. Truman. 2013. "Education and Income—United States, 2009 and 2011." *MMWR* 62 (November 22): 9–19.

Beckwith, Ryan Teague. 2017. "Read President Trump's Remarks at the National Prayer Breakfast." *Time*, February 2. Accessed July 31, 2017 (time.com).

Beech, Hannah. 2013. "The Face of Buddhist Terror." *Time*, July 1, 42–50.

Begley, Sharon. 2010. "Sins of the Grandfathers." *Newsweek*, November 8, 48–50.

Beil, Laura. 2012. "How Much Would You Pay for Three More Months of Life?" *Newsweek*, September 3, 40–44.

Beitsch, Rebecca. 2016. "Tiny Houses Are Trendy, Minimalist and Often Illegal." PBS News Hour, July 6. Accessed August 20, 2017 (pbs.org).

Belkin, Douglas, and Caroline Porter. 2012. "Web Profiles Haunt Students." *Wall Street Journal*, October 4, A3.

Belknap, Joanne. 2007. *The Invisible Woman: Gender, Crime, and Justice,* 3rd edition. Belmont, CA: Wadsworth Press.

Bell, Claes. 2015. "Budgets Can Crumble in Times of Trouble." Bankrate, January 7. Accessed August 26, 2015 (bankrate.com).

Bellini, Jason. 2012. "A Brief History of the NRA." *Wall Street Journal*, December 20. Accessed September 2, 2017 (wsj.com).

Belluck, Pam. 2017. "Pregnancy Prevention Programs Are at Risk of Being Cut." *New York Times*, August 11, 10.

Bendavid, Naftali. 2013. "Countries Expand Recognition for Alternative 'Intersex' Gender." *Wall Street Journal*, October 31, A9.

Bengali, Shashank. 2014. "Toll of Refugees Fleeing Syrian Strife Tops 3 Million, U.N. Says." *Baltimore Sun*, August 30, 10.

Bennett, Jessica. 2009. "Tales of a Modern Diva." *Newsweek,* April 6, 42–43.

Bennett, Jessica, and Jesse Ellison. 2010. "'I Don't': The Case Against Marriage." *Newsweek*, June 21, 42–45.

Bennhold, Katrin, and Rick Gladstone. 2016. "Women Have Led over 70 Nations, but Experts See Pipeline Problem in U.S." *New York Times*, November 4, P11.

Benokraitis, Nijole V. 1997. *Subtle Sexism: Current Practices and Prospects for Change,* 2nd edition. Thousand Oaks, CA: Sage.

Benokraitis, Nijole V. 2015. *Marriages & Families: Changes, Choices, and Constraints*, 8th edition. Upper Saddle River, NJ: Prentice Hall.

Benson, Michael L. 2002. *Crime and the Life Course.* Los Angeles: Roxbury.

Benton, Thomas H. 2011. "A Perfect Storm in Undergraduate Education." *Chronicle of Higher Education,* February 25, A43, A45. Part 2 in April 8, A45–A46.

Berenson, Tessa. 2016. "How Latinos Drive America's Economic Growth." *Time*, August 26, 32–35.

Beres, Damon. 2016. "Here's How Much People Trust Facebook." *Huffington Post*, April 27. Accessed May 1, 2016 (huffingtonpost.com).

Berfield, Susan. 2012. "Levi's Has a New Color for Blue Jeans: Green." *Bloomberg Businessweek*, October 22–28, 26–28.

Berfield, Susan. 2013. "Fast-Food Workers of the World, Unite!" *Bloomberg Businessweek*, December 9–15, 20–22.

Berger, Peter L., and Thomas Luckmann. 1966. *The Social Construction of Reality: A Treatise in the Sociology of Knowledge.* New York: Doubleday.

Berk, Richard A. 1974. *Collective Behavior.* Dubuque, IA: Wm. C. Brown Company.

Berkos, Kristen M., Terre H. Allen, Patricia Kearney, and Timothy G. Plax. 2001. "When Norms Are Violated: Imagined Interactions as Processing and Coping Mechanisms." *Communication Monographs* 68 (September): 289–300.

Berkowitz, Bill. 2011. "Why Is Jerry Falwell's Evangelical University Getting Filthy Rich Off Your Tax Money?" AlterNet, June 29. Accessed June 29, 2011 (www.alternet.org).

Berlin, Overton B., and Paul Kay. 1969. *Basic Color Terms.* Berkeley: University of California Press.

Berrett, Dan. 2014. "Some Elite Colleges Reject AP Credits in Favor of New Core Classes." *Chronicle of Higher Education*, February 21, A13.

Bersani, Bianca E. 2014. "A Game of Catch-Up? The Offending Experience of Second-Generation Immigrants." *Crime & Delinquency* 60 (February): 60–84.

Bertrand, Marianne, and Jessica Pan. 2011. "The Trouble with Boys: Social Influences and the Gender Gap in Disruptive Behavior." National Bureau of Economic Research, October. Accessed April 14, 2012 (www.nber.org).

Berube, Alan, and Natalie Holmes. 2015. "Some Cities Are Still More Unequal Than Others—An Update." Brookings Institution, March 17. Accessed September 31, 2015 (brookings .edu).

Berube, Alan, Audrey Sinter, Jill H. Wilson, and William H. Frey. 2006. "Finding Exurbia: America's Fast-Growing Communities at the Metropolitan Fringe." Brookings Institution. Accessed February 2, 2008 (www.brookings .edu).

Best, Joel. 2006. *Flavor of the Month: Why Smart People Fall for Fads.* Berkeley: University of California Press.

Betcher, R. William, and William S. Pollack. 1993. *In a Time of Fallen Heroes: The Re-Creation of Masculinity.* New York: Atheneum.

Bettinger, Eric P., Bridget T. Long, Philip Oreopoulos, and Lisa Sanbonmatsu. 2012. "The Role of Application Assistance and Information in College Decisions: Results from the H&R Block FAFSA Experiment." *Quarterly Journal of Economics* 127 (August): 1205–1242.

Bharadwaj, Prashant, Mallesh M. Pai, and Agne Suziedelyte. 2015. "Mental Health Stigma." National Bureau of Economic Research, June. Accessed July 22, 2015 (nber.org).

Bhardwaj, Deepika. 2014. "Delhi Takes Forward 'Kiss of Love' Protest Amid Police Barricades and Assaults." *Times of India*, November 10. Accessed September 27, 2015 (timesofindia .indiatimes.com).

Bialik, Carl. 2011. "Irreconcilable Claim: Facebook Causes 1 in 5 Divorces." *Wall Street Journal,* March 12. Accessed November 8, 2011 (www.online.wsj.com).

Bialik, Kristen. 2016. "Most Americans Favor Stricter Environmental Laws and Regulations." Pew Research Center, December 14. Accessed August 25, 2017 (pewresearch.org).

Bialik, Kristen. 2017. "6 Facts About U.S. Mothers." Pew Research Center, May 11. Accessed August 25, 2017 (pewresearch.org).

Bialik, Kristen, and Anthony Cilluffo. 2017. "6 Facts About Black Americans for Black History Month." Pew Research Center, February 22. Accessed June 20, 2017 (pewresearch.org).

Bick, Alexander, Bettina Brüggemann, and Nicola Fuchs-Schundeln. 2017. "Hours Worked in Europe and the US: New Data, New Answers." Institute for the Study of Labor, April 12. Accessed May 25, 2017 (wiwi .uni-frankfurt.de).

Biesecker, Michael. 2017. "Trump Administration Guts EPA Science Panel." *Baltimore Sun*, May 9, 4.

Biggs, Antonia M., Ushma D. Upadhyay, Charles E. McCulloch, and Diana G. Foster. 2017. "Women's Mental Health and Well-being 5 Years After Receiving or Being Denied an Abortion: A Prospective, Longitudinal Cohort Study." *JAMA Psychiatry* 74 (2): 169–178.

Bilefsky, Dan. 2017. "E.U. Court, in 'Bold Step,' Allows Banning Head Scarves at Work." *New York Times*, March 15, A1, A7.

Billingsley, Andrew. 1992. *Climbing Jacob's Ladder: The Enduring Legacy of African-American Families.* New York: Simon & Schuster.

"Biomedical Engineering." 2017. *The Economist*, April 1, 72–73.

Bird, Warren, and Scott Thumma. 2011. "A New Decade of Megachurches: 2011 Profile of Large Attendance Churches in the United States." Leadership and Network. Accessed June 21, 2014 (www.leadnet.org).

Bischoff, Kendra, and Sean F. Reardon. 2013. "Residential Segregation by Income, 1970–2009." Russell Sage Foundation, October 16. Accessed September 19, 2014 (www .russellsage.org).

Bitterman, Amy, Rebecca Goldring, and Lucinda Gray. 2013. "Characteristics of Public and Private Elementary and Secondary School Principals in the United States: Results from the 2011–12 Schools and Staffing Survey." National Center for Education Statistics, August. Accessed November 17, 2014 (www .nces.ed.gov).

Biunno, J. B. 2016. "Inmates Reveal How Smartphones Are Smuggled into Prison." WKRG News, March 16. Accessed March 16, 2017 (wkrg.com).

Bivens, Josh, and Lawrence Mishel. 2015. "Understanding the Historic Divergence Between Productivity and a Typical Worker's Pay." Economic Policy Institute, September 2. Accessed October 7, 2015 (epi.org).

Black, M. C., et al. 2011. "National Intimate Partner and Sexual Violence Survey: 2010 Summary Report." Centers for Disease Control and Prevention. Accessed November 2013 (cdc.gov).

Black, Sandra E., Jane A. Lincove, Jenna Cullinane, and Rachel Veron. 2014. "Can You Leave High School Behind?" National Bureau of Economic Research, January. Accessed June 21, 2014 (www.nber.org).

Black, Sandra E., Paul J. Devereux, Petter Lundborg, and Kaveh Majlesi. 2015. "Poor Little Rich Kids? The Determinants of the Intergenerational Transmission of Wealth." National Bureau of Economic Research, December. Accessed March 1, 2016 (nber.org).

Black, Thomas. 2010. "More Car Jobs Shift to Mexico." *Bloomberg Businessweek*, June 28–July 4, 10–11.

Blake, Mariah. 2014. "Are Any Plastics Safe?" *Mother Jones*, March/April, 19–25, 60–61.

Blanchard, Dallas A. 1994. *The Anti-Abortion Movement and the Rise of the Religious Right: From Polite to Fiery Protest.* New York: Twayne Publishers.

Blau, Francine D., and Christopher Mackie, eds. 2016. *The Economic and Fiscal Consequences of Immigration*. Washington, DC: The National Academies Press.

Blau, Francine D., and Lawrence M. Kahn. 2016. "The Gender Wage Gap: Extent, Trends, and Explanations." National Bureau of Economic Research, March. Accessed March 13, 2017 (nber.org).

Blau, Peter M. 1986. *Exchange and Power in Social Life,* revised edition. New Brunswick, NJ: Transaction.

Blau, Peter M., and Marshall W. Meyer. 1987. *Bureaucracy in Modern Society,* 3rd edition. New York: Random House.

Blay, Zeba. 2015. "How Feminist TV Became the New Normal." *Huffington Post*, June 18. Accessed September 29, 2015 (huffingtonpost.com).

Blee, Kathleen. 2015. "How Grassroots Groups Lose Political Imagination." *Contexts* 14 (1): 32–37.

BLS News Release. 2016. "National Census of Fatal Occupational Injuries in 2015." Bureau of Labor Statistics, December 16. Accessed August 20, 2017 (bls.gov).

BLS News Release. 2016. "Volunteering in the United States—2015." U.S. Department of Labor, February 25. Accessed March 30, 2017 (bls.gov).

BLS News Release. 2017. "The Employment Situation—May 2017." U.S. Bureau of Labor Statistics, June 2. Accessed June 5, 2017 (bls.gov).

BLS News Release. 2017. "Usual Weekly Earnings of Wage and Salary Workers, First Quarter 2017." U.S. Bureau of Labor Statistics, April 18. Accessed May 10, 2017 (bls.gov).

BLS Reports. 2017. "A Profile of the Working Poor, 2015." U.S. Bureau of Labor Statistics, April. Accessed May 5, 2017 (bls.gov).

BLS Reports. 2017. "Women in the Labor Force: A Databook." U.S. Bureau of Labor Statistics, April. Accessed May 10, 2017 (bls.gov).

Blumenstyk, Goldie. 2015. "When a Degree Is Just the Beginning." *Chronicle of Higher Education*, September 18, B4–B7.

Blumer, Herbert. 1946. "Collective Behavior." Pp. 65–121 in *New Outline of the Principles of Sociology*, edited by Alfred M. Lee. New York: Barnes & Noble.

Blumer, Herbert. 1969. *Symbolic Interactionism: Perspective and Method.* Englewood Cliffs, NJ: Prentice Hall.

Blunden, Jessica, and Derek S. Arndt, eds. 2017. "State of the Climate in 2016." Special Supplement to the *Bulletin of the American Meteorological Society* Vol. 98, No. 8, August. Accessed August 20, 2017 (ametsoc.org).

Board of Trustees, The. 2013. *The 2013 Annual Report of the Board of Trustees of the Federal Old-Age and Survivors Insurance and Federal Disability Insurance Trust Funds*. Accessed July 28, 2013 (www.socialsecurity.gov).

Bogage, Joseph. 2015. "Why Companies Are Speaking Up About Gay Marriage." *Washington Post*, June 26. Accessed September 30, 2015 (washingtonpost.com).

Bogle, Kathleen A. 2008. *Hooking Up: Sex, Dating, and Relationships on Campus.* New York: New York University Press.

Bohannon, Paul, ed. 1971. *Divorce and After*. New York: Doubleday.

Bollag, Burton. 2007. "Credential Creep." *Chronicle of Higher Education,* June 22, A10–A12.

Bonacich, Edna. 1972. "A Theory of Ethnic Antagonism: The Split Labor Market." *American Sociological Review* 37 (October): 547–559.

Bookman, Todd, and Brett Neeley. "Labor Unions Appear Set for More State-Level Defeats in 2017." NPR, January 30. Accessed June 19, 2017 (npr.org).

Borgerhoff Mulder, Monique. 2009. "Serial Monogamy as Polygyny or Polyandry?" *Human Nature* 20 (Summer): 130–150.

Borjas, George J. 2017. "The Earnings of Undocumented Immigrants." National Bureau of Economic Research, March. Accessed June 20, 2017 (nber.org).

Bosk, Charles. 1979. *Forgive and Remember: Managing Medical Failure.* Chicago: University of Chicago Press.

Boslaugh, Sarah. 2007. *Secondary Data Sources for Public Health: A Practical Guide*. New York: Cambridge University Press.

Botstein, Leon. 2014. "The SAT Is Part Hoax, Part Fraud." *Time*, March 24, 17.

Bourdieu, Pierre. 1984. *Distinction: A Social Critique of the Judgement of Taste.* Translated by Richard Nice. Cambridge, MA: Harvard University Press.

Bourdieu, Pierre. 1986. "The Forms of Capital." Pp. 241–258 in *Handbook for Theory and Research for the Sociology of Education*, edited by J. G. Richardson. Westport, CT: Greenwood.

Bouvard, Véronique, et al. 2015. "Carcinogenicity of Consumption of Red and Processed Meat." *Lance Oncology*, October 26. Accessed December 15, 2016 (iarc.fr).

Bowie, Liz. 2016. "Schools Teaching Etiquette as an Antidote to Lack of Respect." *Baltimore Sun*, May 16, 1, 11.

Bowles, Samuel, and Herbert Gintis. 1977. *Schooling in Capitalist America: Educational Reform and the Contradictions of Economic Life.* New York: Basic Books.

Boyle, Matthew. 2013. "Yes, Real Men Drink Beer and Use Skin Moisturizer." *Bloomberg Businessweek*, October 7–13, 32–33.

Bradshaw, York W., and Michael Wallace. 1996. *Global Inequalities.* Thousand Oaks, CA: Pine Forge Press.

Braff, Danielle. 2016. "Single Mothers by Choice." *Baltimore Sun*, April 3, 7.

Braga, Anthony A. 2003. "Systematic Review of the Effects of Hot Spots Policing on Crime." Unpublished paper. Accessed May 14, 2005 (www.campbellcollaboration.org).

Braitman, Keli A., Neil K. Chaudhary, and Anne T. McCart. 2011. "Effect of Passenger Presence on Older Drivers' Risk of Fatal Crash Involvement." Insurance Institute for Highway Safety, March. Accessed July 4, 2011 (www.iihs.org).

Breining, Sanni N., Joseph J. Doyle, Jr., David N. Figlio, Krzysztof Karbownik, and Jeffrey Roth. 2017. "Birth Order and Delinquency: Evidence from Denmark and Florida." National Bureau of Economic Research, January. Accessed March 16, 2017 (nber.org).

Brescoll, Victoria L., and Eric L. Uhlmann. 2008. "Can an Angry Woman Get Ahead? Status Conferral, Gender, and Expression of Emotion in the Workplace." *Psychological Science* 19 (March): 268–275.

Brill, Steven. 2013. "Bitter Pill: How Outrageous Pricing and Egregious Profits Are Destroying Our Health Care." *Time*, March 4, 14–55.

Brill, Steven. 2015. "What I Learned from My $190,000 Surgery." *Time*, January 19, 34–43.

Brody, Jane E. 2015. "Limit Children's Screen Time, and Your Own" *New York Times*, July 14, D7.

Brody, Jane E. 2015. "Screens Separate the Obsessed from Life." *New York Times*, July 7, D7.

Brody, Jane E. 2015. "The More We Learn, the More We Ignore." *New York Times*, October 13, D7.

Brooks, Arthur C. 2007. "I Love My Work." *The American* (online), September–October. Accessed September 30, 2007 (www.theamericanmag.com).

Brown, Anna. 2016. "The Challenges of Translating the U.S. Census Questionnaire into Arabic." Pew Research Center, June 3. Accessed June 20, 2017 (pewresearch.org).

Brown, Anna. 2016. "What Americans Say It Takes to Be Middle Class." Pew Research Center, February 4. Accessed July 26, 2016 (pewresearch.org).

Brown, Archie. 2010. "Signposts: Why Did Communism End When It Did?" *History Today*, March. Accessed April 12, 2014 (www.historytoday.com).

Brown, DeNeen L. 2009. "The High Cost of Poverty: Why the Poor Pay More." *Washington Post*, May 18, C1.

Brown, Michael, and Amy Goldin. 1973. *Collective Behavior: A Review and Reinterpretation of the Literature*. Pacific Palisades, CA: Goodyear.

Brown, Susan I. 2005. "How Cohabitation Is Reshaping American Families." *Contexts* 4 (Summer): 33–37.

Bruinius, Harry. 2013. "Whose Holidays Are They?" *Christian Science Monitor*, December 16, 21–23.

Bruinius, Harry. 2016. "A New Twist in Religious Liberty Debate." *Christian Science Monitor Weekly*, April 25 & May 2, 21–23.

Brummelman, Eddie, Sander Thomaes, Stefanie A. Nelemans, Bram Orobio de Castro, Geertjan Overbeek, and Brad J. Bushman. 2015. "Origins of Narcissism in Children." *Proceedings of the National Academy of Sciences* 112 (March 25): 3659–3662.

Brunvand, Jan H. 2001. *The Truth Never Stands in the Way of a Good Story.* Urbana and Chicago: University of Illinois Press.

Bryant, Karl. 2014. "Teaching the Nature-Nurture Debate." *Contexts* 13 (Fall): 22–23.

Buchanan, Larry, Andrew W. Lehren, Jugal K. Patel, and Adam Pearce. 2017. "How Much People in the Trump Administration Are Worth." *New York Times*, April 3. Accessed May 5, 2017 (nytimes.com).

Buchheit, Paul. 2015. "The Corporate Debt to Society: $10,000 Per Household, Per Year." AlterNet, February 22. Accessed August 26, 2015 (alternet.org).

Buchheit, Paul. 2015. "The Number 1 Thing Rich People Get Dead Wrong About Poor People." AlterNet, November 20. Accessed June 4, 2017 (alternet.org).

Buchheit, Paul. 2017. "The Rich Pay Fewer Taxes Than the Poor, and Get More Services." AlterNet, March 20. Accessed June 4, 2017 (alternet.org).

Buckley, Chris. 2015. "China Approves a Two-Child Policy to Help Economy." *New York Times*, October 30, A1, A6.

Buechler, Steven M. 2000. *Social Movements in Advanced Capitalism: The Political Economy and Cultural Construction of Social Activism.* New York: Oxford University Press.

Buffington, Catherine, Benjamin Cerf Harris, Christina Jones, and Bruce A. Weinberg. 2016. "STEM Training and Early Career Outcomes of Female and Male Graduate Students: Evidence from UMETRICS Data Linked to the 2010 Census." *American Economic Review* 106 (May): 333–338.

Bullock, Karen. 2005. "Grandfathers and the Impact of Raising Grandchildren." *Journal of Sociology and Social Welfare* 32 (March): 43–59.

Buntin, John. 2015. "The Myth of Gentrification." *Slate*, January 14. Accessed October 23, 2015 (slate.com).

Burd, Stephen. 2013. "Undermining Pell: How Colleges Compete for Wealthy Students and Leave the Low-Income Behind." New America Foundation, May. Accessed June 21, 2014 (newamerica.org).

Bureau of Labor Statistics. 2017. "Demographics." Accessed July 20, 2017 (bls.gov).

Bureau of Labor Statistics. 2017. "Household Data Annual Averages, Table 39." Accessed April 10, 2017 (bls.gov).

Busteed, Brandon. 2015. "Is College Worth It? That Depends." Gallup, April 8. Accessed July 30, 2017 (gallup.com).

Butler, Kiera. 2014. "Google's Magic Bus." *Mother Jones*, May–June, 70–71.

Butler, Kiera. 2015. "A Tale of Two Diseases." *Mother Jones*, May/June, 68.

C

Cadge, Wendy, and Courtney Bender. 2004. "Yoga and Rebirth in America: Asian Religions Are Here to Stay." *Contexts* 3 (Winter): 45–51.

Cahalan, M., L. W. Perna, M. Yamashita, R. Ruiz, and K. Franklin. 2017. *Indicators of Higher Education Equity in the United States: 2017 Trend Report*. Pell Institute for the Study of Higher Education, Council for Education Opportunity, and Alliance for Higher Education and Democracy. Accessed July 31, 2017 (pellinstitute.org).

Calacal, Celisa. 2017. "Not a Single Black Woman Heads a Top Fortune 500 Company." AlterNet, June 8. Accessed June 20, 2017 (alternet.org).

Calderon, Valerie J., and Preety Sidhu. 2014. "Business Leaders Say Knowledge Trumps College Pedigree." Gallup, February 25. Accessed June 21, 2014 (www.gallup.com).

Calderon, Valerie J., and Susan Sorenson. 2014. "Americans Say College Degree Leads to a Better Life." Gallup, April 7. Accessed June 21, 2014 (www.gallup.com).

Callahan, David. 2004. *The Cheating Culture: Why More Americans Are Doing Wrong to Get Ahead*. Boston: Houghton Mifflin Harcourt.

Cameron, Deborah. 2007. *The Myth of Mars and Venus*. New York: Oxford University Press.

Campion-Vincent, Véronique. 2005. "From Evil Others to Evil Elites: A Dominant Pattern in Conspiracy Theories Today." Pp. 103–122 in *Rumor Mills: The Social Impact of Rumor and Legend*, edited by Gary Alan Fine, Veronique Campion-Vincent, and Chip Heath. New Brunswick, NJ: Transaction Publishers.

Campo-Flores, Arian. 2013. "Religious Dorms Sprout Up." *Wall Street Journal*, September 3, A3.

Cano, Regina Garcia. 2015. "S.D. Tribe to Open Nation's 1st Pot Resort." *Baltimore Sun*, September 30, 10.

Cantor, David, et al. 2015. "Report on the AAU Campus Climate Survey on Sexual Assault and Sexual Misconduct." Westat, September 21. Accessed July 31, 2017 (westat.com).

Cantrell, Steven, and Thomas J. Kane. 2013. "Ensuring Fair and Reliable Measures of Effective Teaching." MET Project, Policy and Practice Brief. Accessed June 21, 2014 (www.metproject.org).

Cantril, Hadley, Hazel Gaudet, and Herta Herzog. 1952. *The Invasion from Mars*. Princeton, NJ: Princeton University Press.

Card, David, and Laura Giuliano. 2016. "Can Tracking Raise the Test Scores of High-Ability Minority Students?" National Bureau of Economic Research, March. Accessed July 31, 2017 (nber.org).

Carey, Anne R., and Paul Trap. 2014. "How Much Are 'Mom Jobs' Worth?" *USA Today*, May 5–9, 1A.

Carey, Benedict. 2010. "Revising Book on Disorders of the Mind." *New York Times*, February 10, 1.

Carey, Kevin. 2013. "Too Much 'Merit Aid' Requires No Merit." *Chronicle of Higher Education*, February 22, A20–A21.

Carl, Traci. 2002. "Amid Latte, Mocha Craze, Coffee Growers Go Hungry in Paradise." Accessed September 19, 2002 (www.yahoo.com).

Carnagey, Nicholas L., and Craig A. Anderson. 2005. "The Effects of Reward and Punishment in Violent Video Games on Aggressive Affect, Cognition, and Behavior." *Psychological Science* 16 (November): 882–889.

Carnevale, Anthony P., and Ban Cheah. 2013. "Hard Times: College Majors, Unemployment, and Earnings." Georgetown Center on Education and the Workforce, May. Accessed June 24, 2014 (www.cew.georgetown.edu).

Carney, Ginny. 1997. "Native American Loanwords in American English." *Wicazo SA Review*, 12 (Spring): 189–203

Carnoy, Martin, and Emma García. 2017. "Five Key Trends in U.S. Student Performance." Economic Policy Institute, January 12. Accessed July 31, 2017 (epi.org).

Carnoy, Martin, and Richard Rothstein. 2013. "What Do International Tests Really Show About U.S. Student Performance?" Economic Policy Institute, January 28. Accessed June 21, 2014 (www.epi.org).

Carpenter, Christopher S., Chandler B. McClellan, and Daniel I. Rees. 2016. "Economic Conditions, Illicit Drug Use, and Substance Use Disorders in the United States." National Bureau of Economic Research, February. Accessed March 16, 2017 (nber.org).

Carrell, Scott E., Mark Hoekstra, and James E. West. 2015. "The Impact of Intergroup Contact on Racial Attitudes and Revealed Preferences." National Bureau of Economic Research, February. Accessed May 1, 2015 (nber.org).

Carson, E. Ann, and Elizabeth Anderson. 2016. "Prisoners in 2015." Bureau of Justice Statistics, December. Accessed March 2, 2017 (bjs.gov).

Case, Anne, and Angus Deaton. 2017. "Mortality and Morbidity in the 21st Century." Brookings, March 23. Accessed May 22, 2017 (brookings.edu).

Cason, Mike. 2017. "Roy Moore Running for Senate, Resigns from Supreme Court to Challenge Luther Strange." *Montgomery News*, April 26. Accessed May 25, 2017 (al.com/news/Montgomery).

Catalano, Shannan. 2013. "Intimate Partner Violence: Attributes of Victimization, 1993–2011." Bureau of Justice Statistics, November. Accessed November 25, 2014 (www.bjs.gov).

Catalyst. 2017. "Pyramid: Women in S&P 500 Companies." March 1. Accessed April 10, 2017 (catalyst.org).

Catalyst. 2017. "Women CEOs of the S&P 500." March 14. Accessed April 2, 2017 (catalyst.org).

"Catching Up." 2013. *The Economist*, January 12, 67–68.

CDC/NCHS National Vital Statistics System. 2017. "National Marriage and Divorce Rate Trends." Centers for Disease Control and Prevention, March 17. Accessed July 8, 2017 (cdc.gov/nchs).

Cellini, Stephanie R., Signe-Mary McKernan, and Caroline Ratcliffe. 2008. "The Dynamics of Poverty in the United States: A Review of Data, Methods, and Findings." *Journal of Policy Analysis and Management* 27 (Summer): 577–605.

Census Bureau News. 2016. "The Majority of Children Live with Two Parents, Census Bureau Reports." News release, November 17. Accessed January 3, 2017 (census.gov).

Center for American Women and Politics. 2017. "Women in Elective Office 2017." Accessed May 22, 2017 (cawp.rutgers.edu).

Center for Behavioral Health Statistics and Quality. 2016. "*Key Substance Use and Mental Health Indicators in the United States: Results from the 2015 National Survey on Drug Use and Health*." Accessed August 15, 2017 (samhsa.gov).

Center for Economic and Policy Research. 2015. "The Rise of Discouraged Workers." March 25. Accessed May 23, 2017 (cepr.net).

Center for Responsive Politics. 2017. "2016 Outside Spending, by Super Pac." Open Secrets, June 25. Accessed June 26, 2017 (opensecrets.org).

Center for Responsive Politics. 2017. "Cost of Election." Open Secrets, June 25. Accessed June 26, 2017 (opensecrets.org).

Center for the Study of Global Christianity. 2007. "Global Table 5: Status of Global Mission, Presence and Activities, AD 1800–2005." Gordon-Conwell Theological Seminary. Accessed August 24, 2007 (www.gcts.edu).

Center on Budget and Policy Priorities. 2016. "Policy Basics: Where Do Our Federal Dollars Go?" March 4. Accessed May 5, 2017 (cbpp.org).

Centers for Disease Control and Prevention. 2016. "Sexually Transmitted Disease Surveillance 2015." Division of STD Prevention, October. Accessed August 15, 2017 (cdc.gov).

Centers for Medicare & Medicaid Services. 2010. "NHE Tables for Selected Calendar 1960–2010." April 11. Accessed April 24, 2010 (www.cms.gov).

Centers for Medicare & Medicaid Services. 2014. "National Health Expenditure Projections 2012–2022." May 6. Accessed July 26, 2014 (www.cms.gov).

Cha, Ariana Eunjung. 2014. "'Smart Pills' with Chips, Cameras and Robotic Parts Raise Legal, Ethical Questions." *Washington Post*, May 24. Accessed October 8, 2014 (www.washingtonpost.com).

Cha, Ariana Eunjung. 2016. "Bad Eating Habits Start Early." *Baltimore Sun*, April 14, 8.

Chambliss, William J. 1969. *Crime and the Legal Process*. New York: McGraw-Hill.

Chambliss, William J., and Robert B. Seidman. 1982. *Law, Order, and Power*, 2nd edition. Reading, MA: Addison-Wesley.

Chamie, Joseph. 2017. "Out-of-Wedlock Births Rise Worldwide." Yale Global Online, March 16. Accessed July 17, 2017 (yaleglobal.yale.edu).

Chandler, Tertius, and Gerald Fox. 1974. *3000 Years of Urban Growth.* New York: Academic Press.

Chandra, Anjani, William D. Mosher, and Casey Copen. 2011. "Sexual Behavior, Sexual Attraction, and Sexual Identity in the United States: Data from the 2006–2008 National Survey of Family Growth." *National Health Statistics Reports*, No. 36, March 3. Accessed March 2, 2014 (www.cdc.gov/nchs).

"Characteristics of Same-Sex Households." 2017. U.S. Census Bureau, Table 1, April 7. Accessed July 20, 2017 (census.gov).

Chasin, Barbara H. 2004. *Inequality & Violence in the United States: Casualties of Capitalism*, 2nd edition. Amherst, NY: Humanity Books.

Chaves, Mark, and Alison Eagle. 2015. "National Congregations Study." Accessed July 31, 2017 (soc.duke.edu/natcong).

Chen, Alice, Emily Oster, and Heidi Williams. 2014. "Why Is Infant Mortality Higher in the US Than in Europe?" National Bureau of Economic Research, September. Accessed December 5, 2014 (www.nber.org).

Chen, Frances S., Julia A. Minson, Maren Schöne1, and Markus Heinrichs. 2013. "In the Eye of the Beholder: Eye Contact Increases Resistance to Persuasion." *Psychological Science* 24 (November): 2254–2261.

Chen, Juan, and David T. Takeuchi. 2011. "Intermarriage, Ethnic Identity, and Perceived Social Standing among Asian Women in the United States." *Journal of Marriage and Family* 73 (August): 876–888.

Chen, Michelle. 2012. "Health Care System Leaves Patients Frustrated—Nurses Work for a Solution." AlterNet, June 4. Accessed June 5, 2012 (www.alternet.org).

Chen, Ronald C., et al. 2016. "Aggressive Care at the End-of-Life for Younger Patients with Cancer: Impact of ASCO's Choosing Wisely Campaign." Paper presented at the American Society of Clinical Oncology meeting, Chicago, Illinois, June 6.

Chesler, Phyllis. 2006. "The Failure of Feminism." *Chronicle of Higher Education,* February 26, B12.

Chesney-Lind, Meda, and Lisa Pasko. 2004. *The Female Offender: Girls, Women, and Crime*, 2nd edition. Thousand Oaks, CA: Sage.

Chetty, Raj, David Grusky, Maximilian Hell, Nathaniel Hendren, Robert Manduca, and Jimmy Narang. 2016. "The Fading American Dream: Trends in Absolute Income Mobility Since 1940." National Bureau of Economic Research, December. Accessed March 15, 2017 (nber.org).

Chetty, Raj, John N. Friedman, Emmanuel Saez, Nicholas Turner, and Danny Yagan. 2017. "Mobility Report Cards: The Role of Colleges in Intergenerational Mobility." Federal Reserve Bank of Boston, January. Accessed May 5, 2017 (bostonfed.org).

Chetty, Raj, and Nathaniel Hendren. 2016. "The Impacts of Neighborhoods on Intergenerational Mobility II: County-Level Estimates." National Bureau of Economic Research, December. Accessed March 15, 2017 (nber.org).

Chetty, Raj, Nathaniel Hendren, Patrick Kline, Emmanuel Saez, and Nicholas Turner. 2014. "Is the United States Still a Land of Opportunity? Recent Trends in Intergenerational Mobility." National Bureau of Economic Research, January. Accessed February 6, 2014 (www.nber.org).

"Child Brides in West Africa: Girls Fight Back." 2014. *The Economist*, August 23, 42.

Child Trends Data Bank. 2013. "Youth Voting." June. Accessed April 20, 2014 (www.childtrendsdatabank.org).

Child Welfare Information Gateway. 2016. "Foster Care Statistics 2014." Children's Bureau, March. Accessed February 28, 2017 (childwelfare.gov).

Children's Defense Fund. 2014. "The State of America's Children 2014." Accessed February 14, 2014 (www.childrensdefense.org).

Chin, Margaret M. 2016. "Asian Americans, Bamboo Ceilings, and Affirmative Action." *Contexts* 15 (Winter): 70–73.

Chira, Susan. 2017. "New Era After O'Reilly? Women Aren't So Sure." *New York Times*, April 21, B7.

Chmielewski, Anna K., and Sean F. Reardon. 2016. "Education." *Pathways*, 45–50.

Choma, Russ. 2015. "One Member of Congress = 18 American Households: Lawmakers' Personal Finances Far from Average." Open Secrets, January 12. Accessed October 7, 2015 (opensecrets.org).

Christakis, Erika. 2012. "The Overwhelming Maleness of Mass Homicide." *Time*, July 24. Accessed November 20, 2013 (www.time.com).

Christina, Greta. 2011. "Wealthy, Handsome, Strong, Packing Endless Hard-Ons: The Impossible Ideals Men Are Expected to Meet." Independent Media Institute, June 20. Accessed June 21, 2011 (www.alternet.org).

Christina, Greta. 2016. "7 of the Less-Noted but Still Very Sexist Attacks on Hillary Clinton." AlterNet, August 19. Accessed May 15, 2016 (alternet.org).

Chronicle of Higher Education. 2014. "Data Point." April 18, A23.

Churchill, Ward. 1997. *A Little Matter of Genocide: Holocaust and Denial in the Americas, 1492 to the Present.* San Francisco: City Lights Books.

Cillizza, Chris. 2014. "Our C-Minus Government (and Why We Deserve It)." *Washington Post*, October 15. Accessed November 23, 2014 (www.brookings.edu).

Citizens Against Government Waste. 2016. "2016 Congressional Pig Book Summary." Accessed June 26, 2017 (cagw.org).

Claxton, Gary, et al. 2016. "Employer Health Benefits: 2016 Annual Survey." Kaiser Family Foundation, September 14. Accessed August 15, 2017 (kff.org).

Clayson, Dennis E., and Debra A. Haley. 2011. "Are Students Telling Us the Truth? A Critical Look at the Student Evaluation of Teaching." *Marketing Education Review* 21 (Summer): 101–112.

Clement, Scott, and Lenny Bernstein. 2016. "One-Third of Long-Term Users Say They're Hooked on Prescription Opioids." *Washington Post*, December 9. Accessed August 15, 2017 (washingtonpost.com).

Clements, David D. 2012. *Corporations Are Not People: Why They Have More Rights Than You Do and What You Can Do About It.* San Francisco: Berrett-Koehler.

Clemmons, Nakia S., Paul A. Gastanaduy, Amy P. Fiebelkorn, Susan B. Redd, and Gregory S. Wallace. 2015. "Measles—United States, January 4 -April 2, 2015." *MMWR* 64 (April 17): 373–6.

Clifford, Stephanie, and Jessica Silver-Greenberg. 2016. "Bank Tellers, with Low Pay and High Access, Pose a Rising Security Risk." *New York Times*, February 2, A22–A23.

Clifford, Stephanie, and Quentin Hardy. 2013. "Attention, Shoppers: Store Is Tracking Your Cell." *New York Times*, July 15, A1.

Clifton, Jim. 2016. "Explaining Trump: Widespread Government Corruption." Gallup, January 6. Accessed June 27, 2017 (gallup.com).

Cloke, Kenneth, and Joan Goldsmith. 2002. *End of Management and the Rise of Organizational Democracy.* San Francisco: Jossey-Bass.

Cloud, John. 2010. "How to Recruit Better Teachers." *Time,* September 20, 47–52.

Cobble, Dorothy Sue, Linda Gordon, and Astrid Henry. 2014. "What 'Lean In' Leaves Out." *Chronicle Review*, September 16, B4–B5.

Cohen, Jere. 2002. *Protestantism and Capitalism: The Mechanisms of Influence.* New York: Aldine de Gruyter.

Cohen, Lawrence J., and Anthony T. DeBenedet. 2012. "Penn State Cover-Up: Groupthink in Action." *Time*, July 17. Accessed April 3, 2017 (ideas.time.com).

Cohen, Patricia. 2015. "The New Shop Class." *New York Times*, March 11, Bi, B4.

Cohen, Patricia. 2015. "When Company Is Fined, Taxpayers Often Share Bill." *New York Times,* February 4, B1, B9.

Cohen, Patricia. 2016. "With Pay Rising, Millions Climb Out of Poverty." *New York Times*, September 26, A1, A3.

Cohen, Patricia. 2017. "In Rural Iowa, a Future Rests on Immigrants." *New York Times*, May 30, A1, A12.

Cohen, Stefanie. 2012. "Why Women Writers Still Take Men's Names." *Wall Street Journal*, December 7, D9.

Cohn, D'Vera. 2012. "Divorce and the Great Recession." Pew Research Center, May 2. Accessed May 22, 2014 (www.pewsocialtrends.org).

Cohn, D'Vera. 2013. "Love and Marriage." Pew Research Social & Demographic Trends, February 13. Accessed May 16, 2014 (www.pewsocialtrends.org).

Cohn, D'Vera, Eileen Patten, and Mark Hugo. 2014. "Puerto Rican Population Declines on Island, Grows on U.S. Mainland." Pew Research Center, August 11. Accessed September 19, 2014 (www.pewresearch.org).

Cohn, D'Vera, Gretchen Livingston, and Wendy Wang. 2014. "After Decades of Decline, a Rise in Stay-at-Home Mothers." Pew Research Center, April 8. Accessed April 12, 2014 (www.pewresearch.org).

Cohn, D'Vera, Jeffrey S. Passel, Wendy Wang, and Gretchen Livingston. 2011. "Barely Half of U.S. Adults Are Married—A Record Low." Pew Research Center, December 14. Accessed May 18, 2014 (www.pewsocialtrends.org).

Cohn, D'Vera, Paul Taylor, Mark H. Lopez, Catherine A. Gallagher, Kim Parker, and Kevin T. Maass. 2013. "Gun Homicide Rate Down 49% Since 1993 Peak; Public Unaware." Pew Research Center, May 7. Accessed October 12, 2013 (www.pewresearch.org).

Cohn, Meredith. 2014. "Kids Are Outfitted with New Hands Made on 3-D Printers." *Baltimore Sun*, September 29, 1, 11.

Colapinto, John. 1997. "The True Story of John/ Joan." *Rolling Stone* (December 11): 54–73, 92–97.

Colapinto, John. 2001. *As Nature Made Him: The Boy Who Was Raised as a Girl.* New York: Harper Perennial.

Colapinto, John. 2004. "What Were the Real Reasons Behind David Reimer's Suicide?" *Slate*, June 3. Accessed April 24, 2008 (www.slate.com).

Colby, Sandra L., and Jennifer M. Ortman. 2015. "Projections of the Size and Composition of the U.S. Population: 2014 to 2060." U.S. Census Bureau, Current Population Reports, March. Accessed July 20, 2017 (census.gov).

"The Collaboration Curse." 2016. *The Economist*, January 23, 63.

College Board. 2013. "2013 College-Bound Seniors: Total Group Profile Report." Accessed June 21, 2014 (www.collegeboard.com).

College Board. 2014. "The 10th Annual AP Report to the Nation." February 11. Accessed June 21, 2014 (apreport.collegeboard.com).

Collins, William J., and Marianne H. Wanamaker. 2017. "Up from Slavery? African American Intergenerational Economic Mobility Since 1880." National Bureau of Economic Research, May. Accessed June 10, 2017 (nber.org).

Colvin, Geoff. 2012. "The Art of the Self-Managing Team." *Fortune*, December 3, 22–23.

Colvin, George. 2014. "In the Future, Will There Be Any Work Left for People to Do?" *Fortune*, June 16, 193–202.

Colvin, Mark, and John Pauly. 1983. "A Critique of Criminology: Toward an Integrated Structural-Marxist Theory of Delinquency Production." *American Journal of Sociology* 89 (November): 513–551.

Commisceo Global. 2016. "Global Guide to Culture, Customs, and Etiquette." Accessed March 13, 2017 (commisceo-global.com).

Congressional Budget Office. 2016. "Trends in Family Wealth, 1989 to 2013." Congress of the United States, August. Accessed May 5, 2017 (cbo.gov).

Conley, Dalton. 1999. *Being Black, Living in the Red: Race, Wealth, and Social Policy in America.* Berkeley: University of California Press.

Conley, Dalton. 2004. *The Pecking Order: Which Siblings Succeed and Why.* New York: Pantheon.

Connell, Dave. 2011. "30 Days of H20: Final Days." The Nature Conservancy, February 22. Accessed June 6, 2012 (blog.nature.org).

Connor, Phillip. 2016. "International Migration: Key Findings from the U.S., Europe and the World." Pew Research Center, December 15. Accessed August 25, 2017 (pewresearch.org).

Connor, Phillip. 2016. "Nearly Half of Refugees Entering the U.S. This Year Are Muslim." Pew Research Center, August 16. Accessed June 20, 2017 (pewresearch.org).

Conrad, Peter, and Kristin K. Barker. 2010. "The Social Construction of Illness: Key Insights and Policy Implications." *Journal of Health and Social Behavior* 51 (Suppl.): S67–S79.

"Consanguineous Marriage." 2016. *Economist*, April 16, 39–40.

Considine, Austin. 2011. "For Asian-American Stars, Many Web Fans." *New York Times*, July 31, ST6.

ConsumerReports.org. 2009. "6 Top Reasons for Not Having Sex." February. Accessed March 4, 2014 (www.consumerreports.org).

Cook, Thomas D., and Donald T. Campbell. 1979. *Quasi-Experimentation: Design and Analysis Issues for Field Settings.* Chicago: Rand McNally.

Cooky, Cheryl, Michael A. Messner, and Michela Musto. 2015. "'It's Dude Time!': A Quarter Century of Excluding Women's Sports in Televised News and Highlight Shows." *Communication & Sport* 3 (September): 261–287.

Cooley, Charles Horton. 1909/1983. *Social Organization: A Study of the Larger Mind.* New Brunswick, NJ: Transaction Books.

Cooper, David. 2015. "Given the Economy's Growth, the Federal Minimum Wage Could Be Significantly Higher." Economic Policy Institute, April 14. Accessed October 7, 2015 (epi.org).

Cooper, O. R., et al. 2010. "Increasing Springtime Ozone Mixing Ratios in the Free Troposphere over Western North America." *Nature* 463 (January 21): 344–348.

Cooperman, Alan, Gregory Smith, and Besheer Mohamed. 2013. "Celebrating Christmas and the Holidays, Then and Now." Pew Research Center, December 18. Accessed June 21, 2014 (www.pewresearch.org).

Copen, Casey E., Anjani Chandra, and Gladys Martinez. 2012. "Prevalence and Timing of Oral Sex with Opposite-Sex Partners Among Females and Males Aged 15–24 Years: United States, 2007–2010." *National Health Statistics Reports*, No. 56, August 16. Accessed March 2, 2014 (www.cdc.gov/nchs).

Copen, Casey E., Anjani Chandra, and Isaedmarie Febo-Vazquez. 2016. "Sexual Behavior Sexual Attraction, and Sexual Orientation Among Adults Ages 18–44 in the United States: Data from the 2011–13 National Survey of Family Growth." *National Health Statistics Reports* 88 (January 7). Accessed May 25, 2017 (cdc.gov/nchs).

Copen, Casey E., Kimberly Daniels, and William D. Mosher. 2013. "First Premarital Cohabitation in the United States: 2006–2010 National Survey of Family Growth." *National Health Statistics Reports*, April 4. Accessed May 16, 2014 (www.cdc.gov/nchs).

Copen, Casey E., Kimberly Daniels, Jonathan Vespa, and William D. Mosher. 2012. "First Marriages in the United States: Data from the 2006–2010 National Survey of Family Growth." *National Health Statistics Reports*, March 22. Accessed May 16, 2014 (www.cdc.gov/nchs).

Corak, Miles. 2013. "Income Inequality, Equality of Opportunity, and Intergenerational Mobility." *Journal of Economic Perspectives* 27 (Summer): 79–102.

Corak, Miles. 2016. "Economic Mobility." *Pathways*, Special Issue, 51–57.

Coren, Arthur. 2011. "Turning a Blind Eye: Faculty Who Ignore Student Cheating." *Journal of Academic Ethics* 9 (4): 291–305.

Corporation for National & Community Service. 2016. "Volunteering and Civic Life in America." Accessed April 2, 2017 (nationalservice.gov).

Couch, Carl J. 1968. "Collective Behavior: An Examination of Some Stereotypes." *Social Problems* 15 (Winter): 310–322.

Couffinhal, Agnès, and Karolina Socha-Dietrich. 2017. "Ineffective Spending and Waste in Health Care Systems: Framework and Findings." Pp. 17–59 in OECD, *Tackling Wasteful Spending on Health.* Accessed August 15, 2017 (oecd.org).

Council for Global Equity. 2014. "The Facts on LGBT Rights in Russia." Accessed March 2, 2014 (www.globalequity.org).

Cox, Cynthia, and Larry Levitt. 2017. "What's the Near-Term Outlook for the Affordable Care Act?" Kaiser Family Foundation, August 4. Accessed August 15, 2017 (kff.org).

Coy, Peter. 2016. "Rising Profits Don't Lift Workers' Boats." *Bloomberg Businessweek*, May 9–15, 19–20.

Crabtree, Steve. 2010. "Religiosity Highest in World's Poorest Nations." Gallup, June 3. Accessed August 31, 2012 (gallup.com).

Craig, Lyn. 2015. "How Mothers and Fathers Allocate Time to Children: Trends, Resources, and Policy Context." *Family Focus*, Spring, F9–F10.

Cremer, Paul C., Yuping Wu, and Haitham M. Ahmed. 2017. "Use of Sex-Specific Clinical and Exercise Risk Scores to Identify Patients at Increased Risk for All-Cause Mortality." *JAMA Cardiology* 2 (1): 15–22.

Cressey, Donald R. 1953. *Other People's Money: A Study in the Social Psychology of Embezzlement.* Glencoe, IL: Free Press.

Cross, Rob, Reb Rebele, and Adam Grant. 2016. "Collaborative Overload." *Harvard Business Review* 94 (January/February): 74–79.

Crowley, Michael. 2014. "Iraq's Eternal War." *Time*, June 30, 28–34.

Cruikshank, Margaret. 2009. *Learning to Be Old: Gender, Culture, and Aging.* Lanham, MD: Rowman & Littlefield.

Cubanski, Juliette, Tricia Newman, Shannon Griffin, and Anthony Damico. 2016. "Medicare Spending at the End of Life: A Snapshot of Beneficiaries Who Died in 2014 and the Cost of Their Care." Kaiser Family Foundation, July 14. Accessed September 4, 2017 (kff.org).

Cudd, Ann E., and Nancy Holstrom. 2010. *Capitalism, For and Against: A Feminist Debate.* New York: Cambridge University Press.

Culbert, Samuel A. 2011. "Why Your Boss Is Wrong About You." *New York Times*, March 3, A25.

Cullen, Lisa T., and Coco Masters. 2008. "We Just Clicked." *Time*, January 28, 86–89.

Cumberworth, Erin. 2010. "Homeboy Industries." *Pathways*, Summer, 22–23.

Cummings, William. 2017. "Millennials Differ from Other Generations in Almost Every Regard. Here's the Data." *USA Today*, April 19. Accessed April 20, 2017 (usatoday.com).

Cunningham, Scott, Jason M. Lindo, Caitlin Myers, and Andrea Schlosser. 2017. "How Far Is Too Far? New Evidence on Abortion Clinic Closures, Access, and Abortions." National Bureau of Economic Research, April. Accessed July 19, 2017 (nber.org).

Cunningham, Solveig A., Michael R. Kramer, and K. M. Venkat Narayan. 2014. "Incidence of Childhood Obesity in the United States." *New England Journal of Medicine* 370 (5): 403–411.

Currie, E. 1985. *Confronting Crime: An American Challenge.* New York: Pantheon.

Curtin, Sally C., Margaret Warner, and Holly Hedegaard. 2016. "Increase in Suicide in the United States, 1999–2014." Centers for Disease Control and Prevention, April. Accessed December 21, 2016 (cdc.gov).

Curtin, Sally C., Stephanie J. Ventura, and Gladys M. Martinez. 2014. "Recent Declines in Nonmarital Childbearing in the United States."

National Center for Health Statistics, August. Accessed November 30, 2014 (www.nchs.gov).

Curtiss, Susan. 1977. *Genie.* New York: Academic Press.

Cyberbullying Research Center. 2016. "2016 Cyberbullying Data." November 26. Accessed March 13, 2017 (cyberbullying.org).

D

D'Innocenzio, Anne. 2015. "Target Stores Aim to Instill Healthier Habits." *Baltimore Sun*, September 17, 10.

Dahl, Robert A. 1961. *Who Governs? Democracy and Power in an American City.* New Haven, CT: Yale University Press.

Dahlin, E. 2011. "There's No 'I' in Innovation." *Contexts* 10 (Fall): 22–27.

Daly, Mary C., and Leila Bengali. 2014. "Is It Still Worth Going to College?" Federal Reserve Bank of San Francisco, May 5. Accessed June 21, 2014 (www.frbsf.org).

Daneshpour, Manijeh. 2015. "A Mini-Narrative About My Praxis as a Muslim Feminist." *Family Focus* 57.4 (Winter): F14–F16.

Daniel, Mary-Alice. 2014. "What White People Need to Learn." AlterNet, February 7. Accessed March 22, 2014 (www.alternet.org).

Daniels, Cynthia R., Janna Ferguson, Grace Howard, and Amada Roberti. 2016. "Informed or Misinformed Consent? Abortion Policy in the United States." *Journal of Health Politics, Policy and Law* 41 (April): 181–209.

Danielson, Melissa L., Susanna N. Visser, Mary Margaret Gleason, Georgina Peacock, Angelika H. Claussen, and Stephen J. Blumberg. 2017. "A National Profile of Attention-Deficit Hyperactivity Disorder Diagnosis and Treatment Among US Children Aged 2 to 5." *Journal of Developmental & Behavioral Pediatrics*, July 14. Accessed August 15, 2017 (journals.lww.com/jrnldbp).

Dank, Meredith, et al. 2014. "Estimating the Size and Structure of the Underground Commercial Sex Economy in Eight Major US Cities." Urban Institute, March. Accessed November 20, 2014 (www.urban.org).

Dao, James. 2013. "In Debate over Military Sexual Assault, Men Are Overlooked Victims." *New York Times*, June 24, A12.

Daskal, Lolly. 2016. "Tangled in Rigid Rules." *Baltimore Sun*, September 25, 19.

Daugherty, Jill, and Casey Copen. 2016. "Trends in Attitudes About Marriage, Childbearing, and Sexual Behavior: United States, 2002, 2006–2010, and 2011–2013." *National Health Statistics Report* 92 (March 17): 1–10.

Davenport, Coral. 2017. "Climate Skeptics Say Rollback at E.P.A. Isn't Going Far Enough." *New York Times*, April 13, A15.

Davenport, Coral. 2017. "Denialists in Charge." *New York Times*, March 29, A19.

Davenport, Coral, and Alissa J. Rubin. 2017. "Trump Signs Rule to Block Efforts on Aiding Climate." *New York Times*, March 29, A1, A19.

Davey, Monica. 2014. "Immigrants Seen a Way to Refill Detroit Ranks." *New York Times*, January 24, A12.

Davidai, Shai, and Thomas Gilovich. 2015. "Building a More Mobile America—One Income Quintile at a Time." *Perspectives on Psychological Science* 10 (1): 60–71.

Davidson, Osha Gray. 1998. *Under Fire: The NRA and the Battle for Gun Control.* Iowa City: University of Iowa Press

Davies, James C. 1979. "The J-Curve of Rising and Declining Satisfaction as a Cause of Revolution and Rebellion." Pp. 413–436 in *Violence in America: Historical and Comparative Perspectives,* edited by Hugh D. Graham and Ted R. Gurr. Beverly Hills, CA: Sage.

Davis, Alyssa. 2016. "In U.S., Concern About Crime Climbs to 15-Year High." Gallup, April 6. Accessed April 15, 2017 (gallup.com).

Davis, Julie Hirschfeld. 2015. "Hacking Exposed 21 Million in U.S., Government Says." *New York Times*, July 10, A1, A3.

Davis, Karen, Kristof Stremikis, David Squires, and Cathy Schoen. 2014. "Mirror, Mirror on the Wall: How the Performance of the U.S. Health Care System Compares Internationally." The Commonwealth Fund, June. Accessed July 26, 2014 (www.commonwealthfund.org).

Davis, Kenneth C. 2010. "America's True History of Religious Tolerance." *Smithsonian*, October. Accessed September 28, 2010 (www.smithsonian.com).

Davis, Kingsley, and Wilbert E. Moore. 1945. "Some Principles of Stratification." *American Sociological Review* 10 (April): 242–249.

Dawson, Steve. 2017. "What Happened to Otto Warmbier—News & Updates." *Gazette Review*, January 16. Accessed February 14, 2017 (gazettereview.com).

de las Casas, Bartolome. 1992. *The Devastation of the Indies: A Brief Account.* Baltimore, MD: Johns Hopkins University Press.

De Long, J. Bradford, and Andrei Shleifer. 1992. "Princes and Merchants: European City Growth Before the Industrial Revolution." National Bureau of Economic Research, December. Accessed January 10, 2008 (www.nper.org).

de Pommereau, Isabelle. 2013. "A Lift for Europe's Women?" *Christian Science Monitor Weekly*, August 5, 16.

de Vogue, Ariane, and Jeremy Diamond. 2015. "Supreme Court Rules in Favor of Same-Sex Marriage Nationwide." CNN, June 27. Accessed September 29, 2015 (cnn.com).

"Deadly Intolerance." 2014. *The Economist*, March 1, 42.

Death Penalty Information Center. 2017. "Facts About the Death Penalty." April 28. The Sentencing Project. Accessed April 29, 2017 (deathpenaltyinfo.org).

Deaton, Angus, and Arthur A. Stone. 2013. "Grandpa and the Snapper: The Wellbeing of the Elderly Who Live with Children." National Bureau of Economic Research, June. Accessed July 10, 2013 (www.nber.org).

Deegan, Mary Jo. 1986. *Jane Addams and the Men of the Chicago School, 1892–1918.* New Brunswick, NJ: Transaction Books.

DeGraw, David. 2014. "How the Ultra-Rich .01% Have Sucked Up Even More of America's Wealth Than You Think." AlterNet, August 22. Accessed November 10, 2014 (www.alternet.org).

DeKeseredy, Walter S. 2011. *Violence Against Women: Myths, Facts, Controversies.* Toronto: University of Toronto Press.

della Porta, Donatella, and Mario Diani. 1999. *Social Movements: An Introduction.* Malden, MA: Blackwell.

Deloitte and Alliance for Board Diversity. 2017. "Missing Pieces Report: The 2016 Board Diversity Census of Women and Minorities on Fortune 500 Boards." February 6. Accessed April 2, 2017 (ww2.deloitte.com/us).

Demographia. 2017. "World Urban Areas, 13th Annual Edition." Accessed August 20, 2017 (demographia.com).

Denny, C. H., J. Tsai, R. L. Floyd, and P. P. Green. 2009. "Alcohol Use Among Pregnant and Nonpregnant Women of Childbearing Age—United States, 1991–2005." *MMWR* 58 (May 22): 529–532.

DeParle, Jason. 2012. "Harder for Americans to Rise from Lower Rungs." *New York Times*, January 5, A1.

Department of Defense. 2016. "Annual Report on Sexual Assault in the Military Fiscal Year 2015." Sexual Assault Prevention and Response, April 21. Accessed April 3, 2017 (sapr.mil).

Deprez, Esmé E., and William Selway. 2014. "Guns Allowed All Over—Except Near Politicians." *Bloomberg Businessweek*, May 19–25, 27–28.

DeSilver, Drew. 2017. "5 Facts About the Minimum Wage." Pew Research Center, January 4. Accessed June 16, 2017 (pewresearch.org).

DeSilver, Drew. 2017. "U.S. Students' Academic Achievement Still Lags that of Their Peers in Many Other Countries." Pew Research Center, February 15. Accessed July 31, 2017 (pewresearch.org).

Desmond-Harris, Jenee. 2009. "Why Michelle's Hair Matters." *Time,* September 7, 55–57.

Deveny, Kathleen. 2008. "They're No Baby Einsteins." *Newsweek*, January 14, 61.

Dewan, Shaila, and Robert Gebeloff. 2012. "Among the Wealthiest One Percent, Many Variations." *New York Times*, January 15, A1.

Diamond, Jeremy. 2017. "Trump Falsely Claims US Murder Rate Is 'Highest' in 47 Years." CNN, February 7. Accessed April 15, 2017 (cnn.com).

Diamond, Milton, and H. Keith Sigmundson. 1997. "Sex Reassignment at Birth: Long-Term Review and Clinical Implications." *Archives of Pediatrics & Adolescent Medicine* 15 (March): 298–304.

Dias, Elizabeth, and Eliana Dockterman. 2016. "The Fine Print." *Time*, October 31, 32–35.

Dickey, Jack. 2015. "Save Our Vacation." *Time*, June 1, 44–49.

Dickman, Samuel L., David U. Himmelstein, and Steffie Woolhandler. 2017. "Inequality and the Health-Care System in the USA." *The Lancet* 389 (April 8): 1431–1441.

"Dietary Inequality." 2016. *The Economist*, August 13, 19

Dilworth-Anderson, Peggye, Linda M. Burton, and William L. Turner. 1993. "The Importance of Values in the Study of Culturally Diverse Families." *Family Relations* 42 (July): 238–242.

Ding, Ming, et al. 2015. "Association of Coffee Consumption with Total and Cause-Specific Mortality in 3 Large Prospective Cohorts." *Circulation* 132 (November): 2305–2315.

DiSalvo, David. 2010. "Are Social Networks Messing with Your Head?" *Scientific American Mind* 20 (January/February): 48–55.

Ditmar, Kelly. 2017. "Finding Gender in Election 2016: Lessons from Presidential Gender Watch." Barbara Lee Foundation. Accessed June 26, 2017 (barbaraleefoundation.org).

"DNA Reveals White Supremacist's Black Heritage." 2013. *Baltimore Sun*, November 14, 8.

Dockterman, Eliana. 2013. "The Digital Parent Trap. Should Your Kids Avoid Tech—or Embrace It?" *Time*, August 19, 54.

Dockterman, Eliana. 2016. "A Barbie for Every Body." *Time*, February 8, 43–51.

Dodson, Lisa, and Wendy Luttrell. 2011. "Families Facing Untenable Choices." *Contexts* 10 (Winter): 38–42.

Doeringer, Peter B., and Michael J. Piore. 1971. *Internal Labor Markets and Manpower Analysis*. Lexington, MA: Heath-Lexington Books.

"Dog-lovers v. Dog-eaters: Pet Food." 2015. *The Economist*, June 20, 46.

Doherty, Carroll. 2015. "Continue Bipartisan Support for Expanded Background Checks on Gun Sales." Pew Research Center, August 13. Accessed August 23, 2015 (pewresearch.org).

Doherty, Carroll, and Alec Tyson. 2015. "Most Say Government Policies Since Recession Have Done Little to Help Middle Class, Poor." Pew Research Center, March 4. Accessed August 26, 2015 (pewresearch.org).

Doherty, Carroll, Juliana M. Horowitz, and Michael Dimock. 2014. "Most See Inequality Growing, but Partisans Differ over Solutions." Pew Research Center, January 23. Accessed April 12, 2014 (www.pewresearch.org).

Dolnick, Sam. 2011. "Dance, Laugh, Drink, Save the Date: It's a Ghanaian Funeral." *New York Times*, April 12, A1.

Domhoff, G. William. 2013. "Interlocking Directorates in the Corporate Community." Who Rules America, October. Accessed April 12, 2014 (whorulesamerica.net).

Dominguez, Kenneth, et al. 2015. "Vital Signs: Leading Causes of Death, Prevalence of Diseases and Risk Factors, and Use of Health Services Among Hispanics in the United States—2009–2013." *MMWR* 64 (17): 469–478.

Domosh, Mona, and Joni Seager. 2001. *Feminist Geographers Make Sense of the World*. New York: Guilford Press.

Donadio, Rachel. 2013. "When Italians Chat, Hands and Fingers Do the Talking." *New York Times*, July 1, A6.

Dorfman, Mark, and Angela Haren. 2014. "Testing the Waters." Natural Resources Defense Council, June. Accessed September 19, 2014 (nrdc.org).

Dorell, Oren. 2017. "Who Will Rebuild Puerto Rico as Young Professionals Leave Island After Hurricane Maria?" *USA Today*, October 12. Accessed October 31, 2017 (usatoday.com).

Douglas, Susan J., and Meredith W. Michaels. 2004. *The Mommy Myth: The Idealization of Motherhood and How It Has Undermined Women*. New York: Free Press.

Dovidio, John F. 2009. "Racial Bias, Unspoken but Heard." *Science* 326 (December 18): 1641–1642.

Dow, Dawn Marie. 2016. "The Deadly Challenges of Raising African American Boys: Navigating the Controlling Image of the 'Thug'." *Gender & Society* 30 (April): 161–188.

Downs, Edward, and Stacy L. Smith. 2010. "Keeping Abreast of Hypersexuality: A Video Game Character Content Analysis." *Sex Roles* 62 (June): 721–733.

Doyle, Francis X. 2011. "Bishops' Missed Opportunity." *Baltimore Sun*, November 14, 15.

Dreid, Nadia. 2016. "Meet the Professor Who's Calling Out Clickbait." *Chronicle of Higher Education*, December 2, A24.

Dreweke, Joerg. 2017. "U.S. Abortion Rate Reaches Record Low Amidst Looming Onslaught Against Reproductive Health and Rights." *Guttmacher Policy Review* 20 (January): 15–19.

Drexler, Peggy. 2012. "The New Face of Infidelity." *Wall Street Journal*, October 21, C3.

Drexler, Peggy. 2013. "The Delicate Protocol of Hugging." *Wall Street Journal*, September 14–15, C3.

Drexler, Peggy. 2014. "Why We Love to Gossip." *Huffington Post*, November 14. Accessed October 25, 2015 (huffingtonpost.com).

"Driven to Distraction." 2017. *The Economist*, February 11, 11.

Dropp, Kyle, and Brendan Nyhan. 2017. "Many Don't Know Obamacare and Affordable Care Act Are the Same." *New York Times*, February 8, A10.

Drucker, Peter F. 2001. *Management Challenges for the 21st Century*. New York: HarperBusiness.

"The Drugs Don't Work." 2014. *The Economist*, May 3, 54.

Druker, Steven M. 2015. *Altered Genes, Twisted Truth*. Salt Lake City, UT: Clear River Press.

Drutman, Lee, and Alexander Furnas. 2014. "K Street Pays Top Dollar for Revolving Door Talent." Sunlight Foundation, January 21. Accessed April 20, 2014 (sunlightfoundation.com).

Du Bois, W. E. B. 1986. *The Souls of Black Folk*. New York: Library of America.

Dugan, Andrew. 2015. "Once Taboo, Some Behaviors Now More Acceptable in U.S." Gallup, June 1. Accessed July 26, 2015 (gallup.com).

Dugan, Andrew. 2017. "US Divorce Rate Dips, but Moral Acceptability Hits New High." Gallup, July 7. Accessed July 20, 2017 (gallup.com).

Dugan, Andrew, and Brad Hoffman. 2014. "Record Low Say Own Representative Deserves Re-Election." Gallup, January 24. Accessed April 20, 2014 (www.gallup.com).

Duhigg, Charles. 2010. "Saving U.S. Water and Sewer Systems Would Be Costly." *New York Times*, March 14, A1.

Dumaine, Brian. 2015. "Can a [Billionaire] Hedge Fund Manager Fix Income Inequality?" *Fortune*, September 1, 87–91.

Duncan, D. T., A. R. Hansen, W. Wang, F. Yan, and J. Zhang. 2015. *"Change* in Misperception of Child's Body Weight among Parents of American Preschool Children." *Childhood Obesity*, National Center for Biotechnology Information, May. Accessed October 15, 2015 (ncbi.nlm.nih.gov).

Dunifon, Rachel, and Lori Kowaleski-Jones. 2007. "The Influence of Grandparents in Single-Mother Families." *Journal of Marriage and Family* 69 (May): 465–481.

Durden, Tyler. 2011. "Here Are the 29 Public Companies with More Cash Than the US Treasury." Zero Hedge, July 15. Accessed January 9, 2012 (www.zerohedge.com).

Durkheim, Émile. 1893/1964. *The Division of Labor in Society*. New York: Free Press.

Durkheim, Émile. 1897/1951. *Suicide: A Study in Sociology*, translated by John A. Spaulding and George Simpson, edited by George Simpson. New York: Free Press.

Durkheim, Émile. 1898/1956. *Education and Sociology*, translated by Sherwood D. Fox. Glencoe, IL: Free Press.

Durkheim, Émile. 1961. *The Elementary Forms of the Religious Life*. New York: Collier Books.

Durose, Matthew R., Alexia D. Cooper, and Howard N. Snyder. 2014. "Recidivism of Prisoners Released in 30 States in 2005: Patterns from 2005 to 2010." Bureau of Justice Statistics, April. Accessed September 2, 2015 (bjs.gov).

Dye, Thomas R., and Harmon Ziegler. 2003. *The Irony of Democracy: An Uncommon Introduction to American Politics*, 12th edition. Belmont, CA: Wadsworth.

E

Eagan, Kevin, Jennifer B. Lozano, Sylvia Hurtado, and Matthew H. Case. 2014. *The American Freshman: National Norms Fall 2013*. Los Angeles: Higher Education Research Institute, UCLA.

Eavis, Peter. 2015. "Judge's Ruling Against 2 Banks Points to Wide Fraud in '08 Crash." *New York Times*, May 12, A1, B2.

Eberstadt, Nicholas. 2009. "Poor Statistics." *Forbes*, March 2, 26.

Eberstadt, Nicholas. 2013. "China's Coming One-Child Crisis." *Wall Street Journal*, November 27, A15.

Eberstadt, Nicholas. 2016. *Men Without Work: America's Invisible Crisis*. West Conshocken, PA: Templeton Press.

Eckenrode, John, Elliott G. Smith, Margaret E. McCarthy, and Michael Dineen. 2014. "Income Inequality and Child Maltreatment in the United States." *Pediatrics* 33 (March): 454–461.

Eckhoff, Vickery. 2015. "Exposing America's Billionaire Welfare Ranchers." AlterNet, March 24. Accessed August 26, 2015 (alternet.org).

Eckholm, Erik. 2014. "'Aid in Dying' Movement Takes Hold in Some States." *New York Times*, February 8, A1.

Eckstein, Rick, Rebecca Schoenike, and Kevin Delaney. 1995. "The Voice of Sociology: Obstacles to Teaching and Learning the Sociological Imagination." *Teaching Sociology* 23 (October): 353–363.

"Economic News Release." 2017. U.S. Bureau of Labor Statistics, June 2. Accessed June 5, 2017 (bls.gov).

Edin, Kathryn, and Timothy J. Nelson. 2013. *Doing the Best I Can: Fatherhood in the Inner City*. Berkeley and Los Angeles: University of California Press.

eDiplomat. 2016. "Cultural Etiquette Around the World." Accessed February 14, 2017 (ediplomat.com).

"Education." 2016. *The Economist*, December 10, 59–61.

"Educational Attainment in the United States: 2016." 2017. U.S. Census Bureau, March 31. Accessed June 20, 2017 (census.gov).

EEOC. 2017. "Age Discrimination and Outdated Views of Older Worker Persist, Experts Tell Commission." U.S. Equal Employment Opportunity Commission, June 14. Accessed July 10, 2017 (eeoc.gov).

Efrati, Amir. 2013. "Google's Data-Trove Dance." *Wall Street Journal*, July 31, B1, B4.

Egan, Mark L., Gregor Matvos, and Amit Seru. 2017. "When Harry Fired Sally: The Double Standard in Punishing Misconduct." National Bureau of Economic Research, March. Accessed March 13, 2017 (nber.org).

Eggebeen, David J. 2005. "Cohabitation and Exchanges of Support." *Social Forces* 83 (May): 1097–1110.

Egley, Arlen, and James C. Howell. 2013. "Highlights of the 2011 National Youth Gang Survey." Office of Juvenile Justice and Delinquency Prevention, September. Accessed January 12, 2014 (www.ojjdp.gov).

Ehrlich, Paul. 1971. *The Population Bomb,* 2nd edition. San Francisco: Freeman.

Ehrlich, Paul R., and Anne H. Ehrlich. 2008. *The Dominant Animal: Human Evolution and the Environment.* Washington, DC: Island Press.

Eilperin, Juliet, and Lenny Bernstein. 2014. "Environmental Groups Say Obama Needs to Address Climate Change More Aggressively." *Washington Post*, January 16. Accessed September 19, 2014 (www.washingtonpost.com).

Eisenberg, Abne M., and Ralph R. Smith, Jr. 1971. *Nonverbal Communication.* Indianapolis: Bobbs-Merrill.

Eisler, Peter. 2013. "What Surgeons Leave Behind Costs Some Patients Dearly." *USA Today*, March 8. Accessed July 26, 2014 (www.usatoday.com).

Ekman, Paul, and Wallace V. Friesen. 1984. *Unmasking the Face: A Guide to Recognizing Emotions from Facial Clues.* Palo Alto, CA: Consulting Psychologists Press.

Elgin, Ben. 2014. "An Oil Giant Dims the Lights on Green Power." *Bloomberg Businessweek*, June 2–8, 21–22.

Eligon, John, and Robert Gebeloff. 2016. "Segregation, the Neighbor That Won't Leave." *New York Times*, August 21, A1.

Eliot, Lise. 2012. *Pink Brain, Blue Brain: How Small Differences Grow into Troublesome Gaps—And What We Can Do About It.* Oxford, England: Oneworld Publications.

Ellis, Bill. 2005. "Legend/AntiLegend: Humor as an Integral Part of the Contemporary Legend Process." Pp. 123–140 in *Rumor Mills: The Social Impact of Rumor and Legend,* edited by Gary Alan Fine, Veronique Campion-Vincent, and Chip Heath. New Brunswick, NJ: Transaction Publishers.

Ellis, Renee. 2013. "Changes in Coresidence of Grandparents and Grandchildren." *Family Focus* (Summer): F13, F15, F17.

Ellis, Renee R., and Tavia Simmons. 2014. "Coresident Grandparents and Their Grandchildren: 2012." *U.S. Census Bureau*, October. Accessed October 25, 2014 (www.census.gov).

Embury-Dennis, Tom. 2017. "US States Consider Laws Allowing Creationism to be Taught by Science Teachers." *Independent*, March 16. Accessed July 31, 2017 (independent.co.uk).

Emery, David. 2015. "Does AriZona Tea Contain Human Urine?" Urban Legends, July 25. Accessed October 25, 2015 (urbanlegends.about.com).

Emmert, Ashley. 2015. "The State of Female Pastors." *Christianity Today*, October 15. Accessed July 31, 2017 (christianitytoday.com).

Emmons, William R., and Bryan J. Noeth. 2015. "Race, Ethnicity, and Wealth." Federal Reserve Bank of St. Louis, February. Accessed September 16, 2015 (stlouisfed.org).

"Employer-Reported Workplace Injuries and Illnesses—2015." 2016. Bureau of Labor Statistics News Release, October 27. Accessed August 15, 2017 (bls.gov).

"Employment Projections." 2017. U.S. Bureau of Labor Statistics, April 14. Accessed June 5, 2017 (bls.gov).

"Ending the Shame." 2013. *The Economist*, September 14, 16, 18.

England, Paula. 2005. "Emerging Theories of Care Work." *Annual Review of Sociology* 31 (August): 381–399.

England, Paula, and Reuben J. Thomas. 2009. "The Decline of the Date and the Rise of the College Hook Up." Pp. 141–152 in *Family in Transition,* 15th edition, edited by Arlene S. Skolnick and Jerome H. Skolnick. Boston: Pearson Higher Education.

Entmacher, Joan, Lauren Frohlich, Katherine G. Robbins, Emily Martin, and Liz Watson. 2014. "Underpaid & Overloaded: Women in Low-Wage Jobs." National Women's Law Center. Accessed November 23, 2014 (www.nwlc.org).

EPA. 2009. "EPA's Endangerment Findings." December 7. Accessed May 1, 2010 (www.epa.gov).

EPA. 2013. "The National Rivers and Streams Assessment 2008–2009: A Collaborative Survey." March. Accessed September 19, 2014 (epa.gov/aquaticsurveys).

EPA. 2014. "Nutrient Pollution." August 26. Accessed September 19, 2014 (epa.gov/nutrientpollution).

EPA. 2015. "Assessment of the Potential Impacts of Hydraulic Fracturing for Oil and Gas on Drinking Water Resources. Executive Summary." June. Accessed October 23, 2015 (epa.gov).

EPA. 2016. "Advancing Sustainable Materials Management: 2014 Fact Sheet." U.S. Environmental Protection Agency, November. Accessed August 20, 2017 (epa.gov).

Epstein, Joseph. 2011. *Gossip: The Untrivial Pursuit.* New York: Houghton Mifflin Harcourt.

Epstein, Robert. 2008. "Same-Sex Marriage Is Too Limiting." *Los Angeles Times*, December 4. Accessed December 6, 2008 (www.latimes.com).

Erbentraut, Joseph. 2016. "If Farm-to-Table Eating Is Cool, What About Toilet-to-Tap Drinking?" *Huffington Post*, March 16. Accessed August 20, 2017 (huffingtonpost.com).

Erikson, Kai T. 1966. *Wayward Puritans: A Study in the Sociology of Deviance.* New York: John Wiley & Sons.

Ernst & Young. 2013. "Navigating Today's Complex Business Risks: Europe, Middle East, India and Africa Fraud Survey 2013." Accessed January 12, 2014 (www.ey.com).

Esipova, Neil, Anita Pagliese, and Julie Ray. 2013. "381 Million Adults Worldwide Migrate Within Countries." Gallup, May 15. Accessed September 19, 2014 (www.gallup.com).

Espelage, Dorothy L., Jun Sung Hong, Sarah Rinehart, and Namrata Doshi. 2016. "Understanding Types, Locations, & Perpetrators of Peer-to-Peer Sexual Harassment in U.S. Middle Schools: A Focus on Sex, Racial, and Grade Differences." *Children and Youth Services Review* 71 (December): 174–183.

Esping-Andersen, Gosta. 1990. *The Three Worlds of Welfare Capitalism.* Cambridge, United Kingdom: Polity Press.

Esterl, Mike. 2013. "The Natural Evolution of Food Labels." *Wall Street Journal*, November 6, B1–B2.

Etter, Lauren, and Shruti Singh. 2017. "Big Meat Braces for a Labor Shortage." *Bloomberg Businessweek*, February 13–19, 19–20.

Etzioni, Amitai. 1975. *A Comparative Analysis of Complex Organizations.* New York: Free Press.

Evans, Becky. 2013. "Bill Gates 'Disrespects' South Korea's Female President by Shaking Hands While Keeping Other in His Pocket." MailOnline, April 23. Accessed November 10, 2013 (dailymail.co.uk).

Evans, Harold, Gail Buckland, and David Lefer. 2006. *They Made America: From the Steam Engine to the Search Engine: Two Centuries of Innovators.* New York: Little, Brown.

Eversley, Melanie. 2017. "Report: Anti-Muslim Groups Triple in U.S. Amid Trump Hate Rhetoric." *USA Today*, February 15. Accessed February 20, 2017 (usatoday.com).

Evich, Helena B. 2016. "Meat Industry Wins Round in War over Federal Nutrition Advice." Politico, January 7. Accessed April 3, 2017 (politico.com).

Ewen, Lynda Ann. 1998. *Social Stratification and Power in America: A View from Below.* Six Hills, NY: General Hall.

Ewert, Stephanie, and Robert Kominski. 2014. "Measuring Alternative Educational Credentials: 2012." U.S. Census Bureau, January. Accessed June 21, 2014 (www.census.gov).

Ewing, Walter A., Daniel E. Martínez, and Rubén G. Rumbaut. 2015. "The Criminalization of Immigration in the United States." American Immigration Council, July. Accessed June 10, 2017 (americanimmigrationcouncil.org).

"Executive Pay." 2016. *The Economist*, June 25, 18–20.

Expedia. 2017. "In-Flight Turbulence: Expedia.com Airplane Etiquette Study Shows Seat-Kicking Edges Bad Parenting as Most Aggravating Behavior." January 17. Accessed March 13, 2017 (viewfinder.expedia.com).

Expedia's Vacation Deprivation Study. 2016. "Work-Life Imbalance: Expedia's 2016 Vacation Deprivation Study Shows Americans Leave Hundreds of Millions of Paid Vacation Days Unused." November 15. Accessed June 22, 2017 (viewfinder.expedia.com).

"Extramarital Affairs." 2014. Pew Research Global Attitudes Project. Accessed April 18, 2014 (www.pewglobal.org).

F

Fahrenthold, David A. 2009. "Environmentalists Seek to Wipe Out Plush Toilet Paper." *Washington Post,* September 24, A1.

Fakhraie, Fatemeh. 2009. "Feminists Don't Understand Muslim Women." Double X, May 20. Accessed April 26, 2010 (www.doublex.com).

"Families and Living Arrangements." 2017. U.S. Census Bureau. Accessed July 4, 2017 (census.gov).

"Family Planning: One-Child Proclivity." 2014. *The Economist*, July 19, 40.

Fang, Ferric C., R. Grant Steen, and Arturo Casadevall. 2012. "Misconduct Accounts for the Majority of Retracted Scientific Publications." *Proceedings of the National Academy of Sciences* 109 (42): 17028–17033.

FAO, IFAD and WFP. 2015. "The State of Food Insecurity in the World. Meeting the 2015 International Hunger Targets: Taking Stock of Uneven Progress." Accessed May 13, 2017 (fao.org).

Farley, Melissa. 2001. "Prostitution: The Business of Sexual Exploitation." Pp. 879–891 in *Encyclopedia of Women and Gender,* volume 2, edited by Judith Worrell. New York: Academic Press.

Farley, Sally D., Amie M. Ashcraft, Mark F. Stasson, and Rebecca L. Nusbaum. 2010. "Nonverbal Reactions to Conversational Interruption: A Test of Complementarity Theory and the Status/Gender Parallel." *Journal of Nonverbal Behavior* 34 (December): 193–206.

"Father's Day: June 18, 2017." 2017. U.S. Census Bureau Newsroom, June 8. Accessed June 20, 2017 (census.gov).

Feagin, Joe R. 2001. "Social Justice and Sociology: Agendas for the Twenty-First Century." *American Sociological Review* 66 (February): 1–20.

Feagin, Joe R., and Clairece Booher Feagin. 2008. *Racial and Ethnic Relations,* 8th ed. Upper Saddle River, NJ: Prentice Hall.

Featherstone, Liza. 2016. "Feminists Misunderstood the Presidential Election from Day One." *The Guardian,* November 15. Accessed January 5, 2017 (theguardian.com).

Federal Bureau of Investigation. 2014. "Crime in the United States, 2013". Fall. Accessed September 2, 2015 (fbi.gov).

Federal Bureau of Investigation. 2016. "2015 Internet Crime Report." Accessed March 13, 2017 (ic3.gov).

Federal Bureau of Investigation. 2016. "Crime in the United States, 2015." Fall. Accessed April 2, 2017 (fbi.gov).

Federal Bureau of Investigation. 2017. "Transnational Organized Crime." Accessed April 2, 2017 (fbi.gov).

Federal Interagency Forum on Aging-Related Statistics. 2016. "Older Americans: 2016 Key Indicators of Well-Being." Accessed July 20, 2017 (agingstats.gov).

Federal Interagency Forum on Child and Family Statistics. 2017. "America's Children: Key National Indicators of Well-Being, 2017." Accessed August 15, 2017 (childstats.gov).

Feldman, Jamie. 2017. "Michelle Obama Reveals Barack 'Wore That Same Tux' for Years." *Huffington Post,* June 7. Accessed June 7, 2017 (huffingtonpost.com).

"Female Genital Cutting." 2016. *The Economist,* June 18, 63–64.

Ferdman, Roberto A. 2015. "France Is Making It Illegal for Supermarkets to Throw Away Edible Food." *Washington Post,* May 22. Accessed July 26, 2015 (washingtonpost .com).

Ferguson, Christopher J., and Cheryl K. Olson. 2013. "Video Game Violence Use Among 'Vulnerable' Populations: The Impact of Violent Games on Delinquency and Bullying Among Children with Clinically Elevated Depression or Attention Deficit Symptoms." *Journal of Youth and Adolescence* 42 (August), published online.

Ferguson, Emily. 2014. "11 Reasons Christianity Needs Feminism." *Huffington Post,* May 21. Accessed June 21, 2014 (www.huffingtonpost .com).

Fernald, Anne, Virginia A. Marchman, and Adriana Weisleder. 2013. "SES Differences in Language Processing Skill and Vocabulary are Evident at 18 Months." *Developmental Science* 16 (March): 234–248.

Ferree, Myra M. 2005. "It's Time to Mainstream Research on Gender." *Chronicle Review,* August 12, B10.

File, Thom. 2017. "Voting in America: A Look at the 2016 Presidential Election." U.S. Census Bureau, May 10. Accessed June 5, 2017 (census.gov).

File, Thom, and Camille Ryan. 2014. "Computer and Internet Use in the United States: 2013." U.S. Census Bureau, American Community Survey Reports, November. Accessed August 22, 2015 (census.gov).

File, Thom, and Sarah Crissey. 2010. "Voting and Registration in the Election of November 2008: Population Characteristics." U.S. Census Bureau, Current Population Reports. Accessed May 15, 2010 (www.census.gov).

Filipovic, Jill. 2016. "Barbie's Problem Is Far Beyond Skin-Deep." *Time,* February 8, 51.

Filipovic, Jill. 2017. "What an All-Male Photo Op Really Says." *New York Times,* March 29, D2.

"Financial Crime." 2015. *The Economist,* May 23, 61.

Finer, Lawrence B., and Mia R. Zolna. 2016. "Declines in Unintended Pregnancy in the United States, 2008–2011." *New England Journal of Medicine* 374 (March 3): 843–852.

Fingerhut, Hannah. 2016. "In Both Parties, Men and Women Differ over Whether Women Still Face Obstacles to Progress." Pew Research Center, August 16. Accessed May 19, 2017 (pewresearch.org).

Fingerhut, Hannah. 2017. "About Seven-in-Ten Americans Oppose Overturning *Roe v. Wade.*" Pew Research Center, January 3. Accessed May 19, 2017 (pewresearch.org).

Finkelhor, David, Heather Turner, Sherry Hamby, and Richard Ormrod. 2011. "Polyvictimization: Children's Exposure to Multiple Types of Violence, Crime, and Abuse." Office of Justice Programs, October. Accessed March 7, 2012 (www.ojp.usdoj .gov).

Finkelstein, Eric A., et al. 2012. "Obesity and Severe Obesity Forecasts Through 2030." *American Journal of Preventive Medicine* 42 (6). Accessed May 14, 2012 (www.ajponline .org).

Fischman, Josh. 2009. "Global Warming Before Smokestacks." *Chronicle Review,* November 6, B11–B12.

Fitch, Kimberly, and Sangeeta Agrawal. 2015. "Female Bosses Are More Engaging Than Male Bosses." Gallup, May 7. Accessed April 4, 2017 (gallup.com).

Flabbi, Luca, Mario Macis, Andrea Moro, and Fabiano Schivardi. 2016. "Do Female Executives Make a Difference? The Impact of Female Leadership on Gender Gaps and Firm Performance." National Bureau of Economic Research, December. Accessed April 10, 2017 (nber.org).

Flasphaler, Julia. 2017. "Watch: Patagonia Ad Calls Out Corporate Takeover of Public Land." AlterNet, August 25. Accessed August 27, 2017 (alternet.org).

Flavin, Jeanne. 2001. "Feminism for the Mainstream Criminologist: An Invitation." *Journal of Criminal Justice* 29 (July/August): 271–285.

Fleegler, Eric W., et al. 2013. "Firearm Legislation and Firearm-Related Fatalities in the United State." *JAMA Internal Medicine,* March 6. Accessed January 15, 2014 (www.archinte .jamanetwork.com).

Fleishman, Jeffrey, and Amro Hassan. 2009. "Gadget to Help Women Feign Virginity Angers Many in Egypt." *Los Angeles Times,* October 7. Accessed October 28, 2009 (www .latimes.com).

Fleming, Charles. 2014. "Ford Greenlights Recycling." *Baltimore Sun,* July 5, 8.

Fleming, John. 2015. "Millennials Most Trusting on Safety of Personal Information." Gallup, May 11. Accessed August 14, 2015 (gallup .com).

Flesher, John, and Michael Biesecker. 2017. "EPA Moves to Withdraw Obama Clean-Water Rule." *Baltimore Sun,* June 28, 6.

Flint Water Advisory Task Force. 2016. "Final Report." Office of Governor Rick Snyder, State of Michigan, March. Accessed August 20, 2017 (michigan.gov).

Flores, Andrew R., Jody L. Herman, Gary J. Gates, and Taylor N. T. Brown. 2016. "How Many Adults Identify as Transgender in the United States?" The Williams Institute, June. Accessed May 10, 2017 (williamsinstitute .law.ucla.edu).

Florida, Richard. 2013. "The Striking Relationship Between Gun Safety Laws and Firearm Deaths." City Lab, March 7. Accessed April 13, 2017 (citylab.com).

Fogg, Piper. 2005. "Don't Stand So Close to Me." *Chronicle of Higher Education,* April 29, A10–A12.

Fomby, Paula, and Angela Estacion. 2011. "Cohabitation and Children's Externalizing Behavior in Low-Income Latino Families." *Journal of Marriage and Family* 73 (February): 46–66.

Fomby, Paula, and Cynthia Osborne. 2017. "Family Instability, Multipartner Fertility, and Behavior in Middle Childhood." *Journal of Marriage and Family* 79 (February): 75–93.

Food and Water Watch. 2013. "Take Back the Tap: Bottled Water Wastes Resources and Money." July. Accessed September 19, 2014 (www .foodandwaterwatch.org).

Food and Water Watch. 2016. "The State of Public Water in the United States." February 16. Accessed August 20, 2017 (foodandwaterwatch .org).

"Food Companies Propose Cutting Back on Junk Food Marketing Aimed at Children." 2011. *Washington Post,* July 14. Accessed July 20, (www.washingtonpost.com).

Ford, Peter. 2011. "Why Japan Will Rebound." *Christian Science Monitor,* March 28, 16–29.

"Foreign Born." 2016. U.S. Census Bureau, July 6. Accessed June 20, 2017 (census.gov).

"Formaldehyde." 2012. Environmental Protection Agency. Accessed September 20, 2014 (www .epa.gov).

Foroohar, Rana. 2011. "Whatever Happened to Upward Mobility?" *Time,* November 14, 27–34.

Fraire, John. 2014. "Why Your College Should Dump the SAT." *Chronicle of Higher Education,* May 2, A44.

Frakt, Austin. 2017. "A Big Culprit of Increased Health Spending? Technology." *New York Times,* January 26, A3.

Frank, Robert. 2017. "Plenty of Billionaires, but Few Are Women." *New York Times,* January 1, BU3.

Frankel, Matthew. 2017. "What's the Average American's Tax Rate?" The Motley Fool, March 4. Accessed May 12, 2017 (fool.com).

Frankel, Robin Saks. 2017. "Worried Sick About Your Health Care? You're Not Alone." Bankrate, July 17. Accessed August 15, 2017 (bankrate.com).

Frayer, Lauren. 2015. "In Spain, Entire Villages Ae Up for Sale—and They're Going Cheap." NPR, August 23. Accessed August 24, 2017 (npr.org).

Freedom House. 2015. "Freedom in the World 2015." Accessed October 12, 2015 (freedomhouse.org).

Freese, Jeremy. 2008. "Genetics and the Social Science Explanation of Individual Outcomes." *American Journal of Sociology* 114 (Suppl.): S1–S35.

Freking, Kevin, and Matthew Daly. 2017. "Congress Scraps Rules on Guns, Coal Mining." *Baltimore Sun*, February 3, 6.

Fremstad, Shawn. 2012. "The Poverty Rate Is Higher Than the Federal Government Says It Is." Center for Economic and Policy Research, September 21. Accessed February 6, 2014 (www.cepr.net).

Frey, William H. 2014. "New Projections Point to a Majority Nation in 2044." Brookings, December 12. Accessed October 4, 2015 (brookings.edu).

Frey, William H. 2014. "The Suburbs: Not Just for White People Anymore." *New Republic*, November 14. Accessed October 23, 2015 (newrepublic.com).

Frey, William H. 2014. "Will This Be the Decade of Big City Growth?" Brookings Institution, May 23. Accessed September 19, 2014 (brookings.edu).

Frey, William H. 2015. *Diversity Explosion: How New Racial Demographics Are Remaking America*. Washington, DC: Brookings Institution.

Frey, William H. 2016. "White Neighborhoods Get Modestly More Diverse, New Census Data Show." Brookings, December 13. Accessed August 20, 2017 (brookings.edu).

Frey, William H. 2017. "Census Shows Pervasive Decline in 2016 Minority Voter Turnout." Brookings, May 18. Accessed May 22, 2017 (brookings.edu).

Friedman, Howard S. 2012. "American Voter Turnout Lower Than Other Wealthy Countries." *Huffington Post*, July 10. Accessed April 20, 2014 (www.huffingtonpost.com).

Friedman, Jaclyn. 2011. *What You Really Really Want: The Smart Girl's Shame-Free Guide to Sex and Safety*. Berkeley, CA: Seal Press.

Friedman, Lisa, and Brad Plumer. 2017. "U.S. Agency Halts Stud of Health Near Mining." *New York Times*, August 22, A12.

Friedrichs, David O. 2009. *Trusted Criminals: White Collar Crime in Contemporary Society*, 4th edition. Belmont, CA: Wadsworth Publishing.

Frohwirth, Lon, Ann M. Moore, and Renata Maniaci. 2013. "Perceptions of Susceptibility to Pregnancy Among Women Obtaining Abortions." *Social Science & Medicine* 99 (December): 18–26.

"From Alpha to Omega." 2015. *The Economist*, August 15, 61.

Fry, Richard. 2012. "No Reversal in Decline of Marriage." Pew Research Center, November 20. Accessed February 9, 2013 (www.pewsocialtrends.org).

Fry, Richard. 2016. "For First Time in Modern Era, Living with Parents Edges Out Other Living Arrangements for 18- to 34-Year-Olds." Pew Research Center, May 24. Accessed March 1, 2017 (pewresearch.org).

Fry, Richard. 2017. "Women May Never Make up Half of the U.S. Workforce." Pew Research Center, September 12. Accessed January 31, 2017 (pewresearch.org).

Fry, Richard, and D'Vera Cohn. 2011. "Living Together: The Economics of Cohabitation." Pew Research Center, June 27. Accessed March 1, 2012 (www.pewsocialtrends.org).

Fry, Richard, and Jeffrey S. Passel. 2014. "In Post-Recession Era, Young Adults Drive Continuing Rise in Multi-Generational Living." Pew Research Center, July 17. Accessed July 25, 2014 (www.pewresearch.org).

Fryar, Cheryl D., Margaret D. Carroll, and Cynthia L. Ogden. 2016. "Prevalence of Overweight and Obesity Among Children and Adolescents Aged 2–19 Years: United States, 1963–1965 Through 2013–2014." National Center for Health Statistics, July. Accessed March 1, 2017 (cdc.gov/nchs).

Fryer, Roland G., Jr., Devah Pager, and Jorg L. Spenkuch. 2011. "Racial Disparities in Job Finding and Offered Wages." National Bureau of Economic Research, September. Accessed December 7, 2011 (www.nber.org).

Fuentes-Nieva, Ricardo, and Nick Galasso. 2014. *Working for the Few*. Oxfam International, January. Accessed February 10, 2014 (www.oxfam.org).

Funk, Cary, and Brian Kennedy. 2016. "The Politics of Climate." Pew Research Center, October 4. Accessed August 25, 2017 (pewresearch.org).

Funk, Cary, Brian Kennedy, and Elizabeth Sciupac. 2016. "U.S. Public Wary of Biomedical Technologies to 'Enhance' Human Abilities." Pew Research Center, July 26. Accessed September 2, 2017 (pewresearch.org).

G

Gallup. 2013. "State of the American Workplace: Employee Engagement Insight for U.S. Business Leaders." Accessed April 12, 2014 (www.gallup.com).

Galperin, Andrew, et al. 2013. "Sexual Regret: Evidence for Evolved Sex Differences." *Archives of Sexual Behavior* 42 (October): 1145–1161.

Galston, William A. 2013. "The Eroding American Middle Class." *Wall Street Journal*, November 13, A15.

Gamson, William. 1999. *Popular Culture & High Culture: An Analysis and Evaluation of Taste*. Revised and updated edition. New York: Basic Books.

Ganem, Joseph. 2016. "Science in Self-Interest." *Baltimore Sun*, August 21, 25.

Gans, Herbert J. 1999. *Popular Culture and High Culture: An Analysis and Evaluation of Taste*. Revised and updated edition. New York: Basic Books.

Gans, Herbert J. 2005. "Wishes for the Discipline's Future." *Chronicle Review*, August 12, B9.

GAO. 2013. "Farm Programs." U.S. Government Accountability Office, June. Accessed May 5, 2017 (gao.gov).

Gardner, Marilyn. 2008. "Happiness Is a Warm 'Thank You.'" *Christian Science Monitor*, January 28, 13, 16.

Gardner, Matthew, Robert S. McIntyre, and Richard Phillips. 2017. "The 35 Percent Corporate Tax Myth: Corporate Tax Avoidance by Fortune 500 Companies, 2008 to 2015." Institute on Taxation and Economic Policy, March. Accessed May 13, 2017 (itep.org).

Garfinkel, Harold. 1967. *Studies in Ethnomethodology*. Englewood Cliffs, NJ: Prentice-Hall.

Garza, Lisa Maria, and Timothy Williams. 2016. "Teenager in 'Affluenza' Case Is Sentenced to Jail." *New York Times*, April 4, A11.

Gates, Gary J. 2014. "LGB Families and Relationships: Analyses of the 2013 National Health Interview Survey." Williams Institute, October. Accessed July 20, 2017 (law.ucla.edu/williamsinstitute).

Gates, Gary J. 2015. "Demographics of Married and Unmarried Same-sex Couples: Analyses of the 2013 American Community Survey." Williams Institute, March. Accessed July 20, 2017 (law.ucla.edu/williamsinstitute).

Gates, Gary J. 2017. "In US, More Adults Identifying as LGBT." Gallup, January 11. Accessed May 20, 2017 (gallup.com).

Gates, Gary J. 2017. "U.S. Satisfaction with Immigration Levels Reaches New High." Gallup, January 18. Accessed June 20, 2017 (gallup.com).

Gavin, Lorrie, et al. 2013. "Vital Signs: Repeat Births among Teens—United States, 2007–2010." *MMWR* 62 (April 5): 249–255.

Gayer, Ted, Austin J. Drukker, and Alexander K. Gold. 2016. "Tax-Exempt Municipal Bonds and the Financial of Professional Sports Stadiums." Brookings, September. Accessed April 10, 2017 (brookings.edu).

"GE Fact Sheet." 2017. Accessed June 19, 2017 (ge.com).

Ge, Yanbo, Christopher R. Knittel, Don MacKenzie, and Stephen Zoepf. 2016. "Racial and Gender Discrimination in Transportation Network Companies." National Bureau of Economic Research, October. Accessed June 20, 2017 (nber.org).

Gee, Lisa Christensen, Matthew Gardner, Misha E. Hill, and Meg Wiehe. 2017. "Undocumented Immigrants' State & Local Tax Contributions." Institute on Taxation and Economic Policy, March. Accessed June 10, 2017 (itep.org).

Geertz, Clifford. 1966. "Religion as a Cultural System." Pp. 1–46 in *Anthropological Approaches to the Study of Religion*, edited by Michael Banton. London: Tavistock.

Geier, Ben. 2014. "Using 3-D Printing to Make Jet Engines." *Fortune*, December 1, 76.

Geiger, Abigail. 2016. "16 Striking Findings from 2016." Pew Research Center, December 21. Accessed June 20, 2017 (pewresearch.org).

Geiger, Abigail. 2016. "Sharing Chores a Key to Good Marriage, Say Majority of Married Adults." Pew Research Center, November 30. Accessed May 19, 2017 (pewresearch.org).

Geiger, Abigail. 2016. "Support for Marijuana Legalization Continues to Rise." Pew Research Center, October 12. Accessed March 1, 2017 (pewresearch.org).

Geiger, Abigail, and Lauren Kent. 2017. "Number of Women Leaders Around the World Has Grown, but They're Still a Small Group." Pew Research Center, March 8. Accessed May 19, 2017 (pewresearch.org).

"Gender, Education and Work." 2015. *The Economist*, March 7, 61–62.

"Gender Equality Universally Embraced, but Inequalities Acknowledged." 2010. Pew Research Center, Global Attitudes Project, July 1. Accessed November 1, 2011 (www.pewglobal.org).

"The Gender Gap in Religion Around the World." 2016. Pew Research Center, March 22. Accessed July 31, 2017 (pewresearch.org).

Gershenson, Seth. 2015. "The Alarming Effect of Racial Mismatch on Teacher Expectations." Brookings, August 18. Accessed March 1, 2017 (brookings.edu).

Gerth, H. H., and C. Wright Mills, eds. 1946. *Max Weber: Essays in Sociology.* New York: Oxford University Press.

Gewertz, Catherine. 2009. "Do Men Deserve a Break in College Admissions?" *Education Week* online, December 9. Accessed December 13, 2010 (www.blogs.edweek.org).

Geyer, Roland, Jenna R. Jambeck, and Kara Lavender Law. 2017. "Production, Use, and Fate of All Plastics Ever Made." *Science Advances* 3 (July 19): e1700782.

Gibson, David. 2012. "Vatican Orders Crackdown on American Nuns." *USA Today*, April 18. Accessed April 20, 2012 (www.usatoday.com).

Gilbert, Dennis. 2011. *The American Class Structure in an Age of Growing Inequality*, 8th edition. Thousand Oaks, CA: Pine Forge Press.

Gillespie, Michael Allen. 2014. "Grade Degradation." *Chronicle Review*, January 24, B2.

Gilliam, Walter S., Angela N. Maupin, Chin R. Reyes, Maria Accavitti, and Frederick Shic. 2016. "Do Early Educators' Implicit Biases Regarding Sex and Race Relate to Behavior Expectations and Recommendations of Preschool Expulsions and Suspensions?" Yale Child Study Center, September 28. Accessed July 31, 2017 (ziglercenter.yale.edu).

Gillis, Justin, and Nadja Popovich. 2017. "The View from Trump Country, Where Renewable Energy Is Thriving." *New York Times*, June 8, A22.

Gilpin, Lindsey. 2014. "The Dark Side of 3D Printing: 10 Things to Watch." TechRepublic, March 5. Accessed October 8, 2014 (www.techrepublic.com).

Gilson, Dave. 2014. "Don't Tread on Me." *Mother Jones*, January/February, 28–33.

Gimpelson, Vladimir, and Daniel Treisman. 2015. "Misperceiving Inequality." National Bureau of Economic Research, May. Accessed September 16, 2015 (nber.org).

Giugni, Marco, Doug McAdam, and Charles Tilley, eds. 1999. *How Social Movements Matter.* Minneapolis: University of Minnesota Press.

Giving USA Foundation. 2017. "Giving USA 2017." June 12. Accessed July 3, 2017 (givingusa.org).

GLAAD. 2017. "Accelerating Acceptance 2017." Accessed May 15, 2017 (glaad.org).

Gladstone, Rick. 2015. "Toll from Hajj Stampede Reaches 2,411 in New Estimate." *New York Times*, December 11, A4.

Glassner, Barry. 2010. "Still Fearful After All These Years." *Chronicle Review,* January 22, B11–B12.

Glauber, Bill. 2011. "Do Monarchies Still Matter?" *Christian Science Monitor,* February 21, 27–31.

Gleiber, Michael A. 2014. "The High Price Our Bodies Pay for High Fashion." *Huffington Post*, October 26. Accessed December 5, 2014 (www.huffingtonpost.com).

Glenn, David. 2011. "For Business Majors, Easy Does It." *Chronicle of Higher Education,* April 22, A1, A3–A5.

Glionna, John M. 2009. "South Korean Kids Get a Taste of Boot Camp." *Los Angeles Times*, August 22, A1.

"Global Christianity: A Report on the Size and Distribution of the World's Christian Population." 2011. Pew Research Center, December 1. Accessed April 8, 2012 (www.pewresearch.org).

"Global Restrictions on Religion Rise Modestly in 2015, Reversing Downward Trend." 2017. Pew Research Center, April 11. Accessed July 31, 2017 (pewresearch.org).

"Global Views on Morality." 2014. Pew Research Center. Accessed September 29, 2015 (www.pewglobal.org).

Global WASH Fast Facts. 2013. "Information on Water, Sanitation, and Hygiene." Centers for Disease Control and Prevention, November 8. Accessed September 19, 2014 (www.cdc.gov).

"Global WASH Fast Facts." 2017. Centers for Disease Control and Prevention, April 11. Accessed August 20, 2017 (cdc.gov).

Goffman, Alice. 2014. *On the Run: Fugitive Life in an American City*. Chicago: University of Chicago Press.

Goffman, Erving. 1959. *The Presentation of Self in Everyday Life.* New York: Doubleday Anchor Books.

Goffman, Erving. 1961. *Asylums: Essays on the Social Situation of Mental Patients and Other Inmates.* Garden City, NY: Anchor Books.

Goffman, Erving. 1963. *Stigma: Notes on the Management of Spoiled Identity.* Englewood Cliffs, NJ: Prentice-Hall.

Goffman, Erving. 1967. *Interaction Ritual: Essays on Face-to-Face Behavior.* New York: Anchor Books.

Goffman, Erving. 1969. *Strategic Interaction.* Philadelphia: University of Pennsylvania Press.

Gold, Rachel Benson, and Elizabeth Nash. 2017. "Flouting the Facts: State Abortion Restrictions Flying in the Face of Science." *Guttmacher Policy Review* 20 (January): 53–58.

Goldberg, Michelle. 2015. "Feminist Writers Are So Besieged by Online Abuse That Some Have Begun to Retire." *Washington Post*, February 20. Accessed September 30, 2015 (washingtonpost.com).

Golden, Daniel. 2006. *The Price of Admission: How America's Ruling Class Buys Its Way into Elite Colleges—and Who Gets Left Outside the Gates.* New York: Crown.

Goldin, Claudia, and Lawrence F. Katz. 2016. "Women Working Longer: Facts and Some Explanations." National Bureau of Economic Research, September. Accessed July 19, 2017 (nber.org).

Goldin, Claudia, Sari Pekkala Kerr, Claudia Olivetti, and Erling Barth. 2017. "The Expanding Gender Earnings Gap: Evidence from the LEHD-2000 Census." *American Economic Review: Papers & Proceedings* 107 (5): 110–114.

Golshan, Tara. 2016. "Why the Stanford Sexual Assault Case Has Become a National Flashpoint, Explained." Vox, December 19. Accessed April 21, 2017 (vox.com).

Goodale, Gloria. 2015. "'Sharing Economy': Helpful or Selfish?" *Christian Science Monitor Weekly*, March 30, 21–23.

Goode, Erica. 2013. "U.S. Prison Populations Decline, Reflecting New Approach to Crime." *New York Times*, July 26, A11.

Goodell, Jeff. 2017. "Scott Pruitt's Crimes Against Nature." *Rolling Stone*, July 27. Accessed August 20, 2017 (rollingstone.com).

Goodnough, Abby. 2010. "Making It Clear That a Clear Parking Space Isn't." *New York Times,* December 28, A10.

Goodnough, Abby. 2017. "Panel Wants Opioid Crisis Declared Emergency." *New York Times*, August 1, A14.

Goodstein, Laurie. 2014. "Presbyterians Vote to Allow Same-Sex Marriages." *New York Times*, June 20, A11.

Goodstein, Laurie, and Michael D. Shear. 2017. "Trump's Order on Religious Liberty Pleases Some, but Lets Down Conservatives." *New York Times*, May 5, A20.

Gopnik, Alison, Andrew N. Meltzoff, and Patricia K. Kuhl. 2001. *The Scientist in the Crib: What Early Learning Tells Us About the Mind.* New York: Perennial.

Gottdiener, Mark, and Ray Hutchison. 2000. *The New Urban Sociology*, 2nd edition. New York: McGraw-Hill.

Gottlieb, Mark. 2011. "Attention to Duty." *Christian Science Monitor,* June 6, 44.

Gottman, John M. 1994. *What Predicts Divorce? The Relationships Between Marital Processes and Marital Outcome.* Hillsdale, NJ: Lawrence Erlbaum Associates.

Goudreau, Jenna. 2012. "A New Obstacle for Professional Women: The Glass Escalator." *Forbes*, May 21. Accessed December 18, 2013 (www.forbes.com).

Gouldner, Alvin W. 1962. "Anti-Minotaur: The Myth of a Value-Free Sociology." *Social Problems* 9 (Winter): 199–212.

Govender, Serusha. 2014. "American Foods the Rest of the World Thinks Are Strange." Fox News, May 13. Accessed February 14, 2017 (foxnews.com).

"Government Efficiency and Effectiveness." 2015. GAO Highlights, GAO-15-440T, March 4. Accessed August 22, 2015 (gao.gov).

"Government Resolves to Start Making Sense Under New Law Forbidding Federal Gibberish." 2011. *Washington Post,* May 20. Accessed May 21, 2011 (www.washingtonpost.com).

Governors Highway Safety Association. 2014. "Distracted Driving Laws." Accessed January 12, 2014 (www.ghsa.org).

Goyette, Braden. 2013. "Oprah Talks to Swiss Newspaper About Alleged Racist Encounter." *Huffington Post*, August 14. Accessed March 22, 2014 (www.huffingtonpost.com).

Graham, Carol. 2015. "The High Costs of Being Poor in America: Stress, Pain, and Worry." Brookings Institution, February 19. Accessed September 31, 2015 (brookings.edu).

Graham, Carol. 2016. "The Rich Even Have a Better Kind of Stress Than the Poor." Brookings, February 10. Accessed December 23, 2016 (brookings.edu).

Graif, Corina, and Robert J. Sampson. 2009. "Spatial Heterogeneity in the Effects of Immigration and Diversity on Neighborhood Homicide Rates." *Homicide Studies* 13 (August): 242–260.

Grall, Timothy. 2013. "Custodial Mothers and Fathers and Their Child Support: 2011." U.S. Census Bureau, October. Accessed May 16, 2014 (www.census.gov).

Gramlich, John. 2016. "Voters' Perceptions of Crime Continue to Conflict with Reality." Pew Research Center, November 16. Accessed December 17, 2016 (pewresearch.org).

Gramlich, John. 2017. "Most Violent and Property Crimes in the U.S. Go Unsolved." Pew Research Center, March 1. Accessed March 2, 2017 (pewresearch.org).

Grandjean, Philippe, and Philip J. Landrigan. 2014. "Neurobehavioral Effects of Developmental Toxicity." *The Lancet Neurology* 13 (March): 330–338.

Grauerholz, Liz, and Sharon Bouma-Holtrop. 2003. "Exploring Critical Sociological Thinking." *Teaching Sociology* 31 (October): 485–496.

Graves, Joseph L., Jr. 2001. *The Emperor's New Clothes: Biological Theories of Race at the Millennium.* New Brunswick, NJ: Rutgers University Press.

Gray, Alex. 2017. "These Are the World's Biggest Corporate Giants." World Economic Forum, January 16. Accessed May 15, 2017 (weforum .org).

Gray, Paul S., John B. Williamson, David R. Karp, and John R. Dalphin. 2007. *The Research Imagination: An Introduction to Qualitative and Quantitative Methods.* New York: Cambridge University Press.

Grazian, David. 2008. *On the Make: The Hustle of Urban Nightlife.* Chicago: University of Chicago Press.

Greeley, Brendan. 2013. "How Inequality Became a Household Word." *Bloomberg Businessweek*, December 16–22, 16–17.

Green, Jeff. 2016. "At Work and Out of the Closet in the Heartland." *Bloomberg Businessweek*, March 3–April 3, 22–23.

Green, Jeff. 2016. "Detroit's Comeback Has an Arabic Accent." *Bloomberg Businessweek*, February 22–26, 18–19.

Green, Jeff, and Tim Higgins. 2016. "LGBT Inc." *Bloomberg Businessweek*, May 2–8, 26–28.

Greenberg, Julie, Arthur McKee, and Kate Walsh. 2013. "Teacher Prep Review: A Review of the Nation's Teacher Preparation Programs 2013." National Council on Teacher Quality. Accessed June 21, 2014 (www.nctq.org).

Greenfield, Rebecca. 2015. "You Take the Vacation. Your Boss Pays the Bill." *Bloomberg Businessweek*, May 18–24, 22–23.

Greenwood, Shannon, Andrew Perrin, and Maeve Duggan. 2016. "Social Media Update 2016." Pew Research Center, November 11. Accessed March 14, 2017 (pewresearch.org).

Gregory, Sean. 2017. "The Jobs That Weren't Saved." *Time*, May 29, 35–41.

Grim, Brian J., and Alan Cooperman. 2014. "Religious Hostilities Reach Six-Year High." Pew Research Center, January 14. Accessed June 21, 2014 (www.pewresearch.org).

Grobart, Sam. 2016. "People Think They Are Middle-Class, Whether They Make $22K or $200K." *Bloomberg Businessweek*, September 19–25, 26.

Gross, Rita M. 1996. *Feminism and Religion: An Introduction.* Boston: Beacon Press.

Guallar, Eliseo, Saverio Stranges, Cynthia Mulrow, Lawrence J. Appel, and Edgar R. Miller III. 2013. "Enough Is Enough: Stop Wasting Money on Vitamin and Mineral Supplements." *Annals of Internal Medicine* 159 (12): 850–851.

Guarino, Mark. 2013. "Auto Jobs Are Back, Too, but at Lower Wages." *Christian Science Monitor Weekly*, May 13, 22–23.

"Gun Laws: A Shot and a Beer." 2014. *The Economist*, May 3, 28.

Gunders, Dana. 2012. "Wasted: How America Is Losing Up to 40 Percent of Its Food from Farm to Fork to Landfill." Natural Resources Defense Council, August. Accessed November 10, 2013 (www.nrdc.org).

Gurkoff, Joe, and Anna Ranieri. 2012. *How Can I Help You? What You Can (and Can't) Do to Counsel a Friend, Colleague or Family Member with a Problem.* North Charleston, SC: CreateSpace Independent Publishing Platform.

Gurney, Joan M., and Kathleen T. Tierney. 1982. "Relative Deprivation and Social Movements: A Critical Look at Twenty Years of Theory and Research." *Sociological Quarterly* 23 (Winter): 33–47.

Guterman, Stanley S. 1969. "In Defense of Wirth's 'Urbanism as a Way of Life.'" *American Journal of Sociology* 74 (March): 492–499.

Guttmacher Institute. 2013. "U.S. Women Who Have Abortions." Infographics. Accessed February 11, 2012 (www.guttmacher.org).

Guttmacher Institute. 2014. "Facts on Induced Abortion in the United States." February. Accessed March 2, 2014 (guttmacher.org).

Guttmacher Institute. 2016. "American Teens' Sexual and Reproductive Health." Fact Sheet, September. Accessed May 5, 2017 (guttmacher.org).

Guttmacher Institute. 2017. "Policy Trends in the States: 2016." January 3. Accessed May 5, 2017 (guttmacher.org).

Guy, Gery P., Jr., et al. 2017. "Vital Signs: Changes in Opioid Prescribing in the United States, 2006–2015." *MMWR* 66 (26): 697–704.

H

Haas, Ann P., Phillip L. Rodgers, and Jody L. Herman. 2014. "Suicide Attempts Among Transgender and Gender Non-Conforming Adults." Williams Institute, January. Accessed March 2, 2014 (www.williamsinstitute.law .ucla.edu).

Haas, Steven A., and David R. Schaefer. 2014. "With a Little Help from My Friends? Asymmetrical Social Influence on Adolescent Smoking Initiation and Cessation." *Journal of Health and Social Behavior* 55 (June): 126–143.

Hacker, Jacob S. 2002. *The Divided Welfare State: The Battle over Public and Private Social Benefits in the United States.* New York: Cambridge University Press.

Hacker, Jacob S., and Paul Pierson. 2010. *Winner-Take-All Politics: How Washington Made the Rich Richer and Turned Its Back on the Middle Class.* New York: Simon and Schuster.

Hagan, Frank E. 2008. *Introduction to Criminology: Theories, Methods, and Criminal Behavior,* 6th edition. Thousand Oaks, CA: Sage.

Hahn, Heather, Julia Isaacs, Sara Edelstein, Ellen Steele, and C. Eugene Steuerle. 2014. "Kids' Share 2014: Report on Federal Expenditures on Children Through 2013." Urban Institute. Accessed July 1, 2017 (urban.org).

Hakim, Danny. 2015. "Banned Abroad: Differences in Rules for Pesticides Point Up U.S.-Europe Trade Bumps." *New York Times*, February 24, B1, B4.

Hales, Emily. 2014. "Incentives for Having Kids: Solution for Falling Birth Rates?" *Deseret News*, June 26. Accessed September 21 (www .newsok.com).

Halikias, Dimitrios, and Richard V. Reeves. 2017. "Ladders, Labs, or Laggards? Which Public Universities Contribute Most." Brookings, July 11. Accessed July 31, 2017 (brookings.edu).

Hall, Aron J., Mary E. Wikswo, Kimberly Pringle, L. Hannah Gould, and Umesh D. Parashar. 2014. "Vital Signs: Foodborne Norovirus Outbreaks—United States, 2009–2012." *MMWR* 63 (22): 491–495.

Hall, Edward T. 1959. *The Silent Language.* New York: Doubleday.

Hall, Edward T. 1966. *The Hidden Dimension.* Garden City, NY: Doubleday.

Hallman, Ben, and Shane Shifflett. 2014. "When Hospices Mistreat the Dying, They Almost Never Get Punished." *Huffington Post*, December 30. Accessed October 16, 2015 (projects.huffingtonpost.com).

Halperin, David M. 2012. "How to Be Gay." *Chronicle Review,* September 7, B13–B17.

Halsey, Ashley, III. 2010. "Study: Older People Are Driving More, Having Fewer Accidents." *Washington Post,* June 22, A10.

Hamby, Sherry, David Finkelhor, Heather Turner, and Richard Ormrod. 2011. "Children's Exposure to Intimate Partner Violence and Other Family Violence." Office of Justice Programs, October. Accessed May 20, 2014 (www.ojp.usdoj.gov).

Hamilton, Brady E., Joyce A. Martin, Michelle J. K. Osterman, Anne K. Driscoll, and Lauren M. Rossen. 2017. "Births: Provisional Data for 2016." National Center for Health Statistics, National Vital Statistics System, June. Accessed July 1, 2017 (cdc.gov/nchs).

Hamilton, Malcolm B. 2001. *The Sociology of Religion,* 2nd edition. New York: Routledge.

Hammadi, Saad. 2013. "Making Asia's Factories Safer." *Christian Science Monitor Weekly*, November 11, 16.

Hanes, Stephanie. 2011. "Pretty in Pink?" *Christian Science Monitor*, September 26, 26–31.

Haney, Craig, Curtis Banks, and Philip Zimbardo. 1973. "Interpersonal Dynamics in a Simulated Prison." *International Journal of Criminology and Psychology* 1: 69–97.

Hansen, Lawrence A. 2010. "Noxious Groupthink." *Chronicle of Higher Education,* November 12, B6–B8.

Hanser, Amy, and Zachary Hyde. 2014. "Foodies Remaking Cities." *Contexts* 13 (Summer): 44–49.

Hanushek, Eric A., and Ludger Woessman. 2005. "Does Educational Tracking Affect Performance and Inequality? Differences-in-Differences Evidence Across Countries." National Bureau of Economic Research, February. Accessed July 25, 2007 (www.nber.org).

Hardoon, Deborah. 2017. "An Economy for the 99%." Oxfam Briefing Paper—Summary, January. Accessed May 13, 2017 (oxfam.org).

Harlow, Harry E., and Margaret K. Harlow. 1962. "Social Deprivation in Monkeys." *Scientific American* 206 (November): 137–146.

Harnish, Verne. 2014. "5 Ways to Tone Your Operations." *Fortune,* October 27, 46.

Harrell, Erika, and Lynn Langton. 2013. "Victims of Identity Theft, 2012." U.S. Department of Justice, Office of Justice Programs, December. Accessed January 12, 2014 (www.ojp .usdoj.gov).

Harrell, Erika, Lynn Langton, Marcus Berzofsky, Lance Couzens, and Hope Smiley-McDonald.

2014. "Household Poverty and Nonfatal Violent Victimization, 2008–2012." Bureau of Justice Statistics, November. Accessed September 11, 2015 (bjs.gov).

Harris, Chauncey D. 1997. "'The Nature of Cities' and Urban Geography in the Last Century." *Urban Geography* 18 (1): 15–35.

Harris, Chauncey D., and Edward L. Ullman. 1945. "The Nature of Cities." *Annals* 242: 7–17.

Harris, Gardiner. 2011. "When the Nurse Wants to Be Called 'Doctor.'" *New York Times*, June 6, A23.

Harris, Gardiner. 2013. "India's New Focus on Rape Shows Only the Surface of Women's Perils." *New York Times*, January 12. Accessed March 4, 2014 (www.nytimes.com).

Harris, Jennifer L., et al. 2013. "Fast Food FACTS 2013: Measuring Progress in Nutrition and Marketing to Children and Teens." Yale Rudd Center for Food Policy & Obesity. Accessed November 10, 2013 (www.fastfoodmarketing.org).

Harter, Jim, and Amy Adkins. 2017. "Alabama Leads US States in Employee Engagement." Gallup, February 20. Accessed June 27, 2017 (gallup.com).

Harter, Jim, and Sangeeta Agrawal. 2013. "'Engaged' Workers Would Keep Jobs upon Winning Lottery." Gallup, December 20. Accessed April 12, 2014 (www.gallup.com).

Harter, Jim, and Sangeeta Agrawal. 2014. "Many Baby Boomers Reluctant to Retire." Gallup, January 20. Accessed May 16, 2014 (www.gallup.com).

Hartig, Hannah, John Lapinski, and Stephanie Psyllos. 2016. "Poll: More Voters Say Trump Doesn't Respect Women After Lewd Tape Surfaces." NBC News, October 10. Accessed January 5, 2017 (nbcnews.com).

Harvard School of Public Health. 2014. "Poll Finds Many in U.S. Lack Knowledge About Ebola and Its Transmission." August 21. Accessed October 8, 2014 (www.hsph.harvard.edu).

Haskins, Ron. 2008. "Immigration: Wages, Education, and Mobility." Pp. 81–90 in *Getting Ahead or Losing Ground: Economic Mobility in America*, edited by Julia B. Isaacs, Isabel V. Sawhill, and Ron Haskins. Washington, DC: Brookings Institution.

Haskins, Ron. 2015. "$2 a Day: A More Complete Picture." Brookings Institution, September 14. Accessed September 31, 2015 (brookings.edu).

Hauer, Mathew E., Jason M. Evens, and Deepak R. Mishra. 2016. "Millions Projected to Be at Risk from Sea-Level Rise in the Continental United States." *Nature Climate Change* 6 (April): 691–695.

Hawai'i Free Press. 2012. "Census Bureau: Hawaii 7th Highest Poverty Rate in US." November 15. Accessed February 14, 2014 (www.hawaiifreepress.com).

Hayden, Dolores. 2002. *Redesigning the American Dream: Gender, Housing, and Family Life*. New York: W. W. Norton

Hayes, Danny, and Jennifer L. Lawless. 2016. *Women on the Run: Gender, Media, and Political Campaigns in a Polarized Era*. New York: Cambridge University Press.

Hayford, Sarah R., and Karen Benjamin Guzzo. 2015. "The Single Mother by Choice Myth." *Contexts* 14 (Fall): 70–72.

"Health Care in America." 2015. *The Economist*, March 7, 63–64.

"The Health Paradox." 2013. *The Economist*, May 11, 27–28.

Health Resources and Services Administration. 2010. "Assuring Access to Essential Health Care." Accessed July 15, 2011 (www.plainlanguage.gov).

Healthcare-NOW! 2014. "Health Insurance CEO Pay Skyrockets in 2013." May 5. Accessed July 26, 2014 (www.healthcare-now.org).

"Health Insurance Coverage." 2017. Centers for Disease Control and Prevention, March 31. Accessed August 15, 2017 (cdc.gov).

Heath, Chip. 2005. "Introduction." Pp. 81–85 in *Rumor Mills: The Social Impact of Rumor and Legend*, edited by Gary Alan Fine, Veronique Campion-Vincent, and Chip Heath. New Brunswick, NJ: Transaction Publishers.

Heath, Jennifer. 2008. *The Veil: Women Writers on Its History, Lore, and Politics*. Berkeley: University of California Press.

Hechter, Michael, and Karl-Dieter Opp. 2001. "Introduction." Pp. xi–xx in *Social Norms*, edited by Michael Hechter and Karl-Dieter Opp. New York: Russell Sage Foundation.

Hedges, Chris. 2011. "Our Public Schools Are Churning Out Drones for the Corporate State." AlterNet, April 11. Accessed April 14, 2011 (www.alternet.org).

Hee Lee, Ye Michelle. 2017. "Fact-Checking Trump's Rhetoric on Crime and the 'American Carnage'." *Washington Post*, January 30. Accessed April 15, 2017 (washingtonpost.com).

Heffer, Simon. 2010. "America Is the Acceptable Face of Cultural Imperialism." *Telegraph Media Group*, July 24. Accessed November 10, 2013 (telegraph.co.uk).

Hegewisch, Ariane, and Heidi Hartmann. 2015. "The Gender Wage Gap: 2014." Institute for Women's Policy Research, September. Accessed September 29, 2015 (iwpr.org).

Heider, Eleanor R., and Donald C. Olivier. 1972. "The Structure of the Color Space in Naming and Memory for Two Languages." *Cognitive Psychology* 3: 337–354.

Heilbroner, R. L., and L. C. Thurow. 1998. *Economics Explained: Everything You Need to Know About How the Economy Works and Where It's Going*. New York: Touchstone.

Heim, Joe. 2015. "Indian Americans Dominate the National Spelling Bee. Why Should They Take Abuse on Social Media for It?" *Washington Post*, May 25. Accessed October 4, 2015 (washingtonpost.com).

Heiman, Julia R., J. Scott Long, Shawna N. Smith, William A. Fisher, Michael S. Sand, and Raymond C. Rosen. 2011. "Sexual Satisfaction and Relationship Happiness in Midlife and Older Couples in Five Countries." *Archives of Sexual Behavior* 40 (4): 741–753.

Heisler, Candace. 2012. "Elder Abuse." Office for Victims of Crime. Accessed May 16, 2014 (www.ovc.ncjrs.gov).

Helliwell, John, Richard Layard, and Jeffrey Sachs (eds). 2013. *World Happiness Report 2013*. United Nations. Accessed February 6, 2014 (www.unsdsn.org).

Hemphill, Thomas A., and Mark J. Perry. 2017. "Wind Power's Future in U.S. Could Be Thwarted by Grassroots Opposition." Real Clear Energy, June 25. Accessed August 20, 2017 (realclearenergy.org).

Henderson, Alex. 2016. "Sick at Work: Staggering Number of Americans Go to Work with Bad Colds or Flu." AlterNet, May 30. Accessed August 15, 2017 (alternet.org).

Henne, Peter. 2015. "Latest Trends in Religious Restrictions and Hostilities." Pew Research Center, February 26. Accessed July 31, 2017 (pewresearch.org).

Henne, Peter, and Katayoun Kishi. 2016. "5 Key Findings About Global Restriction on Religion." Pew Research Center, May 4. Accessed July 31, 2017 (pewresearch.org).

Henry, Meghan, Rian Watt, Lily Rosenthal, and Azim Shivji. 2016. "The 2016 Annual Homeless Assessment Report (AHAR) to Congress." U.S. Department of Housing and Urban Development, November. Accessed May 1, 2017 (hud.gov).

Herman, Bob. 2017. "The Sky-High Pay of Health Care CEOs." Axios, July 24. Accessed August 15, 2017 (axios.com).

Hernandez, David. 2016. "'Pokemon Go' Players Fall off 90-Foot Ocean Bluff." *San Diego Union Tribune*, July 13. Accessed September 2, 2017 (sandiegouniontribune.com).

Herzberg, David. 2009. *Happy Pills in America: From Miltown to Prozac*. Baltimore, MD: Johns Hopkins University Press.

Hibbard, Roberta, Jane Barlow, and Harriet MacMillan. 2012. "Psychological Maltreatment." American Academy of Pediatrics. Accessed May 4, 2013 (www.pediatrics.aappublications.org).

Hicks, Maureen Soyars. 2017. "Flexible Jobs Give Workers Choices." U.S. Bureau of Labor Statistics, Monthly Labor Review, May. Accessed June 5, 2017 (bls.gov).

Higgins, Tim. 2015. "Public Opinion: The Polling Industry Cuts the Cord." *Bloomberg Businessweek*, November 23–29, 30.

Hightower, Jim. 2015. "CEOs Call for Wage Increases for Workers! What's the Catch?" AlterNet, August 26. Accessed August 26, 2015 (alternet.org).

Hilbert, Richard A. 1992. *The Classical Roots of Ethnomethodology: Durkheim, Weber, and Garfinkel*. Chapel Hill: University of North Carolina Press.

Hilger, Nathaniel. 2016. "Upward Mobility and Discrimination: The Case of Asian Americans." National Bureau of Economic Research, October. Accessed June 20, 2017 (nber.org).

Hill, Adam. 2017. "Breaking the Stigma—A Physician's Perspective on Self-Care and Recovery." *New England Journal of Medicine* 376 (March 23): 1103–1105.

Hill, Catey. 2015. "5 Times It's More Expensive to Be a Woman." Market Watch, April 14. Accessed September 29, 2015 (marketwatch.com).

Hill, Catherine, Christianne Corbett, and Andresse St. Rose. 2010. "Why So Few? Women in Science, Technology, Engineering, and Mathematics." American Association of University Women. Accessed October 25, 2011 (www.aauw.org).

Hill, Jason, Christina Stearns, and Chelsea Owens. 2015. "Education and Certification Qualifications of Departmentalized Public High-School-Level Teachers of Selected Subjects: Evidence from the 2011–12 Schools and Staffing Survey." National Center for Education Statistics, June. Accessed July 30, 2017 (nces.ed.gov).

Hillaker, B. D., H. E. Brophy-Herb, F. A. Villarruel, and B. E. Hass. 2008. "The Contributions of Parenting to Social Competencies and Positive Values in Middle School Youth:

Positive Family Communication, Maintaining Standards, and Supportive Family Relationships." *Family Relations* 57 (December): 591–601.

Hillin, Taryn. 2014. "The Divorce Mistakes You Don't Even Know You're Making." *Huffington Post,* March 18. Accessed March 18, 2014 (www.huffingtonpost.com).

Hines, Jonas Z., Steven C. Fiala, and Katrina Hedberg. 2017. "Electronic Cigarettes as an Introductory Tobacco Product Among Eighth and 11th Grade Tobacco Users—Oregon, 2015." *MMWR* 66 (23): 604–606.

Hinkle, Stephen, and John Schopler. 1986. "Bias in the Evaluation of In-Group and Out-Group Performance." Pp. 196 -212 in *Psychology of Everyday Intergroup Relations*, 2nd edition, edited by Stephen Worchel and William G. Austin. Chicago: Nelson-Hall

Hirschi, Travis. 1969. *Causes of Delinquency*. Berkeley: University of California Press.

"Hispanic Heritage Month 2016." 2016. U.S. Census Bureau, October 12. Accessed June 20, 2017 (census.gov).

Hiss, William C., and Valerie W. Franks. 2014. "Defining Promise: Optional Standardized Testing Policies in American College and University Admissions." National Association for College Admission Counseling, February 5. Accessed June 21, 2014 (www.nacacnet .org).

"HIV in the United States: At a Glance." 2017. Centers for Disease Control and Prevention, June. Accessed August 15, 2017 (cdc.gov).

Hlavka, Heather R. 2014. "Normalizing Sexual Violence: Young Women Account for Harassment and Abuse." *Gender & Society* 20 (June): 1–22.

Hoecker-Drysdale, Susan. 1992. *Harriet Martineau: First Woman Sociologist.* Providence, RI: Berg.

Hoffman, Piper. 2014. "7 Countries That Still Kill 'Witches.'" Care2 Causes, October 21. Accessed October 24, 2014 (www.care2.com).

Hokayem, Charles, and Misty L. Heggeness. 2014. "Living in Near Poverty in the United States: 1966–2012." U.S. Census Bureau, May. Accessed May 16, 2014 (www.census.gov).

Holloway, Kali. 2015. "8 Myths That Fuel the Assault on Abortion Rights." AlterNet, July 25. Accessed September 29, 2015 (alternet.org).

Holloway, Kali. 2016. "African-American Women Now Top the List of Most-Educated Group in the Country." AlterNet, June 2. Accessed June 20, 2017 (alternet.org).

Holloway, Kim. 2016. "Yes, Hillary Still Has to Suffer the Sexism of the Dude TV Pundits." AlterNet, February 4. Accessed May 15, 2017 (alternet.org).

Holmes, Elizabeth. 2013. "Grooming Tips from Giant Men." *Wall Street Journal*, November 27, D1.

Homans, George. 1974. *Social Behavior: Its Elementary Forms,* revised edition. New York: Harcourt Brace Jovanovich.

Hondagneu-Sotelo, Pierrette. 2001. *Domestica: Immigrant Workers Cleaning and Caring in the Shadows of Affluence.* Berkeley: University of California Press.

Hong, Gihoon, and John McLaren. 2015. "Are Immigrants a Shot in the Arm for the Local Economy?" National Bureau of Economic Research, February. Accessed June 20, 2017 (nber.org).

Hooke, Alexander E. 2014. "Speak Not, Lest Ye Be Judged." *Baltimore Sun*, May 27, 13.

Hooyman, Nancy R. 2016. "Social and Health Disparities in Aging: Gender Inequities in Long-Term Care." American Society on Aging, August 8. Accessed August 15, 2017 (asaging.org).

Horowitz, Juliana, Kim Parker, Nikki Graf, and Gretchen Livingston. 2017. "Americans Widely Support Paid Family and Medical Leave, but Differ over Specific Policies." Pew Research Center, March 23. Accessed July 20, 2017 (pewresearch.org).

Hotz, V. Joseph, and Juan Pantano. 2013. "Strategic Parenting, Birth Order and School Performance." National Bureau of Economic Research, October. Accessed October 26, 2014 (www.nber.org).

Howard, Beth. 2014. "Should You Have Surgery Abroad?" *AARP Magazine*, October/November, 26–30.

Howard, Jeff. 2003. "Still at Risk: The Causes and Costs of Failure to Educate Poor and Minority Children for the Twenty-First Century." Pp. 81–97 in *A Nation Reformed? American Education 20 Years After A Nation at Risk,* edited by David T. Gordon and Patricia A. Graham. Cambridge, MA: Harvard Education Press.

Hoxie, Josh. 2017. "The Myths Behind Inequality in Our Country." Institute for Policy Studies, May 3. Accessed May 4, 2017 (ips-dc.org).

Hoyt, Homer. 1939. *The Structure and Growth of Residential Neighborhoods in American Cities.* Washington, DC: Federal Housing Administration.

Hsin, Amy, and Yu Xie. 2014. "Explaining Asian Americans' Academic Advantage over Whites." *Proceedings of the National Academy of Sciences*, Early Edition, May 5. Accessed June 21, 2014 (www.pnas.org).

Hsu, A., et al. 2016. "2016 Environmental Performance Index." Yale Center for Environmental Law & Policy. Accessed August 20, 2017 (epi.yale.edu).

Hsu, Angel, et al. 2014. "The 2014 Environmental Performance Index: Full Report and Analysis." Yale Center for Environmental Law & Policy. Accessed September 19, 2014 (www .epi.yale.edu).

Huang, Jon, Samuel Jacoby, Michael Strickland, and K. K. Rebecca Lai. 2016. "Election 2016: Exit." *New York Times*, November 8. Accessed January 5, 2017 (nytimes.com).

Hubbard, Ruth. 1990. *The Politics of Women's Biology.* New Brunswick, NJ: Rutgers University Press.

Huffington Post. 2011. "New American Bible Changes Some Words, Including 'Holocaust.'" March 3. Accessed July 16, 2011 (www.huffingtonpost.com).

Huffman, Matt L. 2013. "Organizations, Managers, and Wage Inequality." *Sex Roles* 68 (February): 216–222.

Hughes, Everett C. 1945. "Dilemmas and Contradictions of Status." *American Journal of Sociology* 50: 353–359.

Huis, Arnold van, et al. 2013. "Edible Insects: Future Prospects for Food and Feed Security." Food and Agriculture Organization of the United Nations. Accessed February 14, 2017 (fao.org).

Huizinga, David, Shari Miller, and the Conduct Problems Prevention Research Group. 2013. "Developmental Sequences of Girls' Delinquent Behavior." Office of Juvenile Justice and Delinquency Prevention, December. Accessed January 12, 2014 (www.ojp.usdoj.gov).

"Human Rights Violations." 2015. Unite to End Violence Against Women. Accessed September 26, 2015 (un.org).

Hummer, Robert A., and Elaine M. Hernandez. 2013. "The Effect of Educational Attainment on Adult Mortality in the United States." *Population Bulletin* 68 (June): 1–16.

Hunnicutt, Gwen. 2009. "Varieties of Patriarchy and Violence Against Women: Resurrecting 'Patriarchy' as a Theoretical Tool." *Violence Against Women* 15 (May): 553–573.

Hunter, Lori M. 2014. "Unmarried Baby Boomers Face Disadvantages as They Grow Older." Population Reference Bureau, February. Accessed May 16, 2014 (www.prb.org).

"Hunting as a Pack." 2014. *Economist Technology Quarterly*, December 6, 6.

Hurwitz, Michael. 2011. "The Impact of Legacy Status on Undergraduate Admissions at Elite Colleges and Universities." *Economics of Education Review* 30 (June): 480–492.

Hutter, Mark. 2007. *Experiencing Cities.* Boston: Allyn & Bacon.

I

Ianzito, Christina. 2013. "Killer Heels." *Baltimore Sun*, August 1, 3.

IBM Journal Staff. 2016. "Cyber Security & Data Protection: A Glimpse into the Dark World of Organized Cybercrime." *IBM Journal*, May 11. Accessed April 15, 2017 (ibmjournal .com).

Iceland, John, Gregory Sharp, and Jeffrey M. Timberlake. 2013. "Sun Belt Rising: Regional Population Change and the Decline in Black Residential Segregation, 1970–2009." *Demography* 50 (February): 97–123.

Identity Theft Resource Center. 2017. "Data Breach Reports: 2016 End of Year Report." January 18. Accessed April 2, 2017 (idtheftcenter .org).

"In First Month, Views of Trump Are Already Strongly Felt, Deeply Polarized." 2017. Pew Research Center, February 16. Accessed June 20, 2017 (pewresearch.org).

Independent Television Service. 2003. "Race the Power of an Illusion: What Is This Thing Called Race?" Accessed March 22, 2014 (www.itvs.org).

"Inequality." 2017. *The Economist*, January 17, 21–22.

Infinit Healthcare. 2017. "The Pros and Cons of the Affordable Care Act." March 2. Accessed August 15, 2017 (infinithealthcare.com).

The Institute for College Access & Success. 2016. "Student Debt and the Class of 2015." October. Accessed July 31, 2017 (ticas.org).

Institute of Medicine. 2015. Beyond Myalgic Encephalomyelitis/Chronic Fatigue Syndrome: Redefining an Illness. Washington, DC: National Academies Press.

Insurance Institute for Highway Safety. 2013. "Status Report." 45 (September 28). Accessed October 18, 2013 (www.iihs.org).

Intergovernmental Panel on Climate Change. 2014. "Climate Change 2014: Impacts, Adaptation and Vulnerability; Summary to Policymakers." Accessed September 19, 2014 (www.ipcc.ch).

International Federation of Health Plans. 2014. "2013 Comparative Price Report: Variation in Medical and Hospice Prices by Country." March 14. Accessed July 26, 2014 (www.ifhp.com).

International Institute for Democracy and Electoral Assistance. 2014. "VAP Turnout." Accessed April 20, 2014 (www.idea.int).

Inter-Parliamentary Union. 2017. "Women in Parliaments: World Classification." May 1. Accessed May 25, 2017 (ipu.org).

Ioannidis, John. 2005. "Contradicted and Initially Stronger Effects in Highly Cited Clinical Research." *JAMA* 294 (July 13): 218–228.

Irving, Shelley K., and Tracy A. Loveless. 2015. "Dynamics of Economic Well-Being: Participation in Government Programs, 2009–2012: Who Gets Assistance?" U.S. Census Bureau, May. Accessed September 11, 2015 (census.gov).

Ivory, Danielle, Ben Protess, and Griff Palmer. 2016. "In American Towns, Pumping Private Profit from Public Works." *New York Times*, December 25, A1.

Ivry, Bob, Bradley Keoun, and Phil Kuntz. 2011. "Secret Fed Loans Gave Banks $13 Billion Undisclosed to Congress." *Bloomberg Businessweek,* November 27. Accessed August 26, 2015 (bloomberg.com).

J

Jackson, D. D. 1998. "'This Hole in Our Heart': Urban Indian Identity and the Power of Silence." *American Indian Culture and Research Journal* 22 (4): 227–254.

Jackson, Janine. 2017. "'It Is Not at All Typical to Stifle Basic Scientific Information.'" FAIR, February 13. Accessed March 3, 2017 (fair.org).

Jackson, Shelly L., and Thomas L. Hafemeister. 2011. "Financial Abuse of Elderly People vs. Other Forms of Elder Abuse: Assessing Their Dynamics, Risk Factors, and Society's Response." U.S. Department of Justice, February. Accessed March 5, 2012 (www.ncjrs.gov).

Jackson, Shelly L., and Thomas L. Hafemeister. 2013. "Understanding Elder Abuse." National Institute of Justice, June. Accessed May 16, 2014 (www.nij.gov).

Jacobe, Dennis. 2013. "U.S. Small Businesses Struggle to Find Qualified Employees." Gallup, February 15. Accessed April 12, 2014 (www.gallup.com).

Jacobs, Anna W., Terrence D. Hill, Daniel Tope, and Laureen K. O'Brien. 2016. "Employment Transitions, Child Care Conflict, and the Mental Health of Low-Income Urban Women with Children." *Women's Health Issues* 26 (4): 366–376.

Jacobs, Jerry A. 2005. "Multiple Methods in *ASR*." *Footnotes* 33 (December): 1, 4.

Jagacinski, Carolyn M. 2013. "Women Engineering Students: Competence Perceptions and Achievement Goals in the Freshman Engineering Course." *Sex Roles* 69 (December): 644–657.

Jamal, Ahmed, Brian A. King, Linda J. Neff, Jennifer Whitmill, Stephen D. Babb, and Corinne M. Graffunder. 2016. "Current Cigarette Smoking Among Adults—United States, 2005–2015." *MMWR* 65 (44): 1205–1211.

Jamal, Ahmed, et al. 2017. "Tobacco Use Among Middle and High School Students—United States, 2011–2016." *MMWR* 66 (23): 597–603.

"Jail Bait." 2016. *The Economist*, December 10, 33.

Jandt, Fred E. 2001. *Intercultural Communication: An Introduction.* Thousand Oaks, CA: Sage.

Janis, Irving L. 1972. *Victims of Groupthink: A Psychological Study of Foreign-Policy Decisions and Fiascoes.* Boston: Houghton Mifflin.

Janis, Irving L. 1982. *Groupthink: A Psychological Study of Policy Decisions and Fiascoes.* Boston: Houghton Mifflin Company.

Janofsky, Michael. 2003. "Young Brides Stir New Outcry on Utah Polygamy." *New York Times,* February 28, 1.

Jaret, Peter. 2015. "Prices Spike for Some Generics." *AARP Bulletin*, July–August, 8, 10.

Jargon, Julie. 2013. "Not-So-Fast Food." *Wall Street Journal*, November 15, B2.

Jargowsky, Paul A. 2015. "The Architecture of Segregation: Civil Unrest, the Concentration of Poverty, and Public Policy." The Century Foundation, August 9. Accessed August 20, 2017 (tcf.org).

Jaspers, Loes, Frederik Feys, Wichor M. Bramer, Oscar H. Franco, Peter Leusink, and Ellen T. M. Laan. 2016. "Efficacy and Safety of Flibanserin for the Treatment of Hypoactive Sexual Desire Disorder in Women: A Systematic Review and Meta-analysis." *JAMA Internal Medicine* 176 (4): 453–462.

Jeffreys, Sheila. 2011. *Man's Dominion: The Rise of Religion and the Eclipse of Women's Rights.* New York: Routledge.

Jenkin, G., N. Madhvani, L. Signal, and S. Bowers. 2014. "A Systematic Review of Persuasive Marketing Techniques to Promote Food to Children on Television." *Obesity Reviews* 15 (April): 281–293.

Jernigan, David. 2010. "Alcohol Marketing and Youth: Why It's a Problem and What You Can Do." Center for Alcohol Marketing and Youth, December 14. Accessed April 12, 2012 (www.camy.org).

Jewkes, Rachel, Emma Fulu, Tim Roselli, and Claudia Garcia-Moreno. 2013. "Prevalence of and Factors Associated with Non-Partner Rape Perpetration: Findings from the UN Multi-Country Cross-Sectional Study on Men and Violence in Asia and the Pacific." *Lancet Global Health* 1 (October): e208–218.

Johnson, Dave. 2014. "8 Ways Being Poor Is Wildly Expensive in America." AlterNet, January 29. Accessed February 14, 2014 (www.alternet.org).

Johnson, Hank, and Bert Klandermans, eds. 1995. *Social Movements and Culture*. Minneapolis: University of Minnesota Press.

Johnson, Heather L. 2016. "Pipelines, Pathways, and Institutional Leadership." American Council on Education. Accessed May 25, 2017 (acenet.edu).

Johnson, Ileana. 2010. "Why Has Communism Failed?" *Orthodoxy Today*, October 30. Accessed April 12, 2014 (www.orthodoxytoday.org).

Johnson, Jenna. 2016. "Trump's Rhetoric on Muslims Plays Well with Fans, but Horrifies Others." *Washington Post*, February 29. Accessed March 1, 2017 (washingtonpost.com).

Johnson, Kevin, and H. Darr Beiser. 2013. "Aging Prisoners' Costs Put Systems Nationwide in a Bind." *USA Today*, July 11. Accessed January 15, 2014 (www.usatoday.com).

Johnson, Kirk. 2011. "Between Young and Old, a Political Collision." *New York Times*, June 4, A10.

Johnson, Paula A., Therese Fitzgerald, Alina Salganicoff, Susan F. Wood, and Jill M. Goldstein. 2014. "Sex-Specific Medical Research: Why Women's Health Can't Wait." Brigham and Women's Hospital. Accessed October 17, 2015 (brighamandwomens.org).

Johnson, Rucker C. 2017. "Health." *Pathways*, Special Issue, 27–31.

Johnston, Pamela. 2005. "Dressing the Part." *Chronicle of Higher Education,* August 10. Accessed August 11, 2005 (www.chronicle.com).

Jones, Ann. 2013. *How the Wounded Return from America's Wars—The Untold Story.* Chicago, IL: Haymarket Books.

Jones, Bradley. 2016. "Americans' Views of Immigrants Marked by Widening Partisan, Generational Divides." Pew Research Center, April 15. Accessed June 20, 2017 (pewresearch.org).

Jones, Charise. 2017. "Los Angeles, New York and San Francisco Are Most Congested U.S. Cities." *USA Today*, February 20. Accessed August 20, 2017 (usatoday.com).

Jones, Jeffrey M. 2012. "Gender Gap in 2012 Vote Is Largest in Gallup's History." Gallup, November 9. Accessed April 20, 2014 (www.gallup.com).

Jones, Jeffrey M. 2013. "U.S. Blacks, Hispanics Have No Preferences on Group Labels." Gallup, July 26. Accessed March 22, 2014 (www.gallup.com).

Jones, Jeffrey M. 2014. "Reports of Alcohol-Related Family Trouble Remain Up in U.S." Gallup, July 29. Accessed November 4, 2014 (www.gallup.com).

Jones, Jeffrey M. 2015. "Majority in U.S. Now Say Gays and Lesbians Born, Not Made." Gallup, May 20. Accessed September 29, 2015 (gallup.com).

Jones, Jeffrey M. 2016. "Americans Still Say Postsecondary Education Very Important." Gallup, April 12. Accessed July 30, 2017 (gallup.com).

Jones, Jeffrey M. 2016. "Americans' Trust in Political Leaders, Public at New Lows." Gallup, September 21. Accessed June 27, 2017 (gallup.com).

Jones, Jeffrey M. 2016. "New Low of 52% 'Extremely Proud' to Be Americans." Gallup, July 1. Accessed July 1, 2016 (gallup.com).

Jones, Jeffrey M. 2017. "Americans Hold Record Liberal Views on Most Moral Issues." Gallup, May 11. Accessed May 20, 2017 (gallup.com).

Jones, Jeffrey M. 2017. "Independent Political ID in U.S. Lowest in Six Years." Gallup, January 6. Accessed June 27, 2017 (gallup.com).

Jones, Jeffrey M. 2017. "In US, 10.2% of LGBT Adults Now Married to Same-Sex Spouse." Gallup, June 22. Accessed July 20, 2017 (gallup.com).

Jones, Jo, and William D. Mosher. 2013. "Fathers' Involvement with Their Children: United States, 2006–2010." *National Health Statistics Reports*, December 20. Accessed May 16, 2014 (www.cdc.gov/nchs).

Jones, Nicholas A., and Jungmiwha Bullock. 2012. "The Two or More Races Population: 2012."

U.S. Census Bureau, September. Accessed March 22, 2014 (www.census.gov).

Jones, R. K., L. Frohwirth, and A. M. Moore. 2013. "More Than Poverty: Disruptive Events Among Women Having Abortions in the USA." *Journal of Family Planning and Reproductive Health Care* 39 (January): 36–43.

Jones, Rachel K., and Jenna Jerman. 2017. "Abortion Incidence and Service Availability in the United States, 2014." *Perspectives on Sexual and Reproductive Health* 49 (1): 17–27.

Jones, Rachel K., and Joerg Dreweke. 2011. "Countering Conventional Wisdom: New Evidence on Religion and Contraceptive Use." Guttmacher Institute, April. Accessed April 24, 2012 (www.guttmacher.org).

Jones, Rachel K., Lori F. Frohwirth, and Nakeisha M. Blades. 2016. "'If I Know I Am on the Pill and I Get Pregnant, It's an Act of God': Women's Views on Fatalism, Agency and Pregnancy." *Contraception* 93 (6): 551–555.

Jones, Robert P., Daniel Cox, E. J. Dionne, Jr., William A. Galston, Betsy Cooper, and Rachel Lienesch. 2016. "How Immigration and Concerns About Cultural Changes Are Shaping the 2016 Election." Public Religion Research Institute, June 23. Accessed June 20, 2017 (publicreligion.org).

Jones, Sophia. 2014. "Turkish Activists Say Their Country Is Sliding Backward on Women's Rights." *Huffington Post*, April 18. Accessed April 20, 2014 (www.huffingtonpost.com).

Jonsson, Jan O., David B. Grusky, Reinhard Pollak, and Matthew Di Carlo. 2009. "Recent Trends in Social Mobility in the United States: A New Approach to Modeling Trend in Big Class, Gradational, and Microclass Reproduction." Stanford Center for the Study of Poverty and Inequality, September. Accessed September 22, 2011 (www.inequality.com).

Jonsson, Patrik. 2007. "In US Justice, How Much Bias?" *Christian Science Monitor*, September 21, 1, 10.

Jordan, William. 2013. "Americans Divided over Blackface Halloween Make Up." YouGov, November 1. Accessed March 18, 2014 (www.today.yougov.com).

Joshi, Suchi P., Jochen Peter, and Patti M. Valkenburg. 2011. "Scripts of Sexual Desire and Danger in US and Dutch Teen Girl Magazines: A Cross-National Content Analysis." *Sex Roles* 64 (April): 463–474.

Juergensmeyer, Mark. 2003. "Thinking Globally About Religion." Pp. 3–13 in *Global Religions: An Introduction*, edited by Mark Juergensmeyer. New York: Oxford University Press.

Julian, Tiffany. 2012. "Work-Life Earnings by Field of Degree and Occupation for People with a Bachelor's Degree: 2011." U.S. Census Bureau, October. Accessed June 21, 2014 (www.census.gov).

K

Kabali, Hilda K., et al. 2015. "Exposure and Use of Mobile Media Devices by Young Children." *Pediatrics* 136 (December): 1044–1050.

Kaeble, Danielle, and Lauren Glaze. 2016. "Correctional Populations in the United States, 2015." Bureau of Justice Statistics, December. Accessed March 2, 2017 (bjs.gov).

Kaeble, Danielle, and Thomas P. Bonczar. 2017. "Probation and Parole in the United States, 2015." Bureau of Justice Statistics, February 2. Accessed March 2, 2017 (bjs.gov).

Kahan, Dan M., Hank Jenkins-Smith, and Donald Braman. 2011. "Cultural Cognition of Scientific Consensus." *Journal of Risk Research* 14 (2): 147–174.

Kahlenberg, Richard D. 2010. "10 Myths About Legacy Preferences in College Admissions." *Chronicle of Higher Education*, October 1, A23, A25.

Kahn, Shulamit, and Donna Ginther. 2017. "Women and STEM." National Bureau of Economic Research, June. Accessed July 31, 2017 (nber.org).

Kaiser Family Foundation. 2016. "Key Facts About the Uninsured Population." Fact Sheet, September. Accessed August 15, 2017 (kff.org).

Kalev, Alexandra, Frank Dobbin, and Erin Kelly. 2006. "Best Practices or Best Guesses? Assessing the Efficacy of Corporate Affirmative Action and Diversity Policies." *American Sociological Review* 71 (August): 589–617.

Kallakuri, Chetana, Shruti Vaidyanathan, Meegan Kelly, and Rachel Cluett. 2016. "The 2016 International Energy Efficiency Scorecard." American Council for and Energy-Efficient Economy, July Accessed August 20, 2017 (aceee.org).

Kalogrides, Demetra, Susanna Loeb, and Tara Beteille. 2013. "Systematic Sorting: Teacher Characteristics and Class Assignments." *Sociology of Education* 86 (April): 103–123.

Kamp, Karin. 2016. "Missing from the Debates: Climate Change, Poverty, Campaign Finance and More." Moyers & Company, October 19. Accessed May 13, 2017 (billmoyers.com).

Kane, Nazneen. 2015. "Stay-at-Home Fatherhoods." *Contexts* 14 (Spring): 74–76.

Kaneda, Toshiko, and Kristin Bietsch. 2015. "2015 World Population Data Sheet." Washington, DC: Population Reference Bureau. Accessed September 13, 2014 (prb.org).

Kaneda, Toshiko, and Kristin Bietsch. 2016. "2016 World Population Data Sheet." Washington, DC: Population Reference Bureau. Accessed May 13, 2017 (prb.org).

Kang, Cecilia. 2015. "Facebook Doubles Black Hires, but They Only Make Up About 1.5% of Staff." *Washington Post*, July 1. Accessed October 4, 2015 (washingtonpost.com).

Kann, Laura, Emily O'Malley Olsen et al. 2016. "Sexual Identity, Sex of Sexual Contacts, and Health-Related Behaviors Among Students in Grades 9–12—United States and Selected Sites, 2015." *MMWR* 65 (9): 1–202.

Kann, Laura, Tim McManus et al. 2016. "Youth Risk Behavior Surveillance—United States, 2015." *MMWR* 65 (6): 1–180.

Kantor, Jodi, and David Streitfeld. 2015. "Amazon's Bruising, Thrilling Workplace." *New York Times*, August 16, A1.

Kantor, Jodi, and Laurie Goodstein. 2014. "From Mormon Women, a Flood of Requests and Questions on Their Role in the Church." *New York Times*, March 7, A2.

Kaplan, David A. 2010. "The Best Company to Work For." *Fortune*, February 8, 57–72.

Kassotis, Christopher D., Donald E. Tillitt, J. Wade Davis, Annette M. Hormann, and Susan C. Nagel. 2014. "Estrogen and Androgen Receptor Activities of Hydraulic Fracturing Chemicals and Surface and Ground Water in a Drilling-Dense Region." *Endocrinology* 155 (March): 897–907.

Kasumovic, Michael M., and Jeffrey H. Kuznekoff. 2015. "Insights into Sexism: Male Status and Performance Moderates Female-Directed Hostile and Amicable Behavior." *PLoS One* 10 (7): e0131613. doi:10.1371/journal.pone.0131613.

Katz, Jackson. 2006. *The Macho Paradox: Why Some Men Hurt Women and How All Men Can Help.* Naperville, IL: Sourcebooks.

Katz, Lawrence F., and Alan B. Krueger. 2016. "The Rise and Nature of Alternative Work Arrangements in the United States, 1995–2015." National Bureau of Economic Research, September. Accessed December 6, 2016 (nber.org).

Katzenbach, Jon. 2003. *Why Pride Matters More Than Money: The Power of the World's Greatest Motivational Force.* New York: Crown Business.

Kaufman, Alexander C. 2015. "Why Marriage Equality Is Great for the Economy." *Huffington Post*, June 26. Accessed September 29, 2015 (huffingtonpost.com).

Kay, Aaron C., Martin V. Day, Mark P. Zanna, and A. David Nussbaum. 2013. "The Insidious (and Ironic) Effects of Positive Stereotypes." *Journal of Experimental Social Psychology* 49 (March): 287–291.

Kearney, Melissa S., and Phillip B. Levine. 2014. "Teen Births Are Falling: What's Going On?" Brookings Institution, March. Accessed May 16, 2014 (www.brookings.edu).

Kearney, Melissa S., Benjamin H. Harris, Elisa Jácome, and Lucie Parker. 2014. "Ten Economic Facts About Crime and Incarceration in the United States." The Hamilton Project, May. Accessed November 4, 2014 (www.hamiltonproject.org).

Kearns, Cristin E., Laura A. Schmidt, and Stanton A. Glantz. 2016. "Sugar Industry and Coronary Heart Disease Research: A Historical Analysis of Internal Industry Documents." *JAMA Internal Medicine* 176 (November): 1680–1685.

Keating, Elizabeth, and Cynthia English. 2015. "U.S. Girls Less Confident They Can Learn Computer Science." Gallup, November 17. Accessed July 30, 2017 (gallup.com).

Keels, Micere. 2014. "Choosing Single Motherhood." *Contexts* 13 (Spring): 70–72.

Keeter, Scott. 2010. "Ask the Expert." Pew Research Center, December 29. Accessed December 30, 2010 (pewresearch.org).

Keeter, Scott, Kyley McGeeney, Ruth Igielnik, Andrew Mercer, and Nancy Mathiowetz. 2015. "From Telephone to the Web: The Challenge of Mode of Interview Effects in Public Opinion Polls." Pew Research Center, May 13. Accessed July 22, 2015 (pewresearch.org).

Kelderman, Eric. 2016. "Chill on Funding Still Puts Limits on Gun-Violence Research." *Chronicle of Higher Education*, January 22, A10–A11.

Kelley, Trista. 2013. "The Big Bucks in Keeping Kids Focused." *Bloomberg Businessweek*, October 14–20, 25–27.

Kelly, William R. 2016. *The Future of Crime and Punishment: Smart Policies for Reducing Crime and Saving Money*. Lanham, MD: Rowman & Littlefield.

Kemp, Alice Abel. 1994. *Women's Work: Degraded and Devalued.* Englewood Cliffs, NJ: Prentice Hall.

Kempner, Joanna, Clifford S. Perlis, and John F. Merz. 2005. "Ethics: Forbidden Knowledge." *Science* 307 (February 11): 854.

Kena, Grace, et al. 2016. *The Condition of Education 2016.* U.S. Department of Education, National Center for Education Statistics, NCES 2016–144, May. Accessed May 25, 2017 (nces.ed.gov).

Kena, Grace, Susan Aud, Frank Johnson, Xiaolei Wang, Jijun Zhang, Amy Rathbun, Sidney Wilkinson-Flicker, and Paul Kristapovich. 2014. "The Condition of Education 2014." National Center for Education Statistics, May. Accessed June 21, 2014 (www.nces .ed.gov).

Kendall, Diana. 2002. *The Power of Good Deeds: Privileged Women and the Social Reproduction of the Upper Class.* Lanham, MD: Rowman & Littlefield.

Kennedy, Courtney, et al. 2016. "Evaluating Online Nonprobability Surveys." Pew Research Center, May 2. Accessed May 8, 2016 (pewresearch.org).

Kessem, Limor. 2016. "2016 Cybercrime Reloaded: Our Predictions for the Year Ahead." Security Intelligence, January 15. Accessed April 18, 2017 (securityintelligence.com).

Kessenides, Dimitra. 2017. "Sustainable Cotton." *Bloomberg Businessweek*, January 6, 62.

Kessler, Aaron M. 2015. "E.P.A. Issues Stiffer Rules on Vehicle Fuel Ratings." *New York Times*, February 24, B1, B8.

Kessler, Glenn. 2016. "Fact Check: How Much Help Did Trump's Father Give His Son?" *Washington Post*, September 26. Accessed May 10, 2017 (washingtonpost.com).

Kettner, Peter M., Robert M. Moroney, and Lawrence L. Martin. 1999. *Designing and Managing Programs: An Effectiveness-Based Approach*, 2nd edition. Thousand Oaks, CA: Sage.

Khan, Shamus R. 2012. *Privilege: The Making of an Adolescent Elite at St. Paul's School.* Princeton, NJ: Princeton University Press.

Kharif, Olga. 2015. "A Technology that Reveals Your Feelings." *Bloomberg Businessweek*, August 10–23, 38–39.

Khaw, Lyndal B. L., and Jennifer L. Hardesty. 2009. "Leaving an Abusive Partner: Exploring Boundary Ambiguity Using the Stages of Change Model." *Journal of Family Theory & Review* 1 (March): 38–53.

Kidder, Katherine, Amy Schafer, Phillip Carter, and Andrew Swick. 2017. "From College to Cabinet: Women in National Security." Center for a New American Security. Accessed May 25, 2017 (cnas.org).

Kiefer, Francine. 2015. "Behind Landmark Sex-Trafficking Bills, 20 Women Senators." *Christian Science Monitor Weekly*, March 30, 14.

Kiel, Paul, and Dan Nguyen. 2015. "Bailout Tracker." ProPublica, September 14. Accessed September 15, 2015 (projects .propublica.org).

Kilgore, James. 2015. "What I Learned in Prison." *Chronicle Review*, August 7, B20.

Killoren, Sarah E., and Andrea L. Roach. 2014. "Sibling Conversations About Dating and Sexuality: Sisters as Confidants, Sources of Support, and Mentors." *Family Relations* 63 (April): 232–243.

Killoren, Sarah E., Edna C. Alfaro, Anna K. Lindell, and Cara Streit. 2014. "Mexican American College Students' Communication with Their Siblings." *Family Relations* 63 (October): 513–525.

Kilmann, R. H. 2011. *Quantum Organizations: A New Paradigm for Achieving Organizational Success and Personal Meaning.* Newport Coast, CA: Kilmann Diagnostics.

Kim, Steven, and James Roland. 2015. "The Affordable Care Act." Healthline, June 15. Accessed August 15, 2017 (healthline.com).

Kindy, Kimberly. 2013. "Tribes' Fight with Government Shakes Contractors." *Baltimore Sun*, December 29, 8.

King, Colbert I. 2015. "The Key Reason Why Racism Remains Alive and Well in America." *Washington Post*, June 26. Accessed October 4, 2015 (washingtonpost.com).

Kinsey Institute. 2011. "Continuum of Human Sexuality." Accessed November 24, 2012 (www.indiana.edu/~kinsey).

Kinsey, Alfred C., Wardell B. Pomeroy, and Clyde E. Martin. 1948. *Sexual Behavior in the Human Male*. Philadelphia: Saunders.

Kirkegaard, Jacob Funk. 2015. "The True Levels of Government and Social Expenditures in Advanced Economies." Peterson Institute for International Economics, March. Accessed September 11, 2015 (piie.org).

Kivisto, Peter, and Dan Pittman. 2001. "Goffman's Dramaturgical Sociology: Personal Sales and Service in a Commodified World." Pp. 311–334 in *Illuminating Social Life: Classical and Contemporary Theory Revisited*, 2nd edition. Thousand Oaks, CA: Pine Forge Press.

Klandermans, Bert. 1984. "Mobilization and Participation: Social Psychological Explanations of Resource Mobilization Theory." *American Sociological Review* 49 (October): 583–600.

Klos, Diana Mitsu. 2013. "The Status of Women in the U.S. Media 3013." Women's Media Center. Accessed March 2, 2014 (www .womensmediacenter.com).

Kluger, Jeffrey. 2014. "Why Mass Killers Are Always Male." *Time*, May 25. Accessed August 6, 2015 (time.com).

Kneebone, Elizabeth. 2016. "Suburban Poverty Is Missing from the Conversation About America's Future." Brookings, September 15. Accessed December 23, 2016 (brookings.edu).

Kneebone, Elizabeth, and Alan Berube. 2013. *Confronting Suburban Poverty in America.* Washington, DC: Brookings Institution.

Kneebone, Elizabeth, and Alan Berube. 2014. "Does the Suburbanization of Poverty Mean the War on Poverty Failed?" Brookings, January 8. Accessed February 6, 2014 (www .brookings.edu).

"Knife-edge Lives." 2016. *The Economist*, September 17, 41.

Koba, Mark. 2013. "Workers Unite! (So You Can Become Capitalists)." CNBC, December 13. Accessed December 19, 2013 (www.nbcnews .com).

Kochhar, Rakesh, and Richard Fry. 2015. "5 Takeaways About the American Middle Class." Pew Research Center, December 10. Accessed March 1, 2017 (pewresearch.org).

Kochhar, Rakesh, and Russ Oates. 2014. "Attitudes About Aging: A Global Perspective." Pew Research Center, January 30. Accessed September 19, 2014 (www.pewresearch.org).

Koenig, David, and Scott Mayerowitz. 2015. "1 or 2 Airlines Controlling Many Airports." *Baltimore Sun*, July 15, 14.

Kohut, Andrew. 2015. "Are Americans Ready for Obama's 'Middle Class' Populism?" Pew Research Center, February 19. Accessed August 26, 2015 (pewresearch.org).

Kolata, Gina. 2017. "Gain May Usher in Pig-to-Human Transplants." *New York Times*, August 11, 3A.

Komando, Kim. 2016. "Facebook Is Watching and Tracking You More Than You Probably Realize." *USA Today*, March 18. Accessed September 2, 2017 (usatoday.com).

Kornhauser, William. 1959. *The Politics of Mass Society.* New York: Free Press.

Korownyk, Christina, et al. 2014. "Televised Medical Talk Shows—What They Recommend and the Evidence to Support Their Recommendations: A Prospective Observational Study." *BMJ Open*, 14 (December): 1–9.

Kotkin, Joel. 2016. *The Human City: Urbanism for the Rest of Us.* Evanston, IL: Agate B2.

Kosova, Weston, and Pat Wingert. 2009. "Crazy Talk." *Newsweek,* June 8, 54–62.

Kowitt, Beth. 2015. "The War on Big Food." *Fortune*, June 1, 61–70.

Kozlowski, Lynn T., and Kenneth E. Warner. 2017. "Adolescents and E-Cigarettes: Objects of Concern May Appear Larger Than They Are." *Drug and Alcohol Dependence* 174 (May): 209–214.

Kozol, Jonathan. 2005. *The Shame of the Nation: The Restoration of Apartheid Schooling in America.* New York: Crown.

Krache, D. 2008. "How to Ground a 'Helicopter Parent.'" CNN, August 19. Accessed July 10, 2009 (www.cnn.com).

Krajicek, David J. 2016. "7 Toxic Assaults on Communities of Color Besides Flint: The Dirty Racial Politics of Pollution." AlterNet, January 23. Accessed March 25, 2016 (alternet.org).

Kramer, Laura, and Ann Beutel. 2015. *The Sociology of Gender: A Brief Introduction*, 4th edition. New York: Oxford University Press.

Krantz, Matt. 2015. "9 CEOs Paid 800 Times More Than Their Workers." *USA Today*, August 5. Accessed October 7, 2015 (usatoday.com).

Kraska, Peter B. 2004. *Theorizing Criminal Justice: Eight Essential Orientations.* Long Grove, IL: Waveland Press.

Krasnova, Hanna, Helena Wenninger, Thomas Widjaja, and Peter Buxmann. 2013. "Envy on Facebook: A Hidden Threat to Users' Life Satisfaction?" Unpublished paper presented at the 11th International Conference on Wirtschaftinformatik, February 27–March 1, Leipzig, Germany. Accessed December 5, 2013 (www.aisel.aisnet.org).

Kraus, Michael W., and Jacinth J. X. Tan. 2015. "Americans Overestimate Social Class Mobility." *Journal of Experimental Social Psychology* 58 (January): 101–111.

Kreager, Derek A., and Jeremy Staff. 2009. "The Sexual Double Standard and Adolescent Peer Acceptance." *Social Psychology Quarterly* 72 (June): 143.

Kreeger, Karen Y. 2002. "Sex-Based Differences Continue to Mount." *The Scientist* 16 (February 18). Accessed November 20, 2013 (www.the-scientist.com).

Kreider, Rose M., and Renee Ellis. 2011. "Number, Timing, and Duration of Marriages and Divorces: 2009." U.S. Census Bureau, Current

Population Reports, May. Accessed March 10, 2012 (www.census.gov).

Kreitner, Ronald. 2016. "Baltimore Subsidies Line Developers' Pockets at the City's Expense." *Baltimore Sun*, March 1, 13.

Krisher, Tom, and Michael Biesecker. 2016. "VW to Spend $10 Billion to Settle Claims." *Baltimore Sun*, June 24, 12.

Krishnamurthy, Prasad, and Aaron Edlin. 2014. "Affirmative Action and Stereotypes in Higher Education Admissions." National Bureau of Economic Research, October. Accessed December 5, 2014 (www.nber.org).

Kristof, Nicholas. 2014. "A Nation of Takers?" *New York Times*, March 26, A31.

Kristof, Nicholas. 2016. "Growing up Poor in America." *New York Times*, October 28, SR1.

Krogstad, Jens Manuel. 2014. "Asian American Voter Turnout Lags Behind Other Groups; Some Non-Voters Say They're 'Too Busy.'" Pew Research Center, April 9. Accessed April 20, 2014 (www.pewresearch.org).

Krogstad, Jens Manuel, and Jynnah Radford. 2017. "Key Facts About Refugees to the U.S." Pew Research Center, January 30. Accessed June 20, 2017 (pewresearch.org).

Kroll, Luisa, and Kerry A. Dolan. 2013. "Forbes Billionaires." *Forbes*, March 25, 85–144, 164–174, 190–200.

Kroll, Luisa, and Kerry A. Dolan. 2017. "Forbes 2017 Billionaires List: Meet the Richest People on the Planet." *Forbes*, March 20. Accessed May 5, 2017 (forbes.com).

Kromer, Braedyn, and David Howard. 2013. "Labor Force Participation and Work Status of People 65 Years and Older." U.S. Census Bureau, January. Accessed May 16, 2014 (www.census.gov).

Kross, Ethan, et al. 2013. "Facebook Use Predicts Declines in Subjective Well-Being in Young Adults." *PLoS One* 8 (8): e69841. Accessed December 5, 2013 (www.plosone.org).

Krueger, Alan B. 2016. "Where Have All the Workers Gone?" Prepared for Federal Reserve Bank of Boston, 60th Economic Conference, October 14. Accessed May 15, 2017 (bostonfed.org).

Krumwiede, Kip. 2017. "IMA Global Salary Survey for 2016." Association of Accountants and Financial Professionals in Business. Accessed June 21, 2017 (imanet.org).

Kueppers, Courtney. 2016. "Should Colleges Ever Disinvite a Controversial Speaker?" *Chronicle of Higher Education*, March 25, A6.

Kuhn, Manford H., and Thomas S. McPartland. 1954. "An Empirical Investigation of Self-Attitudes." *American Sociological Review* 19 (1): 68–76.

Kulczycki, Andrei, and Arun P. Lobo. 2002. "Patterns, Determinants, and Implications of Intermarriage Among Arab Americans." *Journal of Marriage and the Family* 64 (February): 202–210.

Kunkel Dale L., Jessica S. Castonguay, and Christine R. Filer. 2015. "Evaluating Industry Self-Regulation of Food Marketing to Children." *American Journal of Preventive Medicine* 49 (August): 181–187.

Küntzle, Julia, and Paul Blondé. 2013. "The Unhappy Fate of Ghanaian Witches." VICE United States, October 22. Accessed February 14, 2014 (www.vice.com).

Kurzman, Charles, and David Schanzer. 2015. "The Other Terror Threat." *New York Times*, June 16, A27.

Kwok, Alvin C., et al. 2011. "The Intensity and Variation of Surgical Care at the End of Life: A Retrospective Cohort Study." *The Lancet*, October 6. Accessed April 15, 2012 (www.lancet.com).

Kymlicka, Will. 2012. *Multiculturalism: Success, Failure, and the Future*. Washington, DC: Migration Policy Institute.

L

La France, Betty H., David D. Henningsen, Aubrey Oates, and Christina M. Shaw. 2009. "Social-Sexual Interactions? Meta-Analyses of Sex Differences in Perceptions of Flirtatiousness, Seductiveness, and Promiscuousness." *Communication Monographs* 76 (September): 263–285.

La Puma, John. 2014. "Don't Ask Your Doctor About 'Low T.'" *New York Times*, February 4, A21.

Lakoff, Robin. T. 1990. *Talking Power: The Politics of Language*. New York: Basic Books.

Laliberte, Richard. 2016. "12 Ways the Health Care System May Be Harming You." *AARP Bulletin*, September, 18–24.

Lamb, Sharon, and Lyn Mikel Brown. 2007. *Packaging Girlhood: Rescuing Our Daughters from Marketers' Schemes.* New York: St. Martin's Press.

Landgrave, Michelangelo, and Alex Nowrasteh. 2017. "Criminal Immigrants: Their Numbers, Demographics, and Countries of Origin." Cato Institute, March 15. Accessed May 20, 2017 (cato.org).

Landivar, Liana C. 2013. "Disparities in STEM Employment by Sex, Race, and Hispanic Origin." U.S. Census Bureau, September. Accessed June 21, 2014 (www.census.gov).

Landsberger, Henry A. 1958. *Hawthorne Revisited.* Ithaca, NY: Cornell University Press.

Laneri, Raquel. 2011. "Slumdog Millions." *Forbes*, May 9, 100–101.

Lang, James M. 2013. "How Orwell and Twitter Revitalized My Course." *Chronicle of Higher Education*, November 1, A35–A37.

Lang, Kurt, and Gladys Engel Lang. 1961. *Collective Dynamics.* New York: Thomas Y. Crowell.

Langreth, Robert. 2013. "Do Robot Surgeons Do No Harm?" *Bloomberg Businessweek*, March 14, 17–18.

Lankford, Adam. 2015. "Mass Shooters, Firearms, and Social Strains: A Global Analysis of an Exceptionally American Problem." Paper presented at the American Sociological Association, 110th Annual Meeting, August 23, Chicago, Illinois.

Lannutti, Pamela J., Melanie Laliker, and Jerold L. Hale. 2001. "Violations of Expectations and Social-Sexual Communication in Student/Professor Interactions." *Communication Education* 50 (January): 69–82.

Laraña, Enrique, Hank Johnston, and Joseph R. Gusfield, eds. 1994. *New Social Movements: From Ideology to Identity.* Philadelphia: Temple University Press.

Lareau, Annette. 2011. *Unequal Childhoods: Class, Race, and Family Life,* 2nd edition. Berkeley: University of California Press.

LaRossa, Ralph, and Donald C. Reitzes. 1993. "Symbolic Interactionism and Family Studies." Pp. 135–163 in *Sourcebook of Family Theories and Methods: A Contextual Approach,* edited by Pauline G. Boss, William J. Doherty, Ralph LaRossa, Walter R. Schumm, and Suzanne K. Steinmetz. New York: Plenum Press.

Larrimore, Jeff, Sam Dodini, and Logan Thomas. 2016. "Report on the Economic Well-Being of U.S. Households in 2015." Board of Governors of the Federal Reserve System, May. Accessed May 5, 2017 (federalreserve.gov).

Larson, Christina. 2017. "Over the Handlebars." *Bloomberg Businessweek*, August 14, 50.

Lauerman, John. 2013. "Colleges Slow to Investigate Assaults." *Baltimore Sun*, June 20, 10.

Laughlin, Lynda. 2014. "A Child's Day: Living Arrangements, Nativity, and Family Transitions: 2011 (Selected Indicators of Child Well-Being)." U.S. Census Bureau, December. Accessed August 6, 2015 (census.gov).

Lavy, Victor, and Edith Sand. 2015. "On the Origins of Gender Human Capital Gaps: Short and Long Term Consequences of Teachers' Stereotypical Biases." National Bureau of Economic Research, January. Accessed September 16, 2015 (nber.org).

Law Center to Prevent Gun Violence. 2017. "Open Carry." Accessed September 2, 2017 (smartgunlaws.org).

Lawless, Jennifer L., and Richard L. Fox. 2013. "Girls Just Wanna Not Run: The Gender Gap in Young Americans' Political Ambition." Women & Politics Institute, March. Accessed April 21, 2014 (www.american.edu).

Lawn, Richard. 2013. *Experiencing Jazz,* 2nd edition. New York: Routledge.

Lawrence, Lee. 2013. "Was God Expelled?" *Christian Science Monitor*, June 17, 26–32.

Lawrence, Lee. 2015. "The Overbooked Generation." *Christian Science Monitor Weekly*, May 11, 26–32.

Lazear, Edward P., Kathryn L. Shaw, and Christopher T. Stanton. 2013. "The Value of Bosses." National Bureau of Economic Research, June. Accessed December 10, 2013 (www.nber.org).

LeBlanc, Paul. 2017. "The Countries That Allow Transgender Troops to Serve in Their Armed Forces." CNN, July 27. Accessed August 12, 2017 (cnn.com).

Le Bon, Gustave. 1896/1968. *The Crowd: A Study of the Popular Mind,* 2nd edition. Dunwoody, GA: Norman S. Berg.

Lee, David M., James Nazroo, Daryl B. O'Connor, Margaret Blake, and Neil Pendleton. 2015. "Sexual Health and Well-being Among Older Men and Women in England: Findings from the English Longitudinal Study of Ageing." *Archives of Sexual Behavior*, January 27, 1–12. Accessed September 29, 2015 (link.springer.com).

Lee, Jennifer, and Min Zhou. 2015. *The Asian American Achievement Paradox*. New York, NY: Russell Sage Foundation.

Lee, Marlene. 2009. "Aging, Family Structure, and Health." Population Reference Bureau, October. Accessed April 10, 2010 (www.prb.org).

Legatum Prosperity Index. 2013. "The Prosperity Index 2013." Accessed February 6, 2014 (www.prosperity.com).

Leinberger, Christopher B., and Patrick Lynch. 2014. "Foot Traffic Ahead: Ranking Walkable Urbanism in America's Largest Metros." Smart Growth America. Accessed September 19, 2014 (www.smartgrowthAmerica.org).

Lemert, Edwin M. 1951. *Social Pathology: A Systematic Approach to the Theory of Sociopathic Behavior.* New York: McGraw-Hill.

Lemert, Edwin M. 1967. *Human Deviance, Social Problems and Social Control.* Englewood Cliffs, NJ: Prentice Hall.

Lemieux, Frederick. 2016. "6 Things Americans Should Know About Mass Shootings." Scientific American, June 13. Accessed December 3, 2016 (scientificamerican.com).

Lencioni, Patrick. 2002. *The Five Dysfunctions of a Team: A Leadership Fable.* San Francisco: Jossey-Bass.

Lengermann, Patricia Madoo, and Jill Niebrugge-Brantley. 1992. "Contemporary Feminist Theory." Pp. 308–357 in *Contemporary Sociological Theory,* 3rd edition, edited by George Ritzer. New York: McGraw-Hill.

Lenhart, Amanda, Aaron Smith, Monica Anderson, Maeve Duggan, and Andrew Perrin. 2015. "Teens, Technology & Friendships." Pew Research Center, August 6. Accessed August 14, 2015 (pewresearch.org).

Lenhart, Amanda, and Maeve Duggan. 2014. "Couples, the Internet, and Social Media." Pew Research Center, February 11. Accessed May 25, 2017 (pewresearch.org).

Lenhart, Amanda, Michele Ybarra, Kathryn Zickuhr, and Myeshia Price-Feeney. 2016. "Online Harassment, Digital Abuse, and Cyberstalking in America." Data & Society Research Institute and Center for Innovative Public Health Research, November 21. Accessed March 13, 2017 (datasociety.net).

Leonardsen, Dag. 2004. *Japan as a Low-Crime Nation.* New York: Palgrave Macmillan.

Leonhardt, David. 2017. "The Americans Who Saved Health Insurance." *New York Times*, August 1, A23.

Leung, Maxwell. 2013. "Jeremy Lin's Model Minority Problem." *Contexts* 12 (Summer): 53–56.

Leurent, B., et al. 2013. "Spiritual and Religious Beliefs as Risk Factors for the Onset of Major Depression: An International Cohort Study." *Psychological Medicine* 43 (10): 2109–2120.

LeVay, Simon. 2011. *Gay, Straight, and the Reason Why: The Science of Sexual Orientation.* New York: Oxford University Press.

Levin, Diane E., and Jean Kilbourne. 2009. *So Sexy So Soon: The New Sexualized Childhood and What Parents Can Do to Protect Their Kids.* New York: Ballantine Books.

The Levin Institute. 2013. "Pop Culture." State University of New York. Accessed November 10, 2013 (www.globalization101.org).

Levin, Jack, and Jim Nolan. 2017. *The Violence of Hate: Understanding Harmful Forms of Bias and Bigotry*, 4th edition. Lanham, MD: Rowman & Littlefield.

Levine, Hallie. 2014. "5 Reasons Women Live Longer Than Men." *Health*, October 13. Accessed October 16, 2015 (news.health.com).

Levine, Lawrence W. 1998. *Highbrow/Lowbrow: The Emergence of Cultural Hierarchy in America.* Cambridge, MA: Harvard University Press.

Levinson, Lauren. 2016. "What Percentage of White Women Voted for Trump?" Pop Sugar, November 30. Accessed January 5, 2017 (popsugar.com).

Levinthal, Dave, Lateshia Beachum, and Carrie Levine. 2017. "Supreme Court a Millionaire's Club." Public Integrity, June 22. Accessed June 26, 2017 (publicintegrity.org).

Levintova, Hannah. 2016. "Minor Threats." *Mother Jones*, September/October, 36–41, 64.

Lewin, Kurt, Ronald R. Lippit, and Ralph K. White. 1939. "Patterns of Aggressive Behavior in Experimentally Created 'Social Climates.'" *Journal of Social Psychology* 10 (2): 271–301.

Lewin, Tamar. 2013. "New Milestone Emerges: Baby's First iPhone App." *New York Times*, October 26, A17.

Lewis, David Levering. 1993. *W. E. B. Du Bois: Biography of a Race, 1868–1919.* New York: Henry Holt.

Lewis, Oscar. 1966. "The Culture of Poverty." *Scientific American* 115 (October): 19–25.

Lewis, Ricki. 2014. "Noninvasive Prenatal DNA Test Okay for Low-Risk Pregnancies." Medscape, February 27. Accessed October 8, 2014 (www.medscape.com).

Li, Shan. 2015. "Toy-makers Building a New Model for Girls." *Baltimore Sun*, March 1, 6.

Lichter, Daniel T., Zhenchao Qian, and Leanna M. Mellott. 2006. "Marriage or Dissolution? Union Transitions Among Poor Cohabiting Women." *Demography* 43 (May): 223–240.

"Life Size Barbie Shows Young Girls the Dangers of Unrealistic Body Expectations." 2011. Diets in Review, May 3. Accessed October 25, 2011 (www.dietsinreview.com).

Light, Michael T., and Jeffrey T. Ulmer. 2016. "Explaining the Gaps in White, Black, and Hispanic Violence Since 1990: Accounting for Immigration, Incarceration, and Inequality." *American Sociological Review* 81 (April): 290–315.

Lilly, J. Robert, Francis T. Cullen, and Richard A. Ball. 1995. *Criminological Theory: Context and Consequences,* 2nd edition. Thousand Oaks, CA: Sage.

Lindberg, Laura D., Isaac Maddow-Zimet, and Heather Boonstra. 2016. "Changes in Adolescents' Receipt of Sex Education, 2006–2013." *Journal of Adolescent Health* 58 (6): 621–627.

Lindberg, Laura, John Santelli, and Sheila Desai. 2016. "Understanding the Decline in Adolescent Fertility in the United States, 2007–2012." *Journal of Adolescent Health* 59 (November): 577–583.

Lindner, Andrew M. 2012. "An Old Tool with New Promise." *Contexts* 11 (Winter): 70–72.

Lindsey, Linda L. 2005. *Gender Roles: A Sociological Perspective,* 4th edition. Upper Saddle River, NJ: Prentice Hall.

Lino, Mark, Kevin Kuczynski, Nestor Rodriguez, and TusaRebecca Schap. 2017. "Expenditures on Children by Families, 2015." U.S. Department of Agriculture, Center for Nutrition Policy and Promotion. Accessed February 28, 2017 (cnpp.usda.gov).

Linton, Ralph. 1936. *The Study of Man.* New York: Appleton-Century-Crofts.

Linton, Ralph. 1964. *The Study of Man: An Introduction.* New York: Appleton-Century-Crofts.

Lipka, Michael. 2014. "Few in LGBT Community See Pro Sports as Friendly." Pew Research Center, February 10. Accessed May 19, 2017 (pewresearch.org).

Lipka, Michael. 2016. "10 Facts About Atheists." Pew Research Center, May 4. Accessed June 1, 2017 (pewresearch.org).

Lipka, Michael. 2016. "5 Facts About Prayer." Pew Research Center, May 4. Accessed July 31, 2017 (pewresearch.org).

Lipka, Michael. 2016. "Why America's 'Nones' Left Religion Behind." Pew Research Center, May 4. Accessed August 24, 2017 (pewresearch.org).

Lipka, Michael. 2017. "Muslim and Islam: Key Findings in the U.S. and Around the World." Pew Research Center, February 27. Accessed July 31, 2017 (pewresearch.org).

Lipka, Michael, and Sandra Stencel. 2013. "Racial and Ethnic Groups View 'Radical Life Extension' Differently." Pew Research Center, August 6. Accessed May 16, 2014 (www.pewresearch.org).

Lipman, Joanne. 2015. "Let's Expose the Gender Pay Gap." *New York Times*, August 13, A19.

Lipton, Eric, and Julie Creswell. 2016. "Documents Show How Wealthy Hid Millions Abroad." *New York Times*, May 6, A1, A14.

Lipton, Eric, and Rachel Abrams. 2016. "Their Hair Fell Out, and F.D.A. Can't Do a Thing." *New York Times*, August 16, A1, B2.

Liu, Sa, S. Katherine Hammond, and Ann Rojas-Cheatham. 2013. "Concentrations and Potential Health Risks of Metals in Lip Products." *Environmental Health Perspectives* 121 (June): 705–710.

Livingston, Gretchen. 2015. "It's No Longer a 'Leave It to Beaver' World for American Families—but It Wasn't Back Then, Either." Pew Research Center, December 30. Accessed July 20, 2017 (pewresearch.org).

Livingston, Gretchen. 2016. "5 Facts About Immigrant Mothers and U.S. Fertility Trends." Pew Research Center, October 26. Accessed June 20, 2017 (pewresearch.org).

Livingston, Gretchen. 2017. "Among U.S. Cohabiters, 18 Percent Have a Partner of a Different Race or Ethnicity." Pew Research Center, June 8. Accessed July 20, 2017 (pewresearch.org).

Livingston, Gretchen, and Anna Brown. 2014. "Birth Rate for Unmarried Women Declining for First Time in Decades." Pew Research Center, August 11. Accessed August 14, 2014 (www.pewresearch.org).

Livingston, Gretchen, and Anna Brown. 2017. "Intermarriage in the U.S. 50 Years After *Loving v. Virginia*." Pew Research Center, May 18. Accessed June 20, 2017 (pewresearch.org).

Livingston, Gretchen, and D'Vera Cohn. 2013. "Long-Term Trend Accelerates Since Recession: Record Share of New Mothers Are College Educated." Pew Research Center, May 10. Brookings Institution, March. Accessed May 16, 2014 (pewresearch.org).

Livingston, Gretchen, and Kim Parker. 2011. "A Tale of Two Fathers." Pew Social & Demographic Trends, June 15. Accessed May 20, 2014 (pewsocialtrends.org).

Llana, Sara M., and Sibylla Brodzinsky. 2012. "Latin America's Silent Scourge." *Christian Science Monitor,* November 26, 18–20.

"Lobbying Database." 2017. Center for Responsive Politics, Open Secrets. Accessed June 26, 2017 (opensecrets.org).

Locher, David A. 2002. *Collective Behavior.* Upper Saddle River, NJ: Prentice Hall.

Lodge, Amy C., and Debra Umberson. 2012. "All Shook Up: Sexuality of Mid-To-Later Life Married Couples." *Journal of Marriage and Family 74* (June): 428–443.

Lofland, John. 1996. *Social Movement Organizations: Guide to Research on Insurgent Realities.* New York: Aldine De Gruyter.

Logan, John, and Harvey Molotch. 1987. *Urban Fortunes: The Political Economy of Place.* Berkeley: University of California Press.

Lohan, Tara. 2015. "Why America's Deadly Love Affair with Bottles Water Has to Stop." AlterNet, September 17. Accessed October 23, 2015 (alternet.org).

Lohr, Steve. 2011. "Computers That See You and Keep Watch over You." *New York Times,* January 1, A1.

López, Gustavo, and Jynnah Radford. 2017. "Statistical Portrait of the Foreign-Born Population in the United States." Pew Research Center, May 3. Accessed June 20, 2017 (pewhispanic .org).

López, Gustavo, and Kristen Bialik. 2017. "Key Findings About U.S. Immigrants." Pew Research Center, May 3. Accessed June 20, 2017 (pewresearch.org).

Lopez, Mark H., and Ana Gonzalez-Barrera. 2013. "What Is the Future of Spanish in the United States?" Pew Research Center, September 5, 2013. Accessed November 10, 2013 (www .pewresearch.org).

Lopez, Mark Hugo, Rich Morin, and Jens Manuel Krogstad. 2016. "Latinos Increasingly Confident in Personal Finances, See Better Economic Times Ahead." Pew Research Center, June 8. Accessed June 20, 2017 (pewresearch.org).

Lorber, Judith, and Lisa Jean Moore. 2007. *Gendered Bodies: Feminist Perspectives.* Los Angeles: Roxbury.

Louie, J. D. 2015. "10 Examples of Confusing Etiquette in Other Countries." Listverse, February 3. Accessed February 14, 2017 (listverse.com).

Louis, Catherine Saint. 2015. "Many Children Under 5 Are Left to Their Mobile Devices, Survey Finds." *New York Times*, November 2, A16.

Loumarr, Ess. 2015. "How Much Weight Can You Lose by Not Drinking Soda?" Live Strong, January 28. Accessed August 23, 2015 (livestrong.com).

Love, Robert. 2017. "Why Drugs Cost So Much." *AARP Bulletin*, May, 16–23.

Loveless, Tom. 2014. "The 2014 Brown Center Report on American Education: How Well Are American Students Learning?" Brown Center on Education Policy, March. Accessed June 21, 2014 (www.brookings.edu).

Lowenberger, Amanda, Joanna Mauer, Andrew deLaski, Marianne DiMascio, Jennifer Amann, and Steven Nadel. 2012. "The Efficiency Boom: Cashing in on the Savings from Appliance Standards." American Council for an Energy-Efficient Economy, and Appliance Standards Awareness Project, March. Accessed June 22, 2012 (www.aceee.org).

Lowery, Wesley. 2014. "Senate Republicans Unanimously Reject Equal Pay Bill." *Washington Post*, April 9. Accessed April 21, 2014 (www .washingtonpost.com).

Lublin, Joann S. 2013. "Tyco's Former CEO Set for 2014 Parole." *Wall Street Journal*, December 4, B2.

Luckey, John R. 2008. "CRS Report for Congress: The United States Flag: Federal Law Relating to Display and Associated Questions." Congressional Research Service, April 14. Accessed July 20, 2011 (www.senate.gov).

Luff, Tracy, Kristi Hoffman, and Marit Berntson. 2016. "Hooking up and Dating Are Two Sides of a Coin." *Contexts* 15 (Winter): 76–77.

Lugaila, Terry. A. 1998. "Marital Status and Living Arrangements: March 1998 (Update)." U.S. Census Bureau, Current Population Reports. Accessed August 8, 2000 (www.census.gov).

Lugo, Luis, Alan Cooperman, and Cary Funk. 2013. "Living to 120 and Beyond: Americans' Views on Aging, Medical Advances, and Radical Life Extension." Pew Research Center, August 6. Accessed May 16, 2014 (www .pewresearch.org).

Lugo, Luis, Alan Cooperman, Cary Funk, and Gregory A. Smith. 2012. "'Nones' on the Rise: One-in-Five Adults Have No Religious Affiliation." Pew Forum on Religion & Public Life, October 9. Accessed June 21, 2014 (www.pewforum.org).

Lundberg, Shelly, and Robert A. Pollak. 2007. "The American Family and Family Economics." *Journal of Economic Perspectives* 21 (Spring): 3–26.

Lundberg, Shelly, Robert A. Pollak, and Jenna E. Stearns. 2016. "Family Inequality: Diverging Patterns in Marriage, Cohabitation, and Childbearing." National Bureau of Economic Research, March. Accessed July 19, 2017 (nber.org).

Luscombe, Richard. 2017. "Arrival of Puerto Ricans Post-Hurricane Maria Could Have Big Impact on Florida." *The Guardian*, October 12. Accessed November 1, 2017 (theguardian.com).

Lutz, William. 1989. *Doublespeak.* New York: Harper & Row.

Lynas, Mark. 2015. "Even in 2015, the Public Doesn't Trust Scientists." *Washington Post*, January 30. Accessed July 22, 2015 (washingtonpost.com).

Lynch, Eleanor W., and Marci J. Hanson, eds. 1999. *Developing Cross-Cultural Competence: A Guide for Working with Children and Their Families,* 2nd edition. Baltimore, MD: Paul H. Brookes.

Lynn, David B. 1969. *Parental and Sex Role Identification: A Theoretical Formulation.* Berkeley, CA: McCutchen.

Lynn, M., and M. Todoroff. 1995. "Women's Work and Family Lives." Pp. 244–271 in *Feminist Issues: Race, Class, and Sexuality,* edited by Nancy Mandell. Scarborough, Ontario: Prentice Hall Canada.

M

Ma, Hong, et al. 2017. "Correction of a Pathogenic Gene Mutation in Human Embryos." *Science* 548 (August): 413–419.

Maccoby, Eleanor E., and John A. Martin. 1983. "Socialization in the Context of the Family: Parent-Child Interaction." Pp. 1–101 in *Socialization, Personality, and Social Development: Vol. 4. Handbook of Child Psychology,* edited by E. Mavis Hetherington. New York: Wiley.

MacDorman, Marian F., T. J. Mathews, Ashna D. Mohangoo, and Jennifer Zeitlin. 2014. "International Comparisons of Infant Mortality and Related Factors: United States and Europe, 2010." *National Vital Statistics Reports* 63 (5): 1–7.

Mach, Annie L., and Ada S. Cornell. 2013. "Federal Employees Health Benefits Program (FEHBP): Available Health Insurance Options." Congressional Research Service, November, 13. Accessed November 17, 2014 (www.crs.gov).

Macionis, John J., and Vincent N. Parrillo. 2007. *Cities and Urban Life,* 4th edition. Upper Saddle River, NJ: Prentice Hall.

MacKenzie, Donald, and Judy Wajcman, eds. 1999. *The Social Shaping of Technology,* 2nd edition. Philadelphia: Open University Press.

Madden, Mary, Amanda Lenhart, Sandra Cortesi, Urs Gasser, Maeve Duggan, and Aaron Smith. 2013. "Teens, Social Media, and Privacy." Pew Research Center, May 21. Accessed December 4, 2013 (www.pewinternet.org).

Madrick, Jeff. 2013. "The Anti-Economist: The Fall and Rise of Occupy Wall Street." *Harper's Magazine*, March, 9–11.

Magill, Bobby. 2014. "Clean Power Plan Exempts Major CO_2 Emitters." Climate Central, June 17. Accessed September 19, 2014 (www .climatecontrol.org).

Makary, Martin A., and Michael Daniel. 2016. "Medical Error—The Third Leading Cause of Death in the US." *British Medical Journal* 353 (5): i2139.

Makinen, Julie. 2013. "Law on Elder Visits in Effect." *Baltimore Sun*, August 6, 6.

Malone, Clare. 2016. "Clinton Couldn't Win over White Women." FiveThirtyEight, November 9. Accessed January 5, 2017 (fivethirtyeight .com).

Malthus, Thomas Robert. 1798/1965. *An Essay on Population.* New York: Augustus Kelley.

Malthus, Thomas Robert. 1872/1991. *An Essay on the Principle of Population,* 7th edition. London: Reeves & Turner.

Mangan, Katherine. 2013. "Comanche Nation College Tries to Rescue a Lost Tribal Language." *Chronicle of Higher Education*, June 14, A18–A19.

Mangan, Katherine. 2013. "Life After Steel." *Chronicle of Higher Education*, April 26, A33–A35.

Mangan, Katherine. 2015. "Stack Those Credentials." *Chronicle of Higher Education*, September 18, B14–B17.

Mann, Michael E. 2017. "No Apologies, No Regrets." *Chronicle of Higher Education*, February 10, A6.

Manning, Jennifer E. 2017. "Membership of the 115th Congress: A Profile." Congressional Research Service, March 13. Accessed July 31, 2017 (crs.gov).

Manning, Wendy D., and Jessica A. Cohen. 2012. "Premarital Cohabitation and Marital Dissolution: An Examination of Recent Marriages." *Journal of Marriage and Family* 74 (April): 377–387.

Manning, Wendy D., and Pamela J. Smock. 2005. "Measuring and Modeling Cohabitation: New Perspectives from Qualitative Data." *Journal of Marriage and Family* 67 (November): 989–1002.

Manyika, James, Michael Chui, Mehdi Miremadi, Jacques Bughin, Katy George, Paul Willmott, and Martin Dewhurst. 2017. "A Future That Works: Automation, Employment, and Productivity." McKinsey Global Institute, January. Accessed May 25, 2017 (mckinsey .com/mgi).

Manza, Jeff, and Clem Brooks. 2016. "Why Aren't Americans Angrier About Rising Inequality?" *Pathways*, Winter, 22–26.

Marcec, Dan. 2017. "Equilar/Associated Press CEO Pay Study 2017." Equilar, May 23. Accessed June 23, 2017 (equilar.com).

Marcotte, Amanda. 2014. "Is Religion Inherently Oppressive?" AlterNet, April 9. Accessed June 21, 2014 (www.alternet.org).

Marcus, Amy Dockser. 2013. "Genetic Testing Leaves More Patients Living in Limbo." *Wall Street Journal*, November 19, D1–D2.

Marikar, Sheila. 2015. "The Company Boosts Growth by Treating Its Employees Like Royalty." *Baltimore Sun*, June 7, 4.

"Marital Status." 2017. U.S. Census Bureau, Figure MS-2, April 4. Accessed July 20, 2017 (census.gov).

"Marketing in the Digital Age." 2015. *The Economist*, August 29, 51–52.

Markey, Richard M., Charlotte N. Markey, and Juliana E. French. 2014. "Violent Video Games and Real-World Violence: Rhetoric Versus Data." *Psychology of Popular Media Culture*, August 18. APA PsycNet. Accessed August 6, 2015 (psycnet.apa.org).

Marquand, Robert. 2011. "Europe Rejects Multiculturalism." *Christian Science Monitor*, February 21, 14.

"Marriage in India." 2015. *The Economist*, September 5, 43–44.

"Marriage in Japan." 2016. *The Economist*, April 16, 35–36.

"The Marriage Squeeze in India and China." 2015. *The Economist*, April 18, 36–37.

Martin, Andrew. 2015. "Making Washington Fall in Love with Pizza Again." *Bloomberg Businessweek*, March 9–15, 35–36.

Martin, Chris. 2016. "The Greening of Adidas." *Bloomberg Businessweek*, May 9–15, 46–47.

Martin, Joyce A., Brady E. Hamilton, Michelle J. K. Osterman, Sally C. Curtin, and T. J. Mathews. 2013. "Births: Final Data for 2012." *National Vital Statistics Reports* 62 (December 30): 1–87.

Martin, Joyce A., Brady E. Hamilton, Michelle J. K. Osterman, Anne K. Driscoll, and T. J. Matthews. 2017. "Births: Final Data for 2015." National Center for Health Statistics, National Vital Statistics System, January 5. Accessed July 1, 2017 (cdc.gov/nchs).

Martin, Molly A. 2008. "The Intergenerational Correlation in Weight: How Genetic Resemblance Reveals the Social Role of Families." *American Journal of Sociology* 114 (Suppl.): S67–S105.

Martin, Sophie. 2013. "29 Non-Americans on the Food They Find Disgusting." Thought Catalog, October 3. Accessed February 14, 2017 (thoughtcatalog.com).

Martin, Steven, Nan Astone, and Elizabeth Peters. 2014. "Fewer Marriages, More Divergence: Marriage Projections for Millennials to Age 40." Urban Institute, April 29. Accessed July 25, 2014 (www.urban.org).

Martinez, Gladys M., and Joyce C. Abma. 2015. "Sexual Activity, Contraceptive Use, and Childbearing of Teenagers Aged 15–19 in the United States." NCHS Data Brief, No. 209, July. Accessed May 25, 2017 (cdc.gov).

Martins, Vinicius, J. B., et al. 2011. "Long-Lasting Effects of Undernutrition." *International Journal of Environmental Research and Public Health* 8 (June): 1817–1846.

Marx, Gary T., and Douglas McAdam. 1994. *Collective Behavior and Social Movements: Process and Structure*. Upper Saddle River, NJ: Prentice Hall.

Marx, Karl. 1844/1964. *Economic and Philosophic Manuscripts of 1844*. New York: International Publishers.

Marx, Karl. 1845/1972. "The German Ideology." Pp. 110–164 in *The Marx-Engels Reader*, edited by Robert C. Tucker. New York: W. W. Norton.

Marx, Karl. 1867/1967. *Capital*, edited by Friedrich Engels. New York: International Publishers.

Marx, Karl. 1934. *The Class Struggles in France*. New York: International Publishers.

Marx, Karl. 1964. *Karl Marx: Selected Writings in Sociology and Social Philosophy*, translated by T. B. Bottomore. New York: McGraw-Hill.

Masci, David. 2016. "Key Findings About Americans' Views on Religious Liberty and Nondiscrimination." Pew Research Center, September 28. Accessed May 19, 2017 (pewresearch.org).

Masci, David. 2016. "Q&A: Why Millennials Are Less Religious Than Older Americans." Pew Research Center, January 8. Accessed July 31, 2017 (pewresearch.org).

Masci, David. 2017. "For Darwin Day, 6 Facts About the Evolution Debate." Pew Research Center, February 10. Accessed March 3, 2017 (pewresearch.org).

Masci, David, Anna Brown, and Jocelyn Kiley. 2017. "5 Facts About Same-Sex Marriage." Pew Research Center, June 26. Accessed July 1, 2017 (pewresearch.org).

Massey, Douglas S. 2007. *Categorically Unequal: The American Stratification System*. New York: Russell Sage Foundation.

Masucci, Madeline, and Lynn Langton. 2017. "Hate Crime Victimization, 2004–2015." Bureau of Justice Statistics, June. Accessed June 3, 2017 (bjs.gov).

Mata, Douglas A., et al. 2015. "Prevalence of Depression and Depressive Symptoms Among Resident Physicians: A Systematic Review and Meta-Analysis." *JAMA* 314 (22): 2373–2383.

Matanoski, Joseph. 2017. "The Germy Case Against Bottled Water." *Baltimore Sun*, January 8, 21.

"Maternal Mortality." 2015. *The Economist*, July 18, 28.

Mather, Mark, Linda A. Jacobsen, and Kelvin M. Pollard. 2015. "Aging in the United States." Population Reference Bureau, December. Accessed July 20, 2017 (prb.org).

Matjasko, Jennifer L., Phyllis H. Niolon, and Linda A. Valle. 2013. "The Role of Economic Factors and Economic Support in Preventing and Escaping from Intimate Partner Violence." *Journal of Policy Analysis and Management* 32 (October): 122–141.

Matthew, Dayna Bowen, and Richard V. Reeves. 2016. "6 Charts Showing Race Gaps Within the American Middle Class." Brookings, October 21. Accessed June 20, 2017 (brookings.edu).

Matthew, Dayna Bowen, and Richard V. Reeves. 2017. "Trump Won White Voters, but Serious Inequities Remain for Black Americans." Brookings, January 13. Accessed June 28, 2017 (brookings.edu).

Matthews, H. Damon, Tanya L. Graham, Serge Keverian, Cassandra Lamontagne, Donny Seto, and Trevor J Smith. 2014. "National Contributions to Observed Global Warming." *Environmental Research Letters* 9 (January), doi:10.1088/1748-9326/9/1/014010. Accessed October 22, 2015 (iopscience.iop.org).

Matzke, Nicholas J. 2016. "The Evolution of Antievolution Policies After *Kitzmiller v. Dover*." *Science* 351 (6268): 28–30.

Maxwell, Steve. 2012. "Water Is *Still* Cheap: Demonstrating the True Value of Water." *American Water Works Association* 104 (May): 31–36.

Mayo, Elton. 1945. *The Problems of an Industrial Civilization*. Cambridge, MA: Harvard University Press.

Mayo, Michael. 2012. "Home for the Holidays: Convicted Killers Goodman, LeVin." *Sun Sentinel*, December 19. Accessed January 12, 2014 (www.articles.sun-sentinel.com).

McAdam, Doug, and Ronnelle Paulsen. 1994. "Specifying the Relationship Between Social Ties and Activism." *American Journal of Sociology* 99 (November): 640–667.

McCabe, Brian J. 2014. "When Property Values Rule." *Contexts* 13 (Winter): 38–43.

McCabe, Janice. 2016. "Friends with Academic Benefits." *Contexts* 15 (Summer): 22–29.

McCabe, Janice, Emily Fairchild, Liz Grauerholz, Bernice A. Pescosolido, and Daniel Tope. 2011. "Gender in Twentieth-Century Children's Books." *Gender & Society* 25 (April): 197–226.

McCarthy, Julie. 2015. "An Angry Young Man Leads Protests Against India's Affirmative Action." NPR, August 31. Accessed September 11, 2015 (npr.org).

McCarthy, Justin. 2015. "Majority in U.S. Maintain Need for Third Major Party." Gallup, September 25. Accessed October 7, 2015 (gallup.com).

McCarthy, Justin. 2015. "Majority in U.S. Still Say Moral Values Getting Worse." Gallup, June 2. Accessed July 1, 2016 (gallup.com).

McCarthy, Justin. 2015. "More Americans Say Crime Is Rising in U.S." Gallup, October 22. Accessed April 15, 2017 (gallup.com).

McCarthy, Justin. 2016. "Americans Still Split on Government's Healthcare Role." Gallup, December 2. Accessed August 15, 2017 (gallup.com).

McCarthy, Justin. 2016. "Americans' Views Shift on Toughness of Justice System." Gallup, October 20. Accessed April 15, 2017 (gallup.com).

McCarthy, Justin. 2016. "Satisfaction with Acceptance of Gays in U.S. at New High." Gallup, January 18. Accessed May 20, 2017 (gallup.com).

McCarthy, Justin. 2017. "Americans More Positive About Effects of Immigration." Gallup, June 28. Accessed June 27, 2017 (gallup.com).

McCarthy, Justin. 2017. "Americans Split over New LGBT Protections, Restroom Policies." Gallup, May 18. Accessed May 20, 2017 (gallup.com).

McCarthy, Justin. 2017. "US Support for Gay Marriage Edges to New High." Gallup, May 15. Accessed May 20, 2017 (gallup.com).

McClendon, David, and Aleksandra Sandstrom. 2016. "Child Marriage Is Rare in the U.S., Though This Varies by State." Pew Research Center, November 1. Accessed July 20, 2017 (pewresearch.org).

McCoy, Charles, and Roscoe Scarborough. 2015. "Why Do We Watch Trashy TV?" *Washington Post*, May 27. Accessed July 26, 2015 (washingtonpost.com).

McDermott, Rose, James H. Fowler, and Nicholas A. Christakis. 2013. "Breaking Up Is Hard to Do, Unless Everyone Else Is Doing It Too: Social Network Effects on Divorce in a Longitudinal Sample." *Social Forces* 92 (2): 491–519.

McDevitt, Jack, Jack Levin, and Susan Bennett. 2002. "Hate Crime Offenders: An Expanded Typology." *Journal of Social Issues* 58 (Summer): 303–318.

McDonald, Kim A. 1999. "Studies of Women's Health Produce a Wealth of Knowledge on the Biology of Gender Differences." *Chronicle of Higher Education*, June 25, A19, A22.

McDonough, Katie. 2014. "The GOP's Other War on Women: 5 Gender Battlegrounds Beyond Abortion and Contraception." AlterNet, March 10. Accessed April 21, 2014 (www.alternet.org).

McDuff, Daniel, Jeffrey M. Girard, and Rana el Kaliouby. 2017. "Large-Scale Observational Evidence of Cross-Cultural Differences in Facial Behavior." *Journal of Nonverbal Behavior* 41 (March): 1–19.

McEwan, Melissa. 2012. "Unbelievable: Man Beats Wife, Judge Orders Him to Take Her Out to Red Lobster and the Bowling Alley." AlterNet, February 9. Accessed February 10, 2012 (www.alternet.org).

McFadden, Robert D., and Angela Macropoulos. 2008. "Wal-Mart Employee Trampled to Death." *New York Times*, November 29, A16.

McFalls, Joseph A., Jr. 2007. "Population: A Lively Introduction, 5th edition." *Population Bulletin* 62 (March): 1–31.

McFarland et al., 2017. The *Condition of Education 2017*. National Center for Education Statistics, U.S. Department of Education, May. Accessed July 30, 2017 (nces.ed.gov).

McFarland, Mac. 2015. "How a Robot Ended Up Teaching Exercise Classes in a Dutch Retirement Home." *Washington Post*, May 27. Accessed October 25, 2015 (washingtonpost.com).

McGirt, Ellen. 2017. "Google Searches Its Soul." *Fortune*, February 1, 50–56.

McHale, Susan M. 2001. "Free-Time Activities in Middle Childhood: Links with Adjustment in Early Adolescence." *Child Development* 76 (November/December): 1764–1778.

McHale, Susan M., Kimberly A. Updegraff, and Shawn D. Whiteman. 2012. "Sibling Relationships and Influences in Childhood and Adolescence." *Journal of Marriage and Family* 74 (October): 913–930.

McIntosh, Peggy. 1995. "White Privilege and Male Privilege: A Personal Account of Coming to See Correspondences Through Work in Women's Studies." Pp. 76–87 in *Race, Class, and Gender: An Anthology*, 2nd edition, edited by Margaret L. Andersen and Patricia Hill Collins. Belmont, CA: Wadsworth.

McKenna, Phil. 2017. "China's Carbon Emissions Falling, While Trump Points U.S. in Opposite Direction." Inside Climate News, March 1. Accessed August 20, 2017 (insideclimatenews.org).

McKenzie, Brian. 2013. "Out-of-State and Long Commutes: 2011." U.S. Census Bureau, February. Accessed September 19, 2014 (www.census.gov).

McKernan, Signe-Mary, Caroline Ratcliffe, Eugene Steuerle, and Sisi Zhang. 2013. "Less Than Equal: Racial Disparities in Wealth Accumulation." Urban Institute, April. Accessed February 6, 2014 (www.urban.org).

McKinnon, Mark. 2011. "Do We Still Need Unions? No: Let's End a Privileged Class." *Newsweek*, March 7, 19.

McNeill, David. 2012. "Japanese Fraud Case Highlights Weaknesses in Scientific Publishing." *Chronicle of Higher Education*, October 12, A16–A17.

McPhail, Clark. 1991. *The Myth of the Madding Crowd*. New York: deGruyter.

McPhail, Clark, and Ronald T. Wohlstein. 1983. "Individual and Collective Behavior Within Gatherings, Demonstrations, and Riots." *Annual Review of Sociology* 9 (August): 579–600.

McPhate, Mike. 2016. "Record Number of False Convictions Were Overturned in 2015, Study Finds." *New York Times*, February 4, A13.

McRae, Susan. 1999. "Cohabitation or Marriage?" Pp. 172–190 in *The Sociology of the Family*, edited by Graham Allan. Malden, MA: Blackwell Publishers.

Mead, George Herbert. 1934. *Mind, Self, and Society*. Chicago: University of Chicago Press.

Mead, George Herbert. 1964. *On Social Psychology*. Chicago: University of Chicago Press.

"Medical Tourism." 2015. Centers for Disease Control and Prevention, February 23. Accessed October 17, 2015 (cdc.gov).

Meier, Barry, and Eric Lipton, 2013. "F.D.A. Shift on Painkillers Was Years in the Making." *New York Times*, October 28, A1.

Melton, Glennon. 2013. "5 Reasons Social Media Is Dangerous for Me." *Huffington Post*, October 1. Accessed December 4, 2013 (www.huffingtonpost.com).

Melucci, Alberto. 1995. "The New Social Movements Revisited: Reflections on a Sociological Misunderstanding." Pp. 107–119 in *Social Movements and Social Classes: The Future of Collective Action*, edited by Louis Maheu. Thousand Oaks, CA: Sage.

Melzer, Scott. 2009. *Gun Crusaders: The NRA's Culture War*. New York: New York University Press.

Mencher, Steve. 2013. "Visit Your Parents or Be Sued: Is That Un-American?" AARP, July 8. Accessed November 2, 2013 (www.aarp.org).

Mendelsohn, Oliver, and Maria Vicziany. 1998. *The Untouchables, Subordination, Poverty and the State in Modern India*. New York: Cambridge University Press.

Mendoza, Jessica. 2016. "The Price Isn't Right." *Christian Science Monitor Weekly*, April 25 & May 2, 26–31.

Mertens, Richard. 2015. "In Small-Town America, Open Arms." *Christian Science Monitor Weekly*, March 23, 21–23.

Ming, Liu. 2011. "Chinese College Drops Plan to Discourage Kissing and Other 'Uncivilized Behavior' on Campus." *Global Times*, April 22. Accessed October 25, 2011 (www.china.globaltimes.cn).

Menissi, Fatima. 1991. *The Veil and the Male Elite: A Feminist Interpretation of Women's Rights in Islam*. Reading, MA: Addison-Wesley.

Menissi, Fatima. 1996. *Women's Rebellion and Islamic Memory*. Atlantic Highlands, NJ: Zed Books.

Mercer, Andrew, Claudia Deane, and Kyley McGeeney. 2016. "Why 2016 Election Polls Missed Their Mark." Pew Research Center, November 9. Accessed November 11, 2016 (pewresearch.org).

Merelli, Annalisa. 2017. "Only 4.2% of Fortune 500 Companies Are Run by Women." Quartz, March 7. Accessed April 10, 2017 (qz.com).

Merton, Robert K. 1938. "Social Structure and Anomie." *American Sociological Review* 3 (December): 672–682.

Merton, Robert K. 1948/1996. "The Self-Fulfilling Prophecy." Pp. 183–201 in *Robert K. Merton: On Social Structure and Science*, edited by Piotr Sztompka. Chicago: University of Chicago Press.

Merton, Robert K. 1949. "Discrimination and the American Creed." Pp. 99–126 in *Discrimination and National Welfare*, edited by Robert M. MacIver. New York: Harper.

Merton, Robert K. 1968. *Social Theory and Social Structure*. New York: Free Press.

Merton, Robert K., and Alice K. Rossi. 1950. "Contributions to the Theory of Reference Group Behavior." Pp. 40–105 in *Continuities in Social Research*, edited by Robert K. Merton and Paul L. Lazarsfeld. New York: Free Press.

Merz, C. Noel Bairey. 2015. "Improving Gender Equity in Medical Research (Letter to the Editor)." *New York Times*, July 27, A16.

Messerli, Joe. 2012. "Should Affirmative Action Policies, Which Give Preferential Treatment Based on Minority Status, Be Eliminated?" BalancedPolitics.org, January 7. Accessed June 21, 2014 (www.balancedpolitics.org).

Messing, Jill T., Jacquelyn Campbell, Janet Sullivan Wilson, Sheryll Brown, and Beverly Patchell. 2014. "Police Departments' Use of the Lethality Assessment Program: A Quasi-Experimental Evaluation." National Institute of Justice, March 31. Accessed September 2, 2015 (www.nij.gov).

Meston, Cindy M., and David M. Buss. 2007. "Why Humans Have Sex." *Archives of Sexual Behavior* 36 (August): 477–507.

Meteyer, Karen B., and Maureen Perry-Jenkins. 2009. "Dyadic Parenting and Children's Externalizing Symptoms." *Family Relations* 58 (July): 289–302.

"M-Health: Health and Appiness." 2014. *The Economist*, February 1, 56–57.

Mettler, Suzanne. 2014. "Equalizers No More: Politics Thwart Colleges' Role in Upward Mobility." *Chronicle Review*, March 7, B7–B11.

Michels, Robert. 1911/1949. *Political Parties*. Glencoe, IL: Free Press.

Milgram, Stanley. 1963. "Behavioral Study of Obedience." *Journal of Abnormal and Social Psychology* 67 (4): 371–378.

Milgram, Stanley. 1965. "Some Conditions of Obedience and Disobedience to Authority." *Human Relations* 18 (February): 57–76.

"Millennials Outnumber Baby Boomers and Are Far More Diverse, Census Bureau Reports." 2015. U.S. Census Bureau, June 25. Accessed October 4, 2015 (census.gov).

Miller, Claire C. 2013. "Angry over U.S. Surveillance, Tech Giants Bolster Defenses." *New York Times*, November 1, A1.

Miller, D. W. 2001. "DARE Reinvents Itself—With Help from Its Social-Scientist Critics." *Chronicle of Higher Education*, October 16, A12–A14.

Miller, Elizabeth, et al. 2015. "A School Health Center Intervention for Abusive Adolescent Relationships: A Cluster RCT." *Pediatrics* 135 (January): 76–85.

Miller, Lisa. 2010. "A Woman's Place Is in the Church." *Newsweek*, April 12, 34–41.

Miller, Lisa. 2012. "Feminism's Final Frontier? Religion." *Washington Post*, March 8. Accessed March 9, 2012 (www.washingtonpost.com).

Mills, C. Wright. 1956. *The Power Elite*. New York: Oxford University Press.

Mills, C. Wright. 1959. *The Sociological Imagination*. New York: Oxford University Press.

Mills, Theodore M. 1958. "Some Hypotheses on Small Groups from Simmel." *American Journal of Sociology* 63 (May): 642–650.

Millsap, Adam. 2016. "Privatizing Water Facilities Can Help Cash-Strapped Municipalities." *Forbes*, October 5. Accessed August 20, 2017 (forbes.com).

Milman, Oliver, and Jessica Glenza. 2016. "At Least 33 US Cities Used Water Testing 'Cheats' over Lead Concerns." *The Guardian*, June 2. Accessed August 15, 2017 (theguardian.com).

Mischel, Walter. 1966. "A Social Learning View of Sex Differences." Pp. 57–81 in *The Development of Sex Differences*, edited by Eleanor E. Maccoby. Stanford, CA: Stanford University Press.

Mitnik, Pablo A., and David B. Grusky. 2015. "Economic Mobility in the United States." The Pew Charitable Trusts, July. Accessed September 11, 2015 (pewtrusts.org).

Mohdin, Aamna. 2016. "American Women Voted Overwhelmingly for Clinton, Except the White Ones." Quartz, November 9. Accessed January 5, 2017 (qz.com).

Mollborn, Stephanie. 2015. "Mixed Messages About Teen Sex." *Contexts* 14 (Winter): 44–49.

Money, John, and Anke A. Ehrhardt. 1972. *Man & Woman, Boy & Girl: The Differentiation and Dimorphism of Gender Identity from Conception to Maturity*. Baltimore, MD: Johns Hopkins University Press.

Monk, Ellis P., Jr. 2015. "The Cost of Color: Skin Color, Discrimination, and Health among African-Americans." *American Journal of Sociology* 121 (September): 396–444.

Montanaro, Domenico, Rachel Wellford, and Simone Pathe. 2014. "Money Is Pretty Good Predictor of Who Will Win Elections." Public Broadcasting Service, November 11. Accessed November 23, 2014 (www.pbs .org).

Montez, Jennifer Karas, and Anna Zajacova. 2013. "Explaining the Widening Education Gap in Mortality Among U.S. White Women." *Journal of Health and Social Behavior* 54 (June): 165–181.

Montlake, Simon. 2016. "Maine's Minimum Wage Experiment." *Christian Science Monitor Weekly*, May 16, 20–23.

Monto, Martin, and Anna Carey. 2014. "A New Standard of Sexual Behavior? Are Claims Associated with the 'Hookup Culture' Supported by General Social Survey Data?" *Journal of Sex Research* 51(6): 605–615.

Mooney, Chris. 2015. "The Remote Alaskan Village That Needs to Be Relocated Due to Climate Change." *Washington Post*, February 24. Accessed October 23, 2015 (washingtonpost .com).

Moore, Elizabeth S. 2006. "It's Child's Play: Advergaming and the Online Marketing of Food to Children." Henry J. Kaiser Family Foundation, July. Accessed April 12, 2008 (www.kff.org).

Moore, Kathleen. 2008. "Low-Income Homes Green—and Affordable." *Daily Gazette*, July 1. Accessed September 21, 2008 (www .dailygazette.com).

Moore, Peter. 2015. "A Third of Young Americans Say They Aren't 100% Heterosexual." YouGov, August 20. Accessed September 29, 2015 (today.yougov.com).

Morello, Carol. 2014. "Census to Change the Way It Counts Gay Married Couples." *Washington Post*, May 26. Accessed October 23, 2014 (www.washingtonpost.com).

Morgan, Jacob. 2017. "The Gig Economy Explained." INC, April 14. Accessed May 15, 2017 (inc.com).

Morgenson, Gretchen. 2004. "No Wonder C.E.O.'s Love Those Mergers." *New York Times*, July 18, C1.

Morin, Rich. 2016. "Behind Trump's Win in Rural White America: Women Joined Men in Backing Him." Pew Research Center, November 17. Accessed January 5, 2017 (pewresearch.org).

Morin, Rich, and Renee Stepler. 2016. "The Racial Confidence Gap in Police Performance." Pew Research Center, September 29. Accessed March 1, 2017 (pewresearch.org).

Morin, Rich, and Seth Motel. 2012. "A Third of Americans Now Say They Are in the Lower Class." Pew Social & Demographic Trends, September 10. Accessed February 6, 2014 (www.pewsocialtrends.org).

Morin, Rich, Kim Parker, Renee Stepler, and Andrew Mercer. 2017. "Behind the Badge." Pew Research Center, January 11. Accessed April 10, 2017 (pewresearch.org).

Morris, Desmond. 1994. *Bodytalk: The Meaning of Human Gestures*. New York: Crown Trade Paperbacks.

Morris, Edward W., and Brea L. Perry. 2017. "Girls Behaving Badly? Race, Gender, and Subjective Evaluation in the Discipline of African American Girls." *Sociology of Education* 90 (2): 127–148.

Morris, Theresa. 2014. "C-Section Epidemic." *Contexts* 13 (Winter): 70–72.

Morrison, Denton E. 1971. "Some Notes Toward Theory on Relative Deprivation, Social Movements, and Social Change." *American Behavioral Scientist* 14 (May–June): 675–690.

Moss-Racusin, Corrine A., John F. Dovidio, Victoria L. Brescoli, Mark J. Graham, and Jo Handelsman. 2012. "Science Faculty's Subtle Gender Biases Favor Male Students." *Proceedings of the National Academy of Sciences* 109 (October 9): 16474–16479.

Motel, Seth, and Eileen Patten. 2013. "Statistical Portrait of the Foreign-Born Population in the United States, 2011." Pew Research Hispanic Trends Project, January 29. Accessed March 22, 2014 (www.pewhispanic.org).

Motivans, Mark. 2013. "Federal Justice Statistics 2010—Statistical Tables." Bureau of Justice Statistics, December. Accessed January 12, 2014 (www.bjs.gov).

Moyer, Imogene L. 2001. *Criminological Theories: Traditional and Nontraditional Voices and Themes*. Thousand Oaks, CA: Sage.

Moyer, Imogene L. 2003. "Jane Addams: Pioneer in Criminology." *Women & Criminal Justice* 14 (3/4): 1–14.

Mozzafarian, Dariush, and David S. Ludwig. 2015. "Stop Fearing Fat." *New York Times*, July 9, A25.

Mudallal, Zainab. 2015. "Nearly Half of Americans Didn't Take a Vacation Day in 2014." Quartz, January 5. Accessed October 7, 2015 (qz.com).

Mui, Ylan Q. 2015. "Rather Than Raises, Firms Raise Perks." *Baltimore Sun*, July 31, 12.

Mullen, Ann. 2012. "The Not-So-Pink Ivory Tower." *Contexts* 4 (Fall): 34–38.

Mumford, Lewis. 1961. *The City in History: Its Origins, Transformations, and Its Prospects*. New York: Harcourt, Brace.

Munoz-Boudet, Ana Maria, and Ana Revenga. 2017. "Breaking the STEM Ceiling for Girls." Brookings, March 7. Accessed July 31, 2017 (brookings.edu).

Munsell, Christina R., Jennifer L. Harris, Vishnudas Sarda, and Marlene B. Schwartz. 2015. "Parents' Beliefs About the Healthfulness of Sugary Drink Options: Opportunities to Address Misperceptions." *Public Health Nutrition*, March 11, 1–9 published online ahead of print. Accessed October 25, 2015 (http://dx.doi.org/10.1017/S1368980015000397).

Murdock, George P. 1940. "The Cross-Cultural Survey." *American Sociological Review*, 5: 361–370.

Murphey, David, and P. Mae Cooper. 2015. "Parents Behind Bars: What Happens to Their Children?" Child Trends, October. Accessed February 28, 2017 (childtrends.org).

Murphey, David, Megan Barry, and Brigitte Vaughn. 2013. "Mental Health Disorders." Child Trends, January. Accessed July 26, 2014 (www.childtrends.org).

Murphy, Cait. 2005. "Fast-Forward to the Future." *Fortune*, September 19, 271.

Murphy, John. 2004. "S. Africa's New Goal: Economic Equality." *Baltimore Sun*, April 27, 1A, 4A.

Murphy, Kevin. 2013. "Hallmark Responds to 'Gay' Ornament Controversy with Regret." *Huffington Post*, November 1. Accessed November 1, 2013 (www.huffingtonpost.com).

Murray, Charles. 1984. *Losing Ground: American Social Policy, 1950–1980*. New York: Basic Books.

Murray, Charles. 2012. *Coming Apart: The State of White America, 1960–2010*. New York: Crown Forum.

Musick, Kelly, Ann Meier, and Sarah Flood. 2016. "How Parents Fare: Mothers' and Fathers' Subjective Well-Being in Time with Children." *American Sociological Review* 81 (5): 1069–1095.

Myers, Joe. 2016. "This Is What Gun Ownership Looks Like Around the World." World Economic Forum, January 6. Accessed May 1, 2017 (weforum.org).

Myers, John P. 2007. *Dominant–Minority Relations in America: Convergence in the New World*, 2nd edition. Boston: Allyn & Bacon.

N

Nagoski, Emily. 2015. "Nothing Is Wrong with Your Sex Drive." *New York Times*, January 27, A21.

Naili, Hajer. 2011. "Study Details Sex-Traffic in Post-Saddam Iraq." Women's e-News, November 9. Accessed February 25, 2012 (www.womensenews.org).

Najam, Adil. 2016. "America's Nobel Success Is the Story of Immigrants." The Conversation, October 18. Accessed June 8, 2017 (theconversation.com).

"National African-American History Month: February 2017." 2017. U.S. Census Bureau. Accessed June 20, 2017 (census.gov).

National Center for Education Statistics. 2013. "The Nation's Report Card: A First Look: 2013 Mathematics and Reading." U.S. Department of Education, November. Accessed June 21, 2014 (www.nationsreportcard.gov).

National Center for Health Statistics. 2015. "Health, United States, 2014." May. Accessed October 8, 2015 (cdc.gov/nchs).

National Center for Health Statistics. 2017. *Health, United States, 2016: With Chartbook on Long-term Trends in Health*. Centers for Disease Control and Prevention. Accessed July 20, 2017 (cdc.gov/nchs).

National Center for Statistics and Analysis. 2016. "Distracted Driving 2014." National Highway Traffic Safety Administration, April. Accessed December 25, 2016 (nrd.nhtsa.dot.gov).

National Center on Elder Abuse. 2012. "Statistics/Data." Accessed May 16, 2014 (www.ncea.aoa.gov).

National Conference of State Legislatures. 2014. "State Laws Regarding Marriages Between First Cousins." Accessed May 16, 2014 (www.ncsl.org).

National Diabetes Statistics Report. 2017. "Estimates of Diabetes and Its Burden in the United States." National Center for Chronic Disease Prevention and Health Promotion. Accessed August 15, 2017 (cdc.gov).

"National Grandparents Day 2016: Sept. 11." 2016. Census Bureau, Facts for Features, July 29. Accessed February 28, 2017 (census.gov).

"National Grandparents Day 2017: Sept 10." 2017. U.S. Census Bureau, July 10. Accessed July 20, 2017 (census.gov).

National Institute of Mental Health. 2011. "The Teen Brain: Still Under Construction." Accessed November 20, 2013 (www.nimh.nih.gov).

National Institute on Alcohol Abuse and Alcoholism. 2013. "College Drinking." July. Accessed July 26, 2014 (www.niaaa.nih.gov).

National Institute on Alcohol Abuse and Alcoholism. 2014. "Alcohol Facts and Statistics." May. Accessed July 30, 2014 (www.niaaa.nih.gov).

National Institute on Drug Abuse. 2012. "Principles of Drug Addiction Treatment: A Research-Based Guide." Accessed January 15, 2014 (www.drugabuse.gov).

National Institute on Drug Abuse. 2017. "Overdose Death Rates." January. Accessed August 15, 2017 (drugabuse.gov).

National Oceanic and Atmospheric Administration. 2014. "Heat Wave: A Major Summer Killer." NOAA Watch. Accessed September 19, 2014 (www.noaawatch.gov).

National Organization on Fetal Alcohol Syndrome. 2012. "FASD: What Everyone Should Know." Accessed July 25, 2014 (www.nofas.org).

National Public Radio. 2010. "Black Male Privilege?" Interview transcript, March 4. Accessed March 10, 2010 (www.npr.org).

National Retail Federation. 2016. "Retailers See Increase in Organized Retail Crime." October 18. Accessed April 19, 2017 (nrf.com).

National Science Board. 2014. *Science and Engineering Indicators 2014*. Arlington, VA: National Science Foundation.

National Survey of Student Engagement. 2013. "A Fresh Look at Student Engagement—Annual Results 2013." Bloomington, IN: Indiana University Center for Postsecondary Research. Accessed June 21, 2014 (www.nsse.iub.edu).

National Survey of Student Engagement. 2016. "Engagement Insights: Survey Findings on the Quality of Undergraduate Education." Accessed July 31, 2017 (nsse.indiana.edu).

National Women's Law Center. 2014. "Women's Unemployment Rises Despite Job Gains, Growth Concentrated in Low-Wage Sectors, NWLC Analysis Shows." April 4. Accessed April 12, 2014 (www.nwlc.org).

"The Nation's Older Population Is Still Growing, Census Bureau Reports." 2017. June 22. Accessed July 20, 2017 (census.gov).

Nechepurenko, Ivan. 2014. "Russia Reverses Birth Decline—But for How Long?" *The Moscow Times*, June 22. Accessed September 21 (www.themoscowtimes.com).

Neelakantan, Shailaja. 2006. "In India, Conservatives Want Women Under Wraps." *Chronicle of Higher Education*, May 26, A47–A48.

Neff, Roni A., Marie L. Spiker, and Patricia L. Truant. 2015. "Wasted Food: U.S. Consumers' Reported Awareness, Attitudes, and Behaviors." *PLoS One* 10 (6): doi:10.1371/journal.pone.0127881. Accessed October 22, 2015 (journals.plos.org).

Neider, Linda L., and Chester A. Schriesheim, eds. 2005. *Understanding Teams*. Greenwich, CT: Information Age.

Nelson, Libby, and Emily Crockett. 2017. "Sexual Assault Allegations Against Donald Trump: 15 Women Say He Groped, Kissed, or Assaulted Them." Vox, January 19. Accessed March 14, 2017 (vox.com).

Neuhaus, Jessamyn. 2010. "Marge Simpson, Blue-Haired Housewife: Defining Domesticity on *The Simpsons*." *The Journal of Popular Culture* 43 (August): 761–781.

"The New American Father." 2013. Pew Research Center, June 14. Accessed May 16, 2014 (www.pewresearch.org).

New York Times. 2016. "Transcript: Donald Trump's Taped Comments About Women." October 8. Accessed March 14, 2017 (nytimes.com).

Newberry, Sydne. 2013. "What's in a Name? Calling Obesity a Disease Could Help Improve Chronic Disease Outcomes." The RAND Blog, October 16. Accessed July 26, 2014 (www.rand.org).

Newkirk, Margaret, and Gigi Douban. 2012. "Legal Immigrants Wanted for Dirty Jobs." *Bloomberg Businessweek*, October 8–14, 34, 36.

Newmark, David, Ian Burn, and Patrick Button. 2017. "Age Discrimination and Hiring of Older Workers." Federal Reserve Bank of San Francisco, February 27. Accessed July 20, 2017 (frbsf.org).

Newport, Frank. 2014. "In U.S., 42% Believe Creationist View of Human Origins." Gallup, June 4. Accessed June 21, 2014 (www.gallup.com).

Newport, Frank. 2014. "Three-Quarters of Americans Identify as Christian." Gallup, December 24. Accessed July 30, 2017 (gallup.com).

Newport, Frank. 2015. "Most U.S. Smartphone Owners Check Phone at Least Hourly." Gallup, July 9. Accessed August 14, 2015 (gallup.com).

Newport, Frank. 2016. "American Public Opinion, Terrorism and Guns." Gallup, June 13. Accessed April 15, 2017 (gallup.com).

Newport, Frank. 2016. "Americans' Satisfaction with Ability to Get Ahead Edges Up." Gallup, January 21. Accessed January 1, 2017 (gallup.com).

Newport, Frank. 2016. "Five Key Findings on Religion in the U.S." Gallup, April 12. Accessed December 23, 2016 (gallup.com).

Newport, Frank. 2017. "Americans' Confidence in Institutions Edges Up." Gallup, June 26. Accessed July 30, 2017 (gallup.com).

Newport, Frank. 2017. "Middle-Class Identification in US at Pre-Recession Levels." Gallup, June 21. Accessed June 23, 2017 (gallup.com).

Newport, Frank. 2017. "More U.S. Protestants Have No Specific Denominational Identity." Gallup, July 18. Accessed July 30, 2017 (gallup.com).

Newport, Frank, and Igor Himelfarb. 2013. "In U.S., Strong Link Between Church Attendance, Smoking." Gallup, August 5. Accessed June 21, 2014 (www.gallup.com).

Newport, Frank, and Joy Wilke. 2013. "Desire for Children Still Norm in U.S." Gallup, September 25. Accessed May 16, 2014 (www.gallup.com).

Newport, Frank, Dan Witters, and Sangeeta Agrawal. 2012. "In U.S., Very Religious Americans Have Higher Wellbeing Across All Faiths." Gallup, December 1. Accessed April 8, 2012 (www.gallup.com).

Newport, Frank, Jeffrey M. Jones, and Lydia Saad. 2014. "State of the Union: The Public Weighs In on 10 Key Issues." Gallup, January 31. Accessed March 22, 2014 (www.gallup.com).

Ngabirano, Anne-Marcelle. 2017. "'Pink Tax' Forces Women to Pay More Than Men." *USA Today*, March 27. Accessed March 28, 2017 (usatoday.com).

Nguyen, Frances. 2016. "Women Workers Fight Back Against Sexual Harassment and Assault." Women's Media Center, December 20. Accessed December 22, 2016 (womensmediacenter.com).

Nichols, Andrew Howard, and José Luis Santos. 2016. "A Glimpse Inside the Coffers: Endowment Spending at Wealthy Colleges and Universities." The Education Trust, August. Accessed July 31, 2017 (edtrust.org).

Nichols, James. 2014. "Cecil Chao, Hong Kong Billionaire, Doubles Reward for Any Man Who Can Make Lesbian Daughter Straight." *Huffington Post*, January 24. Accessed March 2, 2014 (www.huffingtonpost.com).

Niederdeppe, Jeff, Sahara Byrne, Rosemary J. Avery, and Jonathan Cantor. 2013. "Direct-to-Consumer Television Advertising Exposure, Diagnosis with High Cholesterol, and Statin Use." *Journal of General Internal Medicine* 28 (July): 886–893.

Nielsen. 2016. "The Nielsen Total Audience Report: Q1 2016." Accessed March 13, 2017 (nielsen.com).

Nielsen Company, The. 2008. "College Spring Break Study." February 27. Accessed April 10, 2008 (www.alcoholstats.com).

Niquette, Mark, and Richard Rubin. 2014. "States Target Corporate Cash Stashed Overseas." *Bloomberg Businessweek*, April 21–27, 27–28.

Nir, Sarah Maslin. 2015. "The Price of Nice Nails." *New York Times*, May 7. Accessed October 4, 2015 (nytimes.com).

"No Action on Greenhouse Gases." 2008. *Baltimore Sun*, July 12, 2A.

NOAA National Centers for Environmental Information. 2015. "State of the Climate: Global Analysis for July 2015." Accessed October 23, 2015 (ncdc.noaa.gov).

Nolette, James. 2015. "Using Research to Move Policing Forward." *NIJ Journal* 276 (December): 46–51.

Norman, Jim. 2015. "In U.S., Women, Poor, Urbanites Most Fearful of Walking Alone." Gallup, November 10. Accessed July 15, 2017 (gallup.com).

Norman, Jim. 2015. "Young, Poor, Urban Dwellers Most Likely to Be Crime Victims." Gallup, November 6. Accessed April 15, 2017 (gallup .com).

Norman, Jim. 2016. "Americans' Confidence in Institutions Stays Low." Gallup, June 13. Accessed April 4, 2017 (gallup.com).

Norman, Jim. 2016. "Millennials Like Sanders, Dislike Election Process." Gallup, May 11. Accessed January 9, 2017 (gallup.com/poll).

Norman, Jim. 2016. "Most U.S. Retirees Living Well, but Nonretirees Unconvinced." Gallup, April 29. Accessed July 20, 2017 (gallup .com).

Norris, Tina, Paula L. Vines, and Elizabeth M. Hoeffel. 2012. "The American Indian and Alaska Native Population: 2010." U.S. Census Bureau. 2010 Census Briefs, January. Accessed March 20, 2012 (www.census.gov).

Nossiter, Adam. 2011. "Hinting at an End to a Curb on Polygamy, Interim Libyan Leader Stirs Anger." *New York Times*, October 30, A6.

"Number of Jobs Held, Labor Market Activity, and Earnings Growth Among the Youngest Baby Boomers: Results from a Longitudinal Survey." 2012. Bureau of Labor Statistics, News Release, July 25. Accessed November 18, 2013 (www.bls.gov).

Nyseth, Hollie, Sarah Shannon, Kia Heise, and Suzy Maves McElrath. 2011. "Embedded Sociologists." *Contexts* 10 (Spring): 44–50.

O

O'Brien, Jodi, and Peter Kollock. 2001. *The Production of Reality: Essays and Readings on Social Interaction*, 3rd edition. Thousand Oaks, CA: Pine Forge Press.

O'Connor, Anahad. 2014. "New Concern About Testosterone and Heart Risks." *New York Times*, January 30, A13.

O'Connor, Anahad. 2015. "Study Warns of Diet Supplement Dangers Kept Quiet by F.D.A." *New York Times*, April 8, A1, A12.

O'Connor, Siobhan. 2015. "Farm-to-Fridge Is Heating Up." *Time*, August 24, 28.

O'Donnell, Victoria. 2007. *Television Criticism*. Thousand Oaks, CA: Sage.

O'Keefe, Ed, Matea Gold, and David A. Fahrenthold. 2015. "Bush Tax Forms Show Income of $29 Million Since 2007." *Washington Post*, June 30. Accessed August 31, 2015 (washingtonpost.com).

O'Keefe, Kevin. 2014. "TV's Renaissance for Strong Women Is Happening in a Surprising Place." *The Atlantic*, October 9. Accessed September 29, 2015 (theatlantic.com).

Oakes, Jeannie. 1985. *Keeping Track: How Schools Structure Inequality*. New Haven, CT: Yale University Press.

Obach, Brian K. 2004. *Labor and the Environmental Movement: The Quest for Common Ground*. Cambridge, MA: MIT Press.

Obach, Brian K. 2015. "A Fracking Fracas Demonstrates Movement Potential." *Contexts* 14 (Fall): 72–75.

Oberschall, Anthony. 1995. *Social Movements: Ideologies, Interests, and Identities*. New Brunswick, NJ: Transaction.

"Obesity Prevalence Maps." 2015. Centers for Disease Control and Prevention, September 11. Accessed October 15, 2015 (cdc.gov).

Ochs, Susan M. 2016. "At Banks, the Buck Stops Short." *New York Times*, September 15, A25.

OECD. 2013. "Results from PISA 2012: United States." Accessed June 22, 2014 (www.oecd .org).

OECD. 2014. *OECD Factbook 2014: Economic, Environmental and Social Statistics*. OECD Publishing. Accessed July 26, 2014 (www .oecd.org).

OECD. 2016. "Income Inequality Update." November. Accessed May 5, 2017 (oecd.org).

OECD. 2016. "PISA 2015 Results in Focus." Accessed July 30, 2017 (oecd.org).

OECD. 2017. "Continued Slowdown in Productivity Growth Weighs Down on Living Standards." May 18. Accessed June 22, 2017 (oecd.org).

OECD/European Union. 2015. *Indicators of Immigrant Integration 2015: Settling In*. Paris: OECD Publishing.

Office for Human Research Protections. 2016. "Regulations." U.S. Department of Health and Human Services, February 12. Accessed March 24, 2017 (hhs.gov/ohrp).

Office of Inspector General. 2017. "Semiannual Report to Congress: October 1, 2016 to March 31, 2017." U.S. Department of Health and Human Services. Accessed August 15, 2017 (oig.hhs.gov).

Office of the Deputy Chief of Staff for Intelligence. 2006. "Arab Cultural Awareness: 58 Factsheets." U.S. Army Training and Doctrine Command, Ft. Leavenworth, Kansas, January. Accessed February 15, 2006 (www.fas.org).

Office of the Inspector General. 2015. "Qualifying for Disability Benefits in Puerto Rico Based on an Inability to Speak English." Social Security Administration, April. Accessed October 4, 2015 (oig.ssa.gov).

OfficeTime. 2015. "Top 10 Time Killers." Accessed August 23, 2015 (officetime.net).

Ogburn, William F. 1922. *Social Change with Respect to Culture and Original Nature*. New York: Dell.

Ogunro, Nola. 2012. "10 Great Paying Jobs You Can Get Without a 4 Year College Degree." Alternet, September 21. Accessed June 23, 2014 (www.alternet.org).

Ogunwole, Stella U., Malcolm P. Drewery, Jr., and Merarys Rios-Vargas. 2012. "The Population with a Bachelor's Degree or Higher by Race and Hispanic Origin: 2006–2010." U.S. Census Bureau, May. Accessed March 22, 2014 (www.census.gov).

"Older Americans Month: May 2017." 2017. U.S. Census Bureau, March 27. Accessed July 20, 2017 (census.gov).

Oliphant, Baxter. 2016. "Support for Death Penalty Lowest in More Than Four Decades." Pew Research Center, September 29. Accessed March 1, 2017 (pewresearch.org).

Oliver, Melvin L., and Thomas M. Shapiro. 1995. *Black Wealth/White Wealth: A New Perspective on Racial Inequality*. New York: Routledge.

Olivier, Jos G. J., Greet Janssens-Maenhout, Marilena Muntean, and Jeroen A. H. W. Peters. 2016. "Trends in Global CO_2 Emissions: 2016 Report." PBL Netherlands Environmental Assessment Agency. Accessed August 20, 2017 (edgar.jrc.europa.eu).

Olmstead, Kenneth, and Aaron Smith. 2017. "Americans and Cybersecurity." Pew Research Center, January 26. Accessed March 1, 2017 (pewresearch.org).

Olsen, Lauren D. 2016. "'It's on the MCAT for a Reason': Premedical Students and the Perceived Utility of Sociology." *Teaching Sociology* 44(2): 72–83.

Olson, Jonathan R. 2010. "Choosing Effective Youth-Focused Prevention Strategies: A Practical Guide for Applied Family Professionals." *Family Relations* 59 (April): 207–220.

Olson, Theodore B. 2010. "The Conservative Case for Gay Marriage." *Newsweek*, January 18, 48–53.

"On Views of Race and Inequality, Blacks and Whites Are Worlds Apart." 2016. Pew Research Center, June 27. Accessed June 20, 2017 (pewresearch.org).

Oppel, Richard A., Jr. 2011. "Steady Decline in Major Crime Baffles Experts." *New York Times*, May 4, A17.

Oreopoulos, Philip, and Uros Petronijevic. 2013. "Making College Worth It: A Review of Research on the Returns to Higher Education." Accessed June 21, 2014 (www.ticas.org).

Ortman, Jennifer M., Victoria A. Velkoff, and Howard Hogan. 2014. "An Aging Nation: The Older Population in the United States." U.S. Census Bureau, May. Accessed May 16, 2014 (www.census.gov).

Oudekerk, Barbara, Dara Blachman-Demner, and Carrie Mulford. 2014. "Teen Dating Violence: How Peers Can Affect Risk & Protective Factors." National Institute of Justice, November. Accessed August 6, 2015 (nij.gov).

Owen, Daniela J., Amy M. S. Slep, and Richard E. Heyman. 2012. "The Effect of Praise, Positive Nonverbal Response, Reprimand, and Negative Nonverbal Response on Child Compliance: A Systematic Review." *Clinical Child and Family Psychology Review* 15 (December): 364–385.

Oxfam. 2013. "The Cost of Inequality: How Wealth and Income Extremes Hurt Us All." January 18. Accessed February 6, 2014 (www.oxfam.org).

Oxfam. 2014. "Even It Up: Time to End Extreme Inequality." October. Accessed November 10, 2014 (www.oxfam.org).

Oxfam America. 2017. "Rigged Reform." April 12. Accessed April 15, 2017 (oxfamamerica.org).

P

Packard, Vance. 1959. *The Status Seekers*. New York: David McKay.

Padgett, Tim. 2010. "Robes for Women." *Time*, September 27, 53–55.

Palen, J. John. 2014. *The Urban World*, 10th edition. New York: Oxford University Press.

Paludi, Michele A., and J. Harold Ellens, eds. 2016. *Feminism and Religion: How Faiths View Women and Their Rights*. Santa Barbara, CA: Praeger Press.

Pandit, Sonia G. 2015. "Compete with the NRA?" *Baltimore Sun*, October 27, 13.

Park, Robert, and Ernest Burgess. 1921. *Introduction to the Science of Sociology*. Chicago: University of Chicago Press.

Parker, Kim. 2012. "The Boomerang Generation: Feeling OK About Living with Mom and Dad." Pew Social & Demographic Trends, March 15. Accessed November 20, 2013 (pewsocialtrends.org).

Parker, Kim, and Eileen Patten. 2013. "The Sandwich Generation: Rising Financial Burdens for Middle-Aged Americans." Pew Research Center, January 30. Accessed May 22, 2014 (www.pewsocialtrends.org).

Parker, Kim, and Gretchen Livingston. 2016. "6 Facts About American Fathers." Pew Research Center, June 16. Accessed May 19, 2017 (pewresearch.org).

Parker, Kim, and Juliana Menasce Horowitz. 2015. "Family Support in Graying Societies." Pew Research Center, May 21. Accessed July 22, 2015 (pewresearch.org).

Parker, Kim, and Juliana Menasce Horowitz. 2015. "Women and Leadership." Pew Research Center, January 14. Accessed September 29, 2015 (pewresearch.org).

Parker, Kim, Juliana Horowitz, Ruth Igielnik, Baxter Oliphant, and Anna Brown. 2017. "America's Complex Relationship with Guns." Pew Research Center, June 22. Accessed September 2, 2017 (pewresearch.org).

Parker, Kim, Rich Morin, Juliana Menasce Horowitz, and Mark Hugo Lopez. 2015. "Multiracial in America." Pew Research Center, June 11. Accessed June 13, 2015 (pewresearch.org).

Parker, Kim, Richard Fry, D'Vera Cohn, and Wendy Wang. 2011. "Is College Worth It? College Presidents, Public Assess Value, Quality and Mission of Higher Education." Pew Research Center, May 16. Accessed April 15, 2012 (www.pewsocialtrends.org).

Parker, Laura. 2016. "National Park Service Faces Sex Harassment Scandal." *National Geographic,* September 23. Accessed May 22, 2017 (nationalgeographic.com).

Parrott, Scott, and Caroline Titcomb Parrott. 2015. "U.S. Television's 'Mean World' for White Women: The Portrayal of Gender and Race on Fictional Crime Dramas." *Sex Roles* 73 (June): 70–82.

Parsons, Chelsea, and Eugenio Weigend. 2016. "America Under Fire: An Analysis of Gun Violence in the United States and the Link to Weak Gun Laws." Center for American Progress, October. Accessed March 1, 2017 (americanprogress.org).

Parsons, Talcott, and Robert F. Bales, eds. 1955. *Family, Socialization, and Interaction Process.* New York: Free Press.

Parsons, Talcott. 1951. *The Social System.* Glencoe, IL: Free Press.

Parsons, Talcott. 1954. *Essays in Sociological Theory,* revised edition. New York: Free Press.

Parsons, Talcott. 1959. "The School Class as a Social System: Some of Its Functions in American Society." *Harvard Educational Review* 29 (Fall): 297–313.

Parsons, Talcott. 1960. *Structure and Process in Modern Societies.* New York: Free Press.

Partnership for Public Service. 2011. "Scores by Effective Leadership." Accessed August 13, 2011 (www.bestplacestowork.org).

Passel, Jeffrey S., and D'Vera Cohn. 2016. "Overall Number of U.S. Unauthorized Immigrants Holds Steady Since 2009." Pew Research Center, September 20. Accessed June 20, 2017 (pewresearch.org).

Passel, Jeffrey S., D'Vera Cohn, and Mark Hugo Lopez. 2011. "Hispanics Account for More Than Half of Nation's Growth in Past Decade." Pew Research Center, March 24. Accessed December 7, 2011 (www.pewhispanic.org).

Patel, Sujan. 2017. "7 Deadly Sins of Employee Communication." *Baltimore Sun,* January 5, 19.

Patten, Eileen. 2015. "Who Is Multiracial? Depends on How You Ask." Pew Research Center, November 6. Accessed June 20, 2017 (pewresearch.org).

Patten, Eileen. 2016. "The Nation's Latino Population Is Defined by Its Youth." Pew Research Center, April 20. Accessed June 20, 2017 (pewresearch.org).

Paul, Annie Murphy. 2010. *Origins: How the Nine Months Before Birth Shape the Rest of Our Lives.* New York: Free Press.

Paul, Noel. C. 2002. "The Birth of a Would-Be Fad." *Christian Science Monitor,* September 23, 11, 14–16.

Paul, Richard, and Linda Elder. 2007. *The Miniature Guide to Critical Thinking: Concepts and Tools.* Dillon Beach, CA: Foundation for Critical Thinking.

Paulson, Amanda. 2014. "Changes of Address." *Christian Science Monitor Weekly,* May 26, 12–13.

Payne, David, and Glenn Somerville. 2015. "Where the Jobs Will Be." *Baltimore Sun,* April 19, 3.

Payne, K. K., and J. Copp. 2013. "Young Adults in the Parental Home and the Great Recession." National Center for Family & Marriage Research. Accessed November 20, 2013 (www.hcfmr.bgsu.edu).

Pazol, Karen, Andreea A. Creanga, Kim D. Burley, and Denise J. Jamieson. 2014. "Abortion Surveillance—United States, 2011." *MMWR* 63 (11): 1–41.

Pearlstine, Norman. 2013. "Fix This: Water." *Bloomberg Businessweek,* March 25–31, 46–51.

Pedersen, Paul. 1995. *The Five Stages of Culture Shock: Critical Incidents Around the World.* Westport, CT: Greenwood Press.

Perlmutter, David D. 2001. "Students Are Blithely Ignorant; Professors Are Bitter." *Chronicle of Higher Education,* July 27, B20.

Perlroth, Nicole, Michael Wines, and Matthew Rosenberg. 2017. "Russian Election Hacking Efforts, Wider Than Previously Known, Draw Little Scrutiny." *New York Times,* September 1. Accessed September 2, 2017 (nytimes.com).

Perry, Gina. 2013. *Behind the Shock Machine: The Untold Story of the Notorious Milgram Psychology Experiments.* New York: New Press.

PerryUndem. 2017. "The State of the Union on Gender Equality, Sexism, and Women's Rights." January 17. Accessed January 20, 2017 (perryundem.com).

Persell, Caroline Hodges, and Peter W. Cookson, Jr. 1985. "Chartering and Bartering: Elite Education and Social Reproduction." *Social Problems* 33 (December): 114–129.

Peter, Tom A. 2012. "Mistreatment of Afghan Women Caused by Far More Than Taliban." *Christian Science Monitor,* January 31. Accessed March 4, 2014 (www.csmonitor.com).

Petersen, Andrea. 2012. "Smarter Ways to Discipline Kids." *Wall Street Journal,* December 26, D1, D3.

Peterson, James L., Josefina J. Card, Marvin B. Eisen, and Bonnie Sherman-Williams. 1994. "Evaluating Teenage Pregnancy Prevention and Other Social Programs: Ten Stages of Program Assessment." *Family Planning Perspectives* 26 (May): 116–120, 131.

Peterson, Scott. 2008. "In Iran, Barbie Seen as Cultural Invader." *Christian Science Monitor,* September 15, 4.

Petrosino, Anthony, Carolyn Turpin-Petrosino, Meghan E. Hollis-Peel, and Julia G. Lavenberg. 2013. "Scared Straight and Other Juvenile Awareness Programs for Preventing Juvenile Delinquency: A Systematic Review." Campbell Systematic Reviews. Accessed September 2, 2015 (campbellcollaboration.org).

Pettypiece, Shannon. 2013. "Anything You Can Do, I Can Do Better." *Bloomberg Businessweek,* March 11–17, 27–29.

Pettypiece, Shannon. 2016. "No Cheers When Walmart Packs Up." *Bloomberg Businessweek,* February 1–7, 18–19.

Pew Center on the States. 2010. "Prison Count 2010." April. Accessed August 24, 2011 (www.pewcenteronthestates.org).

Pew Charitable Trusts. 2015. "The Precarious State of Family Balance Sheets." January. Accessed September 11, 2015 (pewtrusts.org).

Pew Hispanic Center. 2012. "When Labels Don't Fit: Hispanics and Their Views of Identity." Pew Research Center, April 4. Accessed March 22, 2014 (www.pewhispanic.org).

Pew Research Center. 2014. "Most Say Religious Holiday Displays on Public Property Are OK." December 14. Accessed July 31, 2017 (pewresearch.org).

Pew Research Center. 2015. "The American Middle Class Is Losing Ground: No Longer the Majority and Falling Behind Financially." December 9. Accessed December 10, 2015 (pewsocialtrends.org).

Pew Research Center. 2015. "America's Changing Religious Landscape." May 4. Accessed July 31, 2017 (pewresearch.org).

Pew Research Center. 2016. "Choosing a New Church or House of Worship." August 23. Accessed July 31, 2017 (pewresearch.org).

Pew Research Center. 2017. "In America, Does More Education Equal Less Religion?" April 26. Accessed July 31, 2017 (pewresearch.org).

Pewewardy, Cornel. 1998. "Fluff and Feathers: Treatment of American Indians in the Literature and the Classroom." *Equity & Excellence in Education* 31 (April): 69–76.

Pflanz, Mike. 2014. "Briefing: Africa's Stance on Gays." *Christian Science Monitor Weekly,* March 17, 13.

Pflaumer, Alicia. 2011. "Texting Bride Video Goes Viral on the Web." *Christian Science Monitor,* November 1. Accessed November 3, 2011 (www.csmonitor.com).

Phelps, Glenn, and Steve Crabtree. 2013. "Worldwide, Median Household Income About $10,000." Gallup, December 16. Accessed February 6, 2014 (www.gallup.com).

Phelps, Glenn, and Steve Crabtree. 2014. "Worldwide, Richest 3% Hold One-Fifth of Collective Income." Gallup, January 3. Accessed February 6, 2014 (www.gallup.com).

Philips, Matthew. 2016. "Regulations Dry Up Wastewater Wells." *Bloomberg Businessweek,* March 21–27, 23–24.

Phillips, L. Taylor, and Brian S. Lowery. 2015. "The Hard-Knock Life? Whites Claim Hardships in Response to Racial Inequity." *Journal of Experimental Social Psychology* 61 (November): 12–18.

Phillips, Matthew. 2014. "Welders, America Needs You." *Bloomberg Businessweek*, March 24–April 6, 19–21.

Philpott, Tom. 2016. "Playing Chicken." *Mother Jones*, May/June, 41–47, 74.

Pianigiani, Gaia. 2016. "Italy's 'Fertility Day' Ads Anger Women Stymied by Lack of Support." *New York Times*, August 14, A4.

Pianta, Robert C., Jay Belsky, Renate Houts, Fred Morrison, and The National Institute of Child Health and Human Development (NICHD) Early Child Care Research Network. 2007. "Teaching: Opportunities to Learn in America's Elementary Classrooms." *Science* 315 (March 30): 1795–1796.

Piazza, Jo. 2016. "Women of Color Hit a 'Concrete Ceiling' in Business." *Wall Street Journal*, September 27. Accessed June 8, 2017 (wsj.com).

Pierce, Lamar, Daniel Snow, and Andrew McAfee. 2013. "Cleaning House: The Impact of Information Technology Monitoring on Employee Theft and Productivity." Social Science Research Network, August 24. Accessed October 24, 2013 (www.ssrn.com).

Piketty, Thomas, Emmanuel Saez, and Gabriel Zucman. 2016. "Distributional National Accounts: Methods and Estimates for the United States." National Bureau of Economic Research, December. Accessed March 15, 2017 (nber.org).

Pipkin, Whitney. 2013. "Are Asian Americans Taking over the Internet?" *Asian Fortune*, November 29. Accessed December 4, 2013 (www.asianfortunenews.com).

Pittz, Will. 2005. "Closing the Gap: Solutions to Race-Based Health Disparities." Applied Research Center & Northwest Federation of Community Organizations, June. Accessed April 20, 2007 (www.arc.org).

Planty, M., W. Hussar, T. Snyder, S. Provasnik, G. Kena, R. Dinkes, A. KewalRamani, and J. Kemp. 2008. *The Condition of Education 2008*. Washington, DC: National Center for Education Statistics, U.S. Department of Education.

Plateris, Alexander A. 1973. *100 Years of Marriage and Divorce Statistics: 1867–1967*. Rockville, MD: National Center for Health Statistics.

Plumer, Brad. 2017. "Energy Dept. Closes Office for Sharing Clean Technology." *New York Times*, June 16, A20.

Plutzer, Eric, Mark McCaffrey, A. Lee Hannah, Joshua Rosenau, Minda Berbeco, and Ann H. Reid. 2016. "Climate Confusion Among U.S. Teachers." *Science* 351 (6274): 664–665.

Polderman, Tinca, J. C., et al. 2015. "Meta-Analysis of the Heritability of Human Traits Based on Fifty Years of Twin Studies." *Nature Genetics* 47 (May): 702–709.

Polgreen, Lydia. 2010. "One Bride for Two Brothers: A Custom Fades in India." *New York Times*, July 16, A4.

"Police Culture." 2015. *The Economist*, April 25, 28–29.

"Policing Philadelphia: Boots on the Street." 2013. *The Economist*, August 24, 33.

Polikoff, Morgan S., and Andrew C. Porter. 2014. "Instructional Alignment as a Measure of Teaching Quality." *Educational Evaluation and Policy Analysis* 36 (December): 399–416.

"The Politics of Financial Insecurity." 2015. Pew Research Center, January 8. Accessed October 7, 2015 (pewresearch.org).

PolitiFact. 2016. "Comparing Hillary Clinton, Donald Trump on the Truth-O-Meter." Accessed January 11, 2017 (politifact.com).

Pollack, Andrew. 2013. "A.M.A. Recognizes Obesity as a Disease." *New York Times*, June 19, B1.

Pollick, Michael. 2014. "Why Didn't Communism Work in Eastern Europe?" Conjecture Corporation, April 1. Accessed April 12, 2014 (www.wisegeek.org).

Polsby, Nelson W. 1959. "Three Problems in the Analysis of Community Power." *American Sociological Review* 24 (December): 796–803.

Pong, Suet-ling, Lingxin Hao, and Erica Gardner. 2005. "The Roles of Parenting Styles and Social Capital in the School Performance of Immigrant Asian and Hispanic Adolescents." *Social Science Quarterly* 86 (December): 928–950.

Porter, Eduardo. 2013. "America's Sinking Middle Class." *New York Times*, September 19, A1.

Porter, Eduardo. 2014. "A Global Boom, but Only for Some." *New York Times*, March 19, B1.

Porter, Michael E., Scott Stern, and Michael Green. 2017. "Social Progress Index 2017." Social Progress Imperative. Accessed August 20, 2017 (socialprogressimperative.org).

Porter, Nicole D. 2016. "The State of Sentencing 2015: Developments in Policy and Practice." The Sentencing Project. Accessed April 20, 2017 (sentencingproject.org).

Portes, Jonathan. 2015. "Immigration Is Good for Economic Growth. If Europe Gets It Right, Refugees Can Be Too." *Huffington Post*, September 15. Accessed October 23, 2015 (huffingtonpost.com).

Postel, Sandra. 2012. "Humanity's Growing Impact on the World's Freshwater." AlterNet, February 23. Accessed June 23, 2012 (alternet.org).

Potok, Mark. 2017. "The Year in Hate and Extremism." Southern Poverty Law Center, February 15. Accessed February 25, 2017 (splcenter.org).

Poushter, Jacob. 2014. "Russia's Moral Barometer: Homosexuality Unacceptable, but Drinking, Less So." Pew Research Center, February 6. Accessed March 2, 2014 (www.pewresearch.org).

Powell, Jon T., Timothy G. Townsend, and Julie B. Zimmerman. 2015. "Estimates of Solid Waste Disposal Rates and Reduction Targets for Landfill Gas Emissions." *Nature Climate Change*, September 21. Accessed October 23, 2015 (nature.com).

Powers, Charles H. 2004. *Making Sense of Social Theory: A Practical Introduction*. Lanham, MD: Rowman & Littlefield.

Pratt, Beverly M., Lindsay Hixson, and Nicholas A. Jones. 2015. "Measuring Race and Ethnicity Across the Decades: 1790–2010." U.S. Census Bureau, November 2. Accessed January 12, 2017 (census.gov).

"Pregnancy Mortality Surveillance System." 2017. Centers for Disease Control and Prevention, June 27. Accessed August 15, 2017 (cdc.gov).

Prein, Andreas, Roy M. Rasmussen, Kyoko Ikeda, Changhai Liu, Martyn P. Clark, and Greg J. Holland. 2017. "The Future Intensification of Hourly Precipitation Extremes." *Nature Climate Change* 7 (December): 48–52.

"Prescription for Change." 2013. *The Economist*, June 29, 61–62.

Preston, Caroline. 2016. "The 20 Most Generous Companies of the Fortune 500." *Fortune*, June 22. Accessed May 15, 2017 (fortune.com).

Preston, Julia. 2015. "States Are Divided by the Lines They Draw on Immigration." *New York Times*, March 30, A10, A12.

Price, Barbara Raffel, and Natalie J. Sokoloff, eds. 2004. *The Criminal Justice System and Women: Offenders, Prisoners, Victims, & Workers*, 3rd edition. New York: McGraw-Hill.

Princiotta, Daniel, Laura Lippman, Renee Ryberg, Hannah Schmitz, David Murphey, and Mae Cooper. 2014. "Social Indicators Predicting Postsecondary Success." Child Trends, April 1. Accessed June 23, 2014 (www.childtrends.org).

"Prisons." 2017. *The Economist*, March 18, 26.

Project on Government Oversight. 2010. "Letter to NIH on Ghostwriting Academics." November 29. Accessed July 3, 2011 (www.pogo.org).

Protect Our Defenders. 2016. "Facts on United States Military Sexual Violence." September. Accessed April 2, 2017 (protectourdefenders.com).

"Pro-natalism." 2015. *The Economist*, July 25, 47–48.

Pryor, John H. 2011. "The Changing First-Year Student: Challenges for 2011." Higher Education Research Institute at UCLA, January 27. Accessed April 10, 2012 (www.heri.ucla.edu).

Przybylski, Andrew K., and Allison F. Mishkin. 2015. "How the Quantity and Quality of Electronic Gaming Relates to Adolescents' Academic Engagement and Psychosocial Adjustment." *Psychology of Popular Media Culture*, March 2. APA PsycNet. Accessed August 6, 2015 (psycnet.apa.org).

"Public Trust in Government Remains Near Historic Lows as Partisan Attitudes Shift." 2017 Pew Research Center, May 3. Accessed June 16, 2017 (pewresearch.org).

"Public Uncertain, Divided over America's Place in the World." 2016. Pew Research Center, May 5. Accessed February 5, 2017 (pewresearch.org).

Puddington, Arch, and Tyler Roylance. 2017. "Populists and Autocrats: The Dual Threat to Global Democracy." Freedom House. Accessed June 25, 2017 (freedomhouse.org).

Putnam, Hannah, Julie Greenberg, and Kate Walsh. 2014. "Easy A's and What's Behind Them." National Council on Teacher Quality, November. Accessed December 5, 2014 (www.nctq.org).

Putnam, Robert D. 2015. *Our Kids: The American Dream in Crisis*. New York: Simon & Schuster.

Q

Qin, Amy. 2015. "'Kingdom of Daughters' Draws Tourists to Its Matrilineal Society." *New York Times*, October 26, A1, A6.

Quinney, Richard. 1980. *Class, State, and Crime*. Boston: Little, Brown.

R

Rackin, Heather, and Christina M. Gibson-Davis. 2012. "The Role of Pre- and Postconception Relationships for First-Time Parents." *Journal of Marriage and Family* 74 (June): 389–398.

Radesky, Jenny, et al. 2014. "Patterns of Mobile Device Use by Caregivers and Children During

Meals in Fast Food Restaurants." *Pediatrics* 133 (4): e843–e849.

Rainie, Lee, and Kathryn Zickuhr. 2015. "Americans' Views on Mobile Etiquette." Pew Research Center, August 26. Accessed August 26, 2015 (pewresearch.org).

Raj, Suhasini. 2015. "Goat Meat, Not Beef, Found in Home of Slain Indian." *New York Times*, December 30, A3.

Ramakrishnan, Karthick, and Farah Z. Ahmad. 2014. "Education: Part of the 'State of Asian Americans and Pacific Islanders' Series." Center for American Progress, April 23. Accessed July 31, 2017 (americanprogress.org).

Rampell, Catherine. 2011. "Companies Spend on Equipment, Not Workers." *New York Times*, June 10, A1.

Rampell, Catherine. 2014. "The Safety Net Catches the Middle Class More Than the Poor." *Washington Post*, April 7. Accessed March 8, 2014 (www.washingtonpost.com).

Rank, Mark R., and Thomas A. Hirschl. 2015. "The Likelihood of Experiencing Relative Poverty over the Life Course." *PLoS One* 10(7): e0133513. doi:10.1371/journal.pone.0133513. Accessed September 16, 2015 (journals.plos.org).

Rano, Jason, and Jane Houlihan. 2012. "Myths on Cosmetic Safety." Skin Deep Cosmetics Database. Accessed May 10, 2012 (www.ewg.org).

Rape Crisis Center. 2014. "Get the Facts." Accessed February 14, 2014 (www.rccmsc.org).

Raphael, Steven, and Michael A. Stoll. 2010. "Job Sprawl and the Suburbanization of Poverty." Brookings Institute, Metropolitan Policy Program, March. Accessed May 1, 2010 (www.brookings.edu).

Ray, Rebecca, Milla Sanes, and John Schmitt. 2013. "No-Vacation Nation Revisited." Center for Economic and Policy Research, May. Accessed April 12, 2014 (cepr.net).

Rayasam, Renuka. 2007. "Immigrants: The Unsung Heroes of the U.S. Economy." *U.S. News & World Report*, February 26, 58.

Reaney, Patricia. 2012. "Average Cost of U.S. Weddings Hits $27,021." Reuters, May 23. Accessed June 2, 2012 (www.reuters.com).

Reardon, Sean F. 2011. "The Widening Academic Achievement Gap Between the Rich and the Poor: New Evidence and Possible Explanations." Pp. 91–116 in *Whither Opportunity? Rising Inequality and the Uncertain Life Chances of Low-Income Children*, edited by R. Murnane and G. Duncan. New York: Russell Sage Foundation Press.

Reardon, Sean F., Lindsay Fox, and Joseph Townsend. 2015. "Neighborhood Income Composition by Household Race and Income, 1990–2009." *Annals of the American Academy of Political and Social Science* 660 (July): 78–99.

Reaves, Brian A. 2017. "Police Response to Domestic Violence, 2006–2015." Bureau of Justice Statistics, May. Accessed July 15, 2017 (bjs.gov).

Redd, Zakia, Tahilin Sanchez Karver, David Murphey, Kristin Anderson Moore, and Dylan Knewstub. 2011. "Two Generations in Poverty: Status and Trends among Parents and Children in the United States, 2000–2010." Child Trends Research Brief, November. Accessed January 12, 2012 (www.childtrends.org).

Reddy, Sumathi. 2013. "'I Don't Smoke, Doc,' and Other Patient Lies." *Wall Street Journal*, February 19, D3.

Reeves, Jay. 2013. "Alabama Woman's Body Removed from Front Yard Grave." *U.S. News & World Report*, November 15. Accessed August 10, 2014 (www.us.news.com).

Reeves, Richard V. 2017. "Race Gaps in SAT Math Scores Are as Big as Ever." Brookings, July 11. Accessed February 1, 2017 (brookings.edu).

Reeves, Richard V., and Joanna Venator. 2015. "Sex, Contraception, or Abortion? Explaining Class Gaps in Unintended Childbearing." Brookings, February. Accessed September 29, 2015 (brookings.edu).

Reeves, Richard V., and Nathan Joo. 2016. "How Much Social Mobility Do People Really Want?" Brookings, January 12. Accessed December 23, 2016 (brookings.edu).

Reeves, Richard V., Isabel V. Sawhill, and Eleanor Krause. 2016. "The Most Educated Women Are the Most Likely to Be Married." Brookings, August 19. Accessed August 25, 2016 (brookings.edu).

Reger, Jo. 2012. *Everywhere and Nowhere: Contemporary Feminism in the United States*. New York: Oxford University Press.

Regnerus, Mark, and Jeremy Uecker. 2011. *Premarital Sex in America: How Young Americans Meet, Mate, and Think About Marrying*. New York: Oxford University Press.

Reich, Robert. 2014. "Why Widening Inequality Is Hobbling Equal Opportunity." AlterNet, February 6. Accessed February 10, 2014 (www.alternet.org).

Reich, Robert. 2015. "Robert Reich Demolishes Myth That College Is Gateway to the Middle Class." AlterNet, March 23. Accessed July 31, 2017 (alternet.org).

Reilly, Steve. 2016. "Teachers Who Sexually Abuse Students Still Find Classroom Jobs." *USA Today*, December 22. Accessed January 5, 2017 (usatoday.com).

Reiman, Jeffrey, and Paul Leighton. 2010. *The Rich Get Richer and the Poor Get Prison: Ideology, Class, and Criminal Justice*, 9th edition. Upper Saddle River, NJ: Prentice Hall.

Rein, Lisa. 2011. "Federal Workers Tell Us What Should Be Cut from the Budget." *Washington Post*, April 14, B4.

Rein, Lisa, and Emily Wax-Thibodeaux. 2015. "Veterans Affairs Improperly Spent $6 Billion Annually, Senior Official Says." *Washington Post*, May 14. Accessed August 23, 2015 (washingtonpost.com).

Reingold, Jennifer. 2016. "The Disappeared." *Fortune*, September 15, 100–108.

Reinhart, R. J. 2017. "More in US Say Government Is the Most Important Problem." Gallup, June 15. Accessed June 27, 2017 (gallup.com).

Reinhold, Steffen 2010. "Reassessing the Link Between Premarital Cohabitation and Marital Instability." *Demography* 47 (August): 719–733.

Reiss, Fraidy. 2015. "America's Child-Marriage Problem." *New York Times*, October 14, A23.

Reiss, Steven. 2004. "The Sixteen Strivings for God." *Zygon* 39 (June): 303–320.

"Religion: The New Strife." 2016. *The Economist*, May 14, 10–12.

"Religions of the World: Number of Adherents, Names of Houses of Worship. . . ." 2007. Religious Tolerance. Accessed August 24, 2007 (www.religioustolerance.org).

Restaurant Opportunities Centers United. 2017. "Secretary of Labor Violations?" January 10. Accessed May 25, 2017 (rocunited.org).

Reuben, Cynthia, and Patricia Pastor. 2015. "Percentage of Children and Adolescents Aged 4–17 Years with Serious Emotional or Behavioral Difficulties, by Poverty Status and Sex—National Health Interview Survey, 2011–2014." *Morbidity and Mortality Weekly Report* 64 (46): 1303.

Reynolds, Kelly A. 2012. "The Price of Drinking Water." *Water Conditioning & Purification* 54 (July): 50–52.

Rheault, Magali, and Kyley McGeeney. 2011. "Education Is a Key Predictor of Emotional Health After 65." Gallup, August 19. Accessed April 3, 2012 (www.gallup.com).

Ricciardelli, Rosemary, Kimberley A. Clow, and Philip White. 2010. "Investigating Hegemonic Masculinity: Portrayals of Masculinity in Men's Lifestyle Magazines." *Sex Roles* 63 (March): 64–78.

Rich, Motoko. 2013. "Subtract Teachers, Add Pupils: Math of Today's Jammed Schools." *New York Times*, December 22, A1.

Rich, Motoko. 2014. "Why Don't More Men Go into Teaching?" *New York Times*, September 6, SR3.

Richburg, Keith B. 2009. "States Seek Less Costly Substitutes for Prison." *Washington Post*, July 13, A1.

Riche, Martha Farnsworth. 2000. "America's Diversity and Growth: Signposts for the 21st Century." *Population Bulletin* 55 (June): 1–41.

Richey, Alexandra S., Brian F. Thomas, Min-Hui Lo, James S. Famiglietti, Sean Swenson, and Matthew Rodell. 2015. "Uncertainty in Global Groundwater Storage Estimates in a Total Groundwater Stress Framework." *Water Resources Research* 51 (July): 5198–5216.

Richie, Christina. 2013. "The Scandal of the (Female) Evangelical Mind." *Chronicle of Higher Education*, June 14, A37–A38.

Richtel, Matt, and Alexei Barrionuevo. 2005. "Wendy's Restaurants." *New York Times*, April 22, A9.

Rideout, Victoria. 2013. "Zero to Eight: Children's Media Use in America 2013." Common Sense Media, Fall. Accessed November 18, 2013 (www.commonsensemedia.org).

Rideout, Victoria. 2014. "Children, Teens, and Reading: A Common Sense Media Research Brief." Accessed March 1, 2017 (commonsensemedia.org).

Rideout, Victoria. 2015. "The Common Sense Census: Media Use by Tweens and Teens." Accessed March 1, 2017 (commonsensemedia.org).

Riesman, David. 1953. *The Lonely Crowd*. New York: Doubleday.

Riffkin, Rebecca. 2014. "New Record Highs in Moral Acceptability." Gallup, May 30. Accessed August 10, 2014 (www.gallup.com).

Riffkin, Rebecca. 2016. "Majority of Americans Dissatisfied with Corporate Influence." Gallup, January 20. Accessed April 4, 2017 (gallup.com).

Rios, Edwin. 2016. "Block the Vote." *Mother Jones*, July/August, 7.

Riosmena, Fernando, Elisabeth Root, Jamie Humphrey, Emily Steiner, and Rebecca Stubbs. 2015. "The Waning Hispanic Health Paradox." *Pathways* (Spring): 25–29.

Ripley, Amanda. 2016. "What the U.S. Can Learn from Other Nations' Schools." *New York Times*, December 8, A3.

Risk Based Security. 2017. "Data Breach QuickView Report." January. Accessed April 2, 2017 (riskbasedsecurity.com).

Ritzer, George. 1992. *Contemporary Sociological Theory*, 3rd edition. New York: McGraw-Hill.

Ritzer, George. 1996. *The McDonaldization of Society: An Investigation into the Changing Character of Contemporary Social Life*. Thousand Oaks, CA: Pine Forge Press.

Ritzer, George. 2008. *The McDonaldization of Society*, 5th edition. Los Angeles: Pine Forge Press.

Rizga, Kristina. 2017. "Betsy DeVos Wants to Use America's Schools to Build 'God's Kingdom'." *Mother Jones* March/April. Accessed July 31, 2017 (motherjones.com).

Robbins, Katherine Gallagher, and Julie Vogtman. 2016. "Low-Wage Jobs Held Primarily by Women Will Grow the Most over the Next Decade." National Women's Law Center, April. Accessed June 22, 2017 (nwlc.org).

Robert Wood Johnson Foundation. 2010. "California Nurse Ratio Law Saves Lives, Improves Nurse Morale, Study Finds." May 16. Accessed June 6, 2012 (www.rwjf.org).

Robert Wood Johnson Foundation. 2015. "Understanding the Uninsured Now." June. Accessed October 16, 2015 (rwjf.org).

Roberts, Andrea L., et al. 2013. "Perinatal Air Pollutant Exposures and Autism Spectrum Disorder in the Children of Nurses' Health Study II Participants." *Environmental Health Perspectives* 121 (August): 978–984.

Roberts, James A., and Meredith E. David. 2016. "My Life Has Become a Major Distraction from My Cell Phone: Partner Phubbing and Relationship Satisfaction Among Romantic Partners." *Computers in Human Behavior* 54 (January): 134–141.

Roberts, Keith A. 2004. *Religion in Sociological Perspective*, 4th edition. Belmont, CA: Wadsworth.

Roberts, Laura M., and Robin J. Ely. 2016. "Why Did So Many White Women Vote for Donald Trump?" *Fortune*, November 18. Accessed January 5, 2017 (fortune.com).

Roberts, Yvonne. 2014. "Addicted to Email? The Germans Have an Answer." AlterNet, September 1, 2015. Accessed September 2, 2014.

Robertson, Campbell. 2016. "Chief Justice in Alabama Is Suspended a Second Time." *New York Times*, October 1, A9.

Robinson, Laurie O., and Jeff Slowikowski. 2011. "Scary—and Ineffective." *Baltimore Sun*, February 1, 11.

Rochman, Bonnie. 2012. "The End of an Epidemic?" *Time*, February 6, 16.

Rochman, Bonnie. 2013. "Hover No More: Helicopter Parents May Breed Depression and Incompetence in Children." *Time*, February 22. Accessed November 20, 2013 (www.healthland.time.com).

Roden, Lee. 2017. "Why Sweden Is NOT the 'Rape Capital of the World'." The Local, February 21. Accessed April 24, 2017 (thelocal.se).

Rodengen, Jeffrey L. 2002. *NRA: An American Legend*. Fort Lauderdale, FL: Write Stuff Enterprises.

Rodgers, Bill. 2017. "Sanctuary Churches in US Mobilize to Help Undocumented." VOA News, March 21. Accessed July 31, 2017 (voanews.com).

Rodgers, Timothy. 2016. "Organized Crime." LinkedIn, December 5. Accessed March 1, 2017 (linkedin.com).

Rodrigue, Edward, and Richard V. Reeves. 2015. "Cutting Poverty by Increasing Program Participation." Brookings Institution, June 11. Accessed September 31, 2015 (brookings.edu).

Roethlisberger, F. J., and William J. Dickson. 1939/1942. *Management and the Worker: An Account of a Research Program Conducted at the Western Electric Company, Hawthorne Works, Chicago*. Cambridge, MA: Harvard University Press.

Rohrlich, Justin. 2010. "Why White-Collar Criminals Don't Fear Getting Caught." Minyanville Media, Inc., November 12. Accessed January 12, 2014 (www.minyanville.com).

Roman, Caterina Gouvis, et al. 2012. "Social Networks, Delinquency, and Gang Membership: Using a Neighborhood Framework to Examine the Influence of Network Composition and Structure in a Latino Community." The Urban Institute, February. Accessed January 12, 2014 (www.urban.org).

Romer, Dan. 2011. "After 11 Years of Setting the Record Straight, Stories About Holiday Suicides Still Outnumber Those Debunking the Myth." Annenberg Public Policy Center, December 13. Accessed December 14, 2011 (www.annenbergpublicpolicycenter.org).

Ronfeldt, Matthew, Hamilton Lankford, Susanna Loeb, and James Wyckoff. 2011. "How Teacher Turnover Harms Student Achievement." National Bureau of Economic Research, June. Accessed April 14, 2012 (nber.org).

Roscigno, Vincent J. 2010. "Ageism in the American Workplace." *Contexts* 9 (Winter): 16–21.

Rose, Fred. 1997. "Toward a Class-Cultural Theory of Social Movements: Reinterpreting New Social Movements." *Sociological Forum* 12 (September): 461–494.

Rose, Peter I. 1997. *They and We: Racial and Ethnic Relations in the United States*, 5th edition. New York: McGraw-Hill.

Rosen, Ruth. 2016. "US Presidential Race: The Feminist Generation Gap." Open Democracy, March 14. Accessed January 9, 2017 (opendemocracy.net).

Rosenberg, Martha. 2016. "How an Army of Pharma Lobbyists in Washington Have Locked in One of the Biggest Corporate Ripoff Schemes in America." AlterNet, February 22. Accessed August 15, 2017 (alternet.org).

Rosenblatt, Lauren. 2017. "Bias Creeps Up on Older Workers." *Baltimore Sun*, July 2, 19.

Rosenfeld, Steven. 2014. "The Supreme Court's Radical Right Wing Majority: Waging War on Women and Boosting Corporate Power." AlterNet, June 30. Accessed July 26, 2014 (www.alternet.org).

Rosenthal, Bill. 2016. "The Ashley Madison Hack—One Year Later." Logical Operations, July 20. Accessed April 13, 2017 (logicaloperations.com).

Rosenthal, Elizabeth. 2014. "Medicine's Top Earners Are Not the M.D.s." *New York Times*, May 18, SR4.

Rosenthal, Robert, and Lenore Jacobsen. 1968. *Pygmalion in the Classroom: Teacher Expectations and Pupils' Intellectual Development*. New York: Holt, Rinehart, and Winston.

Rosin, Hanna. 2012. "Boys on the Side." *The Atlantic*, September. Accessed January 7, 2013 (www.theatlantic.com).

Ross, Brian, Rhonda Schwartz, and Megan Christie. 2016. "Recovered Madoff Money Now over $11 Billion: More Hidden Away?" ABC News, February 1. Accessed March 1, 2017 (abcnews.go.com).

Ross, Martha, Nicole P. Svajlenka, and Jane Williams. 2014. "Part of the Solution: Pre-Baccalaureate Healthcare Workers in a Time of Health System Change." Brookings, July. Accessed December 5, 2014 (www.brookings.edu).

Ross, Terris, Grace Kena, Amy Rathbun, Angelina KewalRamani, Jijun Zhang, Paul Kristapovich, and Eileen Manning. 2012. "Higher Education: Gaps in Access and Persistence Study." National Center for Education Statistics, August. Accessed June 21, 2014 (www.nces.ed.gov).

Rossano, Matt J. 2012. "The Essential Role of Ritual in the Transmission and Reinforcement of Social Norms." *Psychological Bulletin* 138 (May): 529–549.

Rossi, Max, and Lisa Jucca. 2014. "Villages Face 'Slow Death' as Italy's Population Ages." *Baltimore Sun*, March 2, 14.

Rossin-Slater, Maya. 2017. "Maternity and Family Leave Policy." National Bureau of Economic Research, January. Accessed July 19, 2017 (nber.org).

Roth, Steve. 2017. "New Data Reveal the Depressing Truth About How Wealth Is Amassed in America." AlterNet, January 6. Accessed June 4, 2017 (alternet.org).

Rothman, Sheila M. 1978. *Women's Proper Place: A History of Changing Ideals and Practices, 1870 to the Present*. New York: Basic Books.

Rovner, Joshua. 2016. "Racial Disparities in Youth Commitments and Arrests." The Sentencing Project, April. Accessed April 20, 2017 (sentencingproject.org).

Rowe-Finkbeiner, Kristin. 2012. "It's Not a 'Mommy War,' It's a War on Moms." Moms Rising, April 14. Accessed April 12, 2014 (www.momsrising.org).

Rubin, Kenneth, William Bukowski, and Jeffrey G. Parker. 1998. "Peer Interactions, Relationships, and Groups." Pp. 619–700 in *Handbook of Child Psychology: Vol. 3. Social, Emotional, and Personality Development*, edited by William Damon and Nancy Eisenberg. New York: Wiley.

Rubin, Rita. 2004. "'Smart Pills' Make Headway." *USA Today*, July 7, 1D.

Ruetschlin, Catherine, and Dedrick Asante-Muhammad. 2013. "The Challenge of Credit Card Debt for the African American Middle Class." Demos & NAACP, December 4. Accessed March 22, 2014 (www.demos.org).

Rule, James B. 1988. *Theories of Civil Violence*. Berkeley: University of California Press.

Rushe, Dominic, Oliver Milman, Molly Redden, Jamiles Lartey, David Smith, and Oliver Laughland. 2017. "6 Ways Trump Is Dismantling the Country Without Passing a Single Piece of Legislation." AlterNet, July 19. Accessed July 25, 2017 (alternet.org).

Ruth, Jennifer. 2014. "Non-Critical Thinking in China." *Chronicle Review*, February 28, B20.

Rutter, Virginia, and Pepper Schwartz. 2012. *The Gender of Sexuality: Exploring Sexual Possibilities*, 2nd edition. Lanham, MD: Rowman & Littlefield.

Ryan, Camille. 2013. "Language Use in the United States: 2011." U.S. Census Bureau, August. Accessed March 22, 2014 (www.census.gov).

Ryan, Camille L., and Kurt Bauman. 2016. "Educational Attainment in the United States: 2015." U.S. Census Bureau, Current Population Reports, March. Accessed June 20, 2017 (census.gov).

Ryzhkov, Vladimir. 2015. "Russia's New Totalitarianism Depends on Silence." *Moscow Times*, April 23. Accessed June 23, 2017 (themoscowtimes .com).

S

Saad, Lydia. 2001. "Majority Considers Sex Before Marriage Morally Okay." *Gallup Poll Monthly*, No. 428 (May): 46–48.

Saad, Lydia. 2006. "Families of Drug and Alcohol Abusers Pay an Emotional Toll." Gallup News Service, August 25. Accessed August 27, 2006 (www.gallup.com).

Saad, Lydia. 2013. "Half in U.S. Support Publicly Financed Federal Campaigns." Gallup, June 24. (Accessed April 16, 2014 (www .gallup.com).

Saad, Lydia. 2013. "U.S. Support for Euthanasia Hinges on How It's Described." Gallup, May 29. Accessed May 16, 2014 (www.gallup.com).

Saad, Lydia. 2016. "Trump and Clinton Finish with Historically Poor Images." Gallup, November 8. Accessed June 27, 2017 (gallup.com).

Saad, Lydia. 2017. "A Third of Investors Help Parent or Grown Child Financially." Gallup, May 31. Accessed July 20, 2017 (gallup.com).

Saad, Lydia. 2017. "Record Few Americans Believe Bible Is Literal Word of God." Gallup, May 15. Accessed July 30, 2017 (gallup.com).

Sachs, Jeffrey S. 2005. "Confusion over Population: Growth or Dearth?" *Pop!ulation Press* 11 (Winter/Spring): 17.

Sacks, Vanessa, David Murphey, and Kristin Moore. 2014. "Adverse Childhood Experiences: National and State-Level Prevalence." Child Trends Research Brief, July. Accessed November 25, 2014 (www.childtrends.org).

Saez, Emmanuel. 2016. "U.S. Top One Percent of Income Earners Hit New High in 2015 Amid Strong Economic Growth." Washington Center for Equitable Growth, July 1. Accessed May 5, 2017 (equitablegrowth.org).

Sagarin, Edward. 1975. *Deviants and Deviance.* New York: Praeger.

Sagon, Candy. 2017. "Medicine's Gender Issues." *AARP Bulletin*, January–February, 24–27.

Sahgal, Neha. 2014. "Coke, 'America the Beautiful,' and the Language of Diversity." Pew Research Center, February 3. Accessed November 25, 2014 (www.pewresearch.org).

Sahgal, Neha, and Greg Smith. 2009. "A Religious Portrait of African-Americans." Pew Research Center, Pew Forum on Religion & Public Life Project, January 30. Accessed June 21, 2014 (www.pewforum.org).

Salamone, Frank A. 2005. "Jazz and Its Impact on European Classical Music." *The Journal of Popular Culture* 38 (May): 732–743.

Sales, Nancy Jo. 2016. "How Social Media Is Disrupting the Lives of American Girls." *Time*, February 22–29, 26–27.

Salkin, Allen. 2017. "Take a Load Off. The Robots that Fold Laundry Are Coming." *New York Times*, May 25, B5.

Samuelson, Kate. 2016. "7 Ideas from Other Countries That Could Improve U.S. Elections." *Time*, November 14, 11.

Sanburn, Josh. 2015. "The Joy of Less." *Time*, March 23, 44–50.

Sanburn, Josh. 2016. "The Faults of Oklahoma." *Time*, March 21, 36–41.

Sánchez, Erika L. 2013. "The Challenge of Defining Muslim Feminism." *Huffington Post*, February 10. Accessed October 12, 2013 (www.huffingtonpost.com).

Sandbu, Martin. 2013. "Talkin' 'Bout a Revolution." *Financial Times*, April 19. Accessed October 8, 2014 (www.ft.com).

Sandstrom, Aleksandra. 2015. "Nearly All States Allow Religious Exemptions for Vaccinations." Pew Research Center, July 16. Accessed July 31, 2017 (pewresearch.org).

Sandstrom, Aleksandra. 2016. "Most States Allow Religious Exemptions from Child Abuse and Neglect Laws." Pew Research Center, August 12. Accessed July 31, 2017 (pewresearch.org).

Sandstrom, Aleksandra. 2017. "God or the Divine Is Referenced in Every State Constitution." Pew Research Center, August 17. Accessed August 21, 2017 (pewresearch.org).

Sandstrom, Aleksandra, and Angelina E. Theodorou. 2016. "Many Countries Allow Child Marriage." Pew Research Center, September 12. Accessed July 20, 2017 (pewresearch.org).

Sandstrom, Aleksandra, and Becka A. Alper. 2016. "If the U.S. Had 100 People: Charting Americans' Religious Beliefs and Practices." Pew Research Center, May 4. Accessed July 31, 2017 (pewresearch.org).

Sandstrom, Kent L., Daniel D. Martin, and Gary Alan Fine. 2006. *Symbols, Selves, and Social Reality: A Symbolic Interactionist Approach to Social Psychology and Sociology*, 2nd edition. Los Angeles: Roxbury.

Sang-Hun, Choe. 2016. "American Runs Afoul of Nation's Devotion to Slogans." *New York Times*, March 18, A4.

Sapir, Edward. 1929. "The Status of Linguistics as a Science." *Language* 5 (4): 207–214.

Sassler, Sharon, and Amanda J. Miller. 2011. "Class Differences in Cohabitation Processes." *Family Relations* 60 (April): 163–177.

Sassler, Sharon, Fenaba R. Addo, and Daniel T. Lichter. 2012. "The Tempo of Sexual Activity and Later Relationship Quality." *Journal of Marriage and Family* 74 (August): 708–725.

Saulny, Susan. 2011. "Counting by Race Can Throw Off Some Numbers." *New York Times*, February 10, A1.

Sawchuck, Stephen. 2012. "Teacher Quality, Status Entwined Among Top-Performing Nations." *Education Week* 31 (16): 12–16.

Sawhill, Isabel V., and Joanna Venator. 2014. "Families Adrift: Is Unwed Childbearing the New Norm?" Brookings, October 13. Accessed October 14, 2014 (www.brookings.edu).

Scarlett, W. George, Sophie Naudeau, Dorothy Salonius-Pasternak, and Iris Ponte. 2005. *Children's Play.* Thousand Oaks, CA: Sage.

Schachter, Jason P. 2004. "Geographical Mobility: 2002 to 2003." U.S. Census Bureau, Current Population Reports. Accessed May 15, 2007 (www.census.gov).

Schaeffer, Robert K. 2003. *Understanding Globalization: The Social Consequences of Political, Economic, and Environmental Change,* 2nd edition. Lanham, MD: Rowman & Littlefield.

Schanzenbach, Diane Whitmore, Lauren Bauer, Megan Mumford, and Ryan Nunn. 2016. "Money Lightens the Load." Brookings, December 12. Accessed December 23, 2016 (brookings.edu).

Schehr, Robert C. 1997. *Dynamic Utopia: Establishing Intentional Communities as a New Social Movement.* Westport, CT: Bergin & Garvey.

Schieman, Scott. 2010. "Socioeconomic Status and Beliefs About God's Influence in Everyday Life." *Sociology of Religion* 71 (Spring): 25–51.

Schiesel, Seth. 2011. "Supreme Court Has Ruled; Now Games Have a Duty." *New York Times*, July 29, C1.

Schiffrin, Holly H., Miriam Liss, Haley Miles-McLean, Katherine A. Geary, Mindy J. Erchull, and Taryn Tashner. 2014. "Helping or Hovering? The Effects of Helicopter Parenting on College Students' Well Being." *Journal of Child and Family Studies* 23 (April): 548–557.

Schlesinger, Izchak M. 1991. "The Wax and Wane of Whorfian Views." Pp. 7–44 in *Influence of Language on Culture & Thought,* edited by Robert Cooper and Bernard Spolsky. New York: Mounton de Gruyter.

Schlesinger, Robert. 2011. "Two Takes: Collective Bargaining Rights for Public Sector Unions?" *U.S. News Weekly,* February 25, 15–16.

Schmall, Emily. 2007. "The Cult of Chick-fil-A." *Forbes*, July 23, 80, 83.

Schmidt, Peter. 2008. "2 Studies Raise Questions about Research Based on Student Surveys." *Chronicle of Higher Education*, November 6. Accessed November 9, 2008 (www.chronicle .com).

Schmidt, Peter. 2014. "Supreme Court Exposes Affirmative Action at Colleges to Continued Political Assault." *Chronicle of Higher Education*, May 2, A3–A4.

Schmidt, Peter. 2015. "Racism Is Widespread Among College Students, Researcher Says." *Chronicle of Higher Education,* March 20, A4.

Schneider, Mark. 2013. "Higher Education Pays: But a Lot More for Some Graduates Than for Others." College Measures. Accessed June 21, 2014 (www.collegemeasures.org).

Schneider, Monica C., and Angela L. Bos. 2014. "Measuring Stereotypes of Female Politicians." *Political Psychology* 25 (April): 245–266.

Schnittker, Jason. 2009. "Mirage of Health in the Era of Biomedicalization: Evaluating Change in the Threshold of Illness, 1972–1996." *Social Forces* 87 (June): 2155–2182.

"Schools in Finland 2016." *The Economist*, May 14, 46.

Schramm, J. B., Chad Aldeman, Andrew Rotherham, Rachael Brown, and Jordan Cross. 2013. "Smart Shoppers: The End of the 'College for All' Debate." College Summit. Accessed June 21, 2014 (www.collegesummit .org).

Schrobsdorff, Susanna. 2016. "A Distressing Summer of Workplace Sexism Reminds Us How Far We Have to Go." *Time*, September 5, 55.

Schultz, Ellen E. 2011. *Retirement Heist: How Companies Plunder and Profit from the Nest Eggs of American Workers.* New York: Penguin.

Schur, Edwin M. 1968. *Law and Society: A Sociological View.* New York: Random House.

Schurman-Kauflin, Deborah. 2000. *The New Predator: Women Who Kill.* New York: Algora.

Schutz, Alfred. 1967. *The Phenomenology of the Social World.* Evanston, IL: Northwestern University Press.

Schwartz, Larry. 2014. "8 Mistakes We're Making About Ebola That We Also Made When AIDS Appeared." AlterNet, October 17. Accessed October 18, 2014 (www.alternet.org).

Schwartz, Larry. 2017. "The Depressing Trendlines That Are Driving America's Decline in Life Expectancy." AlterNet, March 9. Accessed March 20, 2017 (alternet.org).

Schwartz, Nelson D. 2017. "Workers Needed, but Drug Testing Takes a Toll." *New York Times,* July 25, A1, A15.

Schwarz, Alan. 2014. "Thousands of Toddlers Are Medicated for A.D.H.D., Report Finds, Raising Worries." *New York Times,* May 17, A11.

Schweizer, Peter. 2013. *Extortion: How Politicians Extract Your Money, Buy Votes, and Line Their Own Pockets.* New York: Houghton Mifflin Harcourt.

Schwyzer, Hugo. 2011. "How Our Sick Culture Makes Girls Think They Have to Be Gorgeous to Be Loved." AlterNet, April 12. Accessed October 29, 2012 (www.alternet.org).

Scommegna, Paola. 2011. "U.S. Parents Who Have Children with More Than One Partner." Population Reference Bureau, June. Accessed March 6, 2012 (www.prb.org).

Scommegna, Paola. 2013. "Aging U.S. Baby Boomers Face More Disability." Population Reference Bureau, March. Accessed May 16, 2014 (www.prb.org).

Scott, Robert E. 2012. "The China Toll." EPI Briefing Paper 345, Economic Policy Institute, August 23. Accessed April 12, 2014 (www.epi.org).

"Second-Generation Americans." 2013. Pew Research Center, February 7. Accessed March 22, 2014 (www.pewresearch.org).

"Select-a-Faith." 2014. *The Economist,* May 17, 30.

Selingo, Jeffrey J. 2013. "The Diploma's Vanishing Value." *Wall Street Journal,* April 27–28, C3.

Seltzer, Sarah. 2012. "Skinny Minnie? Our Culture's Bizarre Obsession with Stick-Thin Women." AlterNet, October 16. Accessed May 15, 2013 (www.alternet.org).

Semega, Jessica L., Kayla R. Fontenot, and Melissa A. Kollar. 2017. "Income and Poverty in the United States: 2016." U.S. Census Bureau, Current Population Reports, September. Accessed September 25, 2017 (census.gov).

Semuels, Alana. 2010. "Can a Prison Save a Town?" *Los Angeles Times,* May 3. Accessed May 4, 2010 (www.latimes.com).

Sengupta, Somini. 2013. "What You Didn't Post, Facebook May Still Know." *New York Times,* March 26, B1.

The Sentencing Project. 2013. "Report of the Sentencing Project to the United Nations Human Rights Committee." August. Accessed January 15, 2014 (www.sentencingproject.org).

The Sentencing Project. 2017. "Trends in U.S. Corrections." March 7. Accessed April 20, 2017 (sentencingproject.org).

Senter, Mary S., and Roberta Spalter-Roth. 2016. "Individual Salary Is Not Enough: Measuring the Well-Being of Recent College Graduates in Sociology." American Sociological Association, Bachelors & Beyond Series, August. Accessed December 23, 2016 (asanet.org).

Senter, Mary S., Nicole Van Vooren, and Roberta Spalter-Roth. 2014. "Sociology, Criminology Concentrations, and Criminal Justice: Differences in Reasons for Majoring, Skills, Activities, and Early Outcomes?" American Sociological Association, April. Accessed October 25, 2014 (www.asanet.org).

Seward, Zachary M. 2012. "58 Countries with Better Voter Turnout Than the United States." Quartz, November 6. Accessed April 20, 2014 (www.qz.com).

"Sex Selection." 2017. *The Economist,* January 21, 49–50.

Shannon-Missal, Larry. 2016. "Tattoo Takeover: Three in Ten Americans Have Tattoos, and Most Don't Stop at Just One." Harris Poll, February 10. Accessed April 14, 2017 (theharrispoll.com).

Shapira, Ian. 2014. "In Arizona, a Navajo High School Emerges as a Defender of the Washington Redskins." *Washington Post,* October 26. Accessed October 27, 2014 (www.washingtonpost.com).

Shapiro, Jenessa R., and Amy M. Williams. 2012. "The Role of Stereotype Threats in Undermining Girls' and Women's Performance and Interest in STEM Fields." *Sex Roles* 66 (February): 175–183.

Shariff, Azim F., and Lara B. Aknin. 2014. "The Emotional Toll of Hell: Cross-National and Experimental Evidence for the Negative Well-Being Effects of Hell Beliefs." *PLoS One* 9 (January): e85251.

Sharifzadeh, Virginia-Shirin. 1997. "Families with Middle Eastern Roots." Pp. 441–482 in *Developing Cross-Cultural Competence: A Guide for Working with Children and Families,* edited by Eleanor W. Lynch and Marci J. Hanson. Baltimore, MD: Paul H. Brookes.

Shauk, Zain, and Bradley Olson. 2014. "Where Frackers Are Friendly Neighbors." *Bloomberg Businessweek,* September 8–14, 37–38.

Shea, Christopher. 2014. "The Liar's 'Tell.'" *The Chronicle Review,* October 17, B6–B9.

SHEEO. 2017. "SHEF: FY 2016 State Higher Education Finance." State Higher Education Executive Officers Association. Accessed July 31, 2017 (sheeo.org).

Shellenbarger, Sue. 2013. "Finding the Just-Right Level of Self-Esteem for a Child." *Wall Street Journal,* February 27, D1, D3.

Shepherd, Julianne Escobedo. 2011. "Hard-Partying Rich Boy Kills Two in Hit-and-Run, Buys Himself Out of a Prison Sentence." AlterNet, June 6. Accessed June 10, 2011 (www.alternet.org).

Sheppard, Kate. 2011. "The Hackers and the Hockey Stick." *Mother Jones,* May/June, 33–45.

Sheppard, Kate. 2013. "Scientific Misconceptions: The Junk Science Behind Anti-Abortion Advocates' Wildest Claims." *Mother Jones,* January/February, 14.

Sherman, Erik. 2015. "Report: Disturbing Drop in Women in Computing Field." *Fortune,* March 26. Accessed July 31, 2017 (fortune.com).

Sherwood, Jessica Holden. 2010. *Wealth, Whiteness, and the Matrix of Privilege: The View from the Country Club.* Lanham, MD: Lexington Books.

Shibutani, Tamotsu. 1986. *Social Process: An Introduction to Sociology.* Berkeley: University of California Press.

Shilo, Guy, and Riki Savaya. 2011. "Effects of Family and Friend Support on LGB Youths' Mental Health and Sexual Orientation Milestones." *Family Relations* 60 (July): 318–330.

Shorrocks, Anthony, James B. Davies, Rodrigo Lluberas, and Antonios Koutsoukis. 2016. "Global Wealth Report 2016." Credit Suisse, Research Institute, November. Accessed May 5, 2017 (credit-suisse.com).

Shorto, Russell. 2008. "No Babies?" *New York Times,* June 29, 34.

Shriver, Maria. 2014. *The Shriver Report: A Woman's Nation Pushes Back from the Brink.* New York: Rosetta Books.

Shteir, Rachel. 2013. "Feminism Fizzles." *The Chronicle Review,* February 1, B6–B9.

Siegman, Aron W., and Stanley Feldstein, eds. 1987. *Nonverbal Behavior and Communication.* Hillsdale, NJ: Lawrence Erlbaum Associates.

Sifferlin, Alexandra. 2014. "Rise of the Mumps: What's Behind the New Cases." *Time,* April 7, 24.

Silverman, Rachel E. 2003. "Provisions Boost Rights of Couples Living Together." *Wall Street Journal,* March 5, D1.

Silverman, Rachel E. 2013. "Tracking Sensors Invade the Workplace." *Wall Street Journal,* March 7, B1, B2.

Silverstein, Jason. 2015. "Dylann Roof Obsessed over Trayvon Martin and 'White Race'." *New York Daily News,* June 20. Accessed April 16, 2017 (nydailynews.com).

Simmel, Georg. 1902. "The Number of Members as Determining the Sociological Form of the Group." *American Journal of Sociology* 8 (July): 1–46.

Simon, Paula. 2013. "AP Is Not for Everyone." *Baltimore Sun,* August 24, 17.

Singal, Pooja, Adi Rattner, and Meghana Desale. 2014. "Court Ruling Limits a Doctor's Options." *Baltimore Sun,* July 17, 17.

Singer, Natasha. 2013. "Group Criticizes Learning Apps for Babies." *New York Times,* August 8, B8.

Singer, Natasha. 2013. "They Loved Your G.P.A. Then They Saw Your Tweets." *New York Times,* November 10, BU3.

Singh, Tushar, et al. 2016. "Vital Signs: Exposure to Electronic Cigarette Advertising Among Middle School and High School Students—United States, 2014." *MMWR* 64 (52): 1403–1408.

Sjoberg, Gideon. 1960. *The Preindustrial City: Past and Present.* Glencoe, IL: Free Press.

Skinner, Victor. 2013. "What's the Practice Called 'Passing the Trash' That Allows Teachers to Continue Sexually Targeting Kids?" The Blaze, March 27. Accessed April 3, 2017 (theblaze.com).

Skoning, Gerald D. 2013. "How Congress Puts Itself Above the Law." *Wall Street Journal,* April 16, A15.

Skoumal, Robert J., Michael R. Brudzinski, and Brian S. Currie. 2015. "Earthquakes Induced by Hydraulic Fracturing in Poland Township, Ohio." *Bulletin of the Seismological Society of America* 105 (February): 189–197.

Slater, Dan. 2013. "Darwin Was Wrong About Dating." *New York Times,* January 13, SR1.

Smedley, Audrey. 2007. *Race in North America: Origin and Evolution of a Worldview,* 3rd edition. Boulder, CO: Westview Press.

Smeeding, Tim. 2017. "Cash Matters and Place Matters: A Child Poverty Plan That Capitalizes on New Evidence." *Pathways*, Spring, 14–20.

Smelser, Neil J. 1962. *Theory of Collective Behavior.* New York: Free Press.

Smelser, Neil J. 1988. "Social Structure." Pp. 103–129 in *Handbook of Sociology*, edited by Neil J. Smelser. Newbury Park, CA: Sage.

Smith, Aaron. 2016. "Gig Work, Online Selling and Home Sharing." Pew Research Center, November 17. Accessed June 16, 2017 (pewresearch.org).

Smith, Aaron. 2017. "Record Shares of Americans Now Own Smartphones, Have Home Broadband." Pew Research Center, January 12. Accessed March 14, 2017 (pewresearch.org).

Smith, Aaron, and Monica Anderson. 2015. "5 Facts About Online Dating." Pew Research Center, April 20. Accessed August 14, 2015 (pewresearch.org).

Smith, Aaron, and Monica Anderson. 2016. "5 Facts About Online Dating." Pew Research Center, February 29. Accessed May 19, 2017 (pewresearch.org).

Smith, Aaron, Janna Anderson, and Lee Rainie. 2014. "AI, Robotics, and the Future of Jobs." Pew Research Center, August 6. Accessed October 8, 2014 (www.pewresearch.org).

Smith, Adam. 1776/1937. *An Inquiry into the Nature and Causes of the Wealth of Nations.* New York: Modern Library.

Smith, Dorothy E. 1987. *The Everyday World as Problematic: A Feminist Sociology.* Toronto: University of Toronto Press.

Smith, Jane I. 1994. "Women in Islam." Pp. 303–325 in *Today's Woman in World Religions*, edited by Arvind Sharma. Albany: State University of New York Press.

Smith, N., and A. Leiserowitz. 2013. "American Evangelicals and Global Warming." *Global Environmental Change* 23 (October): 1009–1017.

Smith, Oliver. 2017. "Mapped: The World's Most Polluted Countries." *Telegraph*, January 2. Accessed August 20, 2017 (telegraph.co.uk).

Smith, Ray A. 2013. "A Closet Filled with Regrets." *Wall Street Journal*, April 18, D1, D4.

Smith, Samantha. 2017. "Why People Are Rich and Poor: Republicans and Democrats Have Very Different Views." Pew Research Center, May 2. Accessed July 4, 2017 (pewresearch.org).

Smith, Stacy L., Katherine M. Pieper, Amy Granados, and Marc Choueiti. 2010. "Assessing Gender-Related Portrayals in Top-Grossing G-Rated Films." *Sex Roles* 62 (June): 774–786.

Smith, Stacy L., Marc Choueiti, Katherine Pieper, Traci Gillig, Carmen Lee, and Dylan DeLuca. 2015. "Inequality in 700 Popular Films: Examining Portrayals of Gender, Race, & LGBT Status from 2007 to 2014." USC Annenberg School for Communication and Journalism. Accessed September 29, 2015 (annenberg.usc.edu).

Smith, Tom W., Peter V. Marsden, and Michael Hout. 2011. *General Social Survey, 1972–2010, Cumulative File Codebook.* Ann Arbor, MI: Inter-University Consortium for Political and Social Research. Accessed February 13, 2014 (www.icpsr.umich.edu).

Snider, Laureen. 1993. "Regulating Corporate Behavior." Pp. 177–210 in *Understanding Corporate Criminality*, edited by Michael B. Blankenship. London: Garland Press.

Snipp, C. Matthew. 1996. "A Demographic Comeback for American Indians." *Population Today* 24 (November): 4–5.

Snyder, Thomas D., and Sally A. Dillow. 2013. *Digest of Education Statistics 2012.* U.S. Department of Education, National Center for Education Statistics, December. Accessed June 21, 2014 (www.nces.ed.gov).

Snyder, Thomas D., Cristobal de Brey, and Sally A. Dillow. 2016. *Digest of Education Statistics 2015.* National Center for Education Statistics, December. Accessed May 21, 2017 (nces.ed.gov).

Soble, Jonathan. 2015. "A Sprawl of Ghost Homes in Aging Tokyo Suburbs." *New York Times*, August 24, A1, A3

Soldatov, Andrei. 2016. "Putin Has Finally Reincarnated the KGB." Foreign Policy, September 21. Accessed June 21, 2017 (foreignpolicy.com).

"Social Media Fact Sheet." 2017. Pew Research Center, January 12. Accessed March 14, 2017 (pewresearch.org).

Sonfield, Adam, and Kathryn Kost. 2013. "Public Costs from Unintended Pregnancies and the Role of Public Insurance Programs in Paying for Pregnancy Infant Care: Estimates for 2008." Guttmacher Institute, October. Accessed May 16, 2014 (www.guttmacher.org).

Spalter-Roth, Roberta, Nicole Van Vooren, and Mary S. Senter. 2013. "Social Capital for Sociology Majors: Applied Activities and Peer Networks." American Sociological Association. Accessed October 12, 2013 (www.asanet.org).

Sparks, Sarah D. 2016. "Summing Up Results from TIMSS, PISA." *Education Week*, December 13. Accessed July 31, 2017 (edweek.org).

Speigel, Lee. 2013. "48 Percent of Americans Believe UFOs Could Be ET Visitations." *Huffington Post*, September 11. Accessed September 12, 2013 (www.huffingtonpost.com).

Spencer, Herbert. 1862/1901. *First Principles.* New York: P. F. Collier & Son.

Sperry, Shelley. 2008. "Ozone Defense." *National Geographic* 214 (October).

Spiller, Michael W., Dita Broz, Cyprian Wejnert, Lina Nerlander, and Gabriela Paz-Bailey. 2015. "HIV Infection and HIV-Associated Behaviors Among Persons Who Inject Drugs—20 Cities, United States, 2012." *MMWR* 64 (10): 270–275.

Spitznagel, Eric. 2013. "Men on the Run." *Bloomberg Businessweek*, March 25–31, 82.

Spivak, Howard R., E. Lynn Jenkins, Kristi VanAudenhove, Debbie Lee, Mim Kelly, and John Iskander. 2014. "CDC Grand Rounds: A Public Health Approach to Prevention of Intimate Partner Violence." *MMWR* 63 (January 17): 38–41.

Spotts, Peter N. 2009. "New Climate Change Signal: Oceans Turning Acidic." *Christian Science Monitor*, December 9. Accessed December 12, 2010 (www.csmonitor.com).

Spradley, J. P., and M. Phillips. 1972. "Culture and Stress: A Quantitative Analysis." *American Anthropologist* 74 (3): 518–529.

Sprigg, Peter. 2011. "Marriage's Public Purpose." *Baltimore Sun*, February 2, 27.

Srivastava, Mehul. 2013. "In Bangladesh, Outside Inspectors Are Still MIA." *Bloomberg Businessweek*, November 4–10, 28–30.

Srivastava, Mehul, and Kartikay Mehrotra. 2013. "India's Second-Class Citizens." *Bloomburg Businessweek*, January 14–20, 10–12.

Stager, Curt. 2017. "Sowing Climate Doubt Among Schoolteachers." *New York Times*, April 27, A27.

Stanley, Megan, Ife Floyd, and Misha Hill. 2016. "TANF Cash Benefits Have Fallen by More Than 20 Percent in Most States and Continue to Erode." Center on Budget and Policy Priorities, October 17. Accessed May 5, 2017 (cbpp.org).

Starr, Christine R., and Gail M. Ferguson. 2012. "Sexy Dolls, Sexy Grade-Schoolers? Media and Maternal Influences on Young Girls' Self-Sexualization." *Sex Roles* 67 (October): 463–476.

Startz, Dick. 2016. "Teacher Pay Around the World." Brookings, July 11. Accessed July 20, 2017 (brookings.edu).

"The State of American Jobs." 2016. Pew Research Center, October 6. Accessed June 20, 2017 (pewresearch.org).

"The State of Arab Men." 2017. *The Economist*, May 6, 46.

Statista. 2016. "Daily Media Use Per Capita in the United States in Spring 2016." Accessed March 13, 2017 (statista.com).

Statista. 2017. "Number of Retail Clinics in the United States from 2008 to 2018." The Statistics Portal. Accessed April 3, 2017 (statista.com).

Steel, Emily. 2015. "For Cable Network, Crime Certainly Pays." *New York Times*, January 5, B1.

Steele, Eurídice Martínez, Larissa Galastri Baraldi, Maria Laura da Costa Louzada, Jean-Claude Moubarac, Dariush Mozaffarian, and Carlos Augusto Monteiro. 2016. "Ultra-Processed Foods and Added Sugars in the US Diet: Evidence from a Nationally Representative Cross-Sectional Study." *BMJ Open* 6 (3): 1–9.

Steffensmeier, Darrell J., Jennifer Schwartz, and Michael Roche. 2013. "Gender and Twenty-First-Century Corporate Crime: Female Involvement and the Gender Gap in Enron-Era Corporate Frauds." *American Sociological Review* 78 (June): 448–476.

Stein, Joel. 2015. "Baby, You Can Drive My Car." *Time*, February 9, 32–40.

Stein, Rob. 2017. "Scientists Precisely Edit DNA in Human Embryos to Fix a Disease Gene." NPR, August 2. Accessed September 2, 2017 (npr.org).

Steinberg, Julia R., and Lawrence B. Finer. 2011. "Examining the Association of Abortion History and Current Mental Health: A Reanalysis of the National Comorbidity Survey Using a Common-Risk-Factors Model." *Social Science & Medicine* 72 (1): 72–82.

Steinmetz, Katy. 2014. "Disrupted: The Tech Wealth Transforming San Francisco Has Unemployment Down, Evictions Up and Tensions Flaring." *Time*, February 10, 32–35.

Steinmetz, Katy. 2015. "Help! My Parents Are Millennials." *Time*, October 26, 36–43.

Steinmetz, Katy. 2015. "What the Toy Aisle Can Teach Us About Gender Parity." *Time*, August 24, 25–26.

Steinmetz, Katy. 2017. "Infinite Identities." *Time*, March 27, 48–54.

Stepan-Norris, Judith. 2015. "Social Justice & the Next Upward Surge for Unions." *Contexts* 14 (Spring): 46–51.

Stepler, Renee. 2017. "Led by Baby Boomers, Divorce Rates Climb for America's 50+ Population." Pew Research Center, March 9. Accessed July 20, 2017 (pewresearch.org).

Stepler, Renee. 2017. "Number of U.S. Adults Cohabiting with a Partner Continues to Rise, Especially Among Those 50 and Older." Pew Research Center, April 6. Accessed July 20, 2017 (pewresearch.org).

Stepler, Renee, and Mark Hugo Lopez. 2016. "U.S. Latino Population Growth and Dispersion Has Slowed Since Onset of the Great Recession." Pew Research Center, October 26. Accessed September 8, 2017 (pewresearch.org).

Sternbergh, Adam. 2008. "Why White People Like 'Stuff White People Like.'" *New Republic*, March 17. Accessed November 6, 2008 (www.tnr.com).

Steuerle, C. Eugene, and Caleb Quakenbush. 2013. "Social Security and Medicare Taxes and Benefits over a Lifetime: 2013 Update." Urban Institute, November. Accessed July 20, 2017 (urban.org).

"Steve Jobs: Adopted Child Who Never Met His Biological Father." 2011. *Telegraph*, October 6. Accessed March 22, 2014 (telegraph.co.uk).

Stillars, Alan L. 1991. "Behavioral Observation." Pp. 197–218 in *Studying Interpersonal Interaction*, edited by B. M. Montgomery and S. Duck. New York: Guilford Press.

Stokes, Bruce. 2017. "As Republicans' Views Improve, Americans Give the Economy Its Highest Marks Since Financial Crisis." Pew Research Center, April 3. Accessed June 16, 2017 (pewresearch.org).

Stokes, Bruce. 2017. "What It Takes to Truly Be 'One of Us'." Pew Research Center, February 1. Accessed February 5, 2017 (pewresearch.org).

Stone, Geoffrey R. 2016. "Free Expression in Peril." *Chronicle of Higher Education*, September 16, B9–B11.

Strasburger, Victor C., and Marjorie J. Hogan. 2013. "Policy Statement: Children, Adolescents, and the Media." *Pediatrics* 132 (5): 958–961.

Straus, Murray A. 2011. "Gender Symmetry and Mutuality in Perpetration of Clinical-Level Partner Violence: Empirical Evidence and Implications for Prevention and Treatment." *Aggression and Violent Behavior* 16 (July–August): 279–288.

Straus, Murray A. 2014. "Addressing Violence by Female Partners Is Vital to Prevent or Stop Violence Against Women: Evidence from the Multisite Batterer Intervention Evaluation." *Violence Against Women* 20 (7): 889–899.

"Streaking." 2005. Wikipedia Encyclopedia. Accessed September 8, 2005 (www.wikipedia.org).

Strean, William B. 2009. "Remembering Instructors: Play, Pain, and Pedagogy." *Qualitative Research in Sport and Exercise* 1 (November): 210–220.

Streib, Jessi. 2011. "Class Reproduction by Four Year Olds." *Qualitative Sociology* 34 (June): 337–352.

Stroebe, Margaret, Maarten van Son, Wolfgang Stroebe, Rolf Kleber, Henk Schut, and Jan van den Bout. 2000. "On the Classification and Diagnosis of Pathological Grief." *Clinical Psychology Review* 20 (January): 57–75.

Strom, Stephanie. 2014. "Coca-Cola to Remove an Ingredient Questioned by Consumers." *New York Times*, May 6, B6.

Strom, Stephanie. 2015. "The Fried Chicken Is O.K., but the Cups Are Delicious." *New York Times*, February 26, B3.

Strom, Stephanie. 2016. "At Chobani, It's Not Just the Yogurt That's Rich." *New York Times*, April 27, B1, B5.

The Student Loan Report. 2017. "Student Loan Debt Statistics 2017." Accessed July 31, 2017 (studentloans.net).

"Study Shows Positive Results from Early Head Start Program." 2002. U.S. Department of Health and Human Services, June 3. Accessed July 10, 2002 (www.hhs.gov).

Stullich, Stephanie, Ivy Morgan, and Oliver Schak. 2016. "State and Local Expenditures on Corrections and Education." U.S. Department of Education, July. Accessed April 4, 2017 (ed.gov).

Sturm, Roland, and Ruopeng An. 2014. "Obesity and Economic Environments." *CA: A Cancer Journal for Clinicians* 65 (5): 337–350.

Sudarkasa, Niara 2007. "African American Female-Headed Households: Some Neglected Dimensions." Pp. 172–182 in *Black Families*, 4th edition, edited by Harriette Pipes McAdoo. Thousand Oaks, CA: Sage.

Suddath, Claire. 2013. "So Hard to Say Goodbye." *Bloomberg Businessweek*, November 4–10, 79–81.

Sullivan, Andrew. 2011. "Why Gay Marriage Is Good for Straight America." *Newsweek*, July 25, 12–14.

Sullivan, Deirdre. 2017. "How Much Does a Tiny House Really Cost?" The Spruce, June 1. Accessed August 20, 2017 (thespruce.com).

Summers, Nick. 2010. "Do Fines Ever Make Corporations Change?" Newsweek, November 13, 56.

Sumner, William G. 1906. *Folkways*. New York: Ginn.

"Sun Trust: Many Higher Income Households Living Paycheck-to-Paycheck." 2015. Sun Trust Banks, Inc. April 16. Accessed October 7, 2015 (suntrust.com).

Sunstein, Cass R. 2009. *On Rumors: How Falsehoods Spread, Why We Believe Them, What Can Be Done*. New York: Farrar, Straus and Giroux.

"Super Bull Sunday." 2015. Snopes.com, January 30. Accessed July 22, 2015 (snopes.com).

Superville, Darlene. 2017. "Trump Lets States Block Some Planned Parenthood Money." *Baltimore Sun*, April 14, 4.

Sustainable Packaging Coalition. 2016. "2015–16 Centralized Study on Availability of Plastic Recycling." Accessed August 20, 2017 (sustainablepackaging.org).

Sutherland, Edwin H. 1949. *White Collar Crime*. New York: Holt, Rinehart, and Winston.

Sutherland, Edwin H., and D. R. Cressey. 1970. *Criminology*, 8th edition. Philadelphia: Lippincott.

Sutton, Philip W. 2000. *Explaining Environmentalism: In Search of a New Social Movement*. Burlington, VT: Ashgate.

Swift, Art. 2013. "Honesty and Ethics Rating of Clergy Slides to New Low." Gallup, December 16. Accessed April 20, 2014 (www.gallup.com).

Swift, Art. 2017. "Americans See US World Standing as Worst in a Decade." Gallup, February 10. Accessed March 1, 2017 (gallup.com).

Swift, Art. 2017. "Americans' Worries About Race Relations at Record High." Gallup, March 15. Accessed June 20, 2017 (gallup.com).

Swift, Art. 2017. "More Americans Say Immigrants Help Rather Than Hurt Economy." Gallup, June 29. Accessed June 27, 2017 (gallup.com).

Sykes, Selina. 2015. "Indian Man Has 39 Wives, 94 Children and 33 Grandchildren All Living Under the SAME Roof." Express, November 9. Accessed August 15, 2017 (express.co.uk).

Szarota, Piotr. 2010. "The Mystery of the European Smile: A Comparison Based on Individual Photographs Provided by Internet Users." *Journal of Nonverbal Behavior* 34 (December): 249–256.

T

Tabachnick, Rachel. 2011. "The 'Christian' Dogma Pushed by Religious Schools That Are Supported by Your Tax Dollars." AlterNet, May 23. Accessed July 12, 2011 (www.alternet.org).

Tabuchi, Hiroko. 2015. "A Tiara? No Thanks." *New York Times*, October 29, B1, B8.

Tabuchi, Hiroko, and Neal E. Boudette. 2017. "Four Carmakers Knew of Hazard, Suit Says." *New York Times*, February 28, B1–B2.

"Tackling Poverty." 2015. *The Economist*, September 5, 31–32.

Taibbi, Matt. 2014. *The Divide: American Injustice in the Age of the Wealth Gap*. New York: Spiegel & Grau.

Tajfel, Henri. 1982. "Social Psychology of Intergroup Relations." *Annual Review of Psychology* 33 (February): 1–39.

Tannen, Deborah. 1990. *You Just Don't Understand: Women and Men in Conversation*. New York: Ballantine.

Tannenbaum, Frank. 1938. *Crime and the Community*. New York: Columbia University Press.

Tanur, Judith M. 1994. "The Trustworthiness of Survey Research." *Chronicle of Higher Education*, May 25, B1-B3.

"Tattoos in the Workplace: Ink Blots." 2014. *The Economist*, August 2, 22.

Taub, Amanda, and Brendan Nyhan. 2017. "Why Objectively False Things Continue to Be Believed." *New York Times*, March 22, A18.

Tavris, Carol. 2013. "How Psychiatry Went Crazy." *Wall Street Journal*, May 18, C5.

Tavris, Carol. 2013. "The Experiments That Still Shock." *Wall Street Journal*, September 6, C1, C7.

Taylor, Bruce G., and Elizabeth A. Mumford. 2016. "A National Descriptive Portrait of Adolescent Relationship Abuse: Results from the National Survey on Teen Relationships and Intimate Violence." *Journal of Interpersonal Violence* 3 (6): 963–988.

Taylor, Carl S., and Pamela R. Smith. 2013. "The Attraction of Gangs: How Can We Reduce It?" Pp. 19–29 in *Changing Course: Preventing Gang Membership*, edited by Thomas R. Simon, Nancy M. Witter, and Reshma R. Mahendra. National Institute of Justice and Centers for Disease Control and Prevention. Accessed December 18, 2013 (www.nij.gov).

Taylor, Frederick W. 1911/1967. *The Principles of Scientific Management*. New York: W. W. Norton & Company.

Taylor, Jay. 1993. *The Rise and Fall of Totalitarianism in the Twentieth Century*. New York: Paragon House.

Taylor, Paul, and Mark H. Lopez. 2013. "Skepticism About the Census Voter Turnout Finding." Pew Research Center, May 15. Accessed August 8, 2014 (www.pewresearch.org).

Taylor, Paul, Carroll Doherty, Rich Morin, and Kim Parker. 2014. "Millennials in Adulthood: Detached from Institutions, Networked with Friends." Pew Research Center, March 7. Accessed June 21, 2014 (www.pewresearch.org).

Taylor, Paul, et al. 2012. "The Lost Decade of the Middle Class." Pew Research Center, August 22. Accessed February 14, 2014 (www.pewsocialtrends.org).

Taylor, Paul, et al. 2013. "The Rise of Asian Americans." Pew Research Center, April 4. Accessed May 16, 2014 (www.pewsocialtrends.org).

Taylor, Paul, Rich Morin, Kim Parker, and D'Vera Cohn. 2009. "Growing Old in America: Expectations vs. Reality." Pew Research Center, June 29. Accessed April 30, 2010 (www.pewsocialtrends.org).

Taylor, Susan C. 2003. *Brown Skin: Dr. Susan Taylor's Prescription for Flawless Skin, Hair, and Nails.* New York: HarperCollins.

Tefferi, Andrew. 2015. "In Support of a Patient-Driven Initiative and Petition to Lower the High Price of Cancer Drugs." *Mayo Clinic Proceedings* 90 (8): 996–1000.

Tenbrunsel, Ann, and Jordan Thomas. 2015. "The Street, the Bull and the Crisis: A Survey of the US & UK Financial Services Industry." University of Notre Dame and Lebaton Sucharow, May. Accessed September 2, 2015 (www.secwhistlebloweradvocate.com).

Terando, Adam J., Jennifer Costanza, Curtis Belyea, Robert R. Dunn, Alexa McKerrow, and Jaime A. Collazo. 2014. "The Southern Megalopolis: Using the Past to Predict the Future of Urban Sprawl in the Southeast U.S." *PLoS One* 9 (July): e102261.

Tétreault, Mary Ann. 2001. "A State of Two Minds: State Cultures, Women, and Politics in Kuwait." *International Journal of Middle East Studies* 33 (May): 203–220.

Tharoor, Ishaan. 2015. "A Politician in Finland Declared War on Multiculturalism. This Is How His Country Responded." *Washington Post,* July 31. Accessed July 31, 2015 (washingtonpost.com).

Theodorou, Angelina E. 2014. "64 Countries Have Religious Symbols on Their National Flags." Pew Research Center, November 25. Accessed February 5, 2017 (pewresearch.org).

Theodorou, Angelina, and Peter Henne. 2014. "Religious Police Found in Nearly One-in-Ten Countries Worldwide." Pew Research Center, March 19. Accessed June 21, 2014 (www.pewresearch.org).

Thibaut, John W., and Harold H. Kelley. 1959. *The Social Psychology of Groups.* New York: Wiley.

Thio, Alex D., Jim D. Taylor, and Martin D. Schwartz. 2012. *Deviant Behavior*, 11th edition. Upper Saddle River, NJ: Pearson.

Thomas, Anita Jones, Jason Daniel Hacker, and Denada Hoxha. 2011. "Gendered Racial Identity of Black Young Women." *Sex Roles* 64 (April): 530–542.

Thomas, W. I., and Dorothy Swaine Thomas. 1928. *The Child in America.* New York: Alfred A. Knopf.

Thompson, Arienne. 2010. "16, Pregnant . . . and Famous: Teen Moms Are Newest Stars." *USA Today,* November 23. Accessed November 24, 2011 (www.usatoday.com).

Thorpe-Moscon, Jennifer, and Alixandra Pollock. 2014. "Feeling Different: Being the 'Other' in US Workplaces." Catalyst. Accessed March 22, 2014 (www.catalyst.org).

Tilly, Charles. 1978. *From Mobilization to Revolution.* Reading, MA: Addison-Wesley.

Timberg, Craig, and Ellen Nakashima. 2013. "Amid NSA Spying Revelations, Tech Leaders Call for New Restraints on Agency." *Washington Post,* November 1. Accessed November 10, 2013 (www.washingtonpost.com).

Timberlake, Cotten. 2013. "Don't Even Think About Returning This Dress." *Bloomberg Businessweek*, September 30–October 6, 29–31.

Timmermans, Stefan, and Hyeyoung Oh. 2010. "The Continued Social Transformation of the Medical Profession." *Journal of Health and Social Behavior* 51 (March): S94–S109.

Tobar, Hector. 2009. "Language as a Bridge and an Identity." *Los Angeles Times,* September 22. Accessed September 24, 2009 (www.latimes.com).

Tomasetti, Cristian, and Bert Vogelstein. 2015. "Variation in Cancer Risk Among Tissues Can Be Explained by the Number of Stem Cell Divisions." *Science* 347 (6217): 78–81.

"Top Spenders." 2017. Center for Responsive Politics, Open Secrets. Accessed June 26, 2017 (opensecrets.org).

Tormey, Simon. 1995. *Making Sense of Tyranny: Interpretations of Totalitarianism.* New York: Manchester University Press.

Torpey, Elka. 2014. "Got Skills? Think Manufacturing." U.S. Bureau of Labor Statistics, June. Accessed June 5, 2017 (bls.gov).

Torpey, Elka, and Andrew Hogan. 2016. "Working in a Gig Economy." U.S. Bureau of Labor Statistics, May. Accessed June 5, 2017 (bls.gov).

Toth, Emily. 2011. "No Girls Aloud." *Chronicle of Higher Education,* April 1, A26.

Touraine, Alain. 2002. "The Importance of Social Movements." *Social Movement Studies* 1 (April): 89–96.

Tozzi, John. 2014. "Who Should Pay the Bill for Wonder Drugs?" *Bloomberg Businessweek*, July 10–16, 27–28.

Tozzi, John. 2017. "Health Care Has a Goldilocks Problem." *Bloomberg Businessweek*, July 31, 38–39.

Tracer, Zachary, and Hannah Recht. 2017. "Obamacare's Problems Still Need Solving." *Bloomberg Businessweek*, July 24, 37–38.

Traister, Rebecca. 2016. *All the Single Ladies: Unmarried Women and the Rise of an Independent Nation*. New York: Simon & Schuster.

Tran, Van C. 2016. "Social Mobility Among Second-Generation Latinos." *Contexts* 15 (Spring): 28–33.

"Transgender Rights." 2015. *The Economist*, June 13, 31.

Trottman, Melanie. 2013. "Religious Discrimination Claims on the Rise." *Wall Street Journal*, October 28, B1, B4.

Truman, Jennifer L., and Erica L. Smith. 2012. "Prevalence of Violent Crime Among Households with Children, 1993–2010." Bureau of Justice Statistics, September. Accessed May 16, 2014 (www.bjs.gov).

Truman, Jennifer L., and Lynn Langton. 2014. "Criminal Victimization, 2013." Bureau of Justice Statistics, September. Accessed November 6, 2014 (www.bjs.gov).

Truman, Jennifer L., and Lynn Langton. 2015. "Criminal Victimization, 2014." Bureau of Justice Statistics, August. Accessed September 2, 2015 (bjs.gov).

Truman, Jennifer L., and Rachel E. Morgan. 2014. "Nonfatal Domestic Violence, 2003–2012." Bureau of Justice Statistics, April. Accessed May 16, 2014 (www.bjs.gov).

Truman, Jennifer L., and Rachel E. Morgan. 2016. "Criminal Victimization, 2015." Bureau of Justice Statistics, October. Accessed December 15, 2016 (bjs.gov).

Trumbull, Mark. 2015. "Life in the Squeezed Middle." *Christian Science Monitor Weekly*, November 2, 26–32.

Tsai, Tyjen. 2012. "China Has Too Many Bachelors." Population Reference Bureau, January. Accessed September 19, 2014 (www.prb.org).

Tsukayama, Hayley. 2014. "High-Tech Upgrades May Let Aging Boomers Live Independently in Their Own Homes Longer." *Washington Post*, January 20. Accessed October 8, 2014 (www.washingtonpost.com).

Tucker-Drob, Elliot M., and Timothy C. Bates. 2016. "Large Cross-National Differences in Gene x Socioeconomic Status Interaction on Intelligence." *Psychological Science* 27 (2): 138–149.

Tumin, Melvin M. 1953. "Some Principles of Stratification: A Critical Analysis." *American Sociological Review* 18 (August): 387–393.

Turk, Austin T. 1969. *Criminality and the Legal Order.* Chicago: Rand-McNally.

Turk, Austin T. 1976. "Law as a Weapon in Social Conflict." *Social Problems* 23 (February): 276–291.

Turkle, Sherry. 2015. *Reclaiming Conversation: The Power of Talk in a Digital Age.* New York: Penguin Press.

Turner, Margery A., Rob Santos, Diane K. Levy, Doug Wissoker, Claudia Aranda, and Rob Pitingolo. 2013. "Housing Discrimination Against Racial and Ethnic Minorities 2012." The Urban Institute, June. Accessed March 22, 2014 (www.huduser.org).

Turner, Ralph H., and Lewis M. Killian. 1987. *Collective Behavior,* 3rd edition. Englewood Cliffs, NJ: Prentice Hall.

Twenge, Jean M., Ryne A. Sherman, and Brooke E. Wells. 2017. "Declines in Sexual Frequency Among American Adults, 1989–2014." *Archives of Sexual Behavior* 46 (January): 1–13.

Tynan, Michael A., Jonathan R. Polansky, Kori Titus, Renata Atayeva, and Stanton A. Glantz. 2017. "Tobacco Use in Top-Grossing Movies—United States, 2010–2016." *MMWR* 66 (26): 681–686.

U

Umberson, Debra, Julie S. Olson, Robert Crosnoe, Hui Liu, Tetyana Pudrovska, and Rachel Donnelly. 2017. "Death of Family Members as an Overlooked Source of Racial Disadvantage in the United States." *Proceedings of the National Academy of Sciences* 114 (5): 915–920.

UN Women. 2014. "Facts and Figures: Ending Violence Against Women." Accessed March 2, 2014 (www.unwomen.org).

"Uncle Sam Overpays." 2017. *The Economist*, June 22. Accessed August 15, 2017 (economist.com).

Underhill, Wendy. 2017. "Voter Identification Requirements/ Voter ID Laws." National

Conference of State Legislatures, June 5. Accessed June 26, 2017 (ncsl.org).

UNICEF. 2016. "Female Genital Mutilation and Cutting." Accessed May 25, 2017 (data.unicef.org).

"Union Members—2016." 2017. U.S. Bureau of Labor Statistics, January 26. Accessed June 5, 2017 (bls.gov).

United Human Rights Council. 2004. "History of Genocide." Accessed April 29, 2004 (www.unitedhumanrights.org).

United Nations. 2014. *World Urbanization Prospects: The 2014 Revision.* Department of Economics and Social Affairs, Population Division. Accessed September 19, 2014 (www.esa.un.org).

United Nations. 2015. "The Millennium Development Goals Report 2015." Accessed September 29, 2015 (un.org).

United Nations. 2015. *The World's Women 2015: Trends and Statistics. Department of Economic and Social Affairs.* Accessed May 13, 2017 (un.org).

United Nations Committee on the Rights of the Child. 2014. "Concluding Observations on the Second Periodic Report of the Holy See." February 25. Accessed June 21, 2014 (www.ohchr.org).

United Nations Department of Economic and Social Affairs. 2012. "World Urbanization Prospects: The 2011 Revision, Highlights." March. Accessed June 23, 2012 (www.esa.un.org).

United Nations Development Program. 2013. "Humanity Divided: Confronting Inequality in Developing Countries." November. Accessed February 6, 2014 (www.undp.org).

United Nations High Commissioner for Refugees. 2017. "Global Trends: Forced Displacement in 2016." Accessed August 20, 2017 (unhcr.org).

United Nations Office on Drugs and Crime. 2017. "Statistics." Accessed April 23, 2017 (unodc.org).

United Nations Population Division. 2017. "World Population Prospects: The 2017 Revision." Department of Economic and Social Affairs. Accessed August 20, 2017 (esa.un.org).

U.S. Census Bureau. 2017. "Voting and Registration in the Election of November 2016." May 12. Accessed June 5, 2017 (census.gov).

U.S. Census Bureau. 2008. "An Older and More Diverse Nation by Midcentury." August 14. Accessed August 18, 2008 (www.census.gov).

U.S. Census Bureau. 2012. *Statistical Abstract of the United States: 2012,* 131st edition. Washington, DC: Government Printing Office.

U.S. Census Bureau. 2016. "New Census Data Show Differences Between Urban and Rural Populations." News Release, December 8. Accessed August 20, 2017 (census.gov).

U.S. Census Bureau, Current Population Survey. 2013. "Annual Social and Economic Supplements." November. Accessed October 20, 2014 (www.census.gov).

U.S. Census Bureau News. 2015. "Census Bureau Reports at Least 350 Languages Spoken in U.S. Homes." November 3. Accessed June 20, 2017 (census.gov).

U.S. Census Bureau Newsroom. 2017. "The South Is Home to 10 of the 15 Fastest-Growing Large Cities." May 25. Accessed August 20, 2017 (census.gov).

U.S. Department of Education. 2012. "Degrees and Other Formal Awards Conferred." Higher Education General Information Survey (HEGIS). Accessed July 22, 2017 (nces.ed.gov).

U.S. Department of Education. 2016. "2013–2014 Civil Rights Data Collection: A First Look." Office for Civil Rights, October 28. Accessed July 30, 2017 (ocrdata.ed.gov).

U.S. Department of Health and Human Services. 2013. *Child Maltreatment 2012.* Administration for Children and Families, Children's Bureau. Accessed May 16, 2014 (www.acf.hhs.gov).

U.S. Department of Health and Human Services. 2014. "The Health Consequences of Smoking—50 Years of Progress. A Report of the Surgeon General." Office of the Surgeon General. Accessed July 30, 2014 (www.cdc.gov).

U.S. Department of Health and Human Services. 2016. "E-Cigarette Use Among Youth and Young Adults. Office of the Surgeon General. Accessed August 15, 2017 (surgeongeneral.gov).

U.S. Department of Health and Human Services. 2016. *Child Maltreatment 2014.* Administration for Children and Families, Children's Bureau. Accessed March 2, 2017 (acf.hhs.gov).

U.S. Department of Health and Human Services. 2017. *Child Maltreatment 2016.* Administration for Children and Families, Children's Bureau. Accessed March 2, 2017 (acf.hhs.gov).

U.S. Department of Justice. 1996. "Policing Drug Hot Spots." National Institute of Justice, January. Accessed May 3, 2005 (www.ncjrs.org).

U.S. Department of Justice. 2011. "U.S. Attorney Announces Drug Endangered Children Task Force." Press Notice, May 31. Accessed July 25, 2011 (www.usdoj.gov).

U.S. Department of Justice. 2017. "Health Care Fraud Unit." June 13. Accessed August 15, 2017 (justice.gov).

U.S. Department of State. 2016. "Trafficking in Persons Report." June. Accessed May 22, 2017 (state.gov/j/tip).

U.S. Department of Veteran Affairs. 2017. "Profile of Post–9/11 Veterans: 2015." National Center for Veterans Analysis and Statistics, March. Accessed August 15, 2017 (va.gov/vetdata).

U.S. Office of Personnel Management. 2014. "2014 Federal Employee Viewpoint Survey Results." Accessed November 2, 2016 (opm.gov/fevs).

United States Senate. 2016. "Rigged Justice 2016: How Weak Enforcement Lets Corporate Offenders Off Easy." Office of Senator Elizabeth Warren, January. Accessed April 4, 2017 (warren.senate.gov).

U.S. Senate Special Committee on Aging, American Association of Retired Persons, Federal Council on the Aging, and U.S. Administration on Aging. 1991. *Aging America: Trends and Projections, 1991.* Washington, DC: U.S. Department of Health and Human Services.

U.S. Senate. 2004. *Report on the U.S. Intelligence Community's Prewar Intelligence Assessments on Iraq.* Accessed September 13, 2006 (www.gpoaccess.gov).

U.S. Travel Association. 2014. "Working for Free: U.S. Workforce Forfeits $52.4 Billion in Time Off Benefits Annually." October 21. Accessed November 23, 2014 (www.ustravel.org).

USA Life Expectancy. 2016. "Life Expectancy All Races." Accessed July 14, 2017 (worldlifeexpectancy.com).

"Universities." 2015. *The Economist*, November 28, 25–26.

"Unmarried and Single Americans Week: Sept. 17–23, 2017." 2017. U.S. Census Bureau, August 14. Accessed August 20, 2017 (census.gov).

Upadhyay, Ushma D. 2016. "The Vanishing Abortion Pill." *New York Times*, August 31, A19.

Upton, John. 2014. "Chevron and BP Are Pulling Out of Wind and Solar." Grist, May 30. Accessed August 20, 2017 (grist.org).

Urbina, Ian. 2006. "In Online Mourning, Don't Speak Ill of the Dead." *New York Times*, November 5, 1.

V

"Vacation Habits Around the World." 2013. Expedia 2013 Vacation Deprivation Study, November 18. Accessed April 12, 2014 (www.viewfinder.expedia.com).

Vallotton, Claire, and Catherine Ayoub. 2011. "Use Your Words: The Role of Language in the Development of Toddlers' Self-Regulation." *Early Childhood Research Quarterly* 26 (2): 169–181.

van Agtmael, Peter. 2013. "Laugh After Death." *Time,* November 18, 44–49.

Vanorman, Alicia G., and Paola Scommegna. 2016. "Understanding the Dynamics of Family Change in the United States." Population Reference Bureau, July. Accessed July 1, 2017 (prb.org).

Van Vooren, Nicole, and Roberta Spalter-Roth. 2010. "Tracking Master's Students Through Programs and into Careers." *Footnotes* (September/October): 10–11.

Vassallo, Trae, et al. 2015. "Elephant in the Valley." Accessed January 22, 2016 (elephantinthevalley.com).

Veblen, Thorstein. 1899/1953. *The Theory of the Leisure Class.* New York: New American Library.

Velasco, Schuyler. 2014. "Food Additives: More Fuss Than Harm?" *Christian Science Monitor*, May 26, 43.

Venkatesh, Sudhir. 2008. *Gang Leader for a Day: A Rogue Sociologist Takes to the Streets.* New York: Penguin Press.

Vespa, Jonathan. 2017. "The Changing Economics and Demographics of Young Adulthood: 1975–2016." U.S. Census Bureau, Current Population Reports, April. Accessed July 20, 2017 (census.gov).

Vespa, Jonathan, Jamie M. Lewis, and Rose M. Kerner. 2013. "America's Families and Living Arrangements: 2012." U.S. Census Bureau, August. Accessed May 16, 2014 (www.census.gov).

Vicens, A. J. 2015. "Missing in Action." *Mother Jones*, July/August, 13.

Vitello, Paul. 2006. "The Trouble When Jane Becomes Jack." *New York Times* (August 20): H1, H6.

Vlasic, Bill. 2015. "G.M.'s Ignition Switch Death Toll Hits 100." *New York Times,* May 12, B3.

Vold, George B. 1958. *Theoretical Criminology.* New York: Oxford University Press.

Volz, Matt. 2012. "$3.4 Billion Indian Land Royalty Settlement Upheld." Salon, May 22. Accessed March 22, 2014 (www.salon.com).

Von Drehle, David. 2014. "Breaking the Marriage Bans." *Time*, March 3, 12.

Vongkiatkajorn, Kanyakrit. 2016. "Prison Labor Is Unseen and 'Utterly Exploitative'." *Mother Jones*, October 6. Accessed March 4, 2017 (motherjones.com).

W

Walk Free Foundation. 2014. "The Global Slavery Index 2014." Accessed September 11, 2015 (globalslaveryindex.org).

Wallace, Bruce. 2006. "Japanese Schools to Teach Patriotism." *Los Angeles Times*, December 16. Accessed December 18, 2006 (www.latimes.com).

Wallis, Cara. 2011. "Performing Gender: A Content Analysis of Gender Display in Music Videos." *Sex Roles* 64 (February): 160–172.

Walls, Mark. 2011. "High Court Strikes Down Calif. Law on Violent Video Games." *Education Week*, June 27. Accessed June 28, 2011 (www.edweek.org).

Walmsley, Roy. 2016. "World Prison Population List (Eleventh Edition)." International Center for Prison Studies. Accessed April 15, 2017 (prisonstudies.org).

Walsh, Ben. 2017. "Donald Trump Is Breaking His Promise to Be Tough on Wall Street." *Huffington Post*, January 31. Accessed February 3, 2017 (huffingtonpost.com).

Walsh, Bryan. 2015. "It May Be Too Late to Reverse the Damage of China's One-Child Policy." *Time*, November 16, 23–24.

Wan, William. 2011. "Chinese Dog Eaters and Dog Lovers Spar over Animal Rights." *Washington Post*, May 28, A14.

Wang, Wendy, and Kim Parker. 2014. "Record Share of Americans Have Never Married: As Values, Economic and Gender Patterns Change." Pew Research Center, September 24. Accessed October 26, 2014 (www.pewresearch.org).

Wang, Wendy, Kim Parker, and Paul Taylor. 2013. "Breadwinner Moms." Pew Research Center, May 29. Accessed May 18, 2014 (www.pewresearch.org).

Wang, Wendy. 2012. "The Rise of Intermarriage." Pew Research Center, Social & Demographic Trends, February 16. Accessed March 22, 2014 (www.pewsocialtrends.org).

Wang, Wendy. 2015. "The Link Between a College Education and a Lasting Marriage." Pew Research Center, December 4. Accessed July 20, 2017 (pewresearch.org).

Wang, Y. Claire. 2015. "The Dangerous Silence of Academic Researchers." *Chronicle of Higher Education*. February 27, A48.

Warren, Kenneth R. 2012. "NIH Statement on International FASD Awareness Day." National Institutes of Health, September 5. Accessed November 20, 2013 (www.nih.gov).

Wartik, Nancy. 2005. "The Perils of Playing House." *Psychology Today* (July/August): 42–52.

Wasley, Paula. 2008. "The Syllabus Becomes a Repository of Legalese." *Chronicle of Higher Education* 54, March 14, A1, A8–A11.

Waters, Mary C., and Marisa Gerstein Pineau, eds. 2015. *The Integration of Immigrants into American Society*. Washington, DC: The National Academies Press.

Wealth-X. 2017. "Billionaire Fact File from the Billionaire Census 2017." March 21. Accessed May 13, 2017 (wealthx.com).

Wealth-X and UBS. 2014. "Billionaire Census 2014." Accessed November 10, 2014 (www.billionairecensus.com).

Weathers, Cliff. 2014. "Research Behind Dr. Oz-Endorsed Diet Supplement Was Bogus, Authors Admit." AlterNet, October 22. Accessed December 5, 2014 (www.alternet.org).

Weathers, Cliff. 2015. "It's 2015 and Congress Is Almost Entirely Made Up of White Christian Men." AlterNet, January 6. Accessed September 29, 2015 (alternet.org).

Weaver, Rosanna Landis. 2016. "The 100 Most Overpaid CEOs: Are Fund Managers Asleep at the Wheel?" As You Sow. Accessed June 22, 2017 (asyousow.org).

Webb, Alex. 2013. "Germany's Coalition Looks to Gender Quotas." *Bloomberg Businessweek*, December 16–22, 17–18.

Weber, Lynn, Tina Hancock, and Elizabeth Higginbotham. 1997. "Women, Power, and Mental Health." Pp. 380–396 in *Women's Health: Complexities and Differences,* edited by Sheryl B. Ruzek, Virginia L. Olesen, and Adele E. Clarke. Columbus: Ohio State University Press.

Weber, Max. 1920/1958. *The Protestant Ethic and the Spirit of Capitalism,* translated by Talcott Parsons. New York: Charles Scribner's Sons.

Weber, Max. 1925/1947. *The Theory of Social and Economic Organization.* New York: Free Press.

Weber, Max. 1925/1978. *Economy and Society,* edited by Guenther Roth and Claus Wittich. Berkeley: University of California Press.

Weber, Max. 1946. *From Max Weber: Essays in Sociology,* translated and edited by H. H. Gerth and C. Wright Mills. Berkeley: University of California Press.

Weinhold, Bob. 2012. "A Steep Learning Curve: Decoding Epigenetic Influences on Behavior and Mental Health." *Environmental Health Perspectives* 120 (October): A396–A401.

Weintraub, Arlene. 2010. "Break That Hovering Habit Early." *U.S. News & World Report,* September, 42–43.

Weise, Elizabeth, and Tom Vanden Brook. 2015. "More Ashley Madison Files Published." *USA Today,* August 20. Accessed August 21, 2015 (usatoday.com).

Weisman, Carrie. 2015. "Can a Woman Find Sexual Desire in a Pill?" AlterNet, June 15. Accessed October 15, 2015 (alternet.org).

Weisman, Carrie. 2016. "Avoiding This Behavior Could Help Save Your Relationship." AlterNet, March 4. Accessed December 29, 2016 (alternet.org).

Weiss, Carol H. 1998. *Evaluation: Methods for Studying Programs and Policies,* 2nd edition. Upper Saddle River, NJ: Prentice Hall.

Welch, Ashley. 2015. "GOP Debate Fact Check: Claims About Vaccines and Autism." CBS News, September 17. Accessed October 25, 2015 (cbsnews.com).

Weller, Chris. 2015. "Black Market Sells Human Body Parts for Hundreds of Thousands: What Do You Cost?" *Medical Daily*, April 20. Accessed March 1, 2017 (medicaldaily.com).

Welty, Leah, et al. 2016. "Health Disparities in Drug- and Alcohol-Use Disorders: A 12-Year Longitudinal Study of Youths After Detention." *American Journal of Public Health* 106 (May): 872–880.

Wenneras, Christine, and Agnes Wold. 1997. "Nepotism and Sexism in Peer Review." *Nature* 387 (May 22): 341–343.

West, Candace, and Don H. Zimmerman. 2009. "Accounting for Doing Gender." *Gender & Society* 23 (February): 112–122.

Westcott, Lucy. 2014. "More Americans Moving to Cities, Reversing the Suburban Exodus." The Wire, March 27. Accessed September 19, 2014 (www.thewire.com).

"When Workers Are Owners." 2015. *The Economist*, August 22, 56.

White, Alexander, and Carolyn Witkus. 2013. "How Legal Marijuana May Affect Troubled Families." *Christian Science Monitor Weekly*, March 11, 36.

White, James W. 2005. *Advancing Family Theories.* Thousand Oaks, CA: Sage.

White, Nicole, and Janet L. Lauritsen. 2012. "Violent Crime Against Youth: 1994–2010." Bureau of Justice Statistics, December. Accessed May 4, 2013 (www.bjs.gov).

Whiteaker, Chloe, Laurie Meisler, Yvette Romero, and Karen Weise. 2014. "Voter ID States That Make Voting Super Simple—or Really Hard." *Bloomberg Businessweek*, October 27–November 2, 36.

Whitehead, Jaye C. 2011. "The Wrong Reasons for Same-Sex Marriage." *New York Times,* May 16, A21.

White House Council of Economic Advisers. 2016. "The Long-Term Decline in Prime-Age Male Labor Force Participation." June. Accessed May 15, 2017 (obamawhitehouse.archives.gov).

Whitehurst, Grover J. R. 2016. "Hard Thinking on Soft Skills." Brookings, March 24. Accessed March 1, 2017 (brookings.edu).

Whitelaw, Kevin. 2000. "But What to Call It?" *U.S. News & World Report,* October 16, 42.

Whitesides, John. 2014. "Anti-Union Group Pledges to Remain on the Offensive." *Baltimore Sun*, February 18, 2.

Whitlock, Craig, and Bob Woodward. 2016. "Pentagon Buries Evidence of $125 Billion in Bureaucratic Waste." *Washington Post*, December 5. Accessed April 10, 2017 (washingtonpost.com).

WHO. 2016. "Ambient (Outdoor) Air Quality and Health." Fact sheet, September. Accessed August 20, 2017 (who.int).

WHO. 2016. "Children: Reducing Mortality." Fact Sheet, September. Accessed May 13, 2017 (who.int).

"Who Pays the Bill?" 2013. *The Economist*, July 27, 24–26.

"Who Votes, Who Doesn't, and Why: Regular Voters, Intermittent Voters, and Those Who Don't." 2006. Pew Research Center for the People & the Press, October 18. Accessed May 3, 2007 (www.peoplepress.org).

Whorf, Benjamin Lee. 1956. *Language, Thought, and Reality.* Cambridge, MA: MIT Press.

Whoriskey, Peter, and Dan Keating. 2014. "Dying and Profits: The Evolution of Hospice." *Washington Post*, December 26. Accessed October 16, 2015 (washingtonpost.com).

Wicker, Christine. 2009. "How Spiritual Are We?" *Parade,* October 4, 4–5.

Wiik, Kenneth A., Renske Keizer, and Trude Lappegård. 2012. "Relationship Quality in Marital and Cohabiting Unions Across Europe." *Journal of Marriage and Family* 74 (June): 389–398.

Wilbur, Tabitha G., and Vincent J. Roscigno. 2016. "First-generation Disadvantage and College Enrollment/Completion." *Socius: Sociological Research for a Dynamic World* 2: 1–11.

Wilcox, W. Bradford. 2015. "Knot Now: The Benefits of Marrying in Your Mid-to-Late 20s (Including More Sex!)." *Washington Post*, February 20. Accessed September 30, 2015 (washingtonpost.com).

Wile, Rob. 2015. "Broke Young People Remain Convinced They'll Be Millionaires One Day." Fusion, February 6. Accessed May 3, 2017 (fusion.net).

Wilke, Joy, and Frank Newport. 2014. "Americans' Satisfaction with Economy Sours Most Since 2001." Gallup, January 16. Accessed April 12, 2014 (www.gallup.com).

Wilkinson, Charles. 2006. *Blood Struggle: The Rise of Modern Indian Nations.* New York: W. W. Norton & Company.

Wilkinson, Lindsey, and Jennifer Pearson. 2013. "High School Religious Context and Reports of Same-Sex Attraction and Sexual Identity in Young Adulthood." *Social Psychology Quarterly* 76 (June): 180–202.

Willett, Megan. 2015. "Living in Tiny Homes Was Much Harder Than These People Realized." Business Insider, July 27. Accessed August 20, 2017 (businessinsider.com).

Williams, Claudia, and Barbara Gault. 2014. "Paid Sick Days Access in the United States: Differences by Race/Ethnicity, Occupation, Earnings, and Work Schedule." Institute for Women's Policy Research, March. Accessed April 12, 2014 (www.iwpr.org).

Williams, Frank P., III, and Marilyn D. McShane. 2004. *Criminological Theory,* 4th edition. Upper Saddle River, NJ: Prentice Hall.

Williams, Lauren. 2014. "21 Things You Can't Do While Black." AlterNet, February 12. Accessed March 22, 2014 (www.alternet.org).

Williams, Robin M., Jr. 1970. *American Society: A Sociological Interpretation,* 3rd edition. New York: Knopf.

Williams, Timothy. 2013. "Blighted Cities Prefer Razing to Rebuilding." *New York Times,* November 13, A15.

Wilson, Thomas C. 1993. "Urbanism and Kinship Bonds: A Test of Four Generalizations." *Social Forces* 71 (March): 703–712.

Wilson, Valerie, and William M. Rodgers III. 2016. "Black-White Wage Gaps Expand with Rising Wage Inequality." Economic Policy Institute, September 19. Accessed June 20, 2017 (epi .org).

Wilson, William Julius. 1996. *When Work Disappears: The World of the New Urban Poor.* New York: Knopf.

Wiltz, Teresa. 2016. "Why More Grandparents Are Raising Children." Pew Charitable Trusts, November 2. Accessed July 14, 2017 (pewtrusts.org).

Winerip, Michael. 2013. "Pushed Out of a Job Early." *New York Times*, December 6. Accessed May 16, 2014 (www.nytimes.com).

Wing, Nick, and Carly Schwartz. 2014. "Here's the Painful Truth About What It Means to Be 'Working Poor' in America." *Huffington Post*, May 19. Accessed May 19, 2014 (www .huffingtonpost.com).

WIN-Gallup International. 2013. "Global Index of Religiosity and Atheism, 2012." Accessed June 21, 2014 (www.gallup-international .com).

Wingfield, Adia Harvey. 2010. "Are Some Emotions Marked 'Whites Only'? Racialized Feeling Rules in Professional Workplaces." *Social Problems* 57 (May): 258–268.

Winkler, Adam. 2011. *Gunfight: The Battle over the Right to Bear Arms in America.* New York: W. W. Norton & Co.

Winship, Michael. 2015. "Inside the Cushy, Fabulously Compensated Life of an Ex-Congressman." Alternet, April 8. Accessed October 7, 2015 (alternet.org).

Wirth, Louis. 1938. "Urbanism as a Way of Life." *American Journal of Sociology* 44 (July): 1–24.

Witters, Dan. 2010. "Large Metro Areas Top Small Towns, Rural Areas in Wellbeing." Gallup, May 17. Accessed May 20, 2010 (www.gallup .com).

"Wives Who Earn More Than Their Husbands, 1978–2012." 2014. U.S. Bureau of Labor Statistics, Labor Force Statistics from the Current Population Survey, March 24. Accessed April 9, 2014 (www.bls.gov/cps).

Wolff, Edward N. 2016. "Household Wealth Trends in the United States, 1962 to 2013: What Happened over the Great Recession?" *The Russell Sage Foundation Journal of the Social Sciences* 2 (October): 24–43.

Wolfson, Mark. 2001. *The Fight Against Big Tobacco: The Movement, the State, and the Public's Health.* New York: Aldine de Gruyter.

Wolven, Jacqueline. 2014. "The 3 Things Middle-Aged Women Need to Stop Doing When Speaking in Public." *Huffington Post*, September 18. Accessed September 18, 2014 (huffingtonpost.com).

Wolverton, Brad. 2016. "The New Cheating Economy." *Chronicle of Higher Education*, September 2, A40–A45.

"Women at Work." 2017. U.S. Bureau of Labor Statistics, March. Accessed May 10, 2017 (bls.gov).

Women's Bureau. 2016. "Working Mothers Issue Brief." U.S. Department of Labor, June. Accessed July 20, 2017 (dol.gov/wb).

Wondergem, Taylor R., and Mihaela Friedlmeier. 2012. "Gender and Ethnic Differences in Smiling: A Yearbook Photographs Analysis from Kindergarten Through 12th Grade." *Sex Roles* 67 (October): 403–411.

Wong, Brittany. 2014. "The Top 10 Reasons People Stay in Unhappy Marriages." *Huffington Post*, December 15. Accessed January 1, 2015 (huffingtonpost.com).

Wong, Curtis M. 2013. "Hallmark's 'Holiday Sweater' Keepsake Ornament Omits 'Gay' from 'Deck the Halls' Lyrics." *Huffington Post*, October 28. Accessed November 10, 2013 (www.huffingtonpost.com).

Wood, Jade, and Justin McCarthy. 2017. "Majority of Americans Remain Supportive of Euthanasia." Gallup, June 12. Accessed July 20, 2017 (gallup.com).

Wood, Julia T. 2015. *Gendered Lives: Communication, Gender, & Culture*, 11th edition. Stamford, CT: Cengage.

Woolf, Steven H., and Laudon Aron. 2013. *U.S. Health in International Perspective: Shorter Lives, Poorer Health.* National Research Council and Institute of Medicine of the National Academies. Accessed July 26, 2014 (www.national-academies.org).

World Bank. 2015. "Country and Lending Groups." Accessed October 14, 2015 (data.worldbank .org).

World Bank. 2017. *Atlas of Sustainable Development Goals 2017: World Development Indicators.* Accessed May 13, 2017 (worldbank.org).

World Bank and Collins. 2013. *Atlas of Global Development: A Visual Guide to the World's Greatest Challenges*, 4th edition. Washington, DC, and Glasgow: World Bank and Collins.

World Economic Forum. 2013. "Outlook on the Global Agenda 2014." Accessed February 6, 2014 (www.weforum.org).

World Economic Forum. 2016. "The Global Gender Gap Report 2016." Accessed May 13, 2017 (weforum.org).

World Factbook. 2017. "Sex Ratio." U.S. Central Intelligence Agency. Accessed August 20, 2017 (cia.gov).

World Food Programme. 2017. "Global Report on Food Crises 2017." Food Security Information Network, March. Accessed August 20, 2017 (wfp.org).

World Health Organization and UNICEF. 2014. "Progress on Drinking Water and Sanitation." May. Accessed September 19, 2014 (www .unicef.org).

World Health Organization. 2013. "Global and Regional Estimates of Violence Against Women: Prevalence and Health Effects of Intimate Partner Violence and Non-Partner Sexual Violence." Accessed March 2, 2014 (www.who.int).

World Health Organization. 2013. "Global and Regional Estimates of Violence Against Women: Prevalence and Health Effects of Intimate Partner Violence and Non-Partner Sexual Violence." Accessed August 6, 2015 (who.int).

World Health Organization. 2015. *World Health Statistics 2015.* Accessed October 15, 2015 (who.int).

World Health Organization. 2016. "Violence Against Women." November. Accesses May 25, 2017 (who.int).

World Meteorological Organization. 2014. "Atlas of Mortality and Economic Losses from Weather, Climate and Water Extremes (1970–2012)." Accessed September 19, 2014 (wmo.int).

World Population Review. 2017. "Sex Ratio at Birth by Country 2017." Accessed August 20, 2017 (worldpopulationreview.com).

World Water Assessment Program. 2009. *The United Nations World Water Development Report 3: Water in a Changing World.* Paris: UNESCO and London: Earthscan.

Worley, Heidi. 2014. "Top 10 Countries Closing Gender Gap." Population Reference Bureau, February. Accessed March 2, 2014 (prb.org).

Wright, Wynne, and Elizabeth Ransom. 2005. "Stratification on the Menu: Using Restaurant Menus to Examine Social Class." *Teaching Sociology* 33 (July): 310–316.

WUNRN. 2017. "Women, Slums, & Urbanization." Women's UN Report Network, March 17. Accessed August 20, 2017 (wunrn.com).

WWAP. 2014. *The United Nations World Water Development Report 2014: Water and Energy.* United Nations World Water Assessment Program. Accessed September 19, 2014 (www.unesco.org).

Wysong, Earl, Robert Perrucci, and David Wright. 2014. *The New Class Society: Goodbye American Dream?* 4th edition. Lanham, MD: Rowman & Littlefield.

X

Xie, Yu, James Raymo, Kimberly Goyette, and Arland Thornton. 2003. "Economic Potential and Entry into Marriage and Cohabitation." *Demography* 40 (May): 351–367.

Y

Yan, Holly. 2015. "717 People Dead: What Caused the Haji Stampede?" CNN, September 26. Accessed October 25, 2015 (cnn.com).

Yeager, David. 2016. "The 2015–2020 Dietary Guidelines." *Today's Dietitian* 18 (April): 34.

Yeager, David S., Valerie Purdie-Vaughns, Sophia Yang Hooper, and Geoffrey L. Cohen. 2017. "Loss of Institutional Trust Among Racial and Ethnic Minority Adolescents: A Consequence of Procedural Injustice and a Cause of Life-Span Outcomes." *Child Development* 88 (March/April): 658–676.

Yeung, Ken. 2016. "Yahoo Reveals New Hack: 'Unauthorized Third Party' Stole Data from More than 1 Billion Accounts." Venture Beat, December 14. Accessed September 2, 2017 (venturebeat.com).

Yglesias, Matthew. 2013. "The End of Cheap Airfare." Slate, February 19. Accessed April 12, 2014 (www.slate.com).

Yokota, Fumise, and Kimberly M. Thompson. 2000. "Violence in G-rated Animated Films." *JAMA* 283 (May 24/31): 2716–2720.

Young, Elise. 2015. "A Rough Ride for Renewable Energy." *Bloomberg Businessweek*, May 25-May 31, 28–29.

Z

Zagorsky, Jay L., and Patricia K. Smith. 2017. "The Association Between Socioeconomic Status and Adult Fast-Food Consumption in the U.S." *Economics & Human Biology* 27 (November): 12–25.

Zahedi, Ashraf. 2008. "Concealing and Revealing Female Hair: Veiling Dynamics in Contemporary Iran." Pp. 250–265 in *The Veil: Women Writers on Its History, Lore, and Politics*, edited by Jennifer Heath. Berkeley: University of California Press.

Zarya, Valentina. 2017. "Facebook: Iterating Diversity?" *Fortune*, February 1, 55.

Zhang, Qiang, et al. 2017. "Transboundary Health Impacts of Transported Global Air Pollution and International Trade." *Nature* 543 (March): 705–709.

Zhang, Yuanyuan, Travis L. Dixon, and Kate Conrad. 2010. "Female Body Image as a Function of Themes in Rap Music Videos: A Content Analysis." *Sex Roles* 62 (June): 787–797.

Zibel, Alan. 2017. "These 80 Executives, Lobbyists, Corporate Lawyers and Consultants Are Doing Donald Trump's Bidding." Citizen Vox, June 21. Accessed July 4, 2017 (citizenvox.org).

Ziccarelli, Gabriella E. 2016. "The Price of Pokemon Go." *Baltimore Sun*, July 15, 15.

Zimbardo, Philip G., Christina Maslach, and Craig Haney. 2000. "Reflections on the Stanford Prison Experiment: Genesis, Transformations, Consequences." Pp. 193–237 in *Obedience to Authority: Current Perspectives on the Milgram Paradigm*, edited by Thomas Blass. Mahwah, NJ: Lawrence Erlbaum Associates.

Zimbardo, Philip. G. 1975. "Transforming Experimental Research into Advocacy for Social Change." Pp. 33–66 in *Applying Social Psychology: Implications for Research, Practice, and Training*, edited by Morton Deutsch and Harvey A. Hornstein. Hillsdale, NJ: Erlbaum.

Zimmerman, Jonathan. 2014. "Are E-Cigarettes Ensnaring a New Generation of Smokers?" *Christian Science Monitor Weekly*, March 31, 35.

Zoll, Rachel, and Emily Swanson. 2015. "AP-NORC Poll: Christian-Muslim Split on Religious Freedom." AP-NORC, December 30. Accessed February 17, 2017 (apnorc.org).

Zong, Jie, and Jeanne Batalova. 2017. "Frequently Requested Statistics on Immigrants and Immigration in the United States." Migration Policy Institute, March 8. Accessed June 6, 2017 (migrationpolicy.org).

Zucman, Gabriel. 2016. "Wealth Inequality." *Pathways*, Special Issue, 40–44.

Zweigenhaft, Richard L., and G. William Domhoff. 2006. *Diversity in the Power Elite: How It Happened, Why It Matters*. Lanham, MD: Rowman & Littlefield.

NAME INDEX

A

B

C

D

E

F

H

I

J

K

L

M

N

O

P

Q

R

S

T

U

V

W

Z

SUBJECT INDEX

D

E

I

J

K

L

M

N

O

P

Q

R

S

T

U

V

W

Y

Z

1-1 **What Is Sociology?** *Sociology* is the systematic study of human behavior in society, behavior that takes place among individuals, small groups (such families), large organizations (such as Apple), and entire societies (such as the United States). Sociology goes well beyond common sense and conventional wisdom because it examines claims and beliefs critically, considers many points of view, and enables us to move beyond established ways of thinking.

Sociology the scientific study of human behavior in society. (p. 3)

1-2 **What Is a Sociological Imagination?** The *sociological imagination* emphasizes the connection between personal troubles (biography) and structural (public and historical) issues in understanding the social world. *Microsociology* examines the relationships between individuals, whereas *macrosociology* focuses on large-scale patterns and processes that characterize society as a whole. Macro-level systems shape society, often limiting our personal options on the micro level.

Sociological imagination seeing the relationship between individual experiences and larger social influences. (p. 5)

Microsociology examines the patterns of individuals' social interaction in specific settings. (p. 5)

Macrosociology examines the large-scale patterns and processes that characterize society as a whole. (p. 5)

1-3 **Why Study Sociology?** Regardless of your major, this course will help you (1) make more informed decisions, (2) understand diversity, (3) shape social policies and practices, (4) think critically, and (5) expand your career opportunities.

1-4 **Some Origins of Sociological Theory** Sociologists use *theories* to produce knowledge and to offer solutions to everyday social problems. Some of the most influential theorists have included Auguste Comte, Harriet Martineau, Émile Durkheim, Karl Marx, Max Weber, Jane Addams, and W. E. B. Du Bois. All of these early thinkers shaped contemporary sociological theories.

Theory a set of statements that explains why a phenomenon occurs. (p. 7)

Empirical information that is based on observations, experiments, or other data collection rather than on ideology, religion, intuition, or conventional wisdom. (p. 8)

Social facts aspects of social life, external to the individual, that can be measured. (p. 8)

Social solidarity social cohesiveness and harmony. (p. 9)

Division of labor an interdependence of different tasks and occupations, characteristic of industrialized societies, that produces social unity and facilitates change. (p. 9)

Capitalism an economic system based on the private ownership of property and the means of production. (p. 10)

Alienation feeling separated from one's group or society. (p. 10)

Value free separating one's personal values, opinions, ideology, and beliefs from scientific research. (p. 12)

1-5 **Contemporary Sociological Theories** Sociologists typically use more than one theory to explain human behavior. Four major perspectives help us understand society:

- *Functionalism maintains that society is a complex system of interdependent parts that work together to ensure a society's survival.* Critics contend that functionalism is so focused on order and stability that it often ignores social change, social inequality, and social conflict.
- *Conflict theory sees disagreement and the resulting changes in society as natural, inevitable, and even desirable.* Critics argue that conflict theory ignores the importance of harmony and cooperation.
- *Feminist theories*, which build on conflict theory, *emphasize women's social, economic, and political inequality.* Critics claim that these theories are too narrowly focused.
- *Symbolic interaction focuses on individuals' everyday behavior through the communication of knowledge, ideas, beliefs, and attitudes.* Critics maintain that this theoretical perspective overlooks the impact of macro-level factors on our everyday behavior.

Example: Critical Thinking versus Common Sense and Conventional Wisdom

When thinking critically, it's important to differentiate between common sense/conventional wisdom and data-based facts. Here are a few examples:

Myth: Older people make up the largest group of poor people.

Fact: Children younger than 6, not older people, make up the largest group of poor people.

Myth: Divorce rates are higher today than ever before.

Fact: Divorce rates are lower today than they were between 1980 and 1990.

Now, based on the material you read in this chapter, construct your own "myth" and "fact."

1 Thinking Like a Sociologist

Functionalism (*structural functionalism*) maintains that society is a complex system of interdependent parts that work together to ensure a society's survival. (p. 13)

Dysfunctions social patterns that have a negative impact on a group or society. (p. 13)

Manifest functions purposes and activities that are intended and recognized; they're present and clearly evident. (p. 13)

Latent functions purposes and activities that are unintended and unrecognized; they're present but not immediately obvious. (p. 13)

Conflict theory examines how and why groups disagree, struggle over power, and compete for scarce resources. (p. 14)

Feminist theories examine women's social, economic, and political inequality. (p. 15)

Symbolic interaction theory (*interactionism*) examines people's everyday behavior through the communication of knowledge, ideas, beliefs, and attitudes. (p. 16)

Social interaction a process in which people take each other into account in their own behavior. (p. 17)

Table 1.3 Leading Contemporary Perspectives in Sociology

THEORETICAL PERSPECTIVE	FUNCTIONALISM	CONFLICT	FEMINIST	SYMBOLIC INTERACTION
Level of Analysis	Macro	Macro	Macro and Micro	Micro
Key Points	• Society is composed of interrelated, mutually dependent parts. • Structures and functions maintain a society's or group's stability, cohesion, and continuity. • Dysfunctional activities that threaten a society's or group's survival are controlled or eliminated.	• Life is a continuous struggle between the haves and the have-nots. • People compete for limited resources that are controlled by a small number of powerful groups. • Society is based on inequality in terms of ethnicity, race, social class, and gender.	• Women experience widespread inequality in society because, as a group, they have little power. • Gender, ethnicity, race, age, sexual orientation, and social class—rather than a person's intelligence and ability—explain many of our social interactions and lack of access to resources. • Social change is possible only if we change our institutional structures and our day-to-day interactions.	• People act on the basis of the meaning they attribute to others. Meaning grows out of the social interaction that we have with others. • People continuously reinterpret and reevaluate their knowledge and information in their everyday encounters.
Key Questions	• What holds society together? How does it work? • What is the structure of society? • What functions does society perform? • How do structures and functions contribute to social stability?	• How are resources distributed in a society? • Who benefits when resources are limited? Who loses? • How do those in power protect their privileges? • When does conflict lead to social change?	• Do men and women experience social situations in the same way? • How does our everyday behavior reflect our gender, social class, age, race, ethnicity, sexual orientation, and other factors? • How do macro structures (such as the economy and the political system) shape our opportunities? • How can we change current structures through social activism?	• How does social interaction influence our behavior? • How do social interactions change across situations and between people? • Why does our behavior change because of our beliefs, attitudes, values, and roles? • How is "right" and "wrong" behavior defined, interpreted, reinforced, or discouraged?
Example	• A college education increases one's job opportunities and income.	• Most low-income families can't afford to pay for a college education.	• Gender affects decisions about a major and which college to attend.	• College students succeed or fail based on their degree of academic engagement.

2-1 **How Do We Know What We Know?** Much of our knowledge is based on tradition and authority. In contrast, sociologists rely on *research methods* to get information about a particular topic.

Research methods organized and systematic procedures to gain knowledge about a particular topic. (p. 21)

2-2 **Why Is Sociological Research Important in Our Everyday Lives?** In contrast to tradition and authority, sociological research challenges overgeneralizations, exposes myths, helps explain why people behave as they do, influences social policies, and sharpens critical thinking skills about issues that affect our everyday lives.

2-3 **The Scientific Method** The *scientific method* incorporates careful data collection, exact measurement, accurate recording and analysis of findings, thoughtful interpretation of results, and, when appropriate, a generalization of the findings to a larger group. A research question or a *hypothesis* examines the association between an *independent variable* and a *dependent variable*. Sociologists use both *qualitative* and *quantitative* approaches, and are always concerned about the *reliability* and *validity* of their measures.

Scientific method a body of objective and systematic techniques used to investige phenomena, acquire knowledge, and test hypotheses and theories. (p. 22)

Concept an abstract idea, mental image, or general notion that represents some aspect of the world. (p. 23)

Variable a characteristic that can change in value or magnitude under different conditions. (p. 23)

Independent variable a characteristic that has an effect on the dependent variable. (p. 23)

Dependent variable the outcome that may be affected by the independent variable. (p. 23)

Control variable a characteristic that is constant and unchanged during the research process. (p. 23)

Hypothesis a statement of the expected relationship between two or more variables. (p. 23)

Deductive reasoning begins with a theory, prediction, or general principle that is then tested through data collection. (p. 23)

Inductive reasoning begins with a specific observation, followed by data collection, a conclusion about patterns or regularities, and the formulation of hypotheses that can lead to theory construction. (p. 23)

Reliability the consistency with which the same measure produces similar results time after time. (p. 24)

Validity the degree to which a measure is accurate and really measures what it claims to measure. (p. 24)

Population any well-defined group of people (or things) that researchers want to know something about. (p. 24)

Sample a group of people (or things) that's representative of the population researchers wish to study. (p. 24)

Probability sample each person (or thing) has an equal chance of being selected because the selection is random. (p. 24)

Nonprobability sample there is little or no attempt to get a representative cross section of a population. (p. 24)

Qualitative research examines and interprets nonnumerical material. (p. 25)

Quantitative research focuses on a numerical analysis of people's responses or specific characteristics. (p. 25)

Causation a relationship in which one variable is the direct consequence of another. (p. 25)

Correlation the relationship between two or more variables. (p. 26)

2-4 **Basic Steps in the Research Process** Research involves choosing a topic, summarizing the pertinent research, formulating a hypothesis or asking a research question, describing the data collection methods, collecting the data, presenting the findings, and analyzing and explaining the results.

2-5 **Some Major Data Collection Methods** Five data collection methods are especially useful in sociology (see Table 2.3 on the next page). In designing their studies, sociologists weigh the advantages and limitations of each data collection method. *Evaluation research*, which uses one or more data collection methods, examines whether a social intervention has produced the intended result. Because sociologists don't conduct research in a cultural vacuum, many groups use the findings to change current policies and practices.

Example: The Uneasy Relationship between Research and Practice

Many supporters of the DARE (Drug Abuse Resistance Education) program were unhappy when more than 30 research studies showed that DARE had negligible long-term effects in reducing teen drug use. About 75 percent of U.S. school districts revised the DARE curriculum, but continued to use it because they believed that the program built a positive relationship between police, students, parents, and educators (see text).

Survey a data collection method that includes questionnaires, face-to-face or telephone interviews, or a combination. (p. 27)

Questionnaire a series of written questions that ask for information. (p. 27)

Interview a researcher directly asks respondents a series of questions. (p. 28)

Field research data collected by observing people in their natural surroundings. (p. 29)

Content analysis a data collection method that systematically examines some form of communication. (p. 31)

Experiment a controlled artificial situation that allows researchers to manipulate variables and measure the effects. (p. 32)

Experimental group the participants who are exposed to the independent variable. (p. 32)

Control group the participants who aren't exposed to the independent variable. (p. 32)

Secondary analysis examination of data that have been collected by someone else. (p. 33)

Evaluation research examines whether a social intervention has produced the intended result. (p. 34)

2-6 **Ethics and Social Research** Sociological research demands a strict code of ethics to avoid mistreating participants. For example, participants must give informed consent and must not be harmed, humiliated, abused, or coerced; researchers must honor their guarantees of privacy, confidentiality, and/or anonymity. Still, sociologists often encounter pressure from policy makers and others to limit their research to non-controversial topics.

Table 2.3 Some Data Collection Methods in Sociological Research

Method	Example	Advantages	Disadvantages
Surveys	Sending questionnaires and/or interviewing students on why they succeeded in college or dropped out	Questionnaires are fairly inexpensive and simple to administer; interviews have high response rates; findings are often generalizable	Mailed questionnaires may have low response rates; respondents tend to be self-selected; interviews are usually expensive
Secondary analysis	Using data from the National Center for Education Statistics (or similar organizations) to examine why students drop out of college	Usually accessible, convenient, and inexpensive; often longitudinal and historical	Information may be incomplete; some documents may be inaccessible; some data can't be collected over time
Field research	Observing classroom participation and other activities of first-year college students with high and low grade-point averages (GPAs)	Flexible; offers deeper understanding of social behavior; usually inexpensive	Difficult to quantify and to maintain observer/subject boundaries; the observer may be biased or judgmental; findings are not generalizable
Content analysis	Comparing the transcripts of college graduates and dropouts on variables such as gender, race/ethnicity, and social class	Usually inexpensive; can recode errors easily; unobtrusive; permits comparisons over time	Can be labor intensive; coding is often subjective (and may be distorted); may reflect social class bias
Experiments	Providing tutors to some students with low GPAs to find out if such resources increase college graduation rates	Usually inexpensive; plentiful supply of subjects; can be replicated	Subjects aren't representative of a larger population; the laboratory setting is artificial; findings can't be generalized

3-1 **Culture and Society** *Culture* is learned, transmitted from one generation to another, adaptive, and always changing. A *society* shares a culture and sees itself as a social unit. People construct a *material culture* (e.g., buildings) and *nonmaterial culture* (e.g., rules for behavior) that influence each other (such as forbidding smoking in public buildings).

Culture the learned and shared behaviors, beliefs, attitudes, values, and material objects that characterize a particular group or society. (p. 39)

Society a group of people who share a culture and defined territory. (p. 39)

Material culture the physical objects that people make, use, and share. (p. 40)

Nonmaterial culture the ideas that people create to interpret and understand the world. (p. 40)

3-2 **The Building Blocks of Culture** The following are some of the fundamental building blocks of culture:

- *Symbols* take many forms, distinguish one culture from another, can change over time, and can unify or divide a society.
- *Language* makes us human; it can change over time, and affects perceptions about gender, race, social class, and ethnicity.
- *Values* provide general guidelines for behavior; they sometimes conflict, are usually emotion laden, vary across cultures, and change over time.
- *Norms*—whether folkways, mores, or laws—regulate our behavior; they vary across cultures, and range from mild to severe sanctions.
- *Rituals* transmit and reinforce norms that unite people and strengthen their relationships.

Example: Sanctions for Violating the Dead

Sanctions are more severe for violating laws than folkways. Legacy.com, which carries a death notice or obituary for almost all of the roughly 2.4 million Americans who die each year, dedicates at least 30 percent of its budget to weeding out comments (a relatively mild punishment) that "diss the dead" (Urbina, 2006). In contrast, in many states, vandalizing a tombstone is a property crime that can result in a fine of up to $1,000, a jail sentence of up to a year, or both.

Symbol anything that stands for something else and has a particular meaning for people who share a culture. (p. 40)

Language a system of shared symbols that enables people to communicate with one another. (p. 42)

Values the standards by which people define what is good or bad, moral or immoral, proper or improper, desirable or undesirable, beautiful or ugly. (p. 44)

Norms specific rules of right and wrong behavior. (p. 45)

Folkways norms that involve everyday customs, practices, and interaction. (p. 46)

Mores norms that people consider very important because they maintain moral and ethical behavior. (p. 47)

Laws formally defined norms about what is legal or illegal. (p. 47)

Sanctions rewards for good or appropriate behavior and/or penalties for bad or inappropriate behavior. (p. 47)

Rituals formal and repeated behaviors that unite people. (p. 48)

3-3 **Some Cultural Similarities** Customs and specific behaviors vary across countries, but *cultural universals*, such as food taboos, are common to all societies. *Ideal culture* often differs from *real culture*. *Ethnocentrism* has its benefits but can also lead to conflict and discrimination. *Cultural relativism*, the opposite of ethnocentrism, encourages cross-cultural understanding and respect.

Cultural universals customs and practices that are common to all societies. (p. 48)

Ideal culture the beliefs, values, and norms that people say they hold or follow. (p. 49)

Real culture people's actual everyday behavior. (p. 49)

Ethnocentrism the belief that one's own culture, society, or group is inherently superior to others. (p. 49)

Cultural relativism the belief that no culture is better than another and should be judged by its own standards. (p. 50)

3-4 **Some Cultural Variations** *Subcultures* and *countercultures* account for some of the complexity within a society. Subcultures differ from the larger society in some ways, whereas countercultures oppose or reject some of the dominant culture's basic beliefs, values, and norms. In *multiculturalism*, several cultures coexist without trying to dominate one another. People who encounter an unfamiliar way of life or environment may experience *culture shock* because everyone is culture bound to some degree.

Subculture a group within society that has distinctive norms, values, beliefs, lifestyle, or language. (p. 50)

Counterculture a group within society that openly opposes and/or rejects some of the dominant culture's norms, values, or laws. (p. 50)

Multiculturalism (sometimes called *cultural pluralism*) the coexistence of several cultures in the same geographic area, without one culture dominating another. (p. 51)

Culture shock confusion, disorientation, or anxiety that accompanies exposure to an unfamiliar way of life. (p. 51)

3-5 **High Culture and Popular Culture** *High culture* refers to the cultural expression of a society's highest social classes, whereas *popular culture*—which is widely shared among a population—includes music, social media, sports, hobbies, fashions, the food we eat, the people with whom we spend time, the gossip we share, and the jokes we tell. *Cultural capital* affects social class boundaries, but our everyday life can contain elements of both popular and high culture. Popular culture is typically spread through *mass media*, including television and the Internet, and has enormous power in shaping our attitudes and behavior.

High culture the cultural expression of a society's highest social classes. (p. 52)

Popular culture the beliefs, practices, activities, and products that are widespread among a population. (p. 52)

Cultural capital resources and assets that give a group advantages. (p. 52)

Mass media forms of communication designed to reach large numbers of people. (p. 53)

Cultural imperialism the cultural values and products of one society influence or dominate those of another. (p. 53)

3-6 **Cultural Change and Technology** Some societies are relatively stable because of *cultural integration*, but all societies change over time because of diffusion, innovation, invention, discovery, and external pressures. *Cultural lags* can create confusion, ambiguity about what's right and wrong, conflict, and a feeling of helplessness. They also expose contradictory values and behavior.

Cultural integration the consistency of various aspects of society that promotes order and stability. (p. 54)

Cultural lag the gap that occurs when material culture changes faster than nonmaterial culture. (p. 56)

3-7 **Sociological Perspectives on Culture** See Table 3.3 below.

Table 3.3 Sociological Perspectives of Culture

THEORETICAL PERSPECTIVE	FUNCTIONALIST	CONFLICT	FEMINIST	SYMBOLIC INTERACTIONIST
Level of Analysis	**Macro**	**Macro**	**Macro and Micro**	**Micro**
Key Points	• Similar beliefs bind people together and create stability. • Sharing core values unifies a society and promotes cultural solidarity.	• Culture benefits some groups at the expense of others. • As powerful economic monopolies increase worldwide, the rich get richer and the rest of us get poorer.	• Women and men often experience culture differently. • Cultural values and norms can increase inequality because of sex, race/ethnicity, and social class.	• Cultural symbols forge identities (that change over time). • Culture (like norms and values) helps people merge into a society despite their differences.
Examples	• Speaking the same language (English in the United States) binds people together because they can communicate with one another, express their feelings, and influence one another's attitudes and behaviors.	• Much of the English language reinforces negative images about gender ("slut"), race ("honky"), ethnicity ("jap"), and age ("old geezer") that create inequality and foster ethnocentrism.	• Using male language (e.g., "congressman," "fireman," and "chairman") conveys the idea that men are superior to and dominant over women, even when women have the same jobs.	• People can change the language they create as they interact with others. Many Americans now use "police officer" instead of "policeman," and "single person" instead of "bachelor" or "old maid."

CHAPTER REVIEW

CHAPTER 4 LEARNING OUTCOMES / KEY TERMS

4-1 **Socialization: Its Purpose and Importance** *Socialization*, a lifelong process, fulfills four key purposes: It establishes our social identity, teaches us roles, controls our behavior (through *internalization*), and transmits culture to the next generation. The research on children who are isolated or institutionalized shows that socialization is critical to our social and physical development.

Socialization the lifelong process through which people learn culture and become functioning members of society. (p. 61)

Internalization the process of learning cultural behaviors and expectations so deeply that we accept them without question. (p. 61)

4-2 **Nature and Nurture** Biologists tend to focus on the role of heredity (or genetics) in human development. In contrast, most social scientists, including sociologists, underscore the role of learning, socialization, and culture. This difference of opinion is often called the *nature–nurture debate*. Sociobiologists argue that genetics (nature) can explain much of our behavior, whereas most sociologists maintain that socialization and culture (nurture) shape even biological inputs.

4-3 **Sociological Explanations of Socialization** Sociologists have offered many explanations of socialization, but two of the most influential, both at the micro level, have been social learning and symbolic interaction theories (see Table 4.2 on the next page). The theories help us understand social development processes, but also have limitations.

Social learning theories people learn new attitudes, beliefs, and behaviors through social interaction. (p. 65)

Role models people we admire and whose behavior we imitate. (p. 65)

Self an awareness of one's social identity. (p. 66)

Looking-glass self a self-image based on how we think others see us. (p. 66)

Role taking learning to take the perspective of others. (p. 67)

Significant other someone whose opinions we value and who influences our thinking, especially about ourselves. (p. 67)

Anticipatory socialization learning how to perform a role that a person will occupy in the future. (p. 68)

Generalized other the norms, values, and expectations of society that affect a person's behavior. (p. 68)

Impression management providing information and cues to others to present oneself in a favorable light while downplaying or concealing one's less appealing characteristics. (p. 68)

Reference groups people who shape an individual's self-image, behavior, values, and attitudes in different contexts. (p. 69)

4-4 **Primary Socialization Agents** Parents are the first and most important *socialization agents*, but siblings, grandparents, and other family members also play important roles. Other significant socialization agents include play and peer groups, teachers and schools, popular culture, and the media. Electronic media and advertising are especially powerful socialization sources.

Example: Are Parents Raising Self-Centered Children?

Many parents tell their offspring that they're beautiful, can achieve anything they want, and are very intelligent. In fact, only about 5 percent of American kids can be considered "gifted" (have significantly higher than average intellectual or other abilities), even though many are enrolled in gifted classes. Some educators maintain that telling average—or even above average—children that they're superior does them a disservice: It gives them false expectations on how the world will treat them, encourages being self-centered, and increases anger and unhappiness when they don't succeed in college or the workplace (Deveny, 2008).

Socialization agents the individuals, groups, or institutions that teach us how to participate effectively in society. (p. 69)

Multigenerational households homes in which three or more generations live together. (p. 71)

Peer group people who are similar in age, social status, and interests. (p. 71)

4-5 Socialization Throughout Life As we progress through the life course, we learn culturally approved norms, values, and roles. Infants are born with an enormous capacity for learning that parents and other caregivers can enrich and shape. In adolescence, these and other socialization agents teach children how to form relationships, to get along with others, and to develop their social identity through play and peer groups. In adulthood, people must learn new roles that include singlehood, marriage, parenthood, divorce, work, and experiencing the death of a loved one. Socialization continues in later life when many people learn still new roles such as grandparents, retirees, older workers, and being widowed.

4-6 Resocialization and Total Institutions *Resocialization* can be voluntary or involuntary. Voluntary resocialization includes entering a religious order, joining a religious cult, seeking treatment in a drug abuse rehabilitation facility, or serving in the military. The changes can be short-term and pleasant, but also long, difficult, and intense. Most involuntary resocialization takes place in *total institutions*, where people are isolated from the rest of society, stripped of their former identities, and required to conform to new rules and behavior.

Resocialization unlearning old ways of doing things and adopting new attitudes, values, norms, and behavior. (p. 79)

Total institutions isolated and enclosed social systems that control most aspects of the participants' lives. (p. 80)

Table 4.2 Key Elements of Socialization Theories

Social Learning Theories	Symbolic Interaction Theories
• Social interaction is important in learning appropriate and inappropriate behavior.	• The self emerges through social interaction with significant others.
• Socialization relies on direct and indirect reinforcement.	• Socialization includes role taking and controlling the impression we give to others.
Example: Children learn how to behave when they are scolded or praised for specific behaviors.	*Example:* Children who are praised are more likely to develop a strong self-image than those who are always criticized.

Social Interaction in Everyday Life 5

5-1 **Social Structure** *Social interaction*, central to all social activity, affects people's behavior. Our interaction is part of the *social structure*, which guides our actions and gives us a feeling that life is orderly and predictable. Every society has a social structure that encompasses statuses and roles.

Social interaction acting toward and reacting to people around us. (p. 83)

Social structure an organized pattern of behavior that governs people's relationships. (p. 83)

5-2 **Status** A *status* is a social position that a person occupies in a society. Every person has many statuses that form her or his *status set*, which include both ascribed and achieved statuses. An *ascribed status* is a social position that a person is born into and can't control, change, or choose (e.g., age, race, and ethnicity). An *achieved status* is a social position that a person attains through personal effort or assumes voluntarily (e.g., college student or wife). A *master status* is usually immediately apparent, makes the biggest impression, affects others' perceptions, and, consequently, often shapes a person's entire life.

Because we hold many statuses, some clash. People experience *status inconsistency* when they occupy social positions that are ranked differently (such as being a low-paid college professor).

Status a social position that a person occupies in a society. (p. 83)

Status set a collection of social statuses that a person occupies at a given time. (p. 83)

Ascribed status a social position that a person is born into. (p. 84)

Achieved status a social position that a person attains through personal effort or assumes voluntarily. (p. 84)

Master status overrides other statuses and forms an important part of a person's social identity. (p. 84)

Status inconsistency the conflict that arises from occupying social positions that are ranked differently. (p. 84)

5-3 **Role** A *role* defines how we're expected to behave in a particular status, but people vary considerably in fulfilling the responsibilities associated with their roles. These differences reflect *role performance*, the actual behavior of a person who occupies a status. A *role set* encompasses different roles attached to a single status (e.g., a parent who's a teacher, volunteer, and PTA member). Playing many roles often leads to *role conflict* because it's difficult to meet the requirements of two or more statuses, and to *role strain*, the stress that arises because of incompatible demands among roles within a single status.

To deal with role conflict and role strain, some people deny that there's a problem. More effective ways to minimize role conflict and role strain include compromising, negotiating, setting priorities, compartmentalizing our roles, refusing to take on more roles, and exiting one or more current roles.

Example: Exiting a Marriage

Divorce is a good example of role exit, but it often involves a long process of five stages that may last several decades (Bohannon 1971):

- The "emotional divorce" begins when one or both partners feel disillusioned or unhappy.
- The "legal divorce" is the formal dissolution of the marriage during which the partner who doesn't want the divorce may try to stall the end of the marriage.
- During the "economic divorce" stage, the partners may argue about who should pay past debts and unforeseen expenses (like moving costs).
- The "coparental divorce" stage involves parents' agreeing on issues such as child support and visitation rights.
- During the "community divorce" stage, partners inform friends, family, and others that they are no longer married.
- Finally, the couple goes through a "psychic divorce" in which the partners separate from each other emotionally. One or both spouses never complete this stage because they can't let go of their pain and anger—even if they remarry.

Role the behavior expected of a person who has a particular status. (p. 85)

Role performance the actual behavior of a person who occupies a status. (p. 85)

Role set array of roles attached to a particular status. (p. 85)

Role conflict difficulties in playing two or more contradictory roles. (p. 86)

Role strain difficulties due to conflicting demands within the same role. (p. 86)

5-4 **Explaining Social Interaction** See Table 5.2 on next page.

Self-fulfilling prophecy if we define something as real and act on it, it can, in fact, become real. (p. 88)

Ethnomethodology the study of how people construct and learn to share definitions of reality that make everyday interactions possible. (p. 88)

Dramaturgical analysis examines social interaction as if occurring on a stage where people play different roles and act out scenes for the audiences with whom they interact. (p. 89)

Social exchange theory proposes that individuals seek through their interactions to maximize their rewards and minimize their costs. (p. 90)

Social Interaction in Everyday Life

5-5 **Nonverbal Communication** Our *nonverbal communication* includes gestures, facial expressions, eye contact, and silence. Touching and how we use space are also important forms of nonverbal communication because they send powerful messages about our feelings and power. Nonverbal communication, which varies from society to society, can lead to cross-cultural misinterpretation and misunderstanding.

Nonverbal communication messages that are sent without using words. (p. 91)

5-6 **Online Interaction** Many people interact in *cyberspace*, an online world of computer networks that includes *social media* and *social networking sites*. Internet usage varies by sex, age, race, ethnicity, and social class. Online interaction can be impersonal, socially isolating, and jeopardizes our privacy, but can also save time, foster closer ties among family members and friends, and facilitate working from home.

social media websites that enable users to create, share, and/or exchange information and ideas. (p. 95)

Table 5.2 Sociological Explanations of Social Interaction

Perspective	Key Points
Symbolic Interactionist	• People create and define their reality through social interaction. • Our definitions of reality, which vary according to context, can lead to self-fulfilling prophecies.
Social Exchange	• Social interaction is based on a balancing of benefits and costs. • Relationships involve trading a variety of resources, such as money, youth, and good looks.
Feminist	• Females and males act similarly in many interactions but may differ in communication styles and speech patterns. • Men are more likely to use speech that's assertive (to achieve dominance and goals), whereas women are more likely to use language that connects with others.

6-1 **Social Groups** A *social group* (such as friends or work groups) gives us a common identity and a sense of belonging. Social groups include *primary groups* (such as family members) that shape our social and moral development and *secondary groups* (such as the students in your sociology class) that pursue a specific goal or activity.

Members of an *in-group* share a sense of identity and "we-ness." In contrast, *out-groups* are viewed and treated negatively because they're seen as having values, beliefs, and other characteristics that differ from those of the in-group.

We also have *reference groups* that influence who we are, what we do, and who we'd like to be in the future. Groups often form a *social network* that may be tightly knit and interact every day or may include large numbers of people whom we don't know personally and with whom we interact only rarely or indirectly.

Example: Secondary Groups Can Replace Primary Groups

In 1864, an alcoholic who had ruined a promising career on Wall Street because of his constant drunkenness cofounded Alcoholics Anonymous (AA), a program that would enable people to stop drinking by undergoing a spiritual awakening, and seeking help from a buddy to stay sober. Initially, AA was a secondary group that encouraged its members to attend regular meetings and discuss their accomplishments in staying sober. Over the years, however, AA became a primary group for many members: It offers support to a relatively small group of people who engage in face-to-face interaction over an extended period, especially when their family and friends have rejected them.

Social group two or more people who share some attribute and interact with one another. (p. 101)

Primary group a small group of people who engage in intimate face-to-face interaction over an extended period. (p. 101)

Secondary group a large, usually formal, impersonal, and temporary collection of people who pursue a specific goal or activity. (p. 101)

Ideal types general traits that describe a social phenomenon rather than every case. (p. 101)

In-group people who share a sense of identity and belonging that typically excludes and devalues outsiders. (p. 102)

out-group people who are viewed and treated negatively because they're seen as having values, beliefs, and other characteristics different from those of an in-group. (p. 102)

Reference group people who shape our behavior, values, and attitudes. (p. 102)

Dyad a group with two members. (p. 103)

Triad a group with three members. (p. 103)

Authoritarian leader gives orders, assigns tasks, and makes all major decisions. (p. 103)

Democratic leader encourages group discussion and includes everyone in the decision-making process. (p. 103)

Laissez-faire leader offers little or no guidance to group members and allows them to make their own decisions. (p. 103)

Groupthink in-group members make faulty decisions because of group pressures, rather than critically testing, analyzing, and evaluating ideas and evidence. (p. 106)

Social network a web of social ties that links individuals or groups to one another. (p. 106)

6-2 **Formal Organizations** We depend on a variety of *formal organizations* to provide goods and services in a stable and predictable way. Utilitarian, normative, and coercive organizations differ in their general characteristics and membership, but a single formal organization can fall into all three categories.

Bureaucracies (such as your college), are supposed to accomplish goals and tasks in the most efficient and rational way possible, but many experience shortcomings such as weak reward systems, rigid rules, goal displacement, alienation, communication problems, and dehumanization.

Formal organization a complex and structured secondary group designed to achieve specific goals in an efficient manner. (p. 106)

Bureaucracy a formal organization designed to accomplish goals and tasks in an efficient and rational way. (p. 108)

Goal displacement a preoccupation with rules and regulations rather than achieving the organization's objectives. (p. 108)

Alienation a feeling of isolation, meaninglessness, and powerlessness. (p. 109)

Iron law of oligarchy the tendency of a bureaucracy to become increasingly dominated by a small group of people. (p. 109)

6-3 Sociological Perspectives on Social Groups and Organizations See Table 6.3 below.

Glass ceiling workplace attitudes or organizational biases that prevent women from advancing to leadership positions. (p. 115)

Glass escalator men who enter female-dominated occupations receive higher wages and faster promotions than women. (p. 115)

6-4 Social Institutions A *social institution* meets a society's basic survival needs. Social institutions are abstractions, but they have an organized purpose, weave together norms and values, and, consequently, guide behavior. Because social institutions are linked to one another, they can tell us a lot about how a society functions and how we're connected to one another.

Social institution an organized and established social system that meets one or more of a society's basic needs. (p. 116)

Table 6.3 Sociological Perspectives on Groups and Organizations

Theoretical Perspective	Level of Analysis	Main Points	Key Questions
Functionalist	Macro	Organizations are made up of interrelated parts and rules and regulations that produce cooperation in meeting a common goal.	• Why are some organizations more effective than others? • How do dysfunctions prevent organizations from being rational and effective?
Conflict	Macro	Organizations promote inequality that benefits elites, not workers.	• Who controls an organization's resources and decision making? • How do those with power protect their interests and privileges?
Feminist	Macro and micro	Organizations tend not to recognize or reward talented women and regularly exclude them from decision-making processes.	• Why do many women hit a glass ceiling? • How do gender stereotypes affect women in groups and organizations?
Symbolic Interactionist	Micro	People aren't puppets but can affect what goes on in a group or organization.	• Why do people ignore or change an organization's rules? • How do members of social groups influence workplace behavior?

7-1 **What Is Deviance?** *Deviance*, the violation of social norms, includes *crime*. Perceptions of deviance and crime vary across and within societies, can change over time. People with authority or power decide what's right or wrong, and a *stigma* accompanies deviant behavior.

Example: Deviance and College Drinking

According to many college presidents, alcohol abuse is the most serious problem on campus. Alcohol abuse leads to alcohol poisoning, blackouts, sexual assault, violent behavior, injuries, and academic problems. Because drinking laws are rarely enforced, some college presidents have proposed that the drinking age be lowered from 21 to 18.

Others argue that doing so would increase traffic fatalities and drinking problems. Young people can get a driver's license at 16 and vote and enlist in the military at 18. Should they be the ones, then, to decide whether drinking laws should be changed?

Deviance a violation of social norms. (p. 119)

Crime a violation of society's formal laws. (p. 119)

Stigma a negative label that devalues a person and changes her or his self-concept and social identity. (p. 119)

7-2 **Types of Deviance and Crime** Sociologists study *noncriminal deviance*—such as suicide, alcoholism, lying, mental illness, and adult pornography—and *criminal deviance*, behavior that violates laws. Two of the most important sources of U.S. crime statistics are the FBI's *Uniform Crime Report*, and *victimization surveys*.

The media usually focuses on street crimes, but Americans are much more likely to be victimized by *hate crimes*, *white-collar crimes*, *corporate crimes*, *cybercrime*, and *organized crime*. *Victimless crimes* violate laws, but the parties involved don't consider themselves victims.

Victimization survey interviews people about being crime victims. (p. 121)

Hate crime (also known as *bias crime*) a criminal offense motivated by the fact or perception that the victims differ from the perpetrator. (p. 123)

White-collar crime illegal activities committed by high-status people in the course of their occupations. (p. 123)

Corporate crimes (also called *organizational crimes*) illegal acts committed by executives to benefit themselves and their companies. (p. 124)

Cybercrime (also called *computer crime*) illegal activities that are conducted online. (p. 124)

Organized crime activities of individuals and groups that supply illegal goods and services for profit. (p. 125)

Victimless crimes (also called *public order crimes*) illegal acts that have no direct victim. (p. 125)

7-3 **Functionalist Perspectives on Deviance** See Table 7.1 on the next page.

Anomie the condition in which people are unsure how to behave because of absent, conflicting, or confusing social norms. (p. 127)

Strain theory people may engage in deviant behavior when they experience a conflict between goals and the means available to obtain the goals. (p. 127)

7-4 **Conflict Perspectives on Deviance** See Table 7.1 on the next page.

7-5 **Feminist Perspectives on Deviance** See Table 7.1 on the next page.

Patriarchy a social system in which men control cultural, political, and economic structures. (p. 131)

Rape culture an environment in which sexual violence is prevalent, pervasive, and perpetuated by the media and popular culture. (p. 131)

7 Deviance, Crime, and Social Control

7-6 Symbolic Interaction Perspectives on Deviance
See Table 7.1 below.

Differential association theory asserts that people learn deviance through interaction, especially with significant others. (p. 132)

Labeling theory posits that society's reaction to behavior is a major factor in defining oneself or others as deviant. (p. 132)

Primary deviance the initial act of breaking a rule. (p. 133)

Secondary deviance rule-breaking behavior that people adopt in response to others' reactions. (p. 133)

Medicalization of deviance diagnosing and treating a violation of social norms as a medical disorder. (p. 133)

7-7 Controlling Deviance and Crime The purpose of *social control* is to eliminate, or at least reduce, deviance and crime. Those who have authority or power administer formal social control. Informal social control is internalized from childhood.

The *criminal justice system* relies on three major approaches in controlling crime: prevention and intervention, punishment, and rehabilitation. A *crime control model* supports a tough approach toward criminals in sentencing, imprisonment, and capital punishment. In contrast, many people believe that *rehabilitation* can change offenders into productive and law-abiding citizens.

Social control the techniques and strategies that regulate people's behavior in society. (p. 134)

Control theory proposes that deviant behavior decreases when people have strong social bonds with others. (p. 134)

Criminal justice system government agencies that are charged with enforcing laws, judging offenders, and changing criminal behavior. (p. 134)

Crime control model proposes that crime rates increase when offenders don't fear apprehension or punishment. (p. 135)

Table 7.1 Sociological Explanations of Deviance

Theoretical Perspective	Level of Analysis	Key Points
Functionalist	Macro	• Deviance is both functional and dysfunctional. • Anomie increases the likelihood of deviance. • People are deviant when they experience blocked opportunities to achieve the culturally approved goal of economic success.
Conflict	Macro	• There's a strong association between capitalism, social inequality, power, and deviance. • The most powerful groups define what's deviant. • Laws rarely punish the illegal activities of the powerful.
Feminist	Macro and micro	• There's a large gender gap in deviant behavior. • Women's deviance reflects their general oppression due to social, economic, and political inequality. • Many women are offenders or victims because of patriarchal beliefs and practices and living in a rape culture.
Symbolic Interactionist	Micro	• Deviance is socially constructed. • People learn deviant behavior from significant others such as parents and friends. • If people are labeled or stigmatized as deviant, they're likely to develop negative self-concepts and engage in criminal behavior.

8-1 Social Stratification Systems and Bases *Social stratification* is a society's ranking of people who have different access to valued resources. A *closed stratification system* differs from an *open stratification system* because the latter allows movement from one social class to another. Stratification includes *wealth, prestige,* and *power*. People are more likely to experience status inconsistency if they rank differently on these three dimensions, such as a football player who has great wealth but little power.

Social stratification a society's ranking of people based on their access to valued resources such as wealth, power, and prestige. (p. 139)

Slavery system people own others as property and have almost total control over their lives. (p. 139)

Caste system people's positions are ascribed at birth and largely fixed. (p. 139)

Class system people's positions are based on both birth and achievement. (p. 140)

Social class people who have a similar standing or rank in a society based on wealth, education, power, prestige, and other valued resources. (p. 140)

Wealth economic assets that a person or family owns. (p. 140)

Income the money a person receives, usually through wages or salaries, but can also include other earnings. (p. 140)

Prestige respect or recognition attached to social positions. (p. 141)

Power the ability to influence or control the behavior of others despite opposition. (p. 142)

8-2 Social Class in America A good indicator of social class is *socioeconomic status (SES)*, an overall rank based on a person's income, education, and occupation. Using SES and other variables (such as values, power, and *conspicuous consumption*), most sociologists agree that there are at least four social classes in the United States: upper, middle, working, and lower. These groups can be divided further into upper-upper, lower-upper, upper-middle, lower-middle, and the working class. The lower class includes the *working poor* and the *underclass*. A major outcome of social stratification is *life chances*.

Example: Restaurant Menus and Stratification

Two sociologists—in Iowa and Virginia—asked their students in introductory sociology classes to do a content analysis (see Chapter 1) of 10 menus that represented a sampling of restaurants by social class. The students found that the restaurants that catered to upper-class clientele had higher than average entrée prices, described the entrées in foreign languages, used fancy sauces, recommended expensive wines, and had few illustrations. Middle-class menus emphasized "value for the dollar," presented photos of entrées with "bountiful plates overflowing with appetizing food," and popular items such as quesadillas. Menus at lower-class restaurants featured low prices ($3 to $10 entrées), the items were numbered, none of the entrées had "pretentious names," and the typesetting was simple (Wright and Ransom, 2005). In effect, then, even menus denote social class and social status.

Socioeconomic status (SES) an overall ranking of a person's position in society based on income, education, and occupation. (p. 142)

Working poor people who work at least 27 weeks a year but whose wages fall below the official poverty level. (p. 145)

Underclass people who are persistently poor and seldom employed, residentially segregated, and relatively isolated from the rest of the population. (p. 145)

8-3 Poverty *Absolute poverty* is a more serious social problem than *relative poverty* because millions of Americans live below the *poverty line*. Poverty varies by age, gender, family structure, race, and ethnicity. One explanation for poverty maintains that people are poor because of individual failings; another contends that societal structures create and sustain poverty.

Absolute poverty not having enough money to afford the basic necessities of life. (p. 145)

Relative poverty not having enough money to maintain an average standard of living. (p. 145)

Poverty line the minimal income level that the federal government considers necessary for basic subsistence (also called the *poverty threshold*). (p. 146)

Feminization of poverty the disproportionate number of the poor who are women. (p. 147)

8-4 Social Mobility *Social mobility* can be *horizontal, vertical, intragenerational,* or *intergenerational*. There has been a sharp decline in the share of middle-income adults since 1971 because many have slid into a lower class. Structural, demographic, and individual factors affect a person's social mobility. Much social mobility depends on structural, demographic, and family background factors, all of which are interrelated.

Social mobility movement from one social class to another. (p. 148)

Intragenerational mobility movement up or down a social class over one's lifetime. (p. 148)

Intergenerational mobility movement up or down a social class over two or more generations. (p. 149)

8 Social Stratification: United States and Global

8-5 Global Stratification Global inequality is widespread, and all societies are stratified, but some countries are much wealthier than others, and inequality is increasing across the globe. Historically, currently, and across all nations, women and children experience the greatest poverty. Sociologists use *modernization theory, dependency theory*, and *world-system theory* to explain why inequality is universal.

Global stratification worldwide inequality patterns that result from differences in wealth, power, and prestige. (p. 151)

Infant mortality rate the number of babies under age 1 who die per 1,000 live births in a given year. (p. 152)

8-6 Sociological Explanations: Why There Are Haves and Have-Nots See Table 8.1 below.

Davis–Moore thesis the functionalist view that social stratification benefits a society. (p. 154)

Meritocracy a belief that social stratification is based on people's accomplishments. (p. 155)

Bourgeoisie those who own and control capital and the means of production. (p. 155)

Proletariat workers who sell their labor for wages. (p. 155)

Corporate welfare subsidies, tax breaks, and assistance that the government has created for businesses. (p. 156)

Gender stratification unequal access to wealth, power, status, prestige, and other valued resources because of one's sex. (p. 156)

Table 8.1 Sociological Explanations of Social Stratification

PERSPECTIVE	LEVEL OF ANALYSIS	KEY POINTS
Functionalist	Macro	• Fills social positions that are necessary for a society's survival • Motivates people to succeed and ensures that the most qualified people will fill the most important positions
Conflict	Macro	• Encourages workers' exploitation and promotes the interests of the rich and powerful • Ignores a wealth of talent among the poor
Feminist	Macro and micro	• Constructs numerous barriers in patriarchal societies that limit women's achieving wealth, status, and prestige • Requires most women, not men, to juggle domestic and employment responsibilities that impede upward mobility
Symbolic Interactionist	Micro	• Shapes stratification through socialization, everyday interaction, and group membership • Reflects social class identification through symbols, especially products that signify social status

9-1 Sex, Gender, and Culture *Sex* refers to biological characteristics, whereas *gender* refers to learned attitudes and behaviors. *Gender roles* differ depending on whether people perceive themselves as masculine or feminine and because a society expects women and men to think and behave differently. Many Americans still have *gender stereotypes* about how people will look, act, think, and feel based on their sex.

Our sexual identity incorporates a *sexual orientation* that can be *homosexual*, *heterosexual*, *bisexual*, or *asexual*. Many biological theories maintain that sexual orientation has a strong genetic basis, but social constructionists argue that culture, not biology, plays a large role in forming people's *gender identity*.

Sex the biological characteristics with which we are born. (p. 161)

Gender learned attitudes and behaviors that characterize women and men. (p. 161)

Intersexuals people whose sex at birth isn't clearly either male or female. (p. 161)

Sexual identity an awareness of ourselves as male or female and how we express our sexual values, attitudes, and feelings. (p. 162)

Sexual orientation a preference for sexual partners of the same sex, of the opposite sex, of both sexes, or neither sex. (p. 162)

Homosexuals those who are sexually attracted to people of the same sex. (p. 162)

Heterosexuals those who are sexually attracted to people of the opposite sex. (p. 162)

Bisexuals those who are sexually attracted to more than one gender. (p. 162)

Asexuals those who lack any interest in or desire for sex. (p. 162)

Gender identity a perception of oneself as either masculine or feminine. (p. 163)

Transgender people whose gender identity and behavior don't correspond with their birth sex. (p. 163)

Gender expression how a person communicates gender identity to others. (p. 163)

Gender roles the characteristics, attitudes, feelings, and behaviors that society expects of females and males. (p. 164)

Gender stereotypes expectations about how people will look, act, think, and feel based on their sex. (p. 164)

Sexism an attitude or behavior that discriminates against one sex, usually females, based on the assumed superiority of the other sex. (p. 164)

Heterosexism belief that heterosexuality is the only legitimate sexual orientation. (p. 165)

Homophobia a fear and hatred of lesbians and gays. (p. 165)

9-2 Contemporary Gender Stratification *Sexism* is widespread because of *gender stratification*, which can lead to inequality in the family (particularly child care and housework), education, the workplace (as in *occupational sex segregation* and a *gender pay gap*), and politics. *Sexual harassment* and workplace bullying are also common in the workplace.

Occupational sex segregation (sometimes called *occupational gender segregation*) the process of channeling women and men into different types of jobs. (p. 167)

Gender pay gap the difference between men's and women's earnings (also called the *wage gap*, *pay gap*, and *gender wage gap*). (p. 167)

Sexual harassment any unwanted sexual advance, request for sexual favors, or other conduct of a sexual nature that makes a person uncomfortable and interferes with her or his work. (p. 168)

9-3 Sexuality Sexual attitudes and behavior can vary from situation to situation and change over time, including why we have sex. Contrary to some stereotypes, adolescents aren't sexually promiscuous and older people aren't asexual. All of us have internalized *sexual scripts*, and *sexual double standards* persist.

Example: Gender Roles and Hooking Up—Are Women the Losers?

Hooking up refers to physical encounters, no strings attached, and can mean anything from kissing and genital fondling to oral sex and sexual intercourse. The prevalence of hooking up has increased only slightly since the late 1980s, but is now more common than dating at many high schools and colleges. Hooking up has its advantages. It's much cheaper than dating, and requires no commitment of time or emotion. In addition, hookups remove the stigma from those who can't get dates but can experience sexual pleasure, and they make people feel sexy and desirable. Hooking up also has disadvantages, especially for women, because they generally get a bad reputation as being "easy" (see Chapter 9). In effect, then, and despite the advantages of hooking up, the sexual double standard persists.

Sexual script specifies the formal and informal norms for acceptable or unacceptable sexual behavior. (p. 170)

Sexual double standard a code that permits greater sexual freedom for men than women. (p. 170)

9-4 Some Current Social Issues about Sexuality Two of the most controversial and politically contested issues continue to be *abortion* and *same-sex marriage*. Almost equal percentages of Americans support or condemn abortion. Those who favor same-sex marriage argue that people should have the same legal rights regardless of sexual orientation; those who oppose same-sex marriage contend that such unions are immoral and contrary to religious beliefs.

Abortion expulsion of an embryo or fetus from the uterus. (p. 171)

9 Gender and Sexuality

9-5 Gender and Sexuality across Cultures Worldwide, women have fewer rights and opportunities than men. In all countries and regions, the greatest gender gaps are in economic opportunity and participation and political leadership. People in some countries are more accepting of homosexuality than in the past, but *heterosexism* prevails.

Same-sex marriage (also called *gay marriage*) a legally recognized marriage between two people of the same biological sex and/or gender identity. (p. 173)

9-6 Sociological Explanations of Gender and Sexuality See Table 9.5 on below page.

Table 9.5 Sociological Explanations of Gender and Sexuality

THEORETICAL PERSPECTIVE	LEVEL OF ANALYSIS	KEY POINTS
Functionalist	Macro	• Gender roles are complementary, equally important for a society's survival, and affect human capital. • Agreed-on sexual norms contribute to a society's order and stability.
Conflict	Macro	• Gender roles give men power to control women's lives. • Most societies regulate women's, but not men's, sexual behavior.
Feminist	Macro and micro	• Women's inequality reflects their historical and current domination by men, especially in the workplace. • Many men use violence—including sexual harassment, rape, and global sex trafficking—to control women's sexuality.
Symbolic Interactionist	Micro	• Gender is a social construction that emerges and is reinforced through everyday interactions. • The social construction of sexuality varies across cultures because of societal norms and values.

10-1 U.S. Racial and Ethnic Diversity Perhaps the most multicultural country in the world, the United States includes about 150 distinct ethnic or racial groups among more than 312 million inhabitants. By 2025, only 58 percent of the U.S. population will be white—down from 86 percent in 1950.

10-2 The Social Significance of Race and Ethnicity *Race* refers to physical characteristics, whereas an *ethnic group* identifies with a common national origin or cultural heritage. A *racial-ethnic group* has both distinctive physical and cultural characteristics.

Racial group people who share visible physical characteristics that members of a society consider socially important. (p. 183)

Ethnic group people who identify with a common national origin or cultural heritage. (p. 184)

Racial-ethnic group people who have distinctive physical and cultural characteristics. (p. 184)

10-3 Our Changing Immigration Mosaic In 1900, almost 85 percent of immigrants came from Europe; now immigrants come primarily from Asia and Latin America. Many Americans are ambivalent about immigrants, especially those who are in the country illegally. Many scholars argue, however, that, in the long run, both legal and undocumented immigrants bring more benefits than costs.

10-4 Dominant and Minority Groups A *dominant group* has more economic and political power than a *minority*. The latter may be larger in number than a dominant group but is often subject to differential and unequal treatment because of its physical, cultural, or other characteristics. Patterns of dominant-minority group relations include *genocide*, *segregation*, *acculturation*, *assimilation*, and *pluralism*.

Dominant group a physically or culturally distinctive group that has the most economic and political power, the greatest privileges, and the highest social status. (p. 187)

Minority people who may be treated differently and unequally because of their physical, cultural, or other characteristics. (p. 187)

Genocide the systematic effort to kill all members of a particular ethnic, religious, political, racial, or national group. (p. 187)

Segregation physical and social separation of dominant and minority groups. (p. 188)

Acculturation the process of adopting the language, values, beliefs, and other characteristics of the host culture. (p. 188)

Assimilation conforming to the dominant group's culture, adopting its language and values, and intermarrying with that group. (p. 189)

Pluralism minority groups maintain many aspects of their original culture while living peacefully with the host culture. (p. 189)

10-5 Some Sources of Racial-Ethnic Friction *Racism* justifies and preserves the social, economic, and political interests of dominant groups. *Prejudice* is an attitude; *discrimination* is an act that occurs at both the individual and institutional level. All of us can be prejudiced, but minorities are typically targets of *stereotypes* and *ethnocentrism* that often lead to *scapegoating*.

Example: Stuff White People Like

The popular blog *Stuff White People Like* has generated clones (e.g., *Stuff Educated Black People Like* and *Stuff Asian People Like*). Why are these sites so popular? Many fans say that the descriptions are funny because they're true. According to some critics, however, by poking fun at privileged upper-middle-class white people, the sites fuel stereotypes instead of having painfully frank discussions about U.S. race and racism (Sternbergh, 2008). Do you agree or disagree?

Racism beliefs that one's own racial group is inherently superior to other groups. (p. 189)

Prejudice an attitude that prejudges people, usually in a negative way. (p. 190)

Stereotype an oversimplified or exaggerated generalization about a group of people. (p. 190)

Scapegoats individuals or groups whom people blame for their own problems or shortcomings. (p. 190)

Discrimination behavior that treats people unequally because of some characteristic. (p. 190)

Individual discrimination unequal treatment on a one-to-one basis. (p. 190)

Institutional discrimination unequal treatment because of a society's everyday laws, policies, practices, and customs. (p. 190)

10-6 Major U.S. Racial and Ethnic Groups European Americans, who settled the first colonies, are declining in population, whereas Latinos now comprise 18 percent of the population. Other large racial and ethnic populations are African Americans (12 percent), Asian Americans (6 percent), and American Indians (almost 2 percent). Middle Eastern Americans, who comprise less than 0.5 percent of the population, come from more than 30 countries. All of these groups have experienced prejudice and discrimination, but they have enhanced U.S. society and culture.

10-7 Sociological Explanations of Racial-Ethnic Inequality See Table 10.3 below.

Gendered racism the overlapping and cumulative effects of inequality due to racism *and* sexism. (p. 200)

Contact hypothesis posits that the more people get to know members of a minority group personally, the less likely they are to be prejudiced against that group. (p. 201)

10-8 Interracial and Interethnic Relationships About 97 percent of Americans report being only one race, but the numbers of biracial children are increasing because of interracial dating and marriage. The rise of racial-ethnic intermarriage reflects many micro and macro factors such as greater interethnic and interracial contact, changing attitudes, and acculturation.

Miscegenation marriage or sexual relations between a man and a woman of different races. (p. 203)

Table 10.3 Sociological Explanations of Racial-Ethnic Inequality

THEORETICAL PERSPECTIVE	LEVEL OF ANALYSIS	KEY POINTS
Functionalist	Macro	Immigration provides needed workers; acculturation and assimilation increase social solidarity; racial-ethnic inequality can be dysfunctional, but benefits dominant groups.
Conflict	Macro	There's ongoing strife between dominant and minority groups; powerful groups maintain their advantages primarily through economic exploitation; race is a more important factor than social class in perpetuating racial-ethnic inequality.
Feminist	Macro and micro	Minority women suffer from the combined effects of racism and sexism; gendered racism occurs within and across racial-ethnic groups.
Symbolic Interactionist	Micro	Because race and ethnicity are socially constructed, social interaction can increase or reduce racial and ethnic hostility; antagonistic attitudes toward minorities, which are learned, can be lessened through cooperative interracial and interethnic contacts.

11-1 Global Economic Systems The *economy* determines how a society produces, distributes, and consumes goods and services. *Capitalism* frequently spawns *monopolies* and *oligopolies* that dominate the market and discourage competition. *Socialism* and *communism* promise cooperation and collective ownership of property, but all countries have mixed economies that allow private profits.

Economy determines how a society produces, distributes, and consumes goods and services. (p. 205)

Politics individuals and groups acquire and exercise power and authority and make decisions. (p. 205)

Monopoly domination of a particular market or industry by one person or company. (p. 205)

Oligopoly domination of a market by a few large producers or suppliers. (p. 205)

Socialism an economic system based on the public ownership of the production of goods and services. (p. 205)

Communism a political and economic system in which property is communally owned and all people are considered equal. (p. 206)

Welfare capitalism (also called *state capitalism*) an economic system that combines private ownership of property, market competition, and a government's regulation of many programs and services. (p. 206)

11-2 Corporations and the Economy A *corporation*, usually created for profit, often forms a *conglomerate*. Both corporations and conglomerates are governed by *interlocking directorates* that have become more powerful than ever because of the growth of *transnational corporations* and *transnational conglomerates*.

Corporation an organization that has legal rights, privileges, and liabilities apart from those of its members. (p. 207)

Conglomerate a corporation that owns a collection of companies in different industries. (p. 207)

Interlocking directorate the same people serve on the boards of directors of several companies or corporations. (p. 207)

Transnational corporation (also called a *multinational corporation* or an *international corporation*) a large company that's based in one country but operates across international boundaries. (p. 208)

Transnational conglomerate (also called a *multinational conglomerate*) a corporation that owns a collection of different companies in various industries in a number of countries. (p. 208)

11-3 Work in U.S. Society Today *Work* produces goods or services. Many Americans have been casualties of *deindustrialization* and *globalization*. Others have lost their jobs to *offshoring*. Widespread *downsizing* has resulted in unemployment, *underemployment*, discouraged workers, and low-paid jobs.

Work a physical or mental activity that produces goods or services. (p. 208)

Deindustrialization social and economic change that reduces industrial activity, especially manufacturing. (p. 208)

Globalization the growth and spread of investment, trade, production, communication, and new technology around the world. (p. 209)

Offshoring sending work or jobs to another country to cut a company's costs at home. (p. 209)

Discouraged workers people who stop looking for work because they believe that job hunting is futile. (p. 211)

11-4 Sociological Explanations of Work and the Economy
See Table 11.3 on the next page.

Motherhood penalty (also called *motherhood wage penalty* or *mommy penalty*) a pay gap between women who are and aren't mothers. (p. 217)

11-5 Global Political Systems In a *democracy*, ideally, citizens have a high degree of control over the state. *Totalitarianism* controls people's lives; *authoritarianism* generally permits some degree of individual freedom. A *monarchy* is the oldest type of authoritarian regime.

Government a formal organization that has the authority to make and enforce laws. (p. 217)

Democracy a political system in which, ideally, citizens have control over the state and its actions. (p. 218)

Totalitarianism the government controls almost every aspect of people's lives. (p. 218)

Authoritarianism the state controls the lives of citizens but permits some degree of individual freedom. (p. 219)

Monarchy power is based on heredity and passes from generation to generation. (p. 219)

11-6 Politics, Power, and Authority Politics includes *power* and *authority*. *Legitimate power* comes from having a role, position, or title that people accept as legal and appropriate; *coercive power* relies on force or threat of force to impose one's will on others. Authority can be based on *tradition*, *charisma*, *rational-legal power*, or a combination of these sources.

11 The Economy and Politics

Power the ability of a person or group to influence others, even if they resist. (p. 219)

Authority the legitimate use of power. (p. 219)

Traditional authority power based on customs that justify the ruler's position. (p. 220)

Charismatic authority power based on exceptional individual abilities and characteristics that inspire devotion, trust, and obedience. (p. 220)

Rational-legal authority power based on the belief that laws and appointed or elected political leaders are legitimate. (p. 220)

11-7 **Politics and Power in U.S. Society** A *political party* tries to influence and control government. Whereas political parties include diverse individuals, *special-interest groups* are usually made up of people who are similar in social class and political objectives. The most powerful special-interest groups are *political action committees* and *lobbyists*. The voting rate is much higher among some groups than others. Situational and structural factors, such as voter registration, requiring proof of citizenship, and convenience can also encourage or discourage voting.

Political party an organization that tries to influence and control government by recruiting, nominating, and electing its members to public office. (p. 221)

Special-interest group (also called an *interest group*) a group of people that seeks or receives benefits or special treatment. (p. 222)

Political action committee (PAC) a special-interest group that raises money to elect one or more candidates to public office. (p. 222)

Lobbyist someone hired by a special-interest group to influence legislation on the group's behalf. (p. 222)

11-8 **Sociological Perspectives on Politics and Power**
See Table 11.7 below.

Pluralism a political system in which power is distributed among a variety of competing groups in a society. (p. 225)

Power elite a small group of influential people who make the nation's major political decisions. (p. 226)

Table 11.3 Sociological Explanations of Work and the Economy

Theoretical Perspective	Level of Analysis	Key Points
Functionalist	Macro	Capitalism benefits society; work provides an income, structures people's lives, and gives them a sense of accomplishment.
Conflict	Macro	Capitalism enables the rich to exploit other groups; most jobs are low-paying, monotonous, and alienating; productivity isn't always rewarded.
Feminist	Macro and micro	Gender roles structure women's and men's work experiences differently and inequitably.
Symbolic Interactionist	Micro	How people define and experience work in their everyday lives affects their workplace behavior and relationships with coworkers and employers.

Table 11.7 Sociological Explanations of Political Power

	Functionalism: A Pluralist Model	Conflict Theory: A Power Elite Model	Feminist Theories: A Patriarchal Model
Who has political power?	The people	Rich upper-class people—especially those at top levels in business, government, and the military	White men in Western countries; most men in traditional societies
How is power distributed?	Very broadly	Very narrowly	Very narrowly
What is the source of political power?	Citizens' participation	Wealthy people in government, business corporations, the military, and the media	Being white, male, and very rich
Does one group dominate politics?	No	Yes	Yes
Do political leaders represent the average person?	Yes, the leaders speak for a majority of the people.	No, the leaders are most concerned with keeping or increasing their personal wealth and power.	No, the leaders are rarely women who have decision-making power.

12-1 **What Is a Family?** Among other activities, the members of a *family* care for one another and any children. Worldwide, however, families vary in characteristics such as structure (a *nuclear* or an *extended* family), living arrangements (*patrilocal*, *matrilocal*, or *neolocal*), who has authority (*matriarchal*, *patriarchal*, or *egalitarian*), and how many marriage mates a person can have (*monogamy* or *polygamy*).

Family an intimate group consisting of two or more people who (1) have a committed relationship, (2) care for one another and any children, and (3) share activities and close emotional ties. (p. 231)

Incest taboo cultural norms and laws that forbid sexual intercourse between close blood relatives. (p. 231)

Marriage a socially approved mating relationship. (p. 232)

Endogamy (often used interchangeably with *homogamy*) cultural practice of marrying within one's group. (p. 232)

Exogamy (often used interchangeably with *heterogamy*) cultural practice of marrying outside one's group. (p. 232)

Nuclear family a family form composed of married parents and their biological or adopted children. (p. 232)

Extended family a family form composed of parents, children, and other kin. (p. 233)

Patrilocal residence pattern newly married couples live with the husband's family. (p. 233)

Matrilocal residence pattern newly married couples live with the wife's family. (p. 233)

Neolocal residence pattern a newly married couple sets up its own residence. (p. 233)

Boomerang generation young adults who move back into their parents' home or never leave it in the first place. (p. 233)

Matriarchal family system the oldest females control cultural, political, and economic resources and, consequently, have power over males. (p. 233)

Patriarchal family system the oldest males control cultural, political, and economic resources and, consequently, have power over females. (p. 233)

Egalitarian family both partners share power and authority fairly equally. (p. 234)

Marriage market prospective spouses compare the assets and liabilities of eligible partners and choose the best available mate. (p. 234)

Arranged marriage parents or relatives choose the children's spouses. (p. 234)

Monogamy one person is married exclusively to another person. (p. 234)

Serial monogamy individuals marry several people, but one at a time. (p. 234)

Polygamy a man or woman has two or more spouses. (p. 234)

12-2 **How U.S. Families Are Changing** The United States has one of the highest marriage and divorce rates in the world. *Divorce* is easier to obtain than in the past because all states now have *no-fault divorce* laws, but divorce rates vary by race, ethnicity, and social class. The number of single people has risen greatly, primarily because many people are postponing marriage. There has also been a striking increase in *cohabitation* and nonmarital births. Female-headed single-parent families are increasingly common.

Cohabitation two unrelated and unmarried people live together and are in a sexual relationship. (p. 236)

Fictive kin nonrelatives who are accepted as part of a family. (p. 241)

12-3 **Family Conflict and Violence** We're more likely to experience *intimate partner violence (IPV)* than assault by a stranger. IPV causes are complex and often the product of multiple individual, demographic, and societal factors. *Child maltreatment* is widespread, and parents are the most common offenders. Similarly, in *elder abuse*, most of the offenders are adult children, spouses, or other family members. Both micro- and macro-level reasons help explain IPV, child maltreatment, and elder abuse and neglect.

Intimate partner violence (IPV) abuse that occurs between people in a close relationship. (p. 242)

Child maltreatment (also called *child abuse*) a broad range of behaviors that can result in serious emotional or physical harm. (p. 243)

Elder abuse (also called *elder mistreatment*) any knowing, intentional, or negligent act by a caregiver or other person that causes harm to people age 65 or older. (p. 244)

12-4 **Our Aging Society** How people define "old" varies across societies. There are also significant differences between the *young-old*, the *old-old*, and the *oldest-old* in their health, ability to live independently, and to work. Historically and currently, women live longer than men, but life expectancy varies by race, ethnicity, and social class. Our *old-age dependency ratio* has increased, placing a larger burden on the working-age population. Some important current aging issues include the rise of *multigenerational households*, physician-assisted suicide, and competition for scarce resources.

Example: "He Gets Prettier; I Get Older"
When comparing her own public image with that of her actor husband, the late Paul Newman, actress Joanne Woodward once remarked, "He gets prettier; I get older." Was she right? If aging gracefully is acceptable, why do many companies tout antiaging products? And why are the products targeted primarily at women?

Life expectancy the average expected number of years of life remaining at a given age. (p. 245)

Baby boomers people born between 1946 and 1964. (p. 246)

Old-age dependency ratio (also called the *elderly support ratio*) the number of working age (18 to 64) adults for every person aged 65 and older. (p. 246)

12-5 Sociological Explanations of Family and Aging

Activity theory proposes that many older people remain engaged in numerous roles and activities, including work. (p. 249)

Exchange theory people seek through their social interactions to maximize their rewards and minimize their costs. (p. 251)

Ageism discrimination against older people. (p. 252)

Continuity theory older adults can substitute satisfying new roles for those they've lost. (p. 252)

Table 12.3 Sociological Perspectives on Families and Aging

Theoretical Perspective	Level of Analysis	Key Points
Functionalist	Macro	Families are important in maintaining societal stability and meeting family members' needs. Older people who are active and engaged are more satisfied with life.
Conflict	Macro	Families promote social inequality because of social class differences. Many corporations view older workers as disposable.
Feminist	Macro and micro	Families both mirror and perpetuate patriarchy and gender inequality. Women have an unequal burden in caring for children as well as older family members and relatives.
Symbolic Interactionist	Micro	Families construct their everyday lives through interaction and subjective interpretations of family roles. Many older family members adapt to aging and often maintain previous activities.

13-1 **Education and Society** U.S. *education* and *schooling* have changed in four important ways: Universal education has expanded, community colleges have flourished, public higher education has burgeoned, and student diversity has greatly increased.

Education transmits attitudes, knowledge, beliefs, values, norms, and skills. (p. 255)

Schooling formal training and instruction provided in a classroom setting. (p. 255)

13-2 **Sociological Perspectives on Education** See Table 13.1 on the next page.

Achievement gap a persistent and significant disparity in academic performance between different groups of students. (p. 257)

Hidden curriculum school practices that transmit nonacademic knowledge, values, attitudes, norms, and beliefs. (p. 258)

Credentialism an emphasis on certificates or degrees to show that people have certain skills, educational attainment levels, or job qualifications. (p. 258)

Tracking assigning students to specific educational programs and classes on the basis of test scores, previous grades, or perceived ability. (p. 262)

Implicit bias unconscious prejudices or stereotypes that affect our attitudes, actions, and decisions. (p. 262)

13-3 **Some Current Issues in U.S. Education** Compared with their counterparts in a number of other countries, many U.S. students are performing poorly in elementary and high schools—especially in mathematics and the sciences. U.S. teachers' salaries are low, many are poorly prepared, out of field, and have less control over curricula than ever before. A growing number of critics contend that standardized tests (such as SATs and ACTs) are little more than gatekeeping tools that exclude lower socioeconomic students from access to higher education, and that the tests scores don't predict college success. Despite grade inflation, high school and college dropout rates are high, and cheating is widespread.

Meritocracy a system that rewards people because of their individual accomplishments. (p. 266)

13-4 **Religion and Society** Some form of religion exists in all societies and cultures. *Religion* unites believers into a community. Every known society distinguishes between *sacred* and *profane (secular)* activities. Religion, *religiosity*, and spirituality differ; for example, people who describe themselves as religious may not attend services.

Example: Sacred vs. Secular

The corporate mission of Chick-fil-A (a franchise that prepares sandwiches), as stated on a plaque at company headquarters, is "to glorify God." Chick-fil-A is the only national fast-food chain that closes on Sunday so employees can go to church, and prospective employees are asked about their religious activities. Some franchise operators are delighted with the religious emphasis; others believe that a business should stay out of its workers' personal lives (Schmall, 2007). What do *you* think?

Religion a social institution that involves shared beliefs, values, and practices related to the supernatural. (p. 269)

Sacred anything that people see as awe-inspiring, supernatural, holy, and not part of the natural world. (p. 269)

Profane the ordinary and everyday elements of life that aren't related to religion. (p. 269)

Religiosity how people demonstrate their religious beliefs. (p. 269)

13-5 **Religious Organization and Major World Religions** People express their religious beliefs most commonly through organized groups, including *cults* (also called *new religious movements [NRMs]*), *sects*, *denominations*, and *churches*. Some NRMs, which usually are organized around a *charismatic leader*, have been short-lived, whereas others have become established religions with highly organized bureaucracies. Worldwide, the largest religious group is Christians, followed by Muslims, but no religious group comes close to being a global majority. The third largest group is people who don't have a religious affiliation.

13 Education and Religion

Cult a religious group devoted to beliefs and practices that are outside of those accepted in mainstream society. (p. 269)

New religious movement (NRM) term used instead of *cult* by most sociologists. (p. 269)

Charismatic leader someone that followers see as having exceptional or superhuman powers and qualities. (p. 269)

Sect a religious group that has broken away from an established religion. (p. 270)

Denomination a subgroup within a religion that shares its name and traditions and is generally on good terms with the main group. (p. 270)

Church a large established religious group that has strong ties to mainstream society. (p. 270)

Ecclesia (also called a *state religion*) an official religious organization that claims everyone in society as its members. (p. 270)

13-6 **Religion in the United States** Religion in the United States is complex and diverse. About half of U.S. adults have changed their religion since childhood, many opting for no religion at all. Mainline Protestant groups have declined whereas evangelicals have surged. Religious participation varies by gender, age, race, ethnicity, and social class. Some sociologists maintain that *secularization* is increasing rapidly in the United States; others contend that this claim is greatly exaggerated, especially as witnessed by the growth of *fundamentalism* and the prevalence of *civil religion*.

Secularization religion loses its social and cultural influence. (p. 273)

Fundamentalism belief in the literal meaning of a sacred text. (p. 274)

Civil religion (sometimes called *secular religion*) integrating religious beliefs into secular life. (p. 275)

13-7 **Sociological Perspectives on Religion** See Table 13.4 below.

False consciousness an acceptance of a system of beliefs that prevents people from protesting oppression. (p. 277)

Table 13.1 Sociological Explanations of Education

Theoretical Perspective	Level of Analysis	Key Points
Functionalist	Macro	Contributes to society's stability, solidarity, and cohesion and provides opportunities for upward mobility
Conflict	Macro	Reproduces and reinforces inequality and maintains a rigid social class structure
Feminist	Macro and micro	Produces inequality based on gender
Symbolic Interactionist	Micro	Teaches roles and values through everyday face-to-face interaction and behavior

Table 13.4 Sociological Explanations of Religion

Theoretical Perspective	Level of Analysis	Key Points
Functionalist	Macro	Religion benefits society by providing a sense of belonging, identity, meaning, emotional comfort, and social control over deviant behavior.
Conflict	Macro	Religion promotes and legitimates social inequality, condones strife and violence between groups, and justifies oppression of poor people.
Feminist	Macro and micro	Religion subordinates women, excludes them from decision-making positions, and legitimizes patriarchal control of society.
Symbolic Interactionist	Micro	Religion provides meaning and sustenance in everyday life through symbols, rituals, and beliefs, and binds people together in a physical and spiritual community.

14-1 **Global Health and Illness** *Health* varies among individuals and societies, but all people experience *disease*. In addressing the question of why some people are healthier than others, health and medical practitioners, researchers, and sociologists examine *social epidemiology*, which includes both the incidence and prevalence of illness or health problems within a population during a specific time period. Worldwide, high income populations live longer, have low infant mortality rates, and spend more on health, but experience "diseases of wealth" such as diabetes, heart disease, and various cancers.

Health the state of physical, mental, and social well-being. (p. 283)

Social epidemiology examines how societal factors affect the distribution of disease within a population. (p. 283)

Health care the prevention, management, and treatment of illness. (p. 283)

14-2 **Health and Illness in the United States** Many people live with a *disability*; others die at an earlier age than expected for a variety of macro- and micro-level reasons. The three most important reasons for illness and early death are environmental, demographic, and lifestyle choices.

After age 65, people are more likely to experience *chronic diseases* rather than *acute diseases*. Women tend to live longer than men but have more chronic diseases, such as arthritis, asthma, cancer, and mental illness, and experience higher depression rates than men.

Among U.S. racial-ethnic groups, black babies have the *highest infant mortality rate*, and Latinos are the least likely to die from illicit drugs and prescription drug abuse. Our lifestyle choices also improve or impair health. The top three preventable health hazards, in order of priority, are smoking, obesity, and substance abuse. Sexually transmitted diseases are another important source of preventable health hazards.

Disability any physical or mental impairment that limits a person's ability to perform a basic life activity. (p. 284)

Chronic diseases long-term or lifelong illnesses that develop gradually or are present from birth. (p. 286)

Acute diseases illnesses that strike suddenly and often disappear rapidly. (p. 286)

Alzheimer's disease a progressive, degenerative disorder that attacks the brain and impairs memory, thinking, and behavior. (p. 286)

Substance abuse a harmful overindulgence in or dependence on a drug or other chemical. (p. 290)

14-3 **Health Care: United States and Global** *Medicine* is a vital part of *health care* in diagnosing, treating, and preventing illness, injury, and other health impairments. The United States is one the richest nations in the world, but only the very wealthy don't have to worry about receiving and paying for the best medical care available. About 10 percent of Americans have no health insurance.

Since 2000, employer-based health insurance coverage has deteriorated. Consequently, workers at both large and small firms have been making higher contributions to premiums and paying higher deductibles or copayments. *Medicare* pays many of the medical costs of Americans age 65 and over, regardless of income. *Medicaid*, another government program, provides medical care for those living below the poverty level.

The United States spends more on health care than any other nation in the world, but compared with other high-income countries, covers a smaller percentage of the total population, and has poorer health outcomes. Unlike people in other high-income countries, Americans have fewer and shorter doctor visits, fewer days of inpatient hospital care, and pay considerably more for prescription drugs and routine medical services (such as doctor's office visits and normal birth deliveries).

In 2010, Congress passed the Patient Protection and Affordable Care Act, also called the Affordable Care Act (ACA) or "Obamacare." The ACA's goal is to give more Americans age 64 and under access to affordable, quality health insurance, and to reduce the growth in U.S. health care spending.

The uninsured include people who live in states where Medicaid doesn't cover the "nearly poor," who are undocumented immigrants, who haven't enrolled in Medicaid, who have incomes above the federal poverty level but can't afford the programs offered by employers or private insurers, and those who forgo coverage by choice.

Medicine a social institution that deals with illness, injury, and other health problems. (p. 291)

14-4 Sociological Perspectives on Health and Medicine See Table 14.4 below.

Sick role a social role that excuses people from normal obligations because of illness. (p. 294)

Medical-industrial complex a network of business enterprises that influences medicine and health care. (p. 295)

Medicalization a process that defines a nonmedical condition or behavior as an illness, disorder, or disease that requires medical treatment. (p. 300)

Table 14.4 Sociological Perspectives on Health and Medicine

Perspective	Level of Analysis	Key Points
Functionalism	Macro	• Health and medicine are critical in ensuring a society's survival and are closely linked to other institutions. • Illness is dysfunctional because it prevents people from performing expected roles. • Sick people are expected to seek professional help and get well.
Conflict	Macro	• There are gross inequities in the health care system. • The medical establishment is a powerful social control agent. • A drive for profit ignores people's health needs.
Feminist	Macro and micro	• Women are less likely than men to receive high-quality health care. • Gender stratification in medicine and the health care industry reduces women's earnings. • Men control women's health.
Symbolic Interaction	Micro	• Illness and disease are socially constructed. • Labeling people as ill increases their likelihood of being stigmatized. • Medicalization has increased the power of medical associations, parents, and mental health advocates and the profits of pharmaceutical companies.

15-1

Population Changes *Demography* examines the interplay among *fertility*, *mortality*, and *migration*. The *crude birth rate*, *total fertility rate*, *crude death rate*, and infant mortality rate measure a population's life expectancy and health.

Push and pull factors affect international and internal migration. Demographers also use *sex ratios* and *population pyramids* to understand a population's composition and structure.

Demographers who believe that population growth is a ticking bomb subscribe to *Malthusian theory*, which argues that the world's food supply won't keep up with population growth. *Demographic transition theory*, in contrast, maintains that population growth is kept in check and stabilizes as countries experience greater economic and technological development.

Demography the scientific study of human populations. (p. 303)

Population people who share a geographic territory. (p. 303)

Fertility the number of babies born during a specified period in a particular society. (p. 303)

Crude birth rate (also called the *birth rate*) the number of live births per 1,000 people in a population in a given year. (p. 303)

Total fertility rate (TFR) the average number of children born to a woman during her lifetime. (p. 304)

Mortality the number of deaths in a population during a specified period. (p. 304)

Crude death rate (also called the *death rate*) the number of deaths per 1,000 people in a population in a given year. (p. 304)

Migration movement of people into or out of a specific geographic area. (p. 305)

Sex ratio the proportion of men to women in a population. (p. 306)

Population pyramid a graphic depiction of a population's age and sex distribution at a given point in time. (p. 307)

Malthusian theory the population is growing faster than the food supply needed to sustain it. (p. 308)

Demographic transition theory population growth is kept in check and stabilizes as countries experience economic and technological development. (p. 308)

Zero population growth (ZPG) a stable population level when each woman has no more than two children. (p. 309)

15-2

Urbanization Globally and in the United States, cities and *urbanization* mushroomed during the twentieth century and are expected to increase. As more people move from rural to urban areas, many of the world's largest cities are becoming *megacities*. In the United States, urban growth has led to suburbanization, *edge cities*, *exurbs*, *gentrification*, and *urban sprawl*.

Sociologists offer several perspectives on how and why cities change, and the consequences (see Table 15.3 on the next page).

Urbanization people's movement from rural to urban areas. (p. 310)

Megacities metropolitan areas with at least 10 million inhabitants. (p. 310)

Edge cities business centers that are within or close to suburban residential areas. (p. 311)

Exurbs small, usually prosperous communities beyond a city's suburbs. (p. 311)

Metropolitan statistical area (MSA, also called *metro area*) a central city of at least 50,000 people and urban areas linked to it. (p. 312)

Urban sprawl rapid, unplanned, and uncontrolled spread of development into regions adjacent to cities. (p. 312)

Gentrification upper-middle-class and affluent people buy and renovate houses and stores in downtown urban neighborhoods. (p. 313)

Urban ecology studies the relationships between people and urban environments. (p. 314)

New urban sociology views urban changes as largely the result of decisions made by powerful capitalists and high-income groups. (p. 315)

15 Population, Urbanization, and the Environment

15-3 **The Environment** Population growth and urbanization are changing the planet's *ecosystem* and, many argue, endangering plants, animals, and humans. Water and air pollution and global warming are good examples of threats to the ecosystem in the United States and globally. Clean water has been depleted for many reasons, including pollution, privatization, waste, and mismanagement.

Four of the most common sources and causes of air pollution are burning fossil fuels, manufacturing plants that spew pollutants into the air, winds that carry contaminants across borders and oceans, and lax governmental policies. Air pollution, which can lead to the *greenhouse effect*, is a major cause of *climate change* and *global warming*.

The rise of environmental problems has sparked discussions about *sustainable development*. Those who are pessimistic about achieving sustainable development show that, worldwide, the United States has one of the worst records on environmental performance, largely because of the close ties between government officials and corporations.

Others are optimistic about achieving sustainable development and point to examples such as decreases in the emission of major air pollutants and some large U.S. corporations' switching to practices that decrease pollution and energy consumption.

Ecosystem a community of living and non-living organisms that share a physical environment. (p. 316)

Environmental racism (also called *environmental injustice*) exposure of poor people, especially minorities, to environmental hazards. (p. 318)

Global warming increase in the average temperature of earth's atmosphere. (p. 320)

Greenhouse effect heating of earth's atmosphere because of the presence of certain atmospheric gases. (p. 320)

Climate change a change in overall temperatures and weather conditions over time. (p. 320)

Sustainable development economic activities that don't threaten the environment. (p. 321)

Table 15.3 Sociological Explanations of Urbanization

Perspective	Level of Analysis	Key Points
Functionalist	Macro	Cities serve many important social and economic functions, but urbanization can also be dysfunctional.
Conflict	Macro	Driven by greed and profit, large corporations, banks, developers, and other capitalist groups shape cities' growth or decline.
Feminist	Macro and micro	Whether they live in cities or suburbs, women generally experience fewer choices and more constraints than men.
Symbolic Interactionist	Micro	City residents differ in their types of interaction, lifestyles, and perceptions of urban life.

16-1 Collective Behavior Social scientists offer several explanations of *collective behavior*. Contagion theory proposes that individuals act emotionally and irrationally due to a crowd's almost hypnotic influence. According to convergence theory, crowds consist of like-minded people who deliberately assemble in a place to pursue a common goal. Emergent norm theory emphasizes the importance of social norms in shaping crowd behavior. Structural strain theory proposes that collective behavior occurs only if six conditions are present.

There are many types of collective behavior, some more short-lived or harmful than others. *Rumors*, *gossip*, and *urban legends* are typically untrue, but many people believe and pass them on for a number of reasons, such as anxiety or to reinforce a community's moral standards. In contrast, *panic* and *mass hysteria* can have dire consequences, including death. *Fashions* and *fads* are harmless because they usually last only a short time and change over time. In contrast, a *disaster* is an unexpected event due to social, technological, or natural causes that result in widespread damage and destruction.

Publics, *public opinion*, and *propaganda* also affect large numbers of people. *Crowds* vary in their motives, interests, and emotional level. A casual crowd, for example, has little, if any, interaction, the gathering is temporary, and there is little emotion. On the other hand, protest crowds, especially in *mobs* and *riots*, can wreak considerable havoc on property and result in death.

Example: Crowds Can Be Deadly

On Thanksgiving Day, 2008, crowds started gathering at 9:00 p.m. outside the Walmart store in Valley Stream, New York, for a bargain-hunting ritual known as Black Friday, the day after Thanksgiving. By 4:55 a.m., the crowd had grown to more than 2,000 people and could no longer be held back. Suddenly, according to witnesses, the glass doors shattered and "the shrieking mob surged through in a blind rush for holiday bargains." A 34-year-old male temporary worker, who had been hired for the holiday season, was trampled to death, and four other people, including a 28-year-old woman who was eight months pregnant, were treated for injuries (McFadden and Macropoulos, 2008). Review the types of crowds in this chapter. Which type of crowd do you think this Walmart incident most nearly represents?

Social change transformations of societies and social institutions over time. (p. 327)

Collective behavior spontaneous and unstructured behavior of a large number of people. (p. 327)

Rumor unfounded information that people spread quickly. (p. 329)

Gossip rumors, often negative, about other people's personal lives. (p. 329)

Urban legends (also called *contemporary legends* and *modern legends*) rumors about stories that supposedly happened somewhere. (p. 330)

Panic collective flight, often irrational, from a real or perceived danger. (p. 330)

Mass hysteria an intense, fearful, and anxious reaction to a real or imagined threat by large numbers of people. (p. 330)

Fashion a popular way of dressing during a particular time or among a particular group of people. (p. 331)

Fad a trend that's popular for a short time. (p. 331)

Disaster an unexpected event that causes widespread damage, destruction, distress, and loss. (p. 332)

Public a group of people, not necessarily in direct contact with each other, who are interested in a particular issue. (p. 332)

Public opinion widespread attitudes on a particular issue. (p. 333)

Propaganda spreading information (or misinformation) to influence people's attitudes or behavior. (p. 333)

Crowd a temporary gathering of people who share a common interest or participate in a particular event. (p. 333)

Mob a highly emotional and disorderly crowd that uses force, the threat of force, or violence against a specific target. (p. 334)

Riot a violent crowd that directs its hostility at a wide and shifting range of targets. (p. 334)

16-2 Social Movements Unlike collective behavior, *social movements* are typically organized and have long-lasting effects. Some of the most common social movements are alternative, redemptive, reformative, resistance, and revolutionary (see Table 16.1 on the next page). Sociologists have offered several explanations for the emergence of social movements, including mass society theory, relative deprivation theory, resource mobilization theory, and new social movements theory. Each theory has strengths and weaknesses in explaining social movements.

Social movements generally go through four stages: emergence, organization, institutionalization, and decline. Decline occurs when a social movement is successful and becomes a part of society's fabric; when the members become distracted because the group loses sight of its original goals and/or their enthusiasm wanes; when the membership fragments because the participants disagree about goals, strategies, or tactics; or when a government quashes dissent. Social movements are important because they can create or resist change at the individual, institutional, and societal levels.

Social movement a large and organized group of people who want to promote or resist a particular social change. (p. 335)

Relative deprivation a gap between what people have and what they think they should have compared with others in a society. (p. 336)

16-3 **Technology and Social Change** *Technology* also generates changes. Some of the most important technological advances have included computer technology, biotechnology, and nanotechnology—all of which have changed our lives. Technology has both benefits and costs, however. For example, the Internet and other forms of telecommunication technology can bring people together, but can also intrude on our privacy. In addition, technological advances raise numerous ethical questions, such as their greater availability to higher-income people.

Technology the application of scientific knowledge for practical purposes. (p. 341)

Table 16.1 Five Types of Social Movements

Movement	Goal	Examples
Alternative	Change some people in a specific way	Alcoholics Anonymous, transcendental meditation
Redemptive	Change some people, but completely	Jehovah's Witnesses, born-again Christians
Reformative	Change everyone, but in specific ways	Gay rights advocates, Mothers Against Drunk Driving (MADD)
Resistance	Preserve status quo by blocking or undoing change	Antiabortion groups, white supremacists
Revolutionary	Change everyone completely	Right-wing militia groups, Communism, ISIS in the Middle East